BIOLOGY

^{5e}

TODAY & TOMORROW

Cecie Starr | Christine A. Evers | Lisa Starr

CENGAGE
Learning

Australia • Brazil • Mexico • Singapore • United Kingdom • United States

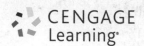

Biology Today and Tomorrow, **Fifth Edition**
Cecie Starr, Christine A. Evers, Lisa Starr

Product Director: Mary Finch

Senior Product Team Manager: Yolanda Cossio

Managing Content Developer: Trudy Brown

Product Assistant: Victor Luu

Content Developer: Lauren Oliveira

Associate Content Developers: Kellie N. Petruzzelli, Casey J. Lozier

Senior Market Development Manager: Tom Ziolkowski

IP Analyst: Christine M. Myaskovsky

IP Project Manager: John N. Sarantakis

Content Project Manager: Hal Humphrey

Senior Art Director: Bethany Casey

Manufacturing Planner: Karen Hunt

Production Service: Grace Davidson & Associates

Photo Researchers: Cheryl DuBois and Megan Cooper, Lumina Datamatics

Text Researcher: Kavitha Balasundaram, Lumina Datamatics

Copy Editor: Anita Wagner Hueftle

Illustrators: Lisa Starr, ScEYEnce Studios, Precision Graphics, Gary Head

Interior Designer: Michael Stratton, Stratton Design

Cover Designer: Michael Stratton, Stratton Design

Cover Image: © Shutterstock, Inc./Felix Rohan

Compositor: Lachina Publishing Services

For product information and technology assistance, contact us at
Cengage Learning Customer & Sales Support, 1-800-354-9706.
For permission to use material from this text or product, submit all requests online at **www.cengage.com/permissions**.
Further permissions questions can be e-mailed to
permissionrequest@cengage.com.

Library of Congress Control Number: 2014958336

Bound Edition:
ISBN: 978-1-305-11735-8

Loose-leaf Edition:
ISBN: 978-1-305-62939-4

Cengage Learning
20 Channel Center Street
Boston, MA 02210
USA

Cengage Learning is a leading provider of customized learning solutions with office locations around the globe, including Singapore, the United Kingdom, Australia, Mexico, Brazil, and Japan. Locate your local office at **www.cengage.com/global**.

Cengage Learning products are represented in Canada by Nelson Education, Ltd.

To learn more about Cengage Learning Solutions, visit **www.cengage.com**.

Purchase any of our products at your local college store or at our preferred online store **www.cengagebrain.com**.

Printed in the United States of America
Print Number: 01   Print Year: 2015

BC

BRIEF CONTENTS

CONTENTS

UNIT 1 HOW CELLS WORK

UNIT 2 GENETICS

UNIT 3 EVOLUTION AND DIVERSITY

15   Animal Evolution

UNIT 4 ECOLOGY

16   Population Ecology

UNIT 5 HOW ANIMALS WORK

19 Animal Tissues and Organs

UNIT 6 HOW PLANTS WORK

27 Plant Form and Function

28 Plant Reproduction and Development

Biology is a huge field, with a wealth of new discoveries being made every day, and biology-related issues such as climate change, stem cell research, and personal genetics often making headlines. This avalanche of information can be intimidating to non-scientists. This book was designed and written specifically for students who most likely will not become biologists and may never again take another science course. It is an accessible and engaging introduction to biology that provides future decision-makers with an understanding of basic biology and the process of science.

A Wealth of Applications This book is packed with everyday applications of biological processes. At every opportunity, we enliven discussions of biological processes with references to their effects on human health and the environment. This edition also continues to focus on real world applications pertaining to the field of biology, including social issues arising from new research and developments. Descriptions of current research, along with photos of scientists who carry it out, underscore the concept that biology is an ongoing endeavor carried out by a diverse community of people. Discussions include not only what was discovered, but also how the discoveries were made, how our understanding has changed over time, and what remains to be discovered. These discussions are provided in the context of an accessible introduction to well-established concepts that underpin modern biology. Every topic is examined from an evolutionary perspective, emphasizing the connections between all forms of life.

Accessible Text Understanding stems from making connections between concepts and details, so a text with too little detail reads as a series of facts that beg to be memorized. However, excessive detail can overwhelm the introductory student. Thus, we constantly strive to strike the perfect balance between level of detail and accessibility. We once again revised the text to eliminate details that do not contribute to a basic understanding of essential concepts. We also know that English is a second language for many introductory students, so we avoid idioms and aim for a clear, straightforward style.

Analogies to familiar objects and phenomena will help students understand abstract concepts. For example, in the discussion of transpiration in Chapter 27 (Plant Form and Function), we explain that a column of water is drawn upward through xylem as a drinker draws fluid up through a straw.

In-Text Learning Tools To emphasize connections between biological topics, each chapter begins with an APPLICATION section that explores a current event or controversy directly related to the chapter's content. For example, a discussion of binge drinking on college campuses introduces the concept of metabolism in Chapter 4. This section presents an overview of the metabolic pathway that breaks down alcohol, linking the function of enzymes in the pathway to hangovers, alcoholism, and cirrhosis. The section is illustrated with a photo of a tailgate party that preceded a recent Notre Dame–Alabama football game, and also a photo of Gary Reinbach just before he died at age 22 of alcoholic liver disease. (In the index, you'll find health-related applications denoted by red squares and environmental applications by green squares.)

To strengthen a student's analytical skills and offer insight into contemporary research, each chapter includes an exercise called DIGGING INTO DATA that is placed in a section with relevant content. The exercise consists of a short text passage—usually about a published scientific experiment—and a table, chart, or other graphic that presents experimental data. A student can use information in the text and graphic to answer a series of questions. For example, the exercise in Chapter 2 asks students to interpret results of a study that examined the effect of dietary fat intake on "good" and "bad" cholesterol levels.

The chapter itself consists of several numbered sections that contain a manageable chunk of information. Every section ends with a boxed TAKE-HOME MESSAGE in which we pose a question that reflects the critical content of the section, and then answer the question in bulleted list format. Every chapter has at least one FIGURE IT OUT QUESTION with an answer immediately following. These questions allow students to quickly check their understanding as they read. Mastering scientific vocabulary challenges many students, so we have included an ON-PAGE GLOSSARY of key terms introduced in each two-page spread, in addition to a complete glossary at the book's end. The end-of-chapter material features a VISUAL SUMMARY that reinforces each chapter's key concepts. A SELF-QUIZ poses multiple choice and other short answer questions for self-assessment (answers are in Appendix I). A set of more challenging CRITICAL THINKING QUESTIONS provides thought-provoking exercises for the motivated student. The end matter of several chapters now includes a VISUAL QUESTION that reinforces learning in a nonverbal style.

Design and Content Revisions Throughout the book, text and art have been revised to help students grasp difficult concepts. The following list highlights some of the revisions to each chapter.

Introduction

1 Invitation to Biology Renewed and updated emphasis on the relevance of new species discovery and the process of science.

Unit 1 How Cells Work

2 Molecules of Life New graphic illustrates radioactive decay.

3 Cell Structure Application section updated with current statistics and 'pink slime' story. Micrograph comparisons now feature *Paramecia* and include a confocal image. Essay about the nature of life expanded to add Gerald Joyce's "life is squishy" concept.

4 Energy and Metabolism Application section now illustrated with a real-life example. Diffusion illustrated with a tea bag in hot water.

5 Capturing and Releasing Energy Application section updated with current statistics and illustrated with a current photo of air pollution in China. Yogurt production added to fermentation section.

Unit 2 Genetics

6 DNA Structure and Function Content reorganized: material on cloning folded into Application section for concept connection, and chromosome structure now appears after DNA structure. New art demonstrates how replication errors become mutations.

7 Gene Expression and Control Ricin discussion revised to include medical applications. New material includes hairlessness mutation (in cats), evolution of lactose tolerance, heritability of DNA methylations, telomeres.

8 How Cells Reproduce New material on telomeres, asexual vs. sexual mud snails. New micrograph shows multiple crossovers.

9 Patterns of Inheritance Epistasis is now illustrated with human skin color. New material about environmentally-triggered hemoglobin production in *Daphnia*; continuous variation in dog face length arising from short tandem repeats foreshadows DNA fingerprinting in chapter 10.

10 Biotechnology Updated coverage of personal genetic testing includes social impact of Angelina Jolie's response to her test. New photos illustrate genetically modified animals. New "who's the daddy" critical thinking question offers students an opportunity to analyze a paternity test based on SNPs.

Unit 3 Evolution and Diversity

11 Evidence of Evolution Photos of 19th century naturalists added to emphasize the process of science that led to natural selection theory. How banded iron formations provide evidence of the evolution of photosynthesis added to fossil section. Plate tectonics art updated to reflect new evidence of lava lamp mantle movements.

12 Processes of Evolution New opening essay on resistance to antibiotics as an outcome of agricultural overuse (warfarin material now exemplifies directional selection). New art illustrates founder effect, and hypothetical example in text replaced with reduced

diversity of ABO alleles in Native Americans. New art illustrates stasis in coelacanths.

13 Early Life Forms and the Viruses New introductory essay about study of the human microbiome, new coverage of Ebola, and new figure depicting mechanisms of gene exchange in prokaryotes.

14 Plants and Fungi Additional coverage of fungal ecology, including information about white-nose syndrome in bats.

15 Animal Evolution New introductory essay about invertebrates as a source of medicines. Updated information about Neanderthals and added coverage of the newly discovered Dennisovans.

Unit 4 Ecology

16 Population Ecology Updated coverage of human demographics.

17 Communities and Ecosystems New photos illustrate species interactions; updated coverage of the increases in greenhouse gases.

18 The Biosphere and Human Effects New essay about dispersion of the radioactive material released at Fukushima and new Digging Into Data about bioaccumulation of this material in tuna.

Unit 5 How Animals Work

19 Animal Tissues and Organs Updated information about stem cell research and tissue regeneration in animals. Improved figures depict epithelial and connective tissues.

20 How Animals Move New information about how different muscle fiber types relate to animal locomotion.

21 Circulation and Respiration Improved coverage of insect respiration, including a new photo.

22 Immunity New photos show skin as a surface barrier, a cytotoxic T cell killing a cancer cell, and victims of HIV. Immune response and lymphatic system illustrations updated.

23 Digestion and Excretion Revised essay about obesity and new comparative information about the ruminant digestive system.

24 Neural Control and the Senses New opening essay about the effects of concussions. Discussion of the human nervous system has been reorganized. New information about echolocation.

25 Endocrine Control Opening essay now focuses on phthalates as endocrine disruptors. New Digging Into Data about BPA's effect on insulin secretion.

26 Reproduction and Development Updated coverage of assisted reproductive technologies. Discussion of human reproductive structure and function has been reorganized.

Unit 6 How Plants Work

27 Plant Form and Function Reorganization consolidates growth into a separate section. Many new photos illustrate stem, leaf, and root structure(s). Material on fire scars added to dendroclimatology.

28 Plant Reproduction and Development Updates reflect current research on colony collapse and ongoing major breakthroughs in the field of plant hormone function. New photos illustrate fruit classification, asexual reproduction, early growth, ABA inhibition of seed germination, and tropisms.

We owe a special debt to the members of our advisory board, listed below. They helped us shape the book's design and to choose appropriate content. We appreciate their guidance.

Andrew Baldwin, *Mesa Community College*
Charlotte Borgeson, *University of Nevada, Reno*
Gregory A. Dahlem, *Northern Kentucky University*
Gregory Forbes, *Grand Rapids Community College*
Hinrich Kaiser, *Victor Valley Community College*
Lyn Koller, *Pierce College*
Terry Richardson, *University of North Alabama*

We also wish to thank the reviewers listed below.

Idris Abdi, *Lane College*
Meghan Andrikanich, *Lorain County Community College*
Lena Ballard, *Rock Valley College*
Barbara D. Boss, *Keiser University, Sarasota*
Susan L. Bower, *Pasadena City College*
James R. Bray Jr., *Blackburn College*
Mimi Bres, *Prince George's Community College*
Randy Brewton, *University of Tennessee*
Evelyn K. Bruce, *University of North Alabama*
Steven G. Brumbaugh, *Green River Community College*
Chantae M. Calhoun, *Lawson State Community College*
Thomas F. Chubb, *Villanova University*
Julie A. Clements, *Keiser University, Melbourne*
Francisco Delgado, *Pima Community College*
Elizabeth A. Desy, *Southwest Minnesota State University*
Brian Dingmann, *University of Minnesota, Crookston*
Josh Dobkins, *Keiser University, online*
Hartmut Doebel, *The George Washington University*
Pamela K. Elf, *University of Minnesota, Crookston*
Johnny El-Rady, *University of South Florida*
Patrick James Enderle, *East Carolina University*
Jean Engohang-Ndong, *BYU Hawaii*
Ted W. Fleming, *Bradley University*
Edison R. Fowlks, *Hampton University*
Martin Jose Garcia Ramos, *Los Angeles City College*
J. Phil Gibson, *University of Oklahoma*
Judith A. Guinan, *Radford University*
Carla Guthridge, *Cameron University*
Laura A. Houston, *Northeast Lakeview–Alamo College*
Robert H. Inan, *Inver Hills Community College*
Dianne Jennings, *Virginia Commonwealth University*
Ross S. Johnson, *Chicago State University*
Susannah B. Johnson Fulton, *Shasta College*
Paul Kaseloo, *Virginia State University*
Ronald R. Keiper, *Valencia Community College West*
Dawn G. Keller, *Hawkeye Community College*
Ruhul H. Kuddus, *Utah Valley State College*
Dr. Kim Lackey, *University of Alabama*
Vic Landrum, *Washburn University*
Lisa Maranto, *Prince George's Community College*
Catarina Mata, *Borough of Manhattan Community College*
Kevin C. McGarry, *Keiser University, Melbourne*
Timothy Metz, *Campbell University*
Ann J. Murkowski, *North Seattle Community College*
Alexander E. Olvido, *John Tyler Community College*
Joshua M. Parke, *Community College of Southern Nevada*
Elena Pravosudova, *Sierra College*
Nathan S. Reyna, *Howard Payne University*
Carol Rhodes, *Cañada College*
Todd A. Rimkus, *Marymount University*
Laura H. Ritt, *Burlington County College*
Lynette Rushton, *South Puget Sound Community College*
Erik P. Scully, *Towson University*

Marilyn Shopper, *Johnson County Community College*
Jennifer J. Skillen, *Community College of Southern Nevada*
Jim Stegge, *Rochester Community and Technical College*
Lisa M. Strain, *Northeast Lakeview College*
Jo Ann Wilson, *Florida Gulf Coast University*

We were also fortunate to have conversations with the following workshop attendees. The insights they shared proved invaluable.

Robert Bailey, *Central Michigan University*
Brian J. Baumgartner, *Trinity Valley Community College*
Michael Bell, *Richland College*
Lois Borek, *Georgia State University*
Heidi Borgeas, *University of Tampa*
Charlotte Borgenson, *University of Nevada*
Denise Chung, *Long Island University*
Sehoya Cotner, *University of Minnesota*
Heather Collins, *Greenville Technical College*
Joe Conner, *Pasadena Community College*
Gregory A. Dahlem, *Northern Kentucky University*
Juville Dario-Becker, *Central Virginia Community College*
Jean DeSaix, *University of North Carolina*
Carolyn Dodson, *Chattanooga State Technical Community College*
Kathleen Duncan, *Foothill College, California*
Dave Eakin, *Eastern Kentucky University*
Lee Edwards, *Greenville Technical College*
Linda Fergusson-Kolmes, *Portland Community College*
Kathy Ferrell, *Greenville Technical College*
April Ann Fong, *Portland Community College*
Kendra Hill, *South Dakota State University*
Adam W. Hrincevich, *Louisiana State University*
David Huffman, *Texas State University, San Marcos*
Peter Ingmire, *San Francisco State*
Ross S. Johnson, *Chicago State University*
Rose Jones, *NW-Shoals Community College*
Thomas Justice, *McLennan Community College*
Jerome Krueger, *South Dakota State University*
Dean Kruse, *Portland Community College*
Dale Lambert, *Tarrant County College*
Debabrata Majumdar, *Norfolk State University*
Vicki Martin, *Appalachian State University*
Mary Mayhew, *Gainesville State College*
Roy Mason, *Mt. San Jacinto College*
Alexie McNerthney, *Portland Community College*
Brenda Moore, *Truman State University*
Alex Olvido, *John Tyler Community College*
Molly Perry, *Keiser University*
Michael Plotkin, *Mt. San Jacinto College*
Amanda Poffinbarger, *Eastern Illinois University*
Johanna Porter-Kelley, *Winston-Salem State University*
Sarah Pugh, *Shelton State Community College*
Larry A. Reichard, *Metropolitan Community College*
Darryl Ritter, *Okaloosa-Walton College*
Sharon Rogers, *University of Las Vegas*
Lori Rose, *Sam Houston State University*
Matthew Rowe, *Sam Houston State University*
Cara Shillington, *Eastern Michigan University*
Denise Signorelli, *Community College of Southern Nevada*
Jennifer Skillen, *Community College of Southern Nevada*
Jim Stegge, *Rochester Community and Technical College*
Andrew Swanson, *Manatee Community College*
Megan Thomas, *University of Las Vegas*
Kip Thompson, *Ozarks Technical Community College*
Steve White, *Ozarks Technical Community College*
Virginia White, *Riverside Community College*
Lawrence Williams, *University of Houston*
Michael L. Womack, *Macon State College*

STUDENT AND INSTRUCTOR RESOURCES

Cengage Learning Testing Powered by Cognero is a flexible, online system that allows you to:

• author, edit, and manage test bank content from multiple Cengage Learning solutions

• create multiple test versions in an instant

• deliver tests from your LMS, your classroom, or wherever you want

Instructor Companion Site   Everything you need for your course in one place! This collection of book-specific lecture and class tools is available online via www.cengage.com/login. Access and download Power-Point presentations, images, instructor's manual, videos, and more.

Cooperative Learning   Cooperative Learning: Making Connections in General Biology, 2nd Edition, authored by Mimi Bres and Arnold Weisshaar, is a collection of separate, ready-to-use, short cooperative activities that have broad application for first year biology courses. They fit perfectly with any style of instruction, whether in large lecture halls or flipped classrooms. The activities are designed to address a range of learning objectives such as reinforcing basic concepts, making connections between various chapters and topics, data analysis and graphing, developing problem solving skills, and mastering terminology. Since each activity is designed to stand alone, this collection can be used in a variety of courses and with any text.

MindTap   A personalized, fully online digital learning platform of authoritative content, assignments, and services that engages students with interactivity while also offering instructors their choice in the configuration of coursework and enhancement of the curriculum via web-apps known as MindApps. MindApps range from ReadSpeaker (which reads the text out loud to students) to Kaltura (which allows you to insert inline video and audio into your curriculum). MindTap is well beyond an eBook, a homework solution or digital supplement, a resource center website, a course delivery platform, or a Learning Management System. It is the first in a new category—the Personal Learning Experience.

New for this edition! MindTap has an integrated Study Guide, expanded quizzing and application activities, and an integrated Test Bank.

Aplia for Biology   The Aplia system helps students learn key concepts via Aplia's focused assignments and active learning opportunities that include randomized, automatically graded questions, exceptional text/art integration, and immediate feedback. Aplia has a full course management system that can be used independently or in conjunction with other course management systems such as MindTap, D2L, or Blackboard.

Acknowledgments

Writing, revising, and illustrating a biology textbook is a major undertaking for two full-time authors, but our efforts constitute only a small part of what is required to produce and distribute this one. We are truly fortunate to be part of a huge team of very talented people who are as committed as we are to creating and disseminating an exceptional science education product.

Biology is not dogma; paradigm shifts are a common outcome of the fantastic amount of research in the field. Ideas about what material should be taught and how best to present that material to students changes from one year to the next. It is only with the ongoing input of our many academic reviewers and advisors (previous page) that we can continue to tailor this book to the needs of instructors and students while integrating new information and models. We continue to learn from and be inspired by these dedicated educators.

On the production side of our team, the indispensable Grace Davidson orchestrated a continuous flow of files, photos, and illustrations while managing schedules, budgets, and whatever else happened to be on fire at the time. Grace, thank you as always for your patience and dedication. Thank you also to Cheryl DuBois, John Sarantakis, and Christine Myaskovsky for your help with photoresearch. Copyeditor Anita Hueftle and proofreader Diane Miller, your valuable suggestions kept our text clear and concise.

Yolanda Cossio, thank you for continuing to support us and for encouraging our efforts to innovate and improve. Thanks also to Cengage Production Manager Hal Humphrey, Marketing Manager Tom Ziolkowski, and to Lauren Oliveira, who creates our exciting technology package, Associate Content Developers Casey Lozier and Kellie Petruzzelli, and Product Assistant Victor Luu.

Lisa Starr and Christine Evers, November 2014

BIOLOGY 5e

TODAY & TOMORROW

1

INVITATION TO BIOLOGY

Application ❯ 1.1 The Secret Life of Earth

A. Paul Oliver discovered this tree frog perched on a sack of rice during a rainy campsite lunch in New Guinea's Foja Mountains. The explorers dubbed the new species "Pinocchio frog" after the Disney character because the male frog's long nose inflates and points upward during times of excitement.

B. Dr. Jason Bond holds a new species of trapdoor spider he discovered in sand dunes of California beaches in 2008. Bond named the spider *Aptostichus stephencolberti*, after TV personality Stephen Colbert.

Figure 1.1 Newly discovered species.
Each of the thousands of species discovered every year is a reminder that we do not yet know all of the organisms living on our own planet. We don't even know how many to look for. Information about the 1.8 million species we do know about is being collected in The Encyclopedia of Life, an online database maintained by collaborative effort (www.eol.org).

(A) Tim Laman/National Geographic Stock; (B) Courtesy East Carolina University.

In this era of detailed satellite imagery and cell phone global positioning systems, could there possibly be any places left on Earth that humans have not yet explored? Actually, there are plenty of them. In 2005, for example, helicopters dropped a team of scientists into the middle of a vast and otherwise inaccessible cloud forest atop New Guinea's Foja Mountains. Within a few minutes, the explorers realized that their landing site, a dripping, moss-covered swamp, had been untouched by humans. Team member Bruce Beehler remarked, "Everywhere we looked, we saw amazing things we had never seen before. I was shouting. This trip was a once-in-a-lifetime series of shouting experiences."

How did the explorers know they had landed in uncharted territory? For one thing, the forest was filled with plants and animals previously unknown even to native peoples that have long inhabited other parts of the region. During the next month, the team members discovered many new species, including a rhododendron plant with flowers the size of a plate and a frog the size of a pea. They also came across hundreds of species that are on the brink of extinction in other parts of the world, and some that supposedly had been extinct for decades. The animals had never learned to be afraid of humans, so they could easily be approached. A few were discovered as they casually wandered through campsites (Figure 1.1A).

New species are discovered all the time, often in places much more mundane than Indonesian cloud forests (Figure 1.1B). How do we know what species a particular organism belongs to? What is a species, anyway, and why should discovering a new one matter to anyone other than a scientist? You will find the answers to such questions in this book. They are part of the scientific study of life, **biology**, which is one of many ways we humans try to make sense of the world around us.

Trying to understand the immense scope of life on Earth gives us some perspective on where we fit into it. For example, hundreds of new species are discovered every year, but about 20 species become extinct every minute in rain forests alone—and those are only the ones we know about. The current rate of extinctions is about 1,000 times faster than normal, and human activities are responsible for the acceleration. At this rate, we will never know about most of the species that are alive on Earth today. Does that matter? Biologists think so. Whether or not we are aware of it, humans are intimately connected with the world around us. Our activities are profoundly changing the entire fabric of life on Earth. These changes are, in turn, affecting us in ways we are only beginning to understand.

Ironically, the more we learn about the natural world, the more we realize we have yet to learn. But don't take our word for it. Find out what biologists know, and what they do not, and you will have a solid foundation upon which to base your own opinions about how humans fit into this world. By reading this book, you are choosing to learn about the human connection—your connection—with all life on Earth.

1.2 Life Is More Than the Sum of Its Parts

What, exactly, is the property we call "life"? We may never actually come up with a good definition, because living things are too diverse, and they consist of the same basic components as nonliving things. When we try to define life, we end up with a long list of properties that differentiate living from nonliving things. These properties often emerge from the interactions of basic components. To understand how that works, take a look at these groups of squares:

A property called "roundness" emerges when the squares are organized one way, but not other ways. The idea that different structures can be assembled from the same basic building blocks is a recurring theme in our world, and also in biology.

Life has successive levels of organization, with new properties emerging at each level (Figure 1.2). This organization begins with interactions between **atoms**, which are fundamental units of matter—the building blocks of all substances ❶. Atoms bond together to form **molecules** ❷. There are no atoms unique to living things, but there are unique molecules. In today's natural world, only living things make the "molecules of life," which are lipids, proteins, DNA, RNA, and complex carbohydrates. The emergent property of "life" appears at the next level, when many molecules of life become organized as a cell ❸. A **cell** is the smallest unit of life. Cells survive and reproduce themselves using energy, raw materials, and information in their DNA.

Some cells live and reproduce independently; others do so as part of a multicelled organism ❹. An **organism** is an individual that consists of one or more cells. In most multicelled organisms, cells are organized as tissues, organs, and organ systems that interact to keep the body working properly.

A **population** is a group of interbreeding individuals of the same type, or species, living in a given area ❺. At the next level, a **community** consists of all populations living in a given area ❻. Communities may be large or small, depending on the area defined.

The next level of organization is the **ecosystem**, which is a community interacting with its physical and chemical environment ❼. The most inclusive level, the **biosphere**, encompasses all regions of Earth's crust, waters, and atmosphere in which organisms live ❽.

Figure 1.2 Levels of organization in nature.

❶ Atoms are fundamental units of matter.

❷ Molecules consist of atoms.

❸ Cells consist of molecules.

❹ Organisms consist of cells.

❺ Populations consist of organisms.

❻ Communities consist of populations.

❼ Ecosystems consist of communities interacting with their environment.

❽ The biosphere consists of all ecosystems on Earth.

Take-Home Message 1.2

How do living things differ from nonliving things?

- All things, living or not, consist of the same building blocks: atoms. Atoms bond together to form molecules.
- In today's natural world, only living things make lipids, proteins, DNA, RNA, and complex carbohydrates. The unique properties of life emerge as these molecules become organized into cells.
- Higher levels of life's organization include multicelled organisms, populations, communities, ecosystems, and the biosphere.

atom Fundamental building block of all matter.

biology The scientific study of life.

biosphere All regions of Earth where organisms live.

cell Smallest unit of life.

community All populations of all species in a given area.

ecosystem A community interacting with its environment.

molecule Two or more atoms bonded together.

organism Individual that consists of one or more cells.

population Group of interbreeding individuals of the same species that live in a given area.

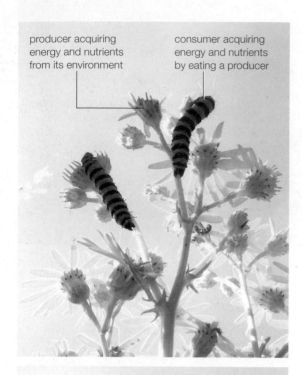

producer acquiring energy and nutrients from its environment

consumer acquiring energy and nutrients by eating a producer

ENERGY IN SUNLIGHT

Producers harvest energy from the environment. Some of that energy flows from producers to consumers.

PRODUCERS
plants and other self-feeding organisms

Nutrients that get incorporated into the cells of producers and consumers are eventually released back into the environment (by decomposition, for example). Producers then take up some of the released nutrients.

CONSUMERS
animals, most fungi, many protists, bacteria

All energy that enters the world of life eventually flows out of it, mainly as heat released back to the environment.

Figure 1.3 The one-way flow of energy and the cycling of materials in the world of life.

Top, © Victoria Pinder, www.flickr.com/photos/vixstarplus.

1.3 How Living Things Are Alike

Even though we cannot precisely define "life," we can intuitively understand what it means because all living things share a particular set of key features. All require ongoing inputs of energy and raw materials; all sense and respond to change; and all pass DNA to offspring.

Organisms Require Energy and Nutrients Not all living things eat, but all require energy and nutrients on an ongoing basis. Inputs of both are essential to maintain the functioning of individual organisms and the organization of life in general. A **nutrient** is a substance that an organism needs for growth and survival but cannot make for itself.

Organisms spend a lot of time acquiring energy and nutrients (Figure 1.3). However, the source of energy and the type of nutrients acquired differ among organisms. These differences allow us to classify living things into two categories: producers and consumers. A **producer** makes its own food using energy and simple raw materials it obtains from nonbiological sources. Plants are producers; by a process called **photosynthesis**, they use the energy of sunlight to make sugars from water and carbon dioxide (a gas in air). Consumers, by contrast, cannot make their own food. A **consumer** obtains energy and nutrients by feeding on other organisms. Animals are consumers. So are decomposers, which feed on the wastes or remains of other organisms. The leftovers from consumers' meals end up in the environment, where they serve as nutrients for producers. Said another way, nutrients cycle between producers and consumers.

Unlike nutrients, energy is not cycled. It flows through the world of life in one direction: from the environment, through organisms, and back to the environment. This flow maintains the organization of every living cell and body, and it also influences how individuals interact with one another and their environment. The energy flow is one-way, because with each transfer, some energy escapes as heat, and cells cannot use heat as an energy source. Thus, energy that enters the world of life eventually leaves it (we return to this topic in Chapter 5).

Organisms Sense and Respond to Change An organism cannot survive for very long in a changing environment unless it adapts to the changes. Thus, every living thing has the ability to sense and respond to change both inside and outside of itself (Figure 1.4). Consider how, after you eat, the sugars from your meal enter your bloodstream. The added sugars set in motion a series of events that causes cells throughout the body to take up sugar faster, so the sugar level in your blood quickly falls. This response keeps your blood sugar level within a certain range, which in turn helps keep your cells alive and your body functioning properly.

All of the fluids outside of cells make up a body's internal environment. That environment must be kept within certain ranges of temperature and other conditions, or the cells that make up the body will die. By sensing and adjusting to change, organisms keep conditions in the internal environment within a range that favors survival. **Homeostasis** is the name for this process, and it is one of the defining features of life.

Organisms Grow and Reproduce With little variation, the same types of molecules perform the same basic functions in every organism. For example, information in an organism's **DNA** (deoxyribonucleic acid) guides ongoing functions that sustain the individual through its lifetime. Such functions include **development**:

Figure 1.4 Organisms sense and respond to stimulation.
This baby orangutan is laughing in response to being tickled. Apes and humans make different sounds when being tickled, but the airflow patterns are so similar that we can say apes really do laugh.

© Dr. Marina Davila Ross, University of Portsmouth.

the process by which the first cell of a new individual becomes a multicelled adult; **growth**: increases in cell number, size, and volume; and **reproduction**: processes by which individuals produce offspring.

Individuals of every natural population are alike in certain aspects of their body form and behavior because their DNA is very similar: Orangutans look like orangutans and not like caterpillars because they inherited orangutan DNA, which differs from caterpillar DNA in the information it carries. **Inheritance** refers to the transmission of DNA to offspring. All organisms receive their DNA from one or more parents.

DNA is the basis of similarities in form and function among organisms. However, the details of DNA molecules differ, and herein lies the source of life's diversity. Small variations in the details of DNA's structure give rise to differences among individuals, and also among types of organisms. As you will see in later chapters, these differences are the raw material of evolutionary processes.

DNA

consumer Organism that gets energy and nutrients by feeding on the tissues, wastes, or remains of other organisms.

development Multistep process by which the first cell of a new multicelled organism gives rise to an adult.

DNA Deoxyribonucleic acid; carries hereditary information that guides development and other activities.

growth In multicelled species, an increase in the number, size, and volume of cells.

homeostasis Process in which an organism keeps its internal conditions within tolerable ranges by sensing and responding to change.

inheritance Transmission of DNA to offspring.

nutrient Substance that an organism needs for growth and survival but cannot make for itself.

photosynthesis Process by which a producer uses light energy to make sugars from carbon dioxide and water.

producer Organism that makes its own food using energy and nonbiological raw materials from the environment.

reproduction Process by which parents produce offspring.

Take-Home Message 1.3

How are all living things alike?

- A one-way flow of energy and a cycling of nutrients sustain life's organization.
- Organisms sense and respond to conditions inside and outside themselves. They make adjustments that keep conditions in their internal environment within a range that favors cell survival, a process called homeostasis.
- All organisms use information in the DNA they inherited from their parent or parents to develop, grow, and reproduce. DNA is the basis of similarities and differences in form and function among organisms.

animal Multicelled consumer that develops through a series of stages and moves about during part or all of its life.

archaea Group of single-celled organisms that lack a nucleus but are more closely related to eukaryotes than to bacteria.

bacteria The most diverse and well-known group of single-celled organisms that lack a nucleus.

biodiversity Scope of variation among living organisms.

eukaryote Organism whose cells characteristically have a nucleus.

fungus Single-celled or multicelled eukaryotic consumer that breaks down material outside itself, then absorbs nutrients released from the breakdown.

plant A multicelled, typically photosynthetic producer.

prokaryote Single-celled organism with no nucleus.

protists A group of diverse, simple eukaryotes.

species Unique type of organism.

taxonomy Practice of naming and classifying species.

A. Bacteria are the most numerous organisms on Earth. Clockwise from upper left, a bacterium with a row of iron crystals that acts like a tiny compass; a common resident of cat and dog stomachs; spiral cyanobacteria; types found in dental plaque.

B. Archaea may resemble bacteria, but they are more closely related to eukaryotes. These are two types of archaea from a hydrothermal vent on the seafloor.

Figure 1.5 A few representative prokaryotes.

1.4 How Living Things Differ

Living things differ tremendously in their observable characteristics. Various classification schemes help us organize what we understand about the scope of this variation, which we call Earth's **biodiversity**.

For example, organisms can be grouped on the basis of whether they have a nucleus, which is a saclike structure containing a cell's DNA. **Bacteria** (singular, bacterium) and **archaea** (singular, archaeon) are organisms whose DNA *is not* contained within a nucleus. All bacteria and archaea are single-celled, which means each organism consists of one cell (Figure 1.5). Collectively, these organisms are the most diverse representatives of life. Different kinds are producers or consumers in nearly all regions of Earth. Some inhabit such extreme environments as frozen desert rocks, boiling sulfurous lakes, and nuclear reactor waste. The first cells on Earth may have faced similarly hostile conditions.

Traditionally, organisms without a nucleus have been called **prokaryotes**, but the designation is now used only informally. This is because, despite the similar appearance of bacteria and archaea, the two types of cells are less related to one another than we once thought. Archaea turned out to be more closely related to **eukaryotes**, which are organisms whose DNA *is* contained within a nucleus. Some eukaryotes live as individual cells; others are multicelled (Figure 1.6). Eukaryotic cells are typically larger and more complex than bacteria or archaea.

Protists are the simplest eukaryotes, but as a group they vary dramatically, from single-celled consumers to giant, multicelled producers.

Fungi (singular, fungus) are eukaryotic consumers that secrete substances to break down food externally, then absorb nutrients released by this process. Many fungi are decomposers. Most fungi, including those that form mushrooms, are multicellular. Fungi that live as single cells are called yeasts.

Plants are multicelled eukaryotes, and the vast majority of them are photosynthetic producers that live on land. Besides feeding themselves, plants also serve as food for most other land-based organisms.

Animals are multicelled eukaryotic consumers that ingest tissues or juices of other organisms. Unlike fungi, animals break down food inside their body. They also develop through a series of stages that lead to the adult form. All animals actively move about during at least part of their lives.

What Is a Species? Each time we discover a new **species**, or unique kind of organism, we name it. **Taxonomy**, the practice of naming and classifying species, began thousands of years ago, but naming species in a consistent way did not become a priority until the eighteenth century. At the time, European explorers who were just discovering the scope of life's diversity started having more and more trouble communicating with one another because species often had multiple names. For example, the dog rose (a plant native to Europe, Africa, and Asia) was alternately known as briar rose, witch's briar, herb patience, sweet briar, wild briar, dog briar, dog berry, briar hip, eglantine gall, hep tree, hip fruit, hip rose, hip tree, hop fruit, and hogseed—and those are only the English names! Species often had multiple scientific names too, in Latin that was descriptive but often cumbersome. The scientific name of the dog rose was *Rosa sylvestris inodora seu canina* (odorless woodland dog rose), and also *Rosa sylvestris alba cum rubore, folio glabro* (pinkish white woodland rose with smooth leaves).

An eighteenth-century naturalist, Carolus Linnaeus, standardized a two-part naming system that we still use. By the Linnaean system, every species is given a

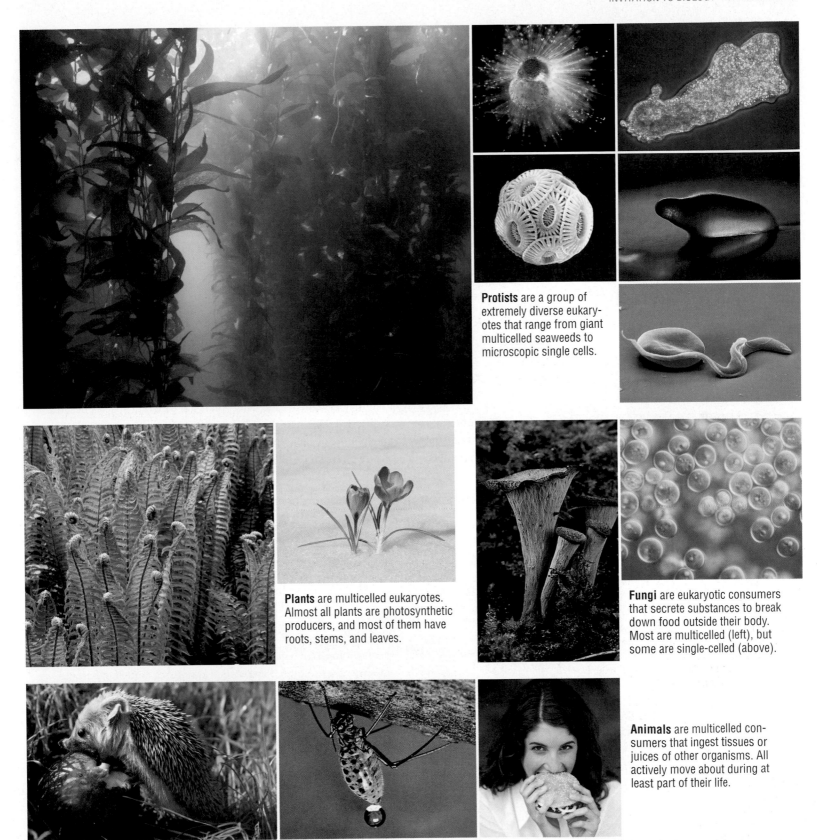

Protists are a group of extremely diverse eukaryotes that range from giant multicelled seaweeds to microscopic single cells.

Plants are multicelled eukaryotes. Almost all plants are photosynthetic producers, and most of them have roots, stems, and leaves.

Fungi are eukaryotic consumers that secrete substances to break down food outside their body. Most are multicelled (left), but some are single-celled (above).

Animals are multicelled consumers that ingest tissues or juices of other organisms. All actively move about during at least part of their life.

Figure 1.6 A few representative eukaryotes.

Protists: from left, © worldswildlifewonders/Shutterstock.com; top middle, Courtesy of Allen W. H. Bé and David A. Caron; bottom middle, © Emiliania Huxleyi photograph, Vita Pariente, scanning electron micrograph taken on a Jeol T330A instrument at Texas A&M University Electron Microscopy Center; top right, M I Walker/Science Source; middle right, © Carolina Biological Supply Company; bottom right, Oliver Meckes/Science Source; Plants: left, © Jag.ca.Shutterstock.com; right, © Martin Ruegner/Radius Images/Getty Images; Fungi, left, Edward S. Ross; right, London Scientific Films/Oxford Scientific/Getty Images; Animals: left, Shironina/Shutterstock.com; middle, © Martin Zimmerman, Science, 1961, 133:73–79, © AAAS; right, © Pixtal/SuperStock.

genus A group of species that share a unique set of traits.

taxon Group of organisms that share a unique set of traits.

> A "species" is a convenient but artificial construct of the human mind.

unique two-part scientific name. The first part of a scientific name is the **genus** (plural, genera), a group of species that share a unique set of features. The second part is the specific epithet. Together, the genus name and the specific epithet designate one species. Thus, the dog rose now has one official name, *Rosa canina*, that is recognized worldwide.

Genus and species names are always italicized. For example, *Panthera* is a genus of big cats. Lions belong to the species *Panthera leo*. Tigers belong to a different species in the same genus (*Panthera tigris*), and so do leopards (*P. pardus*). Note how the genus name may be abbreviated after it has been spelled out once.

A Rose by Any Other Name The individuals of a species share a unique set of inherited traits. For example, giraffes normally have very long necks, brown spots on white coats, and so on. These are morphological (structural) traits. Individuals of a species also share biochemical traits (they make and use the same molecules) and behavioral traits (they respond the same way to certain stimuli, as when hungry giraffes feed on tree leaves). We can rank species into ever more inclusive categories based on some subset of traits it shares with other species. Each rank, or **taxon** (plural, taxa), is a group of organisms that share a unique set of traits. Each category above species—genus, family, order, class, phylum (plural, phyla), kingdom, and domain—consists of a group of the next lower taxon (Figure 1.7). Using this system, we can sort all life into a few categories (Figure 1.8).

It is easy to tell that orangutans and caterpillars are different species because they appear very different. Distinguishing between species that are more closely related may be much more challenging (Figure 1.9). In addition, traits shared by members of a species often vary a bit among individuals, as eye color does among

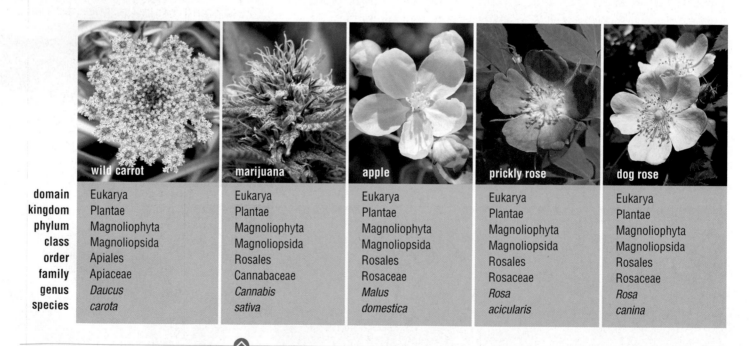

	wild carrot	marijuana	apple	prickly rose	dog rose
domain	Eukarya	Eukarya	Eukarya	Eukarya	Eukarya
kingdom	Plantae	Plantae	Plantae	Plantae	Plantae
phylum	Magnoliophyta	Magnoliophyta	Magnoliophyta	Magnoliophyta	Magnoliophyta
class	Magnoliopsida	Magnoliopsida	Magnoliopsida	Magnoliopsida	Magnoliopsida
order	Apiales	Rosales	Rosales	Rosales	Rosales
family	Apiaceae	Cannabaceae	Rosaceae	Rosaceae	Rosaceae
genus	*Daucus*	*Cannabis*	*Malus*	*Rosa*	*Rosa*
species	*carota*	*sativa*	*domestica*	*acicularis*	*canina*

Figure 1.7 Taxonomic classification of five species that are related at different levels.
Each species has been assigned to ever more inclusive groups, or taxa: in this case, from genus to domain.

Figure It Out: Which of the plants shown here are in the same order?

Answer: Marijuana, apple, prickly rose, and dog rose

Figure 1.8 Two little ways to see the big picture of life. Lines in such diagrams indicate evolutionary connections.

A. Six-kingdom classification system. The protist kingdom includes the most ancient multicelled and all single-celled eukaryotes.

B. Three-domain classification system. The Eukarya domain includes protists, plants, fungi, and animals.

people. How do we decide whether similar-looking organisms belong to the same species? The short answer to that question is that we rely on whatever information we have. Early naturalists studied anatomy and distribution—essentially the only methods available at the time—so species were named and classified according to what they looked like and where they lived. Today's biologists are able to compare traits that the early naturalists did not even know about, including biochemical ones.

The discovery of new information sometimes changes the way we distinguish a particular species or how we group it with others. For example, Linnaeus grouped plants by the number and arrangement of reproductive parts, a scheme that resulted in odd pairings such as castor-oil plants with pine trees. Having more information today, we place these plants in separate phyla.

Evolutionary biologist Ernst Mayr defined a species as one or more groups of individuals that potentially can interbreed, produce fertile offspring, and do not interbreed with other groups. This "biological species concept" is useful in many cases, but it is not universally applicable. For example, we may never know whether two widely separated populations could interbreed if they got together. As another example, populations often continue to interbreed even as they diverge, so the exact moment at which two populations become two species is often impossible to pinpoint. We return to speciation and how it occurs in Chapter 12, but for now it is important to remember that a "species" is a convenient but artificial construct of the human mind.

Figure 1.9 Four butterflies, two species: Which are which?
The top row shows two forms of the species *Heliconius melpomene*; the bottom row, two forms of *H. erato*.

H. melpomene and *H. erato* never cross-breed. Their alternate but similar patterns of coloration evolved as a shared warning signal to predatory birds that these butterflies taste terrible.

© 2006 Axel Meyer, "Repeating Patterns of Mimicry." *PLoS Biology* Vol. 4, No. 10, e341 doi:10.1371/journal.pbio.0040341. Used with Permission.

Take-Home Message 1.4

How do organisms differ from one another?

- Organisms differ in their details; they show tremendous variation in observable characteristics.
- We divide Earth's biodiversity into broad groups based on traits such as having a nucleus or being multicellular.
- Each species is given a unique, two-part scientific name.
- Classification systems group species on the basis of shared traits.

1.5 The Science of Nature

Most of us assume that we do our own thinking, but do we, really? You might be surprised to find out how often we let others think for us. Consider how a school's job (which is to impart as much information to students as quickly as possible)

control group Group of individuals identical to an experimental group except for the independent variable under investigation.

critical thinking Evaluating information before accepting it.

data Experimental results.

experiment A test designed to support or falsify a prediction.

experimental group In an experiment, a group of individuals who have a certain characteristic or receive a certain treatment.

hypothesis Testable explanation of a natural phenomenon.

model Analogous system used for testing hypotheses.

prediction Statement, based on a hypothesis, about a condition that should exist if the hypothesis is correct.

science Systematic study of the observable world.

scientific method Making, testing, and evaluating hypotheses.

variable In an experiment, a characteristic or event that differs among individuals or over time.

How do my own biases affect what I'm learning?

meshes perfectly with a student's job (which is to acquire as much knowledge as quickly as possible). In this rapid-fire exchange of information, it can be very easy to forget about the quality of what is being exchanged. Any time you accept information without questioning it, you let someone else think for you.

Thinking About Thinking **Critical thinking** is the deliberate process of judging the quality of information before accepting it. "Critical" comes from the Greek *kriticos* (discerning judgment). When you use critical thinking, you move beyond the content of new information to consider supporting evidence, bias, and alternative interpretations. How does the busy student manage this? Critical thinking does not necessarily require extra time, just a bit of extra awareness. There are many ways to do it. For example, you might ask yourself some of the following questions while you are learning something new:

> What message am I being asked to accept?
> Is the message based on facts or opinion?
> Is there a different way to interpret the facts?
> What biases might the presenter have?
> How do my own biases affect what I'm learning?

Such questions are a way of being conscious about learning. They can help you decide whether to allow new information to guide your beliefs and actions.

How Science Works Critical thinking is a big part of **science**, the systematic study of the observable world and how it works. A scientific line of inquiry usually begins with curiosity about something observable, such as (for example) a decrease in the number of birds in a particular area. Typically, a scientist will read about what others have discovered before making a **hypothesis**, a testable explanation for a natural phenomenon. An example of a hypothesis would be, "The number of birds is decreasing because the number of cats is increasing."

A **prediction**, or statement of some condition that should exist if the hypothesis is correct, comes next. Making predictions is often called the if–then process, in which the "if" part is the hypothesis, and the "then" part is the prediction: *If* the number of birds is decreasing because the number of cats is increasing, *then* reducing the number of cats should stop the decline.

Next, a researcher will test the prediction. Tests may be performed on a **model**, or analogous system, if working with an object or event directly is not possible. For

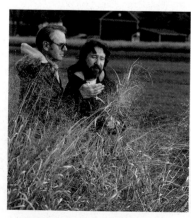

A. Studying the ecological benefits of weedy buffer zones on farms.

B. Measuring how much wood is produced by extremely old trees.

example, animal diseases are often used as models of similar human diseases. Careful observations are one way to test predictions that flow from a hypothesis. So are **experiments**: tests designed to support or falsify a prediction. A typical experiment explores a cause-and-effect relationship using **variables**, which are characteristics or events that can differ among individuals or over time.

Biological systems are typically complex, with many interdependent variables. It can be difficult to study one variable separately from the rest. Thus, biology researchers often test two groups of individuals simultaneously. An **experimental group** is a set of individuals that have a certain characteristic or receive a certain treatment. An experimental group is tested side by side with a **control group**, which is identical to the experimental group except for one independent variable: the characteristic or the treatment being tested. Any differences in experimental results between the two groups is likely to be an effect of changing the variable. Test results—**data**—that are consistent with the prediction are evidence in support of the hypothesis. Data inconsistent with the prediction are evidence that the hypothesis is flawed and should be revised.

A necessary part of science is reporting one's results and conclusions in a standard way, such as in a peer-reviewed journal article. The communication gives other scientists an opportunity to evaluate the information for themselves, both by checking the conclusions drawn and by repeating the experiments. Forming a hypothesis based on observation, and then systematically testing and evaluating the hypothesis, are collectively called the **scientific method** (Table 1.1).

Examples of Experiments in Biology
There are many different ways to do research, particularly in biology (Figure 1.10). Some biologists survey, simply observing without making or testing hypotheses. Others make hypotheses based on observations, and leave the testing to others. However, despite a broad range of approaches, scientific experiments are typically designed in a consistent way, so the effects of changing one variable at a time can be measured. To give you a sense of how biology experiments work, we summarize two published studies here.

In 1996 the U.S. Food and Drug Administration (FDA) approved Olestra®, a fat replacement manufactured from sugar and vegetable oil, as a food additive. Potato chips were the first Olestra-containing food product to be sold in the United States. Controversy about the chip additive soon raged. Many people complained of intestinal problems after eating the chips, and thought that the Olestra was at fault. Two

Table 1.1 The Scientific Method

Observe some aspect of nature.

Think of an explanation for your observation (in other words, form a hypothesis).

Test the hypothesis.
  a. Make a prediction based on the hypothesis.
  b. Test the prediction using experiments or surveys.
  c. Analyze the results of the tests (data).

Decide whether the results of the tests support your hypothesis or not (form a conclusion).

Report your results to the scientific community.

Figure 1.10 A few examples of scientific research in the field of biology.

(A) Photo by Scott Bauer, USDA/ARS; (B) MICHAEL NICHOLS/National Geographic Creative; (C) © Roger W. Winstead, NC State University; (D) National Cancer Institute; (E) Courtesy of Susanna López-Legentil.

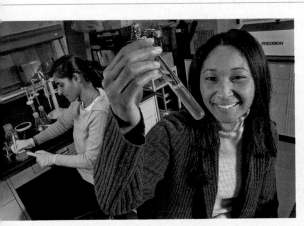

C. Improving efficiency of biofuel production from agricultural waste.

D. Devising a vaccine that helps prevent cancer.

E. Discovering medically active natural products made by marine animals.

Figure 1.11 The steps in a scientific experiment to determine whether Olestra causes intestinal cramps. A report of this study was published in the *Journal of the American Medical Association* in January 1998.

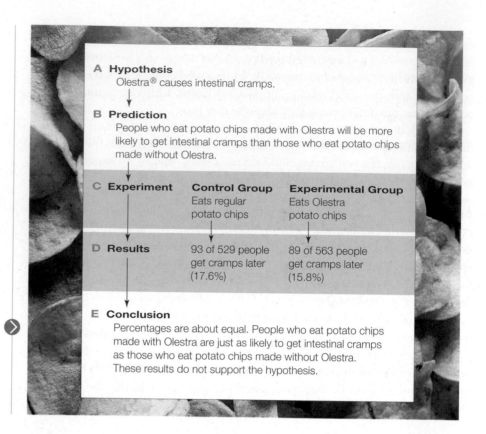

years later, researchers at the Johns Hopkins University School of Medicine designed an experiment to test whether Olestra causes cramps. The researchers made the following prediction: *if* Olestra causes cramps, *then* people who eat Olestra should be more likely to get cramps than people who do not eat it. To test the prediction, they used a Chicago theater as a "laboratory." They asked 1,100 people between the ages of thirteen and thirty-eight to watch a movie and eat their fill of potato chips. Each person received an unmarked bag containing 13 ounces of chips. In this experiment, the individuals who received Olestra-laden potato chips were the experimental group, and the individuals who received regular chips were the control group.

A few days after the movie, the researchers contacted all of the people who participated in the experiment and collected any reports of post-movie gastrointestinal problems. Of 563 people making up the experimental group, 89 (15.8 percent) reported having cramps. However, so did 93 of the 529 people (17.6 percent) making up the control group—who had eaten the regular chips. People were about as likely to get cramps whether or not they ate chips made with Olestra. These results did not support the prediction, so the researchers concluded that eating Olestra does not cause cramps (Figure 1.11).

A different experiment that took place in 2005 investigated whether certain behaviors of peacock butterflies help the insects avoid predation by birds. The researchers performing this experiment began with two observations. First, when a peacock butterfly rests, it folds its wings, so only the dark underside shows (Figure 1.12A). Second, when a butterfly sees a predator approaching, it repeatedly flicks its wings open, while also moving them in a way that produces a hissing sound and a series of clicks (Figure 1.12B).

The researchers were curious about why the peacock butterfly flicks its wings. After they reviewed earlier studies, they came up with two hypotheses that might explain the wing-flicking behavior.

1. Wing-flicking probably attracts predatory birds, but it also exposes brilliant spots that resemble owl eyes. Anything that looks like owl eyes is known to startle small, butterfly-eating birds, so exposing the wing spots might scare off predators.

2. The hissing and clicking sounds produced when the peacock butterfly moves its wings may be an additional defense that deters predatory birds.

The researchers then used their hypotheses to make the following predictions:

1. *If* exposing brilliant wing spots startles butterfly-eating birds, *then* peacock butterflies missing their spots will be more likely to get eaten.

2. *If* hissing and clicking sounds deter birds butterfly-eating birds, *then* peacock butterflies unable to make these sounds will be more likely to get eaten.

The next step was the experiment. The researchers used a black marker to cover up the wing spots of some butterflies, and scissors to cut off the sound-making part of the wings of others. A third group had both treatments, their wings painted and also cut. The researchers then put each butterfly into a large cage with a hungry blue tit (Figure 1.12C) and watched the pair for thirty minutes.

Figure 1.12D lists the results of the experiment. All butterflies with unmodified wing spots survived, regardless of whether they made sounds. By contrast, only half of the butterflies that had spots painted out but could make sounds survived. Most

Figure 1.12 Testing peacock butterfly defenses.

(A) © Matt Rowlings, www.eurobutterflies.com; (B) © Adrian Vallin; (C) © Antje Schulte; (D) *Proceedings of the Royal Society of London,* Series B (2005) 272: 1203–1207.

Figure It Out: What percentage of butterflies with spots painted and wings cut survived the test?
Answer: 20 percent

A. With wings folded, a resting peacock butterfly resembles a dead leaf, so it is appropriately camouflaged from predatory birds.

B. When a predatory bird approaches, a butterfly flicks its wings open and closed, revealing brilliant spots and producing hissing and clicking sounds.

C. Researchers tested whether the wing-flicking behavior of peacock butterflies affected predation by blue tits.

Experimental Treatment	Number of Butterflies Eaten (of Total)
Spots painted out	5 of 10
Wings cut	0 of 8
Spots painted, wings cut	8 of 10
None	0 of 9

D. The researchers painted out the spots of some butterflies, cut the sound-making part of the wings on others, and did both to a third group; then exposed each butterfly to a hungry blue tit for 30 minutes. Results are listed on the right.

Digging Into Data

Peacock Butterfly Predator Defenses The photographs below represent the experimental and control groups used in the peacock butterfly experiment. Identify the experimental groups, and match them up with the relevant control group(s). *Hint:* Identify which variable is being tested in each group (each variable has a control).

Adrian Vallin, Sven Jakobsson, Johan Lind and Christer Wiklund, *Proc. R. Soc. B* (2005: 272, 1203, 1207). Used with permission of The Royal Society and the author.

A. Wing spots painted out   **B.** Wing spots visible; wings silenced   **C.** Wing spots painted out; wings silenced   **D.** Wings painted but spots visible   **E.** Wings cut but not silenced   **F.** Wings painted, spots visible; wings cut, not silenced of the silenced butterflies with painted-out spots were eaten quickly. The test results confirmed both predictions, so they support the hypotheses. Predatory birds are indeed deterred by peacock butterfly wing-flicking behavior.

> ## Take-Home Message 1.5
>
> **How does science work?**
>
> - The scientific method consists of making, testing, and evaluating hypotheses. It is one way of critical thinking—systematically judging the quality of information before allowing it to guide one's beliefs and actions.
> - Natural processes are often very complex and influenced by many interacting variables.
> - Experiments help researchers unravel causes of complex natural processes by focusing on the effects of changing a single variable.

> The scientific community consists of critically thinking people trying to poke holes in one another's ideas.

1.6 The Nature of Science

Bias in Interpreting Experimental Results Experimenting with a single variable apart from all others is not often possible, particularly when studying humans. For example, remember that the people who participated in the Olestra experiment were chosen randomly, which means the study was not controlled for gender, age, weight, medications taken, and so on. These variables may well have influenced the experiment's results.

Humans are by nature subjective, and scientists are no exception. Researchers risk interpreting their results in terms of what they want to find out. That is

Figure 1.13 Example of how generalizing from a subset can lead to a conclusion that is incorrect.

(A) Tim Laman/ National Geographic Stock; (B) © Bruce Beehler/ Conservation International.

A. The cloud forest that covers about 2 million acres of New Guinea's Foja Mountains is extremely remote and difficult to access, even for natives of the region. The first major survey of this forest occurred in 2005.

B. In science, discovery of an error is not always bad news. Kris Helgen holds a golden-mantled tree kangaroo he found during the 2005 Foja Mountains survey. This kangaroo species is extremely rare in other areas, so it was thought to be critically endangered prior to the expedition.

why they typically design experiments that will yield quantitative results, which are counts or some other data that can be measured or gathered objectively. Quantitative results minimize the potential for bias, and also give other scientists an opportunity to repeat the experiments and check the conclusions drawn from them. This last point gets us back to the role of critical thinking in science. Scientists expect one another to recognize and put aside bias in order to test hypotheses in ways that may prove them wrong. If a scientist does not, then others will, because exposing errors is just as useful as applauding insights. The scientific community consists of critically thinking people trying to poke holes in one another's ideas. Ideally, their collective efforts make science a self-correcting endeavor.

Sampling Error Researchers cannot always observe all individuals of a group. For example, the explorers you read about in Section 1.1 did not—and could not—survey every uninhabited part of the Foja Mountains. The cloud forest alone cloaks more than 2 million acres (Figure 1.13A), so surveying all of it would take unrealistic amounts of time and effort.

When researchers cannot directly observe all individuals of a population, all instances of an event, or some other aspect of nature, they may test or survey a subset. Results from the subset are then used to make generalizations about the whole. However, generalizing from a subset is risky because subsets are not necessarily representative of the whole. Consider the golden-mantled tree kangaroo, an animal first discovered in 1993 on a single forested mountaintop in New Guinea. For more than a decade, the species was never seen outside of that habitat, which is getting smaller every year because of human activities. Thus, the golden-mantled tree kangaroo was considered to be one of the most endangered animals on the planet. Then, in 2005, the New Guinea explorers discovered that this kangaroo species is fairly common in the Foja Mountain cloud forest (Figure 1.13B). As a result, biologists now believe its future is secure, at least for the moment.

Sampling error is a difference between results obtained from a subset, and results from the whole (Figure 1.14A). Sampling error may be unavoidable, but knowing how it can occur helps researchers design their experiments to minimize it. For example, sampling error can be a substantial problem with a small subset, so experimenters try to start with a relatively large sample, and they repeat their experiments (Figure 1.14B). To understand why these practices reduce the risk of sampling error, think about flipping a coin. There are two possible outcomes of each flip: The coin lands heads up, or it lands tails up. Thus, the chance that the coin will land heads up is one in two (1/2), or 50 percent. However, when you flip a coin repeatedly, it often lands heads up, or tails up, several times in a row. With just 3 flips, the proportion of times that the coin actually lands heads up may not even be close to 50 percent. With 1,000 flips, however, the overall proportion of times the coin lands heads up is much more likely to approach 50 percent.

Probability is the measure, expressed as a percentage, of the chance that a particular outcome will occur. That chance depends on the total number of possible outcomes. For instance, if 10 million people enter a drawing, each has the same probability of winning: 1 in 10 million, or (an extremely improbable) 0.00001 percent. Analysis of experimental data often includes probability calculations. If there is a very low probability that a result has occurred by chance alone, the result is said to be **statistically significant**. In this context, the word "significant" does not refer to the result's importance. Rather, it means that a rigorous statistical analysis has shown a very low probability (usually 5 percent or less) of the result being incorrect because of sampling error.

A. Natalie chooses a random jelly bean from a jar. She is blindfolded, so she does not know that the jar contains 120 green and 280 black jelly beans.

The jar is hidden from Natalie's view before she removes her blindfold. She sees one green jelly bean in her hand and assumes that the jar must hold only green jelly beans. This assumption is incorrect: 30 percent of the jelly beans in the jar are green, and 70 percent are black. The small sample size has resulted in sampling error.

B. Still blindfolded, Natalie randomly chooses 50 jelly beans from the jar. She ends up choosing 10 green and 40 black beans.

The larger sample leads Natalie to assume that one-fifth of the jar's jelly beans are green (20 percent) and four-fifths are black (80 percent). The larger sample more closely approximates the jar's actual green-to-black ratio of 30 percent to 70 percent.

The more times Natalie repeats the sampling, the greater her chance of guessing the actual ratio.

Figure 1.14 How sample size affects sampling error.
© Gary Head.

probability The chance that a particular outcome of an event will occur; depends on the total number of outcomes possible.

sampling error Difference between results derived from testing an entire group of events or individuals, and results derived from testing a subset of the group.

statistically significant Refers to a result that is statistically unlikely to have occurred by chance alone.

Figure 1.15 Example of error bars in a graph.
This graph was adapted from the peacock butterfly research described in Section 1.5.

The researchers recorded the number of times each butterfly flicked its wings in response to an attack by a bird.

The squares represent average frequency of wing flicking for each sample set of butterflies. The error bars that extend above and below the dots indicate the range of values—the sampling error.

Figure It Out: What was the fastest rate at which a butterfly with no spots or sound flicked its wings?   Answer: 22 times per minute

Variation in data is often shown as error bars on a graph (Figure 1.15). Depending on the graph, error bars may indicate variation around an average for one sample set, or the difference between two sample sets.

Scientific Theories Suppose a hypothesis stands even after years of tests. It is consistent with all data ever gathered, and it has helped us make successful predictions about other phenomena. When a hypothesis meets these criteria, it is considered to be a **scientific theory** (Table 1.2). To give an example, all observations to date have been consistent with the hypothesis that matter consists of atoms. Scientists no longer spend time testing this hypothesis for the compelling reason that, since we started looking 200 years ago, no one has discovered matter that consists of anything else. Thus, scientists use the hypothesis, now called atomic theory, to make other hypotheses about matter.

Scientific theories are our best objective descriptions of the natural world. However, they can never be proven absolutely, because to do so would necessitate testing under every possible circumstance. For example, in order to prove atomic theory, the atomic composition of all matter in the universe would have to be checked—an impossible task even if someone wanted to try.

Like all hypotheses, a scientific theory can be disproven by a single observation or result that is inconsistent with it. For example, if someone discovers a form of matter that does not consist of atoms, atomic theory would have to be revised. The potentially falsifiable nature of scientific theories means that science has a built-in system of checks and balances. A theory is revised until no one can prove it to be incorrect. The theory of evolution, which states that change occurs in a line of descent over time, still holds after a century of observations and testing. As with all other scientific theories, no one can be absolutely sure that it will hold under all possible conditions, but it has a very high probability of not being wrong. Few other theories have withstood as much scrutiny.

You may hear people apply the word "theory" to a speculative idea, as in the phrase "It's just a theory." This everyday usage of the word differs from the way it is used in science. Speculation is an opinion, belief, or personal conviction that is not necessarily supported by evidence. A scientific theory is different. By definition, a scientific theory is supported by a large body of evidence, and it is consistent with all known data.

A scientific theory also differs from a **law of nature**, which describes a phenomenon that has been observed to occur in every circumstance without fail, but for which we do not have a complete scientific explanation. The laws of

Science helps us communicate our experiences without bias.

thermodynamics, which describe energy, are examples. We understand *how* energy behaves, but not exactly *why* it behaves the way it does.

The Scope of Science Science helps us be objective about our observations in part because of its limitations. For example, science does not address many questions, such as "Why do I exist?" Answers to such questions can only come from within as an integration of the personal experiences and mental connections that shape our consciousness. This is not to say subjective answers have no value, because no human society can function for long unless its individuals share standards for making judgments, even if they are subjective. Moral, aesthetic, and philosophical standards vary from one society to the next, but all help people decide what is important and good. All give meaning to our lives.

Neither does science address the supernatural, or anything that is "beyond nature." Science neither assumes nor denies that supernatural phenomena occur, but scientists often cause controversy when they discover a natural explanation for something that was thought to have none. Such controversy arises when a society's moral standards are interwoven with its understanding of nature. Nicolaus Copernicus proposed in 1540 that Earth orbits the sun. Today that idea is generally accepted, but the prevailing belief system had Earth as the immovable center of the universe. In 1610, Galileo Galilei published evidence for the Copernican model of the solar system, an act that resulted in his imprisonment. He was publicly forced to recant his work, spent the rest of his life under house arrest, and was never allowed to publish again.

As Galileo's story illustrates, exploring a traditional view of the natural world from a scientific perspective is often misinterpreted as a violation of morality. As a group, scientists are no less moral than anyone else, but they follow a particular set of rules that do not necessarily apply to others: Their work concerns only the natural world, and their ideas must be testable by other scientists.

Science helps us communicate our experiences of the natural world without bias. As such, it may be as close as we can get to a universal language. We are fairly sure, for example, that the laws of gravity apply everywhere in the universe. Intelligent beings on a distant planet would likely understand the concept of gravity. Thus, we might well use gravity or another scientific concept to communicate with them, or anyone, anywhere. The point of science, however, is not to communicate with aliens. It is to find common ground here on Earth.

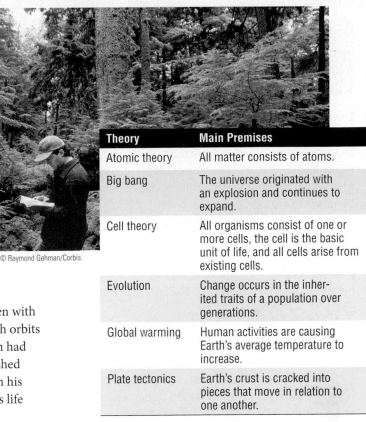

© Raymond Gehman/Corbis.

Theory	Main Premises
Atomic theory	All matter consists of atoms.
Big bang	The universe originated with an explosion and continues to expand.
Cell theory	All organisms consist of one or more cells, the cell is the basic unit of life, and all cells arise from existing cells.
Evolution	Change occurs in the inherited traits of a population over generations.
Global warming	Human activities are causing Earth's average temperature to increase.
Plate tectonics	Earth's crust is cracked into pieces that move in relation to one another.

Table 1.2 Examples of Scientific Theories

Take-Home Message 1.6

Why does science work?

- Researchers minimize sampling error by using large sample sizes and by repeating their experiments. Probability calculations can show whether a result is unlikely to have occurred by chance alone.
- Science is concerned only with testable ideas about observable aspects of nature.
- Ideally, science is a self-correcting process because it is carried out by a community of people who systematically check one another's work and conclusions.
- Because a scientific theory is thoroughly tested and revised until no one can prove it wrong, it is our best way of objectively describing the natural world.

law of nature Generalization that describes a consistent natural phenomenon for which there is incomplete scientific explanation.

scientific theory Hypothesis that has not been disproven after many years of rigorous testing.

Summary

Section 1.1 **Biology** is the scientific study of life. We know about only a fraction of the organisms that live on Earth, in part because we have explored only a fraction of its inhabited regions.

Section 1.2 Biologists think about life at different levels of organization, with new properties emerging at successively higher levels. All matter consists of **atoms**, which bond together to form **molecules**. **Organisms** are individuals that consist of one or more **cells**, the organizational level at which life emerges. A **population** is a group of interbreeding individuals of a species in a given area; a **community** is all populations of all species in a given area. An **ecosystem** is a community interacting with its environment. The **biosphere** includes all regions of Earth that hold life.

Section 1.3 Life has underlying unity in that all living things have similar characteristics: (1) All organisms require energy and **nutrients** to sustain themselves. **Producers** harvest energy from the environment to make their own food by processes such as **photosynthesis**; **consumers** ingest other organisms, or their wastes or remains. (2) Organisms keep the conditions in their internal environment within ranges that their cells tolerate—a process called **homeostasis**. (3) **DNA** contains information that guides an organism's **growth**, **development**, and **reproduction**. The passage of DNA from parents to offspring is **inheritance**.

Section 1.4 The many types of organisms that currently exist on Earth differ greatly in details of body form and function. **Biodiversity** is the sum of differences among living things. **Bacteria** and **archaea** are **prokaryotes**, single-celled organisms whose DNA is not contained within a nucleus. The DNA of single-celled or multicelled **eukaryotes** (**protists**, **plants**, **fungi**, and **animals**) is contained within a nucleus.

Each **species** has a two-part name. The first part is the **genus** name. When combined with the specific epithet, it designates the particular species. With **taxonomy**, species are ranked into ever more inclusive **taxa** on the basis of shared traits.

Section 1.5 **Critical thinking**, the self-directed act of judging the quality of information as one learns, is an important part of **science**. Generally, a researcher observes something in nature, forms a **hypothesis** (testable explanation) for it, then makes a **prediction** about what might occur if the hypothesis is correct. Predictions are tested with observations, **experiments**, or both.

Experiments typically are performed on an **experimental group** as compared with a **control group**, and sometimes on **model** systems. Conclusions are drawn from **data**. A hypothesis that is not consistent with data is modified or discarded. The **scientific method** consists of making, testing, and evaluating hypotheses, and sharing results with the scientific community.

Biological systems are usually influenced by many interacting **variables**. Research approaches differ, but experiments are designed in a consistent way, in order to study a single cause-and-effect relationship in a complex natural system.

Section 1.6 Small sample size increases the potential for **sampling error** in experimental results. In such cases, a subset may be tested that is not representative of the whole. Researchers design experiments carefully to minimize sampling error and bias, and they use **probability** calculations to check the **statistical significance** of their results.

Science helps us be objective about our observations because it is concerned only with testable ideas about observable aspects of nature. Opinion and belief have value in human culture, but they are not addressed by science. A **scientific theory** is a long-standing hypothesis that is useful for making predictions about other phenomena. It is our best way of objectively describing nature. A **law of nature** is a phenomenon that occurs without fail, but has an incomplete scientific explanation.

Self-Quiz

Answers in Appendix I

1. _____ are fundamental building blocks of all matter.
   a. Cells
   b. Atoms
   c. Organisms
   d. Molecules

2. The smallest unit of life is the _____ .
   a. atom
   b. molecule
   c. cell
   d. organism

3. _____ is the transmission of DNA to offspring.
   a. Reproduction
   b. Development
   c. Homeostasis
   d. Inheritance

4. A process by which an organism produces offspring is called _____ .
   a. reproduction
   b. development
   c. homeostasis
   d. inheritance

5. Organisms require _____ and _____ to maintain themselves, grow, and reproduce.
   a. DNA; energy
   b. food; sunlight
   c. nutrients; energy
   d. DNA; cells

6. _____ move around for at least part of their life.

7. By sensing and responding to change, an organism keeps conditions in its internal environment within ranges that its cells can tolerate. This process is called _____ .
   a. sampling error
   b. development
   c. homeostasis
   d. critical thinking

8. DNA _____ .
   a. guides form and function
   b. is the basis of traits
   c. is transmitted from parents to offspring
   d. all of the above

9. A butterfly is a(n) _____ (choose all that apply).
   a. organism
   b. domain
   c. species
   d. eukaryote
   e. consumer
   f. producer
   g. prokaryote
   h. trait

10. A bacterium is _____ (choose all that apply).
    a. an organism
    b. single-celled
    c. an animal
    d. a eukaryote

11. Bacteria, Archaea, and Eukarya are three _____ .

12. A control group is _____ .
    a. a set of individuals that have a characteristic under study or receive an experimental treatment
    b. the standard against which an experimental group is compared
    c. the experiment that gives conclusive results

13. Science addresses only that which is _____ .
    a. alive
    b. observable
    c. variable
    d. indisputable

14. Match the terms with the most suitable description.
    _____ life
    _____ probability
    _____ species
    _____ scientific theory
    _____ hypothesis
    _____ prediction
    _____ producer
    a. if–then statement
    b. unique type of organism
    c. emerges with cells
    d. testable explanation
    e. measure of chance
    f. makes its own food
    g. time-tested hypothesis

15. In one survey, fifteen randomly selected students were found to be taller than 6 feet. This data led to the conclusion that the average height of a student is greater than 6 feet. This is an example of _____ .
    a. experimental error
    b. sampling error
    c. a subjective opinion
    d. experimental bias

Critical Thinking

1. A person is declared to be dead upon the irreversible ceasing of spontaneous body functions: brain activity, or blood circulation and respiration. However, only about 1% of a person's cells have to die in order for all of these things to happen. How can someone be dead when 99% of his or her cells are still alive?

2. Explain the difference between a one-celled organism and a single cell of a multicelled organism.

3. Why would you think twice about ordering from a restaurant menu that lists only the second part of the species name (not the genus) of its offerings? *Hint:* Look up *Ursus americanus*, *Ceanothus americanus*, *Bufo americanus*, *Homarus americanus*, *Lepus americanus*, and *Nicrophorus americanus*.

4. Once there was a highly intelligent turkey that had nothing to do but reflect on the world's regularities. Morning always started out with the sky turning light, followed by the master's footsteps, which were always followed by the appearance of food. Other things varied, but food always followed footsteps. The sequence of events was so predictable that it eventually became the basis of the turkey's theory about the goodness of the world. One morning, after more than 100 confirmations of this theory, the turkey listened for the master's footsteps, heard them, and had its head chopped off.

   Any scientific theory is modified or discarded upon discovery of contradictory evidence. The absence of absolute certainty has led some people to conclude that "theories are irrelevant because they can change." If that is so, should we stop doing scientific research? Why or why not?

5. In 2005, researcher Woo-suk Hwang reported that he had made immortal stem cells from human patients. His research was hailed as a breakthrough for people affected by degenerative diseases, because stem cells may be used to repair a person's own damaged tissues. Hwang published his results in a peer-reviewed journal. In 2006, the journal retracted his paper after other scientists discovered that Hwang's group had faked their data. Does the incident show that results of scientific studies cannot be trusted? Or does it confirm the usefulness of a scientific approach, because other scientists discovered and exposed the fraud?

2 MOLECULES OF LIFE

Application ❯ 2.1 Fear of Frying

The human body requires only about a tablespoon of fat each day to stay healthy, but most people in developed countries eat far more than that. The average American eats about 70 pounds of fat per year, which may be part of the reason why the average American is overweight. Being overweight increases one's risk for many chronic illnesses. However, the total quantity of fat in the diet may have less impact on health than the types of fats. Fats are more than inert molecules that accumulate in strategic areas of our bodies. They are the main constituents of cell membranes, and as such they have powerful effects on cell function.

The typical fat molecule has three fatty acid tails, each a long chain of carbon atoms that can vary a bit in structure. Fats with a certain arrangement of hydrogen atoms around those carbon chains are called *trans* fats. Small amounts of *trans* fats occur naturally in red meat and dairy products, but the main source of these fats in the American diet is an artificial food product called partially hydrogenated vegetable oil. Hydrogenation is a manufacturing process that adds hydrogen atoms to oils in order to change them into solid fats. In 1908, Procter & Gamble Co. developed partially hydrogenated soybean oil as a substitute for the more expensive solid animal fats they had been using to make candles. By 1911, more households in the United States became wired for electricity, so the demand for candles was waning. P & G needed another way to sell its proprietary fat. Partially hydrogenated vegetable oil looks a lot like lard, so the company began aggressively marketing it as a revolutionary new food: a solid cooking fat with a long shelf life, mild flavor, and lower cost than lard or butter.

Figure 2.1 Unhealthy *trans* fats are abundant in partially hydrogenated oils commonly used to make manufactured and fast foods.

© Kentoh/Shutterstock.com.

By the mid-1950s, hydrogenated vegetable oil had become a major part of the American diet. For decades, it was considered to be healthier than animal fats because it was made from plants, but we now know otherwise. *Trans* fats, which are abundant in hydrogenated vegetable oils, raise the level of cholesterol in our blood more than any other fat, and they directly alter the function of our arteries and veins. The effects of such changes are quite serious. Eating as little as 2 grams per day (about 0.4 teaspoon) of hydrogenated vegetable oil measurably increases one's risk of atherosclerosis (hardening of the arteries), heart attack, and diabetes. A small serving of french fries made with hydrogenated vegetable oil contains about 5 grams of *trans* fat (Figure 2.1). At this writing, hydrogenated oil is still a component of many manufactured and fast foods: french fries, stick margarines, ready-to-use frostings, cookies, crackers, cakes and pancakes, peanut butter, pies, doughnuts, muffins, chips, microwave popcorn, pizzas, burritos, chicken nuggets, fish sticks, and so on.

All organisms consist of the same kinds of molecules, but small differences in the way those molecules are put together can have big effects. With this concept, we introduce you to the chemistry of life. This is your chemistry. It makes you far more than the sum of your body's molecules.

2.2 **Start With Atoms**

REMEMBER: Atoms are fundamental units of matter—the building blocks of all substances (Section 1.2).

Even though atoms are about 20 million times smaller than a grain of sand, they consist of even smaller subatomic particles. Positively charged **protons** (p⁺) and uncharged **neutrons** occur in an atom's core, or **nucleus**. Negatively charged **electrons** (e⁻) move around the nucleus (Figure 2.2). **Charge** is an electrical property: Opposite charges attract, and like charges repel. A typical atom has about the same number of electrons and protons. The negative charge of an electron is the same magnitude as the positive charge of a proton, so the two charges cancel one another. Thus, an atom with the same number of electrons and protons carries no charge.

All atoms have protons. The number of protons in the nucleus is called the **atomic number**, and it determines the type of atom, or element. **Elements** are pure substances, each consisting only of atoms with the same number of protons in their nucleus. For example, the element carbon has an atomic number of 6 (Figure 2.3). All atoms with six protons in their nucleus are carbon atoms, no matter how many electrons or neutrons they have. Elemental carbon (the substance) consists only of carbon atoms, and all of those atoms have six protons. Each of the 118 known elements has a symbol that is typically an abbreviation of its Latin or Greek name (see Appendix II). Carbon's symbol, C, is from *carbo*, the Latin word for coal. Coal is mostly carbon.

All atoms of an element have the same number of protons, but they can differ in the number of other subatomic particles. Those that differ in the number of neutrons are called **isotopes**. The total number of neutrons and protons in the nucleus of an isotope is its **mass number**. Mass number is written as a superscript to the left of the element's symbol. For example, the most common isotope of hydrogen has one proton and no neutrons, so it is designated ^1H. Other hydrogen isotopes include deuterium (^2H, one proton and one neutron) and tritium (^3H, one proton and two neutrons).

The most common carbon isotope has six protons and six neutrons (^{12}C). Another naturally occurring carbon isotope, ^{14}C, has six protons and eight neutrons (6 + 8 = 14). Carbon 14 is an example of a **radioisotope**, or radioactive isotope. Atoms of a radioisotope have an unstable nucleus that breaks up spontaneously. As a nucleus breaks up, it emits radiation (subatomic particles, energy, or both), a process called **radioactive decay**. The atomic nucleus cannot be altered by ordinary means, so radioactive decay is unaffected by external factors such as temperature, pressure, or whether the atoms are part of molecules.

Each radioisotope decays at a predictable rate into predictable products. For example, when carbon 14 decays, one of its neutrons splits into a proton and an electron. The nucleus emits the electron as radiation. Thus, a carbon atom with eight neutrons and six protons (^{14}C) becomes a nitrogen atom, with seven neutrons and seven protons (^{14}N):

nucleus of ^{14}C, with 6 protons, 8 neutrons

nucleus of ^{14}N, with 7 protons, 7 neutrons

This process is so predictable that we can say with certainty that about half of the atoms in any sample of ^{14}C will be ^{14}N atoms after 5,730 years. The predictability of

proton
neutron
electron

an atom

Figure 2.2 Atoms consist of subatomic particles. Models such as this do not show what atoms really look like. Electrons move in defined, three-dimensional spaces about 10,000 times bigger than the nucleus. Protons and neutrons occur in the nucleus.

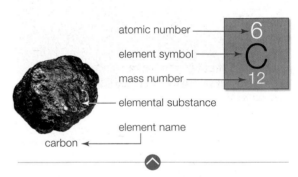

atomic number
element symbol
mass number
elemental substance
element name
carbon
6
C
12

Figure 2.3 Example of an element: carbon.

Left, Theodore Gray/Visuals Unlimited, Inc.

atomic number Number of protons in the atomic nucleus; determines the element.

charge Electrical property; opposite charges attract, and like charges repel.

electron Negatively charged subatomic particle.

element A pure substance that consists only of atoms with the same number of protons.

isotopes Forms of an element that differ in the number of neutrons their atoms carry.

mass number Of an isotope, the total number of protons and neutrons in the atomic nucleus.

neutron Uncharged subatomic particle in the atomic nucleus.

nucleus Core of an atom; occupied by protons and neutrons.

proton Positively charged subatomic particle that occurs in the nucleus of all atoms.

radioactive decay Process by which atoms of a radioisotope emit energy and subatomic particles when their nucleus spontaneously breaks up.

radioisotope Isotope with an unstable nucleus.

Figure 2.4 PET scans.
PET scans use radioactive tracers to form a digital image of a process in the body's interior. These two PET scans reveal the activity of a molecule called MAO-B in the body of a nonsmoker (left) and a smoker (right). The activity is color-coded from red (highest activity) to purple (lowest). Low MAO-B activity is associated with violence, impulsiveness, and other behavioral problems.

Brookhaven National Laboratory.

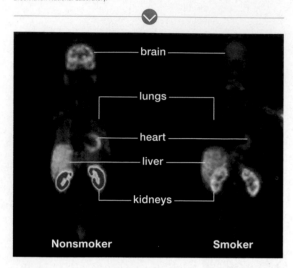

radioactive decay makes it possible for scientists to estimate the age of a rock or fossil by measuring its isotope content (we return to this topic in Section 11.4).

Radioisotopes are often used in **tracers**, which are substances with a detectable component. For example, a molecule in which an atom (such as ^{12}C) has been replaced with a radioisotope (such as ^{14}C) can be used as a radioactive tracer. When delivered into a biological system, a radioactive tracer may be followed as it moves through the system with instruments that detect radiation (Figure 2.4).

Why Electrons Matter The more we learn about electrons, the weirder they seem. Consider that an electron has mass but no size, and its position in space is described as more of a smudge than a point. It carries energy, but only in incremental amounts (this concept will be important to remember when you learn how cells harvest and release energy). An electron gains energy only by absorbing the precise amount needed to boost it to the next energy level. Likewise, it loses energy only by emitting the exact difference between two energy levels.

Imagine that an atom is a multilevel apartment building with a nucleus in the basement. Each "floor" of the building corresponds to a certain energy level, and each has a certain number of "rooms" available for rent. Two electrons can occupy each room. Pairs of electrons populate rooms from the ground floor up. The farther an electron is from the nucleus in the basement, the greater its energy. An electron can move to a room on a higher floor if an energy input gives it a boost, but it immediately emits the extra energy and moves back down.

A. The first shell corresponds to the first energy level, and it can hold up to 2 electrons. Hydrogen has one proton, so it has 1 electron and one vacancy. A helium atom has 2 protons, 2 electrons, and no vacancies.

first shell · hydrogen (H) ←one proton / ←one electron · helium (He)

B. The second shell corresponds to the second energy level, and it can hold up to 8 electrons. Carbon has 6 electrons, so its first shell is full. Its second shell has 4 electrons and four vacancies. Oxygen has 8 electrons and two vacancies. Neon has 10 electrons and no vacancies.

second shell · carbon (C) 6 · oxygen (O) 8 · neon (Ne) 10

C. The third shell corresponds to the third energy level, and it can hold up to 8 electrons. A sodium atom has 11 electrons, so its first two shells are full; the third shell has one electron. Thus, sodium has seven vacancies. Chlorine has 17 electrons and one vacancy. Argon has 18 electrons and no vacancies.

third shell · sodium (Na) 11 · chlorine (Cl) 17 · argon (Ar) 18

Figure 2.5 Shell models.
Each circle (shell) represents one energy level. To make these models, we fill the shells with electrons from the innermost shell out, until there are as many electrons as the atom has protons. The number of protons in each model is indicated.

Figure It Out: Which of these models have unpaired electrons in their outer shell?

Answer: Hydrogen, carbon, oxygen, sodium, and chlorine

A **shell model** helps us visualize how electrons populate atoms (Figure 2.5). In this model, nested "shells" correspond to successively higher energy levels. Thus, each shell includes all of the rooms on one floor (energy level) of our atomic apartment building. We draw a shell model of an atom by filling it with electrons from the innermost shell out, until there are as many electrons as the atom has protons. There is only one room on the first floor, and it fills up first. In hydrogen, the simplest atom, a single electron occupies that room (Figure 2.5A). Helium, with two protons, has two electrons that fill the room—and the first shell. In larger atoms, more electrons rent the second-floor rooms (Figure 2.5B). When the second floor fills, more electrons rent third-floor rooms (Figure 2.5C), and so on.

When an atom's outermost shell is filled with electrons, we say that it has no vacancies, and it is in its most stable state. Helium, neon, and argon are examples of elements with no vacancies. Atoms of these elements are chemically stable, which means they have very little tendency to interact with other atoms. Thus, these elements occur most frequently in nature as solitary atoms. By contrast, when an atom's outermost shell has room for another electron, it has a vacancy. Atoms with vacancies tend to get rid of them by interacting with other atoms; in other words, they are chemically active. For example, the sodium atom (Na) in Figure 2.5C has one electron in its outer (third) shell, which can hold eight. With seven vacancies, we can predict that this atom is chemically active. In fact, this particular sodium atom is not just active, it is extremely so. Why? The shell model shows that a sodium atom has an unpaired electron, but in the real world, electrons really like to be in pairs when they populate atoms. Atoms that have unpaired electrons are called **free radicals**. With a few exceptions, free radicals are very unstable, easily forcing electrons upon other atoms or ripping electrons away from them. This property makes free radicals dangerous to life. A sodium atom with 11 electrons (a sodium radical) quickly evicts the one unpaired electron, so that its second shell—which is full of electrons—becomes its outermost, and no vacancies remain. This is the atom's most stable state. The vast majority of sodium atoms on Earth are like this one, with 11 protons and 10 electrons.

© Michael S. Yamashita/ Corbis.

Atoms with an unequal number of protons and electrons are ions. An **ion** carries a net (or overall) charge. Sodium ions (Na^+) offer an example of how atoms gain a positive charge by losing an electron (Figure 2.6A). Other atoms gain a negative charge by accepting an electron (Figure 2.6B).

Figure 2.6 Ion formation.

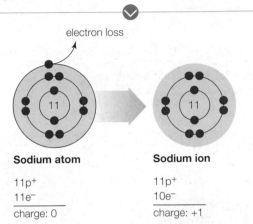

electron loss

Sodium atom	Sodium ion
$11p^+$	$11p^+$
$11e^-$	$10e^-$
charge: 0	charge: +1

A. A sodium atom (Na) becomes a positively charged sodium ion (Na^+) when it loses the single electron in its third shell. The atom's full second shell is now its outermost, so it has no vacancies.

electron gain

Chlorine atom	Chloride ion
$17p^+$	$17p^+$
$17e^-$	$18e^-$
charge: 0	charge: −1

B. A chlorine atom (Cl) becomes a negatively charged chloride ion (Cl^-) when it gains an electron and fills the vacancy in its third, outermost shell.

Figure It Out: Does a chloride ion have an unpaired electron?　　　　**Answer: No**

Take-Home Message 2.2

What are atoms?

- Atoms consist of electrons moving around a nucleus of protons and neutrons. The number of protons determines the element. Isotopes are forms of an element that have different numbers of neutrons.
- Unstable nuclei of radioisotopes emit radiation as they spontaneously break down (decay). Radioisotopes decay at a predictable rate to form predictable products.
- An atom's electrons are the basis of its chemical behavior. When an atom's outermost shell is not full of electrons, it has a vacancy and it is chemically active. Atoms that get rid of vacancies by gaining or losing electrons become ions (charged).

free radical Atom with an unpaired electron.

ion Atom or molecule that carries a net charge.

shell model Model of electron distribution in an atom.

tracer A substance that can be traced via its detectable component.

Figure 2.7 The water molecule.
Each water molecule has two hydrogen atoms bonded to the same oxygen atom.

— one oxygen atom

— two hydrogen atoms

Figure 2.8 Ionic bonds in table salt, or NaCl.

(A) left, Francois Gohier/Science Source; top right, Melica/Shutterstock.

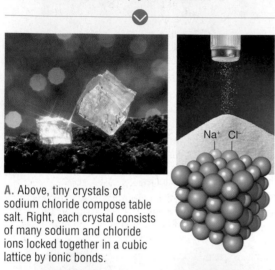

Na⁺ Cl⁻

A. Above, tiny crystals of sodium chloride compose table salt. Right, each crystal consists of many sodium and chloride ions locked together in a cubic lattice by ionic bonds.

ionic bond

11   17

Sodium ion   **Chloride ion**

B. The strong mutual attraction of opposite charges holds a sodium ion and a chloride ion together in an ionic bond.

Na⁺   Cl⁻

positive ←——→ negative
charge   charge

C. Ions taking part in an ionic bond retain their charge, so the molecule itself is polar. One side is positively charged (represented by a blue overlay); the other side is negatively charged (red overlay).

2.3 From Atoms to Molecules

REMEMBER: The same building blocks, arranged different ways, form different products; atoms bond together to form molecules (Section 1.2).

A **chemical bond** is an attractive force that arises between two atoms, and it is one way that atoms rid themselves of vacancies. Chemical bonds make molecules out of atoms. A molecule consists of atoms held together in a particular number and arrangement by chemical bonds. For example, a water molecule consists of three atoms: two hydrogen atoms bonded to the same oxygen atom (Figure 2.7). Because a water molecule has atoms of two or more elements, it is called a **compound**. Other molecules, including molecular oxygen (a gas in air), have atoms of one element only.

The term "bond" applies to a continuous range of atomic interactions. However, we can categorize most bonds into distinct types based on their different properties. Which type forms depends on the atoms taking part in the molecule.

Ionic Bonds Two ions may stay together by the mutual attraction of their opposite charges, an association called an **ionic bond**. Ionic bonds can be quite strong. Ionically bonded sodium and chloride ions make sodium chloride (NaCl), which we know as table salt; a crystal of this substance consists of a lattice of sodium and chloride ions interacting in ionic bonds (Figure 2.8A).

Ions retain their respective charges when participating in an ionic bond (Figure 2.8B). Thus, one "end" of an ionically bonded molecule has a positive charge, and the other "end" has a negative charge. Any such separation of charge into distinct positive and negative regions is called **polarity** (Figure 2.8C).

Covalent Bonds In a **covalent bond**, two atoms share a pair of electrons, so each atom's vacancy becomes partially filled (Figure 2.9). Sharing electrons links the two atoms, just as sharing a pair of earphones links two friends (left). Covalent bonds can be stronger than ionic bonds, but they are not always so.

Table 2.1 shows some of the different ways we represent molecules that are held together with covalent bonds. In structural formulas, a line between two atoms represents a single covalent bond, in which two atoms share one pair of electrons. For example, molecular hydrogen (H_2) has one covalent bond between hydrogen atoms (H—H).

Two, three, or even four covalent bonds may form between atoms when they share multiple pairs of electrons. For example, two atoms sharing two pairs of electrons are connected by two covalent bonds. Such double bonds are represented by a double line between the atoms. A double bond links the two oxygen atoms in molecular oxygen (O=O). Three lines indicate a triple bond, in which two atoms share three pairs of electrons. A triple covalent bond links the two nitrogen atoms in molecular nitrogen (N≡N).

Double and triple bonds are not distinguished from single bonds in structural models, which show positions and relative sizes of the atoms in three dimensions. The bonds are shown as one stick connecting two balls, which represent atoms. Elements are usually coded by color:

carbon   hydrogen   oxygen   nitrogen   phosphorus

Table 2.1 Representing Covalent Bonds in Molecules

Common name	Water	Familiar term
Chemical name	Dihydrogen monoxide	Describes elemental composition.
Chemical formula	H_2O	Indicates unvarying proportions of elements. Subscripts show number of atoms of an element per molecule. The absence of a subscript means one atom.
Structural formula	H—O—H	Represents each covalent bond as a single line between atoms.
Structural model		Shows relative sizes and positions of atoms in three dimensions.
Shell model		Shows how pairs of electrons are shared in covalent bonds.

Atoms share electrons unequally in a polar covalent bond. A bond between an oxygen atom and a hydrogen atom in a water molecule is an example: One atom (the oxygen, in this case) pulls the electrons a little more toward its side of the bond, so that atom bears a slight negative charge. The atom at the other end of the bond (the hydrogen, in this case) bears a slight positive charge. Covalent bonds in most compounds are polar. By contrast, atoms participating in a nonpolar covalent bond share electrons equally. There is no difference in charge between the two ends of such bonds. Molecular hydrogen (H_2), oxygen (O_2), and nitrogen (N_2) are examples.

Take-Home Message 2.3

How do atoms interact in chemical bonds?

- A chemical bond forms between atoms when their electrons interact. Depending on the atoms taking part in it, the bond may be ionic or covalent.
- An ionic bond is a strong mutual attraction between ions of opposite charge.
- Atoms share a pair of electrons in a covalent bond. When the atoms share electrons unequally, the bond is polar.

2.4 Hydrogen Bonds and Water

Life evolved in water. All living organisms are mostly water, many of them still live in it, and all of the chemical reactions of life are carried out in water-based fluids. What makes water so fundamentally important for life?

Water has unique properties that arise from the two polar covalent bonds in each water molecule. Overall, the molecule has no charge, but the oxygen atom

Molecular hydrogen (H—H)
Two hydrogen atoms, each with one proton, share two electrons in a nonpolar covalent bond.

Molecular oxygen (O═O)
Two oxygen atoms, each with eight protons, share four electrons in a double covalent bond.

Water (H—O—H)
Two hydrogen atoms share electrons with an oxygen atom in two covalent bonds. The bonds are polar because the oxygen exerts a greater pull on the shared electrons than the hydrogens do.

Figure 2.9 Covalent bonds, in which atoms fill vacancies by sharing electrons.
Two electrons are shared in each covalent bond. When sharing is equal, the bond is nonpolar. When one atom exerts a greater pull on the electrons, the bond is polar.

chemical bond An attractive force that arises between two atoms when their electrons interact. Links atoms into molecules.

compound Molecule that has atoms of more than one element.

covalent bond Type of chemical bond in which two atoms share a pair of electrons.

ionic bond Type of chemical bond in which a strong mutual attraction links ions of opposite charge.

polarity Any separation of charge into distinct positive and negative regions.

Figure 2.10 Hydrogen bonds and water.

slight negative charge

slight positive charge

A. Polarity of the water molecule. Each of the hydrogen atoms in a water molecule bears a slight positive charge (represented by a blue overlay). The oxygen atom carries a slight negative charge (red overlay).

a hydrogen bond

B. A hydrogen bond is an attraction between a hydrogen atom and another atom taking part in a separate polar covalent bond.

C. The many hydrogen bonds that form among water molecules impart special properties to liquid water.

carries a slight negative charge, and the two hydrogen atoms carry a slight positive charge. Thus, the molecule itself is polar (Figure 2.10A).

The polarity of individual water molecules attracts them to one another. The slight positive charge of a hydrogen atom in one water molecule is drawn to the slight negative charge of an oxygen atom in another. This type of interaction is called a hydrogen bond. A **hydrogen bond** is an attraction between a covalently bonded hydrogen atom and another atom taking part in a separate polar covalent bond (Figure 2.10B). Like ionic bonds, hydrogen bonds form by the mutual attraction of opposite charges. However, unlike ionic bonds, hydrogen bonds do not make molecules out of atoms, so they are not chemical bonds.

Hydrogen bonds lie on the weaker end of the spectrum of atomic interactions; they form and break much more easily than covalent or ionic bonds. Even so, many of them form, and collectively they can be quite strong. As you will see, hydrogen bonds stabilize the characteristic structures of biological molecules such as DNA and proteins. They also form in tremendous numbers among water molecules (Figure 2.10C). Extensive hydrogen bonding among water molecules gives liquid water the special properties that make life possible.

Water Is an Excellent Solvent　The polarity of the water molecule and its ability to form hydrogen bonds make water an excellent **solvent**, which means that many other substances can dissolve in it. Substances that dissolve easily in water are **hydrophilic** (water-loving). Ionic solids such as sodium chloride (NaCl) dissolve in water because the slight positive charge on each hydrogen atom in a water molecule attracts negatively charged ions (Cl^-), and the slight negative charge on the oxygen atom attracts positively charged ions (Na^+). Hydrogen bonds among many water molecules are collectively stronger than an ionic bond between two ions, so the solid dissolves as water molecules tug the ions apart and surround each one (right).

When a substance such as NaCl dissolves, its component ions disperse uniformly among the molecules of liquid, and it becomes a **solute**. Sodium chloride is called a **salt** because it releases ions other than H^+ and OH^- when it dissolves in water (more about this in the next section). A uniform mixture such as salt dissolved in water is called a **solution**. Chemical bonds do not form between molecules of solute and solvent, so the proportions of the two substances in a solution can vary. The amount of a solute that is dissolved in a given volume of fluid is its **concentration**.

Many nonionic solids also dissolve easily in water. Sugars are examples. Molecules of these substances have one or more polar covalent bonds, and atoms participating in a polar covalent bond can form hydrogen bonds with water molecules. Hydrogen bonding with water pulls individual molecules of the solid away from one another and keeps them apart. Unlike ionic solids, these substances retain their molecular integrity when they dissolve, which means they do not dissociate into atoms.

Water does not interact with **hydrophobic** (water-dreading) substances such as oils. Oils consist of nonpolar molecules, and hydrogen bonds do not form between nonpolar molecules and water. When you mix oil and water, the water breaks into small droplets, but quickly begins to cluster into larger drops as new hydrogen bonds form among its molecules. The bonding excludes molecules of oil and pushes them together into drops that rise to the surface of the water. The very same interactions occur at the thin, oily membrane that separates the watery fluid inside cells from the watery fluid outside of them. Such interactions give rise to the structure of cell membranes.

© Herbert Schnekenburger.

Water Has Cohesion Molecules of some substances resist separating from one another, and the resistance gives rise to a property called **cohesion**. Water has cohesion because hydrogen bonds collectively exert a continuous pull on its individual molecules. You can see cohesion in water as surface tension, which means that the surface of liquid water behaves a bit like a sheet of elastic (left).

Cohesion is a part of many processes that sustain multicelled bodies. Consider how sweating helps keep your body cool during hot, dry weather. Sweat, which is about 99 percent water, cools the skin as it evaporates. Why? **Evaporation** is the process in which molecules escape from the surface of a liquid and become vapor. Evaporation of water is resisted by hydrogen bonding among individual water molecules. In other words, overcoming water's cohesion takes energy. Thus, evaporation sucks energy (in the form of heat) from liquid water, and this lowers the water's surface temperature.

Another example of cohesion's importance to life involves plants. Water molecules evaporate from leaves, and replacements are pulled upward from roots. Cohesion makes it possible for columns of liquid water to rise from roots to leaves inside narrow pipelines of vascular tissue. In some trees, these pipelines extend hundreds of feet above the soil (Section 27.4 returns to this topic).

Water Stabilizes Temperature All atoms jiggle nonstop, so the molecules they make up jiggle too. We measure the energy of this motion as degrees of **temperature**. Adding energy (in the form of heat, for example) makes the jiggling faster, so the temperature rises. Hydrogen bonding keeps water molecules from moving as much as they would otherwise, so it takes more heat to raise the temperature of water compared with other liquids. Temperature stability is an important part of homeostasis because most of the molecules of life function properly only within a certain range of temperature.

Below 0°C (32°F), water molecules do not jiggle enough to break hydrogen bonds between them, and they become locked in the rigid, lattice-like bonding pattern of ice (Figure 2.11). Individual water molecules pack less densely in ice than they do in water, which is why ice floats on water. Sheets of ice that form on the surface of ponds, lakes, and streams can insulate the water under them from subfreezing air temperatures. Such "ice blankets" protect aquatic organisms during long, cold winters.

Take-Home Message 2.4

What gives water the special properties that make life possible?

- Extensive hydrogen bonding among water molecules arises from the polarity of the individual molecules.
- Hydrogen bonding among water molecules imparts cohesion to liquid water, and gives it the ability to stabilize temperature and dissolve many substances.

Figure 2.11 Ice.
Top, hydrogen bonds lock water molecules in a rigid lattice in ice. The molecules in this lattice pack less densely than in liquid water (compare Figure 2.10C), so ice floats on water. Bottom, a covering of ice can insulate water underneath it, thus keeping aquatic organisms from freezing during harsh winters.

Bottom, www.flickr.com/photos/roseofredrock.

cohesion Property of a substance that arises from the tendency of its molecules to resist separating from one another.

concentration Amount of solute per unit volume of solution.

evaporation Transition of a liquid to a vapor.

hydrogen bond Attraction between a covalently bonded hydrogen atom and another atom taking part in a separate covalent bond.

hydrophilic Describes a substance that dissolves easily in water.

hydrophobic Describes a substance that resists dissolving in water.

salt Ionic compound that releases ions other than H^+ and OH^- when it dissolves in water.

solute A dissolved substance.

solution Uniform mixture of solute completely dissolved in a solvent.

solvent Liquid in which other substances dissolve.

temperature Measure of molecular motion.

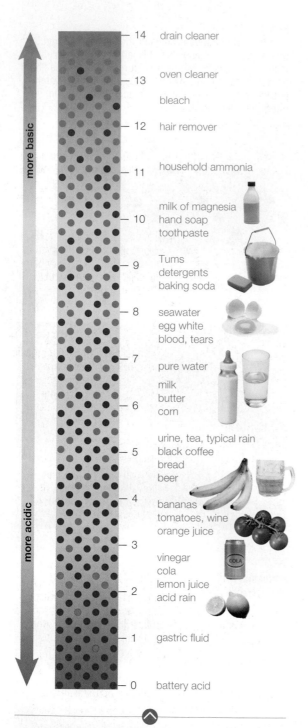

Figure 2.12 A pH scale.
Here, red dots signify hydrogen ions (H⁺) and blue dots signify hydroxyl ions (OH⁻). Also shown are the approximate pH values for some common solutions.

This pH scale ranges from 0 (most acidic) to 14 (most basic). A change of one unit on the scale corresponds to a tenfold change in the amount of H⁺ ions.

Photos, © JupiterImages Corporation.

Figure It Out: What is the approximate pH of cola?

Answer: 2.5

2.5 Acids and Bases

A hydrogen atom, remember, is just a proton and an electron. When a hydrogen atom participates in a polar covalent bond, the electron is pulled away from the proton, just a bit. Hydrogen bonding in water tugs on that proton even more, so much that the proton can be pulled right off of the molecule. The electron stays with the rest of the molecule, which becomes negatively charged (ionic), and the proton becomes a hydrogen ion (H⁺). For example, a water molecule that loses a proton becomes a hydroxyl ion (OH⁻). The loss is more or less temporary, because these two ions easily get back together to form a water molecule (H_2O). With other molecules, the loss of a hydrogen ion in water is essentially permanent.

We use a value called **pH** to measure of the number of hydrogen ions in a water-based fluid. In pure water, the number of H⁺ ions is the same as the number of OH⁻ ions, and the pH is 7, or neutral. The higher the number of hydrogen ions, the lower the pH. A one-unit decrease in pH corresponds to a tenfold increase in the number of H⁺ ions (Figure 2.12). One way to get a sense of the pH scale is to taste dissolved baking soda (pH 9), distilled water (pH 7), and lemon juice (pH 2).

An **acid** is a substance that gives up hydrogen ions in water. Acids can lower the pH of a solution and make it acidic (below pH 7). **Bases** accept hydrogen ions from water, so they can raise the pH of a solution and make it basic (above pH 7). Nearly all of life's chemistry occurs near pH 7. Under normal circumstances, fluids inside cells and bodies stay within a certain range of pH because they are buffered. A **buffer** is a set of chemicals that can keep pH stable by alternately donating and accepting ions that affect pH. For example, two chemicals, carbonic acid and bicarbonate, are part of a homeostatic mechanism that normally keeps your blood pH between 7.3 and 7.5. Carbonic acid forms when carbon dioxide gas dissolves in the fluid portion of blood. It can dissociate into a hydrogen ion and a bicarbonate ion, which in turn recombine to form carbonic acid:

$$H_2CO_3 \longrightarrow H^+ + HCO_3^- \longrightarrow H_2CO_3$$

carbonic acid   bicarbonate   carbonic acid

An excess of OH⁻ ions in the blood causes the carbonic acid in it to release H⁺ ions. These combine with the excess OH⁻ ions to form water, which does not affect pH. Excess H⁺ in blood combines with the bicarbonate, so it does not affect pH either.

Any buffer can neutralize only so many ions. Even slightly more than that limit and the pH of the fluid will change dramatically. Buffer failure can be catastrophic in a biological system because most biological molecules can function properly only within a narrow range of pH. Consider what happens when breathing is impaired suddenly. Carbon dioxide gas accumulates in tissues, and too much carbonic acid forms in blood. If the excess acid reduces blood pH below 7.3, a dangerous level of unconsciousness called coma can be the outcome.

Take-Home Message 2.5

Why are hydrogen ions important in biological systems?

• The number of hydrogen ions in a fluid determines its pH. Most biological systems function properly only within a narrow range of pH. Buffers help keep pH stable.

• Acids release hydrogen ions in water; bases accept them.

2.6 Organic Molecules

The same elements that make up a living body also occur in nonliving things, but their proportions differ. For example, compared to sand or seawater, a human body has a much larger proportion of carbon atoms. Why? Unlike sand or seawater, a body contains a lot of the molecules of life—complex carbohydrates and lipids, proteins, and nucleic acids—and these molecules consist of a high proportion of carbon atoms. Compounds that consist primarily of carbon and hydrogen atoms are said to be **organic**. The term is a holdover from a time when such molecules were thought to be made only by living things, as opposed to the "inorganic" molecules that formed by nonliving processes. We now know that organic compounds were present on Earth long before organisms were.

A. Carbon's versatile bonding behavior allows it to form a variety of structures, including rings.

B. Carbon rings form the backbone of many sugars, starches, and fats, such as those found in foods.

Figure 2.13 Carbon rings.

(B) Getty Images.

Carbon's importance to life arises from its versatile bonding behavior. Carbon has four vacancies in its outer shell, so it can form four covalent bonds with other atoms, including other carbon atoms. Many organic molecules have a chain of carbon atoms, and this backbone often forms rings (Figure 2.13). Small molecular groups that attach to the backbone impart chemical properties to the molecule. For example, carboxyl groups (—COOH) make amino acids and fatty acids acidic; hydroxyl groups (—OH) make sugars polar. Carbon's ability to form chains and rings, and also to bond with many other elements, means that atoms of this element can be assembled into a wide variety of organic compounds.

As you will see shortly, the function of an organic molecule depends on its structure. The structure of even a small organic molecule can be quite complicated (Figure 2.14A), so representations are typically simplified. Hydrogen atoms and some of the bonds may not be shown, but are understood to exist where they should. Carbon rings such as the ones that occur in glucose and other sugars are often depicted as polygons (Figure 2.14B). If no atom is shown at a corner or at the end of a bond, a carbon is implied there. Ball-and-stick models are used to depict an organic molecule's three-dimensional arrangement of atoms (Figure 2.14C); space-filling models are used to show its overall shape (Figure 2.14D). Proteins and nucleic acids are often modeled as ribbon structures, which, as you will see in Section 2.9, show how the molecule folds and twists.

What Cells Do to Organic Compounds All biological systems are based on the same organic molecules, but the details of those molecules differ among organisms. Just as atoms bonded in different numbers and arrangements form different molecules, simple organic building blocks bonded in different numbers and arrangements form different versions of the molecules of life.

Cells assemble complex carbohydrates, lipids, proteins, and nucleic acids from small organic molecules. These small organic molecules—sugars, fatty acids, amino acids, and nucleotides—are called **monomers** when they are used as subunits of larger molecules. A molecule that consists of multiple monomers is a **polymer**. Cells build polymers from monomers, and break down polymers to release

Figure 2.14 Structural models of an organic molecule. All of these models represent the same molecule: glucose.

A. A structural formula for an organic molecule—even a simple one—can be very complicated. The overall structure is obscured by detail.

B. Structural formulas of organic molecules are often simplified by using polygons as symbols for rings, and omitting some bonds and element labels.

C. A ball-and-stick model shows the arrangement of atoms and bonds in three dimensions.

D. A space-filling model can be used to show a molecule's overall shape. Individual atoms are visible in this model. Space-filling models of larger molecules often show only the surface contours.

acid Substance that releases hydrogen ions in water.

base Substance that accepts hydrogen ions in water.

buffer Set of chemicals that can keep the pH of a solution stable by alternately donating and accepting ions that contribute to pH.

monomer Molecule that is a subunit of polymers.

organic Describes a compound that consists mainly of carbon and hydrogen atoms.

pH Measure of the amount of hydrogen ions in a fluid.

polymer Molecule that consists of multiple monomers.

A. Metabolism refers to processes by which cells acquire and use energy as they make and break down molecules. Humans and other consumers break down the molecules in food. Their cells use energy and raw materials from the breakdown to maintain themselves and to build new components.

Exactostock/SuperStock.

B. Cells often build a large molecule from small ones by condensation. In this reaction, an enzyme removes a hydroxyl group from one molecule and a hydrogen atom from another. A covalent bond forms between the two molecules; water forms too.

C. Cells use a water-requiring reaction called hydrolysis to split a large molecule into smaller ones. An enzyme attaches a hydroxyl group and a hydrogen atom (both from water) at the cleavage site.

Figure 2.15 Metabolism.
Two common reactions by which cells build and break down organic molecules are shown.

monomers. These and other processes of molecular change are called **reactions**. Cells constantly run reactions as they acquire and use energy to stay alive, grow, and reproduce—activities collectively called **metabolism** (Figure 2.15A). Metabolism also requires **enzymes**, which are organic molecules (usually proteins) that speed up reactions without being changed by them. For example, in a common reaction called condensation, an enzyme covalently bonds two monomers together (Figure 2.15B). In hydrolysis, the reverse of condensation, an enzyme splits an organic polymer into its component monomers (Figure 2.15C).

Take-Home Message 2.6

How are all of the molecules of life alike?

- The molecules of life (carbohydrates, lipids, proteins, and nucleic acids) are organic, which means they consist mainly of carbon and hydrogen atoms.
- The structure of an organic molecule starts with a chain of carbon atoms (the backbone) that may form one or more rings.
- By processes of metabolism, cells assemble the molecules of life from monomers. They also break apart polymers into component monomers.

2.7 Carbohydrates

A **carbohydrate** is an organic compound that consists of carbon, hydrogen, and oxygen in a 1:2:1 ratio. The term can apply to a sugar molecule or a polymer of them, so these compounds are also called saccharides (saccharide means sugar). Cells use different kinds for fuel, as structural materials, and for storing energy.

Monosaccharides (one sugar) are the simplest carbohydrates, and common types have a backbone of five or six carbon atoms. Glucose, shown in Figure 2.14, is a monosaccharide. Sucrose, which is our table sugar, is a disaccharide (two sugars) that consists of glucose and fructose monomers. Monosaccharides and disaccharides are very soluble in water, so they can move easily through the water-based internal environments of all organisms.

Breaking the bonds of a monosaccharide releases energy that can be harnessed to power other reactions (Chapter 5 returns to this topic). Monosaccharides are also remodeled into other important compounds. For example, cells of plants and many animals make vitamin C from glucose. Human cells are unable to make vitamin C, so we need to get it from our food.

Foods that we call "complex" carbohydrates consist mainly of polysaccharides, which are chains of hundreds or thousands of monosaccharide monomers. The chains may be straight or branched, and can have one or many types of monosaccharides. The most common polysaccharides are cellulose, starch, and glycogen. All consist only of glucose monomers, but as substances their properties are very different. Why? The answer begins with differences in patterns of covalent bonding that link their monomers.

Cellulose, the major structural material of plants, is the most abundant organic molecule on Earth. Hydrogen bonding locks its long, straight chains of covalently bonded glucose monomers into tight, sturdy bundles (Figure 2.16A). The bundles form tough fibers that act like reinforcing rods inside stems and other plant parts, helping these structures resist wind and other forms of mechanical stress. Cellulose

A. Cellulose is the main structural component of plants. Above, in cellulose, hydrogen bonds stabilize chains of glucose monomers in tight bundles that form long fibers. Few organisms can digest this tough, insoluble material.

B. Starch is the main energy reserve in plants, which store it in their roots, stems, leaves, seeds, and fruits. Below, starch consists of long, coiled chains of glucose monomers.

C. Glycogen functions as an energy reservoir in animals, including people. It is especially abundant in the liver and muscles. Above, glycogen consists of highly branched chains of glucose monomers.

Figure 2.16 Three of the most common complex carbohydrates and their locations in a few organisms. Each polysaccharide consists only of glucose units, but different bonding patterns that link the subunits result in substances with very different properties.

Middle photo, © JupiterImages Corporation.

is insoluble (it does not dissolve) in water, and it is not easily broken down. Some bacteria and fungi make enzymes that can break it apart into its component sugars, but humans and other mammals do not. Dietary fiber, or "roughage," usually refers to the indigestible cellulose in our vegetable foods. Bacteria that live in the guts of termites and grazers such as cattle and sheep help these animals digest the cellulose in plants. (Chapter 23 returns to the topic of animal digestion.)

In starch, a different covalent bonding pattern between glucose monomers makes a chain that coils up into a spiral (Figure 2.16B). Like cellulose, starch does not dissolve readily in water, but it is easier to break down. These properties make the molecule ideal for storing sugars in the watery, enzyme-filled interior of plant cells. Most plant leaves make glucose during the day, and their cells store it by building starch. At night, hydrolysis enzymes break the bonds between starch's glucose monomers. The released glucose can be broken down immediately for energy, or converted to sucrose that is transported to other parts of the plant. Humans also have hydrolysis enzymes that break down starch, so this carbohydrate is an important component of our food.

Animals store sugars in the form of glycogen, a polysaccharide that consists of highly branched chains of glucose monomers (Figure 2.16C). Muscle and liver cells contain most of the body's glycogen. When the blood sugar level falls, liver cells break down the glycogen, and the released glucose subunits enter the blood.

Take-Home Message 2.7

What is a carbohydrate?

- Cells use simple carbohydrates (sugars) for energy and to build other molecules.
- Sugar monomers, bonded different ways, form complex carbohydrates such as cellulose, starch, and glycogen.

carbohydrate Molecule that consists primarily of carbon, hydrogen, and oxygen atoms in a 1:2:1 ratio.

cellulose Tough, insoluble carbohydrate that is the major structural material in plants.

enzyme Organic molecule that speeds up a reaction without being changed by it.

metabolism All the enzyme-mediated chemical reactions by which cells acquire and use energy as they build and break down organic molecules.

reaction Process of molecular change.

2.8 Lipids

Lipids are fatty, oily, or waxy organic compounds. They vary in structure, but all are hydrophobic. Many lipids incorporate **fatty acids**, which are small organic molecules that consist of a carbon chain "tail" of variable length, and a carboxyl group "head" (Figure 2.17). The tail is hydrophobic (hence the name "fatty"); the carboxyl group makes the head hydrophilic (and acidic). You are already familiar with the properties of fatty acids because these molecules are the main component of soap. The hydrophobic tails of fatty acids in soap attract oily dirt, and the hydrophilic heads dissolve the dirt in water.

Saturated fatty acids have only single bonds linking the carbons in their tails. In other words, their carbon chains are fully saturated with hydrogen atoms (Figure 2.17A). The tail of a saturated fatty acid is flexible and it wiggles freely. Double bonds between carbons limit the flexibility of the tails of an **unsaturated fatty acid** (Figure 2.17B,C). These bonds are *cis* or *trans*, depending on the way the hydrogens are arranged around them (Figure 2.17D,E).

Fats The carboxyl group head of a fatty acid can easily form a covalent bond with another molecule. When it bonds to a glycerol, a type of alcohol, it loses its hydrophilic character. Three fatty acids bonded to the same glycerol form a **triglyceride**, a molecule that is entirely hydrophobic and therefore does not dissolve in water. Triglycerides are the most abundant and richest energy source in vertebrate bodies; gram for gram, they store more energy than carbohydrates.

A **fat** is a substance that consists mainly of triglycerides. Butter and other fats derived from animals have a high proportion of triglycerides in which all three fatty acid tails are saturated. These triglycerides are commonly called saturated fats, and substances that consist of them are solid at room temperature because floppy saturated fatty acid tails can pack tightly together. Vegetable oils, by contrast, have a high proportion of unsaturated fats, the common term for triglycerides in which at least one of the three fatty acid tails is unsaturated. Each double bond makes a rigid kink, and kinky tails cannot pack tightly. This is why most substances that consist of unsaturated fats are liquid at room temperature. The partially hydrogenated vegetable oils that you learned about in Section 2.1 are an exception. They are solid at room temperature because the special *trans* double bond keeps their fatty acid tails straight, allowing them to pack tightly just like saturated fatty acid tails do.

Phospholipids A **phospholipid** has two fatty acid tails and a head that contains a phosphate group (Figure 2.18A). The tails are hydrophobic, but the phosphate group is highly polar and it makes the head very hydrophilic. These opposing properties give rise to the basic structure of cell membranes,

Figure 2.17 Fatty acids.
Double bonds in the tails are highlighted in red.

A. The tail of stearic acid is fully saturated with hydrogen atoms.

B. Linoleic acid, with two double bonds, is unsaturated. The first double bond occurs at the sixth carbon from the end of the tail, so linoleic acid is called an omega-6 fatty acid. Omega-6 and **C** omega-3 fatty acids are "essential fatty acids," which means your body does not make them and they must come from food.

D. The hydrogen atoms around the double bond in oleic acid are on the same side of the tail. Most other naturally occurring unsaturated fatty acids have these *cis* bonds.

E. Hydrogenation creates abundant *trans* bonds, with hydrogen atoms on opposite sides of the tail.

Figure It Out: Are the double bonds in linolenic acid *cis* or *trans*? Answer: *cis*

A. stearic acid (saturated) B. linoleic acid (omega-6) C. linolenic acid (omega-3) D. oleic acid (*cis*) E. elaidic acid (*trans*)

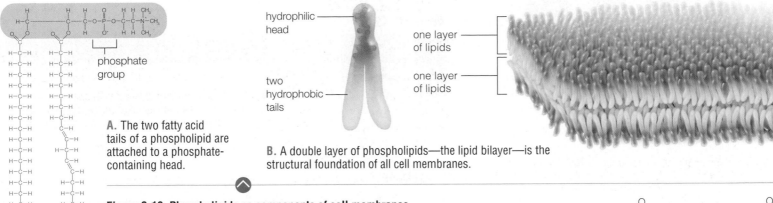

A. The two fatty acid tails of a phospholipid are attached to a phosphate-containing head.

B. A double layer of phospholipids—the lipid bilayer—is the structural foundation of all cell membranes.

Figure 2.18 Phospholipids as components of cell membranes.

which consist mainly of phospholipids. In a cell membrane, phospholipids are arranged in two layers—a **lipid bilayer** (Figure 2.18B). The heads of one layer are dissolved in the cell's watery interior, and the heads of the other layer are dissolved in the cell's fluid surroundings. All of the hydrophobic tails are sandwiched between the hydrophilic heads. Section 3.3 returns to the structure of cell membranes.

Waxes A **wax** is a complex, varying mixture of lipids with long fatty acid tails bonded to carbon rings or other structures. These molecules pack tightly, so waxes are firm and water-repellent. Plants secrete waxes onto their exposed surfaces to restrict water loss and keep out parasites and other pests. Other types of waxes protect, lubricate, and soften skin and hair. Waxes, together with fats and fatty acids, make feathers waterproof. Bees store honey and raise new generations of bees inside a honeycomb of secreted beeswax.

Steroids **Steroids** are lipids with no fatty acid tails; they have a rigid backbone that consists of twenty carbon atoms arranged in a characteristic pattern of four rings (Figure 2.19). As you will see in later chapters, these molecules serve varied and important physiological functions in plants, fungi, and animals. Cholesterol, the most common steroid in animal tissues, is remodeled into other molecules such as bile salts (which help digest fats), vitamin D (required to keep teeth and bones strong), and steroid hormones.

Figure 2.19 Examples of steroids.
Estrogen and testosterone are steroid hormones that govern reproduction and secondary sexual traits in animals. These hormones give rise to the gender-specific traits of many species, including wood ducks.

Bottom, Tim Davis/Science Source.

fat Substance that consists mainly of triglycerides.

fatty acid Organic compound with an acidic carboxyl group "head" and a long carbon chain "tail."

lipid Fatty, oily, or waxy organic compound.

lipid bilayer Double layer of lipids arranged tail-to-tail; structural foundation of all cell membranes.

phospholipid A lipid with a phosphate group in its hydrophilic head, and two nonpolar fatty acid tails.

saturated fatty acid Fatty acid with only single bonds linking the carbons in its tail.

steroid A type of lipid with four carbon rings and no fatty acid tails.

triglyceride A molecule with three fatty acid tails that is entirely hydrophobic; main component of fats.

unsaturated fatty acid Fatty acid with one or more carbon–carbon double bonds in its tail.

wax Water-repellent substance that is a complex, varying mixture of lipids.

Take-Home Message 2.8

What are lipids?

- Lipids are fatty, waxy, or oily organic compounds.
- Fats are substances that consist primarily of triglycerides, which have three fatty acid tails. Triglycerides are called unsaturated fats if there are double bonds in one or more of their fatty acid tails, and saturated fats if there are none.
- Phospholipids arranged in a lipid bilayer are the main component of cell membranes.
- Waxes have complex, varying structures. They are components of water-repelling and lubricating secretions.
- Steroids serve varied and important physiological roles in plants, fungi, and animals.

② A protein's primary structure consists of a linear sequence of amino acids (a polypeptide). Each type of protein has a unique primary structure.

① A condensation reaction joins the carboxyl group of one amino acid and the amine group of another to form a peptide bond. In this example, a peptide bond forms between the amino acids methionine and valine.

Figure 2.20 How protein structure arises.

(3–5) 1BBB, A third quaternary structure of human hemoglobin A at 1.7-A resolution. Silva, M.M., Rogers, P.H., Arnone, A., Journal: (1992) *J.Biol.Chem.* 267: 17248-17256.

2.9 **Proteins**

Cells make the thousands of different **proteins** they need from only twenty kinds of amino acid monomers. An **amino acid** is a small organic compound with an amine group (—NH_2), a carboxyl group (—COOH, the acid), and a side chain called an "R group" that defines the kind of amino acid. In most amino acids, all three groups are attached to the same carbon atom:

The covalent bond that links amino acids in a protein is called a **peptide bond**. During protein synthesis, a peptide bond forms between the carboxyl group of the first amino acid and the amine group of the second (Figure 2.20 **①**). Another peptide bond links a third amino acid to the second, and so on (you will learn more about the details of protein synthesis in Chapter 7). A short chain of amino acids is called a peptide; as the chain lengthens, it becomes a polypeptide. Polypeptides can be hundreds or even thousands of amino acids long.

The idea that structure dictates function is particularly appropriate as applied to proteins, because the diversity in biological activity among these molecules arises from differences in their three-dimensional shape. Protein structure begins with the linear series of amino acids composing a polypeptide **②**. The order of the amino acids, which is called primary structure, defines the type of protein.

The molecule begins to take on three-dimensional shape during protein synthesis, as hydrogen bonds that form between amino acids cause the lengthening polypeptide to twist and fold. Hydrogen bonding holds sections of the polypeptide in loops, helices (coils), or flat sheets, and these patterns constitute the protein's secondary structure **③**. The primary structure of each type of protein is unique, but most proteins have similar patterns of secondary structure.

Much as an overly twisted rubber band coils back upon itself, hydrogen bonding also makes the loops, helices, and sheets of a protein fold up into even more compact domains. These domains are called tertiary structure. Tertiary structure is what makes a protein a working molecule. For example, the helices and loops in a polypeptide called a globin chain fold up together to form a pocket **④**. This pocket

Figure 2.21 An example of a protein domain.
This barrel domain is part of a rotary mechanism in a larger protein. The protein is a molecular motor that pumps hydrogen ions through cell membranes.

pdb ID2W5J, Vollmar, M., Shlieper, D., Winn, M., Buechner, C., Groth, G. "Structure of the C14 rotor ring of the proton translocating chloroplast ATP synthase." (2009) *J. Biol. Chem.* 284:18228.

3 Secondary structure arises as a polypeptide chain twists into a helix (coil), loop, or sheet held in place by hydrogen bonds.

4 Tertiary structure arises when loops, helices, and sheets fold up into a domain. In this example, the helices of a globin chain form a pocket.

5 Many proteins have two or more polypeptides (quaternary structure). Hemoglobin, shown here, consists of four globins (green and blue). The pocket of each globin now holds a heme group (red).

6 Some types of proteins can aggregate into much larger structures. As an example, organized arrays of keratin, a fibrous protein, compose very long filaments that make up your hair.

holds a heme, which is a small compound essential to the function of the finished protein—hemoglobin. In other proteins, sheets, loops, and helices come together as complex structures that resemble barrels, propellers, sandwiches, and so on. Barrel domains often form tunnels through cell membranes, allowing small molecules to cross. Some proteins have barrel domains that rotate like motors in small molecular machines (Figure 2.21). A protein may have several domains, each contributing a particular structural or functional property to the molecule.

Many proteins also have quaternary structure, which means they consist of two or more polypeptides that are closely associated or covalently bonded together. Hemoglobin is like this **5**. So are most enzymes, which have multiple polypeptides that collectively form a roughly spherical shape.

Fibrous proteins aggregate by many thousands into much larger structures, with their polypeptides organized into strands or sheets. The keratin in your hair is an example **6**. Other fibrous proteins are part of the mechanisms that help cells, cell parts, and multicelled bodies move.

Carbohydrates, lipids, or both may get attached to a protein after synthesis. A protein with carbohydrates attached to it is called a glycoprotein. Molecules that allow a body to recognize its own cells are glycoproteins, as are other molecules that help cells interact in immunity. A protein with one or more lipids attached to it is called a lipoprotein. Some lipoproteins are aggregate structures that consist of variable amounts and types of proteins and lipids (Figure 2.22).

The Importance of Protein Structure Protein shape depends on hydrogen bonds and other interactions that heat, some salts, shifts in pH, or detergents can disrupt. Such disruption can cause proteins to lose their three-dimensional shape, or **denature**. Once a protein's shape unravels, so does its function. You can see denaturation in action when you cook an egg. A protein called albumin is a major component of egg white. Cooking does not disrupt the covalent bonds of albumin's primary structure, but it does destroy the hydrogen bonds that maintain the protein's shape. When a translucent egg white turns opaque, the albumin has been denatured. For a very few proteins, denaturation is reversible if normal conditions return, but albumin is not one of them. There is no way to uncook an egg.

Diseases such as bovine spongiform encephalitis (BSE, or mad cow disease) in cattle, Creutzfeldt–Jakob disease in humans, and scrapie in sheep, are the dire aftermath of a protein that changes shape. These diseases may be inherited, but

Figure 2.22 A lipoprotein particle.
The one depicted here (HDL, which is often called "good" cholesterol) consists of thousands of lipids lassoed into a clump by two proteins.

Castrignanò T, De Meo PD, Cozzetto D, Talamo IG, Tramontano A. (2006). The PMDB Protein Model Database. Nucleic Acids Research, 34: D306–D309.

lipids

protein

amino acid Small organic compound that is a subunit of proteins. Consists of a carboxyl group, an amine group, and a characteristic side group (R), all typically bonded to the same carbon atom.

denature To unravel the shape of a protein or other large biological molecule.

peptide bond A bond between the amine group of one amino acid and the carboxyl group of another. Joins amino acids in proteins.

protein Polymer of amino acids; an organic molecule that consists of one or more polypeptides.

Digging Into Data

Effects of Dietary Fats on Lipoprotein Levels

Cholesterol that is made by the liver or that enters the body from food does not dissolve in blood, so it is carried through the bloodstream by lipoproteins. Low-density lipoprotein (LDL) carries cholesterol to body tissues such as artery walls, where it can form deposits associated with cardiovascular disease. Thus, LDL is often called "bad" cholesterol. High-density lipoprotein (HDL) carries cholesterol away from tissues to the liver for disposal, so HDL is often called "good" cholesterol. In 1990, Ronald Mensink and Martijn Katan published a study that tested the effects of different dietary fats on blood lipoprotein levels. Their results are shown in Figure 2.23.

1. In which group was the level of LDL ("bad" cholesterol) highest?
2. In which group was the level of HDL ("good" cholesterol) lowest?
3. An elevated risk of heart disease has been correlated with increasing LDL-to-HDL ratios. Rank the three diets according to their predicted effect on cardiovascular health.

	Main Dietary Fats			
	cis fatty acids	*trans* fatty acids	saturated fats	optimal level
LDL	103	117	121	<100
HDL	55	48	55	>40
ratio	1.87	2.44	2.2	<2

Figure 2.23 Effect of diet on lipoprotein levels.
Researchers placed 59 men and women on a diet in which 10 percent of their daily energy intake consisted of *cis* fatty acids, *trans* fatty acids, or saturated fats. Blood LDL and HDL levels were measured after three weeks on the diet; averaged results are shown in mg/dL (milligrams per deciliter of blood). All subjects were tested on each of the diets. The ratio of LDL to HDL is also shown.

Source, Mensink RP, Katan MB, "Effect of dietary trans fatty acids on high-density and low-density lipoprotein cholesterol levels in healthy subjects." *NEJM* 323(7):439–45.

Figure 2.24 Variant Creutzfeldt–Jakob disease (vCJD).

(A) © Gary I. Rothstein/Reuters/Corbis; (B) Sherif Zaki, MD, PhD, Wun-Ju Shieh, MD, PhD, MPH/CDC.

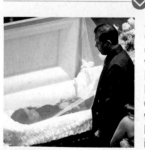

A. Charlene Singh was one of the three people who developed symptoms of vCJD disease while living in the United States. Singh, like the others, most likely contracted the disease elsewhere; she spent her childhood in Britain. Diagnosed in 2001, she died in 2004.

B. Slice of brain tissue from a person with vCJD. Fibers of prion proteins (amyloid fibrils) radiating from several deposits are visible.

more often they arise spontaneously. All are characterized by relentless deterioration of mental and physical abilities that eventually causes death (Figure 2.24A), and all begin with a glycoprotein called PrPC that occurs normally in cell membranes of the mammalian body. This protein is especially abundant in brain cells, but we still know very little about what it does. Sometimes, a PrPC protein misfolds so that part of the molecule forms a sheet instead of a helix. One misfolded molecule should not pose much of a threat, but when this particular protein misfolds it becomes a **prion**, or infectious protein. The shape of a misfolded PrPC protein causes normally folded PrPC proteins to misfold too. Each protein that misfolds becomes infectious, so the number of prions increases exponentially.

The shape of misfolded PrPC proteins allows them to align tightly into long, insoluble fibers that are called amyloid fibrils. Amyloid fibrils grow from their ends as more PrPC proteins misfold (Figure 2.24B). They form patches in the brain that disrupt its function, causing symptoms such as confusion, memory loss, and lack of coordination. Holes form in the brain as its cells die. Eventually, the brain becomes so riddled with holes that it looks like a sponge.

In the mid-1980s, an epidemic of mad cow disease in Britain was followed by an outbreak of a new variant of Creutzfeldt–Jakob disease (vCJD) in humans. Researchers isolated a prion similar to the one in scrapie-infected sheep from cows with BSE, and also from humans affected by the new type of Creutzfeldt–Jakob disease. How did the prion get from sheep to cattle to people? Prions resist denaturation, so treatments such as cooking that inactivate other types of infectious agents have little effect on them. The cattle became infected by the prion after eating feed prepared from the remains of scrapie-infected sheep, and people became infected by eating beef from the infected cattle.

Two hundred people have died from vCJD since 1990. The use of animal parts in livestock feed is now banned in many countries, and the number of cases of BSE and vCJD has since declined. Cattle with BSE still turn up, but so rarely that they pose little threat to human populations.

A. Example of a nucleotide: ATP. ATP is a monomer of RNA, and also a participant in many metabolic reactions.

Take-Home Message 2.9

Why is protein structure important?

- A protein consists of one or more polypeptides, each a chain of amino acids. The order of amino acids in the polypeptide(s) dictates the type of protein.
- During protein synthesis, polypeptides twist and fold into coils, sheets, and loops, which fold and pack further into functional domains.
- A protein's function arises from and depends on its three-dimensional shape.

2.10 Nucleic Acids

Nucleotides are small organic molecules that function as energy carriers, enzyme helpers, chemical messengers, and subunits of DNA and RNA. Each consists of a monosaccharide ring bonded to a nitrogen-containing base and one, two, or three phosphate groups (Figure 2.25A). The monosaccharide is a five-carbon sugar, either ribose or deoxyribose, and the base is one of five compounds with a flat ring structure (we return to the structure of nucleotide bases in Section 6.2). When the third phosphate group of a nucleotide is transferred to another molecule, energy is transferred along with it. You will read about such phosphate-group transfers and their important metabolic role in Section 4.4. The nucleotide **ATP** (adenosine triphosphate) serves an especially important role as an energy carrier in cells.

Nucleic acids are chains of nucleotides in which the sugar of one nucleotide is joined to the phosphate group of the next (Figure 2.25B). An example is **RNA**, or ribonucleic acid, named after the ribose sugar of its component nucleotides. An RNA molecule is a chain of four kinds of nucleotide monomers (one of which is ATP). There are different types of RNA, and they work together to carry out protein synthesis. **DNA**, or deoxyribonucleic acid, is a nucleic acid named after the deoxyribose sugar of its component nucleotides. A DNA molecule consists of two chains of nucleotides twisted into a double helix (Figure 2.25C). Hydrogen bonds hold the chains together (Chapter 6 returns to DNA structure). Each cell starts life with DNA inherited from a parent cell. That DNA contains all of the information necessary to build a new cell and, in the case of multicelled organisms, an entire individual. The cell uses the order of nucleotide bases in DNA—the DNA sequence—to guide production of RNA and proteins (Chapter 7 returns to this topic).

B. A chain of nucleotides is a nucleic acid. The sugar of one nucleotide is covalently bonded to the phosphate group of the next, forming a sugar–phosphate backbone.

C. DNA consists of two chains of nucleotides, twisted into a double helix. Hydrogen bonding maintains the three-dimensional structure of this nucleic acid.

Figure 2.25 Nucleic acid structure.

Take-Home Message 2.10

What are nucleotides and nucleic acids?

- Nucleotides are monomers of the nucleic acids DNA and RNA. Some have additional roles. ATP, for example, is an important energy carrier in cells.
- DNA holds information necessary to build cells and multicelled individuals.
- RNAs carry out protein synthesis.

ATP Nucleotide monomer of RNA; also serves an important role as an energy carrier in cells.

DNA Deoxyribonucleic acid. Nucleic acid that consists of two chains of nucleotides twisted into a double helix; holds hereditary information.

nucleic acid Chain of nucleotides; DNA or RNA.

nucleotide Small molecule with a five-carbon sugar, a nitrogen-containing base, and one, two, or three phosphate groups. Monomer of nucleic acids; some have additional roles.

prion Infectious protein.

RNA Ribonucleic acid. Nucleic acid that consists of a chain of nucleotides. Carries out protein synthesis.

Summary

Section 2.1 All organisms consist of the same kinds of molecules. Seemingly small differences in the way those molecules are put together can have big effects inside a living organism.

Section 2.2 Atoms consist of **electrons**, which carry a negative **charge**, moving about a **nucleus** of positively charged **protons** and uncharged **neutrons**. The number of protons (the **atomic number**) determines the **element**.

Isotopes of an element have the same number of protons but differ in the number of neutrons. The total number of protons and neutrons in the atomic nucleus is called **mass number**.

Tracers can be made with **radioisotopes**, which, by **radioactive decay**, emit particles and energy when their nucleus spontaneously breaks up.

A **shell model** of an atom represents the energy levels of its electrons as concentric circles. Atoms are in their most stable state when all of their shells are full of electrons, so they tend to get rid of vacancies. Many can do so by gaining or losing electrons, thereby becoming **ions**. Atoms with unpaired electrons are **free radicals**. Most free radicals are unstable and dangerous to life.

Section 2.3 A **chemical bond** unites two atoms in a molecule. A **compound** is a molecule that consists of two or more elements. An **ionic bond** is a strong association of two oppositely charged ions. **Polarity** is a separation of charge into positive and negative regions. Atoms share a pair of electrons in a **covalent bond**, which is nonpolar if the sharing is equal, and polar if it is not.

Section 2.4 Two polar covalent bonds give each water molecule an overall polarity. **Hydrogen bonds** that form among water molecules in tremendous numbers are the basis of water's unique life-sustaining properties: **cohesion**, resistance to **temperature** changes, and a capacity to act as a **solvent** for **salts** and other polar **solutes**. The amount of solute in a given volume of a **solution** is the solute's **concentration**. **Hydrophilic** substances dissolve easily in water; **hydrophobic** substances do not. **Evaporation** is the transition of a liquid to a vapor.

Section 2.5 **pH** is a measure of the number of hydrogen ions (H^+) in a liquid. At neutral pH (7), there are an equal number of H^+ and OH^- ions. **Acids** release hydrogen ions in water; **bases** accept them. A **buffer** can keep the pH of solution consistent. Most cell and body fluids are buffered because most molecules of life work only within a narrow range of pH.

Section 2.6 The molecules of life are **organic**, and chains or rings of carbon atoms form their backbones. All are **polymers** of sugar, fatty acid, or amino acid **monomers**. **Enzymes** carry out **reactions** of **metabolism**.

Section 2.7 Cells use **carbohydrates** for energy, and as structural materials. Enzymes assemble polysaccharides such as **cellulose** from monosaccharide monomers.

Section 2.8 All **lipids** are nonpolar. A **fatty acid** is a lipid with an acidic head and a long carbon chain tail. Only single bonds link the carbons in the tail of a **saturated fatty acid**; the tail of an **unsaturated fatty acid** has one or more double bonds. **Fats** consist mostly of **triglycerides**, which have three fatty acid tails and are nonpolar. When all of a triglyceride's fatty acid tails are unsaturated, it is called an unsaturated fat; if one or more of the tails is saturated, it is a saturated fat. A **lipid bilayer** (of **phospholipids**) is the basic structure of all cell membranes. **Waxes** are part of water-repellent and lubricating secretions. Some **steroids** function as hormones.

Section 2.9 The shape of a **protein** is the source of its function. Protein structure begins as a sequence of **amino acids** linked by **peptide bonds** into a polypeptide. Polypeptides twist into loops, sheets, and coils that can pack further into functional domains. Many proteins, including most enzymes, consist of two or more polypeptides. Fibrous proteins aggregate into much larger structures. A protein that **denatures**, or loses its shape, also loses its function. **Prion** diseases are a fatal consequence of misfolded proteins.

Section 2.10 **Nucleotides** consist of a five-carbon sugar, a nitrogen-containing base, and one, two or three phosphate groups. Nucleotides are monomers of **nucleic acids**, and some, especially **ATP**, have additional functions. **DNA** holds heritable information; **RNA** carries out protein synthesis.

Self-Quiz

Answers in Appendix I

1. Which of the following statements is *in*correct?
   a. Isotopes have the same atomic number and different mass numbers.
   b. Atoms have about the same number of electrons as protons.
   c. All molecules consist of atoms.
   d. Free radicals are dangerous because they emit energy.

2. Which element has only one proton?

3. The mutual attraction of opposite charges holds atoms together as molecules in a(n) _____ bond.
   a. ionic
   b. hydrogen
   c. polar covalent
   d. nonpolar covalent

4. A salt does not release _____ in water.
   a. ions
   b. energy
   c. H^+

5. A(n) _____ substance repels water.
   a. acidic
   b. basic
   c. hydrophobic
   d. polar

6. When dissolved in water, a(n) _____ donates H^+ and a(n) _____ accepts H^+.
   a. acid; base
   b. base; acid
   c. buffer; solute
   d. base; buffer

7. _____ is a monosaccharide.
   a. Glucose
   b. Sucrose
   c. Ribose
   d. Starch
   e. a and c
   f. a, b, and c

8. Unlike saturated fatty acids, the tails of unsaturated fatty acids incorporate one or more _____ .
   a. phosphate groups
   b. glycerols
   c. double bonds
   d. single bonds

9. Which of the following is a class of molecules that encompasses all of the other molecules listed?
   a. triglycerides
   b. fatty acids
   c. waxes
   d. steroids
   e. lipids
   f. phospholipids

10. _____ are to proteins as _____ are to nucleic acids.
   a. Amino acids; hydrogen bonds
   b. Amino acids; nucleotides
   c. Sugars; lipids
   d. Sugars; proteins

11. A denatured protein has lost its _____ .
   a. hydrogen bonds
   b. shape
   c. function
   d. all of the above

12. Match the terms with their most suitable description.
    _____ hydrophilic
    _____ atomic number
    _____ hydrogen bonds
    _____ positive charge
    _____ temperature
    _____ negative charge
    _____ solution
    a. protons > electrons
    b. number of protons in nucleus
    c. polar; dissolves easily in water
    d. collectively strong
    e. protons < electrons
    f. measure of molecular motion
    g. solute dissolved in solvent

13. Which of the following are *not* found in DNA?
    a. amino acids
    b. sugars
    c. nucleotides
    d. phosphate groups

14. Match the molecules with the best description.
    _____ wax
    _____ starch
    _____ triglyceride
    _____ DNA
    _____ polypeptide
    _____ ATP
    a. protein primary structure
    b. an energy carrier
    c. water-repellent secretions
    d. carries heritable information
    e. sugar storage in plants
    f. richest energy source in animals

15. Match each molecule with its component(s).
    _____ protein
    _____ phospholipid
    _____ triglyceride
    _____ nucleic acid
    _____ cellulose
    _____ nucleotide
    _____ wax
    _____ glycoprotein
    a. glycerol, fatty acids, phosphate
    b. glycerol, three fatty acids
    c. nucleotides
    d. glucose only
    e. sugar, phosphate, base
    f. amino acid monomers
    g. amino acids, sugars
    h. fatty acids, carbon rings

Critical Thinking

1. Alchemists were the forerunners of modern-day chemists. Many of these medieval scholars and philosophers spent their lives trying to transform lead (atomic number 82) into gold (atomic number 79). Explain why they never succeeded.

2. Draw a shell model of a lithium atom (Li), which has 3 protons, then predict whether the majority of lithium atoms on Earth are uncharged, positively charged, or negatively charged.

3. Polonium is a rare element with 33 radioisotopes. The most common one, ^{210}Po, has 82 protons and 128 neutrons. When ^{210}Po decays, it emits an alpha particle, which is a helium nucleus (2 protons and 2 neutrons). ^{210}Po decay is tricky to detect because alpha particles do not carry very much energy compared to other forms of radiation. For example, they can be stopped by a single sheet of paper or a few inches of air. That is one reason that authorities failed to discover toxic amounts of ^{210}Po in the body of former KGB agent Alexander Litvinenko until after he died suddenly and mysteriously in 2006. What element does an atom of ^{210}Po become after it emits an alpha particle?

4. In the following list, identify the carbohydrate, the fatty acid, the amino acid, and the polypeptide:
   a. NH_2—CHR—COOH
   b. $C_6H_{12}O_6$
   c. (methionine)$_{20}$
   d. $CH_3(CH_2)_{16}COOH$

3 CELL STRUCTURE

Application ⟩ 3.1 Food for Thought

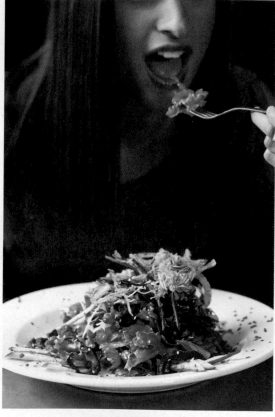

Cell for cell, microorganisms that live in and on a human body outnumber the person's own cells by about ten to one. Most are bacteria that live in the digestive tract, but these cells are not just stowaways. Gut bacteria help with digestion, make vitamins that mammals cannot, prevent the growth of dangerous germs, and shape the immune system. One of the most common intestinal bacteria of warm-blooded animals (including humans) is *Escherichia coli*. Most of the hundreds of types, or strains, of *E. coli*, are helpful, but a few make a toxic protein that can severely damage the lining of the intestine. After ingesting as few as ten cells of a toxic strain, a person may become ill with severe cramps and bloody diarrhea that lasts up to ten days. In some people, complications of infection result in kidney failure, blindness, paralysis, and death. Each year, about 265,000 people in the United States become infected with toxin-producing *E. coli*.

Strains of *E. coli* that are toxic to people live in the intestines of other animals—mainly cattle, deer, goats, and sheep—apparently without sickening them. Humans are exposed to the bacteria when they come into contact with animals that harbor them, for example, by eating fresh fruits and vegetables that have contacted animal feces (Figure 3.1). People also become infected with toxic *E. coli* by eating contaminated ground meat. An animal's feces can contaminate its meat during slaughter. Bacteria in the feces stick to the meat, then get thoroughly mixed into it during the grinding process. Unless the contaminated meat is cooked to at least 71°C (160°F), live bacteria will enter the digestive tract of anyone who eats it.

The United States Department of Agriculture (USDA) recalls food products in which toxic bacteria are discovered. Recalled meat is not necessarily discarded; often, it is cooked and processed into ready-to-eat products such as canned chili. Sterilization by cooking or other means kills bacteria, and it is one way to ensure food safety. Raw beef trimmings, which have a high risk of contact with fecal matter during the butchering process, are effectively sterilized when sprayed with ammonia. Ground to a paste and formed into pellets or blocks, the resulting product is termed "lean finely textured beef" or "boneless lean beef trimmings." This product is routinely used as a filler in prepared food products such as hamburger patties, fresh ground beef, hot dogs, lunch meats, sausages, frozen entrees, canned foods, and other items sold to quick service restaurants, hotel and restaurant chains, institutions, and school lunch programs. In 2012, a series of news reports that nicknamed the product "pink slime" provoked public outrage at its widespread use. Meat industry organizations and the USDA agree that lean finely textured beef, appetizing or not, is perfectly safe to eat because it has been sterilized.

Figure 3.1 Toxin-producing bacteria can contaminate foods.
Top, *Escherichia coli* cells sticking to the surface of a lettuce leaf. Some strains of this bacteria can cause a serious intestinal illness when they contaminate human food (bottom).

Top, © Custom Medical Stock Photo/Getty Images; bottom, Getty Images.

3.2 What, Exactly, Is a Cell?

REMEMBER: The cell is the smallest unit with the properties of life (Section 1.2). A tracer is a substance that can be tracked via a detectable component (2.2). Phospholipids arranged in a lipid bilayer are the main component of cell membranes (2.8).

The Cell Theory Hundreds of years of observations led to the way we now answer the question, What is a cell? Today we know that a cell carries out metabolism and homeostasis, and reproduces either on its own or as part of a larger organism. By this definition, each cell is alive even if it is part of a multicelled body, and all living organisms consist of one or more cells. We also know that cells reproduce by dividing, so it follows that all existing cells must have arisen by division of other cells (later chapters discuss processes by which cells reproduce). As a cell divides, it passes its hereditary material—its DNA—to offspring. Taken together, these generalizations constitute the **cell theory**, which is one of the foundations of modern biology (Table 3.1).

Components of All Cells Cells vary in shape and in what they do, but all share certain organizational and functional features: a plasma membrane, cytoplasm, and DNA (Figure 3.2). A cell's **plasma membrane** separates it from the external environment. Like all other cell membranes, it consists mainly of a phospholipid bilayer, and it is selectively permeable, which means that only certain materials can cross it. Thus, a plasma membrane controls exchanges between the cell and its environment. As you will see in Section 3.3, many different proteins embedded in a lipid bilayer or attached to one of its surfaces carry out membrane functions.

The plasma membrane encloses a jellylike mixture of water, sugars, ions, and proteins called **cytoplasm**. Some or all of a cell's metabolism occurs in cytoplasm, and the cell's internal components, including organelles, are suspended in it. **Organelles** are structures that carry out special metabolic functions inside a cell. Those with membranes compartmentalize substances and activities.

Every cell starts out life with DNA. In nearly all bacteria and archaea, that DNA is suspended directly in cytoplasm. By contrast, the DNA of a eukaryotic cell is contained in a **nucleus** (plural, nuclei), an organelle with a double membrane. All protists, fungi, plants, and animals are eukaryotes. Some of these organisms are independent, free-living cells; others consist of many cells working together as a body.

Constraints on Cell Size A living cell must exchange substances with its environment at a rate that keeps pace with its metabolism. These exchanges occur across the plasma membrane, which can handle only so many exchanges at a time. The rate of exchange across a plasma membrane depends on its surface area: The bigger it is, the more substances can cross it during a given interval. Thus, cell size is limited by a physical relationship called the **surface-to-volume ratio**. By this ratio, an object's volume increases with the cube of its diameter, but its surface area increases only with the square.

Apply the surface-to-volume ratio to a round cell. As Figure 3.3 shows, when a cell expands in diameter, its volume increases faster than its surface area does. Imagine that our round cell expands until it is four times its original diameter. The volume of the cell has increased 64 times (4^3), but its surface area has increased only 16 times (4^2). Each unit of plasma membrane must now handle exchanges for four times as much cytoplasm ($64 \div 16 = 4$). If the cell gets too big, the inward flow of nutrients and the outward flow of wastes across that membrane will not be fast enough to keep the cell alive.

Table 3.1 The Cell Theory

1. Each organism consists of one or more cells.
2. The cell is the structural and functional unit of all organisms. A cell is the smallest unit of life, individually alive even as part of a multicelled organism.
3. All living cells arise by division of preexisting cells.
4. Cells contain hereditary material (DNA), which they pass to their offspring when they divide.

Figure 3.2 General organization of a cell.
All cells start out life with a plasma membrane, cytoplasm, and DNA. This one is a cell from a plant.

cytoplasm

DNA in nucleus

plasma membrane

cell theory Theory that all organisms consist of one or more cells, which are the basic unit of life; all cells come from division of preexisting cells; and all cells pass DNA to offspring.

cytoplasm Semifluid substance enclosed by a cell's plasma membrane.

nucleus Of a cell, an organelle with two membranes that holds the cell's DNA.

organelle Structure that carries out a special metabolic function inside a cell.

plasma membrane Membrane that encloses a cell and separates it from the external environment.

surface-to-volume ratio A relationship in which the volume of an object increases with the cube of the diameter, and the surface area increases with the square.

Figure 3.3 Examples of surface-to-volume ratio. This physical relationship between increases in volume and surface area limits the size and influences the shape of cells.

Diameter (cm)	2	3	6
Surface area (cm²)	12.6	28.2	113
Volume (cm³)	4.2	14.1	113
Surface-to-volume ratio	3:1	2:1	1:1

Surface-to-volume limits also affect the form of colonial organisms such as strandlike algae, in which small cells attach end to end so each one can interact directly with the environment. It also affects the form of cells in multicelled bodies. For example, some types of muscle cells in your thighs run the length of your upper leg. Each of these cells is thin, so it exchanges substances efficiently with fluids in the surrounding tissue.

How Do We See Cells?   No one even knew cells existed until well after the first microscopes were invented. This is because typical cells are in the micrometer range of size—much smaller than the unaided human eye can perceive (Figure 3.4). One micrometer (μm) is one-thousandth of a millimeter, which is one-thousandth of a meter (Table 3.2). In a light microscope, visible light illuminates a sample. As you will learn in Chapter 5, all light travels in waves. This property of light causes it to bend when passing through a curved glass lens. Inside a light microscope, such lenses focus light that passes through a specimen, or bounces off of one, into a magnified image (Figure 3.5A). Microscopes that use polarized light can yield images in which the edges of some structures appear in three-dimensional relief (Figure 3.5B). Photographs of images enlarged with a microscope are called micrographs; those taken with visible light are called light micrographs (LM).

Most cells are nearly transparent, so their internal details may not be visible unless they are first stained, or exposed to dyes that only some cell parts soak up. Parts that absorb the most dye appear darkest. Staining results in an increase in contrast (the difference between light and dark) that allows us to see a greater range of detail. Researchers often use light-emitting tracers to pinpoint the location of a molecule of interest within a cell. When illuminated with a laser, these tracers fluoresce (emit light), and an image of the emitted light can be captured with a fluorescence microscope (Figure 3.5C). Such images are called fluorescence micrographs.

Structures smaller than about 200 nanometers across appear blurry under light microscopes. To observe objects of this size range clearly, we would have to switch to an electron microscope. There are two types of electron microscope; both use magnetic fields as lenses to focus a beam of electrons onto a sample. A transmission electron microscope directs electrons through a thin specimen, and the specimen's internal details appear as shadows in the resulting image, which is called a transmission electron micrograph, or TEM (Figure 3.5D). A scanning electron microscope directs a beam of electrons back and forth across the surface of a specimen that has been coated with a thin layer of gold or other metal. The irradiated metal emits

Table 3.2 Common Units of Length

Unit	Equivalent	
	Meter	Inch
centimeter (cm)	1/100	0.394
millimeter (mm)	1/1000	0.0394
micrometer (μm)	1/1,000,000	0.0000394
nanometer (nm)	1/1,000,000,000	0.0000000394
meter (m)	100 cm 1,000 mm 1,000,000 μm 1,000,000,000 nm	39.4

Figure 3.4 Relative sizes.
Below, the diameter of most cells is between 1 and 100 micrometers. Table 3.2 shows conversions among units of length; also see Units of Measure, Appendix IV.

Louse, Edward S. Ross; Ant, Vladimir Davydov/iStock/360/Getty Images; Frog, © A Cotton Photo/Shutterstock; Rat, © Pakhnyushcha/Shutterstock; Goose, panbazil/Shutterstock.com; Boy, © Piotr Marcinski/Shutterstock; Giraffe, © Valerie Kalyuznnyy/Photos.com; Whale, Dorling Kindersley/Getty Images.

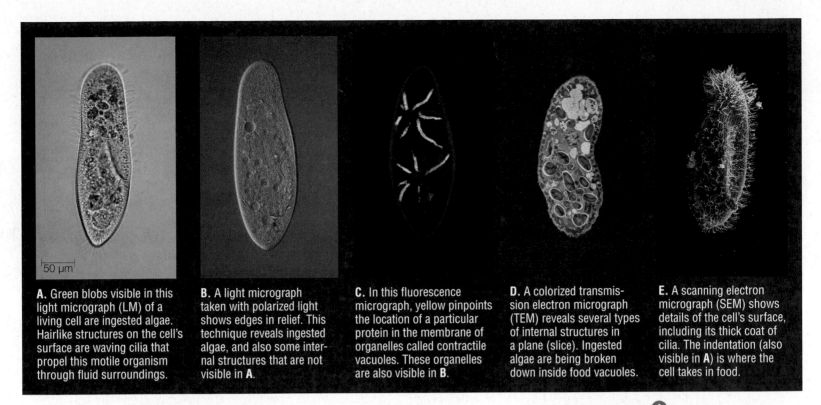

A. Green blobs visible in this light micrograph (LM) of a living cell are ingested algae. Hairlike structures on the cell's surface are waving cilia that propel this motile organism through fluid surroundings.

B. A light micrograph taken with polarized light shows edges in relief. This technique reveals ingested algae, and also some internal structures that are not visible in **A**.

C. In this fluorescence micrograph, yellow pinpoints the location of a particular protein in the membrane of organelles called contractile vacuoles. These organelles are also visible in **B**.

D. A colorized transmission electron micrograph (TEM) reveals several types of internal structures in a plane (slice). Ingested algae are being broken down inside food vacuoles.

E. A scanning electron micrograph (SEM) shows details of the cell's surface, including its thick coat of cilia. The indentation (also visible in **A**) is where the cell takes in food.

electrons and x-rays, which are converted into an image (a scanning electron micrograph, or SEM) of the surface (Figure 3.5E). SEMs and TEMs are always black and white; colored versions have been digitally altered to highlight specific details.

Figure 3.5 Different microscopes reveal different characteristics of the same organism, a single-celled protist called *Paramecium*.

(A) Nancy Nehring/iStockphoto.com; (B) Michael Abbey/Science Source; (C) © Dennis Kunkel Microscopy, Inc./PhototakeUSA.com; (D) © Microworks/PhototakeUSA.com; (E) Steve Gschmeissner/Science Source

Take-Home Message 3.2

How are all cells alike?

- All cells start life with a plasma membrane, cytoplasm, and a region of DNA. In eukaryotic cells only, the DNA is contained within a nucleus.
- The surface-to-volume ratio limits cell size and influences cell shape.
- Different types of microscopes reveal different aspects of cell structure.

human eye (no microscope)

frog eggs · small animals · largest organisms

100 µm · 1 mm · 1 cm · 10 cm · 1 m · 10 m · 100 m

Figure 3.6 Cell membrane structure.
Organization of phospholipids in cell membranes (**A**) and examples of common membrane proteins (**B–E**). For clarity, these proteins are often modeled as blobs or geometric shapes; their structure can be extremely complex.

A. In a watery fluid, phospholipids spontaneously line up into two layers: the hydrophobic tails cluster together, and the hydrophilic heads face outward, toward the fluid. This lipid bilayer forms the framework of all cell membranes. Many types of proteins intermingle among the lipids; a few that are typical of plasma membranes are shown opposite.

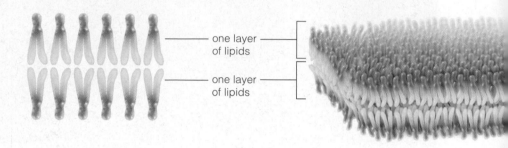

one layer of lipids

one layer of lipids

3.3 Cell Membrane Structure

REMEMBER: Phospholipids have a phosphate-containing head and two fatty acid tails; the tails can vary in length and in saturation (Section 2.8).

> A cell's basic structure is essentially a lipid bilayer bubble filled with fluid.

The foundation of all cell membranes is a lipid bilayer that consists mainly of phospholipids (Figure 3.6A). The head of a phospholipid is highly polar and hydrophilic, which means that it interacts with water molecules. Its two long hydrocarbon tails are very nonpolar and hydrophobic, so they do not interact with water molecules. As a result of these opposing properties, phospholipids swirled into water will spontaneously organize themselves into lipid bilayer sheets or bubbles, with hydrophobic tails together, hydrophilic heads facing the watery surroundings. A cell's basic structure is essentially a lipid bilayer bubble filled with fluid (left).

Other molecules, including cholesterol, proteins, glycoproteins, and glycolipids, are embedded in or attached to the lipid bilayer of a cell membrane. Many of these molecules can move around the membrane more or less freely. We use the term **fluid mosaic** to describe a membrane of a eukaryotic or bacterial cell because it behaves like a two-dimensional liquid of mixed composition. Membrane fluidity occurs because phospholipids in the bilayer are not chemically bonded to one another; they stay organized as a result of collective hydrophobic and hydrophilic attractions. These interactions are, on an individual basis, relatively weak. Thus, individual phospholipids in the bilayer drift sideways and spin around their long axis, and their tails wiggle.

A cell membrane's properties vary depending on the types and proportions of molecules composing it. For example, membrane fluidity decreases with increasing cholesterol content. A membrane's fluidity also depends on the length and saturation of its phospholipids' fatty acid tails. Archaea do not even use fatty acids to build their phospholipids. Instead, they use molecules with reactive side chains, so the tails of archaeal phospholipids form covalent bonds with one another. As a result of this rigid crosslinking, archaeal phospholipids do not drift, spin, or wiggle in a

B. Adhesion proteins fasten cells together or to external proteins. This one connects protein filaments inside the cell with external filaments in animal tissues.

C. Receptor proteins trigger a change in cellular activity in response to a stimulus such as binding to a particular substance. This one occurs on cells of the immune system.

D. Enzymes speed reactions at membranes. This one is part of a membrane-bound set of molecules that together break down drugs and other organic toxins.

E. Transport proteins bind to molecules on one side of the membrane, and release them on the other side. This one transports glucose.

extracellular fluid

lipid bilayer   cytoplasm

bilayer. Thus, membranes of archaea are stiffer than those of bacteria or eukaryotes, a characteristic that may help these cells survive in extreme habitats.

Membrane Proteins A cell membrane physically separates an external environment from an internal one, but that is not its only task. Many types of proteins are associated with a cell membrane, and each type adds a specific function to it. Thus, different cell membranes can carry out different tasks depending on which proteins are associated with them. A plasma membrane incorporates certain proteins that no internal cell membrane has, so it has functions that no other membrane does. For example, cells stay organized in animal tissues because **adhesion proteins** in their plasma membranes fasten them together and hold them in place (Figure 3.6B). Plasma membranes and some internal membranes incorporate **receptor proteins**, which trigger a change in the cell's activities in response to a stimulus (Figure 3.6C). Each type of receptor protein receives a particular stimulus, such as binding to a certain hormone. Each receptor protein also triggers a specific response inside the cell, which may involve metabolism, movement, division, or even cell death.

All cell membranes incorporate enzymes (Figure 3.6D). Some membrane enzymes act on other proteins or lipids that are part of the lipid bilayer. All membranes also have **transport proteins**, which move specific substances across the bilayer (Figure 3.6E). These proteins are important because lipid bilayers are impermeable to ions and polar molecules (we return to this topic in Section 4.5).

Take-Home Message 3.3

What is a cell membrane?

- The foundation of all cell membranes is the lipid bilayer: two layers of phospholipids, tails sandwiched between heads.
- Proteins that associate with lipid bilayers add various functions to a membrane.

adhesion protein Plasma membrane protein that fastens cells together in animal tissues.

fluid mosaic Model of a cell membrane as a two-dimensional fluid of mixed composition.

receptor protein Membrane protein that triggers a change in cell activity in response to a stimulus such as binding a certain substance.

transport protein Protein that moves specific ions or molecules across a membrane.

A. *Escherichia coli* is a common bacterial inhabitant of human intestines. Short, hairlike structures are pili; longer ones are flagella. This one is harmless; others can cause disease in humans.

B. *Oscillatoria* are a type of cyanobacteria, an ancient lineage of photosynthetic bacteria. Photosynthesis occurs at internal membranes (green). The multisided structures (pink) are protein-enclosed organelles called carboxysomes that assist photosynthesis.

C. *Ferroglobus placidus* is an archaeon that thrives in superheated water spewing from the ocean floor. The durable composition of its lipid bilayers (note the gridlike texture) keeps them intact at extreme heat and pH.

Figure 3.7 Some representative prokaryotes.

1. cytoplasm, with ribosomes
2. DNA in nucleoid
3. plasma membrane
4. cell wall
5. capsule
6. pilus
7. flagellum

Figure 3.8 Generalized body plan of a prokaryote.

3.4 Introducing Prokaryotic Cells

REMEMBER: A polysaccharide is a long chain of monosaccharides (Section 2.7); a peptide is a short chain of amino acids (2.9).

All bacteria and archaea are single-celled organisms, although individual cells of many species cluster in filaments or colonies (Figure 3.7). Outwardly, cells of the two groups appear so similar that archaea were once presumed to be an unusual group of bacteria. Both were classified as prokaryotes, a word that means "before the nucleus." By 1977, it had become clear that archaea are more closely related to eukaryotes than to bacteria, so they were given their own separate domain. The term "prokaryote" is now an informal designation only.

Bacteria and archaea are the smallest and most metabolically diverse forms of life that we know about. Chapter 13 revisits them in more detail; here we present an overview of structures shared by both groups (Figure 3.8).

Compared with eukaryotic cells, prokaryotes have little in the way of internal framework, but they do have protein filaments under the plasma membrane that reinforce the cell's shape and act as scaffolding for internal structures. The cytoplasm of these cells ❶ contains many **ribosomes** (organelles upon which polypeptides are assembled), and in some species, additional organelles. The cytoplasm also contains plasmids, which are small circles of DNA that carry a few genes (units of inheritance). The cell's remaining genes typically occur on one large circular molecule of DNA located in an irregularly shaped region of cytoplasm called the nucleoid ❷. In a few species, the nucleoid is enclosed by a membrane. Other internal membranes carry out special metabolic processes such as photosynthesis (Figure 3.7B).

Like all cells, bacteria and archaea have a plasma membrane ❸. In nearly all prokaryotes, a rigid **cell wall** ❹ surrounding the plasma membrane protects the cell and supports its shape. Archaeal cell walls and bacterial cell walls differ, but both types are permeable to water, so dissolved substances easily cross. Many species of bacteria, including the ones shown in Figure 3.7, have a second membrane

D. The square archaeon *Haloquadratum walsbyi* prefers brine pools saltier than soy sauce. Gas-filled organelles (white structures) buoy these highly motile cells, which can aggregate into flat sheets a bit like floor tiles.

E. *Helicobacter pylori*, a bacterium that can cause stomach ulcers when it infects the lining of the stomach, takes on a ball-shaped form (shown) that offers protection from environmental challenges such as antibiotic treatment.

F. The archaeon *Thermococcus gammatolerans* lives under extreme conditions of salt, temperature, and pressure. It is by far the most radiation-resistant organism ever discovered, capable of withstanding thousands of times more radiation than humans can.

around the cell wall. Outside the cell wall (or the second membrane) is a thick, gelatinous capsule ⑤ and/or a loosely attached layer of slime.

Protein filaments called **pili** (singular, pilus) ⑥ project from the wall of some prokaryotes. Pili help these cells move across or cling to surfaces. One kind, a "sex" pilus, attaches to another bacterium and then shortens. The attached cell is reeled in, and DNA is transferred from one cell to the other. Many types of bacteria and archaea also have one or more flagella projecting from their surface ⑦. **Flagella** (singular, flagellum) are long, slender cellular structures used for motion. A prokaryotic flagellum rotates like a propeller that drives the cell through fluid habitats.

Biofilms Bacterial cells often live so close together that an entire community shares a layer of slime. A communal living arrangement in which single-celled organisms occupy a shared mass of slime is called a **biofilm**. A biofilm is often attached to a solid surface, and may include bacteria, algae, fungi, protists, and/or archaea. Participating in a biofilm allows the cells to linger in a favorable spot rather than be swept away by fluid currents, and to reap the benefits of living communally. For example, rigid or netlike secretions of some species serve as permanent scaffolding for others; species that break down toxic chemicals allow more sensitive ones to thrive in habitats that they could not withstand on their own; and waste products of some serve as raw materials for others. Later chapters discuss some medical implications of biofilms, including dental plaque (Figure 3.9).

Figure 3.9 Oral bacteria in dental plaque, a biofilm. This micrograph shows two species of bacteria (tan, green) and a yeast (red) sticking to one another and to teeth via a gluelike mass of shared, secreted polysaccharides (pink). Other secretions of these organisms cause cavities and periodontal disease.

© Dennis Kinkel Microscopy, Inc./Phototake.

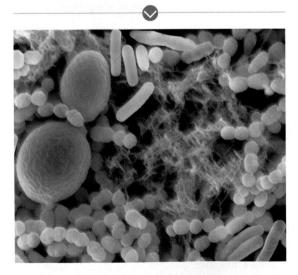

Take-Home Message 3.4

How are bacteria and archaea alike?

- Bacteria and archaea do not have a nucleus. Most kinds have a cell wall around their plasma membrane. The permeable wall reinforces and imparts shape to the cell body.
- The structure of bacteria and archaea is relatively simple, but as a group these organisms are the most diverse forms of life.

biofilm Community of microorganisms living within a shared mass of slime.

cell wall Semirigid but permeable structure that surrounds the plasma membrane of some cells.

flagellum Long, slender cellular structure used for movement.

pilus A protein filament that projects from the surface of some prokaryotic cells.

ribosome Organelle of protein synthesis.

3.5 Introducing Eukaryotic Cells

REMEMBER: Hydrogen bonding makes the loops, helices, and sheets of a polypeptide fold up into functional domains; many proteins consist of two or more polypeptides; fibrous proteins aggregate by many thousands into much larger structures (Section 2.9). ATP has an important metabolic role as an energy carrier (2.10).

Figure 3.10 Common components of eukaryotic cells.
An animal cell is illustrated here.

1 The nucleus protects and controls access to the cell's DNA.

2 Vesicles form by budding from other components of the endomembrane system or from the plasma membrane. Some transport substances among organelles of the ER, and to and from the plasma membrane. Others store or break down substances.

3 Ribosomes attached to rough endoplasmic reticulum (ER) assemble polypeptides that thread into the ER's interior, where they take on tertiary structure and assemble with other polypeptides.

4 Many proteins made in rough ER migrate through the ER compartment to smooth ER. Some of these proteins stay in smooth ER, as enzymes that assemble lipids and break down carbohydrates, wastes, and toxins. Others are packaged in vesicles for transport to Golgi bodies.

5 Golgi bodies modify proteins and lipids, then sort and repackage the finished molecules into new vesicles. Some of the new vesicles become lysosomes. Others carry proteins to the plasma membrane for insertion into the lipid bilayer or secretion.

6 Mitochondria specialize in efficient production of ATP.

In addition to a nucleus, a typical eukaryotic cell has many other membrane-enclosed organelles (Figure 3.10). An enclosing membrane allows an organelle to regulate the types and amounts of substances that enter and exit. Through this control, the organelle maintains a special internal environment that allows it to carry out a particular function—for example, isolating toxic or sensitive substances from the rest of the cell, moving substances through cytoplasm, maintaining fluid balance, or providing a favorable environment for a special process.

The Nucleus A cell nucleus serves two important functions. First, it keeps the cell's genetic material—its one and only copy of DNA—away from metabolic processes that might damage it. Isolated in its own compartment, the DNA stays separated from the bustling activity of the cytoplasm **1**. Second, a nucleus controls the passage of certain molecules across its membrane. The nucleus has a special membrane, the **nuclear envelope**, that carries out this function. A nuclear envelope consists of two lipid bilayers folded together. Proteins embedded in the two lipid bilayers aggregate into thousands of tiny nuclear pores that span the envelope (Figure 3.11). Some bacteria have membranes around their DNA, but we do not consider the bacteria to have nuclei because there are no pores in these membranes.

Large molecules, including RNA and proteins, cannot cross lipid bilayers on their own. Nuclear pores function as gateways for these molecules to enter and exit a nucleus. Protein synthesis offers an example of why this movement is important. Protein synthesis occurs in cytoplasm, and it requires the participation of many molecules of RNA. RNA is produced in the nucleus. Thus, RNA molecules must move from nucleus to cytoplasm, and they do so through nuclear pores. Proteins that carry out RNA synthesis must move in the opposite direction, because this process occurs in the nucleus.

The Endomembrane System The endomembrane system is a series of interacting organelles between the nucleus and the plasma membrane. Its main function is to make lipids, enzymes, and proteins for insertion into the cell's membranes or secretion to the external environment. The endomembrane system also destroys toxins, recycles wastes, and has other special functions. Components of the system vary among different types of cells, but here we present an overview of the most common ones.

Small, membrane-enclosed sacs called **vesicles 2** form by budding from other organelles or when a patch of plasma membrane sinks into the cytoplasm. Many types carry substances from one organelle to another, or to and from the

Figure 3.11
The nucleus.
This organelle is the defining characteristic of eukaryotic cells. TEM shows a liver cell nucleus.

© Kenneth Bart.

nuclear envelope
DNA
nuclear pore

1 µm

outer membrane
inner membrane
inner compartment
outer compartment

0.5 µm

Figure 3.12 The mitochondrion.
Two membranes, one folded inside the other, form the ATP-making machinery of this eukaryotic organelle. The TEM shows a mitochondrion in a cell from bat pancreas.

Keith R. Porter.

Figure It Out: What organelle is visible to the upper right in the micrograph?   **Answer: Rough ER**

plasma membrane. Some are a bit like trash cans that collect and dispose of waste, debris, or toxins. Enzymes in vesicles called **peroxisomes** break down fatty acids, amino acids, and poisons such as alcohol. They also break down hydrogen peroxide, a toxic by-product of fatty acid breakdown. Even more powerful enzymes in vesicles called **lysosomes** break down cellular debris and wastes. Ingested bacteria, cell parts, and other particles are delivered to lysosomes for breakdown.

Vacuoles are sacs that form by the fusion of multiple vesicles. Many isolate or break down waste, debris, toxins, or food. Plant cells have a large central vacuole that collects amino acids, sugars, ions, wastes, and toxins. Fluid pressure in a central vacuole keeps plant cells plump, so stems, leaves, and other plant parts stay firm (a central vacuole is illustrated to the left of the nucleus in Figure 3.2).

Endoplasmic reticulum (**ER**) is a system of tubes and flattened sacs. The membrane of the ER extends from the nuclear envelope, and it encloses a single, continuous compartment. Two kinds of ER, rough and smooth, are named for their appearance in electron micrographs. Thousands of ribosomes attached to the outer surface of rough ER give this organelle its "rough" appearance ❸. These ribosomes make polypeptides that thread into the interior of the ER as they are assembled. Inside the ER, the polypeptides take on their tertiary structure, and many assemble with other polypeptides. Cells that make, store, and secrete proteins have a lot of rough ER.

Some proteins made in rough ER become part of its membrane. Others migrate through the ER compartment to smooth ER. Smooth ER has no ribosomes, so it does not make its own proteins ❹. Some proteins that arrive in smooth ER are immediately packaged into vesicles for delivery elsewhere. Others are enzymes that stay and become part of the smooth ER. Enzymes of the smooth ER make lipids for the cell's membranes, and they break down carbohydrates, fatty acids, and some drugs and poisons.

A **Golgi body** has a folded membrane that often looks like a stack of pancakes ❺. Enzymes inside of Golgi bodies put finishing touches on proteins and lipids that have been delivered from ER. The finished products are sorted and packaged in new vesicles. Some of the vesicles deliver their cargo to the plasma membrane; others become lysosomes.

Mitochondria The **mitochondrion** (plural, mitochondria) ❻ is a eukaryotic organelle that specializes in making ATP by aerobic respiration (Chapter 5 details this metabolic pathway). A mitochondrion has two membranes, one highly folded inside the other (Figure 3.12), that form its ATP-making machinery. The organelle

endoplasmic reticulum (**ER**) Membrane-enclosed organelle that is a continuous system of sacs and tubes extending from the nuclear envelope. Smooth ER makes lipids and breaks down carbohydrates and fatty acids; ribosomes on the surface of rough ER make polypeptides.

Golgi body Organelle that modifies polypeptides and lipids, then packages the finished products into vesicles.

lysosome Enzyme-filled vesicle that breaks down cellular wastes and debris.

mitochondrion Eukaryotic organelle that produces ATP by aerobic respiration.

nuclear envelope A double membrane that constitutes the outer boundary of the nucleus. Nuclear pores in the membrane control the entry and exit of large molecules.

peroxisome Enzyme-filled vesicle that breaks down amino acids, fatty acids, and toxic substances.

vesicle Small, membrane-enclosed organelle; different kinds store, transport, or break down their contents.

Digging Into Data

Organelles and Cystic Fibrosis

A plasma membrane transport protein called CFTR moves chloride ions out of cells lining cavities and ducts of the lungs, liver, pancreas, intestines, and reproductive system. Water that follows the ions creates a thin film that allows mucus to slide easily through these structures.

People with cystic fibrosis (CF) have too few copies of the CFTR protein in the plasma membranes of their cells. Not enough chloride ions leave the cells, and so not enough water leaves them either. The result is thick, dry mucus that clogs the airways to the lungs and other passages. Symptoms include difficulty breathing and chronic lung infections. In 2000, researchers tracked the cellular location of the CFTR protein as it was being produced in cells from people with CF (Figure 3.13).

1. Which organelle contains the least amount of CFTR protein in normal cells? In CF cells?
2. In which organelle is the amount of CFTR protein most similar in both types of cells?
3. Where is the CFTR protein getting held up in cells of people who have CF?

Figure 3.13 Cellular location of the CFTR protein. Graph compares the amounts of CFTR protein found in endoplasmic reticulum, vesicles traveling from ER to Golgi, and Golgi bodies in CF cells and normal cells.

two outer membranes
stroma
inner membrane

Figure 3.14 The chloroplast. Each chloroplast has two outer membranes. Photosynthesis occurs at a third, much-folded inner membrane. TEM shows a chloroplast from a cell in a corn leaf.

Science Source.

resembles a bacterium in size, form, and biochemistry. Mitochondria have their own DNA, which is circular and otherwise similar to bacterial DNA. They divide independently of the cell, and have their own ribosomes. Such clues led to a theory that mitochondria evolved from aerobic bacteria that took up permanent residence inside a host cell (we return to this topic in Section 13.3).

Chloroplasts Photosynthetic cells of plants and many protists have **chloroplasts**, which are organelles specialized for photosynthesis. Most chloroplasts are oval or disk-shaped (Figure 3.14). Each has two outer membranes enclosing a semifluid interior, the stroma, that contains enzymes and the chloroplast's own DNA. In the stroma, a third, highly folded membrane forms a single, continuous compartment. Photosynthesis occurs at this inner membrane. In many ways, chloroplasts resemble the photosynthetic bacteria from which they evolved.

The Cytoskeleton Between the nucleus and plasma membrane of all eukaryotic cells is a system of protein filaments collectively called the **cytoskeleton**. Elements of the cytoskeleton reinforce, organize, and move cell structures, and often the whole cell. Some are permanent; others form only at certain times.

Microtubules are long, hollow cylinders that consist of subunits of the protein tubulin (Figure 3.15A). They form a dynamic scaffolding for many cellular processes, rapidly assembling when they are needed, disassembling when they are not. For example, before a eukaryotic cell divides, microtubules assemble, separate the cell's duplicated DNA molecules, then disassemble. As another example, microtubules that form in the growing end of a young nerve cell support its lengthening in a particular direction (Figure 3.15C).

Microfilaments are fine fibers that consist primarily of subunits of a protein called actin (Figure 3.15B). These fibers strengthen or change the shape of eukaryotic cells, and have a critical function in cell migration, movement, and contraction. Crosslinked, bundled, or gel-like arrays of them make up the cell cortex, a reinforcing mesh under the plasma membrane. Microfilaments also connect plasma membrane proteins to other proteins inside the cell.

Intermediate filaments are the most stable elements of the cytoskeleton, forming a framework that lends structure and resilience to cells and tissues in multicelled organisms. Several types of intermediate filaments are assembled from different proteins. For example, intermediate filaments that make up your hair consist of keratin, a fibrous protein. Intermediate filaments of lamins (another fibrous protein) support the nuclear envelope, and also help regulate processes inside the nucleus such as DNA replication.

Motor proteins that associate with cytoskeletal elements move cell parts when energized by a phosphate-group transfer from ATP. A cell is like a bustling train station, with molecules and structures being moved continuously throughout its interior. Motor proteins are a bit like freight trains, dragging cellular cargo along tracks of microtubules and microfilaments (Figure 3.16). One motor protein, myosin, brings about muscle cell contraction by interacting with microfilaments. Dynein, another motor protein, interacts with microtubules to bring about movement of eukaryotic flagella and cilia. Eukaryotic flagella propel sperm and other motile cells through fluid by whipping back and forth, a motion that differs from the propeller-like rotation of prokaryotic flagella. **Cilia** (singular, cilium) are short, hairlike structures that project from the surface of some eukaryotic cells. The coordinated waving of many cilia can propel a cell through fluid, and stir fluid around a stationary cell. Cilia on thousands of cells lining your airways sweep inhaled particles away from your lungs.

Some eukaryotic cells, including the amoeba at left, form **pseudopods**, or "false feet." As these temporary, irregular lobes bulge outward, they can move the entire cell or engulf a target such as prey. Elongating microfilaments force the lobe to advance in a steady direction. Motor proteins attached to the microfilaments drag the plasma membrane along with them.

Astrid Hanns-FriederMichler/ Science Source

tubulin subunit

actin subunit

25 nm

6–7 nm

A. Microtubule.

B. Microfilament.

10 µm

C. A fluorescence micrograph shows microtubules (yellow) and microfilaments (blue) in the growing end of a nerve cell. These cytoskeletal elements support and guide the cell's lengthening in a particular direction.

Figure 3.15 Cytoskeletal elements.

(C) © Dylan T. Burnette and Paul Forscher.

Figure 3.16 Motor proteins.
Here, kinesin (tan) drags a pink vesicle as it inches along a microtubule.

chloroplast Organelle of photosynthesis in the cells of plants and photosynthetic protists.

cilia Short, movable structures that project from the plasma membrane of some eukaryotic cells.

cytoskeleton Network of protein filaments that support, organize, and move eukaryotic cells and their internal structures.

intermediate filament Stable cytoskeletal element that structurally supports cells and tissues.

microfilament Reinforcing cytoskeletal element that functions in cell movement; a fiber of actin subunits.

microtubule Cytoskeletal element involved in movement; hollow filament of tubulin subunits.

motor protein Type of energy-using protein that interacts with cytoskeletal elements to move the cell's parts or the whole cell.

pseudopod A temporary protrusion that helps some eukaryotic cells move and engulf prey.

Figure 3.17 A plant ECM.
This section through a plant leaf shows the cuticle, a protective covering of deposits secreted by living cells.

George S. Ellmore.

— cuticle

— outer cell of leaf

— photosynthetic cell inside leaf

Figure 3.18 Three types of cell junctions common in animal tissues: tight junctions, gap junctions, and adhering junctions.
The micrograph shows how a profusion of tight junctions (green) seals abutting surfaces of kidney cell membranes to form a waterproof tissue. The DNA in each cell nucleus appears red.

© ADVANCELL (Advanced In Vitro Cell Technologies; S.L.) www.advancell.com.

free surface of epithelial tissue

— tight junctions

— gap junctions

— adhering junctions

basement membrane

Extracellular Matrix Many cells secrete an **extracellular matrix** (**ECM**), a complex mixture of molecules that varies by cell type, but often includes polysaccharides and fibrous proteins. A cell wall is an example of ECM. You learned in Section 3.4 that many prokaryotes have walls. Plants have them too, as do fungi and some protists. Wall composition differs among these groups. Animal cells have no walls, but some types secrete an extracellular matrix called basement membrane. Despite the name, basement membrane is not a cell membrane because it does not consist of lipids. Rather, it is a sheet of fibrous material that structurally supports and organizes tissues. A **cuticle** is a type of ECM secreted by cells at a body surface. In plants, a cuticle of waxes and other hydrophobic compounds helps stems and leaves fend off insects and retain water (Figure 3.17). Crabs, spiders, and other arthropods have a cuticle that consists mainly of chitin, a tough polysaccharide.

Cell Junctions In multicelled species, cells can interact with one another and their surroundings by way of cell junctions. **Cell junctions** are structures that connect a cell directly to other cells or to its environment. Cells send and receive substances and signals through some junctions. Other junctions help cells recognize and stick to each other and to ECM.

Three types of cell junctions are common in animal tissues (Figure 3.18). In tissues that line body surfaces and internal cavities, rows of tight junctions fasten the plasma membranes of adjacent cells and prevent body fluids from seeping between them. For example, the lining of the stomach is leakproof because tight junctions seal its cells together. Adhering junctions consist of adhesion proteins, and they make a tissue quite strong by connecting cytoskeletal elements of adjacent cells, and cytoskeletal elements to basement membrane. Contractile tissues (such as heart muscle) have a lot of adhering junctions, as do tissues subject to abrasion or stretching (such as skin). Gap junctions are closable channels that connect the cytoplasm of adjoining animal cells. When open, they permit water, ions, and small molecules to pass directly from the cytoplasm of one cell to another. These channels allow entire regions of cells to respond to a single stimulus. Heart muscle and other tissues in which the cells perform a coordinated action have many gap junctions.

In plants, plasmodesmata (singular, plasmodesma) are open channels that connect the cytoplasm of adjoining cells. These cell junctions extend across the cell walls, and, like gap junctions, they allow substances to flow quickly from cell to cell.

Take-Home Message 3.5

What structures are common in eukaryotic cells?

- A typical eukaryotic cell has many membrane-enclosed organelles.
- A nucleus protects and controls access to the cell's DNA. Organelles of the endomembrane system make, modify, and transport proteins and lipids. Mitochondria produce ATP; chloroplasts carry out photosynthesis.
- An extensive cytoskeleton of protein filaments reinforces a eukaryotic cell's shape, and is the basis of movement of the cell and its parts.
- Many cells secrete an extracellular matrix (ECM) such as a cell wall. Plant cells, fungi, and some protists are walled, but not animal cells.
- Cell junctions structurally and functionally connect cells in tissues. In animal tissues, cell junctions also connect cells with basement membrane.

3.6 The Nature of Life

Carbon, hydrogen, oxygen, and other atoms of organic molecules are the stuff of you, and us, and all of life. Yet it takes more than organic molecules to complete the picture. Life continues only as long as an ongoing flow of energy sustains its organization, because assembling molecules and cells requires energy. Life is no more and no less than a marvelously complex system for prolonging order. With energy and the hereditary codes of DNA, matter becomes organized, generation after generation.

In this chapter, you learned about the structure of cells, which have at minimum a plasma membrane, cytoplasm, and DNA. We often use differences in other cellular components—the presence or absence of a particular organelle, for example—to categorize life's diversity. What about life's commonality? The cell is the smallest unit with the properties of life, but what is it, exactly, that makes a cell, or an organism that consists of them, alive? According to evolutionary biologist Gerald Joyce, the simplest definition of life might well be "that which is squishy." He says, "Life, after all, is protoplasmic and cellular. It is made up of cells and organic stuff and is undeniably squishy."

Defining life more unambiguously than "squishy" is challenging, if not impossible. Even deciding what sets the living apart from the nonliving can be tricky. For example, living things have a high proportion of the organic molecules of life, but so do the remains of dead organisms in seams of coal. Living things use energy to reproduce themselves, but computer viruses, which are arguably not alive, can do that too.

So how do biologists, who study life as a profession, define it? The short answer is that their best definition is a long list of properties that collectively apply to living things, and not to nonliving things. You already know about two of these properties:

1. Living things make and use the organic molecules of life . . .

2. . . . and they consist of one or more cells.

The remainder of this book details the other properties of life:

3. Living things engage in self-sustaining biological processes such as metabolism and homeostasis . . .

4. . . . and they change over their lifetime, for example by maturing and aging . . .

5. . . . and they use DNA as their hereditary material when they reproduce . . .

6. . . . and they have the collective capacity to change over successive generations, for example by adapting to environmental pressures.

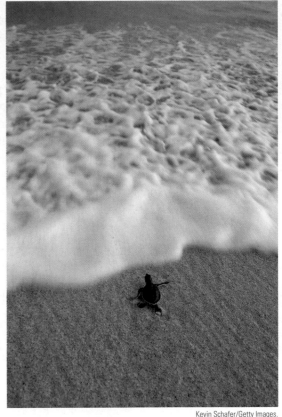

Kevin Schafer/Getty Images.

"Life . . . is made up of cells and organic stuff and is undeniably squishy."

Take-Home Message 3.6

What, exactly, is life?

- We describe the characteristic of "life" in terms of properties that are collectively unique to living things.
- Organisms make and use the organic molecules of life. DNA is their hereditary material.
- In living things, the molecules of life are organized as one or more cells that engage in self-sustaining biological processes.
- Living things change over lifetimes, and also over successive generations.

cell junction Structure that connects a cell to another cell or to extracellular matrix.

cuticle Secreted covering at a body surface.

extracellular matrix (ECM) Complex mixture of cell secretions, the composition and function of which vary by cell type.

Summary

Section 3.1 A huge number of bacteria live in and on the human body. Most of them are helpful; only a few types can cause disease. Contamination of food with disease-causing bacteria can result in illness that is sometimes fatal.

Section 3.2 By the **cell theory**, all organisms consist of one or more cells; the cell is the smallest unit of life; each new cell arises from another, preexisting cell; and a cell passes hereditary material to its offspring. All cells start out life with **cytoplasm**, DNA, and a **plasma membrane** that controls the types and kinds of substances that cross it. Most have many additional components. A eukaryotic cell's DNA is contained within a **nucleus**, which is a membrane-enclosed **organelle**. The **surface-to-volume ratio** limits cell size and influences cell shape. Different types of microscopes and techniques reveal different internal and external details of cells.

Section 3.3 A cell membrane is a lipid bilayer (of mainly phospholipids) with many other molecules attached or embedded in it. A bacterial or eukaryotic cell membrane can be described as a **fluid mosaic**; archaeal membranes are not fluid. Proteins carry out most membrane functions. All cell membranes have enzymes, and all have **transport proteins** that help substances move across the membrane. Plasma membranes also incorporate **adhesion proteins** that lock cells together in tissues. Plasma membranes and some internal membranes have **receptor proteins** that trigger a change in cell activities in response to a stimulus.

Section 3.4 Bacteria and archaea (prokaryotes) are single-celled organisms with no nucleus. All have DNA and **ribosomes**. Many also have a protective, rigid **cell wall** and a sticky capsule, and some have motile structures (**flagella**) and other projections (**pili**). They often have plasmids in addition to the single circular molecule of DNA. Bacteria and other microorganisms may live together in a shared mass of slime as a **biofilm**.

Section 3.5 The nucleus in cells of eukaryotes (protists, fungi, plants, and animals) protects and controls access to the cell's DNA. The nucleus has a **nuclear envelope**, a double membrane studded with pores through which molecules pass into and out of the nucleus. A typical eukaryotic cell has many other membrane-enclosed organelles, including the **endoplasmic reticulum (ER)**, a continuous system of sacs and tubes extending from the nuclear envelope. Polypeptides made in ribosome-studded rough ER pass to smooth ER, which makes lipids and breaks down carbohydrates and fatty acids. **Golgi bodies** modify proteins and lipids before sorting them into **vesicles**. Enzymes in **peroxisomes** break down substances such as amino acids, fatty acids, and toxins. Enzymes in **lysosomes** break down cellular wastes and debris. **Mitochondria** produce ATP by aerobic respiration; **chloroplasts** carry out photosynthesis.

Elements of a **cytoskeleton** reinforce, organize, and move cell structures and often the entire cell. Cytoskeletal elements include microtubules, microfilaments, and intermediate filaments. Interactions between ATP-driven **motor proteins** and **microtubules** bring about movement of **cilia** and eukaryotic flagella. Elongating **microfilaments** bring about movement of **pseudopods**. **Intermediate filaments** lend structural support to cells and tissues, and they help support the nuclear membrane.

A secreted mixture of materials forms **extracellular matrix (ECM)** that has different functions depending on the cell type. In animals, a secreted basement membrane supports and organizes cells in tissues. Among the eukaryotes, plant cells, fungi, and many protists secrete a cell wall around their plasma membrane. Many eukaryotic cell types also secrete a protective **cuticle**.

Cell junctions connect cells to one another and to their environment. Plasmodesmata (in plants) and gap junctions (in animals) connect the cytoplasm of adjacent cells. Also in animal cells, adhering junctions that connect to cytoskeletal elements fasten cells to one another and to basement membrane. Tight junctions form a waterproof seal between cells.

Section 3.6 All living things share a characteristic set of features. They make and use the molecules of life; they consist of one or more cells that engage in self-sustaining biological processes; they change over their lifetime; and they pass their DNA to offspring that can change over generations.

Self-Quiz

Answers in Appendix I

1. All cells have these three things in common:
   a. cytoplasm, DNA, and organelles with membranes
   b. a plasma membrane, DNA, and a nuclear envelope
   c. cytoplasm, DNA, and a plasma membrane
   d. a cell wall, cytoplasm, and DNA

2. Unlike eukaryotic cells, prokaryotic cells _____ .
   a. have no plasma membrane   c. have no nucleus
   b. have RNA but not DNA   d. a and c

3. Every cell is descended from another cell. This idea is part of _____ .
   - a. evolution
   - b. the theory of heredity
   - c. the cell theory
   - d. cell biology

4. The surface-to-volume ratio _____ .
   - a. does not apply to prokaryotic cells
   - b. is part of the cell theory
   - c. constrains cell size
   - d. b and c

5. True or false? Some protists start out life with no nucleus.

6. Cell membranes consist mainly of _____ and _____ .
   - a. lipids; carbohydrates
   - b. phospholipids; proteins
   - c. lipids; carbohydrates
   - d. phospholipids; ECM

7. Which of the following statements is correct?
   - a. Ribosomes are only found in bacteria and archaea.
   - b. Some animal cells are prokaryotic.
   - c. Only eukaryotic cells have mitochondria.
   - d. The plasma membrane is the outermost boundary of all cells.
   - e. Most membrane functions are carried out by phospholipids.

8. In a lipid bilayer, the _____ of all the lipid molecules are sandwiched between all of the _____ .
   - a. hydrophilic tails; hydrophobic heads
   - b. hydrophilic heads; hydrophilic tails
   - c. hydrophobic tails; hydrophilic heads
   - d. hydrophobic heads; hydrophilic tails

9. The main function of the endomembrane system is _____ .
   - a. building and modifying proteins and lipids
   - b. isolating DNA from toxic substances
   - c. secreting extracellular matrix onto the cell surface
   - d. producing ATP by aerobic respiration

10. Enzymes contained in _____ break down worn-out organelles, bacteria, and other particles.
    - a. lysosomes
    - b. mitochondria
    - c. endoplasmic reticulum
    - d. peroxisomes

11. Put the following structures in order according to the pathway of a secreted protein:
    - a. plasma membrane
    - b. Golgi bodies
    - c. endoplasmic reticulum
    - d. post-Golgi vesicles

12. No animal cell has a _____ .
    - a. plasma membrane
    - b. flagellum
    - c. lysosome
    - d. cell wall

13. _____ connect the cytoplasm of plant cells.
    - a. Plasmodesmata
    - b. Adhering junctions
    - c. Tight junctions
    - d. Adhesion proteins

14. Which of the following organelles contains no DNA? Choose all that are correct.
    - a. nucleus
    - b. Golgi body
    - c. mitochondrion
    - d. chloroplast

15. Match each cell component with its main function.
    - _____ mitochondrion
    - _____ chloroplast
    - _____ ribosome
    - _____ nucleus
    - _____ cell junction
    - _____ flagellum
    - _____ cuticle
    - a. connection
    - b. protective covering
    - c. ATP production
    - d. protects DNA
    - e. protein synthesis
    - f. photosynthesis
    - g. movement

Critical Thinking

1. In a classic episode of *Star Trek*, a gigantic amoeba engulfs an entire starship. Spock blows the cell to bits before it has a chance to reproduce. Think of at least one problem a biologist would have with this particular scenario.

2. In plants, the cell wall forms as a young plant cell secretes polysaccharides onto the outer surface of its plasma membrane. Being thin and pliable, this primary wall allows the cell to enlarge and change shape. In mature woody plants, cells in some tissues deposit material onto the primary wall's inner surface. Why doesn't this secondary wall form on the outer surface of the primary wall?

Visual Question

1. What type of micrograph is shown below? Is the organism pictured prokaryotic or eukaryotic? How can you tell?

P. L. Walne and J. H. Arnott, Planta, 77:325–354, 1967.

ENERGY AND METABOLISM

Application ❯ 4.1 A Toast to Alcohol Dehydrogenase

Most college students are under the legal drinking age, but alcohol abuse continues to be the most serious drug problem on college campuses throughout the United States. Recent surveys polled tens of thousands of undergraduates about their drinking habits, and more than half of them reported regularly consuming five or more alcoholic beverages within a two-hour period—a self-destructive behavior called binge drinking. Drinking large amounts of alcohol in a brief period of time is an extremely risky behavior, both for the drinkers and people around them. Every year, around 600,000 students injure themselves while under the influence of alcohol; intoxicated students physically assault 690,000 people and sexually assault 97,000 others. Binge drinking is responsible for killing or causing the death of 1,825 students per year.

Before you drink, consider what you are consuming. All alcoholic drinks—beer, wine, hard liquor—contain the same psychoactive ingredient: ethanol. Almost all ingested ethanol ends up in the liver, a large organ in the abdomen with many important functions. Liver cells make an enzyme, alcohol dehydrogenase (ADH), which is part of a metabolic pathway that detoxifies alcohols (Figure 4.1). This pathway evolved long before humans began to consume alcoholic beverages. Its main function in our bodies is to break down the tiny amount of alcohol that we encounter naturally: Foods such as ripe fruit contain it, and it also forms in our bodies as a metabolic by-product of body cells and gut bacteria.

ADH converts ethanol to acetaldehyde, an organic compound even more toxic than ethanol and the most likely source of various hangover symptoms. A second enzyme converts the toxic acetaldehyde to acetate, which is a nontoxic salt. In the average healthy adult human, these two enzymes can detoxify between 7 and 14 grams of ethanol per hour. The average alcoholic beverage contains between 10 and 20 grams of ethanol, which is why having more than one drink in any two-hour interval may result in a hangover.

Putting more alcohol into your body than your enzymes can detoxify damages it more permanently than a hangover, however. Ethanol breakdown harms liver cells, so the more a person drinks, the fewer liver cells are left to do the breaking down. Ethanol also interferes with normal processes of metabolism. For example, oxygen that would ordinarily take part in breaking down fatty acids is diverted to breaking down the ethanol. As a result, fats tend to accumulate as large globules in the tissues of heavy drinkers. Long-term heavy drinking causes alcoholic hepatitis,

Figure 4.1 Alcohol dehydrogenase. Left, this enzyme helps the body break down toxic molecules such as ethanol, making it possible for humans to drink beer, wine, and other alcoholic beverages.

Right, a tailgate party at a Notre Dame–Alabama football game. Indiana State police arrested 138 Notre Dame students for underage drinking at tailgate parties during 2012.

© Al Diaz/Miami Herald/MCT via Getty Images.

alcohol dehydrogenase

a disease characterized by inflammation and destruction of liver tissue. It also causes cirrhosis, a condition in which the liver becomes so scarred, hardened, and filled with fat that it loses its function. (The term cirrhosis is from the Greek *kirros*, meaning "orange-colored," after the abnormal skin color of people with the disease.) A cirrhotic liver can no longer produce the protein albumin, so the solute balance of body fluids is disrupted, and the legs and abdomen swell with watery fluid. It can no longer remove drugs and other toxins from the blood, so they accumulate in the brain—which impairs mental functioning and alters personality. Restricted blood flow through a cirrhotic liver causes veins to enlarge and rupture, so internal bleeding is a risk. The damage to the body results in a heightened susceptibility to diabetes and liver cancer. Once cirrhosis has been diagnosed, a person has about a 50 percent chance of dying within 10 years (Figure 4.2).

Figure 4.2 Gary Reinbach.
The 22-year-old died from alcoholic liver disease shortly after this photograph was taken, in 2009. The odd color of his skin is a symptom of cirrhosis.

Transplantation is a last-resort treatment for a failed liver, but there are not enough liver donors for everyone who needs a transplant. Reinbach was refused a transplant that may have saved his life because he had not abstained from drinking for the prior 6 months.

Stuart Clark/The Sunday Times/nisyndication.

4.2 Life Runs on Energy

REMEMBER: A one-way flow of energy and a cycling of nutrients sustain life's organization (Section 1.3). A law of nature describes something that occurs without fail, but our explanation of why it occurs is incomplete (1.6).

Energy is formally defined as the capacity to do work, but this definition is not very satisfying. Even the brilliant physicists who study it cannot say exactly what it is. However, we do have an intuitive understanding of energy just by thinking about familiar forms of it, such as light, heat, electricity, and motion. We also understand intuitively that one form of energy can be converted to another. Think about how an automobile changes the chemical energy of gasoline into the energy of motion (kinetic energy), or how a lightbulb changes electricity into light.

The formal study of heat and other forms of energy is called thermodynamics (*therm* means heat; *dynam* means power). By making careful measurements, thermodynamics researchers discovered that the total amount of energy before and after every conversion is always the same. In other words, energy cannot be created or destroyed—a phenomenon that is the **first law of thermodynamics**. Energy also tends to spread out, or disperse, until no part of a system holds more than another part. In a kitchen, for example, heat always flows from a hot pan to cool air until the temperature of both is the same. We never see cool air raising the temperature of a hot pan. The tendency of energy to spread out spontaneously is the **second law of thermodynamics**.

Work occurs as a result of energy transfers. Consider how it takes work to push a box across a floor. In this case, a body (you) transfers energy to another body (the box) to make it move. Similarly, a plant cell works to make sugars. Inside the cell, one set of molecules harvests energy from light, then transfers it to another set of molecules. A second set of molecules uses the energy to build sugars from carbon dioxide and water. This particular energy transfer involves the conversion of light energy to chemical energy. Most other types of cellular work occur by the transfer of chemical energy from one molecule to another.

As you learn about such processes, remember that every time energy is transferred, a bit of it disperses—usually in the form of heat. As a simple example, a typical incandescent lightbulb converts only about 5 percent of the energy of electricity into light. The remaining 95 percent of the energy ends up as heat that disperses from the bulb.

© Piotr Marcinski/Shutterstock.com.

energy The capacity to do work.

first law of thermodynamics Energy cannot be created or destroyed.

second law of thermodynamics Energy tends to disperse spontaneously.

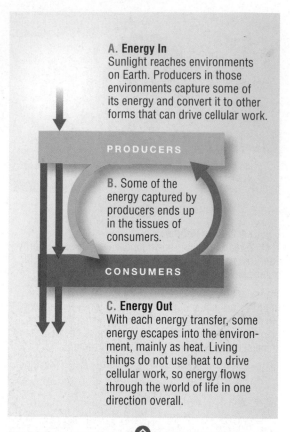

A. Energy In
Sunlight reaches environments on Earth. Producers in those environments capture some of its energy and convert it to other forms that can drive cellular work.

PRODUCERS

B. Some of the energy captured by producers ends up in the tissues of consumers.

CONSUMERS

C. Energy Out
With each energy transfer, some energy escapes into the environment, mainly as heat. Living things do not use heat to drive cellular work, so energy flows through the world of life in one direction overall.

Figure 4.3 Energy flow through the world of life.

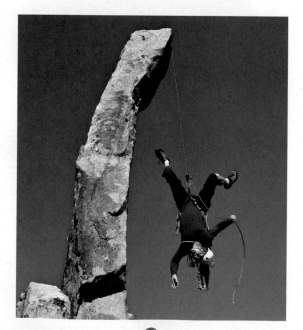

Figure 4.4 Illustration of potential energy.
By opposing gravity's downward pull, the rope keeps the man from falling. Similarly, a chemical bond that attaches two atoms keeps them from flying apart.

© Greg Epperson/Shutterstock.com.

Dispersed heat is not useful for doing work, and it is not easily converted to a more useful form of energy (such as electricity). Because some energy in every transfer disperses as heat, and heat is not useful for doing work, we can say that the total amount of energy in the universe available for doing work is always decreasing.

Is life an exception to this inevitable flow? An organized body is hardly dispersed. Energy becomes concentrated in each new organism as the molecules of life organize into cells. Even so, living things constantly use energy—to grow, to move, to acquire nutrients, to reproduce, and so on—and some energy is lost in every one of these processes (Figure 4.3). Unless the losses are replenished with energy from another source, life's complex organization will end.

The energy that fuels most life on Earth comes from the sun. That energy flows through producers such as plants, then consumers such as animals (Figure 4.3). During this journey, the energy is transferred many times. With each transfer, some energy escapes as heat until, eventually, all of it is permanently dispersed. However, the second law of thermodynamics does not say how quickly the dispersal has to happen. Energy's spontaneous dispersal is resisted by chemical bonds. The energy in chemical bonds is a type of potential energy, which is energy stored in the position or arrangement of objects in a system (Figure 4.4). Think of the bonds in the countless molecules that make up your skin, heart, liver, fluids, and other body parts. Those bonds hold the molecules, and you, together—at least for the time being.

Take-Home Message 4.2

What is energy?

- Energy, which is the capacity to do work, cannot be created or destroyed.
- Energy disperses spontaneously.
- Energy can be transferred between systems or converted from one form to another, but some is lost (as heat, typically) during every exchange.
- Sustaining life's organization requires ongoing energy inputs to counter energy loss. Organisms stay alive by replenishing themselves with energy they harvest from someplace else.

4.3 **Energy in the Molecules of Life**

REMEMBER: A chemical formula (Section 2.3) indicates unvarying proportions of elements in a molecule. Reactions are processes of molecular change (2.6). Cellulose consists of chains of glucose monomers (2.7).

All cells store and retrieve energy in chemical bonds of the molecules of life, and these activities occur by way of reactions. During a reaction, one or more **reactants** (molecules that enter the reaction and become changed by it) become one or more **products** (molecules that are produced by the reaction). Intermediate molecules may form between reactants and products. We show a reaction as an equation in which an arrow points from reactants to products:

$$2H_2 \quad + \quad O_2 \quad \longrightarrow \quad 2H_2O$$
(hydrogen)   (oxygen)   (water)

Figure 4.5 Energy inputs and outputs in chemical reactions.

① Some reactions convert molecules with lower energy to molecules with higher energy, so they require a net energy input in order to proceed.

② Other reactions convert molecules with higher energy to molecules with lower energy, so they end with an energy release.

Figure It Out: Which law of thermodynamics explains energy inputs and outputs in chemical reactions?

Answer: The first law

A number before a chemical formula in such equations indicates the number of molecules; a subscript indicates the number of atoms of that element per molecule. Note that atoms shuffle around in a reaction, but they never disappear: The same number of atoms that enter a reaction remain at the reaction's end.

Every chemical bond holds a certain amount of energy. That is the amount of energy required to break the bond, and it is also the amount of energy released when the bond forms. The particular amount of energy held by a bond depends on which elements are taking part in it. For example, two covalent bonds—one between an oxygen and a hydrogen atom in a water molecule, the other between two oxygen atoms in molecular oxygen (O_2)—both hold energy, but different amounts of it. In most reactions, the energy of the reactants differs from the energy of the products. If the reactants have less energy than the products, the reaction will not proceed without a net energy input (Figure 4.5 ①). If the reactants have more energy than the products, the reaction will end with a net release of energy ②.

Why Earth Does Not Go Up in Flames The molecules of life release energy when they combine with oxygen. Think of how a spark ignites wood in a campfire. Wood is mostly cellulose, which consists of long chains of repeating glucose monomers. A spark starts a reaction that converts cellulose (in wood) and oxygen (in air) to water and carbon dioxide. This reaction releases a lot of energy—enough to initiate the same reaction with other cellulose and oxygen molecules. That is why wood keeps burning after it has been lit (Figure 4.6A).

Earth is rich in oxygen—and in potential energy-releasing reactions. Why doesn't it burst into flames? Luckily, chemical bonds do not break without at least a small input of energy, even in an energy-releasing reaction. We call this input activation energy. **Activation energy**, the minimum amount of energy required to get a chemical reaction started, is a bit like a hill that reactants must climb before they can coast down the other side to become products (Figure 4.6B).

Both energy-requiring and energy-releasing reactions have activation energy, but the amount varies with the reaction. Consider guncotton (nitrocellulose), a highly explosive derivative of cellulose. Christian Schönbein accidentally discovered a way to manufacture it when he used his wife's cotton apron to wipe up a nitric acid spill on his kitchen table, then hung it up to dry next to the oven. The apron exploded, and being a chemist in the 1800s, Schönbein was thrilled. He immediately

A. Wood continues to burn after it has been lit because the combustion reaction between cellulose molecules in wood and oxygen molecules in air releases enough energy to trigger the reaction again with other molecules. Activation energy keeps this and other energy-releasing reactions from starting without an energy input.

B. Most reactions will not begin without an input of activation energy, which is shown as a bump in an energy hill. The graph shows an energy-releasing reaction; energy-requiring reactions also have activation energy.

Figure 4.6 Activation energy.

(A) Tero Hakala/Shutterstock.com.

activation energy Minimum amount of energy required to start a reaction.

product A molecule that is produced by a reaction.

reactant A molecule that enters a reaction and is changed by participating in it.

A. Cells store energy in the chemical bonds of organic compounds.

B. Cells retrieve energy stored in the chemical bonds of organic compounds.

Figure 4.7 Cells store and retrieve energy in the chemical bonds of organic molecules.

A. A glucose molecule meets up with a phosphate in the active site of a hexokinase enzyme.

B. The reaction between glucose and phosphate produced glucose-6-phosphate, shown leaving the active site.

Figure 4.8 Example of an active site.
For simplicity, enzymes are often drawn as blobs or geometric shapes. This model shows the actual contours of an active site in hexokinase, an enzyme that adds a phosphate group to glucose and other six-carbon sugars.

PDB ID: 1GZX; Paoli, M., Liddington, R., Tame, J., Wilinson, A., Dodson, G.; Crystal Structure of T state hemoglobin with oxygen bound at all four haems. *J. Mol.Bio.*, v256, pp. 775–792, 1996.

tried marketing guncotton as a firearm explosive, but it was too unstable to manufacture. So little activation energy is needed to make guncotton react with oxygen that it tends to explode unexpectedly. Several manufacturing plants burned to the ground before guncotton was abandoned for use as a firearm explosive. The substitute: gunpowder, which has a higher activation energy for a reaction with oxygen.

Energy In, Energy Out Cells store energy by running energy-requiring reactions that build organic compounds (Figure 4.7A). For example, light energy drives photosynthesis, a pathway that produces glucose from carbon dioxide and water. Unlike light, glucose can be stored in a cell. Cells harvest energy by running energy-releasing reactions that break the bonds of organic compounds (Figure 4.7B). Most cells do this when they carry out aerobic respiration, a pathway that releases the energy of glucose by breaking the bonds between its carbon atoms. You will see in the next section how cells use energy released from some reactions to drive others (we return to the reactions of photosynthesis and aerobic respiration in Chapter 5).

Take-Home Message 4.3

How do cells use energy?

• Cells store and retrieve energy by making and breaking chemical bonds.
• Some reactions require a net input of energy. Others end with a net release of energy.
• Most chemical reactions require an input of activation energy to begin.

4.4 How Enzymes Work

REMEMBER: Electrons carry energy in incremental amounts (Section 2.2). The number of hydrogen ions in a fluid determines its pH (2.5). Molecules jiggle faster at higher temperatures (2.4). Enzymes speed reactions without being changed by them; metabolism includes all the enzyme-mediated reactions by which cells acquire and use energy as they build and break down organic molecules (2.6). A protein's function arises from and depends on its three-dimensional shape, which is held together by hydrogen bonds; when a protein denatures, it loses its shape and its function (2.9). ATP has an important metabolic role as an energy carrier (2.10).

The Need for Speed Metabolism requires enzymes. Why? Consider that a molecule of glucose can break down to carbon dioxide and water on its own, but the process might take decades. That same conversion takes just seconds inside your cells. Enzymes make the difference. An enzyme makes a reaction run much faster than it would on its own. The enzyme is unchanged by participating in the reaction, so it can work again and again.

Some enzymes are RNAs, but most are proteins. Each kind of enzyme interacts only with specific reactants, or **substrates**, and alters them in a specific way. Such specificity occurs because an enzyme's polypeptide (or nucleotide) chains fold to form a pocket called an **active site**, where substrates bind and a reaction proceeds (Figure 4.8). An active site is complementary in shape, size, polarity, and charge to the enzyme's substrate. That fit is the reason why each enzyme acts in a specific way on a specific substrate (Figure 4.9).

Figure 4.9 How an active site works.

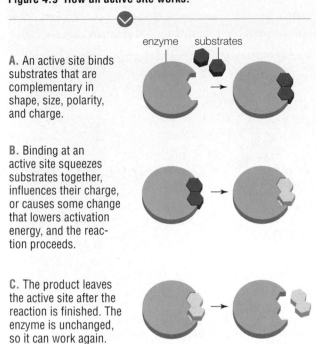

A. An active site binds substrates that are complementary in shape, size, polarity, and charge.

B. Binding at an active site squeezes substrates together, influences their charge, or causes some change that lowers activation energy, and the reaction proceeds.

C. The product leaves the active site after the reaction is finished. The enzyme is unchanged, so it can work again.

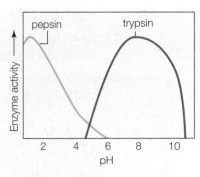

A. The pH-dependent activity of two digestive enzymes. Pepsin acts in the stomach, where the normal pH is 2. Trypsin acts in the small intestine, where the pH is normally around 7.5.

B. Temperature-dependent activity of an enzyme from two bacteria: *E. coli*, which inhabits the human gut (normally 37°C); and *Thermus aquaticus*, which lives in hot springs around 70°C.

Figure 4.10 Enzymes, temperature, and pH.
Each enzyme works best within a characteristic range of conditions—generally, the same environmental conditions in which the enzyme normally occurs.

Figure It Out: At what temperature does the *E. coli* DNA polymerase work fastest?

Answer: About 37°C

An enzyme speeds a reaction by reducing activation energy, so it lowers the barrier that prevents the reaction from proceeding. When we talk about activation energy, we are really talking about the energy required to bring reactant bonds to the breaking point. An active site can bring reactants to this state by (for example) holding them in a certain position, squeezing them, or redistributing their charge.

Factors That Influence Enzyme Activity Environmental factors such as pH, temperature, salt, and pressure influence an enzyme's shape, and so influence its function. Each enzyme works best in a particular range of conditions that reflect the environment in which it evolved.

Consider pepsin, a digestive enzyme that works best at low pH (Figure 4.10A). Pepsin begins the process of protein digestion in the very acidic environment of the stomach (pH 2). During digestion, the stomach's contents pass into the small intestine, where the pH rises to about 7.5. Pepsin denatures (unfolds) above pH 5.5, so this enzyme becomes inactive in the small intestine. Here, protein digestion continues with the assistance of trypsin, an enzyme that functions well at the higher pH.

Adding heat boosts energy, which is why the jiggling motion of atoms and molecules increases with temperature. The greater the energy of reactants, the closer they are to reaching activation energy. Thus, the rate of an enzymatic reaction typically increases with temperature—but only up to a point. An enzyme denatures above a characteristic temperature. Then, the reaction rate falls sharply as the shape of the enzyme changes and it stops working (Figure 4.10B). Body temperatures above 42°C (107.6°F) adversely affect the function of many of your enzymes, which is why such severe fevers are dangerous.

The activity of many enzymes is also influenced by the amount of salt in the surrounding fluid. Too little salt, and polar parts of the enzyme attract one another so strongly that the molecule's shape changes. Too much salt interferes with the hydrogen bonds that hold the enzyme in its characteristic shape, so the enzyme denatures.

Cells use energy released from some reactions to drive others.

active site Pocket in an enzyme where substrates bind and a reaction occurs.

substrate Of an enzyme, a reactant that is specifically acted upon by the enzyme.

Digging Into Data

One Tough Bug

The genus *Ferroplasma* consists of a few species of acid-loving archaea. One species, *F. acidarmanus*, was discovered to be the main component of slime streamers (a type of biofilm) deep inside an abandoned California copper mine (Figure 4.11A).

F. acidarmanus cells use an ancient energy-harvesting pathway that combines oxygen with iron–sulfur compounds in minerals such as pyrite. This reaction dissolves the minerals, so groundwater that seeps into the mine ends up with extremely high concentrations of metal ions such as copper, zinc, cadmium, and arsenic. The reaction also produces sulfuric acid, which lowers the pH of the water around the cells to zero.

Despite living in an environment with a composition similar to hot battery acid, *F. acidarmanus* cells maintain their internal pH at a cozy 5.0. Thus, researchers investigating *Ferroplasma* metabolic enzymes were surprised to discover that most of the cells' enzymes function best at very low pH (Figure 4.11B).

1. What does the dashed line signify?
2. Of the four enzymes profiled in the graph, how many function optimally at a pH lower than 5? How many retain significant function at pH 5?
3. What is the optimal pH for *F. acidarmanus* carboxylesterase?

A. Deep inside one of the most toxic sites in the United States: Iron Mountain Mine, in California. The water in this stream, which is about 1 meter (3 feet) wide in this photo, is hot (around 40°C, or 104°F), heavily laden with arsenic and other toxic metals, and has a pH of zero. Slime streamers growing in it are a biofilm dominated by a species of archaea, *Ferroplasma acidarmanus*.

B. pH profiles of four enzymes isolated from *F. acidarmus*. Researchers had expected these enzymes to function best at the cells' cytoplasmic pH (5.0).

Figure 4.11 pH anomaly of *Ferroplasma acidarmanus* enzymes.

(A) Dr. Katrina J. Edwards; (B) From Golyshina et al., *Environmental Microbiology*, 8(3): 416–425. © 2006 John Wiley and Sons. Used with permission of the publisher.

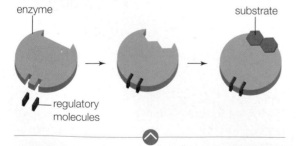

Figure 4.12 Regulatory molecule binding to enzymes. Some types of regulatory molecules (red) bind to an enzyme in a place other than the active site. This binding changes the shape of the enzyme in a way that enhances or inhibits its function.

Regulatory molecules (or ions) enhance or inhibit the activity of many enzymes. Some of these substances exert their effects by binding directly to an active site; others bind elsewhere on the enzyme. In the latter case, binding of the regulatory molecule alters the overall shape of the enzyme (Figure 4.12).

Cofactors Most enzymes (and many other proteins) can function properly only with assistance from metal ions or small organic molecules. These helpers are called **cofactors**. Many dietary vitamins and minerals are essential because they are cofactors or become converted into cofactors. **Coenzymes**, which are organic cofactors, carry chemical groups, atoms, or electrons from one reaction to another, and often into or out of organelles. In some reactions, coenzymes stay tightly bound to the enzyme. In others, they participate as separate molecules.

Unlike enzymes, many coenzymes are modified by taking part in a reaction. They are regenerated in separate reactions. Consider NAD$^+$ (nicotinamide adenine dinucleotide), a coenzyme derived from niacin (vitamin B$_3$). NAD$^+$ can accept electrons and hydrogen atoms, thereby becoming NADH. When electrons and hydrogen atoms are removed from NADH, NAD$^+$ forms again:

$$\text{NAD}^+ + \text{electrons} + \text{H}^+ \longrightarrow \boxed{\text{NADH}} \longrightarrow \text{NAD}^+ + \text{electrons} + \text{H}^+$$

In cells, the nucleotide ATP (adenosine triphosphate) functions as a coenzyme in many reactions. Bonds between phosphate groups hold a lot of energy, and ATP has two of these bonds holding its three phosphate groups together (Figure 4.13A). When a phosphate group is transferred to or from a nucleotide, energy is transferred along with it. Thus, the nucleotide can receive energy from an energy-releasing reaction, and it can also donate energy that contributes to the "energy in" part of an energy-requiring reaction. ATP is such an important currency in a cell's energy economy that we use a cartoon coin to symbolize it.

A reaction in which a phosphate group is transferred from one molecule to another is called a **phosphorylation**. ADP (adenosine diphosphate) forms when an enzyme transfers a phosphate group from ATP to another molecule (Figure 4.13B). Cells constantly run this reaction in order to drive a variety of energy-requiring reactions. Thus, they constantly have to replenish their stockpile of ATP—by running energy-releasing reactions that phosphorylate ADP. This cycle of using and replenishing ATP couples energy-requiring reactions with energy-releasing ones. As you will see in Chapter 5, cells harvest energy from organic compounds by running metabolic pathways that break them down. Energy that cells harvest in these pathways is not released to the environment, but rather stored in the high-energy phosphate bonds of ATP molecules and in electrons carried by coenzymes. The ATP and the coenzymes are then used to drive many of the energy-requiring reactions that a cell runs.

Metabolic Pathways   Building, rearranging, or breaking down an organic substance often occurs stepwise, in a series of enzymatic reactions called a **metabolic pathway**. Some metabolic pathways are linear, meaning that the reactions run straight from reactant to product:

$$\text{reactant} \xrightarrow{\text{enzyme 1}} \text{intermediate} \xrightarrow{\text{enzyme 2}} \text{intermediate} \xrightarrow{\text{enzyme 3}} \text{product}$$

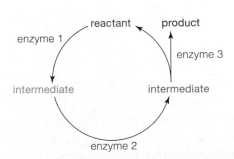

Other reactions are cyclic. In a cyclic pathway, the last step regenerates a reactant of the first step (left).

Controlling Metabolism   Cells conserve energy and resources by making only what they need at any given moment—no more, no less. Several mechanisms help a cell maintain, raise, or lower its production of thousands of different substances. Consider that reactions do not only run from reactants to products. Many also run in reverse at the same time, with some of the products being converted back into reactants. The rates of the forward and reverse reactions often depend on the concentrations of reactants and products: A high concentration of reactants pushes the reaction in the forward direction; a high concentration of products pushes it in reverse.

A. Bonds between phosphate groups hold a lot of energy. ATP has two of these bonds.

B. ADP forms in a reaction that removes a phosphate group from ATP. Energy released in this reaction drives other reactions that are the stuff of cellular work. ATP forms again in reactions that phosphorylate ADP.

Figure 4.13 ATP as an energy carrier.

coenzyme An organic cofactor.

cofactor A molecule or metal ion that associates with a protein and is necessary for its function.

metabolic pathway Series of enzyme-mediated reactions by which cells build, remodel, or break down an organic molecule.

phosphorylation A phosphate-group transfer.

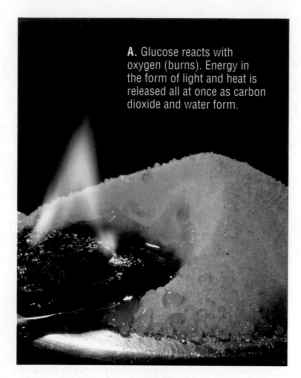

A. Glucose reacts with oxygen (burns). Energy in the form of light and heat is released all at once as carbon dioxide and water form.

B. In cells, glucose reacts with oxygen in a stepwise fashion that involves an electron transfer chain, represented here by a staircase. Energy is released in amounts that cells are able to use.

1 An input of activation energy splits glucose into carbon dioxide, electrons, and hydrogen ions (H⁺).

2 Electrons lose energy ( ) as they pass from one molecule to the next in an electron transfer chain. That energy is harnessed for cellular work.

3 Electrons, hydrogen ions, and oxygen combine to form water.

Figure 4.15 Comparing uncontrolled and controlled energy release.
The overall reaction is the same in both cases.

(A) © 2004 Richard Megna–Fundamental Photographs.

Figure 4.14 Feedback inhibition.
In this example, three different enzymes act in sequence to convert a substrate to a product. The product inhibits the activity of the first enzyme.

Figure It Out: Is this a cyclic or a linear pathway?

Answer: This is a linear pathway.

Other mechanisms more actively regulate pathways. Regulatory substances govern how fast enzyme molecules are made, or influence the activity of enzymes that have already been built. Regulation of a single enzyme can affect an entire metabolic pathway. In some cases, the end product of a series of enzymatic reactions inhibits the activity of one of the enzymes in the series (Figure 4.14). This type of regulatory mechanism, in which a change that results from an activity decreases or stops the activity, is called **feedback inhibition**.

Electron Transfers The bonds of organic molecules hold a lot of energy that can be released in a reaction with oxygen. Burning is one type of reaction with oxygen, and it releases the energy of organic molecules all at once (Figure 4.15A). Cells use oxygen to break the bonds of organic molecules, but they have no way to harvest the sudden burst of energy that occurs during burning. Instead, they break the molecules apart in pathways that release the energy in small, manageable steps. Most of these steps are electron transfers, in which one molecule accepts electrons from another. Energy is harvested in these reactions. In the next chapter, you will learn about the importance of this process in electron transfer chains. An **electron transfer chain** is a series of membrane-bound enzymes and other molecules that give up and accept electrons in turn. Electrons are at a higher energy level when they enter a chain than when they leave. Energy given off by an electron as it drops to a lower energy level is harvested by molecules of the electron transfer chain to do cellular work (Figure 4.15B).

Take-Home Message 4.4

How do enzymes work in metabolic pathways?

- Binding at an enzyme's active site causes substrate bonds to reach their breaking point, and the reaction can run spontaneously to completion.
- Each enzyme works best within a characteristic range of environmental conditions. Many enzymes require the assistance of cofactors.
- ATP often couples reactions that release energy with reactions that require energy.
- A metabolic pathway is a series of enzyme-mediated reactions that builds, breaks down, or remodels an organic molecule. Some pathways involve electron transfer chains.
- Cells conserve energy and resources by producing only what they require at a given time. This metabolic control arises from mechanisms that regulate individual enzymes and often entire pathways.

4.5 Diffusion and Membranes

REMEMBER: An ion carries charge (Section 2.2). Atoms and molecules move faster at higher temperatures (2.4). The amount of a solute per unit volume of solution is its concentration (2.4). The number of hydrogen ions in a fluid determines its pH (2.5). Lipid bilayers are impermeable to ions and polar molecules (3.3).

gresei/Shutterstock.com.

Metabolic pathways require the participation of molecules and ions that must move across membranes and through cells. **Diffusion** is the spontaneous spreading of molecules or atoms (left), and it is an essential way in which substances move into, through, and out of cells. Diffusion occurs because an atom or molecule is always jiggling, and this internal movement causes it to randomly bounce off of nearby objects, including other atoms or molecules. Rebounds from such collisions propel solutes through a liquid, resulting in a gradual and complete mixing. How fast this occurs depends on five factors:

1. Concentration. A difference in solute concentration between adjacent regions of solution is called a concentration gradient. Solutes tend to diffuse "down" their concentration gradient, from a region of higher concentration to one of lower concentration. Why? Consider that moving objects (such as molecules) collide more often when they are more crowded. Thus, during a given interval, more molecules get bumped out of a region of higher concentration than get bumped into it.

2. Temperature. Atoms and molecules jiggle faster at higher temperature, so they collide more often. Thus, diffusion occurs more quickly at higher temperatures.

3. Size. It takes more energy to move a large object than it does to move a small one, so ions and small molecules diffuse more quickly than large molecules.

4. Charge. Each ion or charged molecule in a fluid contributes to the fluid's overall electric charge. A difference in charge between two regions of fluid can affect the rate and direction of diffusion between them. For example, positively charged substances (such as sodium ions) will tend to diffuse toward a region with an overall negative charge.

5. Pressure. Pressure squeezes objects—including atoms and molecules—closer together. Atoms and molecules that are more crowded collide and rebound more frequently. Thus, diffusion occurs faster at higher pressures.

Semipermeable Membranes Lipid bilayers are selectively permeable, which means that they are permeable to some substances and not others. Specifically, water, gases, and hydrophobic molecules can diffuse directly through a lipid bilayer; most solutes—ions and polar molecules, in particular—cannot (Figure 4.16). When a lipid bilayer separates two fluids with differing solute concentrations, water will diffuse across it. The direction of water movement depends on the relative solute concentration of the two fluids, which we describe in terms of tonicity. If the overall solute concentrations of the two fluids differ, the fluid with the lower concentration of solutes is said to be **hypotonic** (*hypo–*, under). The other one, with the higher solute concentration, is **hypertonic** (*hyper–*, over). Water diffuses from a hypotonic fluid into a hypertonic one. The diffusion will continue until the two fluids are **isotonic**, which means they have the same overall solute concentration. The movement of water across a membrane is so important in biology that it is given a special name: **osmosis** (Figure 4.17).

Figure 4.16 Selective permeability of lipid bilayers. Hydrophobic molecules, gases, and water molecules can cross a lipid bilayer on their own. Ions in particular and most polar organic molecules such as glucose cannot.

Figure 4.17 Osmosis. Water moves across a selectively permeable membrane that separates two fluids of differing tonicity (red dots represent solutes). The fluid volume changes in the two compartments as water diffuses across the membrane.

selectively permeable membrane

diffusion The spontaneous spreading of molecules or atoms.

electron transfer chain Series of enzymes and other molecules in a cell membrane that accept and give up electrons, thus releasing the energy of the electrons in steps.

feedback inhibition Mechanism by which a change that results from some activity decreases or stops the activity.

hypertonic Describes a fluid that has a high overall solute concentration relative to another fluid.

hypotonic Describes a fluid that has a low overall solute concentration relative to another fluid.

isotonic Describes two fluids with identical solute concentrations.

osmosis Diffusion of water across a selectively permeable membrane; occurs when there is a difference in solute concentration between the fluids on either side of the membrane.

A. Red blood cells in an isotonic solution (such as the fluid portion of blood) have a normal, indented disk shape.

B. Water diffuses out of red blood cells immersed in a hypertonic solution, so they shrivel up.

C. Water diffuses into red blood cells immersed in a hypotonic solution, so they swell up. Some of these have burst.

Figure 4.18 Effects of tonicity in human red blood cells. These cells have no mechanism to compensate for differences in solute concentration between cytoplasm and extracellular fluid.

(A) Annie Cavanagh/Wellcome Images; (B, C) CMSP/Getty Images.

A. Osmotic pressure keeps plant parts erect. These cells in an iris petal are plump with cytoplasm.

B. Cells from a wilted iris petal. The cytoplasm shrank, and the plasma membrane is pulled away from the cell wall.

Figure 4.19 Turgor, as illustrated in cells of iris petals.

(A,B) Perennou Nuridsany/Science Source; (inset) © Evgenyi/Shutterstock.com.

If a cell's cytoplasm becomes hypertonic with respect to the fluid outside of its plasma membrane, water will diffuse into the cell. If the cytoplasm becomes hypotonic, water will diffuse out. In either case, the solute concentration of the cytoplasm may change. If it changes enough, the cell's enzymes will stop working, with lethal results. Many cells have built-in mechanisms that compensate for differences in solute concentration between cytoplasm and extracellular (external) fluid. In cells with no such mechanism, the volume—and solute concentration—of cytoplasm changes when water diffuses into or out of the cell (Figure 4.18).

The rigid cell walls of plants and many protists, fungi, and bacteria can resist an increase in the volume of cytoplasm even in hypotonic environments. In the case of plant cells, cytoplasm usually contains more solutes than soil water does. Thus, water usually diffuses from soil into a plant—but only up to a point. Stiff walls keep plant cells from expanding very much, so an inflow of water causes pressure to build up inside them. Pressure that a fluid exerts against a structure that contains it is called **turgor**. When enough pressure builds up inside a plant cell, water stops diffusing into its cytoplasm. The amount of turgor that is enough to stop osmosis is called osmotic pressure.

Osmotic pressure keeps walled cells plump, just as high air pressure inside a tire keeps it inflated. A young land plant can resist gravity to stay erect because its cells are plump with cytoplasm (Figure 4.19A). When soil dries out, it loses water, so the concentration of solutes increases in it. If soil water becomes hypertonic with respect to cytoplasm, water will start diffusing out of the plant's cells, causing their cytoplasm to shrink (Figure 4.19B). As turgor inside the cells decreases, the plant wilts.

Take-Home Message 4.5

What influences the movement of ions and molecules?

- Solutes tend to diffuse into an adjoining region of fluid in which they are not as concentrated. The steepness of a concentration gradient as well as temperature, molecular size, charge, and pressure affect the rate of diffusion.

- When two fluids of different solute concentration are separated by a selectively permeable membrane, water diffuses from the hypotonic to the hypertonic fluid. This movement, osmosis, is opposed by turgor.

Extracellular Fluid

glucose

Cytoplasm

A. A glucose molecule (here, in extracellular fluid) binds to a glucose transporter (gray) in the plasma membrane.

B. Binding causes the transport protein to change shape.

C. The transport protein releases the glucose on the other side of the membrane (in cytoplasm) and resumes its original shape.

Figure 4.20 An example of facilitated diffusion.

Figure It Out: In this example, which fluid is hypotonic: extracellular fluid or cytoplasm?

Answer: Cytoplasm

4.6 Membrane Transport Mechanisms

REMEMBER: Phospholipids swirled into water will spontaneously organize themselves into lipid bilayer sheets or bubbles; a transport protein moves specific solutes across a cell membrane (Section 3.3). Some cells use pseudopods to engulf a target (3.5).

Substances that do not diffuse directly through lipid bilayers can cross a cell membrane through transport proteins embedded in it. Each type of transport protein allows a specific substance to cross: Calcium pumps pump only calcium ions; glucose transporters transport only glucose; and so on. This specificity is an important part of homeostasis. Consider how the composition of cytoplasm depends on movement of particular solutes across the plasma membrane, which in turn depends on the transporters in it. Glucose is an important source of energy for most cells, so they normally take up as much as they can from extracellular fluid. They do so with the help of glucose transporters in the plasma membrane. As soon as a molecule of glucose enters cytoplasm, an enzyme (hexokinase, shown in Figure 4.8) phosphorylates it. Phosphorylation traps the molecule in the cell because the transporters are specific for glucose, not phosphorylated glucose. Thus, phosphorylation prevents the molecule from moving back through the transport protein and leaving the cell.

Passive Transport Osmosis is an example of **passive transport**, a membrane-crossing mechanism that requires no energy input. The diffusion of solutes through transport proteins is another example. In this case, the movement of the solute (and the direction of its movement) is driven entirely by the solute's concentration gradient. Some transport proteins form pores: permanently open channels through a membrane. Other channels are gated, which means they open and close in response to a stimulus such as a shift in electric charge or binding to a signaling molecule. With a passive transport mechanism called **facilitated diffusion**, a solute binds to a transport protein, which then changes shape so the solute is released to the other side of the membrane. A glucose transporter is an example of a transport protein that works in facilitated diffusion (Figure 4.20). This protein changes shape when it binds to a molecule of glucose. The shape change moves the glucose to the opposite side of the membrane, where it detaches from the transport protein. Then, the glucose transporter reverts to its original shape.

facilitated diffusion Passive transport mechanism in which a solute follows its concentration gradient across a membrane by moving through a transport protein.

passive transport Membrane-crossing mechanism that requires no energy input.

turgor Pressure that a fluid exerts against a membrane, wall, or other structure that contains it.

A. Two calcium ions (blue) bind to the transport protein (a calcium pump, gray).

B. A phosphate group from ATP causes the protein to change shape so that the calcium ions are ejected to the opposite side of the membrane.

C. After it loses the calcium ions, the transport protein resumes its original shape.

Figure 4.21 Active transport of calcium ions.

Active Transport Maintaining a solute's concentration often means transporting the solute against its gradient, to the side of the membrane where it is more concentrated. This takes energy. In **active transport**, a transport protein uses energy to pump a solute against its gradient across a cell membrane. Typically, an energy input (for example, in the form of a phosphate-group transfer from ATP) changes the shape of the transport protein. The shape change causes the protein to release a bound solute to the other side of the membrane.

A calcium pump moves calcium ions across cell membranes by active transport (Figure 4.21). Calcium ions act as potent messengers inside cells, and they also affect the activity of many enzymes. Thus, their concentration in cytoplasm is tightly regulated. Calcium pumps in the plasma membrane of all eukaryotic cells can keep the concentration of calcium ions in cytoplasm thousands of times lower than it is in extracellular fluid. Another example of active transport involves sodium–potassium pumps (Figure 4.22). Nearly all cells in your body have these transport proteins.

Bear in mind that the membranes of all cells, not just those of animals, have proteins that carry out active transport. In plants, for example, transport proteins in the plasma membranes of photosynthetic cells pump sucrose from cytoplasm into tubes that thread throughout the plant body.

Figure 4.22 The sodium–potassium pump.
This protein (gray) actively transports sodium ions (Na^+) from cytoplasm to extracellular fluid, and potassium ions (K^+) in the other direction. The transfer of a phosphate group () from ATP provides energy required for transporting both ions against their concentration gradient.

Membrane Trafficking Vesicles are constantly carrying materials to and from a cell's plasma membrane (Figure 4.23). Cells use vesicles to take in or expel materials in bulk (as opposed to one molecule or ion at a time via transport proteins). A cell can expel bulk materials by **exocytosis**, a pathway in which a vesicle in the cytoplasm moves to the cell's surface ❶ and fuses with the plasma membrane ❷. As the exocytic vesicle loses its identity, its contents are released to extracellular fluid ❸. The cell can take in bulk materials by **endocytosis**. In one endocytic pathway, a small patch of plasma membrane balloons into the cytoplasm, bringing with it a drop of extracellular fluid (along with solutes and particles suspended in it). As the balloon sinks into the cytoplasm, the hydrophobic tails of the lipids in the bilayer are repelled by the watery fluid on both sides. The fluid "pushes" the phospholipid tails together, which helps round off the bud as a vesicle.

Figure 4.23 Membrane crossings.
A plasma membrane is a hub of activity: Molecules and ions (colored balls) are constantly flowing into and out of a cell via transport proteins embedded in its plasma membrane. Vesicles are also taking in or expelling bulk amounts of solutes and much larger particles.

1 Exocytosis begins as a vesicle moves to the plasma membrane.

2 The vesicle's membrane fuses with the plasma membrane.

3 As the membranes fuse, the vesicle's contents are released to the extracellular fluid.

Phagocytosis (which means "cell eating") is a type of endocytosis in which motile cells engulf microorganisms, cellular debris, or other large particles. Many single-celled protists such as amoebas feed by phagocytosis. Some of your white blood cells use the pathway to engulf viruses and bacteria, cancerous body cells, and other threats to health. (The micrograph on the left shows a phagocytic white blood cell engulfing several tuberculosis bacteria, in red.) Phagocytosis begins when receptor proteins bind to a particular target. The binding causes microfilaments just under the plasma membrane to contract. The contraction forces a lobe of membrane-enclosed cytoplasm to bulge outward as a pseudopod. Pseudopods that merge around a target trap it inside a vesicle that sinks into the cytoplasm.

Science Source.

Take-Home Message 4.6

How do substances and particles that cannot diffuse through lipid bilayers cross cell membranes?

- Transport proteins allow specific ions or molecules to cross a cell membrane. The types and amounts of substances that cross a membrane depend on the transport proteins embedded in it.
- In facilitated diffusion (a type of passive transport), a solute binds to a transport protein that releases it on the opposite side of the membrane. The movement is driven by the solute's concentration gradient.
- In active transport, a transport protein pumps a solute across a membrane against its concentration gradient. The movement requires energy, as from ATP.
- Exocytosis and endocytosis move materials in bulk across a plasma membrane. Some cells can engulf large particles by phagocytosis.

active transport Energy-requiring mechanism in which a transport protein pumps a solute across a cell membrane against the solute's concentration gradient.

endocytosis Process by which a cell takes in a small amount of extracellular fluid (and its contents) by the ballooning inward of the plasma membrane.

exocytosis Process by which a cell expels a vesicle's contents to extracellular fluid.

phagocytosis "Cell eating"; an endocytic pathway by which a cell engulfs large particles such as microbes or cellular debris.

Summary

Section 4.1 Currently the most serious drug problem on college campuses is binge drinking. Drinking more alcohol than the body's enzymes can detoxify damages the body, and it can be lethal in the short term or the long term.

Section 4.2 Energy is the capacity to do work. Energy cannot be created or destroyed (**first law of thermodynamics**), but it can be converted from one form to another and transferred between objects or systems. Energy tends to disperse spontaneously (**second law of thermodynamics**). A bit disperses at each energy transfer, usually in the form of heat.

Living things maintain their organization only as long as they harvest energy from someplace else. Energy flows in one direction through the biosphere, starting mainly from the sun, then into and out of ecosystems. Producers and then consumers use the captured energy to assemble, rearrange, and break down organic molecules that cycle among organisms in an ecosystem.

Section 4.3 Cells store and retrieve energy by making and breaking chemical bonds in chemical reactions, in which **reactants** are converted to **products**. Some reactions require a net energy input; others end with a net energy release. **Activation energy** is the minimum energy input required to start a reaction.

Section 4.4 Enzymes greatly enhance the rate of reactions without being changed by them. Each has an **active site** that is complementary in shape, size, polarity, and charge to the enzyme's **substrate**, and each works best within a characteristic range of conditions, including temperature, salt concentration, and pH. Most enzymes require assistance from **cofactors**; organic cofactors are **coenzymes**. ATP is often used as a coenzyme to carry energy between reactions. When a phosphate group is transferred from ATP to another molecule, energy is transferred along with it. Phosphate-group transfers (**phosphorylations**) to and from ATP couple reactions that release energy with reactions that require energy.

ATP

Metabolic pathways are sequences of enzyme-mediated reactions that build, convert, and break down organic molecules. Regulating metabolic pathways allows cells to conserve energy and resources by making only what they need at a given time. The activity of many enzymes can be regulated, for example by binding of a regulatory molecule. The products of some metabolic pathways inhibit their own production, a regulatory mechanism called **feedback inhibition**. **Electron transfer chains** allow cells to harvest energy in small, manageable steps.

Section 4.5 The rate of **diffusion** is influenced by temperature, solute size, and regional differences in concentration, charge, and pressure. Gases, water, and nonpolar molecules can diffuse directly through a lipid bilayer. Ions and most polar molecules cannot. **Osmosis** is the diffusion of water across a selectively permeable membrane, from a **hypotonic** fluid toward a **hypertonic** fluid. There is no net movement of water between **isotonic** solutions. Osmotic pressure is the amount of **turgor** (fluid pressure against a cell membrane or wall) sufficient to halt osmosis.

Section 4.6 Ions and most polar molecules can cross cell membranes only with the help of a transport protein. With facilitated diffusion, a solute follows its concentration gradient across a membrane through a transport protein. **Facilitated diffusion** is a type of **passive transport** (no energy input is required). With **active transport**, a transport protein uses energy to pump a solute across a membrane against its concentration gradient. A phosphate-group transfer from ATP often supplies energy needed for active transport.

Particles and substances in bulk and large particles are moved across plasma membranes by **exocytosis** and **endocytosis**. In exocytosis, a cytoplasmic vesicle fuses with the plasma membrane, and its contents are released to the outside of the cell. In one endocytic pathway, a patch of plasma membrane balloons into the cell, taking with it a drop of extracellular fluid (along with solutes and particles in it). The balloon forms a vesicle that sinks into the cytoplasm. Some cells can engulf large particles such as other cells by the endocytic pathway of **phagocytosis**.

Self-Quiz

1. _____ is life's primary source of energy.
   a. Food
   b. Water
   c. Sunlight
   d. ATP

2. Which of the following statements is *not* correct?
   a. Energy cannot be created or destroyed.
   b. Energy cannot change from one form to another.
   c. Energy tends to disperse spontaneously.

3. If we liken a reaction to an energy hill, then a reaction that _____ is an uphill run.
   a. requires energy
   b. releases energy
   c. runs from reactants to products
   d. uses an enzyme and a cofactor

4. In an energy-requiring reaction, activation energy is a bit like _____ .
   a. a burst of speed
   b. coasting downhill
   c. a bump at the top of the hill
   d. putting on the brakes

5. _____ are always changed by participating in a reaction. (Choose all that are correct.)
   a. Enzymes
   b. Cofactors
   c. Reactants
   d. Coenzymes

6. An environmental factor that directly influences enzyme function is _____ .
   a. temperature
   b. wind
   c. light
   d. all of the above

7. A metabolic pathway _____ .
   a. may build or break down molecules
   b. generates heat
   c. can include an electron transfer chain
   d. all of the above

8. Which of the following statements is *not* correct?
   a. Some metabolic pathways are cyclic.
   b. Glucose can diffuse directly through a lipid bilayer.
   c. Feedback inhibition controls some metabolic pathways.
   d. All coenzymes are cofactors.
   e. Osmosis is a case of diffusion.

9. Ions or molecules tend to diffuse from a region where they are _____ (more/less) concentrated to another where they are _____ (more/less) concentrated.

10. _____ cannot diffuse directly across a lipid bilayer.
    a. Water
    b. Gases
    c. Ions
    d. all of the above

11. If you immerse a human red blood cell in a hypotonic solution, water will _____ .
    a. diffuse into the cell
    b. diffuse out of the cell
    c. show no net movement
    d. move in by endocytosis

12. Fluid pressure against a wall or cell membrane is called _____ .
    a. osmosis
    b. turgor
    c. diffusion
    d. osmotic pressure

13. A transport protein requires ATP to pump sodium ions across a membrane. This is a case of _____ .
    a. passive transport
    b. active transport
    c. facilitated diffusion
    d. a and c

14. Vesicles are part of _____ .
    a. endocytosis
    b. exocytosis
    c. phagocytosis
    d. all of the above

15. Match each term with its most suitable description.
    ____ reactant
    ____ phagocytosis
    ____ first law of thermodynamics
    ____ product
    ____ cofactor
    ____ concentration gradient
    ____ passive transport
    ____ active transport
    ____ ATP
    ____ cyclic pathway

    a. assists enzymes
    b. forms at reaction's end
    c. enters a reaction
    d. requires energy input
    e. one cell "eats" another
    f. energy cannot be created or destroyed
    g. basis of diffusion
    h. no energy input required
    i. goes in circles
    j. currency in a cell's energy economy

Critical Thinking

1. Beginning physics students are often taught the basic concepts of thermodynamics with two phrases: First, you can never win. Second, you can never break even. Explain.

2. Water molecules tend to diffuse in response to their own concentration gradient. How can water be more or less concentrated?

3. Dixie Bee wanted to make JELL-O shots for her next party, but felt guilty about encouraging her guests to consume alcohol. She tried to compensate for the toxicity of the alcohol by adding pieces of healthy fresh pineapple to the shots, but when she did, the JELL-O never solidified. What happened? *Hint:* JELL-O is mainly sugar and a gelatinous mixture of proteins.

4. The enzyme trypsin is sold as a dietary enzyme supplement. Explain what happens to trypsin taken with food.

5. The enzyme catalase combines two hydrogen peroxide molecules ($H_2O_2 + H_2O_2$) to make two molecules of water ($2H_2O$). A gas also forms. What is the gas?

Application ❯ 5.1 **A Burning Concern**

Your body is about 9.5 percent carbon by weight, which means that you contain an enormous number of carbon atoms. Where did they all come from? Those atoms may have passed through other consumers before you ate them, but at some point they were components of producers. The vast majority of producers get their carbon from carbon dioxide (CO_2), a gas in air. Your carbon atoms—and those of most other organisms on land—were recently part of Earth's atmosphere, in molecules of CO_2.

The main producers in the human food chain are plants. Plants make their own food by photosynthesis, a pathway that harnesses the energy of sunlight to drive the assembly of sugars from carbon dioxide and water. Photosynthesis removes carbon dioxide from the atmosphere, and fixes its carbon atoms in organic compounds. When plants and other organisms break down organic compounds for energy, carbon atoms are released in the form of CO_2, which then reenters the atmosphere. For billions of years, these two processes have constituted a more or less balanced cycle of the biosphere (you will learn more about the carbon cycle in Section 17.6). For now, know that the amount of carbon dioxide that photosynthesis removes from the atmosphere is roughly the same amount that organisms release back into it—at least it was, until humans came along. As early as 8,000 years ago, humans began burning forests to clear land for agriculture. When trees and other plants burn, most of the carbon in their tissues is released into the atmosphere as CO_2. Fires that occur naturally release carbon dioxide the same way.

Today, we burn a lot more than our ancestors ever did. In addition to wood, we are burning fossil fuels—coal, petroleum, and natural gas—to satisfy our greater and greater demands for energy. Fossil fuels are the organic remains of ancient organisms. When we burn fossil fuels, we release the carbon that has been locked in their organic molecules for hundreds of millions of years, mainly as carbon dioxide that reenters the atmosphere.

Our extensive use of fossil fuels has put Earth's atmospheric cycle of carbon dioxide out of balance: We are adding far more CO_2 to the atmosphere than photosynthetic organisms are removing from it, and the excess is fueling global climate change (we return to this topic in Section 18.5). In 2013 alone, humans released over 36 billion tons of CO_2 into the atmosphere—an increase of 61 percent over 1990 and twice as much as photosynthesis removed from the atmosphere during the same year. Most of the CO_2 that humans release comes from burning fossil fuels (Figure 5.1A). How do we know? Researchers can determine how long ago the carbon atoms in a sample of CO_2 were part of a living organism by measuring the ratio of different carbon isotopes in it (you will read more about radioisotope dating techniques in Section 11.4). The results are correlated with global statistics on fossil fuel extraction, refining, and trade.

Tiny pockets of Earth's ancient atmosphere remain in Antarctica, preserved in snow and ice that have been accumulating in layers, year after year, for millions of years (Figure 5.1B). Air and dust trapped in each layer reveal the composition of the atmosphere that prevailed when the layer formed. These layers tell us that the atmospheric CO_2 level was relatively stable for about 10,000 years before the industrial revolution began in the mid-1800s. Since then, the CO_2 level has been steadily rising. Today, the atmospheric CO_2 level is higher than it has been for *15 million years.*

A. Air pollution—smog—blankets Lianyungang, China, in December 2013. The brown color of smog comes from one of its components, a gas (nitric oxide) that is toxic in large amounts. Invisible components include other nitrogen compounds, sulfur compounds, organic molecules, mercury and other heavy metals, and carbon dioxide.

B. Air bubbles trapped in Antarctic ice core slices such as this one are samples of Earth's atmosphere as it was when the ice formed. The deeper the slice, the older the air in the bubbles. From the deepest samples, we know that the level of carbon dioxide in the atmosphere is now higher than it has been for at least 15 million years.

Figure 5.1 Worldwide, most air pollution comes from burning fossil fuels.

(A) ChinaFotoPress/Getty Images; (B) www.photo.antarctica.ac.uk.

Figure 5.2 Properties of light.
Electromagnetic radiation moves through space in waves that we measure in nanometers (nm). Visible light makes up a very small part of this energy. Raindrops or a prism can separate visible light's different wavelengths, which we see as different colors. About 25 million nanometers are equal to 1 inch.

Left, © Robbie George, National Geographic Creative.

5.2 To Catch a Rainbow

REMEMBER: An electron that absorbs energy moves to a higher energy level, then emits the extra energy and moves back down (Section 2.2). Chloroplasts are organelles specialized for photosynthesis (3.5). Sustaining life's organization requires ongoing energy inputs to counter energy loss (4.2). Building, rearranging, or breaking down an organic substance often occurs stepwise, in a series of enzymatic reactions called a metabolic pathway (4.4).

Energy flow through nearly all ecosystems on Earth begins when photosynthesizers capture the energy in sunlight. Harnessing that energy is a complicated business; plants do it by converting it to chemical energy, which they and most other organisms use to drive cellular work. Understanding the conversion process requires a bit of knowledge about the nature of light.

Light is electromagnetic radiation, a type of energy that moves through space in waves, a bit like waves move across an ocean. The distance between the crests of two successive waves is called **wavelength**, and it is measured in nanometers (nm). Light that is visible to the human eye is only a tiny part of the spectrum of electromagnetic radiation emitted by the sun (Figure 5.2). This visible light travels in wavelengths between 380 and 750 nm, and it is the main form of energy that drives photosynthesis. Our eyes perceive all of these wavelengths combined as white light, and particular wavelengths in this range as different colors. White light separates into its component colors when it passes through a prism, or raindrops that act as tiny prisms. A prism bends longer wavelengths more than it bends shorter ones, so a rainbow of colors forms.

Photosynthesizers use pigments to capture the energy of visible light. A **pigment** is an organic molecule that selectively absorbs light of certain wavelengths, a bit like an antenna specialized for receiving light energy. Absorbing energy of the appropriate wavelength excites a pigment's electrons. An excited electron (one that has been boosted to a higher energy level) quickly emits its extra energy and returns to a lower energy level. As you will see, photosynthetic cells can capture energy emitted from an electron returning to a lower energy level.

pigment An organic molecule that can absorb light of certain wavelengths.

wavelength Distance between the crests of two successive waves.

Figure 5.3 A few photosynthetic pigments.
The curves in this graph show the efficiency at which each pigment absorbs the different wavelengths of visible light. Line color indicates the pigment's characteristic color. Using a combination of pigments allows photosynthetic organisms to maximize the range of wavelengths they can capture for photosynthesis.

Top, © Photobac/Shutterstock.

All life is sustained by inputs of energy, but not all forms of energy can sustain life.

Wavelengths of light that are not absorbed by a pigment are reflected, and that reflected light gives each pigment its characteristic color. **Chlorophyll *a*** is the most common photosynthetic pigment in plants and photosynthetic protists. It also occurs in some bacteria. Chlorophyll *a* absorbs violet, red, and orange light, and it reflects green light, so it appears green to us. Accessory pigments, including other chlorophylls, collectively harvest a wide range of additional light wavelengths for photosynthesis and for other purposes. Many accessory pigments protect cells from the damaging effects of ultraviolet (UV) light in the sun's rays. Appealing colors attract animals to ripening fruit or pollinators to flowers. You may already be familiar with some of these molecules: Carrots, for example, are orange because they contain beta-carotene (which is often abbreviated as β-carotene); roses are red and violets are blue because their cells make anthocyanins.

Most photosynthetic organisms maximize the range of wavelengths they can capture for photosynthesis by using a combination of pigments (Figure 5.3). In green plants, chlorophylls are usually so abundant that they mask the colors of the other pigments. Plants that change color during autumn are preparing for a period of dormancy; they conserve resources by moving nutrients from tender parts that would be damaged by winter cold (such as leaves) to protected parts (such as roots). Chlorophylls are not needed during dormancy, so they are disassembled and their components recycled. Yellow and orange accessory pigments are also recycled, but not as quickly as chlorophylls. Their colors begin to show as the chlorophyll content declines in leaves. Anthocyanin synthesis also increases in some plants, adding red and purple tones to turning leaf colors.

Storing Energy in Sugars   All life is sustained by inputs of energy, but not all forms of energy can sustain life. Sunlight, for example, is abundant here on Earth, but it cannot be used to directly power protein synthesis or other energy-requiring reactions that all organisms must run in order to stay alive. Photosynthesis converts the energy of light into the energy of chemical bonds. Unlike light, chemical energy can power the reactions of life, and it can be stored for later use.

In plants and other photosynthetic eukaryotes, photosynthesis takes place in chloroplasts (Figure 5.4A). Plant chloroplasts have two outer membranes, and they are filled with a thick, cytoplasm-like fluid called **stroma** (Figure 5.4B). Suspended in the stroma are the chloroplast's own DNA, some ribosomes, and an inner, much-folded **thylakoid membrane**. The folds of a thylakoid membrane typically form stacks of interconnected disks called thylakoids. The space enclosed by the thylakoid membrane is a single, continuous compartment (Figure 5.4C).

Photosynthesis is often summarized by an equation:

$$CO_2 + water \xrightarrow{\text{light energy}} sugars + O_2$$

This equation means that photosynthesis converts CO_2 and water to sugars and oxygen. However, photosynthesis is not a single reaction. Rather, it is a metabolic pathway with many reactions that occur in two stages. Molecules in the thylakoid membrane carry out the reactions of the first stage, which are driven by light and thus called the light-dependent reactions. The "photo" in photosynthesis means light, and it refers to the conversion of light energy to chemical bond energy of ATP during this stage. In addition to making ATP, the main light-dependent pathway in chloroplasts splits water molecules and releases O_2. Hydrogen ions and electrons from the water molecules end up in the coenzyme NADPH (Figure 5.5A).

The "synthesis" part of photosynthesis refers to the reactions of the second stage, which build sugars from CO_2 and water. These sugar-building reactions run

A. Chloroplasts in leaf cells of a moss.

two outer membranes

one inner membrane

stroma

B. Three membranes of a chloroplast.

C. Part of the thylakoid membrane, cutaway view.

Figure 5.4 Zooming in on a chloroplast.
The first stage of photosynthesis takes place at the thylakoid membrane; the second stage runs in the stroma.

(A) Michael Eichelberger/Visuals Unlimited.

in the stroma. They are collectively called the light-independent reactions because light energy does not power them. Instead, they run on energy delivered by NADPH and ATP that formed during the first stage (Figure 5.5B).

Take-Home Message 5.2

How do cells harvest light for photosynthesis?

- The sun emits light, which travels in waves. We see different wavelengths of visible light as different colors.
- Visible light is the main form of energy that drives photosynthesis. Photosynthetic species use pigments such as chlorophyll *a* to harvest this energy for photosynthesis.
- In eukaryotic cells, the first stage of photosynthesis occurs at the thylakoid membrane of chloroplasts. During these light-dependent reactions, light energy drives the formation of ATP and NADPH.
- The second stage occurs in the stroma. During these light-independent reactions, ATP and NADPH drive the synthesis of sugars from water and carbon dioxide.

ADP
NADP+
H_2O
energy

A. Light-dependent reactions of photosynthesis

ATP
NADPH
O_2

ATP
NADPH
CO_2
H_2O

B. Light-independent reactions of photosynthesis (Calvin–Benson cycle)

ADP
NADP+
sugars

Figure 5.5 Inputs and outputs of the two stages of photosynthesis.

5.3 Light-Dependent Reactions

REMEMBER: Successively higher electron energy levels can be represented as nested shells (Section 2.2). Potential energy is stored in the position or arrangement of objects in a system (4.2). Electron transfer chains can harvest the energy of electrons in small, usable increments; phosphorylation is a reaction in which a phosphate group is transferred from one molecule to another (4.4). Ions and other substances that do not diffuse directly through lipid bilayers can cross a cell membrane through transport proteins embedded in it; in active transport, a transport protein uses energy to pump a solute against its gradient across a cell membrane (4.6).

When a chlorophyll or accessory pigment absorbs light, one of its electrons jumps to a higher energy level (shell). The electron quickly drops back down to a lower shell by emitting its extra energy. In the thylakoid membrane, energy emitted by an electron is not lost to the environment. In this special membrane, photosynthetic pigments occur in clusters held together by proteins. These clusters can hold onto energy by passing it back and forth, a bit like volleyball players pass a ball among team members.

chlorophyll *a* Main photosynthetic pigment in plants.

stroma The cytoplasm-like fluid between the thylakoid membrane and the two outer membranes of a chloroplast.

thylakoid membrane A chloroplast's highly folded inner membrane system; forms a continuous compartment.

Calvin–Benson cycle Light-independent reactions of photosynthesis; cyclic carbon-fixing pathway that forms sugars from CO_2.

carbon fixation Process by which carbon from an inorganic source such as carbon dioxide gets incorporated (fixed) into an organic molecule.

electron transfer phosphorylation Process in which electron flow through electron transfer chains sets up a hydrogen ion gradient that drives ATP formation.

The reactions of photosynthesis begin when energy being passed around the thylakoid membrane reaches and becomes absorbed by a photosystem (Figure 5.6). A photosystem is a grouping of hundreds of chlorophylls, accessory pigments, and other molecules. When a photosystem absorbs light energy ❶, it releases electrons that immediately enter an electron transfer chain in the thylakoid membrane. (With this step, light energy has been converted to chemical energy.) The photosystem must replace its lost electrons, and it does so by pulling them off of water molecules in the thylakoid compartment. Water molecules do not give up electrons very easily; doing so causes them to break apart into hydrogen ions and oxygen atoms ❷. The hydrogen ions remain in the thylakoid compartment; oxygen atoms combine and diffuse out of the cell as oxygen gas (O_2).

Meanwhile, the electrons released by the photosystem have been moving through the electron transfer chain ❸. As the electrons pass from one molecule in the chain to the next, they release a bit of their extra energy. Molecules of the chain use the released energy to actively transport hydrogen ions (H^+) across the membrane, from the stroma into the thylakoid compartment ❹. Thus, the flow of electrons through the electron transfer chain sets up and maintains a hydrogen ion gradient across the thylakoid membrane.

At the end of the electron transfer chain, electrons are accepted by a second photosystem. When this photosystem absorbs light energy, it releases electrons ❺. These electrons immediately enter a second, different electron transfer chain. At the end of this chain, the coenzyme $NADP^+$ accepts the electrons along with H^+, so NADPH forms ❻:

$$NADP^+ \; + \; electrons \; + \; H^+ \; \longrightarrow \; \boxed{\textbf{NADPH}}$$

The hydrogen ion gradient that forms across the thylakoid membrane is a type of potential energy that can be tapped to make ATP. The H^+ ions want to follow their concentration gradient by moving back into the stroma, but ions cannot diffuse

Figure 5.6 Light-dependent reactions in the thylakoid membrane of chloroplasts.
ATP and oxygen gas are produced in this pathway. Electrons that travel through two different electron transfer chains end up in NADPH, which delivers them to sugar-building reactions in the stroma.

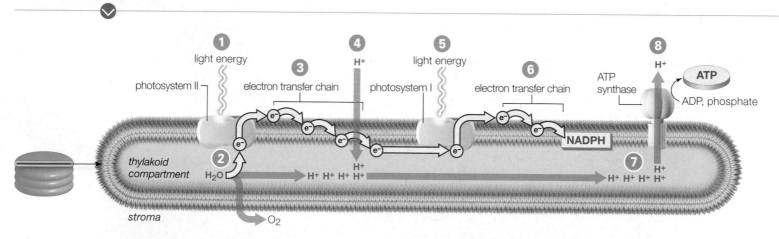

❶ A photosystem absorbs light energy and releases electrons.

❷ The photosystem pulls replacement electrons from water molecules, which then break apart into oxygen atoms and hydrogen ions. The oxygen atoms leave the cell as O_2 gas.

❸ The electrons enter an electron transfer chain in the thylakoid membrane.

❹ Energy lost by the electrons as they move through the chain is used to actively transport hydrogen ions from the stroma into the thylakoid compartment. A hydrogen ion gradient forms across the thylakoid membrane.

❺ Another photosystem absorbs light energy and releases electrons. Replacement electrons come from the first electron transfer chain.

❻ The released electrons move through a second electron transfer chain, then combine with $NADP^+$ and H^+, so NADPH forms.

❼ Hydrogen ions in the thylakoid compartment follow their gradient across the thylakoid membrane by flowing through the interior of ATP synthases.

❽ Hydrogen ion flow causes ATP synthases to phosphorylate ADP, so ATP forms in the stroma.

through the lipid bilayer. H⁺ leaves the thylakoid compartment only by flowing through proteins called ATP synthases embedded in the thylakoid membrane **7**. An ATP synthase is both a transport protein and an enzyme. When hydrogen ions flow through its interior, the protein phosphorylates ADP, so ATP forms in the stroma **8**. The process by which the flow of electrons through electron transfer chains drives ATP formation is called **electron transfer phosphorylation**.

Take-Home Message 5.3

What happens during the first stage of photosynthesis?

- Photosynthetic pigments in the thylakoid membrane transfer the energy of light to photosystems, which release electrons that enter electron transfer chains.
- The flow of electrons through electron transfer chains sets up a hydrogen ion gradient that drives ATP formation.
- Water molecules are split, oxygen is released, and electrons end up in NADPH.

5.4 Light-Independent Reactions

REMEMBER: Plant cells store sugars in the form of starch, a polymer of glucose (Section 2.7). A plant cuticle helps stems and leaves retain water (3.5). An active site is a pocket in an enzyme where substrates bind and a reaction occurs (4.4).

Energy that drives the second stage of photosynthesis is provided not by light; rather, the reactions run on the energy of electrons delivered by NADPH and phosphate-group transfers from ATP. These light-independent reactions, which are collectively called the **Calvin–Benson cycle**, produce sugars in the stroma of chloroplasts (Figure 5.7). This cyclic pathway uses carbon atoms from CO_2 to build the carbon backbones of the sugar molecules. Extracting carbon atoms from an inorganic source (such as CO_2) and incorporating them into an organic molecule is a process called **carbon fixation**.

In most plants, photosynthetic protists, and some bacteria, the enzyme rubisco fixes carbon by attaching CO_2 to a five-carbon compound called RuBP. The six-carbon intermediate that forms by this reaction is unstable, so it splits right away into two three-carbon molecules of PGA, which continue in the cycle. NADPH and ATP are used to convert these molecules to sugars.

It takes six cycles of Calvin–Benson reactions to fix the six carbon atoms necessary to make one molecule of glucose (a six-carbon sugar). Plant cells break down some of the glucose they make to access the energy stored in its bonds. However, most of the glucose they make is converted at once to sucrose or starch by other pathways that conclude the light-independent reactions. Excess glucose is stored as starch grains in chloroplast stroma. When sugars are needed in other parts of the plant, the starch is broken apart into its glucose monomers.

Alternative Carbon-Fixing Pathways The aboveground parts of most plants are covered with a cuticle that limits evaporative water loss. Gases cannot diffuse across the cuticle, but oxygen produced by the light-dependent reactions must escape

Figure 5.7 The Calvin–Benson cycle.
This sketch shows a cross-section of a chloroplast with the reactions cycling in the stroma. Six cycles of the Calvin–Benson reactions produce one six-carbon sugar.

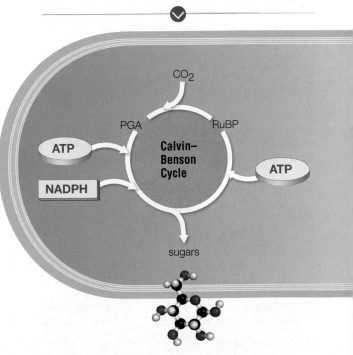

Figure 5.8 Stomata on the surface of a leaf.
When these tiny pores are open, gases can be exchanged between the plant's internal tissues and air. Gas exchange ends when stomata close to conserve water on dry days.

© D. Kucharski & K. Kucharska/Shutterstock.

Figure 5.9 C4 and CAM plants.

(A) Image courtesy msturfweeds.net; (B) © Tamara Kulikova/Shutterstock.

A. Crabgrass "weeds" overgrowing a lawn. Crabgrasses, which are C4 plants, thrive in hot, dry summers, easily outcompeting Kentucky bluegrass and other fine-leaved C3 grasses commonly planted in residential lawns.

B. The jade plant, *Crassula argentea*, and other CAM plants survive in hot deserts by opening stomata to fix carbon only at night. They run the Calvin–Benson cycle during the day, with stomata closed.

the plant, and carbon dioxide needed for the Calvin–Benson cycle must enter it. Thus, most leaves and stems are studded with tiny, closable gaps called **stomata** (singular, stoma; Figure 5.8). When stomata are open, CO_2 diffuses from air into photosynthetic tissues, and O_2 diffuses out of the tissues into air. Stomata close to conserve water on hot, dry days. When that happens, gas exchange comes to a halt.

Both stages of photosynthesis run during the day. With stomata closed, the O_2 level in the plant's tissues rises, and the CO_2 level declines. This outcome can reduce the efficiency of sugar production because both gases are substrates of rubisco, and they compete for its active site. Rubisco starts the Calvin–Benson cycle by attaching CO_2 to RuBP. It also attaches O_2 to RuBP. To convert the product of this alternate reaction to a substrate of the Calvin–Benson cycle, the cell uses additional ATP and NADPH, and it also loses carbon in the form of CO_2. The detrimental effects of the alternate pathway are greatest in **C3 plants**, which use only the Calvin–Benson cycle to fix carbon (they are called C3 plants because a three-carbon molecule, PGA, is the first stable intermediate to form in their light-independent reactions). In a C3 plant, sugar production becomes less and less efficient as daytime temperature rises.

An additional set of carbon-fixing reactions minimizes the effects of rubisco's inefficiency in corn, bamboo, and other **C4 plants** (so named because a four-carbon molecule is the first stable intermediate to form in their light-independent reactions). These plants also close stomata on dry days, but the efficiency of their sugar production does not decline. C4 plants fix carbon twice, in two kinds of cells. In the first cell, carbon is fixed by an enzyme that cannot use oxygen as a substrate. The resulting intermediate is transported to a second cell, where it is converted to CO_2. There, rubisco fixes carbon for the second time as the CO_2 enters the Calvin–Benson cycle. The extra C4 reactions keep the CO_2 level high near rubisco, so sugar production stays efficient even during hot, dry weather (Figure 5.9A).

Succulents, cacti, and other **CAM plants** use a carbon-fixing pathway that allows them to produce sugar efficiently even where typical daytime conditions are extremely hot and dry. CAM stands for crassulacean acid metabolism, after the Crassulaceae family of plants in which this pathway was first studied (Figure 5.9B). Like C4 plants, CAM plants fix carbon twice, but the reactions occur at different times rather than in different cells. Stomata on a CAM plant open only at night, when typically lower temperatures minimize evaporative water loss. The plant's cells use a C4 pathway to fix carbon from CO_2 in the air at this time. The product of the pathway is stored in the cell's central vacuole. When the stomata close the next day, the molecule moves out of the vacuole and becomes broken down to CO_2. Rubisco then fixes carbon for the second time as the CO_2 enters the Calvin–Benson cycle.

Take-Home Message 5.4

What happens during the second stage of photosynthesis?

- During the light-independent reactions (the second stage of photosynthesis), ATP and NADPH drive the synthesis of sugars from CO_2.
- When stomata close on hot, dry days, they also prevent the exchange of gases between plant tissues and the air. This outcome reduces the efficiency of sugar production in C3 plants.
- Plants adapted to hot, dry conditions fix carbon twice. C4 plants separate the two sets of reactions in space; CAM plants separate them in time.

5.5 A Global Connection

REMEMBER: Most free radicals are dangerous to life (Section 2.2). Mitochondria make ATP by aerobic respiration (3.5). Organic molecules release energy when they combine with oxygen (4.3). Most enzymes require assistance from cofactors; coenzymes carry chemical groups, atoms, or electrons between reactions; energy given off by electrons moving through an electron transfer chain drives cellular work (4.4). Ions cross a cell membrane through transport proteins; in active transport, a transport protein uses energy to move a solute across a cell membrane (4.6).

The first cells we know of appeared on Earth about 3.4 billion years ago. Like some modern prokaryotes, these ancient organisms did not tap into sunlight; rather, they extracted the energy they needed from simple molecules such as methane and hydrogen sulfide. Both gases were plentiful in the nasty brew that was Earth's early atmosphere (Figure 5.10). When photosynthesis evolved, sunlight offered cells that used it an essentially unlimited supply of energy, and they were very successful. Oxygen gas released from uncountable numbers of water molecules began seeping out of the photosynthesizers. O_2 reacts easily with metals, so at first, most of it combined with metal atoms in exposed rocks. After the exposed minerals became saturated with oxygen, the gas began to accumulate in the ocean and in the atmosphere. From that time on, the world of life would never be the same.

Before photosynthesis evolved, molecular oxygen had been a very small component of Earth's atmosphere. In what may have been the earliest case of catastrophic pollution, the new abundance of this gas exerted tremendous pressure on all life at the time. Why? Then, as now, enzymes that require metal cofactors were a critical part of metabolism. Oxygen reacts with metal cofactors, and free radicals form during those reactions. Free radicals damage biological molecules, so they are dangerous to life. Most cells had no way to cope with them, and so were wiped out everywhere except deep water, muddy sediments, and other **anaerobic** (oxygen-free) places.

By lucky circumstance, a few types of cells were already making molecules that could detoxify or prevent the formation of free radicals. Cells with these molecules were the first **aerobic** organisms—they could live in the presence of oxygen. As aerobic organisms evolved, their detoxifying molecules became incorporated into new metabolic pathways. One of these pathways put the reactive properties of oxygen to use. Today, this pathway—**aerobic respiration**—is the main ATP-producing sugar breakdown pathway in nearly all eukaryotes (including plants) and some modern bacteria. The pathway requires oxygen, and its products—carbon dioxide and water—are the raw materials used by the majority of photosynthetic organisms to build the sugars in the first place. With this connection, the cycling of carbon, hydrogen, and oxygen through living things connects full circle through the biosphere (right).

Aerobic Respiration in Mitochondria
The bonds of organic molecules hold a lot of energy that can be released in a reaction with oxygen. Aerobic respiration

Figure 5.10 A view of how Earth's atmosphere was permanently changed by photosynthesis.
Earth's early atmosphere was abundant in gases such as methane, sulfur, ammonia, and chlorine. Oxygen released by early photosynthesizers changed the composition of the atmosphere. Photosynthesis is now the main pathway by which energy and carbon enter the world of life.

aerobic Involving or occurring in the presence of oxygen.

aerobic respiration Oxygen-requiring metabolic pathway that breaks down sugars to produce ATP.

anaerobic Occurring in the absence of oxygen.

C3 plant Type of plant that uses only the Calvin–Benson cycle to fix carbon.

C4 plant Type of plant that fixes carbon twice, in two cell types.

CAM plant Type of plant that fixes carbon twice, at different times of day.

stomata Closable gaps on aboveground plant surfaces. When open, they allow the plant to exchange gases with air. When closed, they limit water loss.

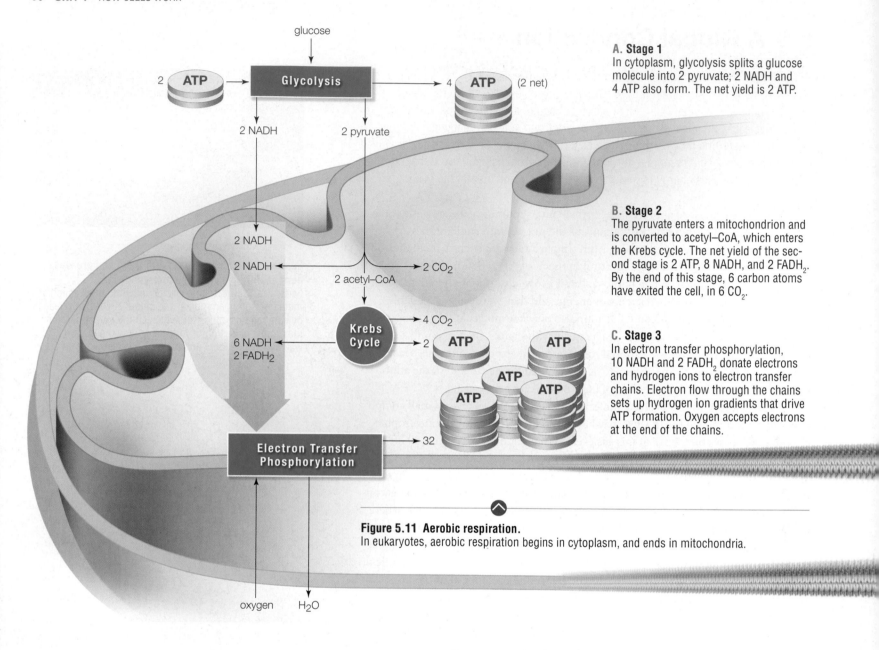

glucose

2 ATP → Glycolysis → 4 ATP (2 net)

A. Stage 1
In cytoplasm, glycolysis splits a glucose molecule into 2 pyruvate; 2 NADH and 4 ATP also form. The net yield is 2 ATP.

2 NADH

2 pyruvate

2 NADH

2 NADH ← 2 acetyl–CoA → 2 CO_2

B. Stage 2
The pyruvate enters a mitochondrion and is converted to acetyl–CoA, which enters the Krebs cycle. The net yield of the second stage is 2 ATP, 8 NADH, and 2 $FADH_2$. By the end of this stage, 6 carbon atoms have exited the cell, in 6 CO_2.

Krebs Cycle → 4 CO_2

6 NADH
2 $FADH_2$ ← → 2 ATP   ATP   ATP   ATP   ATP

C. Stage 3
In electron transfer phosphorylation, 10 NADH and 2 $FADH_2$ donate electrons and hydrogen ions to electron transfer chains. Electron flow through the chains sets up hydrogen ion gradients that drive ATP formation. Oxygen accepts electrons at the end of the chains.

Electron Transfer Phosphorylation → 32

Figure 5.11 Aerobic respiration.
In eukaryotes, aerobic respiration begins in cytoplasm, and ends in mitochondria.

oxygen H_2O

a mitochondrion

completely breaks apart the carbon backbone of a sugar, bond by bond. Energy released as those bonds are broken drives ATP synthesis.

The reactions of aerobic respiration occur in three stages. The first stage, glycolysis, takes place in cytoplasm. **Glycolysis** is a set of reactions that convert one six-carbon sugar into two molecules of pyruvate (Figure 5.11A), an organic compound with a three-carbon backbone. The reactions occur with some variation in the cytoplasm of almost all cells; for clarity, we focus here on those that start with glucose. Two ATP are invested to begin glycolysis, but four ATP form by the end. Thus, we say that the net (overall) yield is two ATP per glucose. Electrons and hydrogen ions released by the reactions combine with the coenzyme NAD⁺, so NADH also forms.

In eukaryotes, aerobic respiration continues in mitochondria. The reactions of the second stage begin when the two pyruvate molecules that formed in glycolysis enter the inner compartment of a mitochondrion (Figure 5.11B). There, each pyruvate reacts with a coenzyme named (rather unimaginatively) coenzyme A. One molecule of CO_2 forms in this reaction and then diffuses out of the cell. The reaction's other product, a molecule called acetyl–CoA, enters a pathway called the **Krebs cycle**. Each cycle of Krebs reactions releases two carbon atoms that leave the cell in molecules of CO_2.

At this point in aerobic respiration, the six-carbon backbone of one glucose molecule has been broken down completely; six carbon atoms have now exited the cell in CO_2. The two ATP that formed during the second stage add to the small net yield (two ATP) of glycolysis. However, six more NADH and two $FADH_2$ (another coenzyme) also formed. Add in the two NADH from glycolysis, and the full breakdown of one glucose molecule has a big potential payoff. Twelve coenzymes now deliver electrons—and the energy they carry—to the third and final stage of aerobic respiration, electron transfer phosphorylation (Figure 5.11C).

1 NADH and $FADH_2$ deliver electrons and hydrogen ions to electron transfer chains in the inner mitochondrial membrane.

2 Molecules of the electron transfer chain use energy released by the electrons to pump the hydrogen ions (H^+) across the membrane, from the matrix to the intermembrane space. This activity sets up and maintains a hydrogen ion gradient across the inner mitochondrial membrane.

3 Hydrogen ion flow back to the matrix through ATP synthases drives the formation of ATP from ADP and phosphate.

4 Electrons at the end of the electron transfer chains combine with oxygen and hydrogen ions, so water forms.

Figure 5.12 Electron transfer phosphorylation in mitochondria.
In eukaryotes, the third and final stage of aerobic respiration, electron transfer phosphorylation, occurs at the inner mitochondrial membrane.

In eukaryotes, electron transfer phosphorylation occurs at the inner mitochondrial membrane (Figure 5.12). The reactions begin when NADH and $FADH_2$ donate their cargo of electrons and hydrogen ions to electron transfer chains embedded in this membrane **1**. As the electrons pass from one molecule in the chain to the next, they release a bit of their extra energy. Molecules of the chain use the released energy to actively transport hydrogen ions (H^+) across the inner mitochondrial membrane, from the matrix to the intermembrane space **2**. Thus, the flow of electrons through the electron transfer chains sets up and maintains a hydrogen ion gradient across the membrane. This gradient attracts the ions back toward the matrix, but ions cannot diffuse through a lipid bilayer. Hydrogen ions cross the inner mitochondrial membrane only by flowing through ATP synthases embedded in the membrane. The flow of hydrogen ions through ATP synthases causes these proteins to attach phosphate groups to ADP, so ATP forms **3**.

At the end of the electron transfer chains, oxygen accepts electrons and H^+ to form water, a product of the third-stage reactions **4**. The term aerobic respiration,

glycolysis Set of reactions in which glucose is broken down to two pyruvate for a net yield of two ATP.
Krebs cycle Cyclic pathway that helps break down pyruvate to carbon dioxide during aerobic respiration.

alcoholic fermentation Anaerobic pathway that breaks down sugars and produces ATP, CO_2, and ethanol.

fermentation An anaerobic pathway that breaks down sugars to produce ATP.

lactate fermentation Anaerobic pathway that breaks down sugars and produces ATP and lactate.

Aerobic respiration literally means "breathing air to live."

A. *Saccharomyces cerevisiae*, a yeast that carries out alcoholic fermentation.

B. Left, one product of *Saccharomyces* alcoholic fermentation (ethanol) makes beer alcoholic; another (CO_2) makes it bubbly. Right, holes in bread are pockets where CO_2 released by fermenting *Saccharomyces* cells accumulated in the dough.

Figure 5.13 Examples of alcoholic fermentation.

(A) © By London Scientific Films/Oxford Scientific/Getty Images; (B) left, © Elena Boshkovska/Shutterstock.com; right, optimarc/Shutterstock.com.

which literally means "breathing air to live," refers to this pathway's requirement for oxygen as the final acceptor of electrons. Every breath you take provides your trillions of aerobically respiring cells with a fresh supply of oxygen.

The following equation summarizes the overall pathway of aerobic respiration:

$$C_6H_{12}O_6 + O_2 \longrightarrow CO_2 + H_2O + \text{ATP}$$
glucose   oxygen   carbon dioxide   water

For each glucose molecule that enters this pathway, four ATP form in the first- and second-stage reactions. Coenzymes deliver enough H^+ and electrons to fuel the synthesis of about thirty-two additional ATP in the third stage. Thus, the breakdown of one glucose molecule typically yields thirty-six ATP. As you will see, other pathways that break down sugars also yield ATP, but not nearly as much as aerobic respiration. You and other large, multicelled eukaryotes could not survive without its higher efficiency.

Take-Home Message 5.5

What happens during aerobic respiration?

- Most cells can make ATP by breaking down sugars in the oxygen-requiring pathway of aerobic respiration.
- Aerobic respiration begins in cytoplasm with glycolysis, and, in eukaryotes, ends in the mitochondrion with electron transfer phosphorylation.
- A typical net yield of aerobic respiration is thirty-six ATP per glucose. Carbon dioxide and water also form.

5.6 Fermentation

REMEMBER: Fungi that live as single cells are called yeasts (Section 1.4). Monomers are subunits of polymers (2.6). Sugar monomers form larger carbohydrates (2.7).

Most types of eukaryotic cells use aerobic respiration exclusively, or they use it most of the time. Many bacteria, archaea, protists, and some eukaryotic cells can harvest energy from carbohydrates by fermentation. **Fermentation** refers to sugar breakdown pathways that produce ATP and do not require oxygen. Like aerobic respiration, fermentation begins with glycolysis in cytoplasm. Unlike aerobic respiration, fermentation's concluding reactions take place in cytoplasm. Electrons do not move through electron transfer chains, so no additional ATP forms, and an organic molecule (instead of oxygen) accepts electrons. In these reactions, pyruvate is converted to other molecules, but it is not fully broken down to CO_2 (as occurs in aerobic respiration). The reactions remove electrons from NADH, so NAD^+ forms; regenerating this coenzyme allows glycolysis—and the ATP it offers—to continue. Thus, the net ATP yield of fermentation consists of the two ATP that form in glycolysis (see Figure 5.11A). Fermentation is inefficient compared with aerobic respiration, but it produces enough ATP to sustain many single-celled species. It also helps cells of multicelled species under anaerobic conditions.

Some yeasts carry out **alcoholic fermentation**, a pathway that converts pyruvate to ethanol. One yeast species, *Saccharomyces cerevisiae*, helps us produce beer, wine, and bread (Figure 5.13). Beer brewers often use barley that has been germinated and dried (a process called malting) as a source of glucose for fermentation by this yeast. As the cells make ATP for themselves, they also produce ethanol (which makes the beer alcoholic) and CO_2 (which makes it bubbly). Flowers of the hop plant add flavor and help preserve the finished product. Winemakers use crushed grapes as a source of sugars for yeast fermentation. The ethanol produced by the cells makes the wine alcoholic, and the CO_2 is allowed to escape to the air.

To make bread, flour is kneaded with water, yeast, and sometimes other ingredients. Flour contains a protein (gluten) and a sugar (maltose) that consists of two glucose subunits. Kneading causes the gluten to form polymers in long, interconnected strands that make the resulting dough stretchy and resilient. The yeast cells in the dough first break down the maltose, then use the released glucose for alcoholic fermentation. The CO_2 they produce accumulates in bubbles that are trapped by the mesh of gluten strands. As the bubbles expand, they cause the dough to rise. The ethanol product of fermentation evaporates during baking.

In **lactate fermentation**, electrons and hydrogen ions carried by NADH are transferred directly to pyruvate, so NAD^+ forms. This reaction converts the pyruvate to lactic acid (also called lactate). Both molecules have three carbons: No carbons are lost, so no CO_2 is produced. We use lactate fermentation by beneficial bacteria to prepare many foods. Yogurt, for example, is made by allowing bacteria such as *Lactobacillus bulgaricus* and *Streptococcus thermophilus* to grow in milk (Figure 5.14A). Milk contains a disaccharide (lactose) and a protein (casein). The cells first break down the lactose into its monosaccharide subunits, then use the monosaccharides for lactate fermentation. The lactate they produce reduces the pH of the milk, which imparts tartness and causes the casein to form a gel.

Cells in animal skeletal muscles are fused as long fibers that carry out aerobic respiration, lactate fermentation, or both (Figure 5.14B). Red fibers have many mitochondria and produce ATP mainly by aerobic respiration. These fibers sustain prolonged activity. They are red because they contain myoglobin, a protein that stores oxygen for aerobic respiration. White muscle fibers contain few mitochondria and no myoglobin; they make most of their ATP by lactate fermentation. This pathway makes ATP quickly, so it is useful for quick, strenuous bursts of activity (Figure 5.14C). The low ATP yield does not support prolonged activity.

Most animal muscles are a mixture of white and red fibers, but the proportions vary. For example, great sprinters tend to have more white fibers in their leg muscles; great marathon runners have more red fibers. Chickens cannot fly far because their flight muscles consist mostly of white fibers (thus, the "white" breast meat). A chicken most often walks or runs. Its leg muscles consist mostly of red muscle fibers, the "dark meat." Section 20.4 returns to the structure and function of skeletal muscle fibers.

Figure 5.14 Examples of lactate fermentation.

(A) left, mexrix/Shutterstock.com; right SCIMAT/Science Source; (B) © William MacDonald, M.D.; (C) © Maxisport/Shutterstock.com.

A. Yogurt is a product of lactate fermentation by bacteria in milk. The micrograph shows *Lactobacillus bulgaricus* (red) and *Streptococcus thermophilus* (purple) in yogurt.

B. Lactate fermentation occurs in white muscle fibers, visible in this cross-section of human thigh muscle. The red fibers, which make ATP by aerobic respiration, sustain endurance activities.

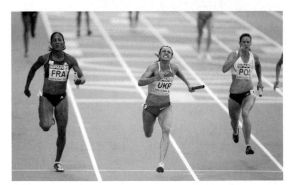

C. Intense activity such as sprinting quickly depletes oxygen in muscles. Under the resulting anaerobic conditions, ATP is produced mainly by lactate fermentation in white muscle fibers. Fermentation does not make enough ATP to sustain this type of activity for long.

Take-Home Message 5.6

What is fermentation?

- Prokaryotes and eukaryotes use fermentation pathways, which are anaerobic, to produce ATP by breaking down carbohydrates.
- Fermentation's small ATP yield (two per molecule of glucose) occurs by glycolysis.

5.7 Food as a Source of Energy

REMEMBER: Animals store sugars in the form of glycogen (Section 2.7). Triglycerides, the main component of fats, have a glycerol head and three fatty acid tails (2.8). Proteins are polymers of amino acids (2.9).

Aerobic respiration produces a lot of ATP by fully dismantling glucose, carbon by carbon. Cells also dismantle other organic molecules to make ATP. Complex carbohydrates, fats, and proteins in food can be converted to molecules that enter glycolysis or the Krebs cycle (Figure 5.15). As in glucose metabolism, the energy of electrons that are transferred to coenzymes ultimately drives the synthesis of ATP.

Complex Carbohydrates In humans and other mammals, the digestive system breaks down starch and other complex carbohydrates to their sugar subunits, which are quickly taken up by cells for glycolysis. When a cell produces more ATP than it uses, ATP accumulates in the cytoplasm. The high concentration of ATP causes glucose to be diverted away from glycolysis and into a pathway that builds glycogen. Liver and muscle cells especially favor the conversion of glucose to glycogen, and these cells contain the body's largest stores of it. Between meals, the liver maintains the blood glucose level by breaking down the stored glycogen to release its glucose monomers.

What happens if you eat too many carbohydrates? When the blood level of glucose gets too high, acetyl–CoA is diverted away from the Krebs cycle and into a pathway that makes fatty acids. This is why excess dietary carbohydrate ends up as fat.

Fats A triglyceride molecule has a glycerol head and three fatty acid tails. Cells dismantle triglycerides in fats by first breaking the bonds that connect fatty acid tails to the glycerol head. Nearly all cells in the body can break down the released fatty acids for energy. First, their long backbones are split into two-carbon fragments. These fragments are converted to acetyl–CoA, which can enter the Krebs cycle. The glycerol released by triglyceride breakdown is converted by liver cells to an intermediate of glycolysis.

On a per carbon basis, fats are a richer source of energy than carbohydrates. This is because typical carbohydrate backbones have many oxygen atoms bonded to them. By contrast, most fatty acid tails have no oxygen atoms bonded to them, so more reactions with oxygen are required to fully break their backbones apart. Coenzymes accept electrons in reactions that break apart carbon backbones. The more

Figure 5.15 A variety of organic compounds from food can enter the reactions of aerobic respiration.

Above, © shabaneiro/Shutterstock.com.

Digging Into Data

Biofuels

A lot of energy is locked up in the chemical bonds of molecules made by plants. That energy can fuel consumers, as when an animal cell powers ATP synthesis by aerobic respiration. It can also fuel our cars, which run on energy released by burning biofuels or fossil fuels. Both processes are fundamentally the same: They release energy by breaking the bonds of organic molecules. Both use oxygen to break those bonds, and both produce carbon dioxide. Unlike fossil fuels, biofuels are a renewable source of energy: We can always make more of them simply by growing more plants. Also unlike fossil fuels, biofuels do not contribute to global climate change, because growing plant matter for fuel recycles carbon that is already in the atmosphere.

Corn, soy, sugarcane, and other food crops are rich in oils, starches, and sugars that can be easily converted to biofuels. The starch in corn kernels, for example, can be enzymatically broken down to glucose, which is fermented to ethanol by bacteria or yeast. However, growing food crops for biofuel production typically requires a lot of energy (in the form of fossil fuels) and it damages the environment. Making biofuels from other plant matter such as weeds or agricultural waste requires additional steps, because these materials contain a higher proportion of cellulose. Breaking down this tough carbohydrate to its glucose monomers adds cost to the biofuel product.

In 2006, David Tilman and his colleagues published the results of a 10-year study comparing the net energy output of various biofuels. The researchers made biofuel from a mixture of native perennial grasses grown without irrigation, fertilizer, pesticides, or herbicides, in sandy soil that was so depleted by intensive agriculture that it had been abandoned. The energy content of this biofuel and the energy it took to produce it were measured and compared with that of biofuels made from food crops (Figure 5.16).

Figure 5.16 Energy inputs and outputs of various biofuels. Input: calculated energy used to grow the crop and produce the biofuel. Output: actual energy content of the biofuel. Corn and soy were grown on fertile farmland; grass, in depleted soil. One hectare is about 2.5 acres.

1. About how much energy did ethanol produced from one hectare of corn yield? How much energy did it take to grow and produce that ethanol?
2. Which of the three crops required the least amount of land to produce a given amount of biofuel energy?
3. The production of which biofuel was most efficient (which had the highest ratio of energy output to energy input)?

coenzymes that accept electrons, the more electrons can be delivered to the ATP-forming machinery of electron transfer phosphorylation.

Proteins Enzymes in the digestive system split dietary proteins into their amino acid subunits, which are absorbed into the bloodstream and used to build proteins or other molecules. When you eat more protein than your body needs for this purpose, the amino acids are broken down. The amino group is removed, and it becomes ammonia, a waste product eliminated in urine. The carbon backbone is split, and acetyl–CoA, pyruvate, or an intermediate of the Krebs cycle forms, depending on the amino acid. These molecules enter aerobic respiration's second stage.

Take-Home Message 5.7

Can the body break down molecules other than sugars for energy?

- Breaking the carbon backbone of any organic molecule releases electrons. Those electrons carry energy that can be harnessed to drive ATP formation in aerobic respiration.
- First the digestive system and then individual cells convert molecules in food (complex carbohydrates, triglycerides, and proteins) into substrates of glycolysis or aerobic respiration's second-stage reactions.

Summary

Section 5.1 By the pathway of photosynthesis, the energy of light is used to build sugars from water and carbon dioxide. Photosynthesis removes CO_2 from the atmosphere, and the metabolic activity of most organisms puts it back. This global cycle was balanced for millions of years. Human activities, especially burning fossil fuels, are currently disrupting the cycle by adding massive amounts of extra CO_2 to the atmosphere. The resulting imbalance is fueling global climate change.

Section 5.2 Light energy travels in waves. Visible light drives photosynthesis, which begins when light energy is absorbed by photosynthetic pigments. A **pigment** absorbs light of particular **wavelengths** only; wavelengths not captured are reflected as its characteristic color. The main photosynthetic pigment, **chlorophyll a**, absorbs violet and red light, so it appears green. Accessory pigments absorb additional wavelengths.

In chloroplasts, the light-dependent reactions of photosynthesis occur at a much-folded **thylakoid membrane.** The light-independent reactions occur in the chloroplast's cytoplasm-like **stroma**. An overview is shown below:

Section 5.3 Clusters of photosynthetic pigments in the thylakoid membrane absorb the energy in light and pass it to photosystems. Receiving energy causes photosystems to release electrons. The electrons flow through electron transfer chains in the thylakoid membrane, and end up in NADPH. Molecules of the electron transfer chain use energy released by the electrons to set up a hydrogen ion gradient across the thylakoid membrane. The ions move back across the membrane through ATP synthases embedded in it. The flow causes these proteins to produce ATP, a process called **electron transfer phosphorylation**.

Photosynthesis releases oxygen because a photosystem replaces lost electrons by pulling them from water molecules, which break apart into hydrogen ions and oxygen as a result.

Section 5.4 Carbon fixation occurs as part of the light-independent reactions of photosynthesis. In the stroma of chloroplasts, the **Calvin–Benson cycle** builds carbon backbones of sugars using carbon atoms from CO_2. This cyclic pathway is driven by phosphate-group transfers from ATP and electrons delivered by NADPH (both of these molecules form during light-dependent reactions).

On hot, dry days, a plant conserves water by closing **stomata**, so carbon dioxide for the light-independent reactions cannot enter it, and oxygen produced by the light-dependent reactions cannot leave. Oxygen buildup in plant tissues reduces the efficiency of sugar production in **C3 plants**. Additional **carbon fixation** reactions in **C4 plants** and **CAM plants** make sugar production more efficient on hot, dry days.

Section 5.5 When photosynthesis evolved, oxygen released by organisms that used it permanently changed the atmosphere, with profound effects on life's evolution. Organisms that could not tolerate the increased atmospheric oxygen persisted only in **anaerobic** habitats. Oxygen-detoxifying pathways allowed other organisms to thrive under **aerobic** conditions.

Most modern organisms convert the chemical energy of carbohydrates to the chemical energy of ATP by oxygen-requiring **aerobic respiration**. In eukaryotes, this pathway starts with **glycolysis** in cytoplasm, and ends in mitochondria. Coenzymes pick up electrons in aerobic respiration's first two stages, glycolysis and the **Krebs cycle**. The energy of those electrons drives ATP synthesis in the third stage, electron transfer phosphorylation. At the end of the electron transfer chains, oxygen accepts electrons and hydrogen ions, so water forms. Aerobic respiration yields about thirty-six ATP per glucose.

Section 5.6 Anaerobic **fermentation** pathways include **alcoholic fermentation** and **lactate fermentation**. Both begin with glycolysis, and they run in the cytoplasm. The electron acceptor at the end of these reactions is an organic molecule. The final steps produce no ATP. Thus, the breakdown of one glucose molecule yields only the two ATP from glycolysis.

Section 5.7 Organic molecules other than sugars can be broken down to make ATP. In humans and other mammals, first the digestive system and then individual cells convert fats, proteins, and complex carbohydrates in food to molecules that are substrates of glycolysis or the second-stage reactions of aerobic respiration.

Self-Quiz

1. Most of the carbon that land plants use for photosynthesis comes from _____ .
   - a. glucose
   - b. the atmosphere
   - c. water
   - d. soil

2. Plants use _____ as an energy source to drive photosynthesis.
   - a. sunlight
   - b. sugars
   - c. O_2
   - d. CO_2

3. Which of the following statements is incorrect?
   - a. Pigments absorb light of certain wavelengths only.
   - b. Many accessory pigments are multipurpose molecules.
   - c. Chlorophyll *a* is green because it absorbs green light.

4. In the light-dependent reactions, _____ .
   - a. carbon dioxide is fixed
   - b. ATP forms
   - c. CO_2 accepts electrons
   - d. sugars form

5. When a photosystem absorbs light, _____ .
   - a. sugar phosphates are produced
   - b. electrons are transferred to ATP
   - c. it ejects electrons

6. The atoms in the oxygen molecules released during photosynthesis come from _____ .
   - a. glucose
   - b. CO_2
   - c. water
   - d. O_2

7. The Calvin–Benson cycle starts when _____ .
   - a. light is available
   - b. carbon is fixed
   - c. electrons leave a photosystem

8. Closed stomata _____ .
   - a. limit gas exchange
   - b. permit water loss
   - c. restrict photosynthesis
   - d. absorb light

9. True or false? Plants make all of their ATP by photosynthesis.

10. In the third stage of aerobic respiration, _____ is the final acceptor of electrons.
    - a. water
    - b. H^+
    - c. O_2
    - d. NADH

11. In eukaryotes, the final reactions of aerobic respiration are completed in _____ .
    - a. the nucleus
    - b. mitochondria
    - c. the plasma membrane
    - d. cytoplasm

12. In eukaryotes, the final reactions of fermentation are completed in _____ .
    - a. the nucleus
    - b. mitochondria
    - c. the plasma membrane
    - d. cytoplasm

13. Your body cells can break down _____ as a source of energy to fuel ATP production.
    - a. fatty acids
    - b. glycerol
    - c. amino acids
    - d. all of the above

14. Which of the following metabolic pathways require(s) molecular oxygen (O_2)?
    - a. aerobic respiration
    - b. lactate fermentation
    - c. alcoholic fermentation
    - d. photosynthesis

15. Match the term with the best description.
    - _____ pyruvate
    - _____ anaerobic
    - _____ mitochondrion
    - _____ pigment
    - _____ carbon dioxide
    - _____ rubisco
    - _____ photosynthesis
    - a. no oxygen required
    - b. converts light to chemical energy
    - c. product of glycolysis
    - d. aerobic respiration in eukaryotes
    - e. carbon-fixing enzyme
    - f. like an antenna
    - g. big in the atmosphere

Critical Thinking

© Martin Shields/Alamy.

1. While looking into an aquarium, you see bubbles coming from an aquatic plant (right). What are the bubbles?

2. How is the function of the thylakoid membrane similar to that of the inner mitochondrial membrane?

3. Which of the following is *not* produced by an animal muscle cell operating under anaerobic conditions?
   - a. heat
   - b. lactate
   - c. ATP
   - d. NAD^+
   - e. pyruvate
   - f. all are produced

4. The bar-tailed godwit is a type of shorebird that makes an annual migration from Alaska to New Zealand and back. The birds make each 11,500-kilometer (7,145-mile) trip by flying over the Pacific Ocean in about nine days. One bird was observed to make the entire journey uninterrupted, a feat that is comparable to a human running a nonstop seven-day marathon at 70 kilometers (43.5 miles) per hour. Would you expect the flight (breast) muscles of bar-tailed godwits to be light or dark colored? Explain your answer.

DNA STRUCTURE AND FUNCTION

6

Application ⊘ 6.1 Cloning

REMEMBER: A cell's DNA contains all of the information necessary to build a new cell and, in the case of multicelled organisms, an entire individual (Section 2.10).

On September 11, 2001, Constable James Symington drove his search dog Trakr from Nova Scotia to Manhattan. Within hours of arriving, the dog led rescuers to the area where the final survivor of the World Trade Center attacks was buried. She had been clinging to life, pinned under rubble from the building where she had worked. Symington and Trakr helped with the search and rescue efforts for three days nonstop, until Trakr collapsed from smoke and chemical inhalation, burns, and exhaustion.

Trakr survived the ordeal, but later lost the use of his limbs from a degenerative neurological disease probably linked to toxic smoke exposure at Ground Zero. The hero dog died in April 2009, but his DNA lives on in his genetic copies—his **clones**. Symington's essay about Trakr's superior nature and abilities as a search and rescue dog won the Golden Clone Giveaway, a contest to find the world's most clone-worthy dog. Trakr's DNA was shipped to Korea, where it was inserted into donor dog eggs, which were then implanted into surrogate mother dogs. Five puppies, all clones of Trakr, were delivered to Symington in July 2009 (Figure 6.1A). Today, Trakr's clones are search and rescue dogs for Team Trakr Foundation, an international humanitarian organization that Symington established in 2010.

Trakr's clones were produced by **somatic cell nuclear transfer** (**SCNT**), a laboratory procedure in which an unfertilized egg's nucleus is replaced with the nucleus of a donor's somatic cell (Figure 6.2). A somatic cell is a body cell, as opposed to a reproductive cell (*soma* is a Greek word that means body). If all goes well, the egg's cytoplasm reprograms the transplanted DNA to direct the development of an embryo, which is then implanted into a surrogate mother. The animal born to the surrogate is a clone of the donor.

SCNT is possible because all cells descended from a fertilized egg inherit the same DNA. That DNA is like a master blueprint that directs the development of the individual's body. Thus, the DNA in each body cell of a multicelled individual

> The DNA in each body cell of a multicelled individual contains all the information necessary to build the individual all over again.

A. Left, James Symington and his dog Trakr assisted in the search for victims at Ground Zero, September 2001. Symington wrote an essay about Trakr's superior abilities in search and rescue operations such as this one, and won an opportunity to have the dog cloned. Right, Symington with Trakr's clones in 2009.

B. Champion dairy cow Liz (right) and her clone, Liz II (left). Liz II, who was produced by SCNT, had already begun to win championships by the time she was one year old.

Figure 6.1 Examples of adult animal cloning.

contains all the information necessary to build the individual all over again. In fact, clones occur all the time in nature. Body cells of many plants and some animal species easily give rise to new individuals that are clones of the parent. Identical twins are the product of embryo splitting, another natural process. The first few divisions of a fertilized egg form a ball of cells that sometimes splits spontaneously. If both halves of the ball continue to develop independently, identical twins result. Humans have long exploited this phenomenon with a technique called artificial embryo splitting. A tiny ball of cells is grown from a fertilized egg in a laboratory. The ball is teased apart into two halves, each of which goes on to develop as a separate embryo. The embryos are implanted in surrogate mothers, who give birth to identical twins.

Twins produced by embryo splitting are identical to one another, but they are not identical to either parent. This is because, in humans and other animals, twins get their DNA from two parents that typically differ in their DNA sequence (Chapter 8 returns to this topic). Thus, animal breeders who want a clone of a specific individual use SCNT. Clones produced by SCNT have the same championship features as their adult donor animals (Figure 6.1B), and there are other benefits. SCNT can yield many more offspring in a given time frame than traditional breeding, and offspring can be produced from a donor animal that is castrated or even dead.

"Cloning" means making an identical copy of something, and it can refer to deliberate interventions in reproduction intended to produce an exact genetic copy of an organism. **Reproductive cloning** refers to any technology, especially SCNT, that yields animal clones. SCNT has been used since 1997, when a lamb called Dolly was cloned from a mammary cell of an adult sheep (the clone was named after voluptuous performer Dolly Parton). At first, Dolly looked and acted like a normal sheep. However, she died early, most likely because she was a clone. SCNT is technically challenging and the outcome is still unpredictable. Depending on the species, few implanted embryos may survive until birth. Until recently, most of the clones that did survive had serious health problems such as enlarged organs and obesity. Cloned mice developed lung and liver problems, and almost all died prematurely. Cloned pigs tended to limp and have heart problems; some developed without a tail or, even worse, an anus.

Why the problems? Even though a somatic cell contains all the DNA required to produce a new individual, it will not automatically start dividing and form an embryo. During early development, an embryo's cells start using different subsets of their DNA. As they do, the cells become different in form and function, a process called **differentiation**. Differentiation is usually a one-way path in animal cells, which means that once a cell has become specialized, all of its descendant cells will be specialized the same way. By the time a liver cell, muscle cell, or other differentiated cell forms, most of its DNA has been turned off, and is no longer used (Chapter 10 returns to this topic). To clone an adult, scientists must transform one of its differentiated cells into an undifferentiated cell by turning the unused DNA back on, reprogramming it to function like the DNA of an egg. Even though we are getting better at doing that, we still have quite a bit to learn. SCNT technology has improved so much in recent years that health problems are much less common in animals cloned today.

Cloning animals brings us closer to the possibility of cloning humans, both technically and ethically. The idea raises uncomfortable ethical questions. For example, if cloning a lost animal for a grieving owner is acceptable, why would it not be acceptable to clone a lost child for a grieving parent? Different people have very different answers to such questions, so controversy over cloning continues to rage even as the technique improves.

A. A cow's egg is held in place by suction through a hollow glass tube called a micropipette. DNA is identified by a purple stain.

B. Another micropipette punctures the egg and sucks out the DNA. All that remains inside the egg's plasma membrane is cytoplasm.

C. A new micropipette prepares to enter the egg at the puncture site. The pipette contains a cell grown from the skin of a donor animal.

D. The micropipette enters the egg and delivers the skin cell to a region between the cytoplasm and the plasma membrane.

E. After the pipette is withdrawn, the donor's skin cell is visible next to the cytoplasm of the egg. The transfer is complete.

F. An electric current causes the foreign cell to fuse with and empty its nucleus into the cytoplasm of the egg. The egg begins to divide, and an embryo forms.

Figure 6.2 Somatic cell nuclear transfer (SCNT) with cells from cattle.
This series of micrographs was taken by scientists at Cyagra, a company that specializes in cloning livestock.

Courtesy of Cyagra, Inc., www.cyagra.com.

clone Genetically identical copy of an organism.
differentiation Process by which cells become specialized during development.
reproductive cloning Any of several laboratory procedures that produce genetically identical animals.
somatic cell nuclear transfer (SCNT) Reproductive cloning method in which the DNA of an adult donor's body cell is transferred into an unfertilized egg.

6.2 Fame, Glory, and DNA Structure

REMEMBER: Radioisotope tracers (Section 2.2) were used in research that led to the discovery that DNA is the hereditary material of all organisms. Protein structure varies greatly, but patterns such as helices are common (2.9). DNA is a polymer of nucleotides that have been linked into a chain (2.10).

Discovery of DNA's Function DNA, the substance (Figure 6.3), was first described in 1869 by Johannes Miescher, a chemist who extracted it from cell nuclei. Miescher determined that DNA is not a protein, and that it is rich in nitrogen and phosphorus, but he never learned its function. Sixty years later, Frederick Griffith unexpectedly found a clue. Griffith was studying pneumonia-causing bacteria in the hope of creating a vaccine. He discovered that these deadly bacteria contained a substance that could transform harmless bacteria into lethal ones. The transformation was permanent and heritable: Even after hundreds of generations, descendants of transformed cells retained the ability to kill.

What was the substance that had caused this transformation? In 1940, Oswald Avery and Maclyn McCarty set out to identify the substance, which they called the "transforming principle." The team extracted lipids, proteins, and nucleic acids from pneumonia-causing bacteria, then used a process of elimination to see which component transformed harmless bacteria into killers. Treating the extract with lipid- and protein-destroying enzymes did not destroy the transforming principle. Thus, the substance could not be lipid or protein. Avery and McCarty realized that the transforming principle must be nucleic acid—DNA or RNA. DNA-degrading enzymes destroyed the extract's ability to transform cells, but RNA-degrading enzymes did not. Thus, DNA had to be the transforming principle.

The result surprised Avery and McCarty, who, along with most other scientists, had assumed that proteins were the material of heredity. After all, traits are diverse, and proteins are the most diverse of all biological molecules. The two scientists were so skeptical that they published their results only after they had convinced themselves, by years of painstaking experimentation, that DNA was indeed hereditary material. They were also careful to point out that they had not proven DNA was the *only* hereditary material.

Avery and McCarty's tantalizing results prompted a stampede of other scientists into the field of DNA research. The resulting explosion of discovery confirmed the molecule's role as carrier of hereditary information. Key in this advance was the realization that any molecule—DNA or otherwise—had to have certain properties in order to function as hereditary material. First, a full complement of hereditary information must be transmitted along with the molecule; second, cells of a given species should contain the same amount of it; third, because the molecule functions as a genetic bridge between generations, it has to be exempt from major change; and fourth, it must be capable of encoding the almost unimaginably huge amount of information required to build a new individual.

In the late 1940s, Alfred Hershey and Martha Chase proved that DNA, and not protein, satisfies the first property of a hereditary molecule: It transmits a full complement of hereditary information. Hershey and Chase specialized in working with bacteriophages, a type of virus that infects bacteria. Like all viruses, these infectious particles carry information about how to make new viruses in their hereditary material. After a virus injects a cell with this material, the cell starts making new virus particles. Hershey and Chase carried out an elegant series of experiments proving that the material a bacteriophage injects into bacteria is DNA, not protein.

Figure 6.3 Human DNA, the substance.
DNA's role as the carrier of hereditary information was uncovered over many years, as scientists built upon one another's discoveries.

Patrick Landmann/Science Source.

Digging Into Data

The Hershey–Chase Experiments

By 1950, researchers had discovered bacteriophages, a type of virus that infects bacteria. Like all viruses, these infectious particles carry hereditary information about how to make new viruses. After a virus infects a cell, the cell starts making new virus particles. Bacteriophages inject genetic material into bacteria, but was that material DNA, protein, or both? Alfred Hershey and Martha Chase carried out experiments to determine the composition of that material. These experiments were based on the knowledge that proteins contain more sulfur (S) than phosphorus (P), and DNA contains more phosphorus than sulfur. Bacteriophage DNA and protein were labeled with radioactive tracers and allowed to infect bacteria. The virus–bacteria mixtures were then whirled in a blender to dislodge any viral components attached to the exterior of the bacteria (Figure 6.4). Afterward, radioactivity from the tracers was measured. The graph shown in Figure 6.4C, which summarizes results from these experiments, is reproduced from an original publication by Hershey and Chase.

Figure 6.4 Hershey–Chase experiments.

A. In one experiment, bacteriophage were labeled with a radioisotope of sulfur (^{35}S), a process that makes their protein components radioactive. The labeled viruses were mixed with bacteria long enough for infection to occur, and then the mixture was whirled in a kitchen blender. Blending dislodged viral parts that remained on the outside of the bacteria. Afterward, most of the radioactive sulfur was detected outside the bacterial cells. The viruses had not injected protein into the bacteria.

B. In another experiment, bacteriophage were labeled with a radioisotope of phosphorus (^{32}P), which makes their DNA radioactive. The labeled viruses were allowed to infect bacteria. After the external viral parts were dislodged from the bacteria, the radioactive phosphorus was detected mainly inside the bacterial cells. The viruses had injected DNA into the cells—evidence that DNA is the genetic material of this virus.

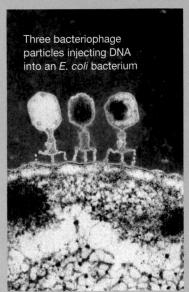

Three bacteriophage particles injecting DNA into an *E. coli* bacterium

Eye of Science/Science Source.

C. Detail from Alfred Hershey and Martha Chase's publication describing their experiments with bacteriophages. In this graph, "infected bacteria" refers to the percentage of bacteria that survived the blender.

Journal of General Physiology, 36(1), Sept. 20, 1952.

1. Before blending, what percentage of each isotope, ^{35}S and ^{32}P, was outside of the bacteria (extracellular)?

2. After 4 minutes in the blender, what percentage of each isotope was extracellular?

3. How did the researchers know that the radioisotopes in the fluid came from outside of the bacterial cells and not from bacteria that had been broken apart by whirling in the blender?

4. The extracellular concentration of which isotope increased the most with blending? Why do these results imply that bacteriophage viruses inject DNA into bacteria?

Variations in the sequence of DNA are the foundation of life's diversity.

Discovery of DNA's Structure DNA is a polymer of four types of nucleotides—adenine (A), guanine (G), thymine (T), and cytosine (C). Each has a five-carbon sugar, three phosphate groups, and a nitrogen-containing base after which it is named (Figure 6.5). Just how those four nucleotides are arranged in a DNA molecule was a puzzle that took over 50 years to solve. As molecules go, DNA is gigantic, and chromosomal DNA has a complex structural organization; both factors made the molecule difficult to work with given the laboratory methods at the time.

Clues about DNA's structure started coming together around 1950, when Erwin Chargaff (one of many researchers investigating its function) made two important discoveries about the molecule. First, the amounts of thymine and adenine are identical, as are the amounts of cytosine and guanine (A = T and G = C). We call this discovery Chargaff's first rule. Chargaff's second discovery, or rule, is that the DNA of different species differs in the proportions of adenine and guanine.

Meanwhile, biologist James Watson and biophysicist Francis Crick had been sharing ideas about the structure of DNA. The helical (coiled) pattern of secondary structure that occurs in many proteins had just been discovered, and Watson and Crick suspected that the DNA molecule was also a helix. The two spent many hours arguing about the size, shape, and bonding requirements of the four DNA nucleotides. They pestered chemists to help them identify bonds they might have overlooked, fiddled with cardboard cutouts, and made models from scraps of metal connected by suitably angled "bonds" of wire.

Biochemist Rosalind Franklin (right) had also been working on the structure of DNA. Like Crick, Franklin was expert in x-ray crystallography, a technique in which x-rays are directed through a purified and crystallized substance. Atoms in the substance's molecules scatter the x-rays in a pattern that can be captured as an image. Researchers can use the pattern to calculate the size, shape, and spacing between any repeating elements of the molecules—all of which are details of molecular structure.

Franklin had been told she would be the only one in her department working on the structure of DNA, so she did not know that Maurice Wilkins was already doing the same thing just down the hall. No one had told Wilkins about Franklin's assignment; he assumed she was a technician hired to do his x-ray crystallography work. And so a clash began. Wilkins thought Franklin displayed an appalling lack of deference that technicians of the era usually accorded researchers. To Franklin, Wilkins seemed prickly and oddly overinterested in her work.

Figure 6.5 The four nucleotides that make up DNA.
Each kind has three phosphate groups, a deoxyribose sugar (orange), and a nitrogen-containing base (blue) after which it is named.

adenine (A)
deoxyadenosine triphosphate

thymine (T)
deoxythymidine triphosphate

cytosine (C)
deoxycytidine triphosphate

guanine (G)
deoxyguanosine triphosphate

Wilkins and Franklin had been given identical samples of DNA. Franklin's meticulous work with hers yielded the first clear x-ray diffraction image of DNA as it occurs in cells. She gave a presentation on this work in 1952. DNA, she said, had two chains twisted into a double helix, with a backbone containing phosphate groups on the outside, and bases arranged in an unknown way on the inside. She had calculated DNA's diameter, the distance between its chains and between its bases, the angle of the helix, and the number of bases in each coil. Crick, with his crystallography background, would have recognized the significance of the work—if he had been there. Watson was in the audience but he was not a crystallographer and did not understand the implications of Franklin's data.

Franklin started to write a research paper on her findings. Meanwhile, and perhaps without her knowledge, Watson reviewed Franklin's x-ray diffraction image with Wilkins, and Watson and Crick read a report detailing Franklin's unpublished data. Crick, who had more experience with molecular modeling than Franklin, immediately understood what the image and the data meant. Franklin's

data provided Watson and Crick with the last piece of the DNA puzzle. In 1953, they put together all of the clues that had been accumulating for the last fifty years and built the first accurate model of the DNA molecule (left). On April 25, 1953, Franklin's paper appeared third in a series of articles about the structure of DNA in the journal *Nature*. It supported with solid experimental evidence Watson and Crick's theoretical model, which appeared in the first article of the series.

Watson and Crick proposed that a DNA molecule consists of two chains (or strands) of nucleotides, running in opposite directions and coiled into a double helix (Figure 6.6). Covalent bonds between the sugar of one nucleotide and a phosphate of the next form the sugar–phosphate backbone of each chain. Hydrogen bonds between the internally positioned bases hold the two strands together. Only two kinds of base pairings form: A to T, and G to C, which explains the first of Chargaff's rules. Most scientists had assumed (incorrectly) that the bases had to be on the outside of the helix, because they would be more accessible to DNA-copying enzymes that way. You will see in Section 6.4 how these enzymes access the nucleotide bases on the inside of the double helix.

Dozens of scientists contributed to the discovery of DNA's structure, but only three received recognition from the general public for their work. Rosalind Franklin died in 1958. Because the Nobel Prize is not given posthumously, she did not share in the 1962 honor that went to Watson, Crick, and Wilkins for the discovery of the structure of DNA.

DNA Sequence A small piece of DNA from a tulip, a human, or any other organism might be:

one base pair

Notice how the two strands of DNA fit together. They are complementary—the base of each nucleotide on one strand pairs with a suitable partner base on the other. This base-pairing pattern (A to T, G to C) is the same in all molecules of DNA. How can just two kinds of base pairings give rise to the incredible diversity of traits we see among living things? Even though DNA is composed of only four

A phosphate group links the sugar of one nucleotide to the sugar of the next. These bonds form each strand's sugar–phosphate backbone.

Hydrogen bonds link internally positioned nucleotide bases.

Figure 6.6 Structure of DNA, as illustrated by a composite of three different models.
The two sugar–phosphate backbones coil in a helix around internally positioned bases.

Figure It Out: What do the yellow balls represent?
Answer: Phosphate groups

nucleotides, the *order* in which one nucleotide follows the next along a strand—the **DNA sequence**—varies tremendously among species (which explains Chargaff's second rule). DNA molecules can be hundreds of millions of nucleotides long, so their sequence can encode a massive amount of information (we return to the nature of that information in the next chapter). DNA sequence variation is the basis of traits that define species and distinguish individuals. Thus DNA, the molecule of inheritance in every cell, is the basis of life's unity. Variations in its sequence are the foundation of life's diversity.

Take-Home Message 6.2

What is DNA?

- DNA is the molecule of inheritance in all organisms.
- A DNA molecule consists of two nucleotide chains (strands) coiled into a double helix. Hydrogen bonding between internally positioned nucleotide bases (A pairs with T, and C with G) hold the two strands together.
- The sequence of bases along a DNA strand—the DNA sequence—varies among species and among individuals. This variation is the basis of life's diversity.

6.3 DNA in Chromosomes

REMEMBER: The DNA of a eukaryotic cell is contained in a nucleus (Section 3.5).

Stretched out end to end, the DNA molecules in a single human cell would be about 2 meters (6.5 feet) long. How can that much DNA cram into a nucleus that is less than 10 micrometers in diameter? Inside a cell, proteins that associate with each DNA molecule twist and pack it into a structure called a **chromosome** (Figure 6.7). In a eukaryotic cell, for example, a DNA molecule ❶ wraps twice at regular intervals around "spools" of proteins called **histones** ❷. These DNA–histone spools look a bit like beads on a string in micrographs. Interactions among histones and other proteins twist the spooled DNA into a tight fiber ❸. This fiber coils, and then it coils again into a hollow cylinder like an old-style telephone cord ❹.

During most of the cell's life, each chromosome consists of one DNA molecule. When the cell prepares to divide, it duplicates its chromosomes (more about this process in the next section). After replication, each chromosome consists of two DNA molecules, or **sister chromatids**, attached to one another at a constricted region called the **centromere**:

The duplicated chromosomes condense into their familiar "X" shapes ❺ just prior to cell division.

The DNA of a eukaryotic cell is divided up among some number of chromosomes that differ in length and shape ❻. That number is called the **chromosome number**, and it is a characteristic of the species. For example, the chromosome number of humans is 46, so our cells have 46 chromosomes.

Figure 6.7 Eukaryotic chromosome structure.
The micrograph shows a duplicated chromosome just before cell division.

Andrew Syred/Science Source.

❶ Two strands of DNA twist into a double helix.

❷ At regular intervals, the DNA (blue) wraps around a core of histone proteins (purple).

❸ The DNA and proteins associated with it twist tightly into a fiber.

❹ The fiber coils and then coils again to form a hollow cylinder.

❺ At its most condensed, a duplicated chromosome has an X shape.

❻ The DNA in the nucleus of a typical eukaryotic cell is divided into a number of chromosomes.

Figure It Out: What is the yellow structure in the upper right corner of the drawing?

Answer: A cell nucleus

A. Karyotype of a female human, with identical sex chromosomes (XX).

B. Karyotype of a female chicken, with nonidentical sex chromosomes (ZW).

Figure 6.8 Karyotypes.
Chromosomes are numbered (as shown).

(A) © University of Washington Department of Pathology; (B) With kind permission from Springer Science+Business Media: *Chromosome Research*, Volume 17, Number 1, 99 113, DOI: 10.1007/s10577-009-9021-6; *Avian comparative genomics: reciprocal chromosome painting between domestic chicken* (*Gallus gallus*) *and the stone curlew* (*Burhinus oedicnemus, Charadriiformes*)—*An atypical species with low diploid number*; Wenhui Nie, Patricia C. M. O'Brien, Bee L. Ng, Beiyuan Fu, Vitaly Volobouev, Nigel P. Carter, Malcolm A. Ferguson-Smith, and Fengtang Yang; Figure 2a.

Actually, human body cells have two sets of 23 chromosomes—two of each type. Having two sets of chromosomes means these cells are **diploid**, or 2*n*. An image of an individual's diploid set of chromosomes is called a **karyotype** (Figure 6.8). To create a karyotype, cells taken from the individual are treated to make the chromosomes condense, and then stained so the chromosomes can be distinguished under a microscope. A micrograph of a single cell is digitally rearranged so the images of the chromosomes are lined up by centromere location, and arranged according to size, shape, and length.

In a human body cell, all but one pair of chromosomes are **autosomes**, which are the same in both females and males. The two autosomes of a pair have the same length, shape, and centromere location. They also hold information about the same traits. Think of them as two sets of books on how to build a house. Your father gave you one set. Your mother had her own ideas about wiring, plumbing, and so on. She gave you an alternate set that says slightly different things about many of those tasks.

Members of a pair of **sex chromosomes** differ between females and males, and the differences determine an individual's sex. The sex chromosomes of humans are called X and Y. The body cells of typical human females have two X chromosomes (XX, Figure 6.8A); those of typical human males have one X and one Y chromosome (XY). This pattern—XX females and XY males—is the rule among fruit flies, mammals, and many other animals, but there are other patterns (Figure 6.8B). Female butterflies, moths, birds, and certain fishes have two nonidentical sex chromosomes; the two sex chromosomes of males are identical. Environmental factors (not chromosomes) determine sex in some species of invertebrates, turtles, and frogs. As an example, the temperature of the sand in which sea turtle eggs are buried determines the sex of the hatchlings.

Take-Home Message 6.3

What is a chromosome?

- A chromosome is a molecule of DNA together with associated proteins that organize it and allow it to pack tightly.
- A eukaryotic cell's DNA is divided among a characteristic number of chromosomes, which differ in length and shape.
- Members of a pair of sex chromosomes differ between males and females. Chromosomes that are the same in both sexes are called autosomes.

autosome A chromosome that is the same in males and females.

centromere Of a duplicated eukaryotic chromosome, constricted region where sister chromatids attach to each other.

chromosome Structure that consists of DNA together with associated proteins; carries part or all of a cell's genetic information.

chromosome number The total number of chromosomes in a cell of a given species.

diploid Having two of each type of chromosome characteristic of the species (2*n*).

DNA sequence Order of nucleotides in a strand of DNA.

histone Type of protein that associates with eukaryotic DNA and structurally organizes chromosomes.

karyotype Image of an individual's complement of chromosomes arranged by size, length, shape, and centromere location.

sex chromosome Member of a pair of chromosomes that differs between males and females.

sister chromatids The two attached DNA molecules of a duplicated eukaryotic chromosome.

Figure 6.9 DNA replication.

1 As replication begins, enzymes begin to unwind and separate the two strands of DNA.

2 Primers base-pair with the exposed single DNA strands.

3 Starting at primers, DNA polymerases (green boxes) assemble new strands of DNA from nucleotides, using the parent strands as templates.

4 DNA ligase seals any gaps that remain between bases of the "new" DNA.

5 Each parental DNA strand (blue) serves as a template for assembly of a new strand of DNA (magenta). Both strands of the double helix serve as templates, so two double-stranded DNA molecules result. One strand of each is parental (old), and the other is new, so DNA replication is semiconservative.

6.4 DNA Replication and Repair

REMEMBER: With a few exceptions, free radicals are dangerous to life (Section 2.2). An enzyme speeds a reaction; bonds between phosphate groups hold a lot of energy (4.4). The wavelength of light is measured in nanometers (5.2).

When a cell reproduces, it divides. The two descendant cells must inherit a complete copy of genetic information or they will not function properly. Thus, in preparation for division, the cell copies its chromosomes so that it contains two sets: one for each of its future offspring. The copying process is an energy-intensive pathway called **DNA replication**. During DNA replication, the double helix of a DNA molecule is opened to expose the internally positioned bases, and an enzyme, **DNA polymerase**, links nucleotides into new strands of DNA according to the sequence of those bases. Each chromosome is copied in its entirety. Two identical molecules of DNA are the result. In eukaryotes, these molecules are sister chromatids that remain attached at the centromere until cell division occurs.

Before DNA replication, a chromosome has one molecule of DNA—one double helix (Figure 6.9). As replication begins, enzymes break the hydrogen bonds that hold the double helix together, so the two DNA strands unwind and separate **1**. Another enzyme starts making **primers**, which are short, single strands of nucleotides that serve as attachment points for DNA polymerase **2**. The nucleotide bases of a primer can form hydrogen bonds with the exposed bases of a single strand of DNA. Thus, a primer can base-pair with a complementary strand of DNA (right). The establishment of base pairing between two strands of DNA (or DNA and RNA) is called hybridization. Hybridization is spontaneous and is driven entirely by hydrogen bonding.

DNA polymerases attach to the hybridized primers and begin DNA synthesis. As a DNA polymerase moves along a strand, it uses the sequence of exposed nucleotide bases as a template, or guide, to assemble a new strand of DNA from free nucleotides **3**. Two of a nucleotide's three phosphate groups are removed when it is added to a DNA strand. Breaking those bonds releases enough energy to drive the attachment.

A DNA polymerase follows base-pairing rules: It adds a T to the end of the new DNA strand when it reaches an A in the template strand; it adds a G when it reaches a C; and so on. Thus, the DNA sequence of each new strand is complementary to the template (parental) strand. The enzyme DNA ligase seals any gaps, so the new DNA strands are continuous **4**. Both strands of the parent molecule are copied at the same time. As each new DNA strand lengthens, it winds up with the template strand into a double helix. So, after replication, two double-stranded molecules of DNA have formed **5**. One strand of each molecule is conserved (parental), and the other is new, so replication is said to be semiconservative. Both double-stranded molecules produced by DNA replication are duplicates of the parent molecule.

How Mutations Arise Sometimes, a new DNA strand is not exactly complementary to its parent strand. A nucleotide may get lost during DNA replication, or an extra one slips in. Occasionally, the wrong nucleotide is added. Most of these replication errors occur simply because DNA polymerases work very fast. Mistakes are inevitable, and some DNA polymerases make a lot of them. Luckily, most DNA polymerases also proofread their work. They can correct a mismatch by reversing the synthesis reaction to remove the mispaired nucleotide, then resuming synthesis in the forward direction. Replication errors also occur after the cell's DNA gets broken or otherwise damaged, because DNA polymerases do not copy damaged DNA very well. In most cases, repair enzymes and other proteins remove and replace damaged or mismatched bases in DNA before replication begins.

When proofreading and repair mechanisms fail to correct an error, it becomes a **mutation**—a permanent change in the DNA sequence of a cell's chromosome(s). Repair enzymes cannot fix a mutation after the altered DNA strand has been replicated, because they do not recognize correctly paired bases (Figure 6.10). Thus, a mutation is passed to one of the cell's offspring and all of its descendants.

Mutations alter DNA's genetic instructions, so they may have a harmful outcome. Cancer begins with mutations. Rosalind Franklin died at the age of 37, of ovarian cancer probably caused by extensive exposure to x-rays during her work. At the time, the link between x-rays, mutations, and cancer was not understood. We now know that electromagnetic energy with a wavelength shorter than 320 nanometers, including x-rays, most ultraviolet (UV) light, and gamma rays, has enough energy to knock electrons out of atoms. Exposure to such ionizing radiation damages DNA, breaking it into pieces that get lost during replication (Figure 6.11). Ionizing radiation can also cause covalent bonds to form between bases on opposite strands of the double helix, an outcome that permanently blocks DNA replication. The nucleotide bases themselves can be irreparably damaged by ionizing radiation. Repair enzymes remove bases damaged in this way, but they leave an empty space in the double helix or even a strand break. Any of these events can result in mutations.

UV light in the range of 320 to 380 nanometers does not have enough energy to knock electrons out of atoms, but it can cause a covalent bond to form between adjacent thymine or cytosine bases. The result is a nucleotide dimer that kinks the DNA strand. DNA polymerase tends to copy the kinked part incorrectly during replication, and mutations are the outcome. Exposing unprotected skin to sunlight increases the risk of cancer because the UV wavelengths cause these nucleotide dimers to form. For every second a skin cell spends in the sun, between 50 and 100 nucleotide dimers form in its DNA.

Exposure to some chemicals also causes mutations. For instance, several of the fifty-five or more cancer-causing chemicals in tobacco smoke transfer methyl groups ($—CH_3$) to the nucleotide bases in DNA. Nucleotides altered in this way do not base-pair correctly. The body converts other chemicals in the smoke to compounds that bind irreversibly to DNA. In both cases, the resulting replication errors can lead to mutation. Cigarette smoke also contains free radicals, which inflict the same damage on DNA as ionizing radiation.

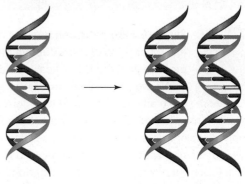

A. Repair enzymes can recognize a mismatched base (yellow), but they sometimes fail to correct it before DNA replication.

B. After replication, both strands base-pair properly. Repair enzymes can no longer recognize the error, which has now become a mutation that will be passed on to the cell's descendants.

Figure 6.10 How replication errors become mutations.

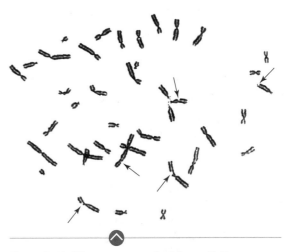

Figure 6.11 Why exposure to ionizing radiation causes mutations.
This micrograph shows major breaks (red arrows) in the chromosomes of human white blood cells that have been exposed to ionizing radiation. Pieces of broken chromosomes often become lost during DNA replication.

Olga Shovman, Andrew C. Riches, Douglas Adamson, and Peter E. Bryant. *An improved assay for radiation-induced chromatid breaks using a colcemid block and calyculin-induced PCC combination.* Mutagenesis (2008) 23(4): 267–270; first published online March 6, 2008. doi:10.1093/mutage/gen009, by permission of Oxford University Press.

Take-Home Message 6.4

What happens during DNA replication?

- When a cell copies its DNA, each strand of the double helix serves as a template for synthesis of a new, complementary strand of DNA. Two double helices result.
- Proofreading and repair mechanisms usually maintain the integrity of a cell's genetic information by correcting mispaired bases and fixing damaged DNA before replication.
- Mismatched or damaged nucleotides that are not repaired become mutations.
- DNA damage by environmental agents such as UV light, chemicals, and free radicals can result in mutations, because damaged DNA is not replicated very well.

DNA polymerase Enzyme that carries out DNA replication.

DNA replication Process by which a cell duplicates its DNA before it divides.

mutation Permanent change in DNA sequence.

primer Short, single strand of DNA that base-pairs with a specific DNA sequence.

Summary

Section 6.1 Somatic cell nuclear transfer (SCNT) and other types of **reproductive cloning** technologies produce genetically identical individuals (**clones**). SCNT works because the DNA in each body cell of an animal contains all the information necessary to build a new individual. However, the outcome of SCNT is still unpredictable. This is because, during development, cells of an embryo become specialized as they begin to use different subsets of their DNA (a process called **differentiation**). Reprogramming an adult cell to behave like an embryonic cell involves rewinding this developmental process, and we are still learning how to do it. Cloning animals continues to raise ethical questions, particularly about the possibility of cloning humans.

Section 6.2 It took decades of research by many scientists to determine that deoxyribonucleic acid (DNA) is the hereditary material of all life, and to unravel its structure.

Each DNA nucleotide has a five-carbon sugar, three phosphate groups, and one of four nitrogen-containing bases after which the nucleotide is named: adenine, thymine, guanine, or cytosine. DNA is a polymer that consists of two strands of these nucleotides coiled into a double helix. Hydrogen bonding between the internally positioned bases holds the strands together. The bases pair in a consistent way: adenine with thymine (A–T), and guanine with cytosine (G–C). The order of bases along a strand of DNA—the **DNA sequence**—varies among species and among individuals, and this variation is the basis of life's diversity.

Section 6.3 The DNA of eukaryotes is divided among a characteristic number of **chromosomes** that differ in length and shape. In eukaryotic chromosomes, the DNA wraps around **histones**. When duplicated, a eukaryotic chromosome has an X shape and consists of two **sister chromatids** attached at a **centromere**.

Diploid cells have two of each type of chromosome. **Chromosome number** is the sum of all chromosomes in a cell of a given species. A human body cell has twenty-three pairs of chromosomes. Members of a pair of **sex chromosomes** differ among males and females. Chromosomes that are the same in males and females are **autosomes**. Autosomes of a pair have the same length, shape, and centromere location. A **karyotype** is an image of an individual's complete set of chromosomes.

Section 6.4 Before a cell divides, it copies all of its DNA (by **DNA replication**) so both of its cellular offspring inherit a complete set of chromosomes. For each molecule of DNA that is copied, two DNA molecules are produced; each is a duplicate of the parent. DNA replication is said to be semiconservative because one strand of each molecule is new, and the other is parental.

During DNA replication, enzymes unwind the double helix. **Primers** form and base-pair (hybridize) with the exposed single strands of DNA. Starting at the primers, **DNA polymerase** enzymes use each strand as a template to assemble new, complementary strands of DNA from free nucleotides. DNA ligase seals any gaps.

Proofreading by DNA polymerases corrects most DNA replication errors as they occur. DNA damage by environmental agents, including ionizing and nonionizing radiation, free radicals, and some chemicals, can lead to replication errors because DNA polymerase does not copy damaged DNA very well. Most DNA damage is repaired before replication begins. Uncorrected replication errors become **mutations**: permanent changes in the nucleotide sequence of a cell's DNA.

Self-Quiz

Answers in Appendix I

1. _____ is an example of reproductive cloning.
   a. Somatic cell nuclear transfer (SCNT)
   b. Multiple offspring from the same pregnancy
   c. Artificial embryo splitting
   d. a and c
   e. all of the above

2. Which is *not* a nucleotide base in DNA?
   a. adenine   c. glutamine   e. cytosine
   b. guanine   d. thymine   f. All are in DNA.

3. What are the base-pairing rules for DNA?
   a. A–G, T–C   c. A–T, G–C
   b. A–C, T–G   d. A–A, G–G, C–C, T–T

4. Variation in _____ is the basis of variation in traits.
   a. karyotype   c. the double helix
   b. DNA sequence   d. chromosome number

5. One species' DNA differs from others in its _____ .
   a. nucleotides
   b. DNA sequence
   c. double helix
   d. replication process
   e. sugar–phosphate backbone
   f. all of the above

6. In eukaryotic chromosomes, DNA wraps around _____ .
   a. histone proteins
   b. sister chromatids
   c. centromeres
   d. nucleotides

7. Chromosome number _____ .
   a. refers to a particular chromosome in a cell
   b. is a characteristic feature of a species
   c. is the number of autosomes in cells of a given type
   d. is the same in all species

8. Human body cells are diploid, which means they _____ .
   a. are complete
   b. have two sets of chromosomes
   c. contain sex chromosomes
   d. divide to form two cells

9. When DNA replication begins, _____ .
   a. the two DNA strands unwind from each other
   b. the two DNA strands condense for base transfers
   c. old strands move to find new strands
   d. mutations occur

10. DNA replication requires _____ .
    a. DNA polymerase
    b. nucleotides
    c. primers
    d. all are required

11. Energy that drives the attachment of a nucleotide to the end of a growing strand of DNA comes from _____ .
    a. phosphate-group transfers from ATP
    b. DNA polymerase
    c. the nucleotide itself
    d. a and c

12. After DNA replication, a eukaryotic chromosome _____ .
    a. consists of two sister chromatids
    b. has a characteristic X shape
    c. is constricted at the centromere
    d. all of the above

13. Exposure to _____ can lead to mutations.
    a. UV light
    b. cigarette smoke
    c. chemicals
    d. x-rays
    e. sunlight
    f. all of the above

14. All mutations _____ .
    a. arise from DNA damage
    b. lead to evolution
    c. are caused by radiation
    d. change the DNA sequence

15. Match the terms appropriately.
    ____ nucleotide
    ____ clone
    ____ autosome
    ____ DNA polymerase
    ____ mutation
    ____ bacteriophage
    ____ semiconservative replication

    a. replication enzyme
    b. does not determine sex
    c. copy of an organism
    d. nitrogen-containing base, sugar, phosphate
    e. injects DNA
    f. can cause cancer
    g. something old, something new

Critical Thinking

1. Mutations are the original source of genetic variation. How can mutations accumulate in DNA, given that cells have repair systems that fix mispaired nucleotides or breaks in DNA strands?

2. Woolly mammoths have been extinct for about 10,000 years, but we often find their well-preserved remains in Siberian permafrost. Research groups are now planning to use SCNT to resurrect these huge elephant-like mammals. No mammoth eggs have been recovered so far, so elephant eggs would be used instead. An elephant would also be the surrogate mother for the resulting embryo. The researchers may try a modified SCNT technique used to clone a mouse that had been dead and frozen for sixteen years. Ice crystals that form during freezing break up cell membranes, so cells from the frozen mouse were in bad shape. Their DNA was transferred into donor mouse eggs, and cells from the resulting embryos were fused with mouse stem cells. Four healthy clones were born from the hybrid embryos. What are some of the pros and cons of cloning an extinct animal?

Visual Question

1. Determine the complementary strand of DNA that forms on this template DNA fragment during replication:

G G T T T C T T C A A G A G A

7

GENE EXPRESSION AND CONTROL

Application ❯ 7.1 Ricin, RIP

REMEMBER: A fat consists mainly of triglycerides (Section 2.8). A domain contributes a structural or functional property to a protein (2.9). A few strains of *E. coli* make a toxic protein that can severely damage the lining of the intestine (3.1). Ribosomes assemble polypeptides (3.4). Enzymes are unchanged by participating in a reaction (4.4). Cells can take in materials across their membrane by endocytosis (4.6).

Castor-oil plants grow wild in tropical regions, and they are widely cultivated for their seeds (Figure 7.1A). Castor-oil seeds are rich in ricin, a toxic protein that effectively deters beetles, birds, mammals, and other animals from eating them. They are also rich in castor oil, a fat that we extract from the seeds to use as an ingredient in plastics, cosmetics, paints, soaps, and many other items. Ricin-containing seed pulp left over from the extraction process is usually—but not always—discarded.

A dose of ricin as small as a few grains of salt can kill an adult human, and there is no antidote. Ricin's lethal effect was known as long ago as 1888, but using it as a weapon is banned by most countries under the Geneva Protocol. Controlling production of the toxin is impossible because no special skills or equipment are required to manufacture it from easily obtained raw materials. Thus, ricin appears periodically in the news as a tool of criminals. Perhaps the most famous example occurred in 1978 at the height of the Cold War, when defectors from countries under Russian control were targets for assassination. Bulgarian journalist Georgi Markov had defected to England and was working for the BBC. As he made his way to a bus stop on a London street, an assassin used a modified umbrella (Figure 7.1B) to fire a tiny pellet of ricin into Markov's leg. Markov died in agony three days later.

Ricin is called a ribosome-inactivating protein (RIP) because it inactivates ribosomes, the organelles that assemble amino acids into proteins. Other RIPs are made by some bacteria, mushrooms, algae, and many plants (including food crops such as tomatoes, barley, and spinach). Most of these proteins are not particularly toxic in humans because they do not cross intact cell membranes very well. Those that do, including ricin, have a domain that binds to molecules on our plasma membranes. Binding causes the cell to take up the RIP by endocytosis. Once inside the cell, a second domain of the RIP—an enzyme—begins to inactivate ribosomes. One molecule of ricin can inactivate more than 1,000 ribosomes per minute. If enough ribosomes are affected, protein synthesis grinds to a halt. Proteins are critical to all life processes, so cells that cannot make them die quickly.

Fortunately, few people actually encounter ricin. Contact with other toxic RIPs is much more common. Bracelets made from beautiful seeds were recalled from stores in 2011 after a botanist recognized the seeds as jequirity beans. These beans contain abrin, an RIP even more toxic than ricin. Shiga toxin, an RIP made by *Shigella dysenteriae* bacteria, causes a severe bloody diarrhea (dysentery) that can be lethal. Some strains of *E. coli* make Shiga-like toxin, an RIP that causes symptoms associated with food poisoning.

Despite their toxicity, the main function of RIPs may not be destroying ribosomes. Many are part of plant immunity, and have antiviral and anticancer activity. Plants that make toxic RIPs have been used as traditional medicines for many centuries; now, Western scientists are exploiting the unique properties of these proteins in drug design. For example, researchers have attached ricin's toxic enzyme domain to antibodies that can find specific types of cancer cells in a person's body. Other RIPs have also been modified to specifically target cancer cells (Figure 7.1C). The intent of both strategies: to assassinate the cancer cells without harming normal ones.

A. Seeds of the castor-oil plant, source of ribosome-busting ricin. Eating just eight of these seeds can kill an adult human.

B. Bulgarian spy's weapon: an umbrella modified to fire a tiny pellet of ricin into a victim. An umbrella like this one was used to assassinate Georgi Markov on the streets of London in 1978.

C. Researchers incorporated a peptide that binds to skin cancer cells into the enzymatic domain of Shiga-like toxin. The model shows the active site of the toxin in red; the changed bit, in blue. This engineered RIP specifically kills skin cancer cells.

Figure 7.1 RIPs, examples of source and delivery.

(A) Vaughan Fleming/SPL/Science Source; (B) Cary Wolinsky/National Geographic Creative; (C) Cheung et al. Molecular Cancer, 9:28, 2010.

nucleotides in DNA—adenine, thymine, guanine, and cytosine. In this chapter, we investigate "words" that can be made from those letters, and "sentences" that can be made from the words.

The nature of information represented by the sequence of nucleotides in a DNA molecule occurs in hundreds or thousands of units called genes. The DNA sequence of a **gene** encodes (contains instructions for building) an RNA or protein product. Converting the information encoded by a gene into a product starts with RNA synthesis, which is called transcription. During **transcription**, enzymes use a strand of DNA as a template to assemble a strand of RNA.

Most of the RNA inside cells occurs in single strands that are similar in structure to single strands of DNA (Figure 7.2). For example, both are chains of four kinds of nucleotides. Like a DNA nucleotide, an RNA nucleotide has three phosphate groups, a sugar, and one of four bases. However, the sugar in an RNA nucleotide (ribose) is slightly different from the sugar in a DNA nucleotide (deoxyribose). Three bases (adenine, cytosine, and guanine) occur in both RNA and DNA nucleotides, but the fourth base differs between the two molecules. In DNA, the fourth base is thymine (T); in RNA, it is uracil (U). These small differences in structure give rise to very different functions. DNA's important but only role is to store a cell's genetic information. By contrast, a cell makes several kinds of RNAs on an ongoing basis, and the different types have different functions. **Messenger RNA (mRNA)** was named for its function as the "messenger" between DNA and protein. By the process of **translation**, the protein-building information in an mRNA is decoded (translated) into a sequence of amino acids. The result is a polypeptide that twists and folds into a protein.

Transcription and translation are both part of **gene expression**, the multistep process by which information in a gene guides the assembly of an RNA or protein product. Expression of a gene that encodes an RNA product (such as an mRNA) involves transcription. Expression of a gene that encodes a protein product involves both transcription and translation:

$$\text{DNA} \xrightarrow{\text{transcription}} \text{mRNA} \xrightarrow{\text{translation}} \text{protein}$$

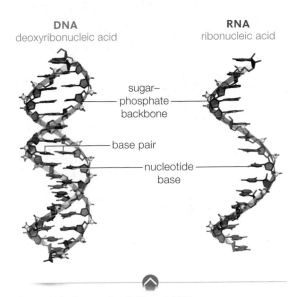

DNA
deoxyribonucleic acid

RNA
ribonucleic acid

sugar–phosphate backbone

base pair

nucleotide base

Figure 7.2 Comparing DNA and RNA.
DNA permanently stores a cell's genetic information. Cells continually make different types of RNAs that have various functions.

Take-Home Message 7.2

What is the nature of the information carried by DNA?

- Information in a DNA sequence occurs in units that are called genes.
- A cell uses the information encoded by a gene to make an RNA or protein product, a process called gene expression.
- Transcription converts information in a gene to RNA; translation converts information in an mRNA to protein.

gene A part of a chromosome that encodes an RNA or protein product in its DNA sequence.

gene expression Process by which the information in a gene guides assembly of an RNA or protein product. Includes transcription and translation.

messenger RNA (mRNA) RNA that has a protein-building message.

transcription Process by which enzymes assemble an RNA using a strand of DNA as a template.

translation Process by which the protein-building instructions in an mRNA guide the assembly of a polypeptide.

Figure 7.3 Base pairing in DNA replication and RNA synthesis.

A. During DNA replication, a nucleotide is added to the end of a growing DNA strand only if it base-pairs with the corresponding nucleotide on the template DNA strand. Cytosine (C) pairs with guanine (G), and adenine (A) pairs with thymine (T).

B. During RNA synthesis, a nucleotide is added to the end of a growing RNA strand only if it base-pairs with the corresponding nucleotide on the template DNA strand. The same base-pairing rules apply, except that uracil in RNA base-pairs with adenine (A).

Figure 7.4 DNA Christmas trees.
Typically, many RNA polymerases simultaneously transcribe the same gene, producing a structure called a Christmas tree after its shape. Here, several genes next to one another on the same chromosome are being transcribed simultaneously.

© O. L. Miller.

7.3 Transcription: DNA to RNA

REMEMBER: Only two kinds of base pairings occur in DNA: G pairs with C, and A pairs with T (Section 6.2). Each molecule of DNA produced during DNA replication is a duplicate of the parent (6.4).

The process of transcription is similar to DNA replication. Base-pairing rules are followed, for example. In DNA replication, cytosine (C) pairs with guanine (G), and adenine (A) pairs with thymine (T). The very same base-pairing rules also govern RNA synthesis in transcription, except that RNA, remember, has uracil instead of thymine (Figure 7.3). Uracil (U) base-pairs with adenine (A). During transcription, a strand of DNA acts as a template upon which a strand of RNA is assembled from nucleotides. A nucleotide can be added to a growing RNA only if it is complementary to the corresponding nucleotide of the template DNA. As in DNA replication, each nucleotide provides the energy for its own attachment to the end of a growing strand. Transcription is also similar to DNA replication in that one strand of a nucleic acid serves as a template for synthesis of another. However, in contrast to DNA replication, only part of one DNA strand, not the whole molecule, is used as a template for transcription. The enzyme **RNA polymerase**, not DNA polymerase, adds nucleotides to the end of a growing RNA. Also, transcription produces a single strand of RNA, not two DNA double helices.

In eukaryotic cells, transcription occurs in the nucleus; in prokaryotes, it occurs in cytoplasm. In both cases, the process begins at a regulatory site called a **promoter**, a short DNA sequence that serves as a binding site for RNA polymerase. Generally, a gene's promoter is close to it and a bit upstream:

Binding at a promoter positions an RNA polymerase close to the gene that will be transcribed. The polymerase starts moving along the DNA, over the gene region.

Figure It Out: Are the polymerases transcribing this DNA molecule moving from left to right or from right to left?

Answer: Left to right (the RNAs get longer as the polymerases move along the DNA)

As it moves, the polymerase unwinds the double helix just a bit so it can "read" the nucleotide sequence of the template DNA strand:

RNA

DNA winding up   DNA unwinding

An RNA polymerase moving over a gene region joins free RNA nucleotides into a chain, in the order dictated by the DNA sequence of the gene:

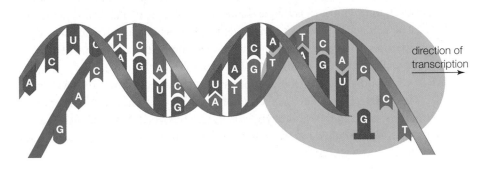

direction of transcription

When the polymerase reaches the end of the gene region, it releases the DNA and the new RNA. The new RNA strand is complementary to the DNA strand from which it was transcribed. It is an RNA copy of a gene, in the same way that a paper transcript of a conversation carries the same information in a different format. Typically, many polymerases transcribe a particular gene region at the same time, so many new RNA strands can be produced very quickly (Figure 7.4).

RNA Modifications Just as a dressmaker may snip off loose threads or add bows to a dress before it leaves the shop, so do eukaryotic cells tailor their RNA before it leaves the nucleus. Consider that most eukaryotic genes contain intervening sequences called **introns**. Introns are removed in chunks from a newly transcribed RNA before it leaves the nucleus. Sequences that remain in the RNA after this process are called **exons** (Figure 7.5). Exons can be rearranged and spliced together in different combinations, so one gene may encode two or more versions of the same product. A newly transcribed RNA that will become an mRNA is further tailored after splicing. For example, a tail of 50 to 300 adenines is added to the end of a new mRNA. Among other functions, this poly-A tail is a signal that allows an mRNA to be exported from the nucleus.

> An mRNA is ... of a gene, in ... same way th... paper trans... conversation carries the same information in a different format.

Figure 7.5 Post-transcriptional modification of RNA. Introns are removed and exons spliced together. Messenger RNAs also get a poly-A tail.

Take-Home Message 7.3

How is RNA assembled?

• During transcription, RNA polymerase uses a gene region in a chromosome as a template to assemble a strand of RNA. The new strand is an RNA copy of the gene from which it was transcribed.

• Post-transcriptional modification of RNA occurs in the nucleus of eukaryotes.

exon Nucleotide sequence that remains in an RNA after post-transcriptional modification.

intron Nucleotide sequence that intervenes between exons and is removed during post-transcriptional modification.

promoter DNA sequence that is a site where RNA polymerase binds for transcription.

RNA polymerase Enzyme that carries out transcription.

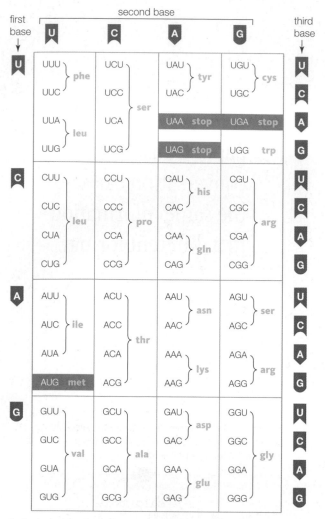

first base → second base → third base

A. Codon table. Each codon in mRNA is a set of three nucleotide bases. The left column lists a codon's first base, the top row lists the second, and the right column lists the third.

Sixty-one of the triplets encode amino acids; one of those, AUG, both codes for methionine and serves as a signal to start translation. Three codons are signals that stop translation.

ala alanine (A)	gly glycine (G)	pro proline (P)
arg arginine (R)	his histidine (H)	ser serine (S)
asn asparagine (N)	ile isoleucine (I)	thr threonine (T)
asp aspartic acid (D)	leu leucine (L)	trp tryptophan (W)
cys cysteine (C)	lys lysine (K)	tyr tyrosine (Y)
glu glutamic acid (E)	met methionine (M)	val valine (V)
gln glutamine (Q)	phe phenylalanine (F)	

B. Names and abbreviations of the 20 naturally occurring amino acids specified by the genetic code (**A**).

Figure 7.6　The genetic code.

Figure It Out: Which codons specify the amino acid lysine (lys)?

Answer: AAA and AAG

7.4 The Genetic Code

REMEMBER: A polypeptide is a linear sequence of amino acids joined by peptide bonds (Section 2.9).

An mRNA is essentially a disposable copy of a gene; its job is to carry the gene's protein-building information to types of RNA during translation. The protein-building message is encoded by nucleotides, "genetic words" that occur one after another along its length. Like the words of a sentence, a series of these genetic words can form a meaningful parcel of information—in this case, the sequence of amino acids of a protein.

Each of the genetic "words" carried by an mRNA is three nucleotide bases long, and each is a code—a **codon**—for a particular amino acid. The sequence of bases in a triplet determines which amino acid the codon specifies (Figure 7.6A). For instance, the codon UUU codes for the amino acid phenylalanine (phe), and UUA codes for leucine (leu).

Codons occur one after another along the length of an mRNA. When the mRNA is translated, the order of its codons determines the order of amino acids in the resulting polypeptide. Thus, the DNA sequence of a gene is transcribed into the nucleotide sequence of an mRNA, which is in turn translated into an amino acid sequence:

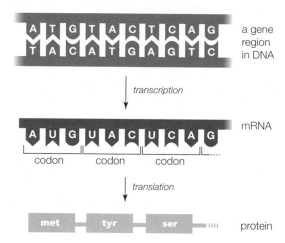

With four possible bases (G, A, U, or C) in each of the three positions of a codon, there are a total of sixty-four (or 4^3) mRNA codons. Collectively, the sixty-four codons constitute the **genetic code**. These codons specify a total of twenty naturally occurring amino acids (Figure 7.6B), so some amino acids have more than one codon. For instance, the amino acid tyrosine (tyr) is specified by two codons: UAU and UAC.

Other codons signal the beginning and end of a protein-coding sequence. In eukaryotes, the first AUG in an mRNA usually serves as the signal to start translation. AUG is the codon for methionine, so methionine is the first amino acid in new polypeptides. The codons UAA, UAG, and UGA do not specify an amino acid. These are signals that stop translation, so they are called stop codons. A stop codon marks the end of the protein-coding sequence in an mRNA.

The genetic code is highly conserved, which means that most organisms use the same code and probably always have. Bacteria, archaea, and some protists have a few codons that differ from the eukaryotic code, as do mitochondria and chloroplasts—a clue that led to a theory of how these two organelles evolved (we return to this topic in Section 13.3).

Take-Home Message 7.4

What is the genetic code?

- The genetic code consists of sixty-four codons (base triplets).
- Three codons are signals that stop translation; the remaining codons specify an amino acid. In eukaryotic mRNAs, the first occurrence of the codon that specifies methionine is a signal to begin translation.

7.5 Translation: RNA to Protein

REMEMBER: Ribosomes are organelles that carry out protein synthesis. Some enzymes are RNAs (Section 4.4).

A ribosome has two subunits, one large and one small (Figure 7.7). Each subunit consists mainly of **ribosomal RNA (rRNA)**, with associated structural proteins. During translation, a large and a small ribosomal subunit converge as an intact ribosome on an mRNA. **Transfer RNAs (tRNAs)** then deliver amino acids to the intact ribosome. Ribosomal RNA is one example of RNA with enzymatic activity: During translation, the rRNA components of a ribosome (not the protein components) cause peptide bonds to form between amino acids.

Each tRNA has two attachment sites. The first is an **anticodon**, which is a triplet of nucleotides that base-pairs with an mRNA codon (Figure 7.8). The other attachment site binds to an amino acid—the one specified by the codon. Transfer RNAs with different anticodons carry different amino acids. During translation, these tRNAs deliver amino acids to a ribosome, one after another in the order specified by the codons in an mRNA. As the amino acids are delivered, the ribosome joins them via peptide bonds into a new polypeptide. Thus, the order of codons in an mRNA—DNA's protein-building message—is translated into a new protein.

Figure 7.8 tRNA structure.
Each tRNA's anticodon is complementary to an mRNA codon. Each also carries the amino acid specified by that particular codon. These models depict the tRNA that carries the amino acid tryptophan.

anticodon

trp

amino acid attachment site

Figure 7.7 Ribosome structure.
An intact ribosome consists of a large and a small subunit. Protein components of both subunits are shown in the ribbon models in green; rRNA components, in brown.

large subunit   +   small subunit   =   intact ribosome

anticodon In a tRNA, set of three nucleotides that base-pairs with an mRNA codon.

codon In an mRNA, a nucleotide base triplet that codes for an amino acid or stop signal during translation.

genetic code Complete set of sixty-four mRNA codons.

ribosomal RNA (rRNA) RNA that becomes part of ribosomes.

transfer RNA (tRNA) RNA that delivers amino acids to a ribosome during translation.

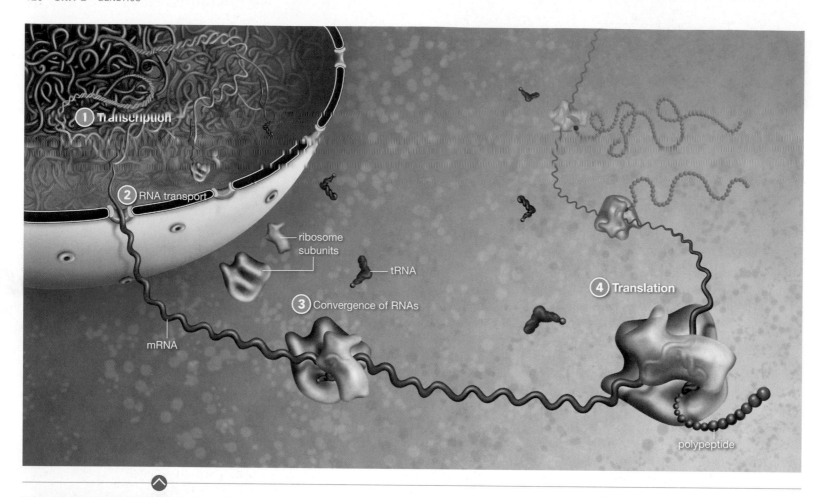

Figure 7.9 Overview of translation in a eukaryotic cell.
In eukaryotes, RNAs are transcribed in the nucleus, then transported into cytoplasm. Translation begins when ribosomal subunits and tRNA converge on an mRNA in cytoplasm. Then, tRNAs deliver amino acids in the order dictated by successive codons in the mRNA. The ribosome links the amino acids together as it moves along the mRNA, so a polypeptide forms and elongates.

In all cells, translation occurs in cytoplasm. In a eukaryotic cell, RNAs that carry out transcription are produced in the nucleus (Figure 7.9 **1**), then transported through nuclear pores into cytoplasm **2**. Translation is initiated when ribosomal subunits and tRNAs converge on an mRNA **3**. The complex of molecules is now ready to carry out protein synthesis. The intact ribosome begins to move along the mRNA and assemble a polypeptide **4**.

The first tRNA carries the amino acid methionine, the first amino acid of the new polypeptide. Another tRNA joins the complex when its anticodon base-pairs with the second codon in the mRNA. This tRNA brings with it the second amino acid. The ribosome joins the first two amino acids by way of a peptide bond:

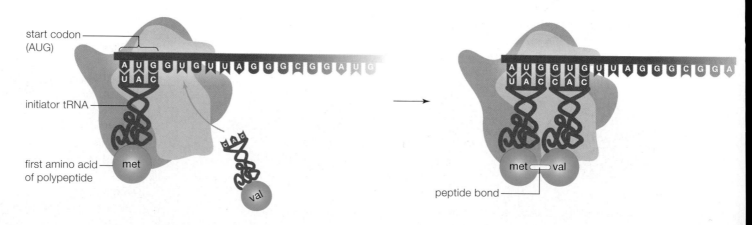

The first tRNA is released and the ribosome moves to the next codon. Another tRNA brings the third amino acid to the ribosome as its anticodon base-pairs with the third codon of the mRNA. The ribosome joins the second and third amino acids by way of a peptide bond:

The second tRNA is released and the ribosome moves to the next codon. Another tRNA brings the fourth amino acid to the complex as its anticodon base-pairs with the fourth codon of the mRNA. The ribosome joins the third and fourth amino acids by way of a peptide bond:

The new polypeptide continues to elongate as the ribosome joins successive amino acids delivered by tRNAs. Translation ends when the ribosome reaches a stop codon in the mRNA. The mRNA and the polypeptide detach from the ribosome, and the ribosomal subunits separate from each other.

Most of the energy that fuels translation is provided by GTP, an RNA nucleotide. Phosphate-group transfers from GTP help the ribosome move from one codon to the next along an mRNA. The RIPs you learned about in Section 7.1 are toxic because they remove a particular adenine base from one of the rRNAs in the ribosome's large subunit. The adenine is part of a binding site for proteins involved in the GTP-requiring steps of elongation. After the base has been removed, the ribosome can no longer bind to these proteins, and elongation stops.

Take-Home Message 7.5

How is mRNA translated into protein?

• Translation is an energy-requiring process in which a polypeptide is synthesized based on the sequence of codons in an mRNA.

• During elongation, amino acids are delivered to the ribosome by tRNAs in the order dictated by successive mRNA codons. As amino acids arrive, the ribosome joins each to the end of the polypeptide.

A. Hemoglobin, an oxygen-binding protein in red blood cells. This protein consists of four polypeptides: two alpha globins (blue) and two beta globins (green). Each globin forms a pocket that cradles a cofactor called a heme (red). Oxygen gas binds to the iron atom at the center of each heme.

16 17 18 19 20 21 22 23 24 25 26 27 28 29 30

pro	glu	glu	lys	ser

B. Part of the DNA (blue), mRNA (brown), and amino acid sequence of human beta globin. Numbers indicate nucleotide position in the mRNA.

pro	val	glu	lys	ser

C. A base-pair substitution replaces a thymine with an adenine. When the altered mRNA is translated, valine replaces glutamic acid as the sixth amino acid. Hemoglobin with this form of beta globin is called HbS, or sickle hemoglobin.

pro	gly	arg	ser	leu

D. A base-pair deletion shifts the reading frame for the rest of the mRNA, so a completely different protein product forms. The mutation shown results in a defective beta globin. The outcome is beta thalassemia, a genetic disorder in which a person has an abnormally low amount of hemoglobin.

pro	glu	glu	lys	val

E. An insertion of one nucleotide causes the reading frame for the rest of the mRNA to shift. The protein translated from this mRNA is too short and does not assemble correctly into hemoglobin molecules. As in **D**, the outcome is beta thalassemia.

Figure 7.10 Examples of mutations.

(A) 1BBB, A third quaternary structure of human hemoglobin A at 1.7-A resolution. Silva, M.M., Rogers, P.H., Arnone, A., Journal: (1992) *J.Biol.Chem*. 267: 17248–17256.

7.6 Products of Mutated Genes

REMEMBER: Hydrophilic substances dissolve easily in water; hydrophobic substances do not (Section 2.4). Hemoglobin has multiple polypeptides that fold around hemes; fibrous proteins such as keratin aggregate by many thousands into much larger structures (2.9). Many proteins can function properly only with assistance from a cofactor (4.4). A diploid cell has two sets of chromosomes (6.3). A mutation is a permanent change in the DNA sequence of a cell's chromosome (6.4).

Mutations are relatively uncommon events in a normal cell. Consider that the chromosomes in a diploid human cell collectively consist of about 6.5 billion nucleotides, any of which may become mutated each time that cell divides. However, the mutation rate in human somatic cells is about 10^{-8}, which means only one nucleotide changes every 10^8 times DNA replication occurs in these cells, on average. On top of that, less than 2 percent of human DNA encodes products, so there is an extremely low probability that any mutation will occur in a coding region. When a nucleotide in a protein-coding region does change, the redundancy of the genetic code offers the cell a margin of safety. For example, a mutation that changes a codon from CCU to CCC may have no further effect, because both of these codons specify proline.

Very rarely, a mutation changes an amino acid in a protein, or results in a premature stop codon that shortens it. Such mutations can have drastic effects on an organism. Consider hemoglobin, an oxygen-transporting protein in your red blood cells. Hemoglobin's structure allows it to bind and release oxygen. In adult humans, a hemoglobin molecule consists of four polypeptides called globins: two alpha globins and two beta globins (Figure 7.10A). Each globin folds around a cofactor called heme. Oxygen molecules bind to hemoglobin at those hemes.

As red blood cells circulate through the lungs, the hemoglobin inside of them bind to oxygen molecules. The cells then travel to other regions of the body, and the hemoglobin releases its oxygen cargo wherever the oxygen level is low. When the red blood cells return to the lungs, the hemoglobin binds to more oxygen.

Mutations that alter hemoglobin's structure can greatly impact health. For example, mutations in either the alpha or beta globin genes can cause a condition called anemia, in which a person's blood is deficient in red blood cells or in hemoglobin. Both outcomes limit the blood's ability to carry oxygen, and the resulting symptoms can range from mild to life-threatening. Sickle-cell anemia arises because of a particular mutation in the beta globin gene. The mutation changes one nucleotide to another, so it is called a **base-pair substitution**. In this case, the substitution results in a version of beta globin that has valine instead of glutamic acid as its sixth amino acid (Figure 7.10B,C). Hemoglobin assembled with this altered beta globin chain is called sickle hemoglobin, or HbS.

Unlike glutamic acid, which carries a negative charge, valine carries no charge. As a result of that one base-pair substitution, a tiny patch of the beta globin polypeptide that is normally hydrophilic becomes hydrophobic. This change slightly alters the behavior of hemoglobin. Under certain conditions, HbS molecules stick together and form large, rodlike

Figure 7.11 A sickled red blood cell compared with a normal one.
A single base-pair substitution gives rise to an abnormal beta globin chain that, when assembled in hemoglobin molecules, forms HbS. The sixth amino acid in these abnormal beta globin chains is valine, not glutamic acid. In the body, the difference causes HbS molecules to form rod-shaped clumps that distort normally round blood cells (red) into the sickle shape (tan) characteristic of sickle-cell anemia. Sickled cells clog small blood vessels.

EM Unit, UCL Medical School, Royal Free Campus/Wellcome Images.

clumps. Red blood cells that contain the clumps become distorted into a crescent (sickle) shape (Figure 7.11). Sickled cells clog tiny blood vessels, thus disrupting blood circulation throughout the body. Over time, repeated episodes of sickling can damage organs and eventually cause death.

A different type of anemia called beta thalassemia is caused by a **deletion**, which is a mutation in which one or more nucleotides is lost from the DNA. In this case, the twentieth nucleotide in the coding region of the beta globin gene is lost (Figure 7.10D). Like most other deletions, this one causes the reading frame of the mRNA codons to shift. A frameshift usually has drastic consequences because it garbles the genetic message, just as incorrectly grouping a series of letters garbles the meaning of a sentence:

> *The fat cat ate the sad rat.*
> *T hef atc ata tet hes adr at.*
> *Th efa tca tat eth esa dra t.*

The frameshift caused by the beta globin deletion results in a polypeptide that is very different from normal beta globin in amino acid sequence and in length. This outcome is the source of the anemia. Beta thalassemia can also be caused by an **insertion**, which is a mutation in which nucleotides are added to the DNA. Insertions, like deletions, often cause frameshifts (Figure 7.10E).

Some mutations can affect a gene's expression without changing any of its codons. Special nucleotide sequences in DNA affect the expression of nearby genes. A promoter is one example; an intron–exon splice site is another. Consider a mutation that causes the hairless appearance of sphynx cats (Figure 7.12). In this case, a base-pair substitution disrupts an intron–exon splice site in a gene for keratin, a fibrous protein. The intron is not correctly removed during post-transcriptional processing. The altered protein translated from the resulting mRNA cannot properly assemble into filaments that make up hair. Cats that have this mutation still make hair, but it falls out before it gets very long.

Figure 7.12 A sphynx cat.
The hairless appearance of a sphynx cat arises from a single base-pair mutation in a gene for keratin, a fibrous protein that makes up hair. The altered keratin that results from the mutation does not assemble correctly into filaments. Sphynx cats are not truly hairless; they produce hair, but it is easily dislodged.

Glennis Siverson/National Geographic Creative.

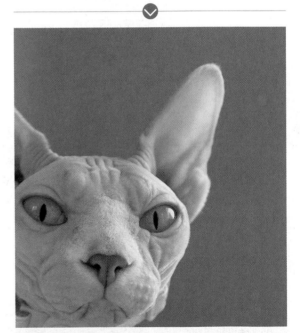

Take-Home Message 7.6

What happens after a gene becomes mutated?

- Mutations that change a protein can have drastic consequences on an organism's form and function.
- A base-pair substitution may change an amino acid in a protein, or it may introduce a premature stop codon.
- Frameshifts that occur after an insertion or deletion can change an mRNA's codon reading frame, thus garbling its protein-building instructions.

base-pair substitution Mutation in which a single base pair changes.

deletion Mutation in which one or more nucleotides are lost.

insertion Mutation in which one or more nucleotides become inserted into DNA.

Which genes a cell uses determines the molecules it will produce, which in turn determines what kind of cell it will be.

Figure 7.13 Example of a homeotic gene.
Most homeotic genes, including this one, encode transcription factors that work by binding directly to DNA.

(B) left, © Jürgen Berger, Max-Planck-Institute for Developmental Biology, Tübingen; right, Science VU/Dr. F. Rudolph Turner/Visuals Unlimited, Inc.

A. A model of the protein product (in gold) of the homeotic gene *antennapedia* attached to a promoter. Expression of *antennapedia* in embryonic tissues of the insect thorax causes legs to form.

antenna   leg

B. The head of a normal fruit fly (left) has two antennae. Right, a mutation that triggers expression of the *antennapedia* gene in embryonic tissues of the head causes legs to form instead of antennae.

7.7 Control of Gene Expression

REMEMBER: Free radicals are dangerous to life (Section 2.2). Estrogen and testosterone are steroid hormones that govern reproduction and secondary sexual traits in animals (2.8). Bacteria can produce ATP by breaking down sugars in fermentation (5.6). As cells in an embryo start using different subsets of their DNA, they become different in form and function, a process called differentiatiation (6.1). The DNA of eukaryotic cells wraps around histones; body cells of human females typically have two X chromosomes (6.3).

A typical cell in your body uses only about 10 percent of its genes at a time. Some of the active genes affect structures and metabolic pathways common to all cells; others are expressed only by certain subsets of cells. For example, most body cells express genes that encode the enzymes of glycolysis, but only immature red blood cells express globin genes. Differentiation occurs as different cell lineages begin to express different subsets of their genes during development. Which genes a cell uses determines the molecules it will produce, which in turn determines what kind of cell it will be. Thus, control over gene expression is necessary for proper development of complex, multicelled bodies. It also allows individual cells to respond appropriately to changes in their internal and external environments.

All steps of gene expression are regulated, starting with transcription and ending with delivery of an RNA or protein product to its final destination in the cell. The "switches" that turn a gene on or off are molecules or processes that affect individual steps of its expression. For example, proteins called **transcription factors** affect whether and how fast a gene is transcribed by binding directly to the DNA. Some transcription factors prevent RNA polymerase from attaching to a gene's promoter, which in turn prevents the gene's transcription. Others help RNA polymerase bind to a promoter, which speeds up transcription. The rest of this section introduces some specific examples of gene expression control in eukaryotes.

Master Genes As an animal embryo develops, its cells differentiate and form tissues, organs, and body parts. The entire process of development is driven by cascades of master gene expression. The products of **master genes** affect the expression of many other genes. Expression of a master gene causes other genes to be expressed, which in turn cause other genes to be expressed, and so on.

Gene expression that orchestrates animal development begins when different maternal mRNAs are delivered to different regions of cytoplasm in an egg as it forms. These mRNAs are translated only after the egg is fertilized. Then, their protein products diffuse away, forming gradients that span the developing embryo. The position of a cell within the embryo determines how much of these proteins it is exposed to. This in turn determines which master genes it turns on. The products of those master genes also form in gradients that diffuse away from cells expressing them. Still other master genes are transcribed depending on where a cell falls within these gradients, and so on. Eventually, the products of master genes cause undifferentiated cells to differentiate and become specialized.

A homeotic gene is a type of master gene whose expression directs the formation of a specific body part such as an eye, leg, or wing. Most homeotic genes encode transcription factors, and the function of many of them has been discovered by deliberately manipulating the genes' expression (Figure 7.13). For example, in a common experiment called a **knockout**, researchers inactivate a gene by introducing a mutation that prevents its expression, or by deleting it entirely. A knockout organism (one that has had a gene knocked out) may differ from normal individuals, and the differences are clues to the function of the missing gene product.

Homeotic genes control development by the same mechanisms in all multicelled eukaryotes, and many are interchangeable among different species. Thus, we can infer that they evolved in the most ancient eukaryotic cells. Consider the *eyeless* gene. Eyes form in embryonic fruit flies wherever this gene is expressed, which is normally in tissues of the head only. Flies with a mutated *eyeless* gene lack eyes (Figure 7.14A). If the *eyeless* gene is expressed in another part of the developing embryo, eyes form there too (Figure 7.14B). Humans, squids, mice, fishes, and many other animals have a gene called *PAX6*, which is very similar in DNA sequence to the *eyeless* gene of flies. In humans, mutations in *PAX6* cause eye disorders such as aniridia, in which the irises are underdeveloped or missing (Figure 7.14C). If a *PAX6* gene from a human or a mouse is inserted into a fly, it has the same effect as the *eyeless* gene: An eye forms wherever it is expressed. (Because *PAX6* is just a switch, the eye that forms is a fly eye, not a human or mouse eye.) The same principle applies in reverse: The *eyeless* gene from flies switches on eye formation in frogs. Such studies are evidence of shared ancestry among these evolutionarily distant animals.

Sex Chromosome Genes

In humans and other mammals, a female's cells have two X chromosomes, one inherited from her mother, the other one from her father. In each cell, one X chromosome is always tightly condensed (Figure 7.15). We call the condensed X chromosomes **Barr bodies**, after Murray Barr, who discovered them. Condensation inhibits transcription, so most of the genes on a Barr body are not expressed. This X chromosome inactivation ensures that only one of the two X chromosomes in a female's cells is active, thus equalizing expression of X chromosome genes between the sexes—a mechanism called dosage compensation. The body cells of male mammals (XY) have one set of X chromosome genes. Body cells of female mammals (XX) have two sets, but female embryos do not develop properly when both sets are expressed.

At this writing, researchers have discovered 1,805 genes on the human X chromosome. Only a few of them are associated with traits that differ between males and females; most govern nonsexual traits such as blood clotting and color perception. The Y chromosome has only 458 known genes, but one of them is *SRY*—the master gene for male sex determination in mammals. Its expression in XY embryos triggers the formation of testes, which are male reproductive organs. Some of the cells in testes make testosterone, a steroid hormone that controls the emergence of male secondary sexual traits such as facial hair, increased musculature, and deepened voice. We know that *SRY* is the master gene that controls emergence of male sexual traits because mutations in this gene cause XY individuals to develop external genitalia that appear female. An XX embryo has no Y chromosome, no *SRY* gene, and much less testosterone, so primary female reproductive organs (ovaries) form instead of testes. Ovaries make estrogens and other sex hormones that will govern the development of female secondary sexual traits, such as enlarged, functional breasts, and fat deposits around the hips and thighs.

Lactose Tolerance

Humans and other mammals break down lactose, a carbohydrate in milk, but most do so only when young. An individual's ability to digest lactose ends at a certain age that depends on the species. In the majority of humans worldwide, this switch occurs at about age five, when transcription of the gene for lactase slows. The resulting decline in production of the enzyme causes a common condition known as lactose intolerance.

A. A normal fruit fly (left) has large, round eyes. A fruit fly with a mutation in its *eyeless* gene (right) develops without eyes.

B. Eyes form wherever the *eyeless* gene is expressed in fly embryos—here, on the head and also on the wing.

The *PAX6* gene of humans, mice, squids, and some other animals is so similar to *eyeless* that it similarly triggers eye development in fruit flies.

C. A normal human eye has a colored iris surrounding the pupil (dark area where light enters). Mutations in *PAX6* cause eyes to develop without an iris, a condition called aniridia.

Figure 7.14 Eyes and *eyeless*.

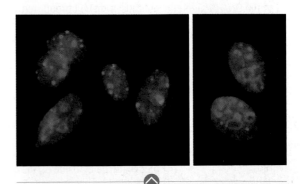

Figure 7.15 X chromosome inactivation.
Barr bodies are visible as red spots in the nucleus of the four XX cells on the left. Compare the nucleus of two XY cells to the right.

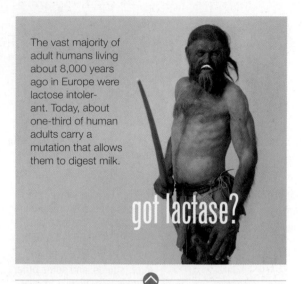

The vast majority of adult humans living about 8,000 years ago in Europe were lactose intolerant. Today, about one-third of human adults carry a mutation that allows them to digest milk.

got lactase?

Figure 7.16 Milk wasn't on the Stone Age menu.

South Tyrol Museaum of Archaeology/A. Ochsenreiter, as altered by Lisa Starr.

cytosine

guanine

A. In the DNA of differentiated cells, a methyl group (red) is most often attached to a cytosine (C) that is followed by a guanine (G).

B. A model of DNA shows methyl groups (red) attached to a cytosine–guanine pair on complementary DNA strands. When the cytosine on one strand is methylated, enzymes methylate the cytosine on the other strand. This is why a methylation tends to persist in a cell's descendants.

Figure 7.17 DNA methylation.

Cells in the intestinal lining secrete lactase into the small intestine, where the enzyme cleaves lactose into its monosaccharide monomers. Monosaccharides are absorbed directly by the small intestine, but lactose is not. Thus, when lactase production slows, lactose consumed in food passes undigested through the small intestine. The lactose ends up in the large intestine, which hosts huge numbers of bacteria. These cells respond to the presence of lactose by switching on genes that break it down. Carbon dioxide and other gaseous products of their various fermentation reactions accumulate quickly in the large intestine, distending its wall and causing pain. Other products of their metabolism disrupt the solute–water balance inside the large intestine, and diarrhea results.

Not everybody is lactose intolerant. About one-third of human adults carry a mutation that allows them to digest milk; this mutation is more common in some populations than in others. Recent analyses of DNA from well-preserved skeletons show that the vast majority of adult humans living about 8,000 years ago in Europe were lactose intolerant (Figure 7.16). Around that time, a mutation appeared in the DNA of prehistoric people inhabiting a region between what is now central Europe and the Balkans. This mutation allowed its bearers to continue digesting milk as adults, and it spread rapidly to the rest of the continent along with the practice of dairy farming. Today, most adults of northern and central European ancestry are able to digest milk because they carry this mutation, a single base-pair substitution in a region of DNA that helps control the lactase gene promoter. Other mutations in the same region of DNA arose independently in North Africa, southern Asia, and the Middle East. Some people descended from these populations can continue to digest milk as adults.

DNA Methylation In a eukaryotic cell, only regions of DNA that have been unwound from histones are accessible to RNA polymerase for transcription. Modifications to histone proteins change the way they interact with DNA wrapped around them, thus affecting transcription. Some modifications make histones release their grip on DNA; others make them tighten it. For example, adding methyl groups ($-CH_3$) to a histone tightens the DNA around it, so enzymes that methylate histones can slow or shut down transcription.

Direct methylation of DNA nucleotides also suppresses transcription, often more permanently than histone modifications. Once a particular nucleotide has become methylated in a cell's DNA, it will usually stay methylated in the DNA of the cell's descendants. Methylation and other heritable modifications to DNA that affect its function but do not alter the nucleotide sequence are said to be **epigenetic**.

DNA methylation is a part of differentiation, so it begins very early in embryonic development: Genes actively expressed in a zygote (the first cell of a new individual) become silenced as their promoters get methylated. This silencing is the basis of selective gene expression that drives differentiation. During development, and also during the remainder of the individual's life, each cell's DNA continues to acquire methylations. Between 3 and 6 percent of the DNA has been methylated in a normal, differentiated body cell.

Methyl groups are usually added to a cytosine that is followed by a guanine (Figure 7.17), but which of these cytosines get methylated varies by the individual. This is because methylation is influenced by environmental factors. For instance, humans conceived during a famine end up with an unusually low number of methyl groups in the DNA of certain genes. The product of one of those genes is a hormone that promotes prenatal growth and development. The resulting increase in expression of this gene may offer a survival advantage in a poor nutritional environment.

Digging Into Data

Effect of Grandmother's Food Supply on Infant Mortality

Researchers are investigating long-reaching epigenetic effects of starvation, in part because historical data on periods of famine are widely available. Before the industrial revolution, a failed harvest in one autumn typically led to severe food shortages the following winter. A retrospective study has correlated female infant mortality at certain ages with the abundance of food during the paternal grandmother's childhood. Figure 7.18 shows some of the results of this study.

1. Compare the mortality risk of girls whose paternal grandmothers ate well at age 2 with that of girls whose grandmothers were starving at the same age. Which girl was more likely to die early? About how much more likely was she to die?

2. Children have a period of slow growth around age 9. What trend in this data can you see around that age?

3. There was no correlation between early death of a male child and eating habits of his paternal grandmother, but there was a strong correlation with the eating habits of his paternal grandfather. What does this tell you about the probable location of epigenetic changes that gave rise to these data?

Figure 7.18 An epigenetic effect.
The graph shows relative risk of early death of a female child correlated with the age at which her paternal grandmother experienced a winter with a food supply that was scarce (blue) or abundant (red) during childhood. The dotted line represents no difference in risk of mortality. A value above the line means an increased risk; one below the line indicates a reduced risk.

Source: Pembrey et al., *European Journal of Human Genetics* (2006) 14, 159–166.

Nucleotides also become methylated by chance during DNA replication, so cells that divide a lot tend to have more methyl groups in their DNA than inactive cells. Free radicals and toxic chemicals add more methyl groups.

Factors that influence DNA methylation can have multigenerational effects. When an organism reproduces, it passes its DNA to offspring. Methylation of parental DNA is normally "reset" in the zygote, with new methyl groups being added and old ones being removed. This reprogramming does not remove all of the parental methyl groups, however, so some methylations acquired during an individual's lifetime are passed to future offspring. Inheritance of epigenetic modifications can adapt offspring to an environmental challenge much more quickly than evolution (we return to evolutionary processes in Chapter 12). Epigenetic modifications are not considered to be evolutionary because the underlying DNA sequence does not change. Even so, they may persist for generations after an environmental challenge has faded.

> Factors that influence DNA methylation can have multigenerational effects.

Take-Home Message 7.7

How is gene expression controlled?

- Gene expression can be switched on or off, or speeded or slowed, by molecules and processes that operate at each step.
- This control allows cells to respond appropriately to environmental change. It is also critical for differentiation and development in multicelled eukaryotes.
- Epigenetic modifications of DNA acquired during an individual's lifetime can be passed to offspring, so gene expression patterns can persist for generations.

epigenetic Refers to heritable changes in gene expression that are not the result of changes in DNA sequence.

Summary

Section 7.1 The ability to make proteins is critical to all life processes. Ricin and other ribosome-inactivating proteins (RIPs) have an enzyme domain that alters ribosomes, thus destroying the cell's ability to make proteins. A cell that cannot make proteins dies quickly.

Section 7.2 Information encoded within the nucleotide sequence of DNA occurs in subsets called **genes**. **Gene expression** is the conversion of information in a gene to an RNA or protein product. RNA is produced during **transcription**. During **translation**, protein-building information in an **mRNA (messenger RNA)** is converted into a sequence of amino acids in a polypeptide:

$$DNA \xrightarrow{\text{transcription}} mRNA \xrightarrow{\text{translation}} protein$$

Section 7.3 During transcription, **RNA polymerase** binds to a **promoter** near a gene region in a chromosome, then links RNA nucleotides in the order dictated by the nucleotide base sequence of the DNA strand. The resulting RNA strand is an RNA copy of the gene. In eukaryotes, RNA is modified before it leaves the nucleus. **Intron** sequences are removed, and the remaining **exon** sequences may be rearranged and spliced in different combinations. Messenger RNAs are further modified.

Section 7.4 An mRNA's protein-building information consists of a series of **codons**, sets of three nucleotides. All sixty-four codons, most of which specify amino acids, constitute the **genetic code**.

Section 7.5 Each **tRNA (transfer RNA)** has an **anticodon** that can base-pair with a codon, and it binds to the amino acid specified by that codon. Proteins and enzymatic **rRNA (ribosomal RNA)** make up the two subunits of a ribosome. During translation, codons in an mRNA direct synthesis of a polypeptide. First, the mRNA, a tRNA, and two ribosomal subunits converge. Next, successive amino acids are delivered by tRNAs in the order specified by the codons in the mRNA. The ribosome causes a peptide bond to form between the amino acids as they arrive, so a polypeptide forms. Translation ends when the ribosome encounters a stop codon in the mRNA.

Section 7.6 **Insertions**, **deletions**, and **base-pair substitutions** are mutations. A mutation that changes a gene's product may have harmful effects. In humans, an example is sickle-cell anemia, a disorder caused by a single base-pair substitution in the gene for the beta globin chain of hemoglobin.

Section 7.7 Gene expression control is the basis of cell differentiation in multicelled eukaryotes, and it also allows individual cells to respond to changes in their environment. Molecules and processes that influence gene expression operate at every step between transcription and delivery of the gene product to its final destination. Proteins called **transcription factors** influence transcription by binding directly to DNA.

All cells of an embryo share the same genes; different cells in it become specialized as they begin to use different subsets of those genes, a process called differentiation. Various **master genes** are expressed locally in different parts of the developing embryo. Their products, which diffuse through the embryo in gradients, affect expression of other master genes, which in turn affect the expression of others, and so on. Cells differentiate according to their exposure to these gradients. Eventually, master gene expression induces the expression of homeotic genes, the products of which trigger development of specific body parts. The function of many homeotic genes was revealed by **knockouts**.

In cells of female mammals, one of the two X chromosomes is condensed as a **Barr body**. The condensation makes most of this chromosome's genes permanently inaccessible. In humans, the *SRY* gene determines male sex. A mutation that abolishes control over expression of the lactase gene allows some people to continue digesting milk in adulthood.

DNA modifications (such as methylations) that affect gene expression but do not involve changes to the DNA sequence are said to be **epigenetic**. An individual acquires these modifications during its lifetime, as a consequence of cell division and also by exposure to environmental challenges. Epigenetic modifications are passed to the cell's descendants; some have multigenerational effects because they can be passed to the individual's offspring.

Self-Quiz

Answers in Appendix I

1. A chromosome contains many genes that are transcribed into different _____ .
   a. proteins   c. RNAs
   b. polypeptides   d. a and b

2. RNAs form by _____ ; proteins form by _____ .
   a. replication; translation
   b. translation; transcription
   c. transcription; translation
   d. replication; transcription

3. In cells, most RNA molecules are _____ , and DNA molecules are _____ .
   a. single-stranded; double-stranded
   b. double-stranded; single-stranded

4. The main function of an mRNA molecule is to _____ .
   a. store heritable information
   b. carry a translatable message
   c. form peptide bonds between amino acids

5. Where does transcription take place in a eukaryotic cell?
   a. the nucleus   c. the cytoplasm
   b. ribosomes   d. b and c are correct

6. What is the maximum number of amino acids that can be encoded by a gene with 45 bases plus a stop codon?
   a. 15   c. 90
   b. 45   d. 135

7. Most codons specify a(n) _____ .
   a. protein   c. amino acid
   b. polypeptide   d. mRNA

8. Where does translation take place in a eukaryotic cell?
   a. the nucleus   c. the cytoplasm
   b. ribosomes   d. b and c are correct

9. A mutation called a _____ often results in a frameshift that garbles the genetic message.
   a. deletion   c. base-pair substitution
   b. insertion   d. a and b

10. Muscle cells differ from bone cells because _____ .
    a. they carry different genes   c. they are eukaryotic
    b. they use different genes   d. both a and b

11. The expression of a gene may depend on _____ .
    a. the type of organism   c. the type of cell
    b. environmental conditions d. all of the above

12. A gene that is knocked out is _____ .
    a. deleted   c. expressed
    b. inactivated   d. either a or b

13. A cell with a Barr body is _____ .
    a. prokaryotic   c. from a female mammal
    b. a sex cell   d. infected by Barr virus

14. True or false? Some gene expression patterns are heritable.

15. Put the following processes in order of their occurrence during expression of a eukaryotic gene:
    a. mRNA processing   c. transcription
    b. translation   d. RNA leaves nucleus

16. Match each term with the most suitable description.
    _____ methylation   a. cells become specialized
    _____ insertion   b. cascades of control
    _____ promoter   c. can be epigenetic
    _____ genetic message   d. assembles amino acids
    _____ differentiation   e. read in threes
    _____ *SRY*   f. makes a man out of you
    _____ master gene   g. extra nucleotides
    _____ ribosome   h. binding site for
                                   RNA polymerase

Critical Thinking

1. An anticodon has the sequence GCG. What amino acid does this tRNA carry? What would be the effect of a mutation that changed the C of the anticodon to a G?

2. Each position of a codon can be occupied by one of four (4) nucleotides. What is the minimum number of nucleotides per codon necessary to specify all 20 of the naturally occurring amino acids that are assembled into proteins?

3. Why are some genes expressed and some not?

4. Bacteria use the same stop codons as eukaryotes. However, bacterial transcription is also terminated in places where the mRNA folds back on itself to form a hairpin-looped structure like the one shown on the right. How do you think that this structure stops transcription?

Visual Question

1. Use Figure 7.6 to translate the following sequence of bases in an mRNA into an amino acid sequence, starting at the first base. Use the one-letter abbreviations for the amino acids.
   GGUGAAAAUGAGACCAUUUGUAGU

2. Translate the base sequence in the previous question, starting at the second base.

8

HOW CELLS REPRODUCE

Application ❯ 8.1 Henrietta's Immortal Cells

Cells are individually alive, even as part of a multicelled body.

Each human starts out as a fertilized egg. By the time of birth, that single cell has given rise to about a trillion other cells, all organized as a human body. Even in the adult, billions of cells divide every day as new cells replace worn-out ones. However, despite the ability of human cells to continue dividing as part of a body, they tend to divide a limited number of times when grown in the laboratory. As early as the mid-1800s, researchers were trying to coax human cells to keep dividing outside of the body because they realized immortal cell lineages—cell lines—would allow them to study human diseases (and potential cures) without experimenting on people. The quest to create a human cell line continued unsuccessfully until 1951. By this time, George and Margaret Gey had been trying to culture human cells for nearly thirty years. Then their assistant, Mary Kubicek, prepared a new sample of human cancer cells. Mary named the cells HeLa, after the first and last names of the patient from whom the cells had been taken. The HeLa cells began to divide, again and again. The cells were astonishingly vigorous, quickly coating the inside of their test tube and consuming their nutrient broth. Four days later, there were so many cells that the researchers had to transfer them to more tubes. The cell populations increased at a phenomenal rate. The cells were dividing every twenty-four hours and coating the inside of the tubes within days.

Sadly, cancer cells in the patient were dividing just as fast. Only six months after she had been diagnosed with cervical cancer, malignant cells had invaded tissues throughout her body. Two months after that, Henrietta Lacks, a young woman from Baltimore, was dead.

Even after Henrietta had passed away, her cells lived on in the Geys' laboratory. The Geys discovered how to grow poliovirus in HeLa cells, a practice that enabled them to determine which strains of the virus cause polio. That work was a critical step in the development of polio vaccines, which have since saved millions of lives.

Henrietta Lacks was just thirty-one, a wife and mother of five, when runaway cell divisions of cancer killed her. Her cells, however, are still dividing, again and again, more than sixty years after she died. Frozen away in tiny tubes packed in Styrofoam boxes, HeLa cells continue to be shipped among laboratories all over the world. They are still widely used to investigate cancer (Figure 8.1), viral growth, protein synthesis, the effects of radiation, and countless other processes important in medicine and research. HeLa cells helped several researchers win Nobel Prizes, and they even traveled into space for experiments on satellites.

Understanding why cancer cells are immortal—and why we are not—begins with learning about the structures and mechanisms that cells use to divide.

Figure 8.1 HeLa.
This micrograph shows two dividing cells of the HeLa line—cellular legacy of cancer victim Henrietta Lacks (left). Blue and green tracers identify the location of proteins that help attach chromosomes (white) to microtubules (red) during the division process. The location of the proteins is abnormal, which means the chromosomes are not properly attached to microtubules that are supposed to distribute them evenly into descendant cells. Defects in these and other proteins that orchestrate cell division result in cells with too many or too few chromosomes, an outcome associated with cancer.

Top, Dr. Paul D. Andrews/University of Dundee; inset, courtesy of the family of Henrietta Lacks.

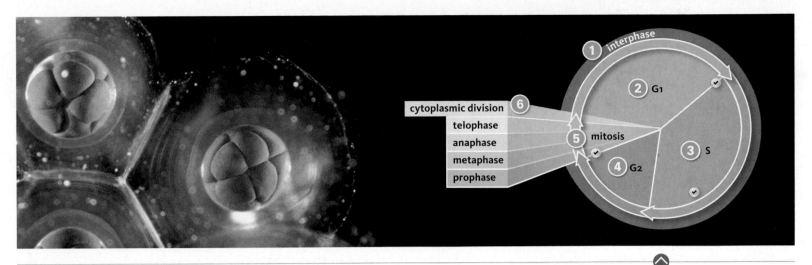

Carolina Biological Supply Company/ Phototake.

8.2 Multiplication by Division

REMEMBER: Processes by which individuals produce offspring are called reproduction (Section 1.3). Microtubules consist of tubulin subunits; microfilaments form a cell cortex that reinforces the plasma membrane; a rigid cell wall surrounds the plasma membrane of a plant cell; the motor protein myosin interacts with microfilaments to bring about contraction (3.5). When a phosphate group is transferred from ATP, energy is transferred along with it (4.4). Sister chromatids attach to one another at the centromere (6.3). A cell copies its chromosomes by DNA replication (6.4).

A life cycle is a sequence of recognizable stages that occur during an organism's lifetime, from the first cell of the new individual until its death. Multicelled organisms and free-living cells have life cycles, but what about cells that make up a multicelled body? Biologists consider such cells to be individually alive, each with its own lifetime. A cell's life passes through a series of recognizable intervals and events collectively called the **cell cycle** (Figure 8.2). A typical body cell spends most of its life in interphase ➊. During **interphase**, the cell increases its mass, roughly doubles the number of its cytoplasmic components, and copies its chromosomes in preparation for division. Interphase has three major stages: G1, S, and G2. G1 and G2 were named "Gap" phases because outwardly they seem to be periods of inactivity, but they are not. Most cells going about their metabolic business are in G1 ➋. Cells preparing to divide enter S ➌, when they undergo DNA replication. During G2 ➍, cells make the proteins that will carry out division.

The rest of the cell cycle consists of the division process itself. When the cell divides, both of its two cellular offspring end up with a blob of cytoplasm and some DNA. Each of the offspring of a eukaryotic cell inherits its DNA packaged inside a nucleus. Thus, a eukaryotic cell's nucleus has to divide before its cytoplasm does. **Mitosis** is a nuclear division mechanism that maintains the chromosome number ➎. In multicelled organisms, mitosis and cytoplasmic division ➏ are the basis of increases in body size and tissue remodeling during development, as well as ongoing replacements of damaged or dead cells in the adult. Mitosis and cytoplasmic division are also part of **asexual reproduction**, a reproductive mode by which offspring are produced by one parent only. Some multicelled eukaryotes and many single-celled ones use this mode of reproduction. (Prokaryotes do not have a nucleus and do not undergo mitosis. We discuss their reproduction in Section 13.5.)

Figure 8.2 The eukaryotic cell cycle.
The photo shows early frog embryos, each a product of three mitotic divisions of one fertilized egg.

➊ A cell spends most of its life in interphase, which includes three stages: G1, S, and G2.

➋ G1 is the phase of growth before DNA replication. The cell's chromosomes are unduplicated.

➌ S is the phase of synthesis, during which the cell makes copies of its chromosome(s) by DNA replication.

➍ G2 is the phase after DNA replication and before mitosis. The cell prepares to divide during this stage.

➎ The nucleus divides during mitosis.

➏ After mitosis, the cytoplasm may divide. Each descendant cell begins the cycle anew, in interphase.

✔ Built-in checkpoints stop the cycle from proceeding until certain conditions are met (see Section 8.3).

Figure It Out: Each of the embryos in the photo consists of how many cells?   **Answer:** Eight

asexual reproduction Reproductive mode of eukaryotes by which offspring arise from a single parent.

cell cycle The collective series of intervals and events of a cell's life, from the time it forms until it divides.

interphase In a eukaryotic cell cycle, the interval between mitotic divisions when a cell grows, roughly doubles the number of its cytoplasmic components, and replicates its DNA.

mitosis Nuclear division mechanism that maintains the chromosome number.

anaphase Stage of mitosis during which sister chromatids separate and move toward opposite spindle poles.

homologous chromosomes In the nucleus of a diploid eukaryotic cell, chromosomes with the same length, shape, and set of genes.

metaphase Stage of mitosis at which all chromosomes are aligned in the middle of the cell.

prophase Stage of mitosis during which chromosomes condense and become attached to a newly forming spindle.

spindle Temporary structure that moves chromosomes during nuclear division; consists of microtubules.

telophase Stage of mitosis during which chromosomes arrive at opposite ends of the cell. Two new nuclei form as the chromosomes loosen.

A. Pair of homologous chromosomes in a cell during G1.

B. By the end of G2, each chromosome has been duplicated, so it has two sister chromatids.

C. Mitosis and cytoplasmic division package one copy of each chromosome into each of two new cells.

Figure 8.3 How mitosis maintains the chromosome number in a diploid cell.
For clarity, only one pair of homologous chromosomes is illustrated. Pink indicates a chromosome of maternal origin; blue, a chromosome of paternal origin.

When a cell divides by mitosis, it produces two descendant cells, each with the chromosome number of the parent. However, if only the total number of chromosomes mattered, then one of the descendant cells might get, say, two pairs of chromosome 22 and no chromosome 9. A cell cannot function properly without a full complement of DNA, which means it needs to have *a copy of each* chromosome. Thus, the two cells produced by mitosis have the same number and types of chromosomes as the parent. Figure 8.3 shows how mitosis maintains the chromosome number. Consider how your body's cells are diploid, which means their nuclei contain pairs of chromosomes—two of each type. One chromosome of a pair was inherited from your father; the other, from your mother. Except for a pairing of nonidentical sex chromosomes (XY) in males, the chromosomes of each pair are homologous. **Homologous chromosomes** have the same length, shape, and genes (*hom–* means alike). When a cell is in G1, each of its chromosomes consists of one double-stranded DNA molecule. The cell replicates its DNA in S, so by G2, each of its chromosomes consists of two double-stranded DNA molecules. These molecules stay attached to one another at the centromere as sister chromatids until mitosis is almost over, and then they are pulled apart and packaged into two separate nuclei. When sister chromatids are pulled apart, each becomes an individual chromosome that consists of one double-stranded DNA molecule. Thus, each of the two new nuclei that form in mitosis contains a full complement of chromosomes. When the cytoplasm divides, these nuclei are packaged into separate cells. Each new cell has the parental chromosome number, and starts its life in G1 of interphase.

Figure 8.4 shows the details of mitosis. When a cell is in interphase, its chromosomes are loosened to allow transcription and DNA replication. Loosened chromosomes are spread out, so they are not easily visible under a light microscope ❶. In preparation for nuclear division, the chromosomes begin to pack tightly ❷. Transcription and DNA replication stop as the chromosomes condense into their most compact "X" forms. Tight condensation keeps the chromosomes from getting tangled and breaking during nuclear division. A cell reaches **prophase**, the first stage of mitosis, when its chromosomes have condensed so much that they are visible under a light microscope ❸. "Mitosis" is from *mitos*, the Greek word for thread, after the threadlike appearance of the chromosomes during nuclear division.

Also during prophase, microtubules assemble and extend from two regions on opposite sides of the cell. These microtubules form a **spindle**, a temporary structure that moves chromosomes during nuclear division. The spindle penetrates the nuclear region as the nuclear envelope breaks up. Some of the microtubules stop lengthening when they reach the middle of the cell. Others lengthen until they reach a chromosome and attach to it at the centromere. By the end of prophase, one sister chromatid of each chromosome has become attached to microtubules extending from one end of the cell, and the other sister has become attached to microtubules extending from the other end.

The opposing sets of microtubules then begin a tug-of-war by adding and losing tubulin subunits. As the microtubules lengthen and shorten, they push and pull the chromosomes. When all the microtubules are the same length, the chromosomes are aligned in the middle of the cell ❹. The alignment marks **metaphase**. During **anaphase**, the sister chromatids of each duplicated chromosome separate, so each becomes an individual, unduplicated chromosome. The spindle moves the chromosomes toward opposite sides of the cell ❺.

Telophase begins when the two clusters of chromosomes reach opposite ends of the cell ❻. Each cluster has the same number and kinds of chromosomes as the parent cell nucleus had: two of each chromosome, if the parent cell was diploid.

Mitosis in a plant cell   **Mitosis in an animal cell**

1 **Interphase**

Interphase cells are shown for comparison, but inter-phase is not part of mitosis. The red spots in the plant cell nucleus are areas where ribosome subunits are being transcribed and assembled.

2 **Early Prophase**

Mitosis begins. Transcription stops, and the DNA begins to appear grainy as it starts to condense.

nuclear envelope breaking up

3 **Prophase**

The duplicated chromosomes become visible as they condense. The nuclear envelope breaks up. Spindle microtubules assemble and bind to chromosomes at the centromere. Sister chromatids become attached to microtubules extending from opposite ends of the cell.

spindle

4 **Metaphase**

All of the chromosomes are aligned in the middle of the cell.

5 **Anaphase**

Spindle microtubules separate the sister chromatids and move them toward opposite sides of the cell. Each sister chromatid has now become an individual, unduplicated chromosome.

6 **Telophase**

The chromosomes reach opposite sides of the cell and loosen up. Mitosis ends when a new nuclear envelope forms around each cluster of chromosomes.

Figure 8.4 Mitosis.
Micrographs show nuclei of plant cells (onion root, left), and animal cells (fertilized egg of a roundworm, right). A diploid (2*n*) animal cell with two chromosome pairs is illustrated.

Figure 8.5 Cytoplasmic division of an animal cell.

1 In a dividing animal cell, the spindle disassembles as mitosis ends.

2 At the midpoint of the former spindle, a ring of actin and myosin filaments attached to the plasma membrane contracts.

3 This contractile ring pulls the cell surface inward as it shrinks. The indentation is called a cleavage furrow.

4 The ring contracts until it pinches the cell in two.

Figure 8.6 Cytoplasmic division of a plant cell.

5 Vesicles containing wall-building materials cluster at the future plane of division before mitosis ends.

6 The vesicles fuse with each other, forming a cell plate.

7 The cell plate expands to the plasma membrane. When it attaches to the membrane, it partitions the cytoplasm.

8 The cell plate matures as two new cell walls. These walls join with the parent cell wall, so each descendant cell becomes enclosed by its own wall.

A new nucleus forms around each cluster as the chromosomes loosen up again, after which telophase—and mitosis—are finished.

Cytoplasmic Division In most eukaryotes, the cell cytoplasm divides between late anaphase and the end of telophase, so two cells form, each with their own nucleus. The mechanism of cytoplasmic division differs between plants and animals.

Typical animal cells pinch themselves in two after nuclear division ends (Figure 8.5). How? The cell cortex, which is the mesh of cytoskeletal elements just under the plasma membrane, includes a band of microfilaments and motor proteins that wraps around the cell's midsection. The band is called a contractile ring because it contracts when the motor proteins are energized by phosphate-group transfers from ATP. Contraction occurs after telophase, when the spindle disassembles **1**.

Because the contractile ring is attached to the plasma membrane, it pulls the membrane inward as it contracts **2**. The sinking plasma membrane is visible on the outside of the cell as an indentation (left), which is called a **cleavage furrow 3**. The cleavage furrow advances around the cell and deepens until the cytoplasm—and the cell—is pinched in two **4**. Each of the two cells formed by this division has its own nucleus and some of the parent cell's cytoplasm, and each is enclosed by a plasma membrane.

Dividing plant cells face a particular challenge because a stiff cell wall surrounds their plasma membrane. Accordingly, plant cells have their own mechanism of cytoplasmic division (Figure 8.6). By the end of anaphase, a set of short microtubules has formed on either side of the future plane of division. These microtubules now guide vesicles from Golgi bodies and the cell surface to the division plane **5**. There, the vesicles and their wall-building contents start to fuse into a disk-shaped cell plate **6**. The plate expands at its edges until it reaches the plasma membrane and attaches to it, thus partitioning the cytoplasm **7**. In time, the cell plate will develop into two new cell walls, so each of the descendant cells will be enclosed by its own plasma membrane and wall **8**.

Take-Home Message 8.2

How do eukaryotic cells reproduce?

- A eukaryotic cell reproduces by division: nucleus first, then cytoplasm.
- Mitosis (a nuclear division mechanism that maintains the chromosome number) and cytoplasmic division are part of the cell cycle, a series of events and stages through which a cell passes during its lifetime.
- DNA replication occurs before mitosis begins, so each chromosome consists of two DNA molecules attached as sister chromatids.
- During mitosis, a spindle assembles, then separates all of the sister chromatids and moves them to opposite sides of the cell. A new nuclear envelope forms around each of the two clusters of chromosomes. The two new nuclei have the same chromosome number as the parent cell.

8.3 Mitosis and Cancer

REMEMBER: A receptor protein triggers a change in cell activity in response to a stimulus such as binding to a certain substance; adhesion proteins help cells stick together in animal tissues (Section 3.3). Dolly the sheep died early, most likely because she was a clone (6.1). An image of a cell's chromosomes is a karyotype (6.3). Proofreading and repair mechanisms usually maintain the integrity of a cell's genetic information by correcting mispaired bases and fixing DNA damage (6.4). Gene expression can be switched on or off by molecules or processes that operate at each step; a knockout is an organism with a gene that has been deliberately inactivated (7.7).

When a cell divides—and when it does not—is determined by mechanisms of gene expression control. Like the accelerator of a car, some of these mechanisms cause the cell cycle to advance. Others are like brakes, preventing the cycle from proceeding. In the adult body, brakes on the cell cycle normally keep the vast majority of cells in G1. Most of your nerve cells, skeletal muscle cells, heart muscle cells, and fat-storing cells have been in G1 since you were born, for example.

Control over the cell cycle also ensures that a dividing cell's descendants receive intact copies of its chromosomes. Built-in checkpoints ensure the cell's DNA has been copied completely, that it is not damaged, and even that enough nutrients are available to support division (a few checkpoints are indicated in Figure 8.2). Protein products of "checkpoint genes" interact to carry out this control process. For example, a checkpoint that operates in S monitors whether the cell's chromosomes have been damaged during DNA replication. Checkpoint proteins recognize damaged DNA and bind to it. This binding puts the brakes on the cell cycle, and also enhances expression of genes involved in DNA repair. After the damage has been fixed, the brakes are lifted and the cell cycle proceeds. If the DNA remains unrepaired, other checkpoint proteins trigger events that cause the cell to self-destruct.

Cell Division Gone Wrong Sometimes a checkpoint gene mutates so that its protein product no longer works properly. In other cases, the controls that regulate its expression fail, and a cell makes too much or too little of the gene's product. When enough checkpoint mechanisms fail, a cell loses control over its cell cycle. Interphase may be skipped, so division occurs over and over with no resting period. Signaling mechanisms that cause abnormal cells to die may stop working. The problem is compounded because checkpoint malfunctions are inherited by the cell's descendants, which form a **neoplasm**—an accumulation of abnormally dividing cells.

A neoplasm that makes a lump in the body is called a **tumor**, but the two terms are sometimes used interchangeably. Once a tumor-causing mutation has occurred, the gene it affects is called an oncogene. An **oncogene** is any gene that helps transform a normal cell into a tumor cell. Oncogene mutations in reproductive cells can be passed to offspring, which is a reason that some types of tumors run in families.

Not every gene can become an oncogene; those that can are called protooncogenes. Consider how most of your body cells have receptors for growth factors, which are molecules that stimulate cell division and differentiation. When a growth factor binds to its receptor on a cell, a series of events is triggered that advances the cell cycle from interphase into mitosis. Mutations can result in a receptor that stimulates mitosis even when the growth factor is not present, and in fact most neoplasms carry mutations resulting in an overactivity or overabundance of growth factor receptors (Figure 8.7).

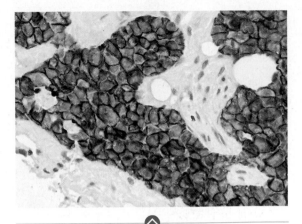

Figure 8.7 An oncogene causing a neoplasm.
In this section of human breast tissue, a brown-colored tracer shows the active form of a growth factor receptor. Normal cells are lighter in color.

The dark cells have an overactive receptor that is constantly stimulating mitosis; these cells have formed a neoplasm. Cells of most neoplasms have mutations that cause this receptor to be overproduced or overactive.

© From *Expression of the epidermal growth factor receptor (EGFR) and the phosphorylated EGFR in invasive breast carcinomas.* http://breast-cancer-research.com/content/10/3/R49.

cleavage furrow In a dividing animal cell, the indentation where cytoplasmic division will occur.

neoplasm An accumulation of abnormally dividing cells.

oncogene Gene that helps transform a normal cell into a tumor cell.

tumor A neoplasm that forms a lump.

Figure 8.8 Neoplasms and malignancy.

A. Benign neoplasms grow slowly and stay in their home tissue.

B. Cells of a malignant neoplasm can break away from their home tissue.

C. The malignant cells become attached to the wall of a lymph vessel or blood vessel (as shown here). They release digestive enzymes that create an opening in the wall, then enter the vessel.

D. The cells creep or tumble along in the vessel, then exit the same way they got in. Migrating cells may start growing in other tissues, a process called metastasis.

Figure 8.9 Skin cancer can be detected and treated early with periodic screening.

(A) Dr. Allan Harris/Phototake; (B) Biophoto Associates/Science Source; (C) James Stevenson/Science Source.

A. Basal cell carcinoma is the most common type of skin cancer. This slow-growing, raised lump may be uncolored, reddish-brown, or black.

B. Squamous cell carcinoma is the second most common form of skin cancer. This pink growth, firm to the touch, grows under the skin's surface.

C. Melanoma spreads fastest. Cells form dark, encrusted lumps that may itch or bleed easily.

Checkpoint gene products that inhibit mitosis are called tumor suppressors because tumors form when they are missing. Two examples are *BRCA1* and *BRCA2*: Mutations in these genes give rise to neoplasms in the breast, prostate, ovary, and other tissues. The multiple functions of the *BRCA* gene products are still being unraveled, but we do know they help maintain the structure and number of chromosomes in a dividing cell, and they participate directly in DNA repair. *BRCA* gene products also bind to hormone receptors that are particularly abundant on cells of breast and ovarian tissues; the binding suppresses transcription of growth factor genes in these cells. When a mutation alters a *BRCA* gene so that its product cannot bind to these hormone receptors, the cells overproduce growth factors—an outcome associated with neoplasm formation.

Viruses such as HPV (human papillomavirus) cause a cell to make proteins that interfere with its own tumor suppressors. Infection with HPV causes skin growths called warts, and some kinds are associated with neoplasms that form on the cervix.

Cancer Benign neoplasms such as warts are not usually dangerous. They grow very slowly, and their cells retain the plasma membrane adhesion proteins that keep them properly anchored in their home tissue. A malignant neoplasm is one that gets progressively worse, and is dangerous to health. Malignant cells typically display the following three characteristics:

First, like cells of all neoplasms, malignant cells grow and divide abnormally. Controls that usually keep cells from getting overcrowded in tissues are lost in malignant cells, so their populations may reach extremely high densities with cell division occurring very rapidly. The number of small blood vessels that transport blood to the growing cell mass also increases abnormally.

Second, the cytoplasm and plasma membrane of malignant cells are altered. Both are indications of cellular malfunction. The cytoskeleton may be shrunken, disorganized, or both. Malignant cells typically have an abnormal chromosome number, with some chromosomes present in multiple copies, and others missing or damaged. The balance of metabolism is often shifted, as in an amplified reliance on ATP formation by fermentation rather than aerobic respiration.

Altered or missing proteins impair the function of the plasma membrane of malignant cells. For example, these cells do not stay anchored properly in tissues because their plasma membrane adhesion proteins are defective or missing. Malignant cells can slip easily into and out of vessels of the circulatory and lymphatic systems (Figure 8.8). By migrating through these vessels, the cells can establish neoplasms elsewhere in the body. The process in which malignant cells break loose from their home tissue and invade other parts of the body is called **metastasis**. Metastasis is the third hallmark of malignant cells.

The disease called **cancer** occurs when the abnormally dividing cells of a malignant neoplasm disrupt body tissues, both physically and metabolically. Unless chemotherapy, surgery, or another procedure eliminates malignant cells from the body, they can put an individual on a painful road to death. Each year, cancer causes 15 to 20 percent of all human deaths in developed countries. The good news is that mutations in multiple checkpoint genes are required to transform a normal cell into a malignant one, and such mutations may take a lifetime to accumulate. Lifestyle choices such as not smoking and avoiding exposure of unprotected skin to sunlight reduce one's risk of acquiring mutations in the first place. Some neoplasms can be detected with periodic screening such as gynecology or dermatology exams (Figure 8.9). If detected early enough, many types of malignant neoplasms can be removed before metastasis occurs.

Digging Into Data

HeLa Cells Are a Genetic Mess

HeLa cells can vary in chromosome number. Defects in proteins that orchestrate cell division result in descendant cells with too many or too few chromosomes, an outcome that is one of the hallmarks of cancer cells. The karyotype in Figure 8.10, originally published in 1989, shows all of the chromosomes in a single metaphase HeLa cell.

1. What is the chromosome number of this HeLa cell?
2. How many extra chromosomes does this cell have, compared with a normal human body cell?
3. Can you tell that this cell came from a female? How?

Figure 8.10 Karyotype of a HeLa cell.

© Dr. Thomas Ried, NIH and the Association for Cancer Research.

Telomeres Remember that Dolly the cloned sheep died early. The life expectancy of a sheep is normally about 10 to 12 years; by the time Dolly was five, however, she was as fat and arthritic as a twelve-year-old sheep. The following year, she contracted a lung disease typical of much older sheep and had to be euthanized.

Dolly's early demise may have been the result of abnormally short telomeres. Telomeres are noncoding DNA sequences that occur at the ends of eukaryotic chromosomes (Figure 8.11). Vertebrate telomeres consist of a short, noncoding DNA sequence repeated perhaps thousands of times. These "junk" repeats provide a buffer against the loss of more valuable DNA internal to the chromosomes.

A telomere buffer is particularly important because, under normal circumstances, a eukaryotic chromosome shortens by about 100 nucleotides with each DNA replication. When a cell's offspring receive chromosomes with too-short telomeres, checkpoint gene products halt the cell cycle, and the descendant cells die shortly thereafter. Most body cells can divide only a certain number of times before this happens. This cell division limit may be a fail-safe mechanism in case a cell loses control over the cell cycle and begins to divide again and again. A limit on the number of divisions keeps the resulting neoplasm from overrunning the body.

The cell division limit varies by species, and it may be part of the mechanism that sets an organism's life span. Dolly's DNA came from the nucleus of a mammary gland cell donated by an adult sheep. When Dolly was only two years old, her telomeres were as short as those of a six-year-old sheep—the exact age of the adult animal that had been her genetic donor.

A few normal cells in an adult retain the ability to divide indefinitely. These cells are called stem cells, and their descendants replace other cell lineages that eventually die out when they reach their division limit. Stem cells are immortal because they continue to make telomerase, an enzyme that reverses the telomere shortening associated with DNA replication (differentiated body cells produce no telomerase).

Mice that have had their telomerase enzyme knocked out age prematurely, with a life expectancy about half that of a normal mouse. When one of these knockout mice is close to the end of its shortened life span, rescuing the function of its telomerase enzyme results in lengthened telomeres. The rescued mouse also regains

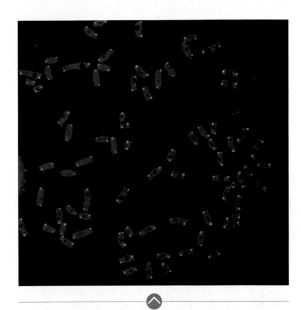

Figure 8.11 Telomeres.
The bright dots at the end of each DNA strand in these (duplicated) chromosomes are telomere sequences.

© ISM/Phototakeusa.

cancer Disease that occurs when a malignant neoplasm physically and functionally disrupts body tissues.

metastasis The process in which cells of a malignant neoplasm spread from one part of the body to another.

vitality: Worn-out tissue in the brain and other organs repairs itself and begins to function normally, and the once-geriatric individual even begins to reproduce again.

Researchers are careful to point out that shortening telomeres may be an effect of aging rather than a cause. Also, while telomerase holds therapeutic promise for rejuvenating aged tissues, it can be dangerous. Cancer cells—including the HeLa cells you learned about in Section 8.1—characteristically express high levels of this molecule, which is why, like stem cells, they can divide indefinitely.

Take-Home Message 8.3

What is cancer?

- Gene expression controls advance, delay, or block the cell cycle in response to internal and external conditions. Checkpoints built into the cell cycle allow problems to be corrected before the cycle proceeds.
- Neoplasms form when cells lose control over their cell cycle and begin dividing abnormally.
- Mutations in multiple checkpoint genes can give rise to a progressively worsening malignant neoplasm.
- Cancer is a disease that occurs when the abnormally dividing cells of a malignant neoplasm physically and metabolically disrupt body tissues.
- In most cases, lifestyle choices and early intervention can reduce one's risk of cancer.

8.4 Sex and Alleles

REMEMBER: All organisms inherit their DNA from parents by processes of reproduction (Section 1.3). Traits shared by members of a species often vary a bit among individuals, as eye color does among people (1.4). A somatic cell is a body cell (6.1). DNA sequence variation is the basis of traits that define species and distinguish individuals (6.2). The two autosomes of a pair hold information about the same traits (6.3). Mutations (6.4) that alter a gene's product can be harmful (7.6).

Your homologous chromosomes carry the same genes, but their DNA sequence may not be identical (Figure 8.12). This is because you inherited your chromosomes from two parents who are, most likely, not closely related. Unique mutations accumulated in their separate lines of descent over time. Thus, the DNA sequence of any of your genes may differ a bit from the corresponding gene on the homologous chromosome. Different forms of the same gene are called **alleles**.

Alleles may encode slightly different forms of a gene's product, and such differences influence the details of shared, inherited traits. Members of a species have the same traits because they have the same genes. However, almost every shared trait varies a bit among individuals of a sexually reproducing species. Alleles of the shared genes are the basis of this variation. Consider just one human gene, *HBB*, which encodes beta globin. *HBB* has has more than 700 alleles; a few cause sickle-cell anemia, several cause beta thalassemia, and so on. There are approximately 20,000 human genes, and most of them have multiple alleles.

On the Advantages of Sex You learned earlier in this chapter that mitosis and cytoplasmic division are part of asexual reproduction in eukaryotes, but only a few

A. Corresponding colored patches in this fluorescence micrograph indicate corresponding DNA sequences in a homologous chromosome pair. These chromosomes carry the same set of genes.

Genes occur in pairs on homologous chromosomes.

The members of each pair of genes may be identical, or they may differ slightly, as alleles. (Color variations represent sequence differences.)

B. Homologous chromosomes carry the same series of genes, but the DNA sequence of any one of those genes might differ just a bit from that of its partner on the homologous chromosome.

Figure 8.12 Genes on chromosomes.
Different forms of a gene are called alleles.

(A) Courtesy of Carl Zeiss MicroImaging, Thornwood, NY.

species use this reproductive mode (Figure 8.13). Most eukaryotes reproduce sexually. **Sexual reproduction** is the process in which offspring arise from two parents and inherit genes from both.

If the function of reproduction is the perpetuation of one's genes, then an asexual reproducer would seem to win the evolutionary race. When it reproduces, it passes all of its genes to every one of its offspring. Only about half of a sexual reproducer's genes are passed to each offspring. So why is sex so widespread?

All offspring of an asexual reproducer are clones: Barring new mutations, they have exactly the same alleles as their one parent. Consistency is a good evolutionary strategy in a favorable, unchanging environment; alleles that help an organism survive and reproduce in the environment do the same for its descendants. However, most environments are constantly changing, and change is not always favorable. Individuals that are identical are equally vulnerable to challenges. In a changing environment, sexual reproducers have the evolutionary edge because they carry different alleles. Sexual reproduction mixes up the genetic information of two parents that differ in the alleles they carry, so offspring typically inherit new combinations of alleles—ones that do not occur in either parent (or other offspring). Some may inherit a particular combination of alleles that suits them perfectly to a new environmental challenge. As a group, their diversity offers them a better chance of surviving environmental change than clones.

To understand why environments constantly change, think about one example: interactions between a predatory species and its prey. To the prey species, the predators are an environmental challenge. Prey individuals that can best escape predation tend to leave more offspring, so alleles that help them be best tend to become more common in the species over generations. At the same time, prey that can better escape predation are an environmental challenge to the predator species. Predator individuals best able to capture prey tend to leave more offspring, so alleles that help them be best tend to become more common in the predator species over generations (Chapter 12 returns to evolutionary processes). The two species are locked in a constant race, with each genetic improvement in one countered by a genetic improvement in the other. This idea is called the Red Queen hypothesis, a reference to Lewis Carroll's book *Through the Looking Glass*. In the book, the Queen of Hearts tells Alice, "It takes all the running you can do, to keep in the same place."

Another advantage of sexual reproduction involves the inevitable occurrence of mutations that are harmful but not lethal. A population of sexual reproducers has a better chance of weathering the effects of such mutations. With asexual reproduction, individuals bearing a harmful mutation necessarily pass it to all of their offspring. This outcome would be rare in sexual reproduction, because each offspring of a sexual union has a 50 percent chance of inheriting a parent's mutation. Thus, all else being equal, harmful mutations can spread through an asexually reproducing population more quickly than a sexually reproducing one.

Figure 8.13 Tiny New Zealand mud snails can reproduce on their own (asexually) or with a partner (sexually).
Like most sexual organisms, the sexual snails have two chromosome sets; like most animals that cannot reproduce sexually, the asexual snails have at least three. DNA has a high phosphorus content; having the extra sets of chromosomes multiplies each organism's requirement for this nutrient. Fertilizers and detergents contain phosphorus, so agricultural runoff and other types of water pollution may be fueling the gigantic populations of asexual snails currently invading ecosystems worldwide.

© Bart Zijistra.

Almost every shared trait varies a bit among individuals of a sexually reproducing species.

Take-Home Message 8.4

Why do species that reproduce sexually vary in shared traits?

- Paired genes on homologous chromosomes may vary in DNA sequence as alleles. Alleles arise by mutation.
- Alleles of shared genes are the basis of differences in traits shared by a species.
- Offspring of sexual reproducers inherit new combinations of parental alleles.

alleles Forms of a gene with slightly different DNA sequences; may encode slightly different versions of the gene's product.

sexual reproduction Reproductive mode by which offspring arise from two parents and inherit genes from both.

haploid Having one of each type of chromosome characteristic of the species.

meiosis Nuclear division process that halves the chromosome number. Basis of sexual reproduction.

8.5 Meiosis in Sexual Reproduction

Meiosis, a nuclear division mechanism that halves the chromosome number, is the process inherent to sexual reproduction that gives rise to new combinations of alleles in offspring. It parcels chromosomes into new nuclei two times (Figure 8.14 shows the stages of meiosis in a diploid (2n) cell). DNA replication occurs before meiosis begins, so each chromosome has two sister chromatids.

The first stage of meiosis I is prophase I **1** . During this phase, the chromosomes condense, and homologous chromosomes align tightly and swap segments (more about segment-swapping shortly). The nuclear envelope breaks up. A spindle forms, and by the end of prophase I, microtubules attach one chromosome of each homologous pair to one spindle pole, and the other to the opposite spindle pole. These microtubules grow and shrink, pushing and pulling the chromosomes as they do. At metaphase I **2** , all of the microtubules are the same length, and the chromosomes are aligned midway between the spindle poles. During anaphase I **3** , the spindle pulls the homologous chromosomes of each pair away from one another

MEIOSIS I One diploid nucleus to two haploid nuclei

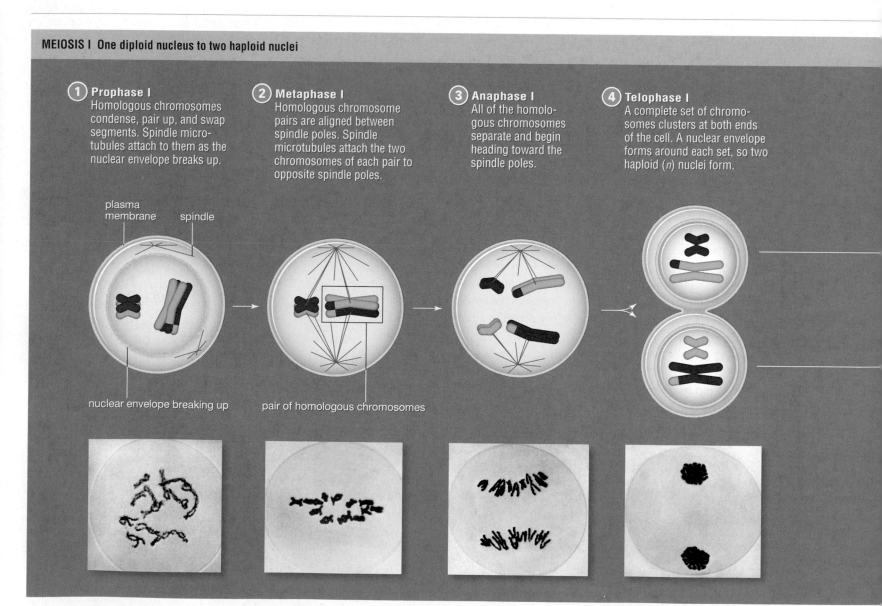

1 Prophase I
Homologous chromosomes condense, pair up, and swap segments. Spindle microtubules attach to them as the nuclear envelope breaks up.

2 Metaphase I
Homologous chromosome pairs are aligned between spindle poles. Spindle microtubules attach the two chromosomes of each pair to opposite spindle poles.

3 Anaphase I
All of the homologous chromosomes separate and begin heading toward the spindle poles.

4 Telophase I
A complete set of chromosomes clusters at both ends of the cell. A nuclear envelope forms around each set, so two haploid (n) nuclei form.

plasma membrane   spindle

nuclear envelope breaking up

pair of homologous chromosomes

and toward opposite spindle poles. The two sets of chromosomes reach the spindle poles during telophase I ❹ , and a new nuclear envelope forms around each cluster of chromosomes as the DNA loosens up. The two new nuclei are **haploid** (n); each contains one set of chromosomes—half of the diploid ($2n$) number. The cytoplasm often divides at this point. Each chromosome is still duplicated (it consists of two sister chromatids). Meiosis may pause at this point, but no DNA replication occurs before meiosis II.

Meiosis II proceeds simultaneously in both nuclei that formed in meiosis I. During prophase II ❺ , the chromosomes condense and the nuclear envelope breaks up. A new spindle forms. By the end of prophase II, spindle microtubules attach each chromatid to one spindle pole, and its sister chromatid to the opposite spindle pole. These microtubules push and pull the chromosomes, aligning them midway between spindle poles at metaphase II ❻ . During anaphase II ❼ , the spindle microtubules pull the sister chromatids apart and toward opposite spindle poles. Each chromosome is now unduplicated (it consists of one molecule of DNA). During telophase II ❽ , these chromosomes reach the spindle poles.

Figure 8.14 Meiosis.
The micrographs show meiosis in a lily cell. The illustrations show two pairs of chromosomes in a diploid ($2n$) animal cell; homologous chromosomes are indicated in blue and pink.

Bottom photos, With thanks to the John Innes Foundation Trustees, computer enhanced by Gary Head.

Figure It Out: Which phase of meiosis reduces the chromosome number?   Answer: Anaphase I

MEIOSIS II Two haploid nuclei to four haploid nuclei

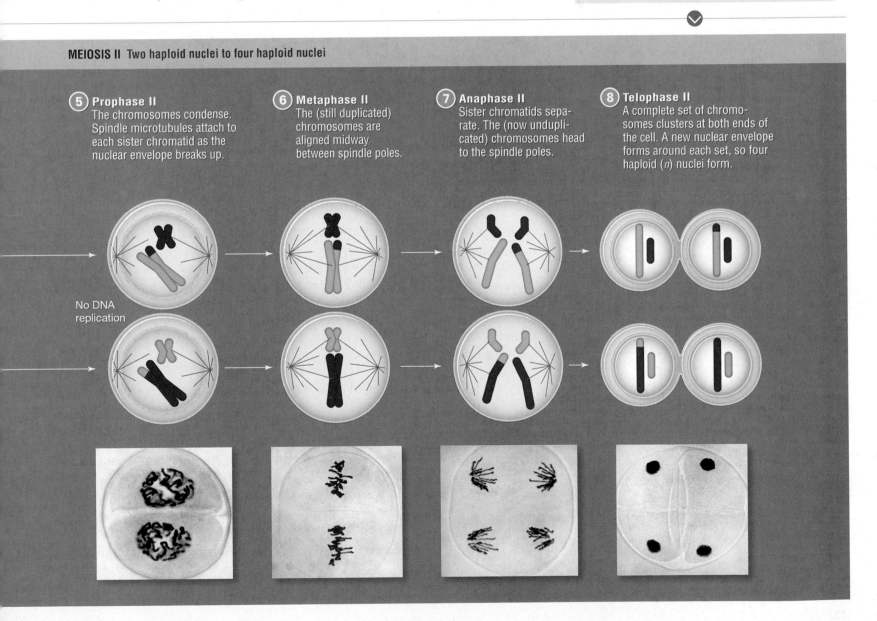

❺ **Prophase II**
The chromosomes condense. Spindle microtubules attach to each sister chromatid as the nuclear envelope breaks up.

❻ **Metaphase II**
The (still duplicated) chromosomes are aligned midway between spindle poles.

❼ **Anaphase II**
Sister chromatids separate. The (now unduplicated) chromosomes head to the spindle poles.

❽ **Telophase II**
A complete set of chromosomes clusters at both ends of the cell. A new nuclear envelope forms around each set, so four haploid (n) nuclei form.

No DNA replication

New nuclear envelopes form around the four clusters of chromosomes as the DNA loosens up. Each of the four nuclei that form are haploid (*n*), with one set of (unduplicated) chromosomes. The cytoplasm often divides at this point.

How Meiosis Mixes Alleles   We mentioned briefly that duplicated chromosomes swap segments with their homologous partners during prophase I. As the chromosomes condense, each is drawn close to its homologous partner, so that the chromatids align along their length:

This tight, parallel orientation favors **crossing over**, a process by which a chromosome and its homologous partner exchange corresponding pieces of DNA during meiosis (Figure 8.15A–C). Homologous chromosomes may swap any segment of DNA along their length, although crossovers tend to occur more frequently in certain regions.

Swapping segments of DNA shuffles alleles between homologous chromosomes. It breaks up the particular combinations of alleles that occurred on the parental chromosomes, and makes new ones on the chromosomes that end up in offspring. Thus, crossing over introduces novel combinations of alleles—and new combinations of traits—among offspring. It is a normal and frequent process in meiosis, but the rate of crossing over varies among species and among chromosomes. In humans, between 46 and 95 crossovers occur per meiosis, so on average each chromosome crosses over at least once (Figure 8.15D).

From Gametes to Offspring   Sexual reproduction involves the fusion of mature reproductive cells—**gametes**—from two parents. All gametes are haploid, and they arise by division of germ cells, which are immature reproductive cells that form in organs set aside for reproduction. Animals and plants make gametes somewhat differently. In animals, meiosis in diploid germ cells gives rise to eggs (female gametes) or sperm (male gametes). In plants, haploid germ cells form by meiosis. Gametes form when these cells divide by mitosis. We leave details of sexual reproduction in animals and plants for later chapters, but you will need to know a few concepts before you get there.

At fertilization, two haploid gametes fuse and produce a diploid **zygote**, which is the first cell of a new individual. Thus, meiosis halves the chromosome number, and fertilization restores it. If meiosis did not precede fertilization, the chromosome number would double with every generation. If the chromosome number changes, so does the individual's set of genetic instructions. An individual's set of chromosomes is like a fine-tuned blueprint that must be followed exactly, in order to build a body that functions normally. As you will see in Chapter 9, chromosome number changes can have drastic consequences for health, particularly in animals.

Fertilization also contributes to the variation that we see among offspring of sexual reproducers. Think about it in terms of human reproduction. Cells that give rise to human gametes have twenty-three pairs of homologous chromosomes. Each time a human germ cell undergoes meiosis, the four gametes that form end up with one of 8,388,608 (or 2^{23}) possible combinations of homologous chromosomes. In

Figure 8.15 Crossing over.
Blue signifies a paternal chromosome, and pink, its maternal homologue. For clarity, only one pair of homologous chromosomes is shown.

(D) James Kezer/Courtesy of Dr. Sessions.

A. Here, we focus on only two of the many genes on a chromosome. In this example, one gene has alleles *A* and *a*; the other has alleles *B* and *b*.

B. Close contact between homologous chromosomes promotes crossing over between nonsister chromatids. Paternal and maternal chromatids exchange corresponding pieces.

C. Crossing over mixes up paternal and maternal alleles on homologous chromosomes.

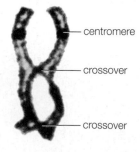

D. Each pair of homologous chromosomes can cross over multiple times. This is a normal and common process of meiosis.

centromere

crossover

crossover

Digging Into Data

BPA and Abnormal Meiosis

In 1998, researchers at Case Western University were studying meiosis in mouse oocytes when they saw an unexpected and dramatic increase in abnormal meiosis events (Figure 8.16). The improper sorting of chromosomes into gametes is one of the main causes of human genetic disorders, which we will discuss in Chapter 9.

The researchers discovered that the spike in abnormal meiosis events began immediately after the mouse facility started washing the animals' plastic cages and water bottles in a new, alkaline detergent. The detergent had damaged the plastic, which as a result was leaching bisphenol A (BPA). BPA is a synthetic chemical that mimics estrogen, the main female sex hormone in animals. Though it has been banned for use in baby bottles, BPA is still widely used to manufacture other plastic items and epoxies (such as the coating on the inside of metal cans of food). BPA-free plastics are often manufactured with a related compound, bisphenol S (BPS), that has effects similar to BPA.

Caging materials	Total number of oocytes	Abnormalities
Control: New cages with glass bottles	271	5 (1.8%)
Damaged cages with glass bottles		
Mild damage	401	35 (8.7%)
Severe damage	149	30 (20.1%)
Damaged bottles	197	53 (26.9%)
Damaged cages with damaged bottles	58	24 (41.4%)

1. What percentage of mouse oocytes displayed abnormalities of meiosis with no exposure to damaged caging?
2. Which group of mice had the most meiotic abnormalities?
3. What is abnormal about metaphase I as it is occurring in the oocytes shown in Figure 8.16B, C, and D?

Figure 8.16 **Meiotic abnormalities associated with exposure to plastic.** Fluorescence micrographs show nuclei of single mouse oocytes (female germ cells) in metaphase I. (**A**) Normal metaphase; (**B–D**) examples of abnormal metaphase. Chromosomes are stained red; spindle fibers, green.

(A–D) Reprinted from *Current Biology*, Vol 13, (Apr 03), Authors Hunt, Koehler, Susiarjo, Hodges, Ilagan, Voight, Thomas, Thomas, and Hassold, Bisphenol A Exposure Causes Meiotic Aneuploidy in the Female Mouse, pp. 546–553, © 2003 Cell Press. Published by Elsevier Ltd. With permission from Elsevier.

addition, any number of genes may occur as different alleles on the maternal and paternal chromosomes, and crossing over makes mosaics of that genetic information. Then, out of all the male and female gametes that form, which two actually get together at fertilization is a matter of chance. Are you getting an idea of why such fascinating combinations of traits show up among the generations of your own family tree?

Take-Home Message 8.5

How does meiosis give rise to new combinations of parental alleles among the offspring of sexual reproducers?

- During meiosis, the nucleus of a diploid (2*n*) cell divides twice. DNA replication does not occur between the two divisions, so the chromosome number is reduced to the haploid number (*n*) for forthcoming gametes.
- Crossing over—recombination between nonsister chromatids of homologous chromosomes—occurs during meiosis. It makes new combinations of parental alleles.
- The union of two haploid gametes at fertilization results in a diploid zygote, the first cell of a new individual.

Meiosis halves the chromosome number, and fertilization restores it.

crossing over Process in which homologous chromosomes exchange corresponding segments during meiosis.

gamete Mature, haploid reproductive cell; e.g., an egg or a sperm.

zygote Diploid cell that forms when two gametes fuse; the first cell of a new individual.

Summary

Section 8.1 An immortal line of human cells (HeLa) is a legacy of cancer victim Henrietta Lacks. Researchers all over the world continue to work with these cells in their efforts to unravel the mechanisms of cancer.

Section 8.2 A **cell cycle** starts when a new cell forms, and ends when the cell reproduces. Most of a cell's activities occur during **interphase**. A eukaryotic cell reproduces by dividing: nucleus first, then cytoplasm. **Mitosis**, a nuclear division mechanism that maintains the chromosome number, is the basis of growth and tissue repair in multicelled species. Mitosis is also the basis of **asexual reproduction** in many species.

DNA replication occurs before mitosis begins, so each of the cell's **homologous chromosomes** consists of two molecules of DNA (sister chromatids) attached at the centromere. Mitosis occurs in four stages. During **prophase**, the chromosomes condense, the nuclear envelope breaks up, and microtubules assemble into a **spindle**. Spindle microtubules attach to the chromosomes at the centromere. At **metaphase**, all of the chromosomes are aligned in the middle of the cell. During **anaphase**, the spindle separates the sister chromatids of each chromosome and moves them toward opposite sides of the cell. During **telophase**, two new nuclei form, each with the parental chromosome number.

In most cases, nuclear division is followed by cytoplasmic division. In animal cells, a contractile ring pulls the plasma membrane inward, forming a **cleavage furrow** that pinches the cytoplasm in two. In plant cells, vesicles merge as a cell plate that expands and fuses with the plasma membrane, thus becoming a cross-wall that partitions the cytoplasm.

Section 8.3 The products of checkpoint genes work together to control the cell cycle. These molecules monitor the integrity of the cell's DNA, and can pause the cycle until breaks or other problems are fixed. When checkpoint mechanisms fail, a cell loses control over its cell cycle, and its abnormally dividing descendants form a **neoplasm**. Neoplasms may form lumps called **tumors**.

Mutations can turn some genes into tumor-causing **oncogenes**. Mutations in multiple checkpoint genes can transform benign neoplasms into malignant ones. Cells of malignant neoplasms can break loose from their home tissues and colonize other parts of the body, a process called **metastasis**. **Cancer** occurs when cells of a malignant neoplasm physically and metabolically disrupt body tissues. Telomere length limits the number of times that a normal cell can divide, a fail-safe mechanism in case the cell loses control over its cell cycle.

Section 8.4 **Sexual reproduction** mixes up the genetic information of two parents. The offspring of sexual reproducers typically vary in shared, inherited traits. Particularly in changing environments, this variation can offer an evolutionary advantage over genetically identical offspring produced by asexual reproduction.

Sexual reproduction produces offspring whose body cells contain pairs of chromosomes, one of each homologous pair from the mother and the other from the father. The two chromosomes of a homologous pair carry the same genes. Paired genes on homologous chromosomes may vary in DNA sequence, in which case they are called **alleles**. Alleles are the basis of differences in shared traits. They arise by mutation.

Section 8.5 **Meiosis**, the basis of sexual reproduction in eukaryotes, is a nuclear division mechanism that halves the chromosome number. DNA replication occurs before meiosis begins, so each chromosome consists of two molecules of DNA (sister chromatids). Two nuclear divisions occur during meiosis. In the first nuclear division (meiosis I), all of the homologous chromosomes line up and exchange corresponding segments. This **crossing over** mixes up the alleles on maternal and paternal chromosomes. The homologous chromosomes are then moved apart and packaged in separate nuclei. This stage reduces the chromosome number, from diploid ($2n$) to **haploid** (n). The second nuclear division (meiosis II) occurs in both haploid nuclei that formed in meiosis I. Sister chromatids separate in this stage, so at the end of meiosis each chromosome consists of one molecule of DNA. Four haploid nuclei typically form. Meiosis is necessary for the production of haploid **gametes**. The fusion of two gametes at fertilization restores the diploid parental chromosome number in the **zygote**, the first cell of the new individual.

Self-Quiz

1. Mitosis and cytoplasmic division function in _____ .
   a. asexual reproduction of single-celled eukaryotes
   b. growth and tissue repair in multicelled species
   c. asexual reproduction in prokaryotes
   d. both a and b

2. A duplicated chromosome has how many chromatids?
   a. one   c. three
   b. two   d. four

3. A cell with two of each type of chromosome has a chromosome number that is _____ .
   a. diploid   c. tetraploid
   b. haploid   d. abnormal

4. Homologous chromosomes _____ .
   a. originate with two parents
   b. are sister chromatids

5. Interphase is the part of the cell cycle when _____ .
   a. a cell ceases to function
   b. the spindle forms prior to nuclear division
   c. DNA replication occurs
   d. mitosis proceeds

6. After mitosis, the chromosome number of a descendant cell is _____ the parent cell's.
   a. the same as   c. rearranged compared to
   b. one-half of   d. doubled compared to

7. One evolutionary advantage of sexual over asexual reproduction may be that it produces _____ .
   a. more offspring per individual
   b. more variation among offspring
   c. healthier offspring

8. Alternative forms of the same gene are _____ .
   a. gametes   c. alleles
   b. homologous   d. oncogenes

9. Meiosis is a necessary part of sexual reproduction because it _____ .
   a. divides two nuclei into four new nuclei
   b. reduces the chromosome number for gametes
   c. produces clones that can cross over

10. Crossing over mixes up _____ .
    a. chromosomes   c. zygotes
    b. alleles   d. gametes

11. Sexual reproduction in animals requires _____ .
    a. meiosis   c. gametes
    b. fertilization   d. all of the above

12. Which of the following is one of the very important differences between mitosis and meiosis?
    a. Chromosomes align in the middle of the cell only in meiosis.
    b. Homologous chromosomes swap segments only in meiosis.
    c. Sister chromatids separate only in meiosis.

13. The cell illustrated on the right is in anaphase I, not anaphase II. I know this because _____ .
    a. crossing over has already occurred
    b. the chromosomes are still duplicated
    c. a spindle has formed.
    d. sister chromatids have separated

14. Match each stage with the events listed.
    _____ prophase   a. sister chromatids move apart
    _____ metaphase   b. chromosomes condense
    _____ anaphase   c. new nuclei form
    _____ telophase   d. DNA replication
    _____ interphase   e. all chromosomes are aligned in the middle of the cell

15. Match each term with the best description.
    _____ cell plate   a. lump of abnormal cells
    _____ spindle   b. forms at fertilization
    _____ tumor   c. divides plant cells
    _____ contractile ring   d. mash-up time
    _____ gamete   e. dangerous metastatic cells
    _____ cancer   f. made of microtubules
    _____ zygote   g. haploid
    _____ prophase I   h. makes an indentation

Critical Thinking

1. When a cell reproduces by mitosis and cytoplasmic division, does its life end?
3. Make a sketch of meiosis in a cell with a diploid chromosome number of 4. Now try it when the chromosome number is 3.
4. The diploid chromosome number for the body cells of a frog is 26. What would that number be after three generations if meiosis did not occur before gamete formation?
5. Which nuclear division, meiosis I or meiosis II, is most similar to mitosis?

Visual Question

The eukaryotic cell in the photo on the left is in the process of cytoplasmic division. Is this cell from a plant or an animal? How do you know?

© Michel Delarue © ISM/Phototake.

9

Application ⊙ 9.1 Menacing Mucus

REMEMBER: Polypeptide chains twist, fold, and pack into functional domains; a protein's function arises from and depends on its shape (Section 2.9). Vesicles move substances among organelles of the ER, and to and from the plasma membrane (3.5). Water moves across a membrane by osmosis, from a hypotonic to a hyper-tonic fluid (4.5). In active transport, a transport protein uses energy to pump a solute against its gradient across a cell membrane; in endocytosis, a small patch of plasma membrane balloons into cytoplasm and becomes a vesicle (4.6). During protein synthesis, amino acids are specified by three-nucleotide codons (7.4). A deletion is a mutation in which one or more nucleotides are lost from a chromosome (7.6).

In 1988, researchers discovered a gene that, when mutated, causes the most common fatal genetic disorder in the United States: cystic fibrosis (CF). The gene they discovered, *CFTR*, encodes an active transport protein that moves chloride ions out of epithelial cells. Sheets of epithelial cells line the passageways and ducts of the lungs, liver, pancreas, intestines, and reproductive system. When the CFTR protein pumps chloride ions out of these cells, water follows the solute by osmosis. The two-step process maintains a thin, watery film on the surface of epithelial cell sheets. Mucus slides easily over the wet sheets of cells.

The mutation most commonly associated with cystic fibrosis is a deletion that removes a single codon from the *CFTR* gene. The protein product of the mutated gene is missing one amino acid, and as a result it misfolds in a tiny region. This small defect interferes with cellular processes that would otherwise finish the protein and install it in the plasma membrane. Normally, a newly translated CFTR polypeptide is modified by the endoplasmic reticulum (ER) and exported to a Golgi body, which attaches carbohydrates to it. The finished protein is then packaged in vesicles routed to the plasma membrane. CFTR polypeptides with the missing amino acid are produced properly, but a cellular quality control mechanism recognizes the misfolded region and destroys most of them before they leave the ER. The few that make it to the plasma membrane are quickly taken back into the cell by endocytosis and destroyed.

Epithelial cell membranes that lack the CFTR protein cannot transport chloride ions. Too few chloride ions leave these cells. Not enough water leaves them either, so the surfaces of epithelial cell sheets are too dry. Mucus that normally slips and slides through the body's tubes sticks to their walls instead. Thick globs of mucus accumulate and clog passageways and ducts throughout the body. Breathing becomes difficult as the mucus obstructs the smaller airways of the lungs. Digestive problems arise as ducts that lead to the gut become clogged with mucus. Males are typically infertile because their sperm flow is hampered.

In addition to its role in chloride ion transport, the CFTR protein also helps alert the immune system to the presence of disease-causing bacteria in the lungs. It functions as a receptor by binding directly to these bacteria and causing them to be taken into the cell by endocytosis. In epithelial cells lining the respiratory tract, endocytosis of bacteria triggers an immune response. Bacteria-fighting molecules that are produced in this response keep microbial populations at bay. When the cells lack CFTR, the early alert system fails, so bacteria have time to multiply before being detected by the immune system. Thus, chronic bacterial infections of the lungs are a hallmark of cystic fibrosis. Antibiotics help control infections, but there is no cure for the disorder. Most affected people die before age thirty, when their tormented lungs fail (Figure 9.1).

Figure 9.1 A few of the many victims of cystic fibrosis. At least one young person dies every day in the United States from complications of this disease, which occurs most often in people of northern European ancestry.

9.2 Tracking Traits

REMEMBER: Inheritance is transmission of DNA to offspring (Section 1.3). DNA was proven to be hereditary material in the 1950s (6.2). Information encoded in a DNA sequence occurs in units called genes; a cell uses the information in a gene to make an RNA or protein product (7.2). Diploid body cells of sexual reproducers have two sets of chromosomes, one inherited from each of two parents; homologous chromosomes carry the same genes, which can differ as alleles; alleles may encode different forms of the gene's product; almost every shared trait varies a bit among individuals of a species, and alleles of shared genes are the basis of this variation (8.4). At fertilization, two gametes fuse to produce a zygote, the first cell of a new individual (8.5).

In the nineteenth century, people thought that hereditary material must be some type of fluid, with fluids from both parents blending at fertilization like milk into coffee. However, the idea of "blending inheritance" failed to explain what people could see with their own eyes. Children sometimes have traits such as freckles that do not appear in either parent. A cross between a black horse and a white one does not produce gray offspring. The naturalist Charles Darwin had no hypothesis to explain such phenomena, even though inheritance was central to his theory of natural selection (Chapter 11 returns to this theory). At the time, no one knew

The Moravian Museum, Brno

that hereditary information is divided into discrete units (genes), an insight that is critical to understanding how traits are inherited. However, even before Darwin presented his theory, someone had been gathering evidence that would support it. Gregor Mendel (left), an Austrian monk, had been carefully breeding thousands of pea plants. By keeping detailed records of how traits passed from one generation to the next, Mendel had been collecting evidence of how inheritance works.

Mendel's Experiments Mendel cultivated the garden pea plant (Figure 9.2). This species is naturally self-fertilizing, which means its flowers produce male and female gametes ❶ that form viable embryos when they meet up. To prevent an individual pea plant from self-fertilizing, Mendel removed the pollen-bearing parts (anthers) from its flowers. He then cross-fertilized the flowers by brushing their egg-bearing parts (carpels) with pollen from another plant ❷. He collected seeds ❸ from the cross-fertilized individual, and recorded the traits of the new pea plants that grew from them ❹.

Many of Mendel's experiments, which are called crosses, started with plants that "breed true" for particular traits such as white flowers or purple flowers. Breeding true for a trait means that, new mutations aside, all offspring have the same form of the trait as the parent(s), generation after generation. For example, all offspring of pea plants that breed true for white flowers also have white flowers. As you will see in the next section, Mendel discovered that crossing pea plants yields offspring with traits that often appear in predictable patterns. Mendel's meticulous work tracking pea plant traits led him to conclude (correctly) that hereditary information passes from one generation to the next in distinct units.

Inheritance in Modern Terms Mendel discovered hereditary units, which we now call genes, almost a century before the discovery of DNA. Today, we know that individuals of a species share certain traits because their chromosomes carry the

Figure 9.2 Breeding experiments with the garden pea.

Top, Valentina Razumova/Shutterstock; (#1) Jean M. Labat/ardea.com.

❶ In the flowers of a garden pea plant (above), pollen grains that form in anthers (right) produce male gametes. Female gametes form in carpels.

carpel

anther

❷ Experimenters control the transfer of hereditary material from one pea plant to another by cutting off a flower's pollen-producing anthers (to prevent it from self-fertilizing), then brushing pollen from another flower onto its egg-producing carpel.

In this example, pollen from a plant with purple flowers is brushed onto the carpel of a white-flowered plant.

❸ Later, seeds develop inside pods of the cross-fertilized plant. An embryo in each seed develops into a mature pea plant.

❹ Every plant that arises from the cross has purple flowers. Predictable patterns such as this are evidence of how inheritance works.

Figure 9.3 Locations of a few genes on two human chromosomes.
Genetic disorders arising from mutations in the genes are shown in parentheses. The number or letter below a chromosome is its name; staining reveals characteristic banding patterns that are illustrated here. Appendix III has a similar map of all 23 human chromosomes.

Figure 9.4 Genotype gives rise to phenotype.
In this example, the dominant allele *P* specifies purple flowers; the recessive allele *p*, white flowers.

© Tamara Kulikova/Shutterstock.

Figure It Out: Which individual is a hybrid?

Answer: The heterozygous one

same genes, and that each gene occurs at a specific location on a particular chromosome (Figure 9.3). The diploid body cells of humans and other animals contain pairs of genes, on pairs of homologous chromosomes. In most cases, both genes of a pair are expressed. Genes at the same location on a pair of homologous chromosomes may be identical, or they may vary as alleles. Organisms breed true for a trait because they carry identical alleles of genes governing that trait. An individual with two identical alleles of a gene is **homozygous** for the allele (*homo–* means the same). By contrast, an individual with two different alleles of a gene is **heterozygous** (*hetero–*, mixed). A hybrid is a heterozygous individual produced by a cross or mating between individuals that breed true for different forms of a trait. Homozygous and heterozygous are examples of **genotype**, the particular set of alleles an individual carries. Genotype is the basis of **phenotype**, which refers to the individual's observable traits. "White-flowered" and "purple-flowered" are examples of pea plant phenotypes that arise from differences in genotype.

The phenotype of a heterozygous individual depends on how the products of its two different alleles interact. In many cases, the effect of one allele influences the effect of the other, and the outcome of this interaction is reflected in the individual's phenotype. An allele is **dominant** when its effect masks that of a **recessive** allele paired with it. A dominant allele is often represented by an italic capital letter such as *A*; a recessive allele, with a lowercase italic letter such as *a*. Consider the purple- and white-flowered pea plants that Mendel studied. In these plants, the allele that specifies purple flowers (let's call it *P*) is dominant over the allele that specifies white flowers (*p*). Thus, a pea plant homozygous for the dominant allele (*PP*) has purple flowers; one homozygous for the recessive allele (*pp*) has white flowers (Figure 9.4). A heterozygous plant (*Pp*) has purple flowers.

Take-Home Message 9.2

How do alleles contribute to traits?

• Genotype (an individual's set of alleles) is the basis of phenotype (the individual's observable traits).

• A homozygous individual has two identical alleles of a gene. A heterozygous individual has two nonidentical alleles.

• A dominant allele masks the effect of a recessive allele in a heterozygous individual.

9.3 Mendelian Inheritance Patterns

REMEMBER: During the nuclear division process of meiosis, homologous chromosomes cross over and then separate; crossing over makes new combinations of parental alleles for forthcoming gametes; at fertilization, two gametes fuse and produce a zygote, which is the first cell of a new individual (Section 8.5).

When homologous chromosomes separate during meiosis, the gene pairs on those chromosomes separate too. Each gamete that forms carries only one of the two genes of a pair. Let's use our alleles for purple and white flowers in an example (Figure 9.5). Plants homozygous for the dominant allele (*PP*) can only make gametes that carry allele *P* **1**. Plants homozygous for a recessive allele (*pp*) can only make gametes that carry allele *p* **2**. If these homozygous plants are crossed (*PP* × *pp*),

Figure 9.5 Segregation of genes on homologous chromosomes into gametes.
Homologous chromosomes separate during meiosis, so the pairs of genes they carry separate too. Each of the resulting gametes carries one of the two members of each gene pair.

1 All gametes made by a parent homozygous for a dominant allele carry that allele.

2 All gametes made by a parent homozygous for a recessive allele carry that allele.

3 If these two parents are crossed, the union of any of their gametes at fertilization produces a zygote with both alleles. All offspring of this cross will be heterozygous.

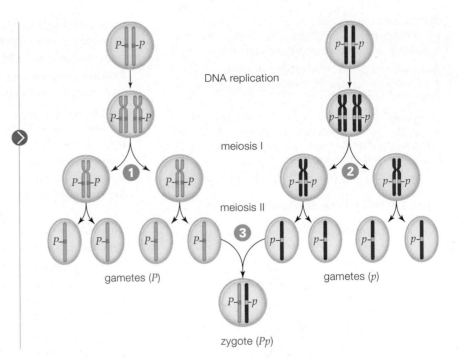

Figure 9.6 Making a Punnett square.
Parental gametes are listed in circles on the top and left sides of a grid. Each square is filled with the combination of alleles that would result if the gametes in the corresponding row and column met up.

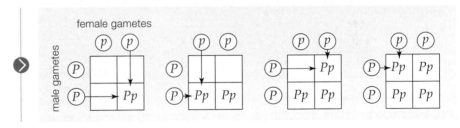

only one outcome is possible: A gamete carrying allele *P* meets up with a gamete carrying allele *p* **3**. All of the offspring of this cross will have both alleles—they will be heterozygous (*Pp*). Because all of the offspring will carry the dominant allele *P*, all will have purple flowers. A grid called a **Punnett square** is helpful for predicting the genetic and phenotypic outcomes of such crosses (Figure 9.6).

Our example illustrated a pattern so predictable that it can be used as evidence of a dominance relationship between alleles, as the following experiments illustrate.

Monohybrid Crosses Dominance relationships between alleles determine the phenotypic outcome of a **monohybrid cross**, in which individuals with the same two alleles of a gene are crossed (*Pp* × *Pp*, for example). The frequency at which traits associated with the alleles appear among the offspring depends on whether one of the alleles is dominant over the other.

To make a monohybrid cross, we would start with individuals that breed true for two different forms of a trait. In garden pea plants, flower color (purple and white) is one example of a trait with two distinct forms, but there are many others. A cross between individuals that breed true for different forms of the trait yields hybrid offspring, all with the same set of alleles governing the trait. A cross between two of these F₁ (first generation) hybrids is the monohybrid cross. The frequency at which the two traits appear in the F₂ (second generation) offspring offers information about a dominance relationship between the two alleles. (F is an abbreviation for filial, which means offspring.)

dominant Refers to an allele that masks the effect of a recessive allele on the homologous chromosome.

genotype The particular set of alleles that is carried in an individual's chromosomes.

heterozygous Having two different alleles of a gene.

homozygous Having identical alleles of a gene.

monohybrid cross Cross between two individuals identically heterozygous for alleles of one gene; for example *Aa* × *Aa*.

phenotype An individual's observable traits.

Punnett square Diagram used to predict the genetic and phenotypic outcomes of a cross.

recessive Refers to an allele with an effect that is masked by a dominant allele on the homologous chromosome.

codominance Inheritance pattern in which the full and separate phenotypic effects of two alleles are apparent in heterozygous individuals.

dihybrid cross Cross between two individuals identically heterozygous for alleles of two genes; for example *AaBb × AaBb*.

incomplete dominance Inheritance pattern in which one allele is not fully dominant over another, so the heterozygous phenotype is an intermediate blend between the two homozygous phenotypes.

Figure 9.7 Example of a monohybrid cross.

parent plant homozygous for purple flowers

parent plant homozygous for white flowers

PP × *pp*

Pp hybrid

P *p* two types of gametes

A. All of the F₁ (first generation) offspring of a cross between two plants that breed true for different forms of a trait are identically heterozygous (*Pp*). These offspring make two types of gametes: *P* and *p*.

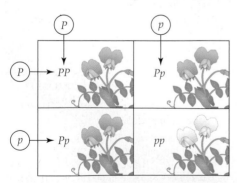

	P	*p*
P	*PP*	*Pp*
p	*Pp*	*pp*

B. A monohybrid cross is a cross between the F₁ offspring. In this example, the phenotype ratio among the F₂ (second generation) offspring is 3:1 (three purple to one white).

Figure It Out: In this example, how many possible genotypes are there in the F₂ generation?

Answer: Three (*PP*, *Pp*, and *pp*)

A cross between two purple-flowered heterozygous plants (*Pp*) offers an example of a monohybrid cross. Each individual can make two types of gametes: ones that carry a *P* allele, and ones that carry a *p* allele (Figure 9.7A). So, in a monohybrid cross between *Pp* plants (*Pp × Pp*), the two types of gametes can meet up in four possible ways at fertilization:

Possible Event	Offspring Genotype	Resulting Phenotype
Sperm *P* meets egg *P* →	zygote genotype is *PP* →	individual has purple flowers
Sperm *P* meets egg *p* →	zygote genotype is *Pp* →	individual has purple flowers
Sperm *p* meets egg *P* →	zygote genotype is *Pp* →	individual has purple flowers
Sperm *p* meets egg *p* →	zygote genotype is *pp* →	individual has white flowers

Three of four possible outcomes of this cross include at least one copy of the dominant allele *P*. In other words, each time fertilization occurs, there are 3 chances in 4 that the resulting zygote will have a *P* allele (and the individual will make purple flowers). There is 1 chance in 4 that the zygote will have two *p* alleles (and the individual will make white flowers). Thus, the probability that a particular offspring of this cross will have purple or white flowers is 3 purple to 1 white—a ratio of 3:1 (Figure 9.7B). The 3:1 pattern is evidence that purple and white flower color are specified by alleles with a clear dominance relationship: Purple is dominant; white, recessive. If the probability of an individual inheriting a particular genotype is difficult to imagine, think about it in terms of many offspring: In this example, there will be roughly three purple-flowered plants for every white-flowered one.

Dihybrid Crosses A monohybrid cross allows us to study a dominance relationship between alleles of one gene. What about alleles of two genes? An individual heterozygous for alleles of two genes (*AaBb*, for example) is called a dihybrid, and a cross between two such individuals is a **dihybrid cross**. As with a monohybrid cross, the frequency of traits appearing among the offspring of a dihybrid cross depends on the dominance relationships between the alleles.

To make a dihybrid cross, we would start with individuals that breed true for two different traits. Let's use a gene for flower color (*P*, purple; *p*, white) and one for plant height (*T*, tall; *t*, short) in an example. Figure 9.8 shows a dihybrid cross starting with one parent plant that breeds true for purple flowers and tall stems (*PPTT*), and one that breeds true for white flowers and short stems (*pptt*). The *PPTT* plant only makes gametes with the dominant alleles (*PT*); the *pptt* plant only makes gametes with the recessive alleles (*pt*) **1**. So, all offspring from a cross between these two plants (*PPTT × pptt*) will be dihybrids (*PpTt*) with purple flowers and tall stems **2**.

Four combinations of *P* and *T* alleles are possible in the gametes of *PpTt* dihybrids **3**. If two of these plants are crossed (a dihybrid cross, *PpTt × PpTt*), the four types of gametes can combine in sixteen possible ways at fertilization **4**. Nine of the sixteen genotypes would give rise to tall plants with purple flowers; three, to short plants with purple flowers; three, to tall plants with white flowers; and one, to short plants with white flowers. Thus, the ratio of phenotypes among the offspring of this dihybrid cross would be 9:3:3:1.

How two gene pairs get sorted into gametes depends partly on whether the two genes are on the same chromosome. When homologous chromosomes separate during meiosis, either member of the pair can end up in either of the two new nuclei that form. Thus, genes on one chromosome pair assort into gametes independently of genes on the other chromosome pairs. What about genes on the same chromosome? Pea plants have seven chromosomes. Mendel studied seven pea genes, and all of them assorted into gametes independently of one another. Was he lucky enough to choose one gene on each of those chromosomes? As it turns out, some of the genes Mendel studied *are* on the same chromosome. These genes are far enough apart that crossing over occurs between them very frequently—so frequently that they tend to assort into gametes independently, just as if they were on different chromosomes. By contrast, genes that are very close together on a chromosome usually do not assort independently into gametes, because crossing over does not happen between them very often. Thus, gametes usually end up with parental combinations of alleles of these genes.

Take-Home Message 9.3

How are alleles distributed into gametes?

- Diploid cells have pairs of genes, on pairs of homologous chromosomes. The two genes of a pair (which may be identical or not) are separated from each other during meiosis, so they end up in different gametes.
- In most cases, the two genes of a pair are distributed into gametes independently of other gene pairs on other chromosomes.
- Gene pairs that are far apart on the same chromosome also tend to be distributed independently into gametes.

9.4 Beyond Simple Dominance

REMEMBER: Sickle-cell anemia is the outcome of a base-pair substitution in the beta globin gene (Section 7.6).

In the Mendelian inheritance patterns discussed in the previous section, the effect of a dominant allele on a trait fully masks that of a recessive one. Other inheritance patterns are more common, and more complex.

Incomplete Dominance In an inheritance pattern called **incomplete dominance**, one allele is not fully dominant over the other, so the heterozygous phenotype is an intermediate blend of the two homozygous phenotypes. A gene that affects flower color in snapdragon plants is an example. One allele of the gene (*R*) encodes an enzyme that makes a red pigment. The enzyme encoded by an allele with a mutation (*r*) cannot make any pigment. Plants homozygous for the *R* allele (*RR*) make a lot of red pigment, so they have red flowers. Plants homozygous for the *r* allele (*rr*) make no pigment, so their flowers are white. Heterozygous plants (*Rr*) make only enough pigment to tint their flowers pink. A cross between two pink-flowered heterozygous plants yields red-, pink-, and white-flowered offspring in a 1:2:1 ratio.

Codominance In an inheritance pattern called **codominance**, both alleles are fully expressed in heterozygous individuals; neither allele is dominant or recessive. Alleles of the *ABO* gene offer an example. This gene encodes an enzyme that modifies a carbohydrate on the surface of human red blood cells.

Figure 9.8 An example of a dihybrid cross.
P and *p* are dominant and recessive alleles for flower color; *T* and *t* are dominant and recessive alleles for plant height.

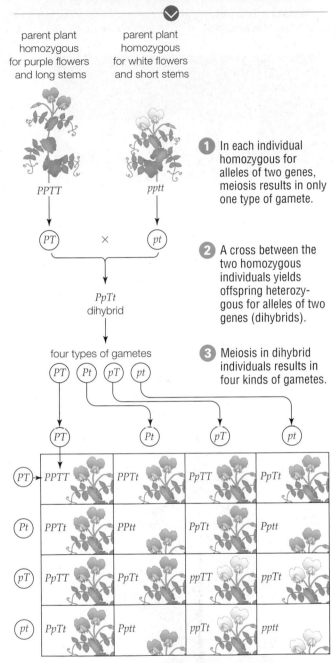

parent plant homozygous for purple flowers and long stems

parent plant homozygous for white flowers and short stems

PPTT   *pptt*

1 In each individual homozygous for alleles of two genes, meiosis results in only one type of gamete.

PT × *pt*

2 A cross between the two homozygous individuals yields offspring heterozygous for alleles of two genes (dihybrids).

PpTt dihybrid

four types of gametes
PT *Pt* *pT* *pt*

3 Meiosis in dihybrid individuals results in four kinds of gametes.

PT *Pt* *pT* *pt*

	PT	*Pt*	*pT*	*pt*
PT	*PPTT*	*PPTt*	*PpTT*	*PpTt*
Pt	*PPTt*	*PPtt*	*PpTt*	*Pptt*
pT	*PpTT*	*PpTt*	*ppTT*	*ppTt*
pt	*PpTt*	*Pptt*	*ppTt*	*pptt*

4 If two of the dihybrid individuals are crossed, the four types of gametes can meet up in 16 possible ways. Of 16 possible offspring genotypes, 9 will result in plants that are purple-flowered and tall; 3, purple-flowered and short; 3, white-flowered and tall; and 1, white-flowered and short. Thus, the ratio of phenotypes is 9:3:3:1.

epistasis Polygenic inheritance, in which a trait is influenced by multiple genes.

pleiotropy Inheritance pattern in which a single gene affects multiple traits.

Figure 9.9 Combinations of alleles that are the basis of blood type.

Photo, Annie Cavanagh/Wellcome Images.

Figure 9.10 An example of epistasis.
Interactions among products of two gene pairs affect fur color in Labrador retrievers. Dogs with alleles *E* and *B* have black fur. Those with an *E* and two recessive *b* alleles have brown fur. Dogs homozygous for the recessive *e* allele have yellow fur.

Top, Susan Schmitz/Shutterstock.

Two alleles, *A* and *B*, encode slightly different versions of this enzyme, which in turn modify the carbohydrate differently. A third allele, *O*, has a mutation that prevents the enzyme product from becoming active, so the carbohydrate remains unmodified.

The alleles of the *ABO* gene that you carry determine the form of the carbohydrate on your blood cells, so they are the basis of your ABO blood type (Figure 9.9). The *A* and the *B* allele are codominant when paired. If your genotype is *AB*, then you have both versions of the enzyme, and your blood type is AB. The *O* allele is recessive when paired with either the *A* or *B* allele. If your genotype is *AA* or *AO*, your blood type is A. If your genotype is *BB* or *BO*, it is type B. If you are *OO*, it is type O.

The immune system attacks any cell bearing molecules that do not occur in one's own body, so receiving incompatible blood in a transfusion can be dangerous. An immune attack causes red blood cells to clump or burst, with potentially fatal results. Almost everyone makes the red blood cell carbohydrate (which is later modified in people with an *A* or *B* allele), so type O blood does not usually trigger an immune response in transfusion recipients. People with type O blood are called universal donors because they can donate blood to anyone. However, because their body is unfamiliar with the modified form of the carbohydrate made by people with type A or B blood, they can receive type O blood only. People with type AB blood can receive a transfusion of any ABO blood type, so they are called universal recipients.

Pleiotropy and Epistasis In an inheritance pattern called **pleiotropy**, a single gene influences multiple traits. Mutations that affect the gene's product or its expression affect all of the traits. Many complex genetic disorders, including sickle-cell anemia and cystic fibrosis, are caused by mutations in single genes. Marfan syndrome, another example, is a result of mutations that affect fibrillin. Long fibers of this protein are part of elastic tissues that make up the heart, skin, blood vessels, tendons, and other body parts. Mutations can cause tissues to form with defective fibrillin or none at all. The largest blood vessel leading from the heart, the aorta, is particularly affected. Without a proper scaffold of fibrillin, the aorta's thick wall is not as elastic as it should be, and it eventually stretches and becomes leaky. Thinned and weakened, the aorta can rupture during exercise—an abruptly fatal outcome. About 1 in 5,000 people have Marfan syndrome, and there is no cure. Its effects—and risks—are manageable with early diagnosis, but symptoms are easily missed. Affected people are tall and loose-jointed, but there are plenty of tall, loose-jointed people who do not have the syndrome. Thus, people with Marfan syndrome may die suddenly and early without ever knowing they had the disorder.

In a common inheritance pattern called polygenic inheritance or **epistasis**, one trait is affected by multiple gene products. Consider how fur color in dogs and other animals arises from pigments called melanins. A dark brown form of melanin gives rise to brown or black fur; a reddish melanin, to yellow fur. The products of several genes interact to carry out melanin synthesis and deposition in fur. In Labrador retriever dogs, alleles of two of these genes determine whether the individual has black, brown, or yellow fur (Figure 9.10).

Human skin color offers another example of epistasis. At least 100 gene products affect this trait, which begins with melanosomes (organelles that make melanins). Most people have about the same number of melanosomes in their skin cells. Variations in skin color arise from differences in the size, shape, and cellular distribution of melanosomes in the skin, as well as in the kinds and amounts of melanins they make. These variations have a genetic basis. Consider one gene that encodes a transport protein in melanosome membranes. Nearly all people of African, Native American, or east Asian descent carry the same allele of this gene. A mutation that

Digging Into Data

The Cystic Fibrosis Mutation and Typhoid Fever

The CF allele that causes most cases of cystic fibrosis is at least 50,000 years old and very common: 1 in 25 people carry it in some populations. Why is this allele so common if it is so dangerous? Consider that the CF allele is eventually lethal in homozygous individuals, but not in those who are heterozygous. This allele is codominant with the normal allele. Heterozygous individuals typically have no symptoms of cystic fibrosis because their cells make enough of the normal CFTR protein to have normal chloride ion transport.

Researchers think the CF allele has persisted because it offers heterozygous individuals an advantage in surviving certain deadly infectious diseases. The unmutated CFTR protein triggers endocytosis when it binds to bacteria. This process is an essential part of the body's immune response to bacteria in the respiratory tract. However, the same function of CFTR allows bacteria to enter cells of the gastrointestinal tract, where they can be deadly. For example, internalization of *Salmonella typhi* bacteria by epithelial cells in the gut causes a common worldwide disease called typhoid fever. Symptoms include extreme fever and diarrhea, and the resulting dehydration causes delirium that may last several weeks. If untreated, it kills up to 30 percent of those infected. Around 600,000 people, most of whom are children, die from the disease each year.

In 1998, Gerald Pier and his colleagues compared the uptake of *S. typhi* by different types of epithelial cells: those heterozygous for the CF allele, and those homozygous for the normal allele. (Cells homozygous for the CF allele do not take up any *S. typhi* bacteria.) Some of the results are shown in Figure 9.11.

Figure 9.11 **Effect of the CF mutation on uptake of three strains of *Salmonella typhi* bacteria by epithelial cells.**

1. Regarding the Ty2 strain of *S. typhi*, about how many more bacteria were able to enter normal cells (those expressing unmutated *CFTR*) than cells expressing the gene with the CF mutation?

2. Which strain of bacteria entered normal epithelial cells most easily?

3. Entry of all three *S. typhi* strains into the heterozygous epithelial cells was inhibited. Is it possible to tell from this graph which strain was most inhibited?

occurred between 6,000 and 12,000 years ago gave rise to a different allele. This mutation, a single base-pair substitution, changed the 111th amino acid of the transport protein from alanine to threonine. The change results in less melanin—and lighter skin color—than the original African allele does. Today, nearly all people of European descent are homozygous for the mutated allele.

A person of mixed ethnicity may make gametes that contain different combinations of alleles for dark and light skin. It is fairly rare that one of those gametes contains all of the alleles for dark skin, or all of the alleles for light skin, but it happens (Figure 9.12). Skin color is only one of many human traits that vary as a result of single nucleotide mutations. The small scale of such changes offers a reminder that all of us share the genetic legacy of common ancestry.

Figure 9.12 Variation in human skin color begins with differences in alleles.
Twins Kian and Remee are shown with their parents. Both of the children's grandmothers are of European descent, and have pale skin. Both of their grandfathers are of African descent, and have dark skin. The twins inherited different alleles of some genes that affect skin color from their parents, who, given the appearance of their children, must be heterozygous for those alleles.

© Gary Roberts/worldwidefeatures.com.

Take-Home Message 9.4

Do all traits appear in a Mendelian inheritance pattern?

- Some alleles are not dominant or recessive when paired. The heterozygous phenotype may include both homozygous phenotypes (codominance), or it may be a blend of the two homozygous phenotypes (incomplete dominance).
- In some cases, one gene influences multiple traits. In other cases, multiple genes influence the same trait.

A. Under low-oxygen conditions, a water flea switches on genes involved in producing hemoglobin. Producing this red protein enhances the individual's ability to take up oxygen from water. The flea on the left has been living in water with a normal oxygen content; the one on the right, in water with a low oxygen content.

B. The color of the snowshoe hare's fur varies by season. In summer, the fur is brown (left); in winter, white (right). Both forms offer seasonally appropriate camouflage from predators.

C. The height of a mature yarrow plant depends on the elevation at which it grows.

Figure 9.13 Some environmental effects on phenotype.

9.5 Complex Variation in Traits

REMEMBER: Mutations are uncommon in a normal cell (Section 7.6). Expression of a homeotic gene directs formation of a specific body part; environmentally triggered changes in DNA methylation can be permanent and heritable (7.7). Growth factors stimulate cell division and differentiation (8.3). Genetic diversity offers sexual reproducers an advantage in a changing environment (8.4).

The pea plant phenotypes that Mendel studied appeared in two or three forms, which made them easy to track through generations. However, many other traits do not appear in distinct forms. Such traits are often the result of complex genetic interactions—multiple genes, multiple alleles, or both—with added environmental influences. Tracking traits with complex variation presents a special challenge, which is why the genetic basis of many of them has not yet been completely unraveled.

The phrase "nature versus nurture" refers to a centuries-old debate about whether human behavioral traits arise from one's genetics (nature) or from environmental factors (nurture). We now know that both play a substantial role. The environment affects the expression of many genes, which in turn affects phenotype—including behavioral traits. We can summarize this thinking with an equation:

$$\text{genotype} + \textbf{environment} \longrightarrow \text{phenotype}$$

Epigenetics research is revealing that the environment has an even greater contribution to this equation than most biologists had suspected. Environmental cues trigger internal pathways that methylate particular regions of DNA, so they suppress gene expression in those regions. In humans and other animals, DNA methylation patterns can be permanently and heritably affected by diet, stress, and exercise, and also by exposure to drugs and toxins such as tobacco, alcohol, arsenic, and asbestos.

Mechanisms that adjust phenotype in response to external cues are part of an individual's normal ability to adapt to environmental change. Consider the water flea, a tiny aquatic animal that lives in standing pools of fresh water such as seasonal ponds and ditches. In these pools, water conditions such as temperature, oxygen content, and salinity vary dramatically over time and between different parts of the pool. Water fleas have a lot of genes—many more than humans and other animals. Environmental cues trigger adjustments in gene expression that change a flea's form and function to suit its current environment. For example, a water flea that swims to the bottom of a pond can survive the low oxygen conditions there by turning on expression of seven genes involved in the production of hemoglobin—and turning red (Figure 9.13A). The newly produced hemoglobin improves the individual's ability to absorb oxygen from the water. In the presence of insect predators, a water flea makes a protective pointy helmet, a long tail spine, and teeth on its neck. Water fleas can also switch between asexual and sexual modes of reproduction. During early spring, food and space are typically abundant, and competition for these resources is scarce. Under these conditions, water fleas reproduce rapidly by asexual means, giving birth to large numbers of female offspring that quickly fill the pool. Later in the season, competition intensifies as the pool's water becomes warmer, saltier, and more crowded. Then, some of the water fleas start giving birth to males, and the population begins to reproduce sexually. The increased genetic diversity of sexually produced offspring may offer an advantage in the more challenging environment.

In many mammals, seasonal changes in temperature and the length of day affect the production of melanin and other pigments that color skin and fur. These species have different color phases in different seasons (Figure 9.13B). Hormonal signals triggered by the seasonal changes cause fur to be shed, and new fur grows back with different types and amounts of pigments deposited in it. The resulting change in phenotype provides these animals with seasonally appropriate camouflage from predators.

In plants, a flexible phenotype gives immobile individuals an ability to thrive in diverse habitats. For example, genetically identical yarrow plants grow to different heights at different altitudes (Figure 9.13C). More challenging temperature, soil, and water conditions are typically encountered at higher altitudes. Differences in altitude are also correlated with changes in the reproductive mode of yarrow: Plants at higher altitude tend to reproduce asexually, and those at lower altitude tend to reproduce sexually.

Researchers recently discovered several mutations associated with five human psychiatric disorders: autism, depression, schizophrenia, bipolar disorder, and attention deficit hyperactivity disorder (ADHD). However, there must be environmental components to these disorders too, because the majority of people who carry these mutations never end up with a psychiatric disorder. Moreover, one person with the mutations might get one type of disorder, while a relative with the same mutations might get another: two different results from the same genetic underpinnings.

Animal models are helping us unravel some of the mechanisms by which environment can influence mental state. For example, we have discovered that learning and memory are associated with dynamic and rapid DNA modifications in animal brain cells. Mood is, too. Stress-induced depression causes methylation-based silencing of a particular nerve growth factor gene; some antidepressants work by reversing this methylation. As another example, rats whose mothers are not very nurturing end up anxious and having a reduced resilience for stress as adults. The difference between these rats and ones who had nurturing maternal care is traceable to epigenetic DNA modifications that result in a lower than normal level of another nerve growth factor. Drugs can reverse these modifications—and their effects. We do not yet know all of the genes that influence human mental state, but the implication of such research is that future treatments for many psychiatric disorders will involve deliberate modification of methylation patterns in an individual's DNA.

Continuous Variation Some traits occur in a range of small differences that is called **continuous variation**. Continuous variation can be an outcome of epistasis, in which multiple genes affect a single trait. The more genes and environmental factors that influence a trait, the more continuous is its variation. Traits that arise from genes with a lot of alleles may also vary continuously. Some genes have regions of DNA in which a series of 2 to 6 nucleotides is repeated many times in a row. These **short tandem repeats** can spontaneously expand or contract very quickly compared with the typical rate of mutation, and the resulting changes in the gene's DNA sequence may be preserved as alleles. For example, in dogs, short tandem repeats have given rise to 12 alleles of a homeotic gene that influences the length of the face, with longer repeats associated with longer faces (Figure 9.14A).

Human skin color varies continuously, as does human eye color (Figure 9.14B). The colored part of the eye is a doughnut-shaped structure called the iris. Iris color, like skin color, is the result of interactions among gene products that make and distribute melanins. The more melanin deposited in the iris, the less light is reflected from it. Dark irises have dense melanin deposits that absorb almost all light, and reflect almost none. Green and blue eyes have the least amount of melanin, so they reflect the most light.

The environment affects the expression of many genes, which in turn affects phenotype.

A. Face length is a trait that varies continuously in dogs. A gene with 12 alleles influences this trait; all arose by the spontaneous insertion of short tandem repeats. The longer the alleles, the longer the face.

B. Iris color is a trait that varies continuously in humans.

Figure 9.14 Examples of continuous variation.

(A) WilleeCole/Shutterstock.com; (B) from left: first row, © szefel/Shutterstock.com; © Aaron Amat/Shutterstock.com; © Villedieu Christophe/Shutterstock.com; David Goodin; second row, jayfish/Shutterstock.com; © J. Helgason/Shutterstock.com; © Tischenko Irina/Shutterstock.com; © Tatiana Makotra/Shutterstock.com; third row, © Vaaka/Shutterstock.com; © evantravels/Shutterstock.com; © rawcaptured/Shutterstock.com; Anemone/Shutterstock.com; fourth row, © Andrey Armyagov/Shutterstock.com; © Tatiana Makatra/Shutterstock.com; © lightpoet/Shutterstock.com; © Anemone/Shutterstock.com.

continuous variation A range of small differences in a shared trait.

short tandem repeat In chromosomal DNA, a sequence of a few nucleotides repeated multiple times in a row.

bell curve Bell-shaped curve; typically results from graphing frequency versus distribution for a trait that varies continuously.

pedigree Chart of family connections that shows the appearance of a trait through generations.

A. To see if human height varies continuously, male biology students at the University of Florida were divided into categories of one-inch increments in height and counted.

B. Graphing the data that resulted from the experiment in **A** produces a bell-shaped curve, which is evidence that height does vary continuously in humans.

Figure 9.15 How to determine whether a particular trait varies continuously.

(A) Courtesy of Ray Carson, University of Florida News and Public Affairs.

How do we know if a particular trait varies continuously? Let's use another human trait, height, in an example. First, the total range of phenotypes is divided into measurable categories—inches, in this case (Figure 9.15A). The individuals in each category are counted; these counts reveal the relative frequencies of phenotypes across the range of values. Finally, this data is plotted as a bar chart (Figure 9.15B). A graph line around the top of the bars shows the distribution of values for the trait. If the line is a bell-shaped curve, or **bell curve**, then the trait varies continuously.

Take-Home Message 9.5

Do all traits occur in distinct forms?

- Many traits do not occur in distinct forms. The more genes and other factors that influence a trait, the more continuous is its range of variation.
- The environment influences gene expression, and therefore can alter phenotype.

9.6 Human Genetic Analysis

REMEMBER: Sampling error is a difference between results obtained from a subset, and results from the whole (Section 1.6). Environmentally driven epigenetic modifications of DNA—and the resulting gene expression patterns—can be inherited (7.7).

Some organisms, including pea plants and fruit flies, are ideal for genetic studies. They have relatively few chromosomes, they reproduce quickly under controlled conditions, and breeding them in controlled settings poses few ethical problems. It does not take long to follow a trait through many generations. Humans, however, are a different story. We live under variable conditions, in different places, and we live as long as the geneticists who study our inheritance patterns. Most of us select our own mates and reproduce if and when we want to. Our families tend to be on the small side, so sampling error is a major factor in studying them. Because of these and other challenges, geneticists often use historical records to track traits through many generations of a family. They make and study **pedigrees**, standardized charts that illustrate the phenotypes of family members and genetic connections among them (Figure 9.16). Analysis of a pedigree can reveal whether a trait is associated with a dominant or recessive allele, and whether the allele is on an autosome or a sex chromosome. Pedigree analysis also allows geneticists to determine the probability that a trait will reappear in future generations of a family or a population.

Types of Genetic Variation Few easily observed human traits follow Mendelian inheritance patterns. Like the flower color of Mendel's pea plants, these traits arise from a single gene with alleles that have a clear dominance relationship. Consider the gene called *MC1R*. In humans, dogs, and other animals, *MC1R* encodes a protein that triggers production of the brownish melanin. Mutations can result in a defective protein; an allele with one of these mutations is recessive when paired with an unmutated allele. A person who is homozygous for a mutated allele does not make the brownish melanin—only the reddish type—so this individual has red hair. Most other human traits are polygenic, and many have epigenetic contributions that can originate in parents or even grandparents. Environmental effects make these traits even harder to study.

Most of what we know about human genetics comes from research on inherited abnormalities and disorders, because this information helps us develop treatments for affected people. A genetic abnormality is a rare or uncommon version of a trait, such as having six fingers on a hand or a web between two toes. Such abnormalities are not inherently life-threatening, and how you view them is a matter of opinion. By contrast, a genetic disorder sooner or later causes medical problems that may be severe. A genetic disorder is often characterized by a specific set of symptoms (a syndrome). The next section discusses a few genetic disorders and abnormalities that are caused by mutations in single genes. Alleles that give rise to severe genetic disorders are generally rare in populations because they compromise the health and reproductive ability of their bearers. Why do they persist? Mutations periodically reintroduce them. In some cases, a codominant allele offers a survival advantage in a particular environment. You already learned about one example, the CF allele that causes cystic fibrosis. You will see additional examples in later chapters. Keep in mind that single-gene disorders are the least common kind. Far more people are affected by disorders that arise from a complex interplay of multiple genes and environmental factors. Diabetes, asthma, obesity, cancer, heart disease, and multiple sclerosis are like this. Despite intense research, our understanding of these disorders remains incomplete.

Take-Home Message 9.6

How do we study inheritance patterns in humans?

- Human inheritance patterns are often studied by tracking genetic abnormalities or disorders through family trees.
- A genetic disorder is an inherited condition that causes medical problems.
- A genetic abnormality is a rare but harmless version of an inherited trait.
- Only a few genetic disorders are governed by single genes and inherited in a Mendelian pattern. Most human traits are polygenic, and many have epigenetic contributions.

9.7 Human Genetic Disorders

REMEMBER: Thousands of tiny pores span the nuclear envelope; intermediate filaments of lamins support the nuclear envelope, and also help regulate processes inside the nucleus such as DNA replication; basement membrane supports and organizes animal tissues (Section 3.5). Autosomes are the same in both females and males; sex chromosomes differ between the sexes (6.3). An insertion is a mutation in which nucleotides are added to DNA; anemia is a deficiency in red blood cells or hemoglobin (7.6). One of the two X chromosomes in each cell of a female is an inactivated Barr body (7.7). Growth factors stimulate cell division (8.3).

Human genetic disorders inherited in a Mendelian pattern are typically categorized by the chromosome of origin (autosome or sex chromosome) and whether alleles associated with them are dominant or recessive.

A. Standard symbols used in pedigrees.

B. Above, a pedigree for polydactyly, a genetic abnormality in which a person has extra fingers (right), toes, or both. The number of fingers on each hand is indicated in black; toes on each foot, in red. Polydactyly that appears on its own is typically inherited in an autosomal dominant pattern. When part of a syndrome (such as Ellis–van Creveld syndrome), it can be inherited in an autosomal recessive pattern.

C. For more than 30 years, researcher Nancy Wexler has studied the genetic basis of Huntington's disease, an inherited disorder that causes progressive degeneration of the nervous system. Wexler and her team constructed an extended family tree for nearly 10,000 Venezuelans. Their pedigree analysis of relationships among unaffected and affected individuals revealed that a dominant allele on human chromosome 4 is the culprit. Wexler has a personal interest in the disorder: It runs in her family.

Figure 9.16 Pedigrees.

(B) Courtesy of Irving Buchbinder, DPM, DABPS, Community Health Services, Hartford CT; (C) Acey Harper/ Time & Life Pictures/Getty Images.

Table 9.1 Some Autosomal Dominant Traits in Humans

Disorder/Abnormality	Main Symptoms
Achondroplasia	One form of dwarfism
Aniridia	Defects of the eyes
Camptodactyly	Rigid, bent fingers
Familial hypercholesterolemia	High cholesterol, clogged arteries
Huntington's disease	Degeneration of the nervous system
Marfan syndrome	Abnormal or missing connective tissue
Polydactyly	Extra fingers, toes, or both
Progeria	Drastic premature aging
Neurofibromatosis	Tumors of nervous system, skin

The Autosomal Dominant Pattern　A trait associated with a dominant allele on an autosome appears in people who are heterozygous for it as well as those who are homozygous. Table 9.1 lists a few examples. Such traits appear in every generation of a family, and they occur with equal frequency in both sexes. When one parent is heterozygous, and the other is homozygous for the recessive allele, each of their children has a 50 percent chance of inheriting the dominant allele and having the associated trait (Figure 9.17A).

A form of hereditary dwarfism called achondroplasia offers an example of an autosomal dominant disorder (one caused by a dominant allele on an autosome). Mutations associated with achondroplasia occur in a gene for a growth factor receptor. The mutations cause the receptor, which normally slows bone development, to be overly active. About 1 in 10,000 people is heterozygous for one of these mutations. As adults, affected people are, on average, about 4 feet 4 inches (1.3 meters) tall, with arms and legs that are short relative to torso size (Figure 9.17B). An allele that causes achondroplasia can be passed to children because its expression does not interfere with reproduction, at least in heterozygous people. The homozygous condition results in severe skeletal malformations that cause early death.

Huntington's disease is also inherited in an autosomal dominant pattern. Mutations that cause this disorder alter a gene for a cytoplasmic protein whose function is still unknown. The mutations are insertions that occur when a three-nucleotide short tandem repeat expands spontaneously. The oversized protein product of the altered gene gets chopped into pieces inside nerve cells of the brain. The pieces accumulate in cytoplasm as large clumps that eventually prevent the cells from functioning properly. Brain cells involved in movement, thinking, and emotion are particularly affected. Dramatic, involuntary jerking and writhing movements that are symptoms of the most common form of Huntington's appear after age thirty. Affected people die during their forties or fifties. With this and other late-onset disorders, people may reproduce before symptoms appear, so the allele can be passed unknowingly to children.

Figure 9.17 Autosomal dominant inheritance.

(B) © Newcastle Photos and Ivy & Violet Broadhead and family; (C) Photo courtesy of The Progeria Research Foundation and John Hurley.

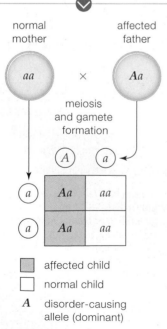

A. A dominant allele on an autosome (red) is fully expressed in heterozygous people.

B. Achondroplasia affects Ivy Broadhead (left), her brother, father, and grandfather.

C. Symptoms of Hutchinson–Gilford progeria are already evident in Megan Nighbor at age 5.

carrier mother   carrier father

Aa × *Aa*

meiosis and gamete formation

A   *a*

	A	*a*
A	*AA*	*Aa*
a	*Aa*	*aa*

☐ affected child
☐ carrier child
☐ normal child

a disorder-causing allele (recessive)

A. Only people homozygous for a recessive allele on an autosome have the associated trait. In this example, both parents are carriers. Each of their children has a 25 percent chance of inheriting two recessive alleles, and being affected by the trait.

B. The albino phenotype is associated with autosomal recessive alleles that cause a deficiency in melanin.

Figure 9.18 Autosomal recessive inheritance.

(B) © Rick Guidotti, Positive Exposure.

Hutchinson–Gilford progeria is an autosomal dominant disorder characterized by drastically accelerated aging. It is usually caused by a mutation that affects lamin A, a protein component of intermediate filaments that support the nuclear envelope. In cells that carry this mutation, the nucleus is grossly abnormal, with improperly assembled nuclear pore complexes and membrane proteins on the wrong side of the nuclear envelope. The function of the nucleus as protector of chromosomes and gateway for transcription is severely impaired, and DNA damage accumulates quickly. The effects are pleiotropic. Outward symptoms begin to appear before age two, as skin that should be plump and resilient starts to thin, muscles weaken, and bones soften. Premature baldness is inevitable (Figure 9.17C). Most people with the disorder die in their early teens as a result of a stroke or heart attack brought on by hardened arteries, a condition typical of advanced age. Progeria does not run in families because affected people do not live long enough to reproduce.

The Autosomal Recessive Pattern A recessive allele on an autosome is expressed only in homozygous individuals, so traits associated with the allele tend to skip generations. They also appear in both sexes at equal frequency. Table 9.2 lists a few examples. Heterozygous individuals are called carriers because they have the allele but not the trait. Any child of two carriers has a 25 percent chance of inheriting the allele from both parents—and developing the trait (Figure 9.18A).

Albinism, a phenotype characterized by an abnormally low level of melanin, is inherited in an autosomal recessive pattern. Mutations associated with albinism affect proteins involved in melanin synthesis. Skin, hair, or eye pigmentation may be reduced or missing. In the most dramatic form, the skin is very white and does not tan, and the hair is white (Figure 9.18B). The irises of the eyes appear red because the lack of pigment allows underlying blood vessels to show through. Melanin also plays a role in the retina, so vision problems are typical. In skin, melanin acts as a sunscreen; without it, the skin is defenseless against UV radiation. Thus, people with the albino phenotype have a very high risk of skin cancer.

With late-onset disorders, people may reproduce before symptoms appear, so the allele can be passed unknowingly to children.

Table 9.2 Some Autosomal Recessive Traits in Humans

Trait	Description
Albinism	Absence of pigmentation
Cystic fibrosis	Abnormally thick mucus damages tissues and organs
Ellis–van Creveld syndrome	Dwarfism, heart defects, polydactyly
Friedreich's ataxia	Progressive loss of motor and sensory function
Hereditary methemoglobinemia	Blue skin coloration
Phenylketonuria (PKU)	Mental impairment
Sickle-cell anemia	Red blood cells can sickle and disrupt circulation
Tay–Sachs disease	Deterioration of mental and physical abilities; early death

polyploid Having three or more of each type of chromosome.

Figure 9.19 Tay–Sachs disease.
Conner Hopf was diagnosed with Tay–Sachs when he was 7 months old. He died before his second birthday.

Courtesy of © Conner's Way Foundation, www.connersway.com.

Inheriting more than two full sets of chromosomes is fatal in humans.

Table 9.3 Some X-Linked Traits in Humans

Disorder or Abnormality	Main Symptoms
Androgen insensitivity syndrome	XY individual has the traits of a female; sterility
Red–green color blindness	Inability to distinguish red from green
Fragile X syndrome	Intellectual, emotional disability
Hemophilia	Impaired blood clotting
Incontinentia pigmenti	Abnormalities of skin, hair, teeth, nails, eyes
Muscular dystrophies	Progressive loss of muscle function
SCID-X1	Severe immune system deficiency
X-linked anhidrotic dysplasia	Mosaic skin; other ill effects

Alleles associated with Tay–Sachs disease are inherited in an autosomal recessive pattern. In the general population, about 1 in 300 people is a carrier for one of these alleles, but the incidence is ten times higher in some groups, such as Jews of eastern European descent. The gene altered in Tay–Sachs encodes a lysosomal enzyme responsible for breaking down a particular type of lipid. Mutations result in an enzyme that misfolds and becomes destroyed, so cells make the lipid but cannot break it down. Typically, newborns homozygous for a Tay–Sachs allele seem normal, but within three to six months they become irritable, listless, and may have seizures as the lipid accumulates in their nerve cells. Blindness, deafness, and paralysis follow. Affected children usually die by age five (Figure 9.19).

Figure 9.20 The human X chromosome.
This chromosome carries about 1,900 genes—almost 10 percent of the total. Most X chromosome alleles that cause genetic disorders are inherited in a recessive pattern. A few disorders are listed (in parentheses).

The X-Linked Recessive Pattern Many genetic disorders are associated with alleles on the X chromosome (Figure 9.20 and Table 9.3). Almost all of them are inherited in a recessive pattern, probably because those caused by dominant X chromosome alleles tend to be lethal in male embryos.

A recessive allele on an X chromosome leaves two clues when it causes a genetic disorder. First, an affected father never passes the disorder to a son, because all children who inherit their father's X chromosome are female (Figure 9.21A). Thus, a heterozygous female is always the bridge between an affected male and his affected grandson. Second, the disorder appears in males more often than in females. This is because all males who carry the allele have the disorder, but not all heterozygous females do. One of the two X chromosomes in each cell of a female is an inactive Barr body, so only about half of a heterozygous female's cells express the recessive allele. The other half of her cells express the dominant, normal allele that she carries on her other X chromosome, and this expression can mask the phenotypic effects of the recessive allele.

Duchenne muscular dystrophy (DMD) is a genetic disorder characterized by progressive muscle degeneration. It is caused by mutations in the X chromosome gene for dystrophin, a protein that links actin microfilaments in cytoplasm to a complex of proteins in the plasma membrane. This complex structurally and functionally links the cell to basement membrane. When dystrophin is absent, the entire protein complex is unstable. Muscle cells, which are subject to stretching, are particularly affected. Their plasma membrane is easily damaged, and they become flooded with calcium ions. Eventually, the cells die and become replaced by fat cells and connective tissue. DMD affects about 1 in 3,500 people; almost all are boys. Symptoms begin between ages three and seven. Anti-inflammatory drugs can slow the progression of the disorder, but there is no cure. When an affected boy is about twelve years old, he will begin to use a wheelchair and his heart will start to fail. Even with the best care, he will probably die before the age of thirty, from a heart disorder or respiratory failure (suffocation).

Hemophilias are genetic disorders in which the blood does not clot properly. Most of us have a blood clotting mechanism that quickly stops bleeding from minor injuries. That mechanism involves two proteins, clotting factor VIII and IX, both products of X chromosome genes. Mutations in these two genes cause two types of hemophilia (A and B, respectively). Males who carry one of these mutations have prolonged bleeding, as do homozygous females (heterozygous females make enough clotting factor to have a clotting time that is close to normal). Affected people bruise easily, but internal bleeding is their most serious problem. Repeated bleeding inside the joints disfigures them and causes chronic arthritis. Today, about 1 in 7,500 people in the general population is affected. That number may be rising because the disorder is now treatable, and more affected people now live long enough to transmit a mutated allele to children.

Color blindness refers to a range of conditions in which an individual cannot distinguish among colors in the spectrum of visible light. These conditions are typically inherited in an X-linked recessive pattern, because most of the genes involved in color vision are on the X chromosome. Humans can sense the differences among 150 colors, and this perception depends on pigment-containing receptors in the eyes. Mutations that result in altered or missing receptors affect color vision. For example, people who have red–green color blindness see fewer than 25 colors because receptors that respond to the red and green wavelengths of light are weakened or absent (Figure 9.21B,C). Some people with red–green color blindness confuse red and green, and others see green as gray.

Take-Home Message 9.7

How do we know whether a trait is associated with a dominant or recessive allele on an autosome or sex chromosome?

- With an autosomal dominant inheritance pattern, anyone with the allele (whether homozygous or heterozygous) has the associated trait. The trait characteristically appears in every generation.
- With an autosomal recessive inheritance pattern, only persons who are homozygous for an allele have the associated trait. The trait tends to skip generations.
- Men who have an X-linked allele have the associated trait, but not all heterozygous women do. Thus, the trait appears more often in men. Men transmit an X-linked allele to their daughters, but not to their sons.

9.8 Chromosome Number Changes

REMEMBER: Researchers risk interpreting their results in terms of what they want to find out; sampling error can be a substantial problem when testing a small group (Section 1.6). Amyloid fibrils form insoluble deposits that disrupt brain function (2.9). Cells that are diploid (2*n*) have two sets of chromosomes (6.3). Meiosis reduces the chromosome number during gamete formation; the fusion of two haploid (*n*) gametes at fertilization restores the diploid chromosome number in the resulting zygote (8.5).

A **polyploid** individual has three or more complete sets of chromosomes. About 70 percent of flowering plant species are polyploid, as are some insects, fishes, and other animals—but not humans. In our species, inheriting more than two full sets of chromosomes is invariably fatal, although some somatic cells are normally polyploid in adult tissues. Aneuploidy—having too many or too few copies of a particular

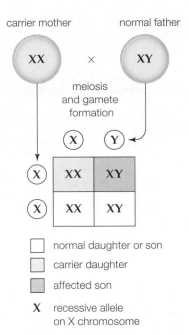

A. In this example of X-linked inheritance, the mother carries a recessive allele on one of her two X chromosomes (red).

B. A view of color blindness. The image on the left shows how a person with red–green color blindness sees the image on the right. The perception of blues and yellows is normal; red and green appear similar.

C. Part of a standardized test for color blindness. A set of 38 of these circles is commonly used to diagnose deficiencies in color perception. You may have one form of red–green color blindness if you see a 7 instead of a 29 in the circle on the left. You may have another form of red–green color blindness if you see a 3 instead of an 8 in the circle on the right.

Figure 9.21 X-linked recessive inheritance.

(B) Photos by Gary L. Friedman, www.FriedmanArchives.com; (C) Life Nature Library, The Primates, 1965, Sarel Eimerl and Irven DeVore.

chromosome—does occur in humans. A few babies (less than 1 percent) are born with a diploid chromosome number that differs from the normal 46. Changes in chromosome number are usually an outcome of **nondisjunction**, the failure of chromosomes to separate properly during mitosis or meiosis. Nondisjunction during meiosis (Figure 9.22) can affect chromosome number at fertilization. For example, if a normal gamete (n) fuses with a gamete that has an extra chromosome ($n+1$), the resulting zygote will have three copies of one type of chromosome and two of every other type ($2n+1$), an aneuploid condition called trisomy. If a normal gamete (n) fuses with a gamete missing a chromosome ($n-1$), the new individual will have one copy of one chromosome and two of every other type ($2n-1$), an aneuploid condition called monosomy. A few disorders associated with aneuploidy are listed in Table 9.4.

Autosomal Change and Down Syndrome In most cases, inheriting the wrong number of autosomes is fatal in humans before birth or shortly thereafter. An important exception is trisomy 21. A person born with three chromosomes 21 will have Down syndrome and a high likelihood of surviving infancy. Mild to moderate mental impairment and health problems such as heart disease are hallmarks of this syndrome. Other effects may include a somewhat flattened facial profile, a fold of skin that starts at the inner corner of each eyelid, white spots on the iris (Figure 9.23), and one deep crease (instead of two shallow creases) across each palm. The skeleton grows and develops abnormally, so older children have short body parts, loose joints, and misaligned bones of the fingers, toes, and hips. Muscles and reflexes are weak, and motor skills such as speech develop slowly. Early training can help these individuals learn to care for themselves and to take part in normal activities. By age 40, all persons with Down syndrome will begin to have symptoms of Alzheimer's disease, a condition of progressive mental deterioration. A gene on chromosome 21 is the culprit. Having three of these chromosomes, a person with Down syndrome makes an excess of the gene's product, a protein that forms the main component of amyloid fibrils characteristic of Alzheimer's. Down syndrome occurs in about 1 of 700 live births, and the risk increases with maternal age.

Change in the Sex Chromosome Number About 1 in 400 human babies is born with an atypical number of sex chromosomes. Most often, such alterations lead to mild difficulties in learning and impaired motor skills such as a speech delay, but these problems may be very subtle.

About 1 in 2,500 girls is born with Turner syndrome, an outcome of having an X chromosome and no corresponding X or Y chromosome (XO). This aneuploid condition is thought to arise most frequently with an unstable Y chromosome inherited from the father. The zygote starts out being genetically male, with an X and a Y chromosome. The Y chromosome breaks up and is lost during early development, so the embryo continues to develop as a female. Affected individuals grow up well proportioned but a bit short, with an average height of 4 feet 8 inches (1.4 meters). Extra folds of skin on the neck, mild skeletal abnormalities, and heart defects are common. The ovaries do not develop properly, so these individuals do not make enough sex hormones to become sexually mature or develop secondary sexual traits such as enlarged breasts. Hormone therapy can trigger sexual development, but even with treatment most girls affected by the syndrome are infertile.

A female may inherit multiple X chromosomes, a condition called triple X syndrome or trisomy X. This syndrome occurs in about 1 of 1,000 births; as with Down syndrome, risk increases with maternal age. Only one X chromosome is typically

Figure 9.22 An example of nondisjunction during meiosis.
Of the two pairs of homologous chromosomes shown here, one fails to separate during anaphase I. The chromosome number is altered in the resulting gametes.

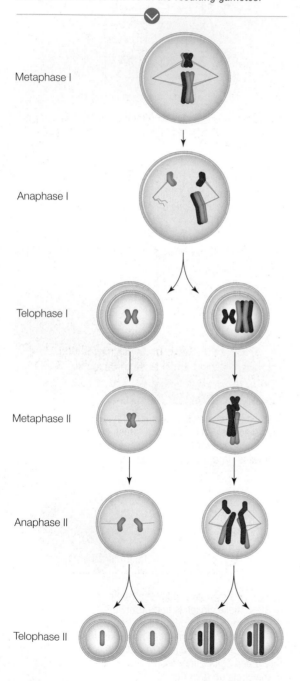

Metaphase I

Anaphase I

Telophase I

Metaphase II

Anaphase II

Telophase II

Figure It Out: During which stage of meiosis does nonjunction occur in this example?

Answer: Anaphase I

A. Example of a Down syndrome genotype.

B. Example of a Down syndrome phenotype. Excess tissue deposits on the iris give rise to a ring of starlike white speckles, a lovely effect of the chromosome number change that causes Down syndrome.

Figure 9.23 Down syndrome, genotype and phenotype.

(A) L. Willatt, East Anglian Regional Genetics Service/Science Source; (B) Ciarra, photo by © Michelle Harmon.

active in female cells, so having extra X chromosomes usually does not cause physical or medical problems, but mild mental impairment may occur.

About 1 out of every 500 males has two or more X chromosomes (XXY, XXXY, and so on). The resulting disorder, Klinefelter syndrome, develops at puberty. As adults, affected males tend to be overweight and tall, with small testes. Underproduction of the hormone testosterone interferes with sexual development and can result in sparse facial and body hair, a high-pitched voice, enlarged breasts, and infertility. Testosterone injections during puberty can minimize some of these traits.

About 1 in 1,000 males is born with an extra Y chromosome (XYY), a result of nondisjunction of the Y chromosome during sperm formation. Adults affected by the resulting XYY syndrome tend to be taller than average, but are within a normal range of phenotype. Sexual development occurs normally, and fertility is normal. Having XYY syndrome was once thought to predispose an individual to a life of crime. This misguided view was based on sampling error (too few cases in narrowly chosen groups such as prison inmates) and bias (the researchers who gathered the karyotypes also took the personal histories of the participants). That view has since been disproven: Men with XYY syndrome are only slightly more likely to be convicted for crimes than other men. Researchers believe this slight increase can be explained by poor socioeconomic conditions related to mild mental impairment of individuals with the syndrome.

Table 9.4 Some Traits Associated With Chromosome Number Changes

Disorder or Abnormality	Main Symptoms
Down syndrome	Mental impairment; heart defects
Turner syndrome (XO)	Sterility; abnormal ovaries, abnormal sexual traits
Klinefelter syndrome (XXY)	Sterility, mental impairment
XXX syndrome	Minimal abnormalities
XYY syndrome	Mild mental impairment

Take-Home Message 9.8

What are the effects of chromosome number changes in humans?

- Polyploidy is fatal in humans, but not in some other organisms.
- Aneuploidy can arise from nondisjunction during meiosis. In humans, most cases of aneuploidy are associated with some degree of mental impairment.

nondisjunction Failure of chromosomes to separate properly during nuclear division.

9.9 Genetic Screening

REMEMBER: Cells make the thousands of different proteins they need from only twenty kinds of amino acid monomers (Section 2.9). During early development, an embryo's cells start using different subsets of their DNA, thus becoming different in form and function—a process called differentiation (6.1).

Studying human inheritance patterns has given us many insights into how genetic disorders arise and progress, and how to treat them. Surgery, prescription drugs, hormone replacement therapy, and dietary controls can minimize and in some cases eliminate the symptoms of a genetic disorder. Some disorders can be detected early enough to start countermeasures before symptoms develop. For these reasons, most hospitals in the United States now screen newborns for mutations that cause phenylketonuria, or PKU. The mutations affect an enzyme that converts one amino acid (phenylalanine) to another (tyrosine). Without this enzyme, the body becomes deficient in tyrosine, and phenylalanine accumulates to high levels. The imbalance inhibits protein synthesis in the brain, which in turn results in severe neurological symptoms. Restricting all intake of phenylalanine can slow the progression of PKU, so routine early screening has resulted in fewer individuals suffering from the symptoms of the disorder.

Parents can also benefit from human genetics studies. The probability that a future child will inherit a genetic disorder can be estimated by testing prospective parents for alleles known to be associated with genetic disorders. Karyotypes and pedigrees are also useful in this type of screening, which can help people make informed decisions about family planning.

Genetic screening is also done post-conception, in which case it is called prenatal diagnosis (prenatal means before birth). Prenatal diagnosis checks an embryo or fetus for physical and genetic abnormalities. Early diagnosis of these conditions gives parents time to prepare for the birth of an affected child, and an opportunity to decide whether to continue with the pregnancy or terminate it. Dozens of conditions are detectable prenatally, including aneuploidy, hemophilia, Tay–Sachs disease, sickle-cell anemia, muscular dystrophy, and cystic fibrosis. If a disorder is treatable, early detection can allow the newborn to receive prompt and appropriate treatment. A few conditions are even surgically correctable before birth.

As an example of how prenatal diagnosis works, consider a woman who becomes pregnant at age thirty-five. Her doctor will probably perform a procedure called obstetric sonography, in which ultrasound waves directed across the woman's abdomen form images of the fetus's limbs and internal organs (Figure 9.24A,B). If the images reveal a physical defect that may be the result of a genetic disorder, a more invasive technique such as fetoscopy would be recommended for further diagnosis. With fetoscopy, sound waves pulsed from inside the mother's uterus yield images much higher in resolution than ultrasound (Figure 9.24C). Samples of tissue or blood are often taken at the same time, and some corrective surgeries can be performed.

Human genetics studies show that our thirty-five-year-old woman has about a 1 in 80 chance that her baby will be born with a chromosomal abnormality, a risk more than six times greater than when she was twenty years old. Thus, even if no abnormalities are detected by ultrasound, she probably will be offered an additional diagnostic procedure, amniocentesis, in which a small sample of fluid is drawn from the amniotic sac enclosing the fetus (Figure 9.25A). The fluid contains cells shed by the fetus, and those cells can be tested for genetic disorders. Chorionic villus sampling (CVS) can be performed earlier than amniocentesis. With this technique,

A. Conventional ultrasound.

B. 4D ultrasound.

C. Fetoscopy.

Figure 9.24 Three ways of imaging a human fetus.

(A) Mediscan/Corbis; (B) Dr. Benoit/Mona Lisa/LooksatSciences/Phototake; (C) © Neil Bromhall/Science Source.

a few cells from the chorion are removed and tested (the chorion is a membrane that surrounds the amniotic sac and helps form the placenta, an organ that allows substances to be exchanged between mother and embryo).

An invasive procedure often carries a risk to the fetus. The risks vary by the procedure. Amniocentesis has improved so much that, in the hands of a skilled physician, it no longer increases the risk of miscarriage. CVS occasionally disrupts the placenta's development and thus causes underdeveloped or missing fingers and toes in 0.3 percent of newborns. Fetoscopy raises the miscarriage risk by a whopping 2 to 10 percent, so it is rarely performed unless surgery or another medical procedure is required before the baby is born.

Couples who discover they are at high risk of having a child with a genetic disorder may opt for reproductive interventions such as *in vitro* fertilization. With this procedure, sperm and eggs taken from the prospective parents are mixed in a test tube. If an egg becomes fertilized, the resulting zygote will begin to divide. In about forty-eight hours, it will have become an embryo that consists of a ball of eight cells (Figure 9.25B). All of the cells in this ball have the same genes, but none has yet committed to being specialized one way or another. Doctors can remove one of these undifferentiated cells and analyze its genes, a procedure called preimplantation diagnosis. The withdrawn cell will not be missed. If the embryo has no detectable genetic defects, it is inserted into the woman's uterus to develop. Most of the resulting "test-tube babies" are born in good health.

amniotic sac

chorion

A. With amniocentesis, a tiny bit of the fluid inside the amniotic sac is removed, and fetal cells that have been shed into the fluid are tested for genetic disorders. Chorionic villus sampling tests cells of the chorion, which is part of the placenta.

B. About 48 hours after *in vitro* fertilization, a human embryo is a ball of eight identical cells. If one cell is removed for genetic analysis, the remaining seven can continue development.

Figure 9.25 Cells tested for prenatal and preimplantation diagnosis.

(A) © Lennart Nilsson/Bonnierforlagen AB; (B) Fran Heyl Associates © Jacques Cohen, computer-enhanced by © Pix Elation.

Take-Home Message 9.9

How do we use what we know about human inheritance?

• Studying inheritance patterns for genetic disorders has helped researchers develop treatments for some of them.

• Genetic testing can provide prospective parents with information about the health of their future children.

Summary

Section 9.1 Symptoms of cystic fibrosis are pleiotropic effects of mutations in one gene. The allele associated with most cases persists at high frequency despite its devastating effects in homozygous people. Carrying the allele may offer heterozygous individuals protection from dangerous gastrointestinal tract infections.

Section 9.2 Gregor Mendel indirectly discovered the role of genes and alleles in inheritance by breeding pea plants and carefully tracking traits of the offspring over many generations. Each gene occurs at a particular location on a chromosome. Individuals with

identical alleles are **homozygous** for the allele. **Heterozygous** individuals have two nonidentical alleles.

A **dominant** allele masks the effect of a **recessive** allele partnered with it on the homologous chromosome. **Genotype** (an individual's particular set of alleles) gives rise to **phenotype** (the individual's observable traits).

Section 9.3 Diploid cells have pairs of genes on homologous chromosomes. The two genes of a pair (which may differ as alleles) are separated from each other during meiosis, so they end up in different gametes. Crossing individuals that breed

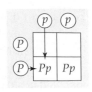

true for two forms of a trait yields offspring that are identically heterozygous for alleles governing the trait. A cross between such offspring is a **monohybrid cross**. The frequency at which the two forms of the trait appear among the offspring of a monohybrid cross can reveal a dominance relationship between the alleles.

In most cases, the two genes of a pair on homologous chromosomes are distributed into gametes independently of other gene pairs on other chromosomes. Crossing individuals that breed true for two forms of two traits yields offspring that are identically heterozygous for alleles governing those traits. A cross between such offspring is a **dihybrid cross**. The frequency at which the two forms of the two traits appear among the offspring of such crosses can reveal dominance relationships between the alleles.

A **Punnett square** can be useful for determining the probability that certain genotypes (and phenotypes) will appear among the offspring of a monohybrid cross or a dihybrid cross.

 Section 9.4 Not all traits have a Mendelian inheritance pattern. With **incomplete dominance**, the phenotype of heterozygous individuals is an intermediate blend of the two homozygous phenotypes. With **codominance**, heterozygous individuals have both homozygous phenotypes. With **epistasis**, two or more genes affect the same trait. With **pleiotropy**, one gene affects two or more traits.

 Section 9.5 Environmental factors can influence phenotype by altering gene expression. A trait that is influenced by multiple genes often occurs in a range of phenotype called **continuous variation**. Continuous variation typically occurs as a **bell curve** in the range of values. Multiple alleles such as those that arise in regions of **short tandem repeats** can give rise to continuous variation.

 Section 9.6 Geneticists study inheritance patterns in humans by tracking genetic disorders and abnormalities through generations of families. A genetic abnormality is an uncommon version of a heritable trait that does not result in medical problems. A genetic disorder sooner or later causes mild or severe medical problems (which often occur in a syndrome). **Pedigrees** can reveal inheritance patterns for alleles that can be predictably associated with specific phenotypes.

 Section 9.7 An allele is inherited in an autosomal dominant pattern if the trait it specifies appears in everyone who carries it, and both sexes are affected with equal frequency. Such traits appear in every generation of families that have the allele. An allele is inherited in an autosomal recessive pattern if the trait it specifies

appears only in homozygous people. Such traits also appear in both sexes equally, but they can skip generations.

An allele is inherited in an X-linked pattern when it occurs on the X chromosome. Most X-linked disorders are inherited in a recessive pattern, and these tend to appear in men more often than in women. Heterozygous women have a dominant, normal allele that can mask the effects of the recessive one; men do not. Men can transmit an X-linked allele to their daughters, but not to their sons. Only a woman can pass an X-linked allele to a son.

 Section 9.8 Occasionally, new individuals end up with the wrong chromosome number. Consequences of such changes range from minor to lethal alterations in form and function.

Chromosome number change is usually an outcome of **nondisjunction**, in which chromosomes fail to separate properly during nuclear division. **Polyploid** individuals have three or more of each type of chromosome. Polyploidy is lethal in humans, but not in flowering plants and some insects, fishes, and other animals.

In humans, most cases of autosomal aneuploidy are lethal. Trisomy 21, which causes Down syndrome, is an exception. Some changes in the number of sex chromosomes result in an impairment in learning and motor skills.

Section 9.9 Prospective parents can use genetic screening to estimate their risk of transmitting a harmful allele to offspring. The procedure involves analysis of parental pedigrees and genotype by a genetic counselor. Amniocentesis and other methods of prenatal testing can reveal a genetic disorder before birth.

Self-Quiz

Answers in Appendix I

1. A heterozygous individual has a _____ for a trait being studied.
   a. pair of identical alleles
   b. pair of nonidentical alleles
   c. haploid condition, in genetic terms

2. An organism's observable traits constitute its _____ .
   a. phenotype   c. genotype
   b. variation   d. pedigree

3. The offspring of the cross *AA* × *aa* are _____ .
   a. all *AA*
   b. all *aa*
   c. all *Aa*
   d. 1/2 *AA* and 1/2 *aa*

4. The probability of a crossover occurring between two genes on the same chromosome _____ .
   a. is unrelated to the distance between them
   b. decreases with the distance between them
   c. increases with the distance between them

5. If one parent is heterozygous for a dominant allele on an autosome and the other parent does not carry the allele, any child of theirs has a _____ chance of being heterozygous.
   a. 25 percent
   b. 50 percent
   c. 75 percent

6. True or false? All traits are inherited in a Mendelian pattern.

7. One gene that affects three traits is an example of _____ .
   a. dominance
   b. codominance
   c. pleiotropy
   d. epistasis

8. _____ in a trait is indicated by a bell curve.
   a. An epigenetic effect
   b. Nondisjunction
   c. Incomplete dominance
   d. Continuous variation

9. Pedigree analysis is necessary when studying human inheritance patterns because _____ .
   a. humans have approximately 20,000 genes
   b. of ethical problems with experimenting on humans
   c. inheritance in humans is more complicated than it is in other organisms
   d. genetic disorders occur only in humans
   e. all of the above

10. A female child inherits one X chromosome from her mother and one from her father. What sex chromosome does a male child inherit from each of his parents?

11. Nondisjunction at meiosis can result in _____ .
    a. base-pair substitutions
    b. aneuploidy
    c. crossing over
    d. pleiotropy

12. True or false? An individual with three or more complete sets of chromosomes is polyploid.

13. Klinefelter syndrome (XXY) is most easily diagnosed by _____ .
    a. pedigree analysis
    b. aneuploidy
    c. karyotyping
    d. a Punnett square

14. Match each example with the best description.
    ____ dihybrid cross   a. *bb*
    ____ monohybrid cross   b. *AaBb* × *AaBb*
    ____ homozygous   c. *Aa*
    ____ heterozygous   d. *Aa* × *Aa*

15. Match the terms appropriately.
    ____ polyploid   a. symptoms of a genetic disorder
    ____ syndrome   b. extra sets of chromosomes
    ____ aneuploidy   c. caused by a short tandem repeat
    ____ Mendelian   d. one extra chromosome
    ____ genotype   e. dominant > recessive
    ____ Huntington's   f. an individual's alleles
         disease

Critical Thinking

1. Mendel crossed a true-breeding pea plant with green pods and a true-breeding pea plant with yellow pods. All offspring had green pods. Which color is recessive?

2. Assuming that independent assortment occurs during meiosis, what type(s) of gametes will form in individuals with the following genotypes?
   a. *AABB*
   b. *AaBB*
   c. *Aabb*
   d. *AaBb*

3. Refer to problem 2. Determine the frequencies of each genotype among offspring from an *AABB* × *aaBB* cross.

4. Duchenne muscular dystrophy (DMD), which is inherited in an X-linked recessive pattern, occurs almost exclusively in males. Suggest why.

5. Heterozygous individuals perpetuate some alleles that have lethal effects in homozygous individuals. A mutated allele (M^L)

Leslie Falteisek/Clacritter Manx.

associated with taillessness in Manx cats is an example (left). Cats homozygous for this allele ($M^L M^L$) typically die before birth due to severe spinal cord defects. In a case of incomplete dominance, cats heterozygous for the M^L allele and the normal, unmutated allele (M) have a short, stumpy tail or none at all. Two $M^L M$ cats mate. What is the probability that any one of their surviving kittens will be heterozygous?

10

BIOTECHNOLOGY

Application ❯ 10.1 **Personal Genetic Testing**

REMEMBER: Some lipoproteins consist of variable amounts and types of proteins and lipids (Section 2.9). A base-pair substitution is a mutation in which one nucleotide replaces another (7.6). *BRCA1* mutations give rise to neoplasms (8.3). Alleles may encode slightly different forms of a gene's product, and such differences influence the details of traits shared by a species (8.4). A heterozygous individual has two nonidentical alleles of a gene (9.2). A person who is homozygous for a mutated allele of the *MC1R* gene makes only the reddish type of melanin, so this individual has red hair; most other human traits are polygenic, and many have epigenetic contributions (9.6). Alzheimer's disease is a condition of progressive mental deterioration (9.8).

Personalized genetic testing is already beginning to revolutionize medicine.

About 99 percent of your DNA is exactly the same as everyone else's. The shared part is what makes you human; the differences make you a unique member of the species. If you compared your DNA with your neighbor's, about 2.97 billion nucleotides of the two sequences would be identical; the remaining 30 million nonidentical nucleotides are sprinkled throughout your chromosomes. The sprinkling is not entirely random because some regions of DNA vary less than others; these conserved regions are of particular interest to researchers because they are most likely to have an essential function. If a conserved sequence does vary among people, the variation tends to be in single nucleotides at a particular location. A base-pair substitution that is carried by a measurable percentage of a population, usually above 1 percent, is called a **single-nucleotide polymorphism**, or **SNP** (pronounced "snip").

Alleles of most genes differ by single nucleotides, and differences in alleles are the basis of the variation in human traits that makes each individual unique. Thus, SNPs account for many of the differences in the way humans look, and they also have a lot to do with differences in the way our bodies work— how we age, respond to drugs, weather assaults by pathogens and toxins, and so on. Finding out which ones you carry has never been easier. Genetic testing companies can extract your DNA from a few drops of spit or a cheek swab, then analyze it using a tiny glass plate called a SNP-chip (Figure 10.1A). Results may include the predicted likelihood of having traits associated with your particular SNPs. For example, the test will probably determine whether you are homozygous for one allele of the *MC1R* gene. If you are, then you have red hair. Few SNPs have such a clear effect, however. Most human traits arise from a complex interplay of genes and environmental factors that we are still unraveling. Thus, although a DNA test can reliably determine an individual's SNPs, it cannot reliably predict the effect of most of those SNPs on the individual.

Consider the lipoprotein particles that carry fats and cholesterol through our bloodstreams. These particles consist of variable amounts and types of lipids and proteins, one of which is specified by the gene *APOE*. About one in four people carries an allele of this gene, *E4*, that increases the risk of developing Alzheimer's disease later in life. If you are heterozygous for this allele, a DNA testing company

Figure 10.1 Personal genetic tests.

(A) Image courtesy of Illumina, Inc.; (B) © Oli Scarff/Getty Images.

A. Only about 1 percent of the 3 billion bases in a person's DNA are unique to the individual. Personal genetic testing companies use chips like this one to analyze their customers' chromosomes for SNPs. This chip reveals which versions of 4,301,331 SNPs occur in the DNA of four individuals at a time.

B. Celebrity Angelina Jolie chose preventive treatment after genetic testing showed she had a very high risk of breast cancer. She carries a *BRCA1* mutation associated with an 87% lifetime risk of developing breast cancer. Even though Jolie did not yet have cancer, she underwent a double mastectomy, thereby reducing her risk of breast cancer to 5%.

cannot tell you whether you will develop Alzheimer's. However, it may report your lifetime risk of developing the disease, which is about 30 percent, as compared with about 14 percent for someone who has no *E4* allele.

What, exactly, does a 30 percent lifetime risk of Alzheimer's disease mean? The number is a probability statistic; it means, on average, 3 of every 10 people who have one *E4* allele eventually get the disease. However, a risk is just that. Not everyone who has an *E4* allele develops Alzheimer's, and not everyone who develops the disease has the allele. Other unknown factors, including epigenetic modifications of DNA, contribute to the disease. We still have a limited understanding of how genes contribute to many health conditions, particularly age-related ones such as Alzheimer's disease. Geneticists believe that it will be at least five to ten more years before genotyping can be used to accurately predict an individual's future health problems. Nonetheless, we are at a tipping point; personalized genetic testing is already beginning to revolutionize medicine. Cancer treatments are now being tailored to fit the genetic makeup of individual patients. People who discover they carry alleles associated with a heightened risk of a medical condition are being encouraged to make lifestyle changes that could delay the condition's onset or prevent it entirely. Preventive treatments based on personal genetics are becoming more common—and more mainstream (Figure 10.1B).

10.2 Finding Needles in Haystacks

REMEMBER: A tracer has a detectable component (Section 2.2). Typical prokaryotes have plasmids that carry a few genes (3.4). Cloning technologies produce identical copies of an organism (6.1). A bacteriophage infects a bacterium by injecting DNA into it; hydrogen bonds between bases hold the two strands of DNA together in a double helix (6.2). Hybridization is the spontaneous establishment of base-pairing between nucleic acid strands; primers are short, single strands of nucleotides; DNA ligase seals gaps in DNA strands during DNA replication (6.4).

Cutting and Pasting DNA In the 1950s, excitement over the discovery of DNA's structure gave way to frustration: No one could determine the order of nucleotides in a molecule of DNA. Identifying a single nucleotide among thousands or millions of others turned out to be a huge technical hurdle. Research in a seemingly unrelated field yielded a solution when Werner Arber, Hamilton Smith, and their coworkers discovered how some bacteria resist infection by bacteriophages. These bacteria have enzymes that chop up any injected viral DNA before it has a chance to integrate into the bacterial chromosome. The enzymes restrict viral growth; hence their name, restriction enzymes. A **restriction enzyme** cuts DNA wherever a specific nucleotide sequence occurs (Figure 10.2). For example, the enzyme *Eco*RI (named after *E. coli*, the bacteria from which it was isolated) cuts DNA at the nucleotide sequence GAATTC ①. Other restriction enzymes cut at different sequences.

The discovery of restriction enzymes allowed researchers to cut chromosomal DNA into manageable chunks. It also allowed them to combine DNA fragments from different organisms. How? Many restriction enzymes, including *Eco*RI, leave single-stranded tails on DNA fragments ②. Researchers realized that complementary tails will base-pair, regardless of the source of the DNA ③. The tails are called "sticky ends," because two DNA fragments stick together when their matching tails base-pair. DNA ligase can be used to seal the gaps between base-paired sticky ends, so continuous DNA strands form ④. Thus, using appropriate restriction enzymes

① The restriction enzyme *Eco*RI recognizes the nucleotide sequence GAATTC in DNA from two sources.

② The enzyme cuts the DNA into fragments. *Eco*RI leaves single-stranded tails ("sticky ends").

③ When the DNA fragments from the two sources are mixed together, matching sticky ends base-pair.

④ DNA ligase joins the base-paired DNA fragments to produce molecules of recombinant DNA.

Figure 10.2 Making recombinant DNA.

restriction enzyme Type of enzyme that cuts DNA at a specific nucleotide sequence.

single-nucleotide polymorphism (SNP) A one-nucleotide DNA sequence variation carried by a measurable percentage of a population.

cloning vector A DNA molecule that can accept foreign DNA and be replicated inside a host cell.

DNA cloning Set of methods that uses living cells to mass-produce targeted DNA fragments.

DNA library Collection of cells that host different fragments of foreign DNA, often representing an organism's entire genome.

genome An organism's complete set of genetic material.

PCR Polymerase chain reaction. Method that rapidly generates many copies of a specific section of DNA.

probe Fragment of DNA or RNA labeled with a tracer; can hybridize with a nucleotide sequence of interest.

recombinant DNA A DNA molecule that contains genetic material from more than one organism.

and DNA ligase, researchers can cut and paste DNA from different sources. The result, a hybrid molecule that consists of genetic material from two or more organisms, is called **recombinant DNA**.

Making recombinant DNA is the first step in **DNA cloning**, a set of laboratory methods that uses living cells to mass-produce specific DNA fragments. Researchers clone a fragment of DNA by inserting it into a **cloning vector**, which is a molecule that can carry foreign DNA into host cells. Bacterial plasmids may be used as cloning vectors, for example (Figure 10.3). When a bacterium reproduces, its offspring inherit a full complement of genetic information—one chromosome plus plasmids. If a plasmid carries a fragment of foreign DNA, that fragment gets copied and distributed to descendant cells along with the plasmid DNA. A host cell into which a recombinant cloning vector has been inserted can be grown in the laboratory (cultured) to yield a huge population of genetically identical cells. Each of these clones contains a copy of the vector and the inserted DNA fragment. The hosted DNA fragment can be harvested in large quantities from the clones.

DNA Libraries The entire set of genetic material—the **genome**—of most organisms consists of thousands of genes. To study or manipulate a single gene, researchers first find it, and then separate it from all of the other genes in a genome. They often begin by cutting an organism's DNA into fragments, and then cloning all the fragments. The result is a set of clones that collectively contain all of the DNA in a genome. A set of cells that hosts various cloned DNA fragments is a **DNA library**.

In DNA libraries, a cell that contains a particular DNA fragment of interest is mixed up with thousands or millions of others that do not—a needle in a genetic haystack. One way to find that one clone among the others involves the use of a **probe**, a fragment of DNA or RNA labeled with a tracer. For example, to find a targeted gene, researchers may use radioactive nucleotides to synthesize a short strand of DNA complementary in sequence to a similar gene. Because the nucleotide sequences of the probe and the gene are complementary, the two can hybridize. When the probe is mixed with DNA from a library, it will hybridize with the gene, but not with other DNA. Researchers can pinpoint a cell that hosts the gene by detecting the label on the probe. That cell is isolated and cultured, and DNA can be extracted in bulk from the cultured cells for research or other purposes.

Figure 10.3 An example of cloning.
Here, a fragment of chromosomal DNA is inserted into a plasmid.

chromosomal DNA

fragments of chromosomal DNA

plasmid cloning vector

cut plasmid

recombinant plasmid

A. A restriction enzyme (gold triangles) cuts a specific nucleotide sequence in chromosomal DNA and also in a plasmid cloning vector.

B. A fragment of chromosomal DNA and the cut plasmid base-pair at their sticky ends. DNA ligase joins the two pieces of DNA, so a recombinant plasmid forms.

C. The recombinant plasmid is inserted into a host bacterial cell. When the cell reproduces, it copies the plasmid along with its chromosome. Each descendant cell receives a plasmid.

Figure 10.4 Two rounds of PCR.
Each cycle of this reaction can double the number of copies of a targeted sequence of DNA. Thirty cycles can make a billion copies.

targeted section

1 DNA (blue) with a targeted sequence is mixed with primers (pink), nucleotides, and heat-tolerant *Taq* DNA polymerase.

2 When the mixture is heated, the double-stranded DNA separates into single strands. When the mixture is cooled, some of the primers base-pair with the DNA at opposite ends of the targeted sequence.

3 *Taq* polymerase begins DNA synthesis at the primers, so it produces complementary strands of the targeted DNA sequence.

4 The mixture is heated again, so all double-stranded DNA separates into single strands. When it is cooled, primers base-pair with the targeted sequence in the original template DNA and in the new DNA strands.

5 Each cycle of heating and cooling can double the number of copies of the targeted DNA section.

PCR The polymerase chain reaction (**PCR**) is a technique used to mass-produce copies of a particular section of DNA without having to clone it in living cells (Figure 10.4). The reaction can transform a needle in a haystack—that one-in-a-million fragment of DNA—into a huge stack of needles with a little hay in it.

The starting material for PCR is any sample of DNA with at least one molecule of a targeted sequence. It might be extracted from a mixture of 10 million different clones, a sperm, a hair left at a crime scene, or a mummy—essentially any sample that has DNA in it.

The PCR reaction is similar to DNA replication. It requires two synthetic DNA primers, each designed to base-pair with one end of the section of DNA to be amplified, or mass-produced **1**. Researchers mix these primers with the starting (template) DNA, nucleotides, and DNA polymerase, then expose the reaction mixture to repeated cycles of high and low temperatures. A few seconds at high temperature disrupts the hydrogen bonds that hold two strands of DNA together, so every molecule of DNA unwinds and becomes single-stranded. As the temperature of the reaction mixture is lowered, the single DNA strands hybridize with the primers **2**.

The DNA polymerases of most organisms denature at the high temperature required to separate DNA strands. The kind that is used in PCR reactions, *Taq* polymerase, is from *Thermus aquaticus*. This bacterial species lives in hot springs and hydrothermal vents, so its DNA polymerase necessarily tolerates heat. *Taq* polymerase, like other DNA polymerases, recognizes hybridized primers as places to start DNA synthesis **3**. Synthesis proceeds along the template strand until the temperature rises and the DNA separates into single strands **4**. The newly synthesized DNA is a copy of the targeted section. When the mixture is cooled, the primers rehybridize, and DNA synthesis begins again. Each cycle of heating and cooling takes only a few minutes, but it can double the number of copies of the targeted section of DNA **5**. Thirty PCR cycles may amplify that number a billionfold.

Take-Home Message 10.2

What techniques allow researchers to study DNA?

- DNA cloning uses living cells to mass-produce particular DNA fragments. Restriction enzymes cut DNA into fragments, then DNA ligase seals the fragments into cloning vectors. Recombinant DNA molecules result.
- A cloning vector that holds foreign DNA can be introduced into a living cell. When the host cell divides, it gives rise to huge populations of genetically identical cells (clones), each of which contains a copy of the foreign DNA.
- Researchers can isolate one gene from the many others in a genome by making a DNA library. A probe can be used to identify one clone that hosts a targeted DNA fragment among many other clones in the library.
- PCR quickly mass-produces copies of a targeted section of DNA.

DNA profiling Identifying an individual by analyzing the unique parts of his or her DNA.

electrophoresis Technique that separates DNA fragments by size.

genomics The study of genomes.

sequencing Method of determining the order of nucleotides in DNA.

10.3 Studying DNA

REMEMBER: A fat is a substance that consists mainly of triglycerides (Section 2.8). James Watson and Francis Crick built the first accurate model of the DNA molecule in 1953 (6.2). In DNA replication, DNA polymerases begin assembling new strands of DNA at hybridized primers; for each molecule of DNA that is copied, two DNA molecules are produced, each a duplicate of the parent—one strand is new, and the other is parental (6.4). A knockout is an organism in which a gene has been deliberately inactivated (7.7). A short tandem repeat is a series of a few nucleotides repeated several times in a row in chromosomal DNA (9.5).

Sequencing the Human Genome Once a fragment of DNA has been isolated (for example by cloning or PCR), researchers can use a technique called **sequencing** to determine the order of nucleotides in it. The most common method uses DNA polymerase. This enzyme is mixed with a primer, nucleotides, and the DNA to be sequenced (the template). Starting at the primer, the polymerase joins the nucleotides into a new strand of DNA, in the order dictated by the sequence of the template. The DNA fragments are then separated by length. In a technique called **electrophoresis**, an electric field pulls the fragments through a semisolid gel. Fragments of different sizes move through the gel at different rates. The shorter the fragment, the faster it moves, because shorter fragments slip through the tangled molecules of the gel faster than longer fragments do. All fragments of the same length move through the gel at the same speed, so they gather into bands. The order of the bands in the gel reflects the sequence of the template DNA (Figure 10.5A).

The sequencing method we have just described was invented in 1975. Ten years later, it had become so routine that scientists began to consider sequencing the entire human genome—all 3 billion nucleotides. Proponents of the idea said it could provide huge payoffs for medicine and research. Opponents said this daunting task would divert attention and funding from more urgent research. It would require 50 years to sequence the human genome given the techniques of the time. However, the techniques continued to improve rapidly, and with each improvement more nucleotides could be sequenced in less time. Automated (robotic) DNA sequencing and PCR had just been invented. Both were still too cumbersome and expensive to be useful in routine applications, but they would not be so for long. Waiting for faster, cheaper technologies seemed the most efficient way to sequence the genome, but just how fast did they need to be before the project should begin?

A few privately owned companies decided not to wait, and started sequencing. One of them intended to determine the genome sequence in order to patent it. The idea of patenting the human genome provoked widespread outrage, but it also spurred commitments in the public sector. In 1988, the National Institutes of Health (NIH) essentially took over the project by hiring James Watson (of DNA structure fame) to head an official Human Genome Project, and providing $200 million per year to fund it. A partnership formed between the NIH and international institutions that were sequencing different parts of the genome. Watson set aside 3 percent of the funding for studies of ethical and social issues arising from the work. He later resigned over a patent disagreement, and geneticist Francis Collins took his place.

Amid ongoing squabbles over patent issues, Celera Genomics formed in 1998. With biologist Craig Venter at its helm, the company intended to commercialize human genetic information. Celera invented faster techniques for sequencing genomic DNA because the first to have the complete sequence had a legal basis for patenting it. The competition motivated the international partnership to accelerate

A. Human DNA sequence data. On this computer screen, the four nucleotides (adenine, thymine, guanine, and cytosine) are color-coded green, red, yellow, and blue. The order of colored bands in each vertical "lane" represents part of the sequence of the template DNA.

B. Today's automated DNA sequencing machines can sequence an individual's entire genome in 2–4 hours, for a cost of about $1,000.

Figure 10.5 Sequencing, a method of determining the nucleotide sequence of a DNA molecule.

(A) Patrick Landmann/Science Source; (B) © Michelle McLoughlin/Reuters/Corbis.

```
758  GATAATCCTGTTTTGAACAAAAGGTCAAATTGCTGAATAGAAA-GTCTTGATTAACTAAAAGATGTACAAAGTGGAATTA  836  Human
752  GATAATCCTGTTTTGAACAAAAGGTCAAATTGCTGAATAGAAA-GTCTTGATTAACTAAAAGATGTACAAAGTGGAATTA  830  Mouse
751  GATAATCCTGTTTTGAACAAAAGGTCAAATTGCTGAATAGAAA-GTCTTGATTAACTAAAAGATGTACAAAGTGGAATTA  829  Rat
754  GATAATCCTGTTTTGAACAAAAGGTCAAATTGCTGAATAGAAA-GTCTTGATTAACTAAAAGATGTACAAAGTGGAATTA  832  Dog
782  GATAATCCTGTTTTGAACAAAAGGTCAAATTGCTGAATAGAAA-GTCTTGATTAACTAAAAGATGTACAAAGTGGAATTA  860  Chicken
758  GATAATCCTGTTTTGAACAAAAGGTCAAATTGCTGAATAGAAA-GTCTTGATTAAGTAAAAGATGTACAAAGTGGAATTA  836  Frog
823  GATAATCCTGTTTTGAACAAAAGGTCAGATTGCTGAATAGAAAAGGCTTGATTAAAGCAGAGATGTACAAAGTGGACGCA  902  Zebrafish
763  GATAATCCTGTTTTGAACAAAAGGTCAAATTGTTGAATAGAGACGCTTTGATAAAGCGGAGGAGGTACAAAGTGGGACC-  841  Pufferfish
```

Figure 10.6 Alignment of a section of genomic DNA from various species.
This is a region of the gene for a DNA polymerase. Nucleotides that differ from those in the human sequence are highlighted. The chance that any two of these sequences would randomly match is 1 in 10^{46}.

its efforts. Then, in 2000, U.S. President Bill Clinton and British Prime Minister Tony Blair jointly declared that the sequence of the human genome could not be patented. Celera kept sequencing anyway, and, in 2001, the competing governmental and corporate teams published about 90 percent of the sequence. In 2003, fifty years after the discovery of the structure of DNA, the sequence of the human genome was officially completed.

Genomics It took 15 years to sequence the human genome for the first time, but the technology has improved so much that sequencing an entire genome now takes a few hours (Figure 10.5B). Anyone can now pay to have their genome sequenced. However, despite our ability to determine the sequence of an individual's genome, it will be a long time before we understand all the information coded within that sequence. The human genome contains a massive amount of seemingly cryptic data. We can decipher some of this data by comparing genomes of different species, the premise being that all organisms are descended from shared ancestors, so all genomes are related to some extent. We see evidence of such genetic relationships simply by comparing the raw sequence data, which, in some regions of DNA, is extremely similar across many species (Figure 10.6).

The study of genomes is called **genomics**, a broad field that encompasses whole-genome comparisons, structural analysis of gene products, and surveys of small-scale variations in sequence. Genomics is providing powerful insights into evolution, and it has many medical benefits. We have learned the function of many human genes by studying their counterparts in other species. For instance, researchers comparing human and mouse genomes discovered a human version of a mouse gene, *APOA5*, that encodes a lipoprotein. Mice with an *APOA5* knockout have four times the normal level of triglycerides in their blood. The researchers then looked for—and found—a correlation between *APOA5* mutations and high triglyceride levels in humans. High triglycerides are a risk factor for coronary artery disease.

DNA Profiling As you learned in Section 10.1, about 99 percent of the human genome sequence is identical in every member of the species. The differences you carry in your DNA make you unique. In fact, those differences are so unique that they can be used to identify you. Identifying an individual by his or her DNA is a method called **DNA profiling**. One DNA profiling method uses SNP-chips (an example is shown in Figure 10.1). A SNP-chip is a tiny glass plate with microscopic spots of DNA stamped on it. The DNA sample in each spot is a short, synthetic single strand with a unique SNP sequence. When an individual's genomic DNA is washed over a SNP-chip, it hybridizes only with DNA spots that have a matching SNP sequence. Probes reveal where the genomic DNA has hybridized—and which of the SNPs are carried by the individual (Figure 10.7).

Figure 10.7 A SNP-chip analysis.
This SNP chip (inset), shown actual size, tests 550,000 SNPs. The small white box indicates the magnified portion shown. Each spot is a region where the individual's genomic DNA has hybridized with one SNP sequence. A red or green dot means that the individual is homozygous for a SNP; a combined signal (yellow dot) indicates heterozygosity.

(Inset) The Sanger Institute. Wellcome Images; (background) Wellcome Trust Sanger Institute.

genetic engineering Process by which deliberate changes are introduced into an individual's genome.

genetically modified organism (GMO) Organism whose genome has been modified by genetic engineering.

transgenic Refers to a genetically modified organism that carries a gene from a different species.

Another method of DNA profiling involves analysis of short tandem repeats in an individual's chromosomes. Short tandem repeats usually occur in the same location in human chromosomes, but the number of times a sequence is repeated in each location differs among individuals. For example, one person's DNA may have fifteen repeats of the nucleotides TTTTC at a certain spot on one chromosome. Another person's DNA may have four repeats of this sequence in the same location. Short tandem repeats slip spontaneously into DNA during replication, and their numbers grow or shrink over generations. Unless two people are identical twins, the chance that they have identical short tandem repeats in even three regions of DNA is 1 in a quintillion (10^{18}), which is far more than the number of people who have ever lived. Thus, an individual's array of short tandem repeats is, for all practical purposes, unique.

Analyzing a person's short tandem repeats begins with PCR, which is used to copy ten to thirteen particular regions of chromosomal DNA known to have repeats. The lengths of the copied DNA fragments differ among most individuals, because the number of tandem repeats in those regions also differs. Thus, electrophoresis can be used to reveal an individual's unique array of short tandem repeats (Figure 10.8).

A. Gray boxes indicate the regions of the individual's DNA that were tested.

B. The number of repeats is shown in a box below each peak. A peak's location on the x-axis corresponds to the length of the DNA fragment amplified (a measure of the number of repeats). Peak size reflects amount of DNA.

Figure 10.8 An individual's (partial) short tandem repeat profile.
Remember, human body cells are diploid. Double peaks appear on a profile when the two members of a chromosome pair carry a different number of repeats.

Figure It Out: How many repeats does this individual have at the Penta D region?

Answer: 12 on one chromosome, and 14 on the other

Short tandem repeat analysis will soon be replaced by full genome sequencing, but for now it continues to be a common DNA profiling method. Geneticists compare short tandem repeats on Y chromosomes to determine relationships among male relatives, and to trace an individual's ethnic heritage. They also track mutations that accumulate in populations over time by comparing DNA profiles of living humans with those of ancient ones. Such studies are allowing us to reconstruct population dispersals that happened in the ancient past.

Short tandem repeat profiles are routinely used to resolve kinship disputes, and as evidence in criminal cases. Within the context of a criminal or forensic investigation, DNA profiling is called DNA fingerprinting. As of January 2014, the database of DNA fingerprints maintained by the Federal Bureau of Investigation (the FBI) contained the short tandem repeat profiles of 10.7 million convicted offenders, and had been used in more than 200,000 criminal investigations. DNA fingerprinting is also used to identify human remains, including the individuals who died in the World Trade Center on September 11, 2001.

Take-Home Message 10.3

How do we use what researchers discover about DNA?

- Improvements in DNA sequencing techniques allowed the human genome sequence to be determined. The technology has improved so much that an individual's genome can now be sequenced in a few hours.
- Analysis of the human genome sequence is yielding new information about our genes and how they work.
- DNA profiling identifies individuals by the unique parts of their DNA.

10.4 Genetic Engineering

REMEMBER: Carrots are orange because they contain the photosynthetic accessory pigment β-carotene (Section 5.2). The DNA sequence of a gene encodes an RNA or protein product; gene expression is the process by which information in a gene guides the assembly of an RNA or protein product (7.2).

Traditional cross-breeding methods can alter genomes, but only if individuals with the desired traits will interbreed. Genetic engineering takes gene-swapping to an entirely different level. **Genetic engineering** is a process by which an individual's genome is deliberately modified. A gene from one species may be transferred to another to produce an organism that is **transgenic**, or a gene may be altered and reinserted into an individual of the same species. Both methods yield a **genetically modified organism**, or **GMO**.

Genetically Modified Microorganisms Most genetic engineering involves yeast and bacteria (Figure 10.9). Both types of cells have the metabolic machinery to make complex organic molecules, and they are easily engineered to produce, for example, medically important proteins. People with diabetes were among the first beneficiaries of such organisms. Insulin for their injections was once extracted from animals, but it provoked an allergic reaction in some people. Human insulin, which does not provoke allergic reactions, has been produced by transgenic *E. coli* since 1982. Slight modifications of the gene have yielded fast-acting and slow-release forms of human insulin.

Genetically engineered microorganisms also make proteins used in foods. For example, enzymes produced by modified microorganisms improve the taste and clarity of beer and fruit juice, slow bread staling, or modify certain fats. Cheese is traditionally made with an enzyme, chymosin, extracted from calf stomachs. Today, almost all cheese is made with calf chymosin produced by transgenic yeast.

Designer Plants As crop production expands to keep pace with human population growth, it places unavoidable pressure on ecosystems everywhere. Irrigation leaves mineral and salt residues in soils. Tilled soil erodes, taking topsoil with it. Runoff clogs rivers, and fertilizer in it causes algae to grow so fast that fish suffocate. Pesticides can be harmful to humans and other animals, including beneficial insects such as bees.

Pressured to produce more food at lower cost and with less damage to the environment, many farmers have begun to rely on genetically engineered crop plants. Genes can be introduced into plant cells by way of electric or chemical shocks, by blasting them with microscopic DNA-coated pellets, or by using *Agrobacterium tumefaciens* bacteria. *A. tumefaciens* carries a plasmid with genes that cause tumors to form on infected plants; hence the name Ti plasmid (for Tumor-inducing). Researchers replace the tumor-inducing genes with foreign or engineered genes, then use the plasmid as a vector to deliver the desired genes into plant cells. Whole plants can be grown from plant cells that integrate a recombinant plasmid into their chromosomes (Figure 10.10).

Many genetically modified crops carry genes that impart resistance to devastating plant diseases and pests. GMO crops such as Bt corn and soy help farmers use smaller amounts of toxic pesticides. Organic farmers often spray their crops with spores of Bt (*Bacillus thuringiensis*), a bacterial species that makes a protein toxic only to some insect larvae. Researchers transferred the gene encoding the Bt protein

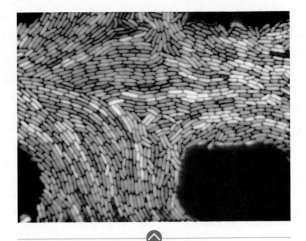

Figure 10.9 Genetically modified bacteria.
These *E. coli* bacteria are transgenic for a fluorescent jellyfish protein. The cells are genetically identical, so the visible variation in fluorescence among them reveals differences in gene expression. Such differences may help us discover why some bacteria of a population become dangerously resistant to antibiotics, and others do not.

Courtesy of Systems Biodynamics Lab, P. I. Jeff Hasty, UCSD Department of Bioengineering, and Scott Cookson.

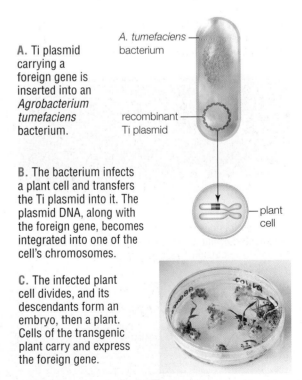

A. Ti plasmid carrying a foreign gene is inserted into an *Agrobacterium tumefaciens* bacterium.

A. tumefaciens bacterium

recombinant Ti plasmid

B. The bacterium infects a plant cell and transfers the Ti plasmid into it. The plasmid DNA, along with the foreign gene, becomes integrated into one of the cell's chromosomes.

plant cell

C. The infected plant cell divides, and its descendants form an embryo, then a plant. Cells of the transgenic plant carry and express the foreign gene.

Figure 10.10 How to make a transgenic plant.

(C) Pascal Goetgheluck/Science Source.

A. The genetically modified plants that produced the corn on the left carry a gene from the bacteria *Bacillus thuringiensis* (Bt) that conferred insect resistance. Compare the corn from unmodified plants, right. No pesticides were used on either crop.

B. Mice transgenic for multiple pigments ("brainbow mice") are allowing researchers to map the complex neural circuitry of the brain. Individual nerve cells in the brain stem of a brainbow mouse are visible in this fluorescence micrograph.

C. Zebrafish genetically modified to glow in places where BPA, an endocrine-disrupting chemical, is present. The fish are literally illuminating where this pollutant acts in the body—and helping researchers discover what it does when it gets there.

Figure 10.11 Some useful GMOs.

(A) The Bt and Non-Bt corn photos were taken as part of field trial conducted on the main campus of Tennessee State University at the Institute of Agricultural and Environmental Research. The work was supported by a competitive grant from the CSREES, USDA titled *Southern Agricultural Biotechnology Consortium for Underserved Communities*, (2000–2005). Dr. Fisseha Tegegne and Dr. Ahmad Aziz served as Principal and Co-principal Investigators respectively to conduct the portion of the study in the State of Tennessee; (B) Courtesy of © Dr. Jean Levit. The Brainbow technique was developed in the laboratories of Jeff W. Lichtman and Joshua R. Sanes at Harvard University. This image has received the Bioscape imaging competition 2007 prize; (C) © Charles Taylor/University of Exeter.

into plants. The engineered plants produce the Bt protein, but otherwise they are essentially identical with unmodified plants. Larvae die shortly after eating their first and only GMO meal. Farmers can use much less pesticide on crops that make their own (Figure 10.11A).

Genetic modifications can make food plants more nutritious. For example, rice plants have been engineered to make β-carotene, an orange photosynthetic pigment that is remodeled by cells of the small intestine into vitamin A. These rice plants carry two genes in the β-carotene synthesis pathway: one from corn, the other from bacteria. One cup of their seeds—grains of Golden Rice—has enough β-carotene to satisfy a child's daily need for vitamin A.

The USDA Animal and Plant Health Inspection Service (APHIS) regulates the introduction of GMOs into the environment. At this writing, APHIS has deregulated ninety-two crop plants, which means the plants are approved for unrestricted use in the United States. Worldwide, more than 330 million acres are currently planted in GMO crops, the majority of which are corn, sorghum, cotton, soy, canola, and alfalfa genetically engineered for resistance to the herbicide glyphosate. Rather than tilling the soil to control weeds, farmers can spray their fields with glyphosate, which kills the weeds but not the GMO crops.

Crops genetically engineered to resist glyphosate have been used in conjunction with the herbicide since the mid-1970s. Genes that confer glyphosate resistance are now appearing in weeds and other wild plants, as well as in unmodified crops—which means that recombinant DNA can (and does) escape into the environment. Glyphosate resistance genes are probably being transferred from transgenic plants to nontransgenic ones via pollen carried by wind or insects.

Biotech Barnyards Genetically modified animals can be produced by injecting DNA into a zygote, and then implanting the resulting embryo in a surrogate to complete its development. Genetically modified mice are invaluable in research (Figure 10.11B). We have discovered the function of many human genes (including the *APOA5* gene discussed in Section 10.3) by inactivating their counterparts in mice. Genetically modified mice are also used as models of human diseases. For example, researchers inactivated the molecules involved in the control of glucose metabolism, one by one. Studying the effects of the knockouts in mice has resulted in much of our current understanding of how diabetes works in humans.

Genetically engineered animals other than mice are also useful in research (Figure 10.11C), and some make molecules that have medical and industrial applications. Various transgenic goats produce proteins used to treat cystic fibrosis, heart attacks, blood clotting disorders, and even nerve gas exposure. Milk from goats transgenic for lysozyme, an antibacterial protein in human milk, may protect infants and children in developing countries from acute diarrheal disease. Goats transgenic for a spider silk gene produce the silk protein in their milk; researchers can spin this protein into nanofibers that have medical and electronics applications. Rabbits make human interleukin-2, a protein that triggers immune cells to divide and is used as a cancer drug. We also engineer food animals. Genetic engineering has given us pigs with heart-healthy fat and environmentally friendly low-phosphate feces, musclebound trout, chickens that do not transmit bird flu, and cows that do not get mad cow disease, among other examples.

Many people think that genetically engineering livestock is unconscionable. Others see it as an extension of thousands of years of acceptable animal husbandry practices. The techniques have changed, but not the intent: We humans continue to have an interest in improving our livestock. Either way, tinkering with the genes of

Digging Into Data

Enhanced Spatial Learning Ability in Mice With an Autism Mutation

Autism is a neurobiological disorder with symptoms that include impaired social interactions and stereotyped patterns of behavior. Around 10 percent of autistic people also have an extraordinary skill or talent such as greatly enhanced memory. Mutations in the gene for neuroligin 3, an adhesion protein that connects brain cells to one another, have been associated with autism. One of these mutations is called *R451C* because the altered gene encodes a protein with an amino acid substitution: a cysteine (C) instead of an arginine (R) in position 451.

In 2007, Katsuhiko Tabuchi and his colleagues introduced the *R451C* mutation into the neuroligin 3 gene of mice. The researchers discovered that the genetically modified mice had impaired social behavior, and also that spacial learning ability was affected.

Spatial learning in mice is tested with a water maze, which consists of a small platform submerged a bit below the surface of a pool of water so it is invisible to swimming mice. Mice do not particularly enjoy swimming, so they try to locate the hidden platform as quickly as they can. When tested again, they remember the platform's location by checking visual cues around the edge of the pool. How quickly they remember is a measure of their spatial learning ability. Figure 10.12 shows some of Tabuchi's results.

a water maze

Figure 10.12 **Spatial learning ability in mice genetically modified to have a mutation associated with autism.**
Performance of genetically modified mice (*R451C*) in a water maze was compared with that of unmodified (wild-type) mice.

1. In the first test, how many days did unmodified mice need to learn to find the location of a hidden platform within 10 seconds?
2. Did the modified or the unmodified mice learn the location of the platform faster in the first test?
3. Which mice learned faster the second time around?
4. Which mice had the greatest improvement in memory?

animals raises a host of ethical dilemmas. Consider animals genetically engineered to carry mutations associated with human diseases—multiple sclerosis, cystic fibrosis, diabetes, cancer, or Huntington's disease, for example. Researchers study these animals in order to understand the diseases, and to test potential treatments, without experimenting on humans. However, the modified animals often suffer the same terrible symptoms of the condition as humans do.

Some worry that our ability to tinker with genetics has surpassed our ability to understand the impact of the tinkering. Controversy raised by GMO use invites you to read the research and form your own opinions. The alternative is to be swayed by media hype (the term "Frankenfood," for instance), or by reports from possibly biased sources (such as herbicide manufacturers).

> Some worry that our ability to tinker with genetics has surpassed our ability to understand the impact of the tinkering.

Take-Home Message 10.4

What is genetic engineering?

- Genetic engineering is the directed alteration of an individual's genome, and it results in a genetically modified organism (GMO).
- A transgenic organism carries a gene from a different species. Transgenic organisms, including bacteria and yeast, are used in research, medicine, and industry.
- Genetically modified crop plants can help farmers be more productive. However, widespread use of these crops is having unintended environmental effects.
- Genetic engineering creates animals that would be impossible to produce using traditional cross-breeding methods.

10.5 Modifying Humans

REMEMBER: A gene that helps transform a normal cell into a tumor cell is an onco-gene; a proto-oncogene can become an oncogene (Section 8.3). An X-linked genetic disorder is associated with an allele on the X chromosome (9.7).

Gene Therapy We know of more than 15,000 serious genetic disorders. Collectively, they cause 20 to 30 percent of infant deaths each year, and account for half of all mentally impaired patients and a fourth of all hospital admissions. They also contribute to many age-related disorders, including cancer, Parkinson's disease, and diabetes. Drugs and other treatments can minimize the symptoms of some genetic disorders, but gene therapy is the only cure. **Gene therapy** is the transfer of recombinant DNA into an individual's body cells, with the intent to correct a genetic disorder or treat a disease. Typically, the transfer inserts an unmutated gene into the individual's chromosomes.

DNA can be introduced into human cells in many ways, for example by direct injection, electrical pulses, lipid clusters, nanoparticles, or genetically engineered viruses. Viruses have molecular machinery that delivers their genomes into cells they infect. Those used as vectors have DNA that splices itself into the infected cells' chromosomes, along with foreign DNA that has been inserted into it.

Human gene therapy is a compelling reason to embrace genetic engineering research. It is now being tested as a treatment for AIDS, muscular dystrophy, heart attack, sickle-cell anemia, cystic fibrosis, hemophilia A, Parkinson's disease, Alzheimer's disease, several types of cancer, and inherited diseases of the eye, the ear, and the immune system.

People have already benefited from gene therapy. Consider SCID-X1, a severe X-linked genetic disorder that stems from a mutated allele of the *IL2RG* gene. The gene encodes a receptor for an immune signaling molecule. Without treatment, people affected by this disorder can survive only in germ-free isolation tents because they cannot fight infections (a diminished life in a sterile isolation tent was the source of the term "bubble boy"). In the late 1990s, researchers used a genetically engineered virus to insert unmutated copies of *IL2RG* into cells taken from the bone marrow of twenty boys with SCID-X1. Each child's modified cells were infused back into his bone marrow. Within months of their treatment, eighteen of the boys left their isolation tents (Figure 10.13A). Gene therapy had repaired their immune systems.

Gene therapy is now being tested as a treatment for acute lymphocytic leukemia, a typically fatal cancer of bone marrow cells. A viral vector is used to insert a gene into immune cells extracted from patients. When the resulting genetically modified cells are reintroduced into the patients' bodies, the inserted gene directs the destruction of cancer cells (Figure 10.13B). The therapy seems to work astonishingly well: In one patient, all traces of the leukemia vanished in eight days.

Despite the successes, manipulating a gene within the context of a living individual is unpredictable even when we know its sequence and location on a chromosome. No one, for example, can predict with absolute certainty where a virus-injected gene will become integrated into a chromosome. Its insertion might disrupt other genes. If it interrupts a gene that is part of the controls over cell division, then cancer might be the outcome. Consider that five of the twenty boys first treated with gene therapy for SCID-X1 developed leukemia, and one of them died. Developers of the gene therapy had wrongly predicted that cancer related to it would be rare. Research now implicates the very gene targeted for repair, especially

A. Rhys Evans, shown here at age 10, was born with SCID-X1. His immune system has been permanently repaired by gene therapy.

B. Six-year-old Emily Whitehead was just days from death when she underwent an experimental gene therapy for acute lympho-cytic leukemia. Two years later, she remains cancer-free.

Figure 10.13 Gene therapy: success stories.

(A) © Huw Evans Agency, Cardiff; (B) © Kari Whitehead Photography.

when combined with the virus that delivered it. The viral DNA preferentially inserted itself into the children's chromosomes at a site near a proto-oncogene. The insertion activated the gene by triggering its transcription, and that is how the leukemia began. Since that time, researchers have used PCR to detect viral integration sites and improve the design of the vector. The development of more efficient and specific viral vectors has reduced the risk associated with all types of gene therapy.

Eugenics Eliminating undesirable human traits is part of **eugenics**, the philosophy of deliberately improving the genetic qualities of the human race. Eugenics has been used as a justification for some of the most horrific episodes in human history, including the genocide of 6 million Jews during World War II; thus, it continues to be a hotly debated social issue. For example, using gene therapy to cure genetic disorders seems like an acceptable goal to most people, but imagine taking this idea a bit further. Would it also be acceptable to engineer the genome of an individual who is within a normal range of phenotype in order to modify a particular trait? Researchers have already produced mice that have improved memory, enhanced learning ability, bigger muscles, and longer lives. Why not people?

Given the pace of genetics research, the debate is no longer about how we would engineer desirable traits, but how we would choose the traits that are desirable. Realistically, cures for many severe but rare genetic disorders will not be found, because the financial return would not cover the cost of the research. Eugenics, however, may be profitable. How much would potential parents pay to be sure that their child will be tall or blue-eyed, with breathtaking strength or intelligence? What about a treatment that can help you lose that extra weight, and keep it off permanently? The gray area between interesting and abhorrent can be very different depending on who is asked. In a survey conducted in the United States, more than 40 percent of those interviewed said they would be fine with using gene therapy to make smarter and cuter babies. In one poll of British parents, 10 percent would use it to keep a child from growing up to be homosexual, and 18 percent would be willing to use it to keep a child from being aggressive.

Some people are concerned that gene therapy puts us on a slippery slope that may result in irreversible damage to ourselves and to the biosphere. We as a society may not have the wisdom to know how to stop once we set foot on that slope; one is reminded of our peculiar human tendency to leap before we look. And yet, something about the human experience allows us to dream of such things as wings of our own making, a capacity that carried us into space. In this brave new world, the questions before you are these: What do we stand to lose if serious risks are not taken? And, do we have the right to impose the potential consequences on people who would choose not to take those risks?

Corbis/SuperStock

The debate is no longer about how we would engineer desirable traits, but how we would choose the traits that are desirable.

Take-Home Message 10.5

Can people be genetically modified?

- Genes can be transferred into a person's cells to correct a genetic disorder or treat a disease. However, the outcome of altering a person's genome remains unpredictable given our current understanding of how the genome works.
- We as a society continue to work our way through the ethical implications of applying DNA technologies.

eugenics Idea of deliberately improving the genetic qualities of the human race.

gene therapy Treating a genetic defect or disorder by transferring a normal or modified gene into the affected individual.

Summary

Section 10.1 Personal genetic testing, which reveals a person's unique array of **single-nucleotide polymorphisms**, or **SNPs**, is beginning to revolutionize the way medicine is practiced.

Section 10.2 In **DNA cloning**, researchers use **restriction enzymes** to cut a sample of DNA into fragments, and then use DNA ligase to splice the fragments into plasmids or other **cloning vectors**. The resulting molecules of **recombinant DNA** are inserted into host cells such as bacteria. Division of host cells produces huge populations of genetically identical descendant cells (clones), each with a copy of the cloned DNA fragment.

A **DNA library** is a collection of cells that contain different fragments of DNA, often representing an organism's entire **genome**. Researchers can use **probes** to identify cells that carry a specific fragment of DNA. The polymerase chain reaction (**PCR**) uses primers and a heat-resistant DNA polymerase to rapidly increase the number of copies of a targeted section of DNA.

Section 10.3 Advances in **sequencing**, which reveals the order of nucleotides in DNA, allowed the DNA sequence of the entire human genome to be determined. DNA polymerase is used to partially replicate a DNA template. The reaction produces a mixture of DNA fragments of all different lengths; **electrophoresis** separates the fragments by length into bands.

Genomics gives us insights into the function of the human genome. Similarities between genomes of different organisms are evidence of evolutionary relationships, and can be used as a predictive tool in research. **DNA profiling** identifies a person by the unique parts of his or her DNA. An example is the determination of an individual's unique array of SNPs or short tandem repeats. Within the context of a criminal investigation, a DNA profile is called a DNA fingerprint.

Section 10.4 Recombinant DNA technology is the basis of **genetic engineering**, the directed modification of an organism's genetic makeup with the intent to modify its phenotype. A gene from one species is inserted into an individual of a different species to make a **transgenic** organism, or a gene is modified and reinserted into an individual of the same species. The result of either process is a **genetically modified organism** (**GMO**).

Bacteria and yeast, the most common genetically engineered organisms, produce proteins that have medical value. The majority of the animals that are being created by genetic engineering are used for medical applications or research. Most transgenic crop plants, which are now in widespread use worldwide, were created to help farmers produce food more efficiently. Some crops are genetically modified to have enhanced nutritional value.

Section 10.5 With **gene therapy**, a gene is transferred into body cells to correct a genetic defect or treat a disease. Potential benefits of genetically modifying humans must be weighed against potential risks. The practice raises ethical issues such as whether **eugenics** would be desirable in some circumstances.

Self-Quiz

1. _____ cut(s) DNA molecules at specific sites.
   a. DNA polymerase   c. Restriction enzymes
   b. DNA probes   d. DNA ligase

2. A _____ is a molecule that can be used to carry a fragment of DNA into a host organism.
   a. cloning vector   c. GMO
   b. chromosome   d. cDNA

3. For each species, all _____ in the complete set of chromosomes is/are the _____ .
   a. genomes; genotype   c. SNPs; genome
   b. DNA; genome   d. transgenics; GMOs

4. A set of cells that host various DNA fragments collectively representing an organism's entire set of genetic information is called a _____ .
   a. genome   d. GMO
   b. DNA library   e. single-nucleotide
   c. clone   polymorphism

5. PCR can be used _____ .
   a. to increase the number of specific DNA fragments
   b. in DNA profiling
   c. to modify a human genome
   d. as a cloning vector
   e. a and b are correct
   f. all of the above

6. Fragments of DNA can be separated by electrophoresis according to _____ .
   a. sequence   c. species
   b. length   d. SNPs

7. *Taq* polymerase is used for PCR because it _____ .
   a. tolerates the high temperature needed to separate DNA strands
   b. is an enzyme from a bacterium
   c. does not require primers
   d. is genetically modified

8. _____ is a technique to determine the order of nucleotides in a fragment of DNA.
   a. PCR
   b. Sequencing
   c. Electrophoresis
   d. Nucleic acid hybridization

9. Which of the following can be used to carry foreign DNA into host cells? Choose all of the correct answers.
   a. RNA
   b. viruses
   c. PCR
   d. plasmids
   e. lipid clusters
   f. blasts of microscopic pellets
   g. a DNA library
   h. sequencing

10. Put the following tasks in the order they would occur during a cloning experiment.
    a. using DNA ligase to seal DNA fragments into vectors
    b. using a probe to identify a clone in the library
    c. using DNA polymerase to sequence the DNA of the clone
    d. making a DNA library of clones
    e. cutting genomic DNA with restriction enzymes

11. A transgenic organism _____ .
    a. carries a gene from another species
    b. has been genetically modified
    c. both a and b

12. True or false? A transgenic organism can pass its foreign gene to offspring.

13. _____ can correct a genetic defect in an individual.
    a. Cloning vectors
    b. Gene therapy
    c. Xenotransplantation
    d. a and b

14. True or false? Some humans are genetically modified.

15. Match the terms with the best description.
    ____ DNA profile
    ____ Ti plasmid
    ____ eugenics
    ____ SNP
    ____ transgenic
    ____ GMO
    a. GMO with a foreign gene
    b. alleles commonly have them
    c. a person's unique collection of short tandem repeats
    d. selecting "desirable" traits
    e. genetically modified
    f. used in plant gene transfers

Critical Thinking

1. Restriction enzymes in bacterial cytoplasm cut injected bacteriophage DNA wherever certain sequences occur. Why do you think these enzymes do not chop up the bacterial chromosome, which is exposed to the enzymes in cytoplasm?

2. In 1918, an influenza pandemic that originated with avian flu killed 50 million people. Researchers isolated samples of that virus from bodies of infected people preserved in Alaskan permafrost since 1918. From the samples, they sequenced the viral genome, then reconstructed the virus. The reconstructed virus is 39,000 times more infectious than modern influenza strains, and 100 percent lethal in mice.

   Understanding how this virus works can help us defend ourselves against other strains that may arise. For example, discovering what makes it so infectious and deadly would help us design more effective vaccines. Critics of the research are concerned: If the virus escapes the containment facilities (even though it has not done so yet), it might cause another pandemic. Worse, the published DNA sequence and methods to make the virus could be used for criminal purposes. Do you think this research makes us more or less safe?

Visual Question

Table 10.1 shows the results of a paternity test using short tandem repeats. Who's the daddy? How sure are you?

Table 10.1 Paternity Test Results

Marker	DNA Samples Tested			
	Mother	Baby	Alleged Father #1	Alleged Father #2
CSF1PO	15, 17	17, 23	23, 27	17, 15
FGA	9, 9	9, 9	9, 12	9, 12
THO1	29, 29	29, 27	27, 28	29, 28
TPOX	14, 18	18, 20	15, 20	17, 22
VWA	14, 14	14, 14	14, 14	14, 16
D3S1358	11, 14	14, 16	12, 16	14, 20
D5S818	11, 13	10, 13	8, 10	18, 18
D7S820	7, 13	13, 13	13, 19	13, 13
D8S1179	13, 13	13, 15	12, 15	10, 12
D13S317	12, 12	10, 12	8, 10	12, 17
D16S539	12, 14	14, 12	14, 14	18, 25
D18S51	5, 6	6, 22	22, 6	5, 22
D21S11	15, 17	17, 22	15, 22	22, 22

EVIDENCE OF EVOLUTION

Application ◉ 11.1 **Reflections of a Distant Past**

How do you think about time? Perhaps you can conceive of a few hundred years of human events, maybe a few thousand, but how about a few million? Envisioning the very distant past requires an intellectual leap from the familiar to the unknown. One way to make that leap involves, surprisingly, asteroids. Asteroids are small planets hurtling through space. They range in size from 1 to 1,500 kilometers (roughly 0.5 to 1,000 miles) across. Millions of them orbit the sun between Mars and Jupiter—cold, stony leftovers from the formation of our solar system. Asteroids do not emit light, so they are difficult to see even with the best telescopes. Thus, asteroids can pass very close to Earth undetected, and many are discovered as they do. Some have not passed us by at all.

Consider the mile-wide Barringer Crater in Arizona (Figure 11.1A). An asteroid 45 meters (150 feet) wide made this impressive pockmark in the desert sandstone when it slammed into Earth 50,000 years ago, its impact 150 times more powerful than the bomb that leveled Hiroshima. No humans were in North America at the time of the impact. If there were no witnesses, how is it possible to know anything about what happened? We often reconstruct history by studying physical evidence of past events. Geologists were able to infer the most probable cause of the Barringer Crater by analyzing tons of meteorites, melted sand, and other rocky clues at the site.

Similar evidence points to even larger impacts in the more distant past. For example, fossil hunters have long known about a mass extinction, or permanent loss of major groups of organisms, that occurred 66 million years ago. The event is marked by an unusual, worldwide formation of sedimentary rock (Figure 11.1B) called the K–Pg boundary sequence (it was formerly known as the K–T bound-

A. What made the Barringer Crater? Rocky evidence points to a 300,000-ton asteroid that collided with Earth 50,000 years ago.

ary). There are plenty of dinosaur fossils below this formation. Above it, in layers of rock that were deposited more recently, there are no dinosaur fossils, anywhere. The formation consists of an unusual clay that is rich in iridium, an element rare on Earth's surface but common in asteroids. It also contains shocked quartz (left) and small glass spheres called tektites—rocks that form when quartz or sand (respectively) undergoes a sudden, violent application of extreme pressure. The only processes on Earth known to produce these minerals are nuclear explosions and asteroid impacts.

Geologists concluded that the K–Pg boundary layer must have originated with extraterrestrial material, and began looking for evidence of an asteroid that hit Earth 66 million years ago—one big enough to cover the entire planet with its debris. Twenty years later, they found it: an impact crater the size of Ireland off the coast of the Yucatán Peninsula. To make a crater this big, an asteroid 20 km (12 miles) wide would have slammed into Earth with a force *40 million times* more powerful than the one that made the Barringer Crater—enough to cause an ecological disaster of sufficient scale to wipe out almost all life on Earth.

B. The K–Pg boundary sequence is an unusual, world-wide formation of sedimentary rock that formed 66 million years ago. This rock formation marks an abrupt transition in the fossil record that implies a mass extinction of the dinosaurs. It also contains materials consistent with an asteroid impact. The impact of an asteroid big enough to blanket the Earth with its debris would have had catastrophic aftereffects on life at the time. The red pocketknife gives an idea of scale.

Figure 11.1 From evidence to inference.

You are about to make an intellectual leap through time, to places that were not even known a few centuries ago. We invite you to launch yourself from this premise: Natural phenomena that occurred in the past can be explained by the same physical, chemical, and biological processes that operate today. That premise is the foundation for scientific research into the history of life. The research represents a shift from experience to inference—from the known to what can only be surmised—and it gives us astonishing glimpses into the distant past.

A. Ostrich, native to Africa.

B. Rhea, native to South America.

C. Emu, native to Australia.

Figure 11.2 Similar-looking, related species native to distant geographic realms.
These birds are unlike most others in several unusual features, including long, muscular legs and an inability to fly. All are native to open grassland regions about the same distance from the equator.

(A) Rebecca Yale/Getty Images; (B) © Nico Stengert/Novarc Images/Alamy; (C) © Earl & Nazima Kowall/Corbis.

11.2 Confusing Discoveries

REMEMBER: Naming species in a consistent way became a priority in the 18th century (Section 1.4). Scientific discoveries may provoke controversy when a society's moral standards are interwoven with its understanding of nature (1.6).

About 2,300 years ago, the Greek philosopher Aristotle described nature as a continuum of organization, from lifeless matter through complex plants and animals. Aristotle's work greatly influenced later European thinkers, who adopted his view of nature and modified it in light of their own beliefs. By the fourteenth century, Europeans generally believed that a "great chain of being" extended from the lowest form of life (plants), up through animals, humans, and spiritual beings. Each link in the chain was a species, and each was said to have been forged at the same time, in one place, and in a perfect state. The chain was complete. Because everything that needed to exist already did, there was no room for change.

In the 1800s, European naturalists embarked on globe-spanning survey expeditions and brought back tens of thousands of plants and animals from Asia, Africa, North and South America, and the Pacific Islands. Each newly discovered species was carefully catalogued as another link in the chain of being. The explorers began to see patterns in where species live and similarities in body plans, and had started to think about natural forces that shape life. These explorers were pioneers in **biogeography**, the study of patterns in the geographic distribution of species and communities. Some of the patterns raised questions that could not be answered within the framework of prevailing belief systems. For example, globe-trotting explorers had discovered plants and animals living in extremely isolated places. The isolated species looked suspiciously similar to species living on the other side of impassable mountain ranges, or across vast expanses of open ocean. Consider the emu, rhea, and ostrich, three types of bird native to three different continents. These birds share a set of unusual features (Figure 11.2). Alfred Wallace, an explorer particularly interested in the geographical distribution of animals, thought that the shared traits might mean that the birds descended from a common ancestor (and he was correct), but he had no idea how they could have ended up on different continents.

Naturalists of the time also had trouble classifying organisms that are very similar in some features, but different in others. For example, both plants shown in Figure 11.3 live in hot deserts where water is seasonally scarce. Both have rows of sharp spines that deter herbivores, and both store water in their thick, fleshy stems.

Figure 11.3 Similar-looking, unrelated species.
On the left, saguaro cactus (*Carnegiea gigantea*), native to the Sonoran Desert of Arizona. On the right, an African milk barrel cactus (*Euphorbia horrida*), native to the Great Karoo desert of South Africa.

Left, © Marka/SuperStock; right, © Richard J. Hodgkiss, www.succulent-plant.com.

biogeography Study of patterns in the geographic distribution of species and communities.

Figure 11.4 A vestigial structure: human tailbones.
Nineteenth-century naturalists were well aware of—but had trouble explaining—body structures such as human tailbones that had apparently lost most or all function.

© Zephyr/Science Photo Library/Science Source.

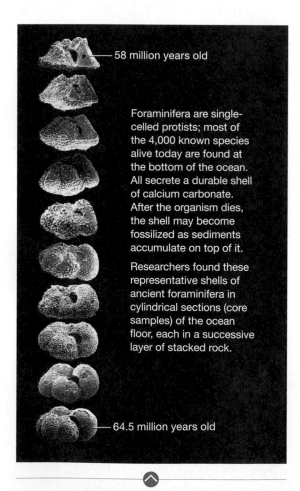

58 million years old

Foraminifera are single-celled protists; most of the 4,000 known species alive today are found at the bottom of the ocean. All secrete a durable shell of calcium carbonate. After the organism dies, the shell may become fossilized as sediments accumulate on top of it.

Researchers found these representative shells of ancient foraminifera in cylindrical sections (core samples) of the ocean floor, each in a successive layer of stacked rock.

64.5 million years old

Figure 11.5 Sequence of ten fossil foraminifera.

Courtesy of Daniel C. Kelley, Anthony J. Arnold, and William C. Parker, Florida State University Department of Geological Science.

However, their reproductive parts are very different, so these plants cannot be (and are not) as closely related as their outward appearance might suggest.

Observations such as these are examples of **comparative morphology**, the study of anatomical patterns: similarities and differences among the body plans of organisms. Today, comparative morphology is only one branch of taxonomy, but in the nineteenth century it was the only way to distinguish species. In some cases, comparative morphology revealed anatomical details (body parts with no apparent function, for example) that added to the mounting confusion. If every species had been created in a perfect state, then why were there useless parts such as wings in birds that do not fly, eyes in moles that are blind, or remnants of a tail in humans (Figure 11.4)?

Fossils were puzzling too. A **fossil** is physical evidence—remains or traces—of an organism that lived in the ancient past. Geologists mapping rock formations exposed by erosion or quarrying had discovered identical sequences of rock layers in different parts of the world. Deeper layers held fossils of simple marine life. Layers above those held similar but more complex fossils (Figure 11.5). In higher layers, fossils that were similar but even more complex resembled modern species. What did these sequences mean? Fossils of many animals unlike any living ones were also being unearthed. If these animals had been perfect at the time of creation, then why had they become extinct?

Taken as a whole, the accumulating discoveries from biogeography, comparative morphology, and geology did not fit with prevailing beliefs of the nineteenth century. If species had not been created in a perfect state (and extinct species, fossil sequences, and "useless" body parts implied that they had not), then perhaps species had indeed changed over time.

Take-Home Message 11.2

Why did observations of nature change our thinking in the nineteenth century?

- Increasingly extensive observations of nature in the nineteenth century did not fit with prevailing belief systems.
- Cumulative findings from biogeography, comparative morphology, and geology led naturalists to question traditional ways of interpreting the natural world.

11.3 A Flurry of New Ideas

REMEMBER: Epigenetic modifications of DNA that alter gene expression may persist for generations (Section 7.7). Paired genes on homologous chromosomes may vary as alleles; alleles are the basis of differences in traits shared by a species (8.4). The environment influences gene expression, and therefore affects phenotype (9.5).

Squeezing New Evidence Into Old Beliefs In the nineteenth century,

naturalists were faced with increasing evidence that life on Earth, and even Earth itself, had changed over time. Around 1800, Georges Cuvier (left), an expert in zoology and paleontology, was trying to make sense of the new information. He knew that many fossil species seemed to have no living counterparts. Given this evidence, he proposed an idea

---

Figure 11.6 Charles Darwin and the _Beagle_.
With Darwin (left) aboard as ship's naturalist, the _Beagle_ (middle) originally set sail to map the coast of South America, but ended up circumnavigating the globe. Right, the path of the voyage is shown from red to blue.

Darwin's detailed observations of the geology, fossils, plants, and animals he encountered on this expedition changed the way he thought about evolution.

Left, painting by George Richmond; middle, © Gordon Chancellor.

startling for the time: Many species that had once existed were now extinct. Cuvier also knew about evidence that Earth's surface had changed. For example, he had seen fossilized seashells in rocks at the tops of mountains far from modern seas. Like most others of his time, he assumed Earth's age to be in the thousands, not billions, of years. He reasoned that geologic forces unlike any known at the time would have been necessary to raise seafloors to mountaintops in such a short time span. These catastrophic geological events would have caused extinctions, after which surviving species repopulated the planet. Cuvier's idea came to be known as catastrophism. We now know that catastrophism is incorrect; geologic processes have not changed over time.

Another naturalist, Jean-Baptiste Lamarck, was thinking about processes that might drive **evolution**, or change in a line of descent. A line of descent is also called

a **lineage**. Lamarck (left) thought that a species gradually improved over generations because of an inherent drive toward perfection, up the chain of being. The drive directed an unknown "fluida" into body parts needing change. By Lamarck's hypothesis, environmental pressures cause an internal requirement for change in an individual's body, and the resulting change is inherited by offspring. Imagine using Lamarck's hypothesis to explain why a giraffe's neck is very long. You might predict that some short-necked ancestor of the modern giraffe stretched its neck to browse on leaves beyond the reach of other animals. The stretches may have even made its neck a bit longer. By Lamarck's hypothesis, the animal's offspring would inherit a longer neck. The modern giraffe would have been the result of many generations that strained to reach ever loftier leaves. Lamarck was correct in thinking that environmental factors affect traits, but his understanding of how inheritance works was incomplete.

Darwin and the HMS _Beagle_   Lamarck's ideas about evolution influenced the thinking of Charles Darwin, who, at the age of 22, joined a survey expedition to South America on a ship called _Beagle_. Since he was eight years old, Darwin had wanted to hunt, fish, collect shells, or watch insects and birds—anything but sit in school. After a failed attempt to study medicine in college, he earned a degree in theology from Cambridge. During his studies, Darwin had spent most of his time with faculty members and other students who embraced natural history.

The _Beagle_ set sail for South America in December 1831 (Figure 11.6). The young man who had hated school and had no formal training in science quickly became an enthusiastic naturalist. During the _Beagle_'s five-year voyage, Darwin found many unusual fossils, and saw diverse species living in environments that ranged from the sandy shores of remote islands to plains high in the Andes. Along

> Accumulating discoveries in the nineteenth century did not fit with prevailing beliefs.

comparative morphology The scientific study of similarities and differences in body plans.

evolution Change in a line of descent.

fossil Physical evidence of an organism that lived in the ancient past.

lineage Line of descent.

the way, he read the first volume of a new and popular book, Charles Lyell's *Principles of Geology*. Lyell (left) was a proponent of what became known as the theory of uniformity, the idea that gradual, repetitive change had shaped Earth. For many years, geologists had been chipping away at the sandstones, limestones, and other types of rocks that form from accumulated sediments at the bottom of lakes, rivers, and oceans. These rocks held evidence that the gradual processes of geologic change operating in the present were the same ones that operated in the distant past.

By the theory of uniformity, strange catastrophes were not necessary to explain Earth's surface. Gradual, everyday geologic processes such as erosion by wind and water could have sculpted Earth's current landscape over great spans of time. This theory challenged the prevailing belief that Earth was 6,000 years old. According to traditional scholars, people had recorded everything that happened in those 6,000 years—and in all that time, no one had mentioned seeing a species evolve. However, by Lyell's calculations, it must have taken millions of years to sculpt Earth's surface. Darwin's exposure to Lyell's ideas gave him insights into the geologic history of the regions he would encounter on his journey. Was millions of years enough time for species to evolve? Darwin thought that it was.

A Key Insight—Variation in Traits   Among the thousands of specimens Darwin collected on his voyage and sent to England were fossil glyptodons from Argentina. These armored mammals are extinct, but they have many traits in common with modern armadillos. Like glyptodons, armadillos have helmets and protective shells that consist of unusual bony plates (Figure 11.7). Armadillos also live only in places where glyptodons once lived. Could the shared traits and distribution mean that glyptodons were ancient relatives of armadillos? If so, perhaps traits of their common ancestor had changed in the line of descent that led to armadillos. But why would such changes have occurred?

After Darwin returned to England, he pondered his notes and fossils, and read an essay by one of his contemporaries, economist Thomas Malthus. Malthus (left) had correlated increases in the size of human populations with episodes of famine, disease, and war. He proposed the idea that humans run out of food, living space, and other resources because they tend to reproduce beyond the capacity of their environment to sustain them. When that happens, the individuals of a population must either compete with one another for the limited resources, or develop new technologies to increase productivity. Darwin realized that Malthus's ideas had wider application: All populations, not just human ones, must have the capacity to produce more individuals than their environment can support.

Reflecting on his journey, Darwin started thinking about how individuals of a species often vary a bit in the details of shared traits such as size, coloration, and so on. He saw such variation among finch species on isolated islands of the Galápagos archipelago. This island chain is separated from South America by 900 kilometers (550 miles) of open ocean, so most species living on the islands did not have the opportunity for interbreeding with mainland populations. The Galápagos island birds resembled finch species in South America, but many had unique traits that suited their particular island habitat.

Darwin was familiar with dramatic variations in traits that selective breeding could produce in pigeons, dogs, and horses. He recognized that a natural environment could similarly select traits that make individuals of a population suited to it. It dawned on Darwin that having a particular form of a shared trait might give an

A. A modern armadillo, about a foot long.

B. Fossil of a glyptodon, an automobile-sized mammal that existed from 2 million to 15,000 years ago.

Figure 11.7 Ancient relatives: armadillo and glyptodon. Even though these animals are widely separated in time, they share a restricted distribution and unusual traits, including a shell and helmet of keratin-covered bony plates—a material similar to crocodile and lizard skin. (The fossil in **B** is missing its helmet.) Their similarities were a clue that helped Darwin develop the theory of evolution by natural selection.

(A) © John White; (B) 2004 Arent.

individual an advantage over competing members of its species. In any population, some individuals have forms of shared traits that make them better suited to their environment than others. In other words, individuals of a natural population vary in fitness. Today, we define **fitness** as the degree of adaptation to a specific environment, and measure it by relative genetic contribution to future generations. A form of a heritable trait that enhances an individual's fitness is called an **adaptive trait**, or evolutionary **adaptation**.

Over many generations, individuals that have adaptive traits tend to survive longer and reproduce more than their less fit rivals. Darwin understood that this process, which he called **natural selection**, could be a mechanism by which evolution occurs. If an individual has a form of a trait that makes it better suited to an environment, then it is better able to survive. If an individual is better able to survive, then it has a better chance of living long enough to produce offspring. If individuals with an adaptive form of a trait produce more offspring than those that do not, then the frequency of that form will tend to increase in the population over successive generations. Table 11.1 summarizes this reasoning in modern terms.

Great Minds Think Alike Darwin wrote out his ideas about natural selection,

but let ten years pass without publishing them. In the meantime, Alfred Wallace (left), who had been studying wildlife in the Amazon basin and the Malay Archipelago, wrote an essay and sent it to Darwin for advice. Wallace's essay outlined evolution by natural selection—the very same theory as Darwin's. Wallace had written earlier letters to Darwin and Lyell about patterns in the geographic distribution of species, and had come to the same conclusion.

In 1858, just weeks after Darwin received Wallace's essay, the theory of evolution by natural selection was presented at a scientific meeting. Both Darwin and Wallace were credited as authors. Wallace was still in the field and knew nothing about the meeting, which Darwin did not attend. The next year, Darwin published *On the Origin of Species*, which laid out detailed evidence in support of the theory. Many people had already accepted the idea of descent with modification (evolution). However, there was a fierce debate over the idea that natural selection drives evolution. Decades would pass before experimental evidence from the field of genetics led to its widespread acceptance as a theory by the scientific community. As you will see in the remainder of this unit, the theory of evolution by natural selection is supported by and helps explain the fossil record as well as similarities and differences in the form, function, and biochemistry of living things.

Observations about populations
- Natural populations have an inherent capacity to increase in size over time.
- As a population expands, resources that are used by its individuals (such as food and living space) eventually become limited.
- When resources are limited, the individuals of a population compete for them.

Observations about genetics
- Individuals of a species share certain traits.
- Individuals of a natural population vary in the details of those shared traits.
- Shared traits have a heritable basis, in genes. Slightly different forms of those genes (alleles) give rise to variation in shared traits.

Inferences
- A certain form of a shared trait may make its bearer better able to survive.
- Individuals of a population that are better able to survive tend to leave more offspring.
- Thus, an allele associated with an adaptive trait tends to become more common in a population over time.

Table 11.1 Principles of Natural Selection

Take-Home Message 11.3

How did new evidence change the way people in the 19th century thought about the history of life?

- Evidence found in the 1800s led to the idea that Earth and the species on it had changed over very long spans of time. This idea set the stage for Darwin's theory of evolution by natural selection.
- Natural selection is a process that drives evolutionary change: Individuals of a population survive and reproduce with differing success depending on the details of their shared, heritable traits.
- Adaptive traits enhance fitness.

adaptive trait (adaptation) A form of a heritable trait that enhances an individual's fitness.

fitness Degree of adaptation to an environment, as measured by an individual's relative genetic contribution to future generations.

natural selection Differential survival and reproduction of individuals of a population based on differences in shared, heritable traits.

A. Fossil skeleton of an ichthyosaur that lived about 200 million years ago. These marine reptiles were about the same size as modern porpoises, breathed air like them, and probably swam as fast, but the two groups are not closely related.

B. Extinct wasp encased in amber, which is ancient tree sap. This 9-mm-long insect lived about 20 million years ago.

C. Fossilized leaf from a 260-million-year-old *Glossopteris,* a type of plant called a seed fern.

D. Ancient footprints of a theropod, a type of carnivorous dinosaur that arose 250 million years ago. *Tyrannosaurus rex* was a theropod.

E. Coprolite (fossilized feces). Fossilized food remains and parasitic worms in coprolites offer clues about the diet and health of extinct species. A foxlike animal excreted this one.

Figure 11.8 Examples of fossils.

(A) Jonathan Blair; (B) © Dr. Michael Engel, University of Kansas; (C) Martin Land/Science Source; (D) © Pixtal/SuperStock; (E) Courtesy of Stan Celestian/Glendale Community College Earth Science Image Archive; (in text) Natural History Museum, London/Science Photo Library/Science Source.

11.4 Fossil Evidence

REMEMBER: Atoms of a radioisotope have an unstable nucleus that breaks up spontaneously, emitting radiation in a process called radioactive decay (Section 2.2). Molecular oxygen had been a very small component of Earth's atmosphere before photosynthesis evolved (5.5).

Even before Darwin's time, fossils were recognized as stone-hard evidence of earlier forms of life. Most fossils consist of mineralized bones, teeth, shells, seeds, spores, or other durable body parts (Figure 11.8). Trace fossils such as footprints and other body impressions, nests, burrows, trails, eggshells, or feces, are evidence of an ancient organism's activities.

The process of fossilization typically begins when an organism or its traces become covered by sediments, mud, or ash. Groundwater then seeps into the remains, filling spaces around and inside of them. Minerals dissolved in the water gradually replace minerals in bones and other hard tissues. Mineral particles that settle out of the groundwater and crystallize inside cavities and impressions form detailed imprints of internal and external structures. Sediments that slowly accumulate on top of the site exert increasing pressure, and, after a very long time, extreme pressure transforms the mineralized remains into rock.

Most fossils are found in layers of sedimentary rock (Figure 11.9). Sedimentary rocks form as rivers wash silt, sand, volcanic ash, and other materials from land to sea. Mineral particles in the materials settle on seafloors in horizontal layers that often vary in thickness and composition. After hundreds of millions of years, the layers of sediments become compacted into layers of rock. Even though most sedimentary rock forms at the bottom of a sea, geologic processes can tilt the rock and lift it far above sea level, where the layers may become exposed by the erosive forces of water and wind.

Biologists study sedimentary rock formations in order to understand the historical context of ancient life. Usually, the deepest layers in a particular formation of rock were the first to form, and those closest to the surface formed most recently. Thus, in general, the deeper the layer, the older the fossils it contains. Features of the formations can provide information about local and global events that were occur-

ring as they formed. Consider banded iron, a unique formation named after its distinctive striped appearance (left). Huge deposits of this sedimentary rock are the source of most iron we mine for steel today, but they also hold a record of how the evolution of photosynthesis changed the chemistry of Earth.

Banded iron started forming about 2.4 billion years ago, right after photosynthesis evolved. At that time, Earth's atmosphere and ocean contained very little oxygen, and ocean water contained a lot of dissolved iron. Oxygen released into the ocean by early photosynthetic bacteria quickly combined with the dissolved iron. The resulting iron compounds are completely insoluble in water, and they began to rain down on the ocean floor in massive quantities, forming layers of sediment that compacted into banded iron formations. This process continued for about 600 million years, until ocean water no longer contained very much dissolved iron.

The Fossil Record We have fossils for more than 250,000 known species. Considering the current range of biodiversity, there must have been many millions more, but we will never know all of them. Why not? The odds are against finding evidence of an extinct species, because fossils are relatively rare. Typically, when an organism dies, its body is quickly eaten by predators, scavengers, or both. Unconsumed remains decompose in the presence of moisture and oxygen, so they can only endure if they dry out, freeze, or become encased in an air-excluding material such as sap, tar, or mud. Fossils that do form are often crushed or scattered by erosion and other geologic assaults. In order for us to know about an extinct species that existed long ago, we have to find a fossil of it. At least one specimen had to escape being eaten, and be preserved before it decomposed. Its burial site also had to persist intact, and end up in a place that we can find today.

Most ancient species had no hard parts to fossilize, so we do not find much evidence of them. For example, there are many more fossils of bony fishes and hard-shelled mollusks than fossils of soft-bodied animals (such as jellyfishes and worms) that were probably much more numerous in life. Also think about relative numbers of organisms. Fungal spores and pollen grains are typically released by the billions. By contrast, the earliest humans lived in small bands and few of their offspring survived. The odds of finding even one fossilized human bone are much smaller than the odds of finding a fossilized fungal spore. Finally, imagine two species, one that existed only briefly and the other for billions of years. Which is more likely to be represented in the fossil record?

Despite these challenges, the fossil record is substantial enough to help us reconstruct large-scale patterns in the history of life.

Radiometric Dating Atoms of a radioisotope become atoms of other elements—daughter elements—as their nucleus disintegrates. This radioactive decay is not influenced by temperature, pressure, chemical bonding state, or moisture; it is influenced only by time. Thus, like the ticking of a perfect clock, each type of

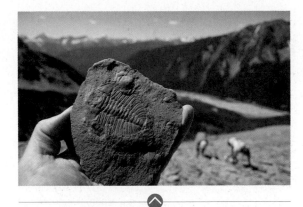

Figure 11.9 Fossil hunting.
Fossil hunters found this fossilized trilobite (an ancient marine relative of centipedes) in a shale formation in Yoho National Park, British Columbia. Fossils are most often found in layered sedimentary rock. This type of rock forms over hundreds of millions of years, often at the bottom of a sea. Geologic processes can tilt the rock and lift it far above sea level, where the layers become exposed by the erosive forces of water and wind.

© Michael Melford/National Geographic Creative.

Digging Into Data

Abundance of Iridium in the K–Pg Boundary Layer

In the late 1970s, geologist Walter Alvarez was investigating the composition of the K–Pg boundary sequence in different parts of the world. He asked his father, Nobel Prize–winning physicist Luis Alvarez, to help him analyze the elemental composition of the layer.

The Alvarezes and their colleagues tested samples of the layer taken from formations in Italy and Denmark. The researchers discovered that the K–Pg boundary sequence contains a much higher iridium content than the surrounding rock layers. Some of their results are shown in Figure 11.10.

Iridium belongs to a group of elements that are much more abundant in asteroids and other solar system materials than they are in Earth's crust, so the Alvarez group concluded that the K–Pg boundary sequence must have originated with extraterrestrial material. They calculated that an asteroid about 14 kilometers (8.7 miles) in diameter would contain enough iridium to account for the extra iridium in the K–Pg boundary sequence.

1. What was the iridium content of the K–Pg boundary layer?
2. What was the difference in iridium content between the boundary layer and the sample taken 0.7 meters above the layer?

Sample Depth	Average Abundance of Iridium (ppb)*
+ 2.7 m	< 0.3
+ 1.2 m	< 0.3
+ 0.7 m	0.36
boundary layer	41.6
− 0.5 m	0.25
− 5.4 m	0.30

Figure 11.10 Abundance of iridium in and near the K–Pg boundary sequence.
Left, Luis and Walter Alvarez with a section of the boundary. The researchers tested the iridium content of many rock samples above, below, and at this boundary in different parts of the world. Right, some of the Alvarezes' results from Stevns Klint, Denmark. Sample depths are given as meters above or below the boundary.
* ppb, parts per billion.

The iridium content of an average Earth rock is 0.4 ppb. The average meteorite contains about 550 ppb iridium.

Lawrence Berkeley National Laboratory.

radioisotope decays at a constant rate. The time it takes for half of the atoms in a sample of radioisotope to decay is called **half-life** (Figure 11.11). Half-life is a characteristic of each radioisotope. For example, radioactive uranium 238 decays into thorium 234, which decays into something else, and so on until it becomes lead 206. The half-life of the decay of uranium 238 to lead 206 is 4.5 billion years.

The predictability of radioactive decay can be used to find the age of a volcanic rock (the date it solidified). Rock forms from magma, which is a hot, molten material deep under Earth's surface. Atoms swirl and mix in this material. When magma cools, for example after reaching the surface as lava, it hardens and becomes rock. Different kinds of minerals crystallize in the rock as it hardens, each with a characteristic structure and composition. For example, the mineral called zircon consists mainly of orderly arrays of zirconium silicate molecules ($ZrSiO_4$). Some of the molecules in a newly formed zircon crystal have uranium atoms substituted for zirconium atoms, but never lead atoms. However, uranium decays into lead at a predictable rate. Thus, over time, uranium atoms disappear from a zircon crystal, and lead atoms accumulate in it. The ratio of uranium atoms to lead atoms in a zircon crystal can be measured precisely, and that ratio can be used to calculate how long ago the crystal formed (its age).

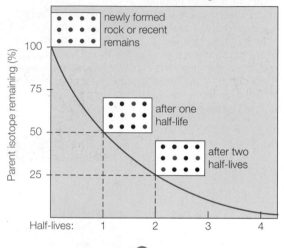

- parent isotope
- daughter elements

newly formed rock or recent remains

after one half-life

after two half-lives

Parent isotope remaining (%)

Half-lives: 1 2 3 4

Figure 11.11 Half-life.

Figure 11.12 The oldest Earth rock.
Dr. Simon Wilde holds a 4.4 billion-year-old speck of zircon embedded in protective plastic.

AP Images/Andy Manis.

Figure 11.13 Example of how radiometric dating can be used to find the age of a carbon-containing fossil.
Carbon 14 (^{14}C) is a radioisotope of carbon that decays into nitrogen. It forms in the atmosphere and combines with oxygen to become CO_2, which enters food chains by way of photosynthesis.

(A) © PhotoDisc/Getty Images.

A. Long ago, ^{14}C and ^{12}C were incorporated into the tissues of a nautilus. Both carbon isotopes were part of organic molecules in the animal's food. ^{12}C is stable and ^{14}C decays, but the proportion of the two isotopes in the nautilus's tissues remained the same. Why? The nautilus continued to gain both types of carbon atoms in the same proportions from its food.

B. The nautilus stopped eating when it died, so its body stopped gaining carbon. The ^{12}C atoms in its tissues were stable, but the ^{14}C atoms (represented as red dots) were decaying into nitrogen atoms. Thus, over time, the amount of ^{14}C decreased relative to the amount of ^{12}C. After 5,730 years, half of the ^{14}C had decayed; after another 5,730 years, half of what was left had decayed, and so on.

C. Fossil hunters discover the fossil and measure its content of ^{14}C and ^{12}C. They use the ratio of these isotopes to calculate how many half-lives have passed since the organism died. For example, if its ^{14}C to ^{12}C ratio is one-eighth of the ratio in living organisms, then three half-lives $(1/2)^3$ must have passed since it died. Three half-lives of ^{14}C is 17,190 years.

We have just described **radiometric dating**, a method that can reveal the age of a material by measuring its content of a radioisotope and daughter elements. The oldest known terrestrial rock, a tiny zircon crystal found in Australia, formed 4.4 billion years ago (Figure 11.12).

Fossils that still contain carbon can be radiometrically dated by measuring the ratio of carbon isotopes in them (Figure 11.13). Most of the ^{14}C in a fossil will have decayed after about 60,000 years. The age of fossils older than that can be estimated by dating volcanic rock formations above and below the fossil-containing layer.

Missing Links The discovery of intermediate forms of cetaceans (an order of animals that includes whales, dolphins, and porpoises) offers an example of how fossil finds and radiometric dating can be used to reconstruct evolutionary history. For some time, evolutionary biologists had thought that the ancestors of modern cetaceans walked on land, then took up life in the water. Evidence in support of this line of thinking includes a set of distinctive features of the skull and lower jaw that cetaceans share with some kinds of ancient carnivorous land animals. DNA sequence comparisons indicate that the ancient land animals were probably artiodactyls, hooved mammals with an even number of toes (two or four) on each foot (Figure 11.14A). Modern representatives of the artiodactyl lineage include camels, hippopotamuses, pigs, deer, sheep, and cows.

Until recently, no one had discovered fossils demonstrating gradual changes in skeletal features that would have accompanied a transition of whale lineages from terrestrial to aquatic life. These intermediate forms had to exist, because a representative fossil skull of an ancient whalelike animal had been discovered, but without a complete skeleton the rest of the story remained speculative.

Then, in 2000, Philip Gingerich and his colleagues unearthed complete fossilized skeletons of two ancient whales. They found *Rodhocetus kasrani* (Figure 11.14B) embedded in a 47-million-year-old rock formation in Pakistan, and *Dorudon atrox* (Figure 11.14C), in a 37-million-year-old formation in Egypt. Both fossil skeletons had whalelike skull bones, as well as intact ankle bones. The ankle bones of both fossils have distinctive features in common with those of extinct and modern artiodactyls. Modern cetaceans do not have even a remnant of an ankle bone (Figure 11.14D).

The proportions of limbs, skull, neck, and thorax indicate *Rodhocetus* swam with its feet, not its tail. Like modern whales, *Dorudon* was clearly a fully aquatic tail-swimmer: The entire hind limb was only about 12 centimeters (5 inches) long, much too small to have supported the animal's 5-meter (16-foot) body out of water.

A. *Elomeryx*, a small terrestrial animal that lived about 30 million years ago. This is a member of the same artiodactyl group (even-toed hooved mammals) that gave rise to modern representatives, including hippopotamuses. *Elomeryx* is thought to resemble a 60-million-year-old ancestor that it shares with whales.

B. *Rodhocetus kasrani*, an ancient whale that lived about 47 million years ago. Its distinctive ankle bones are evidence of a close evolutionary connection to artiodactyls. Artiodactyls are defined by the unique "double-pulley" shape of the bone (right) that forms the lower part of their ankle joint.

ankle bones
Rodhocetus   antelope

C. *Dorudon atrox*, an ancient whale that lived about 37 million years ago. Its tiny, artiodactyl-like ankle bones were much too small to have supported the weight of its body on land, so this animal had to be fully aquatic.

D. Modern cetaceans such as the sperm whale have remnants of a pelvis and leg, but no ankle bones.

Figure 11.14 Comparison of cetacean skeletons.
The ancestor of whales was an artiodactyl that walked on land. Over millions of years, the lineage transitioned from life on land to life in water, and as it did, bones of the hind limb (highlighted in blue) became smaller.

(B left and right) © Phillip Gingerich, University of Michigan.

Take-Home Message 11.4

What do rocks and fossils have to do with biology?

- Sedimentary rock holds evidence of the historical context of fossils embedded in it.
- Fossils are a stone-hard historical record of ancient life. The fossil record will never be complete, but even so it is substantial enough to help us reconstruct patterns and trends in the history of life.
- The predictability of radioisotope decay can be used to estimate the age of rock layers and fossils in them. Radiometric dating helps evolutionary biologists retrace changes in ancient lineages.

geologic time scale Chronology of Earth's history.

Gondwana Supercontinent that existed before Pangea, more than 500 million years ago.

Pangea Supercontinent that began to form about 300 million years ago; broke up 100 million years later.

plate tectonics theory Theory that Earth's outermost layer of rock is cracked into plates, the slow movement of which conveys continents to new locations over geologic time.

Figure 11.15 Plate tectonics.
Huge pieces of Earth's outer layer of rock slowly drift apart and collide. As these plates move, they convey continents around the globe.

Right, © Kevin Schafer/Corbis.

11.5 **Drifting Continents**

REMEMBER: A hypothesis is a testable explanation for a natural phenomenon (Section 1.5). The scientific community consists of critically thinking people trying to poke holes in one another's ideas (1.6).

Wind, water, and other erosive forces continuously sculpt Earth's surface, but they are only part of a much bigger picture of geological change. Earth itself also changes dramatically. Consider that all continents on Earth today were once part of a bigger supercontinent—**Pangea**—that split into fragments and drifted apart. The idea that continents move around, originally called continental drift, was proposed in the early 1900s to explain why the Atlantic coasts of South America and Africa seem to "fit" like jigsaw puzzle pieces, and why the same types of fossils occur in identical rock formations on both sides of the Atlantic Ocean. It also explained why the magnetic poles of gigantic rock formations point in different directions on different continents. As magma solidifies into rock, some iron-rich minerals in it become magnetic, and their magnetic poles align with Earth's poles when they do. If the continents never moved, then all of these ancient rocky magnets should be aligned

1 At oceanic ridges, magma (red) welling up from Earth's interior drives the movement of tectonic plates. New crust spreads outward as it forms on the surface, forcing adjacent tectonic plates away from the ridge and into trenches elsewhere.

2 At trenches, the advancing edge of one plate plows under an adjacent plate and buckles it.

3 Faults are ruptures in Earth's crust where plates meet. The diagram shows a rift fault, in which two plates move apart. The aerial photo on the right shows a strike-slip fault, in which two abutting plates slip against one another in opposite directions.

4 Magma ruptures a tectonic plate at what are called "hot spots." The Hawaiian Islands have been forming from magma that continues to erupt at a hot spot under the Pacific Plate.

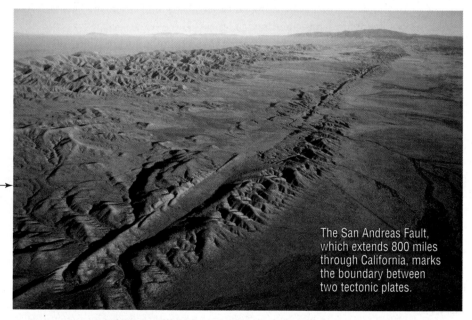

The San Andreas Fault, which extends 800 miles through California, marks the boundary between two tectonic plates.

fault **3**   trench **2**   ridge **1**   hot spot **4**   trench north-to-south, like compass needles. Indeed, the magnetic poles of rocks in each formation are aligned with one another, but the alignment is not always north-to-south. Either Earth's magnetic poles veer dramatically from their north–south axis, or the continents wander.

Continental drift was initially greeted with intense skepticism because there was no known mechanism for continents to move. Then, in the late 1950s, deep-sea explorers found immense ridges and trenches stretching thousands of kilometers across the seafloor. The discovery led to the **plate tectonics theory**, which explains how continents move (Figure 11.15). By this theory, Earth's outer layer of rock is cracked into huge plates, like a gigantic cracked eggshell. Magma welling up at an undersea ridge ❶ or continental rift at one edge of a plate pushes old rock at the opposite edge into a trench ❷. The movement is like that of a colossal conveyor belt that slowly transports continents on top of it to new locations. The plates move no more than 10 centimeters (4 inches) a year—about half as fast as your toenails grow—but it is enough to carry a continent all the way around the world after 40 million years or so.

Evidence of tectonic movement is all around us, in faults ❸ and other geological features of our landscapes. For example, volcanic island chains (archipelagos) form as a plate moves across an undersea hot spot. These hot spots are places where magma ruptures a tectonic plate ❹.

The fossil record also provides evidence in support of plate tectonics. Consider an unusual geological formation that occurs in a belt across Africa. The sequence of rock layers in this formation is so complex that it is quite unlikely to have formed more than once, but identical sequences also occur in huge belts that span India, South America, Madagascar, Australia, and Antarctica. Across all of these continents, the layers are the same ages. They also hold fossils found nowhere else, including remains of the seed fern *Glossopteris* (pictured in Figure 11.8C), which lived 299–252 million years ago, and an early reptile called *Lystrosaurus* that existed 270–225 million years ago. The unusual layered rock formation that contains fossils of these organisms probably formed in one long belt on a single continent that later broke up.

We have evidence of at least five supercontinents that formed and split up again since Earth's outer layer of rock solidified 4.55 billion years ago. One of them, a supercontinent called **Gondwana**, formed about 600 million years ago. Over the next 300 million years, Gondwana wandered across the South Pole, then drifted north until it merged with another supercontinent to form Pangea 300 million years ago (Figure 11.16). Most of the landmasses currently in the Southern Hemisphere as well as India and Arabia were once part of Gondwana. Some modern species, including the birds pictured in Figure 11.2, live only in these places.

Geologic changes brought on by plate tectonics would have had a profound impact on life. For example, when two continents collided into one, they brought together organisms that had been living apart on the separate landmasses, and physically separated organisms living in an ocean. When one continent broke up, organisms living on it would have been separated, and those living in different parts of the ocean would have come together. Events like these have been a major driving force of evolution, as you will see in the next chapter.

Putting Time Into Perspective Similar sequences of sedimentary rock layers occur around the world. Transitions between the layers mark boundaries between great intervals of time in the **geologic time scale**, which is a chronology of Earth's history. Each layer's composition offers clues about conditions on Earth during the

Figure 11.16 Reconstructions of ancient Earth.
mya: million years ago.

© Ron Blakey and Colorado Plateau Geosystems, Inc.

Eon	Era	Period	Epoch	mya*	Major Geologic and Biological Events
Phanerozoic	Cenozoic	Quaternary	Holocene	0.01	Modern humans evolve. Major extinction event is now under way.
			Pleistocene	2.6	
		Neogene	Pliocene	5.3	Tropics, subtropics extend poleward. Climate cools; dry woodlands and grasslands emerge. Adaptive radiations of mammals, insects, birds.
			Miocene	23.0	
		Paleogene	Oligocene	33.9	
			Eocene	56.0	
			Paleocene	66.0	◄ **Major extinction event**
	Mesozoic	Cretaceous	Upper		Flowering plants diversify; sharks evolve. All dinosaurs and many marine organisms disappear at the end of this epoch.
				100.5	
			Lower		Climate very warm. Dinosaurs continue to dominate. Important modern insect groups appear (bees, butterflies, termites, ants, and herbivorous insects including aphids and grasshoppers). Flowering plants originate and become dominant land plants.
				145.0	
		Jurassic			Age of dinosaurs. Lush vegetation; abundant gymnosperms and ferns. Birds appear. Pangea breaks up.
				201.3	◄ **Major extinction event**
		Triassic			Recovery from the major extinction at end of Permian. Many new groups appear, including turtles, dinosaurs, pterosaurs, and mammals.
				252	◄ **Major extinction event**
	Paleozoic	Permian			Supercontinent Pangea and world ocean form. Adaptive radiation of conifers. Cycads and ginkgos appear. Relatively dry climate leads to drought-adapted gymnosperms and insects such as beetles and flies.
				299	
		Carboniferous			High atmospheric oxygen level fosters giant arthropods. Spore-releasing plants dominate. Age of great lycophyte trees; vast coal forests form. Ears evolve in amphibians; penises evolve in early reptiles (vaginas evolve later, in mammals only).
				359	◄ **Major extinction event**
		Devonian			Land tetrapods appear. Explosion of plant diversity leads to tree forms, forests, and many new plant groups including lycophytes, ferns with complex leaves, seed plants.
				419	
		Silurian			Radiations of marine invertebrates. First appearances of land fungi, vascular plants, bony fishes, and perhaps terrestrial animals (millipedes, spiders).
				443	◄ **Major extinction event**
		Ordovician			Major period for first appearances. The first land plants, fishes, and reef-forming corals appear. Gondwana moves toward the South Pole and becomes frigid.
				485	
		Cambrian			Earth thaws. Explosion of animal diversity. Most major groups of animals appear (in the oceans). Trilobites and shelled organisms evolve.
				541	
Precambrian	Proterozoic				Oxygen accumulates in atmosphere. Origin of aerobic metabolism. Origin of eukaryotic cells, then protists, fungi, plants, animals. Evidence that Earth mostly freezes over in a series of global ice ages between 750 and 600 mya.
				2,500	
	Archean and earlier				3,800–2,500 mya. Origin of bacteria and archaea.
					4,600–3,800 mya. Origin of Earth's crust, first atmosphere, first seas. Chemical, molecular evolution leads to origin of life (from protocells to anaerobic single cells).

Figure 11.17 The geologic time scale (above) correlated with sedimentary rock exposed by erosion in the Grand Canyon (opposite).
Red triangles mark times of great mass extinctions. "First appearance" refers to appearance in the fossil record, not necessarily first appearance on Earth.
* mya: million years ago. Dates are from the International Commission on Stratigraphy, 2014.

Photo, © Michael Pancier.

Figure It Out: Which formation is marked with the in the photo?
Answer: Tapeats sandstone

time the layer was deposited. Fossils in the layers are a record of life during that period of time (Figure 11.17).

Take-Home Message 11.5

How has Earth changed over geological time?

• Over geologic time, movements of Earth's crust have caused dramatic changes in continents and oceans. The changes profoundly influenced the course of life's evolution.

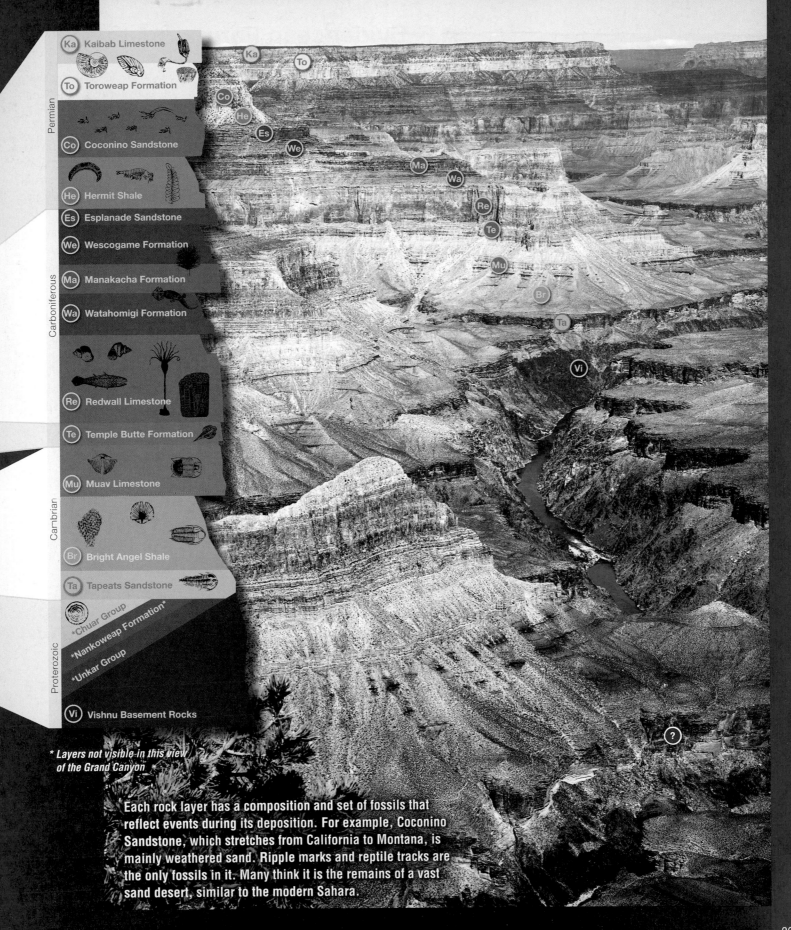

Ka Kaibab Limestone

To Toroweap Formation

Permian

Co Coconino Sandstone

He Hermit Shale

Es Esplanade Sandstone

We Wescogame Formation

Carboniferous

Ma Manakacha Formation

Wa Watahomigi Formation

Re Redwall Limestone

Te Temple Butte Formation

Mu Muav Limestone

Cambrian

Br Bright Angel Shale

Ta Tapeats Sandstone

*Chuar Group

Nankoweap Formation

*Unkar Group

Proterozoic

Vi Vishnu Basement Rocks

*Layers not visible in this view
of the Grand Canyon*

Each rock layer has a composition and set of fossils that reflect events during its deposition. For example, Coconino Sandstone, which stretches from California to Montana, is mainly weathered sand. Ripple marks and reptile tracks are the only fossils in it. Many think it is the remains of a vast sand desert, similar to the modern Sahara.

203

analogous structures Similar body structures that evolved separately in different lineages.

homologous structures Body structures that are similar in different lineages because they evolved in a common ancestor.

morphological convergence Evolutionary pattern in which similar body parts evolve separately in different lineages.

morphological divergence Evolutionary pattern in which a body part of an ancestor changes differently in its different descendants.

11.6 Evidence in Form

Evolutionary biologists are a bit like detectives, using clues to piece together history that no human witnessed. Fossils provide some clues. The body form and function of organisms that are alive today provide others.

Morphological Divergence Comparative morphology can be used to unravel evolutionary relationships in many cases. Body parts that appear similar in separate lineages because they evolved in a common ancestor are called **homologous structures** (*hom–* means "the same"). Homologous structures may be used for different purposes in different groups, but the same genes direct their development.

A body part that outwardly appears very different in separate lineages may be homologous in underlying form. Vertebrate forelimbs, for instance, vary in size, shape, and function. However, they are alike in the structure and positioning of internal elements such as bones, nerves, blood vessels, and muscles.

Populations that are not interbreeding tend to diverge genetically, and in time these genetic divergences give rise to changes in body form. Change from the body form of a common ancestor is an evolutionary pattern called **morphological divergence**. Consider the limb bones of modern vertebrate animals. Fossil evidence suggests that many vertebrates are descended from a family of ancient "stem reptiles" that crouched low to the ground on five-toed limbs. Descendants of this ancestral group diversified over millions of years, and eventually gave rise to modern reptiles,

Figure 11.18 Morphological divergence among vertebrate forelimbs.
The number and position of many skeletal elements were preserved when these diverse forms evolved from a stem reptile; notice the bones of the forearms. Certain bones were lost over time in some of the lineages (compare the digits numbered 1 through 5). Drawings are not to scale.

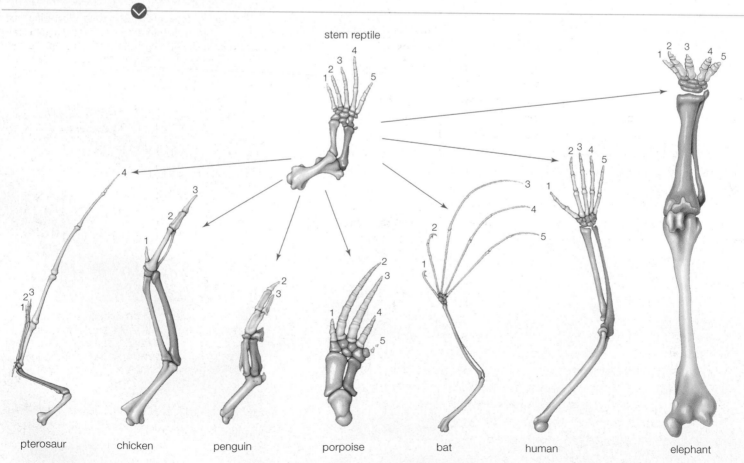

birds, and mammals. As you learned in Section 11.4, a few lineages even returned to life in the seas. As these lineages diversified, their five-toed limbs became adapted for appropriate purposes (Figure 11.18). The limbs became modified for flight in extinct reptiles called pterosaurs and in bats and most birds. In penguins and cetaceans, they are now flippers useful for swimming. Human forelimbs are arms and hands with four fingers and an opposable thumb. Elephant limbs are strong and pillarlike, capable of supporting a great deal of weight. The five-toed limb degenerated to nubs in pythons and boa constrictors, and disappeared entirely in other snakes.

Morphological Convergence Body parts that appear similar in different species are not always homologous; they sometimes evolve independently in lineages subject to the same environmental pressures. The independent evolution of similar body parts in different lineages is an evolutionary pattern called **morphological convergence**. Structures that are similar as a result of morphological convergence are **analogous structures**, which look alike but did not evolve in a shared ancestor; they evolved independently after the lineages diverged. For example, bird, bat, and insect wings all perform the same function, which is flight. However, several clues tell us that the wing surfaces are not homologous. All of the wings are adapted to the same physical constraints that govern flight, but each is adapted in a different way. In the case of birds and bats, the limbs themselves are homologous, but the adaptations that make those limbs useful for flight differ. The surface of a bat wing is a thin, membranous extension of the animal's skin. By contrast, the surface of a bird wing is a sweep of feathers, which are specialized structures derived from skin. Insect wings differ even more. An insect wing forms as a saclike extension of the body wall. Except at forked veins, the sac flattens and fuses into a thin membrane. The sturdy, chitin-reinforced veins structurally support the wing. Unique adaptations for flight are evidence that wing surfaces of birds, bats, and insects are analogous structures that evolved after the ancestors of these modern groups diverged (Figure 11.19A).

As another example of morphological convergence, consider the similar external structures of the African euphorbia and American cactus shown in Figure 11.3. These structures adapt the plants to similarly harsh desert environments where rain is scarce. Accordion-like pleats allow the plant body to swell with water when rain does come; water stored in the plants' tissues allows them to survive long dry periods. As the stored water is used, the plant body shrinks, and the folded pleats provide some shade in an environment that typically has none. Despite these similarities, a closer look reveals differences that indicate the two types of plants are not closely related (Figure 11.19B). For example, cactus spines have a simple fibrous structure; they are modified leaves that arise from dimples on the plant's surface. Euphorbia spines project smoothly from the plant surface, and they are not modified leaves: In many species the spines are dried flower stalks.

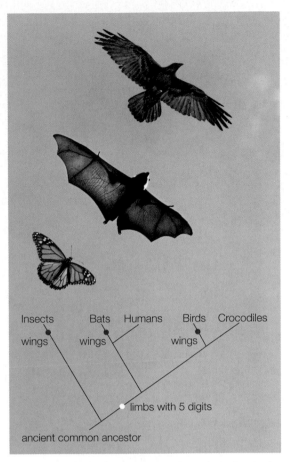

A. The flight surfaces of an insect wing, a bat wing, and a bird wing are analogous structures. The diagram shows how the evolution of wings (red dots) occurred independently in the three separate lineages. You will read more about diagrams that show evolutionary relationships in Section 12.8.

B. Spines of a saguaro cactus (left) differ from those of an African milk barrel cactus (a type of *Euphorbia*, right). This and other differences indicate the two plants are not closely related. Their similar appearances (see Figure 11.3) are an outcome of morphological convergence.

Figure 11.19 Examples of morphological convergence.

(A) Top, © iStockphoto.com/DanCardiff; middle, © Taro Taylor, www.flickr.com/photos/tjt195; bottom, © Alberto J. Espiñeira Francés - Alesfra/Getty Images; (B) Left, George Burba/Shutterstock.com; right, © James C. Gaither, www.flickr.com/people/jim-sf/.

Take-Home Message 11.6

What evidence does evolution leave in body form?

- Body parts are often modified differently in different lines of descent.
- Body parts that appear alike may have evolved independently in lineages that have faced similar environmental pressures.

11.7 Evidence in Function

REMEMBER: A statistically significant result is very unlikely to have occurred by chance alone (Section 1.6). Mitochondria have their own DNA and divide independently of the cell (3.5). In eukaryotes, the third stage of aerobic respiration—electron transfer phosphorylation—occurs at the inner mitochondrial membrane (5.5). Most codons specify an amino acid (7.4). Some mutations have no effect; those that alter a gene product can have drastic consequences (7.6). Embryonic development is orchestrated by layers of master gene expression; expression of a homeotic gene directs the formation of a specific body part (7.7). Members of a species have the same traits because they have the same genes (8.4). All organisms are descended from shared ancestors, so all genomes (10.2) are related to some extent (10.3).

Over time, inevitable mutations change a genome's DNA sequence. Most of these mutations are neutral—they have no effect on an individual's survival or reproduction—and they alter the DNA of each lineage independently of all other lineages. The more recently two lineages diverged, the less time there has been for unique mutations to accumulate in the DNA of each one. That is why the genomes of closely related species tend to be more similar than those of distantly related ones—a general rule that can be used to estimate relative times of divergence. Thus, similarities in the nucleotide sequence of a shared gene (or in the amino acid sequence of a shared protein) are often used as evidence of an evolutionary relationship. Biochemical comparisons like these are often combined with morphological comparisons, in order to provide data for hypotheses about shared ancestry.

Two species with very few similar proteins probably have not shared an ancestor for a long time—long enough for many mutations to have accumulated in the DNA of their separate lineages. Evolutionary biologists often compare a protein's sequence among several species, and use the number of amino acid differences as a measure of relative relatedness (Figure 11.20).

Among species that diverged relatively recently, many proteins have identical amino acid sequences. Nucleotide sequence differences may be instructive in such cases. Even if the amino acid sequence of a protein is identical among species, the nucleotide sequence of the gene that encodes the protein may differ because of redundancies in the genetic code. The DNA from nuclei, mitochondria, or chloroplasts can be used in nucleotide comparisons. Mitochondrial DNA accumulates mutations faster than nuclear DNA, so it can even be used to compare different individuals of the same sexually reproducing animal species. In most animals, mitochondria are inherited intact from a single parent (usually the mother). Thus, in most cases, differences in mitochondrial DNA sequences between maternally related individuals are due to mutations.

The genomes of closely related species tend to be more similar than those of distantly related ones.

Figure 11.20 Example of a protein comparison. Here, part of the amino acid sequence of the same protein from 19 species is aligned. This protein, cytochrome *b*, is a component of mitochondrial electron transfer chains. The honeycreeper sequence is identical in ten species; amino acids that differ in the other species are shown in red. Dashes are gaps in the alignment.

Figure It Out: By this comparison, which species is most closely related to honeycreepers?

Answer: The song sparrow

```
honeycreepers (10) ...CRDVQFGWLIRNLHANGASFFFICIYLHIGRGIYYGSYLNK--ETWNIGVILLLTLMATAFVGYVLPWGQMSFWG...
        song sparrow ...CRDVQFGWLIRNLHANGASFFFICIYLHIGRGIYYGSYLNK--ETWNVGIILLLALMATAFVGYVLPWGQMSFWG...
   Gough Island finch ...CRDVQFGWLIRNIHANGASFFFICIYLHIGRGIYYGSYLYK--ETWNVGIVLLLTLMATAFVGYVLPWGQMSFWG...
          deer mouse ...CRDVNYGWLIRYMHANGASMFFICLFLHVGRGMYYGSYTFT--ETWNIGIVLLFAVMATAFMGYVLPWGQMSFWG...
    Asiatic black bear ...CRDVHYGWIIRYMHANGASMFFICLFMHVGRGLYYGSYLLS--ETWNIGIILLFTVMATAFMGYVLPWGQMSFWG...
        bogue (a fish) ...CRDVNYGWLIRNLHANGASFFFICIYLHIGRGLYYGSYLYK--ETWNIGVVLLLLVMGTAFVGYVLPWGQMSFWG...
              human ...TRDVNYGWIIRYLHANGASMFFICLFLHIGRGLYYGSFLYS--ETWNIGIILLLATMATAFMGYVLPWGQMSFWG...
    thale cress (a plant) ...MRDVEGGWLLRYMHANGASMFLIVVYLHIFRGLYHASYSSPREFVWCLGVVIFLLMIVTAFIGYVLPWGQMSFWG...
         baboon louse ...ETDVMNGWMVRSIHANGASWFFIMLYSHIFRGLWVSSFTQP--LVWLSGVIILFLSMATAFLGYVLPWGQMSFWG...
          baker's yeast ...MRDVHNGYILRYLHANGASFFFMVMFMHMAKGLYYGSYRSPRVTLWNVGVIIFTLTIATAFLGYCCVYGQMSHWG...
```

Figure 11.21 Comparing vertebrate embryos.
All vertebrates go through an embryonic stage in which they have four limb buds, a tail, and divisions called somites along their back. From left to right: human, mouse, bat, chicken, alligator.

From left, © Lennart Nilsson/Bonnierforlagen AB; Courtesy of Anna Bigas, IDIBELL-Institut de Recerca Oncologica, Spain; From *Embryonic staging system for the short-tailed fruit bat, Carollia perspicillata, a model organism for the mammalian order Chiroptera, based upon timed pregnancies in captive-bred animals.* C.J. Cretekos et al., Developmental Dynamics Volume 233, Issue 3, July 2005, Pages: 721–738. Reprinted with permission of Wiley-Liss, Inc. a subsidiary of John Wiley & Sons, Inc.; Courtesy of Prof. Dr. G. Elisabeth Pollerberg, Institut für Zoologie, Universität Heidelberg, Germany; USGS.

Getting useful information from comparing DNA requires a lot more data than comparing proteins. This is because coincidental homologies are statistically more likely to occur with DNA comparisons—there are only four nucleotides in DNA versus twenty amino acids in proteins. However, DNA sequencing has become so fast that there is a lot of data available to compare. Genomics studies with such data have shown us (for example) that about 88 percent of the mouse genome sequence is identical with the human genome, as is 73 percent of the zebrafish genome, 47 percent of the fruit fly genome, and 25 percent of the rice genome.

Patterns in Animal Development In general, the more closely related animals are, the more similar is their development. For example, all vertebrates go through a stage during which a developing embryo has four limb buds, a tail, and a series of somites—divisions of the body that give rise to the backbone and associated skin and muscle (Figure 11.21). Animals have similar patterns of embryonic development because the very same master genes direct the process. Because a mutation in a master gene can unravel development completely, these genes tend to be highly conserved. Even among lineages that diverged a very long time ago, many master genes retain similar sequences and functions.

If the same genes direct development in all vertebrate lineages, how do the adult forms end up so different? Part of the answer is that there are differences in the timing of early steps in development. These differences are brought about by variations in master gene expression patterns. Consider homeotic genes called *Hox*, which, like other homeotic genes, help sculpt details of the body's form during embryonic development. Insects have a *Hox* gene called *antennapedia* that determines the identity of the thorax (the body part with legs). Humans and other vertebrates have a version of the same gene, *Hoxc6*, which determines the identity of the back (as opposed to the neck or tail). Expression of the *Hoxc6* gene in a developing embryo triggers the formation of ribs on a vertebra (Figure 11.22).

Figure 11.22 How differences in body form arise from differences in master gene expression.
Expression of the *Hoxc6* gene is indicated by purple stain in two vertebrate embryos, chicken (left) and garter snake (right). Expression of this homeotic gene causes a vertebra to develop ribs as part of the back. Chickens have 7 vertebrae in their back and 14 to 17 vertebrae in their neck; snakes have upwards of 450 back vertebrae and essentially no neck. *Hoxc6* is a vertebrate version of the *antennapedia* gene in insects.

Courtesy of Ann C. Burke, Wesleyan University.

Take-Home Message 11.7

What evidence does evolution leave in body function?

- Lineages that diverged long ago generally have more differences between their DNA (and their proteins) than do lineages that diverged more recently.
- Similarities in patterns of embryonic development are the result of master genes that have been conserved over evolutionary time.

Summary

Section 11.1 Events of the ancient past can be explained by the same physical, chemical, and biological processes that operate today. An asteroid impact 66 million years ago may have wiped out the dinosaurs and most other life on Earth.

Section 11.2 Expeditions by nineteenth-century explorers yielded increasingly detailed observations of nature. Geology, **biogeography**, and **comparative morphology** of organisms and their **fossils** led to new ways of thinking about the natural world.

Section 11.3 In the 19th century, attempts to reconcile traditional beliefs with physical evidence of **evolution**, which is change in a **lineage** over time, led to new ways of thinking about the natural world. Charles Darwin and Alfred Wallace came up with a theory of how environments select traits. A population tends to grow until it exhausts environmental resources. As that happens, competition for those resources intensifies among the population's members. Individuals with forms of shared, heritable traits that make them more competitive for the resources tend to produce more offspring. Thus, **adaptive traits** (**adaptations**) imparting greater **fitness** tend to become more common in the population over generations. The process in which environmental pressures result in the differential survival and reproduction of individuals of a population is called **natural selection**. It is one of the processes that drives evolution.

Section 11.4 Fossils are typically found in stacked layers of sedimentary rock. Fossils of many organisms are relatively scarce, so the fossil record will always be incomplete. A radioisotope's characteristic **half-life** can be used to determine the age of rocks and fossils. This technique, **radiometric dating**, helps us understand the ancient history of many lineages.

Section 11.5 According to the **plate tectonics theory**, Earth's crust is cracked into giant plates that carry landmasses to new positions as they move. Earth's landmasses have periodically converged as supercontinents such as **Gondwana** and **Pangea**. Transitions in the fossil record are the boundaries of great intervals of the **geologic time scale**.

Section 11.6 Comparative morphology can reveal evidence of evolutionary connections among lineages. **Homologous structures** are similar body parts that, by **morphological divergence**, became modified differently in different lineages. Such parts are evidence of a common ancestor. **Analogous structures** are body parts that look alike in different lineages but did not evolve in a common ancestor. By the process of **morphological convergence**, they evolved separately after the lineages diverged.

Section 11.7 We can discover and clarify evolutionary relationships by comparing amino acid sequences of proteins, or DNA sequences, from different organisms. In general, these sequences are more similar among lineages that diverged more recently. Master genes that affect development tend to be highly conserved, so similarities in patterns of embryonic development reflect shared ancestry that can be evolutionarily ancient.

Self-Quiz

Answers in Appendix I

1. The number of species on an island usually depends on the size of the island and its distance from a mainland. This statement would most likely be made by _____ .
   a. an explorer
   b. a biogeographer
   c. a geologist
   d. a philosopher

2. The bones of a bird's wing are similar to the bones in a bat's wing. This observation is an example of _____ .
   a. uniformity
   b. evolution
   c. comparative morphology
   d. a lineage

3. Evolution _____ .
   a. is natural selection
   b. is change in a line of descent
   c. can occur by natural selection
   d. b and c are correct

4. A trait is adaptive if it _____ .
   a. arises by mutation
   b. increases fitness
   c. is passed to offspring
   d. occurs in fossils

5. In which type of rock are you more likely to find a fossil?
   a. basalt, a dark, fine-grained volcanic rock
   b. limestone, composed of sedimented calcium carbonate
   c. slate, a volcanically melted and cooled shale
   d. granite, which forms by crystallization of molten rock below Earth's surface

6. True or false? Wrinkly textures in rock that formed from ancient biofilms living in marine sediments are considered trace fossils.

7. If the half-life of a radioisotope is 20,000 years, then a sample in which three-quarters of that radioisotope has decayed is _____ years old.
   a. 15,000
   b. 26,667
   c. 30,000
   d. 40,000

8. Forces that cause geologic change include _____ (select all that are correct).
   a. water movement
   b. natural selection
   c. volcanic activity
   d. tectonic plate movement
   e. wind
   f. asteroid impacts

9. Did Pangea or Gondwana form first?

10. The dinosaurs disappeared about _____ million years ago.
    a. 10
    b. 16.5
    c. 66
    d. 200

11. Through _____ , a body part of an ancestor is modified differently in different lines of descent.
    a. homologous evolution
    b. morphological convergence
    c. adaptive divergence
    d. morphological divergence

12. Homologous structures among major groups of organisms may differ in _____ .
    a. size
    b. shape
    c. function
    d. all of the above

13. A mutation that alters the embryonic expression pattern of a _____ may lead to major differences in the adult form.
    a. derived trait
    b. master gene
    c. homologous structure
    d. all of the above

14. Match each term with the most suitable description.
    _____ fitness
    _____ fossils
    _____ natural selection
    _____ homeotic genes
    _____ half-life
    _____ catastrophism
    _____ uniformity
    _____ analogous structures
    _____ homologous structures
    _____ sedimentary rock
    _____ neutral mutation

    a. does not affect fitness
    b. geological change occurs continuously
    c. geological change occurs in unusual major events
    d. good for finding fossils
    e. survival of the fittest
    f. characteristic of a radioisotope
    g. insect wing and bird wing
    h. human arm and bird wing
    i. evidence of life in distant past
    j. measured by reproductive success
    k. similar across diverse taxa

15. All of the following data types can be used as evidence of shared ancestry except similarities in _____ .
    a. amino acid sequences
    b. DNA sequences
    c. fossil morphologies
    d. embryonic development
    e. form due to convergence
    f. all are appropriate

Critical Thinking

1. In the late 1800s, a biologist studying animal embryos coined the phrase "ontogeny recapitulates phylogeny," meaning that the physical development of an animal embryo (ontogeny) seemed to retrace the changing form of the species during its evolutionary history (phylogeny). Why would embryonic development retrace evolutionary steps?

2. Radiometric dating does not measure the age of an individual atom. It is a measure of the age of a quantity of atoms—a statistic. As with any statistical measure, its values may deviate around an average (see sampling error, Section 1.6). Imagine that one sample of rock is dated ten different ways. Nine of the tests yield an age close to 225,000 years. One test yields an age of 3.2 million years. Do the nine consistent results imply that the one that deviates is incorrect, or does the one odd result invalidate the nine that are consistent?

Visual Question

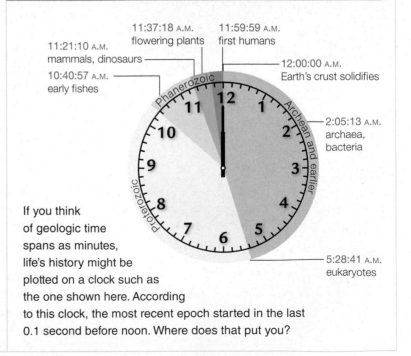

11:21:10 A.M. mammals, dinosaurs
11:37:18 A.M. flowering plants
11:59:59 A.M. first humans
10:40:57 A.M. early fishes
12:00:00 A.M. Earth's crust solidifies
2:05:13 A.M. archaea, bacteria
5:28:41 A.M. eukaryotes

Phanerozoic
Archean and earlier
Proterozoic

If you think of geologic time spans as minutes, life's history might be plotted on a clock such as the one shown here. According to this clock, the most recent epoch started in the last 0.1 second before noon. Where does that put you?

12

PROCESSES OF EVOLUTION

Application ❯ 12.1 Superbug Farms

REMEMBER: Contamination of food with disease-causing bacteria can result in food poisoning that is sometimes fatal (Section 3.1). Mutations can be introduced during DNA replication (6.4). Genetic diversity in a population is an advantage in a changing environment (8.4). An allele associated with an adaptive trait tends to become more common in a population over time (11.3).

Scarlet fever, tuberculosis, and pneumonia once caused one-fourth of the annual deaths in the United States. Since the 1940s, we have been relying on antibiotics to fight these and other dangerous bacterial diseases. We have also been using them in other, less dire circumstances. For an unknown reason, antibiotics promote growth in cattle, pigs, poultry, and even fish. The agricultural industry uses a lot of antibiotics, mainly for this purpose. In 2011, 13.7 million kilograms (about 30 million pounds) of antibiotics were used for agriculture in the U.S.—more than four times the amount used to treat people in the same year.

A natural population of bacteria is diverse, and it can evolve astonishingly fast. Consider how each cell division is an opportunity for mutation. The common intestinal bacteria *E. coli* can divide every 17 minutes, so even if a population starts out as clones, its cells diversify quickly. Bacteria share DNA even among distantly related species, and this adds even more genetic diversity to their populations. When a natural population of bacteria is exposed to a selection pressure such as an antibiotic, some cells in the population are likely to survive because they carry an allele that offers an advantage—antibiotic resistance, in this case. As susceptible cells die and the survivors reproduce, the frequency of antibiotic-resistance alleles in the population increases. A typical two-week course of treatment with antibiotics can exert selection pressure on over a thousand generations of bacteria. The pressure drives genetic change in bacterial populations so they become composed mainly of antibiotic-resistant cells. Thus, using antibiotics on an ongoing basis effectively guarantees the production of antibiotic-resistant bacterial populations (Figure 12.1).

Farms where antibiotics are used to promote growth are hot spots for the evolution of antibiotic-resistant bacteria and their spread to humans. Veterinarians and other people who work with the animals on these farms tend to carry more antibiotic-resistant bacteria in their bodies. So do neighbors who live within a mile. The bacteria spread much farther than the farm, however. Bacteria on an animal's skin or in its digestive tract can easily contaminate its meat during slaughter, and contaminated meat ends up in restaurant and home kitchens. A 2013 investigation found "worrisome" amounts of bacteria in 97% of the chicken meat in stores across the United States. About half of the samples tested were contaminated with superbugs—bacteria that are resistant to multiple antibiotics—and one in ten contained multiple types of superbugs. An earlier study found antibiotic-resistant bacteria in more than half of supermarket ground beef and pork chops, and in over 80 percent of ground turkey. Bacteria can be killed by the heat of cooking, but it is almost impossible to prevent them from spreading from contaminated meat to kitchen surfaces—and to people—during the process.

We have only a limited number of antibiotic drugs, and developing new ones is much slower than bacterial evolution. As resistant bacteria become more common, the number of antibiotics that can be used to effectively treat infections in humans dwindles. Using a particular antibiotic only in animals, or only in humans, is not a solution to this problem because there are only a few mechanisms by which these drugs kill bacteria; resistance to one antibiotic often confers resistance to others.

Figure 12.1　A breeding ground for antibiotic-resistant bacteria.
The vast majority of chickens raised for meat in the United States spend their lives in gigantic flocks that crowd huge buildings like this one. Growth-promoting antibiotics are given to the entire flock in food, a practice that pressures normal bacterial populations to become antibiotic resistant.

Bob Nichols/USDA photo.

For example, bacteria that become resistant to Flavomycin® (an antibiotic used only in animals) also resist vancomycin (an antibiotic used only in humans). Superbugs that are resistant to most currently available antibiotics are turning up at a very alarming rate.

All of this amounts to bad news. An infection with antibiotic-resistant bacteria tends to be longer, more severe, and more likely to be deadly than one more easily treatable with antibiotics. Superbugs cause more than 2 million cases of serious illness each year in the United States alone; they outright kill 23,000 of these people. Many, many more die because the infection complicates another, preexisting illness.

12.2 Alleles in Populations

REMEMBER: A population is a group of interbreeding individuals of the same species living in a given area (Section 1.2). Individuals of a species share a unique set of inherited traits (1.4) because they have the same genes; and alleles of the shared genes are the basis of variation in shared traits (8.4). Sexual reproduction mixes up genetic information from two parents, giving rise to offspring that vary in shared traits (8.5). In a Mendelian inheritance pattern, the phenotypic effect of a dominant allele fully masks that of a recessive allele (9.3). The more genes and environmental factors that influence a trait, the more continuous is its range of variation (9.4–9.6). Neutral mutations have no effect on an individual's survival or reproduction (11.7).

The individuals of a population (and a species) share certain features. Humans, for example, normally have a short neck, a thumb on each hand, and so on. These are examples of morphological traits (*morpho*– means form). Individuals of a species also share physiological traits, such as details of metabolism. They also respond the same way to certain stimuli, as when hungry humans eat food. These responses are behavioral traits.

Some traits appear in two distinct forms, or morphs, in which case the forms are called a dimorphism (*di*– means two). Flower color in the pea plants that Gregor Mendel studied is a dimorphic trait. The interaction of two alleles that have a clear dominance relationship gives rise to the dimorphism—purple or white flowers—in these plants. Other traits appear in three or more distinct forms, in which case the forms are called polymorphisms (*poly*–, many). ABO blood type in humans, which is determined by the codominant alleles of the *ABO* gene, is an example. Most other traits are complex (Figure 12.2), as is their genetic basis. Any or all of the genes that influence such traits may have multiple alleles.

In earlier chapters, you learned that alleles arise by mutation, and other events shuffle them among individuals of a population (Table 12.1). To understand the potential scope of variation that results from these events, consider alleles in our species. There are more than $10^{100,000,000}$ potential combinations of human alleles at fertilization. Not even 10^{10} people are living today. Unless you have an identical twin, it is extremely unlikely that another person with your precise genetic makeup has ever lived, or ever will.

An Evolutionary View of Mutations Being the original source of new alleles, mutations are worth another look, this time in the context of populations. We cannot predict when or in which individual a particular gene will mutate. We can, however, predict the average mutation rate of a species, which is the probability that

Figure 12.2 Sampling morphological variation among zigzag snails (top) and humans (bottom).
Variation in shared traits is an outcome of differences in alleles that influence those traits.

Top, © David McIntyre/Photographer's Direct; bottom, clockwise from top left, © Roderick Hulsbergen/http://www.photography.euweb.nl; Olga Reutska/Shutterstock; NinaMalyna/Shutterstock; Lane Oatey/Blue Jean Images/Getty Images; Djomas/Shutterstock; Paul Matthew Photography/Shutterstock.

Table 12.1 Some Sources of Variation in Shared Traits

Genetic Event	Effect
Mutation	Original source of new alleles
Crossing over at meiosis I	Introduces new combinations of alleles into chromosomes
Independent assortment at meiosis I	Mixes maternal and paternal chromosomes
Fertilization	Combines alleles from two parents

allele frequency Abundance of a particular allele among members of a population.

directional selection Mode of natural selection in which a phenotype at one end of a range of variation is favored.

disruptive selection Mode of natural selection in which traits at the extremes of a range of variation are adaptive, and intermediate forms are not.

gene pool All the alleles of all the genes in a population; a pool of genetic resources.

microevolution Change in allele frequency.

stabilizing selection Mode of natural selection in which an intermediate form of a trait is favored over extreme forms.

Evolution is not purposeful. It simply fills the nooks and crannies of opportunity.

a mutation will occur in a given interval. In the human species, that rate is about 2.2×10^{-9} mutations per base pair per year. In other words, about 70 nucleotides in the human genome sequence change every decade.

In humans at least, most mutations are neutral. For instance, a mutation that results in your earlobes being attached to your head instead of swinging freely should not in itself stop you from surviving and reproducing as well as anybody else. So, natural selection would not affect the frequency of this mutation in the human population. Other mutations give rise to structural, functional, or behavioral alterations that reduce an individual's chances of surviving and reproducing. Consider collagen, a protein component of the skin, bones, tendons, lungs, blood vessels, and other vertebrate organs. If one of the genes for collagen mutates in a way that changes the protein's function, the entire body may be affected. A mutation such as this can change phenotype so drastically that it results in death, in which case it is called a lethal mutation.

Occasionally, a change in the environment favors a mutation that had previously been neutral or even somewhat harmful. Even if a beneficial mutation bestows only a slight advantage, its frequency tends to increase in a population over time. This is because natural selection operates on traits with a genetic basis. With natural selection, remember, environmental pressures result in an increase in the frequency of an adaptive form of a trait in a population over generations. Mutations have been altering genomes for billions of years, and they continue to do so. Cumulatively, mutations have given rise to Earth's staggering biodiversity. Think about it: The reason you do not look like an avocado or an earthworm or even your next-door neighbor began with mutations that occurred in different lines of descent.

Allele Frequency Together, all the alleles of all the genes of a population constitute a pool of genetic resources—a **gene pool**. Members of a population breed with one another more often than they breed with members of other populations, so their gene pool is more or less isolated. The abundance of a particular allele in a gene pool is called **allele frequency**, and it is expressed as a percentage. For example, if a particular allele occurs on half of the chromosomes carried by the population, then the frequency of that allele is 50 percent.

Allele frequency can change, and this change is called **microevolution**. Microevolution is always occurring in natural populations because, as you will see in the next sections, processes that drive it—mutation, natural selection, and genetic drift—are always operating in nature. As you learn about these patterns, remember an important point: Evolution is not purposeful; it simply fills the nooks and crannies of opportunity.

Take-Home Message 12.2

What is microevolution?

• Individuals of a natural population share morphological, physiological, and behavioral traits characteristic of the species. Alleles are the basis of differences in the details of those shared traits.

• All alleles of all individuals in a population make up the population's gene pool. An allele's abundance in the gene pool is called its allele frequency.

• Microevolution is change in allele frequency. It is always occurring in natural populations because processes that drive it are always operating.

12.3 Modes of Natural Selection

REMEMBER: Many vitamins and minerals are essential in the diet because they are cofactors or become converted to them; coenzymes modified in reactions are regenerated in separate reactions (Section 4.4). A homozygous individual has two identical alleles of a gene, and a heterozygous individual has two nonidentical alleles; the effect of a dominant allele masks that of a recessive allele paired with it (9.2). Blood clotting in humans involves proteins called clotting factors (9.7). With natural selection, individuals of a population survive and reproduce with differing success depending on the details of their shared, heritable traits (11.3).

Natural selection influences allele frequency by operating on phenotypes that have a genetic basis. How phenotype is affected depends on the species and the selection pressures in the environment. With **directional selection**, forms at one end of a range of phenotypic variation become more common over time (Figure 12.3A, B). With **stabilizing selection**, an intermediate form of a trait is favored, and extreme forms are selected against (Figure 12.3C). With **disruptive selection**, forms of a trait at both ends of a range of variation are favored, and intermediate forms are selected against (Figure 12.3D).

Directional Selection Antibiotic use that fosters resistant bacterial populations is one example of directional selection; another example involves rats. Rats thrive in urban centers, where garbage is plentiful and natural predators are not. Part of their success stems from an ability to reproduce very quickly: Rat populations can expand within weeks to match the amount of garbage available for them to eat. For decades, people have been fighting rats with poisons. Baits laced with warfarin, an organic compound that interferes with blood clotting, became popular in the 1950s. Rats that ate the poisoned baits died within days after bleeding internally or losing blood through cuts or scrapes. Warfarin was extremely effective, and its impact on harmless species was much lower than that of other rat poisons. It quickly became the rat poison of choice. By 1980, however, about 10 percent of rats in urban areas were resistant to warfarin. What happened?

Warfarin interferes with blood clotting because it inhibits the function of an enzyme called VKORC1. This enzyme regenerates vitamin K, which functions as a coenzyme in the production of blood clotting factors. When vitamin K is not regenerated, the clotting factors are not properly produced, and clotting cannot occur. Rats resistant to warfarin have a mutated version of the *VKORC1* gene. The enzyme encoded by this allele is insensitive to warfarin. "What happened" was evolution by natural selection. Rats with the normal allele died after eating warfarin; the lucky ones with a mutated allele survived and passed it to offspring. The rat populations recovered quickly, and a higher proportion of individuals in the next generation carried a mutation. With each onslaught of warfarin, the frequency of the mutation in rat populations increased. Exposure to warfarin had exerted directional selection.

The mutation that results in warfarin resistance also reduces the activity of the VKORC1 enzyme, so rats that have it require a lot of extra vitamin K. However, being vitamin K deficient is not so bad when compared with being dead from rat poison. In the absence of warfarin, though, rats with the allele are at a serious disadvantage because they cannot easily obtain enough vitamin K from their diet to sustain normal blood clotting and bone formation. Thus, the frequency of a warfarin resistance allele in a rat population declines quickly after warfarin exposure ends—another example of directional selection.

A. Population before selection occurs.

B. With directional selection, forms of a trait at one end of a range of variation is adaptive.

C. With stabilizing selection, extreme forms of a trait are eliminated, and an intermediate form is maintained.

D. With disruptive selection, a midrange form of a trait is eliminated, and extreme forms are maintained.

Figure 12.3 Comparing three modes of natural selection.
In these examples, red arrows indicate which forms are being selected against; blue, forms that are adaptive.

Digging Into Data

Resistance to Rodenticides in Wild Rat Populations

Beginning in 1990, rat infestations in northwestern Germany started to intensify despite continuing use of rat poisons. In 2000, Michael H. Kohn and his colleagues tested wild rat populations around Münster. For part of their research, they trapped wild rats in five towns, and tested those rats for resistance to warfarin and the more recently developed poison bromadiolone. The results are shown in Figure 12.4.

1. In which of the five towns were most of the rats susceptible to warfarin?
2. Which town had the highest percentage of poison-resistant wild rats?
3. What percentage of rats in Olfen were warfarin resistant?
4. In which town do you think the application of bromadiolone was most intensive?

○ not resistant to warfarin or bromadiolone

◐ resistant to warfarin

○ resistant to warfarin and bromadiolone

Figure 12.4 Poison resistance in wild rats in Germany, 2000.

Photo, © Rollin Verlinde/Vilda.

Directional selection also affects the color of rock pocket mice in Arizona's Sonoran Desert. Rock pocket mice are small mammals that spend the day sleeping in underground burrows, emerging at night to forage for seeds. Light brown granite dominates their environment, but there are also patches of dark basalt: the remains of ancient lava flows. Most of the mice in populations that inhabit the dark rock have dark coats. Most of the mice in populations that inhabit the light brown rock have light brown coats. The difference arises because mice that match the rock color in each habitat are camouflaged from their natural predators. Night-flying owls more easily see mice that do not match the rocks, and they preferentially eliminate these mice from each population. Thus, in both habitats, selective predation has resulted in a directional shift in the frequency of alleles that affect coat color.

Another well-documented case of directional selection involves coloration changes in peppered moths. These moths feed and mate at night, then rest on trees during the day. In preindustrial England, the vast majority of peppered moths were white with black speckles, and a small number were much darker. At the time, the air was clean, and light-gray lichens grew on the trunks and branches of most trees. Light-colored moths that rested on lichen-covered trees were well camouflaged, but darker moths were not (Figure 12.5A). By the 1850s, the industrial revolution had begun, and smoke emitted by coal-burning factories was killing the lichens. Dark moths, which were better camouflaged on lichen-free, soot-darkened trees, had become more common (Figure 12.5B).

Figure 12.5 Adaptive value of two color forms of the peppered moth.

J. A. Bishop, L. M. Cook.

A. Light-colored moths on a nonsooty tree trunk (left) are hidden from predators; dark ones (right) stand out.

B. Where soot darkens tree trunks, the dark color (left) provides more camouflage than the light color (right).

Scientists suspected that predation by birds was the selective pressure that shaped moth coloration, and in the 1950s, H. B. Kettlewell set out to test this hypothesis. He bred dark and light moths in captivity, marked them for easy identification, then released them in several areas. His team recaptured more of the dark moths in the polluted areas and more light ones in the less polluted ones. The researchers also observed predatory birds eating more light-colored moths in soot-darkened forests, and more dark-colored moths in cleaner, lichen-rich forests. Dark-colored moths were clearly at a selective advantage in industrialized areas.

Pollution controls went into effect in 1952. As a result of improved environmental standards, tree trunks gradually became free of soot, and lichens made a comeback. Kettlewell observed that moth phenotypes shifted too: Wherever pollution decreased, the frequency of dark moths decreased as well. Recent research has confirmed Kettlewell's results implicating birds as selective agents of peppered moth coloration, and also that this selection causes a shift in the frequency of alleles underlying the coloration. Peppered moth color is determined by a single gene; individuals with a dominant allele of this gene are dark, and those homozygous for a recessive allele are light.

Stabilizing Selection With stabilizing selection, an intermediate form of a trait is favored, and extreme forms are selected against. Consider how environmental pressures maintain an intermediate body mass in populations of sociable weavers (Figure 12.6). These birds live in the African savanna, and their body mass has a genetic basis. Between 1993 and 2000, Rita Covas and her colleagues investigated selection pressures that operate on sociable weaver body mass by capturing and weighing thousands of birds before and after the breeding seasons. The results of this study indicated that optimal body mass in sociable weavers is a trade-off between the risks of starvation and predation. Birds that carry less fat are more likely to starve than fatter birds. However, birds that carry more fat spend more time eating, which in this species means foraging in open areas where they are easily accessible to predators. Fatter birds are also more attractive to predators, and not as agile when escaping. Thus, predators are agents of selection that eliminate the fattest individuals. Birds of intermediate weight have the selective advantage, and they make up the bulk of sociable weaver populations.

Disruptive Selection With disruptive selection, forms of a trait at both ends of a range of variation are favored, and intermediate forms are selected against. Disruptive selection maintains a dimorphism in black-bellied seedcrackers. These birds are native to Cameroon, Africa, and the size of their beaks (bills) has a genetic basis. The bill of a typical black-bellied seedcracker, male or female, is either 12 millimeters wide, or wider than 15 millimeters (Figure 12.7). Birds with a bill size between 12 and 15 millimeters are uncommon. It is as if every human adult were 4 feet or 6 feet tall, with no one of intermediate height.

Seedcrackers with the large and small bill forms inhabit the same geographic range, and they breed randomly with respect to bill size. The dimorphism arises from (and is maintained by) environmental factors that affect feeding performance. The finches feed mainly on the seeds of two types of sedge, which is a grasslike plant. One sedge produces hard seeds; the other produces soft seeds. Small-billed birds are better at opening the soft seeds, but large-billed birds are better at cracking the hard ones. Both hard and soft sedge seeds are abundant during Cameroon's semiannual wet seasons. At these times, all seedcrackers feed on both seed types. The seeds become scarce during the region's dry seasons. As competition for food

Figure 12.6 Stabilizing selection in sociable weavers. The graph shows the number of birds (out of 977) that survived a breeding season. Compare Figure 12.3C.

Top, Peter Chadwick/Science Source.

Figure It Out: According to these data, what is the optimal weight of a sociable weaver?

Answer: About 29 grams

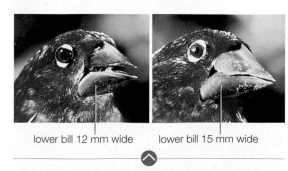

lower bill 12 mm wide   lower bill 15 mm wide

Figure 12.7 Disruptive selection in African seedcracker populations maintains a dimorphism in bill size. Competition for scarce food during dry seasons favors birds with bills that are either 12 millimeters wide (left) or 15 to 20 millimeters wide (right). Birds with bills of intermediate size are selected against.

Thomas Bates Smith.

intensifies, each bird focuses on eating the seeds that it opens most efficiently: Small-billed birds feed mainly on soft seeds, and large-billed birds feed mainly on hard seeds. Birds with intermediate-sized bills cannot open either type of seed as efficiently as the other birds, so they are less likely to survive the dry seasons.

A. Male elephant seals engaged in combat. Males of this species typically compete for access to clusters of females.

B. A male bird of paradise engaged in a courtship display has caught the eye (and, perhaps, the sexual interest) of a female.

C. Female stalk-eyed flies prefer to mate with males that have the longest eyestalks. In this species, long eyestalks provide no known survival advantage.

Figure 12.8 Sexual selection in action.

(A) © Ingo Arndt/Nature Picture Library; (B) Tim Laman/National Geographic Creative;
(C) Minden Pictures/SuperStock.

> ## Take-Home Message 12.3
>
> **How does natural selection drive evolution?**
>
> - Natural selection influences allele frequency by operating on traits that have a genetic basis.
> - With directional selection, forms at one end of a range of variation in a trait are favored.
> - With stabilizing selection, an intermediate form of a trait is favored, and extreme forms are selected against.
> - With disruptive selection, an intermediate form of a trait is selected against, and extreme forms are favored.

12.4 Natural Selection and Diversity

REMEMBER: A particular mutation in the beta globin gene gives rise to sickle-cell anemia (Section 7.6). A homozygous individual has two identical alleles of a gene, and a heterozygous individual has two nonidentical alleles; the effect of a dominant allele masks that of a recessive allele paired with it (9.2). In some cases, a codominant allele offers a survival advantage in a particular environment (9.6).

Survival of the Sexiest Not all evolution is driven by selection for traits that enhance survival. Competition for mates is another selective pressure that can shape form and behavior. Consider how individuals of many sexually reproducing species have a distinct male or female phenotype (a trait that differs between males and females is called a sexual dimorphism). Individuals of one sex are more colorful, larger, or more aggressive than individuals of the other sex. These traits can seem puzzling because they take energy and time away from activities that enhance survival, and some actually hinder an individual's ability to survive. Why, then, do they persist? The answer is **sexual selection**, in which the evolutionary winners outreproduce others of a population because they are better at securing mates. With this mode of natural selection, the most adaptive forms of a trait are those that help individuals defeat rivals for mates, or are most attractive to the opposite sex.

For example, the females of some species cluster in defensible groups when they are sexually receptive, and males compete for sole access to the groups. Competition for the ready-made harems favors brawny, combative males (Figure 12.8A).

Males or females that are choosy about mates act as selective agents on their own species. The females of some species shop for a mate among males that display species-specific cues such as a highly specialized appearance or courtship behavior (Figure 12.8B). The cues often include flashy body parts or movements, traits that tend to attract predators and in some cases are a physical hindrance. However, to a female member of the species, a flashy male's survival despite his obvious handicap may imply health and vigor, two traits that are likely to improve her chances of bearing healthy, vigorous offspring. Selected males pass alleles for their attractive traits to the next generation of males, and females pass alleles that influence mate preference to the next generation of females. Highly exaggerated traits can be an evolutionary outcome (Figure 12.8C).

Maintaining Multiple Alleles Any mode of natural selection may keep two or more alleles circulating at relatively high frequency in a population's gene pool, a state called balanced polymorphism. For example, sexual selection maintains multiple alleles that govern eye color in populations of *Drosophila* fruit flies. Female flies prefer to mate with rare white-eyed males, until the white-eyed males become more common than red-eyed males, at which point the red-eyed flies are again preferred.

Balanced polymorphism can also arise in environments that favor heterozygous individuals. Consider the gene that encodes the beta globin chain of hemoglobin. *HbA* is the normal allele; the codominant *HbS* allele carries a mutation that causes sickle-cell anemia. Even with medical care, about 15 percent of individuals homozygous for the *HbS* allele die by age 18 from complications of the disorder. Despite being so harmful, the *HbS* allele persists at very high frequency among the human populations in tropical and subtropical regions of Asia, Africa, and the Middle East (Figure 12.9A). Why? Populations with the highest frequency of the *HbS* allele also have the highest incidence of malaria (Figure 12.9B). Mosquitoes transmit *Plasmodium*, the parasitic protist that causes malaria, to human hosts. *Plasmodium* multiplies in the liver and then in red blood cells. The cells rupture and release new parasites during recurring bouts of severe illness. People who make both normal and sickle hemoglobin are more likely to survive malaria than people who make only normal hemoglobin. In *HbA*/*HbS* heterozygous individuals, *Plasmodium*-infected red blood cells sometimes sickle. The abnormal shape brings the cells to the attention of the immune system, which destroys them along with the parasites they harbor. By contrast, *Plasmodium*-infected red blood cells of individuals homozygous for the *HbA* allele do not sickle, so the parasite may remain hidden from the immune system.

In areas where malaria is common, the persistence of the *HbS* allele is a matter of relative evils. Malaria and sickle-cell anemia are both potentially deadly. Heterozygous individuals may not be completely healthy, but they do have a better chance of surviving malaria than people homozygous for the normal allele (*HbA*/*HbA*). With or without malaria, people who have both alleles (*HbA*/*HbS*) are more likely to live long enough to reproduce than individuals homozygous for the sickle allele (*HbS*/*HbS*). The result is that nearly one-third of people living in the most malaria-ridden regions of the world carry the *HbS* allele.

Figure 12.9 Malaria and sickle-cell anemia.
After Ayala and others.

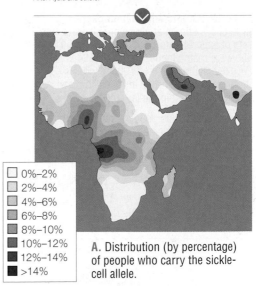

0%–2%
2%–4%
4%–6%
6%–8%
8%–10%
10%–12%
12%–14%
>14%

A. Distribution (by percentage) of people who carry the sickle-cell allele.

B. Distribution of malaria cases (orange) in Africa, Asia, and the Middle East in the 1920s, before the start of programs to control mosquitoes, which transmit the parasitic protist that causes the disease. Notice the correlation with the distribution of the sickle-cell allele in **A.**

Take-Home Message 12.4

How does natural selection maintain diversity?

• With sexual selection, adaptive forms of a trait are those that give an individual an advantage in securing mates.

• Sexual selection can reinforce phenotypic differences between males and females, and sometimes it results in exaggerated traits.

• Balanced polymorphism can be an outcome of sexual selection, or of environmental pressures that favor heterozygous individuals.

sexual selection Mode of natural selection in which some individuals outreproduce others of a population because they are better at securing mates.

Figure 12.10 Genetic drift in flour beetles.
A flour beetle is shown at above on a flake of cereal. In two sets of experiments, beetles heterozygous for alleles b^+ and b were maintained for 20 generations in populations of 10 individuals (**A**) or 100 individuals (**B**). Graph lines in **B** are smoother than in **A**, which means that less genetic drift occurred in the larger populations.

Allele b^+ was lost in one population (one graph line ends at 0). Notice that the average frequency of allele b^+ rose at the same rate in both groups, an indication that natural selection was at work too: Allele b^+ was weakly favored.

(A,B) Adapted from S. S. Rich, A. E. Bell, and S. P. Wilson, "Genetic drift in small populations of Tribolium," *Evolution* 33:579–584, Fig. 1, p. 580, © 1979 by John Wiley and Sons. Used by permission of the publisher; above, Photo by Peggy Greb/USDA.

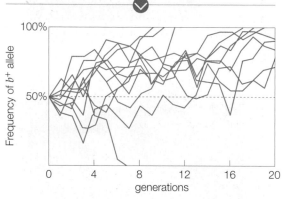

A. In these experiments, population size was maintained at 10 beetles. Several populations were tested; notice that allele b^+ was lost in one (one graph line ends at 0).

B. In these experiments, population size was maintained at 100 beetles. Genetic drift was less pronounced in these populations than in the 10-beetle populations in **A**.

Figure It Out: In how many populations did allele b^+ become fixed?

Answer: Six

12.5 Genetic Drift and Gene Flow

REMEMBER: Members of a sexually reproducing population vary in the details of their shared traits because they carry different alleles; this diversity offers an evolutionary advantage in a changing environment (Section 8.4). Codominant alleles of the *ABO* gene are the basis of ABO blood type in humans (9.4). Engineered genes are being transferred in pollen from genetically modified plants to wild plants (10.4).

In a natural population, individuals reproduce with differing success, and the difference is not always an outcome of natural selection. By chance, a perfectly fit and healthy individual may not pass its alleles to offspring, for example by dying in a random event before it has a chance to reproduce. Such events can change a population's allele frequencies. Change in allele frequency brought about by chance alone is called **genetic drift**.

Genetic drift occurs in all natural populations of sexual reproducers, but its effects are more pronounced in small ones (Figure 12.10). For example, genetic drift makes small populations particularly vulnerable to the loss of genetic diversity. To understand why, imagine a hypothetical gene with two alleles, neither of which confers a selective advantage. These alleles (let's call them *A* and *a*) occur in a population's gene pool at a frequency of 95 percent and 5 percent, respectively. If the population consists of 10 members, then one individual is heterozygous (*Aa*) and the remaining nine are homozygous (*AA*). A random event that eliminates the heterozygous individual from the population before it reproduces also eliminates allele *a* from the population's gene pool. If all individuals of a population are homozygous for an allele (*A*, in this example), we say that the allele is **fixed**. The frequency of a fixed allele will not change unless a new mutation occurs, or an individual bearing another allele enters the population. Now imagine that our population with alleles *A* and *a* consists of 100 members instead of 10. Five individuals in this larger population would be heterozygous (*Aa*). In order for allele *a* to be lost from the population's gene pool, all five would have to be eliminated before they reproduce. The chance of random events eliminating all heterozygous individuals is smaller in the larger population. This is a simplified example of a general effect: The loss of genetic diversity is possible in all populations, but it is more likely to occur in small ones.

Bottlenecks and the Founder Effect A drastic reduction in population size, which is called a **bottleneck**, can greatly reduce diversity. Consider the northern elephant seal (pictured in Figure 12.8A). Overhunting during the late 1890s left only about twenty individuals of this species alive. Since then, hunting restrictions have allowed the population to recover, but genetic diversity among its members has been greatly reduced. The bottleneck and subsequent genetic drift eliminated many alleles that had previously been present in the population.

A loss of genetic diversity can also occur when a small group of individuals establishes a new population. If the founding group is not representative of the original population in terms of allele frequencies, then the new population will not be representative of it either. This outcome is called the **founder effect** (Figure 12.11A). Consider that all three *ABO* alleles for blood type are common in most human populations. Native Americans are an exception, with the majority of individuals being homozygous for the *O* allele. Native Americans are descendants of early humans that migrated from Asia between 14,000 and 21,000 years ago, across a narrow land bridge that once connected Siberia and Alaska. Analysis of DNA from ancient skeletal remains reveals that most early Americans were also homozygous for the *O* allele. Modern Siberians have all three alleles. Thus, the humans who first populated the Americas were probably members of a small group that had reduced genetic diversity compared with the general population.

Founding populations are often necessarily inbred. **Inbreeding** is nonrandom breeding or mating between close relatives. Closely related individuals tend to share more alleles than nonrelatives do, so inbred populations often have unusually high numbers of individuals homozygous for recessive alleles, some of which are harmful. This outcome is minimized in human populations that discourage or forbid incest (mating between parents and children or between siblings).

The Old Order Amish in Lancaster County, Pennsylvania, offer an example of the effects of inbreeding. Amish people marry only within their community. Intermarriage with other groups is not permitted, and no "outsiders" are allowed to join the community. As a result, Amish populations are moderately inbred, and many of their individuals are homozygous for harmful recessive alleles. The Lancaster community has an unusually high frequency of a recessive allele that causes Ellis–van Creveld syndrome, a genetic disorder characterized by dwarfism, heart defects, and polydactyly (extra fingers or toes), among other symptoms. This allele has been traced to a man and his wife, two of a group of 400 Amish people who immigrated to the United States in the mid-1700s. As a result of the founder effect and inbreeding since then, about 1 of 8 people in the Lancaster community is now heterozygous for the allele, and 1 in 200 is homozygous for it (Figure 12.11B).

Gene Flow Individuals tend to mate or breed most frequently with other members of their own population. However, not all populations of a species are completely isolated from one another, and nearby populations may occasionally interbreed. Also, individuals sometimes leave one population and join another. **Gene flow**, the movement of alleles between populations, occurs in both cases, and it can change or stabilize allele frequencies.

Gene flow is common among populations of animals, but it also occurs in less mobile organisms. Consider the acorns that jays disperse when they gather nuts for the winter (left). Every fall, these birds visit acorn-bearing oak trees repeatedly, then bury the acorns in the soil of territories as much as a mile away. The jays transfer acorns (and the alleles carried by these seeds) among populations of oak trees that may otherwise be genetically isolated. Gene flow also occurs when wind or an animal transfers pollen from one plant to another, often over great distances. Many opponents of genetic engineering cite the movement of engineered genes from transgenic crop plants into wild populations via pollen.

© Ashok Khosla, www.seeingbirds.com.

A. The founder effect: a group that founds a new population is not genetically representative of the original population, so allele frequencies differ between the new and the old populations.

B. An allele that causes Ellis–van Creveld syndrome occurs at high frequency in the gene pool of the Lancaster Amish—a result of the founder effect. Outward indications of the disorder (polydactyly and dwarfism) appear in this Amish baby.

Figure 12.11 The founder effect and one outcome.

(B) © Dr. Victor A. McKusick.

Take-Home Message 12.5

What mechanisms other than natural selection affect allele frequencies?

- Genetic drift, or change in allele frequency that occurs by chance alone, makes small populations particularly vulnerable to losing genetic diversity.
- Loss of genetic diversity can occur after a bottleneck or because of the founder effect.
- Gene flow, which is the movement of alleles between populations, can stabilize or change allele frequencies.

bottleneck Reduction in population size so severe that it reduces genetic diversity.

fixed Refers to an allele for which all members of a population are homozygous.

founder effect After a small group of individuals found a new population, allele frequencies in the new population differ from those in the original population.

gene flow The movement of alleles between populations.

genetic drift Change in allele frequency due to chance alone.

inbreeding Mating among close relatives.

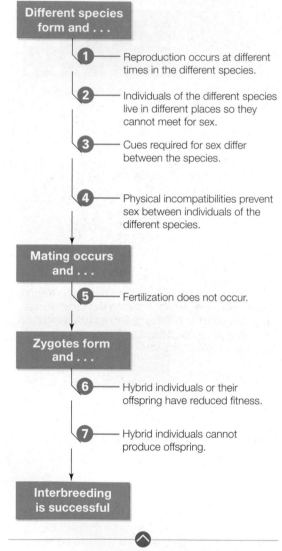

Different species form and . . .

1 — Reproduction occurs at different times in the different species.

2 — Individuals of the different species live in different places so they cannot meet for sex.

3 — Cues required for sex differ between the species.

4 — Physical incompatibilities prevent sex between individuals of the different species.

Mating occurs and . . .

5 — Fertilization does not occur.

Zygotes form and . . .

6 — Hybrid individuals or their offspring have reduced fitness.

7 — Hybrid individuals cannot produce offspring.

Interbreeding is successful

Figure 12.12 How reproductive isolation prevents interbreeding.

12.6 Speciation

REMEMBER: A "species" is a convenient but artificial construct of the human mind (Section 1.4). Mitochondria have their own DNA and divide independently of the cell (3.5). Wavelengths of light that are not absorbed by a pigment are reflected, and that reflected light gives each pigment its characteristic color (5.2). A somatic cell is a body cell (6.1). Mitosis, a nuclear division mechanism that maintains the chromosome number, is the basis of growth and tissue repair in multicelled organisms (8.2). Homologous chromosomes undergo crossing over during meiosis; at fertilization in sexual reproducers, an egg (a female gamete) fuses with a sperm (a male gamete) to produce a zygote, the first cell of a new individual (8.5). A polyploid individual has three or more complete sets of chromosomes (9.8). The slow movement of Earth's crustal plates conveys continents to new locations over geologic time (11.5). Lineages that diverged long ago generally have more differences between their DNA (and their proteins) than do lineages that diverged more recently (11.7).

When two populations of a species do not interbreed, the number of genetic differences between them increases because mutation, natural selection, and genetic drift occur independently in each one. Over time, the populations may become so different that we consider them to be different species. Evolutionary processes in which new species arise are called **speciation**.

Evolution is a dynamic, extravagant, messy, and ongoing process that can be challenging for people who like neat categories. Speciation offers a perfect example, because it rarely occurs at a precise moment in time: Individuals often continue to interbreed even as populations are diverging, and populations that have already diverged may come together and interbreed again.

Reproductive Isolation Every time speciation happens, it happens in a unique way, which means that each species is a product of its own unique evolutionary history. However, there are recurring patterns. For example, **reproductive isolation**, the end of gene flow between populations, is always part of the process by which sexually reproducing species attain and keep their separate identities. Several mechanisms of reproductive isolation prevent successful interbreeding, and thus reinforce differences between diverging populations (Figure 12.12). For example, some closely related species cannot interbreed because the timing of their reproduction differs 1. Consider the periodical cicada (right). Larvae of these insects feed on roots as they mature underground, then the adults emerge to reproduce. Three cicada species reproduce every 17 years. Each has a sibling species with nearly identical form and behavior, except that the siblings emerge on a 13-year cycle instead of a 17-year cycle. Sibling species have the potential to interbreed, but they can only get together once every 221 years!

Alvin E. Staffan/Science Source.

Adaptation to different environmental conditions may prevent closely related species from interbreeding 2. For example, two species of manzanita, a plant native to the Sierra Nevada mountain range, rarely hybridize. One species that lives on dry, rocky hillsides is better adapted for conserving water. The other species, which requires more water, lives on lower slopes where water stress is not as intense. The physical separation makes cross-pollination unlikely.

Differences in behavior can prevent mating between related animal species 3. For example, males and females of many animal species engage in courtship displays

Figure 12.13 An example of reproductive isolation: courtship displays in peacock spiders. This male peacock spider is signaling his intent to mate with the female by raising and waving colorful flaps, and gesturing his legs in time with abdominal vibrations. If his species-specific courtship display fails to impress her, she will kill him.

© Jürgen Otto.

A. Black sage is pollinated mainly by honeybees and other small insects.

B. The flowers of black sage are too delicate to support larger insects. Big insects, including this carpenter bee, access the nectar of small sage flowers only by piercing from the outside. When they do so, they avoid touching the flower's reproductive parts.

C. The reproductive parts (anthers and stigma) of white sage flowers are too far away from the petals to be brushed by honeybees, so honeybees are not efficient pollinators of this species. White sage is pollinated mainly by larger bees and hawkmoths, which brush the flower's stigma and anthers as they pry apart the petals to access nectar.

Figure 12.14 An example of reproductive isolation: pollinator specialization in sage.

(A) Courtesy of Dr. James French; (B) Courtesy of © Ron Brinkmann, www.flickr.com/photos/ronbrinkmann; (C) © David Goodin.

before sex (Figure 12.13). In a typical pattern, the female recognizes the sounds and movements of a male of her species as an overture to sex; females of different species do not.

The size or shape of an individual's reproductive parts may prevent it from mating with members of closely related species ❹. For example, plants called black sage and white sage grow in the same areas, but hybrids rarely form because the flowers of these two related species have become specialized for different pollinators (Figure 12.14).

Even if gametes of different species do meet up, they often have molecular incompatibilities that prevent a zygote from forming ❺. For example, the molecular signals that trigger pollen germination in flowering plants are species-specific (we return to pollen germination and other aspects of flowering plant reproduction in Section 28.2). Gamete incompatibility may be the primary speciation route among animals that release their eggs and free-swimming sperm into water.

Genetic changes are the basis of divergences in form, function, and behavior. Even chromosomes of species that diverged relatively recently may be different enough that a hybrid zygote ends up with extra or missing genes, or genes with incompatible products. Such outcomes typically disrupt embryonic development ❻. Hybrid individuals that do survive embryonic development often have reduced fitness. For example, hybrid offspring of lions and tigers have more health problems and a shorter life expectancy than individuals of either parent species. If hybrids live long enough to reproduce, their offspring often have lower and lower fitness with each successive generation. Incompatible nuclear and mitochondrial DNA may be the cause (mitochondrial DNA is inherited from the mother only).

Some interspecies crosses produce robust but sterile hybrid offspring ❼. For example, mating between a female horse (64 chromosomes) and a male donkey (62 chromosomes) produces a mule (63 chromosomes: 32 from the horse, and 31 from the donkey). Mules are healthy but their chromosomes cannot pair up properly for crossing over, so this animal makes few viable gametes.

Allopatric Speciation Genetic changes that lead to a new species can begin with physical separation between populations. With **allopatric speciation**, a physical barrier arises and separates two populations, ending gene flow between them (*allo–* means different; *patria*, fatherland). Then, reproductive isolating mechanisms evolve that prevent interbreeding even if the diverging populations meet again.

Gene flow between populations separated by distance is often inconsistent. Whether a geographic barrier can completely block that gene flow depends on how the species travels (such as by swimming, walking, or flying), and how it reproduces (for example, by internal fertilization or by pollen dispersal).

A geographic barrier can arise in an instant, or over an eon. The Great Wall of China is an example of a barrier that arose relatively quickly. As it was being built, the wall interrupted gene flow among nearby populations of insect-pollinated plants; DNA sequence comparisons show that trees, shrubs, and herbs on either side of the wall are diverging genetically. Geographic barriers usually arise much more slowly. For example, it took millions of years of tectonic plate movements to bring the two continents of North and South America close enough to collide. The land bridge where the two continents now connect is called the Isthmus of Panama. When this isthmus formed about 4 million years ago, it cut off the flow of water— and gene flow among populations of aquatic organisms—as it separated one large ocean into what are now the Pacific and Atlantic Oceans (Figure 12.15).

Sympatric Speciation In **sympatric speciation**, populations in the same geographic region speciate in the absence of a physical barrier between them (*sym–* means together). Sympatric speciation can occur in a single generation when the chromosome number multiplies. Polyploidy typically arises when an abnormal nuclear division during meiosis or mitosis doubles the chromosome number. For example, if the nucleus of a somatic cell in a flowering plant fails to divide during mitosis, the resulting cell—which is polyploid—may proliferate and give rise to shoots and flowers. If the flowers can self-fertilize, a new polyploid species may be the result. Common bread wheat originated after related species hybridized, and then the chromosome number of the hybrid offspring doubled (Figure 12.16).

Sympatric speciation can also occur with no change in chromosome number. The sage plants you just learned about speciated with no physical barrier to gene flow. As another example, more than 500 species of cichlid fishes arose by sympatric

Figure 12.15 Example of allopatric speciation.
When the Isthmus of Panama formed 4 million years ago, it cut off gene flow among ocean-dwelling populations of snapping shrimp. Today, shrimp species on opposite sides of the isthmus are so similar that they might interbreed, but they are reproductively isolated. Instead of mating when they are brought together, they snap their claws at one another aggressively. The photos show two of the many closely related species that live on opposite sides of the isthmus.

Right, © Arthur Anker.

Alpheus nuttingi (Atlantic)

Alpheus millsae (Pacific)

Figure 12.16 Sympatric speciation in wheat.
The wheat genome, which consists of seven chromosomes, occurs in slightly different forms called A, B, C, D, and so on. Many wheat species are polyploid, carrying more than two copies of the genome. For example, modern bread wheat (*Triticum aestivum*) is hexaploid, with six copies of the wheat genome: two each of genomes A, B, and D (or 42 AABBDD).

Photos © J. Honegger, courtesy of S. Stamp, E. Merz, www.sortengarten/ethz.ch.

Triticum urartu (wild einkorn)   *Aegilops* (wild goatgrass, unknown species)   *Triticum* (hybrid)   *Triticum turgidum* (emmer)   *Aegilops tauschii* (goatgrass)   *Triticum aestivum* (bread wheat)

14 AA × 14 BB → 14 AB → 28 AABB × 14 DD → 42 AABBDD

A. About 11,000 years ago, a diploid wheat (einkorn) hybridized with a diploid species of wild goatgrass.

B. Tetraploid (4*n*) emmer arose when the chromosome number of the resulting hybrid doubled.

C. Common bread wheat is the result of a hybridization between emmer and a diploid goatgrass.

speciation in the shallow waters of Lake Victoria. This large freshwater lake sits isolated from river inflow on an elevated plain in Africa's Great Rift Valley. Since Lake Victoria formed about 400,000 years ago, it has dried up three times. DNA sequence comparisons indicate that almost all of the cichlid species in this lake arose since the last dry spell, which was 12,400 years ago. How could hundreds of species arise so quickly? In this case, the answer begins with differences in the color of ambient light in different parts of the lake. The light in the lake's shallower, clear water is mainly blue; light that penetrates the deeper, muddier water is mainly red. The cichlid species vary in color (Figure 12.17), and female cichlids prefer to mate with brightly colored males. Their preference has a genetic basis, in alleles that encode light-sensitive pigments of the retina (part of the eye). Retinal pigments made by species that live mainly in shallow areas of the lake are more sensitive to blue light. The males of these species are also the bluest. Retinal pigments made by species that prefer deeper areas of the lake are more sensitive to red light. Males of these species are redder. In other words, the colors that a female cichlid sees best are the same colors displayed by males of her species. Thus, mutations that affect color perception are likely to affect a female's choice of mates. Such mutations are probably the way sympatric speciation occurs in these fishes.

Figure 12.17 Males of four closely related species of cichlid native to Lake Victoria, Africa.
Hundreds of cichlid species arose by sympatric speciation in this lake. Mutations that affect female cichlids' perception of the color of ambient light in deeper or shallower regions of the lake also affect their choice of mates. Female cichlids prefer to mate with the most brightly colored males.

Kevin Bauman, www.african-cichlid.com

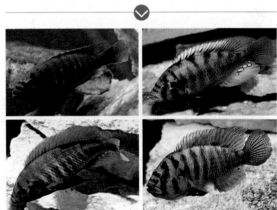

Figure It Out: Which form of natural selection is driving sympatric speciation in these cichlids?

Answer: Sexual selection

Take-Home Message 12.6

How do species attain and maintain separate identities?

- Speciation is an evolutionary process by which new species form. It varies in its details and duration, but always includes reproductive isolation.
- A physical barrier that intervenes between populations of a species prevents gene flow among them. When gene flow ends, genetic divergences give rise to new species.
- Divergence within a population can lead to new species that inhabit the same geographical area, with no physical barrier to gene flow.

adaptive radiation A lineage undergoes a burst of genetic divergences that gives rise to many species.

extinct Refers to a species that no longer has any living members.

macroevolution Large-scale evolutionary patterns and trends.

12.7 Macroevolution

REMEMBER: The scientific community consists of critically thinking people trying to poke holes in one another's ideas (Section 1.6). Predator and prey are locked in a constant race, with each genetic improvement in one species countered by a genetic improvement in the other (8.4). The dinosaurs died out in a mass extinction that occurred 66 million years ago (11.1). Transitions in the fossil record are correlated with boundaries between rock layers in the geologic time scale (11.5).

Microevolution is change in allele frequency within a single species or population. We also see evolutionary patterns on a larger scale than microevolution, and these large-scale patterns are called **macroevolution**. Macroevolution includes trends such as land plants evolving from green algae, all dinosaurs disappearing in a mass extinction, a burst of divergences from a single species, and so on.

Very little change may occur over a very long period of time. Consider coelacanths, an order of ancient lobe-finned fish that had been assumed extinct for at least 70 million years until a fisherman caught one in 1938. In its unique form and other traits, modern coelacanth species are similar to fossil specimens hundreds of millions of years old (Figure 12.18).

Major evolutionary novelties often stem from the adaptation of an existing structure for a completely new purpose. For example, the feathers that allow modern birds to fly are derived from feathers that first evolved in some dinosaurs. Those dinosaurs could not have used their feathers for flight, but they probably did use them for insulation.

By current estimates, more than 99 percent of all species that ever lived are now **extinct**, which means they no longer have living members. In addition to continuing extinctions of individual species, the fossil record indicates that there have been more than twenty mass extinctions, which are simultaneous losses of many lineages. These include five catastrophic events in which the majority of species on Earth disappeared.

Figure 12.18 The coelacanth: a living fossil.
Photos on the left compare a 320-million-year-old coelacanth fossil found in Montana with a live coelacanth. The diagram on the right shows a few of the coelacanth's unusual ancestral features that have been lost in almost all other fish lineages over evolutionary time.

Top left, Courtesy of The Virtual Fossil Museum, www.fossilmuseum.net; bottom left, AllessandroZocc/Shutterstock.com; right, Raul Martin Domingo/National Geographic Creative.

Notochord
This tough, elastic tube, which is partially hollow and filled with fluid, is ancestral to the spinal cord.

Lobed fins
These fleshy fins retain a few of the ancestral bones that gave rise to legs and arms in other lineages.

Long gestation
Coelacanths give birth to litters of up to 26 fully developed "pups" after gestation of more than a year.

Rostral organ
This sensory organ perceives electrical impulses in water. It probably helps the fish locate prey in dark ocean depths.

Akepa (*Loxops coccineus*)   Akikiki (*Oreomystis bairdi*)   Nihoa finch (*Telespiza ultima*)   Akekee (*Loxops caeruleirostris*)

Iiwi (*Vestiaria coccinea*)   Akohekohe (*Palmeria dolei*)   Hawaii Amakihi (*Hemignathus virens*)   Akiapolaau (*Hemignathus munroi*)

Maui parrotbill (*Pseudonestor xanthophrys*)   Maui Alauahio (*Paroreomyza montana*)   Kauai Amakihi (*Hemignathus kauaiensis*)   Apapane (*Himatione sanguinea*)

Figure 12.19 A few Hawaiian honeycreepers.

The bills of these birds are adapted to feed on insects, seeds, fruits, nectar in floral cups, and other foods. All Hawaiian honeycreepers descended from a Eurasian rosefinch that probably resembled modern rosefinches (left).

(Akepa, Nihoa finch, Maui Alauahio) © Jack Jeffrey Photography; (Akekee, Akohekohe, Apapane, Akiapolaau, Maui parrotbill, Kauai Amakihi) © Eric VanderWerf/Pacific Rim Photos; (Iiwi) Michael Ord/Science Source; (Hawaii Amakihi) James A. Hancock/Science Source; (Akikiki) U.S. Geological Survey/photo by Carter Atkinson; (rosefinch) Andrzej Sliwinski/Shutterstock.

With **adaptive radiation**, one lineage rapidly diversifies into several new species. Adaptive radiation can occur after a population colonizes a new environment that has a variety of different habitats and few competitors. Speciation occurs along with adaptation to the different habitats. The Lake Victoria cichlids that you learned about in the previous section arose this way, as did the Hawaiian honeycreepers (Figure 12.19). Some finch species migrate far outside of their normal range when food becomes scarce in a preferred overwintering spot, traveling in flocks of thousands or even tens of thousands of individuals. About 5.8 million years ago in southern Asia, one of these migratory flocks was caught up in the winds of a huge storm. The birds—rosefinches—were blown at least 7,000 miles (11,000 kilometers) across the open ocean to the islands of the Hawaiian archipelago (right). Enough individuals survived the journey to found a new population. The birds' arrival had been preceded by insects and plants, but no predators, and their descendants spread into habitats along the coasts, through dry lowland forests, and into highland rain forests. Isolation from gene flow with mainland finch populations allowed the island colonizers to diverge. Over many generations, populations living in the different habitats became hundreds of separate species—the honeycreepers—as unique forms and behaviors evolved in them. These unique traits helped the birds exploit special opportunities presented by their particular island habitats.

Hawaiian archipelago

Figure 12.20 Example of coevolved species.

(A) © Brian Raine, www.flickr.com/people/25801055@N00. (B) © Jeremy Thomas/Natural Visions.

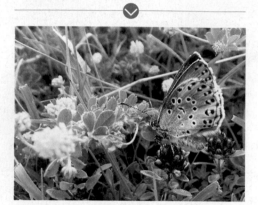

A. A *Maculinea arion* butterfly emerges from a pupa to mate and lay eggs on wild thyme flowers. Larvae that emerge from the eggs will survive only if a colony of *Myrmica sabuleti* ants adopts them.

B. A *Maculinea arion* caterpillar interacting with a *Myrmica sabuleti* ant. This beguiled ant is preparing to carry the honey-exuding, hunched-up caterpillar back to its nest, where the caterpillar will feed on ant larvae for the next 10 months until it becomes a pupa.

Adaptive radiation may also occur after a key innovation evolves. A **key innovation** is a trait that allows its bearer to exploit a habitat more efficiently or in a novel way. The evolution of lungs offers an example, because lungs were a key innovation that opened the way for an adaptive radiation of vertebrates on land. A geologic or climatic event that eliminates some species from a habitat can spur adaptive radiation; species that survive the event then have access to resources from which they had previously been excluded. This is the way mammals were able to undergo an adaptive radiation after the dinosaurs disappeared 66 million years ago.

Two species that have close ecological interactions may evolve jointly, a pattern called **coevolution**. One species acts as an agent of selection on the other, and each adapts to changes in the other. Over evolutionary time, the two species may become so interdependent that they can no longer survive without one another. Relationships between coevolved species can be quite intricate. Consider the large blue butterfly (*Maculinea arion*), a parasite of ants. After hatching, the butterfly larvae (caterpillars) feed on wild thyme flowers (Figure 12.20A) and then drop to the ground. An ant that finds a caterpillar strokes it, which makes the caterpillar exude honey. The ant eats the honey and continues to stroke the caterpillar, which secretes more honey. This interaction continues for hours, until the caterpillar suddenly hunches itself up (Figure 12.20B). The ant then picks up the caterpillar and carries it back to its nest, where, in most cases, other ants kill it—except if the ants are of the species *Myrmica sabuleti*. The caterpillar secretes the same chemicals as *Myrmica sabuleti* larvae, and makes the same sounds as their queen—behaviors that deceive the ants into adopting the caterpillar and treating it better than their own larvae. The adopted caterpillar feeds on ant larvae for about 10 months, then undergoes metamorphosis, changing into a butterfly that emerges from the ground to mate. Eggs are deposited on wild thyme near another *M. sabuleti* nest, and the cycle starts anew. This relationship between ant and butterfly is typical of coevolved relationships in that it is extremely specific. Any increase in the ants' ability to identify a caterpillar in their nest selects for caterpillars that better deceive the ants, which in turn select for ants that can better identify the caterpillars. Each species exerts directional selection on the other.

Evolutionary Theory Biologists have no doubt that macroevolution occurs, but many disagree about how it occurs. However we choose to categorize evolutionary processes, the very same genetic change may be at the root of all evolution—fast or slow, large-scale or small-scale. Dramatic jumps in form, if they are not artifacts of gaps in the fossil record, may be the result of mutations in homeotic or other regulatory genes. Macroevolution may include more processes than microevolution, or it may not. It may be an accumulation of many microevolutionary events, or it may be an entirely different process. Evolutionary biologists may disagree about these and other hypotheses, but all of them are trying to explain the same thing: how all species are related by descent from common ancestors.

Take-Home Message 12.7

What is macroevolution?

• Large-scale patterns of evolutionary change such as adaptive radiation, the origin of major groups, and mass extinctions are called macroevolution.

12.8 Phylogeny

REMEMBER: A "species" is a convenient but artificial construct of the human mind; with taxonomy, we systematically name species and rank them into higher taxa based on shared traits (Section 1.4). A homeotic gene is a type of master gene, and its expression directs the formation of a specific body part during embryonic development (7.7). Humans reconstruct history by studying physical evidence of past events (11.1). Body parts that appear alike may have evolved independently in lineages that have faced similar environmental pressures (11.6). Similarities in the nucleotide sequence of a shared gene (or in the amino acid sequence of a shared protein) are often used as evidence of an evolutionary relationship (11.7).

Classifying life's tremendous diversity into a series of taxonomic ranks is a useful endeavor, in the same way that it is useful to organize a telephone book or contact list in alphabetical order: The result is convenient. Traditional (Linnaean) classification schemes rank species into higher taxa based on shared traits—birds have feathers, cacti have spines, and so on—but these rankings do not necessarily reflect evolutionary relationships.

Today's biologists work from the premise that every living thing is related if you just look back far enough in time. Grouping species according to evolutionary relationships is a way to fill in the details of this bigger picture of evolution. Thus, reconstructing **phylogeny**, the evolutionary history of a species or a group of species, is a priority. Phylogeny is a kind of genealogy that follows evolutionary relationships through time. Instead of ranking species by shared traits, evolutionary biologists spend their time pinpointing what makes the species share the traits in the first place: a common ancestor. They determine common ancestry by looking for a derived trait—one that is present in a group under consideration, but not in any of the group's ancestors.

A group whose members share one or more defining derived traits is called a **clade**. By this definition, each species is a clade. Many higher taxonomic rankings are also equivalent to clades—flowering plants, for example, are both a phylum and a clade—but some are not. For example, the traditional Linnaean class Reptilia ("reptiles") includes crocodiles, alligators, tuataras, snakes, lizards, turtles, and tortoises. While it is convenient to classify these animals together, they would not constitute a clade unless birds are also included, as you will see in Chapter 15.

All species are interconnected in the big picture of evolution; an evolutionary biologist's job is to figure out where the connections are. All species bear traces of their evolutionary history in their traits. For example, humans and bacteria use some of the same proteins to repair DNA. Even so, humans and bacteria are not close relatives. It is the relative newness of a shared trait that defines a clade. Consider how alligators look a lot more like lizards than birds. In this case, the similarity in appearance does indicate shared ancestry, but it is a more distant relationship than alligators have with birds. A unique set of traits that include a gizzard and a four-chambered heart evolved in the lineage that gave rise to alligators and birds, but not in the lineage that gave rise to lizards.

Making hypotheses about evolutionary relationships among clades is called **cladistics**. A **cladogram** is an evolutionary tree diagram that visually summarizes a hypothesis about how a group of clades are related (Figure 12.21). Data from an outgroup (a species not closely related to any member of the group under study) may be included in order to "root" the tree. Each line is a lineage, which may branch into two lineages. The branch point represents a common ancestor of two lineages.

A. Evolutionary connections are represented as lines on a cladogram. Each line is a lineage, and each branch point represents a common ancestor.

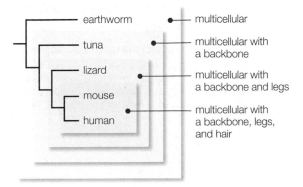

B. A cladogram can be viewed as "sets within sets" of derived traits. Each set (an ancestor together with all of its descendants) is a clade.

Figure 12.21 An example of a cladogram.

© National Geographic/SuperStock.

clade A group whose members share one or more defining derived traits.

cladistics Making hypotheses about evolutionary relationships among clades.

cladogram Evolutionary tree diagram that summarizes hypothesized relationships among a group of clades.

coevolution The joint evolution of two closely interacting species; each species is a selective agent for traits of the other.

key innovation An evolutionary adaptation that gives its bearer the opportunity to exploit a particular environment more efficiently or in a new way.

phylogeny Evolutionary history of a species or group of species.

A. The palila has an adaptation that allows it to feed on seeds of a native Hawaiian plant that are toxic to most other birds. The one remaining palila population is declining because these plants are being trampled by cows and eaten by goats and sheep. Only about 1,200 palila remained in 2010.

B. The lower bill of the akekee points to one side, allowing this bird to pry open buds that harbor insects. Avian malaria is wiping out the last population of this species. Between 2000 and 2007, the number of akekee dropped from 7,839 birds to 3,536.

C. This poouli—rare, old, and missing an eye—died in 2004 from avian malaria. There were two other poouli alive at the time, but neither has been seen since then.

Figure 12.22 Three honeycreeper species: going, going, and gone.
The genetic diversity of Hawaiian honeycreepers is dwindling along with their continued extinctions. Deciphering their evolutionary connections may help us preserve the remaining species.

(A) © Eric VanderWerf/Pacific Rim Photos; (B) Courtesy of © Lucas Behnke; (C) Bill Sparklin/
Ashley Dayer.

Evolutionary history does not change because of events in the present: A species' ancestry remains the same no matter how it evolves. However, as with traditional taxonomic rankings, we can make mistakes grouping organisms based on incomplete information. Thus, a clade or cladogram may change when new discoveries are made. As with all hypotheses, the more data in support of an evolutionary grouping, the less likely it is to require revision.

Applications of Phylogeny   Studies of phylogeny reveal how species relate to one another and to species that are now extinct. In doing so, they inform our understanding of how shared ancestry interconnects all species—including our own.

The story of the Hawaiian honeycreepers offers an example of how finding ancestral connections can help species that are still living. The first Polynesians arrived on the Hawaiian islands sometime before 1000 A.D.; Europeans followed in 1778. Hawaii's rich ecosystem was hospitable to the newcomers and their domestic animals and crops. Entire forests were cleared to grow imported crops, and plants that escaped cultivation began to crowd out native plants. Escaped livestock ate and trampled rain forest plants that had provided the honeycreepers with food and shelter. Mosquitoes accidentally introduced in 1826 spread diseases such as avian malaria from imported chickens to native bird species. Stowaway rats ate their way through populations of native birds and their eggs; mongooses deliberately imported to eat the rats preferred to eat birds and bird eggs.

The isolation that had allowed honeycreepers to arise by adaptive radiation also made them vulnerable to extinction. Divergence from the ancestral species had led to the loss of unnecessary traits such as defenses against mainland predators and diseases. Traits that had been adaptive—such as a long, curved beak matching the flower of a particular plant—became hindrances when habitats suddenly changed or disappeared. Thus, at least 43 Hawaiian honeycreeper species that had thrived on the islands before humans arrived were extinct by 1778. Conservation efforts began in the 1960s, but another 43 species have since disappeared.

Today, the few remaining Hawaiian honeycreepers are still being heavily pressured by established populations of nonnative species of plants and animals (Figure 12.22A). Rising global temperatures are also allowing mosquitoes to invade high-altitude habitats that had previously been too cold for the insects, so honey-creeper species remaining in these habitats are now succumbing to mosquito-borne diseases (Figure 12.22B). Of the 18 remaining honeycreeper species, only two are not in danger of extinction.

As more and more honeycreeper species become extinct, the group's reservoir of genetic diversity dwindles. The lowered diversity means the group as a whole is less resilient to change, and more likely to suffer catastrophic losses. Deciphering their phylogeny can tell us which honeycreeper species are most different from the others—and those are the ones most valuable in terms of preserving the group's genetic diversity. Such research allows us to concentrate our resources and conservation efforts on those species whose extinction would mean a greater loss to biodiversity. For example, we now know the poouli (Figure 12.22C) to be the most distant relative in the Hawaiian honeycreeper family. Unfortunately, the knowledge came too late; the poouli is probably extinct now. Its extinction means the loss of a large part of evolutionary history of the group: One of the longest branches of the honeycreeper family tree is gone forever.

Cladistics analyses are also used to correlate past evolutionary divergences with behavior and dispersal patterns of existing populations. Such studies are useful in conservation efforts. For example, a decline in antelope populations in African savannas is at least partly due to competition with domestic cattle. A cladistic analysis of mitochondrial DNA sequences suggested that current populations of blue wildebeest (Figure 12.23) are genetically less similar than they should be, based on other antelope groups of similar age. Combined with behavioral and geographic data, the analysis helped conservation biologists realize that a patchy distribution of preferred food plants is preventing gene flow among blue wildebeest populations. The absence of gene flow can lead to a catastrophic loss of genetic diversity in populations under pressure. Restoring appropriate grasses in intervening, unoccupied areas of savanna would allow isolated wildebeest populations to reconnect.

Researchers often study the evolution of viruses and other infectious agents by grouping them into clades based on biochemical traits. Even though viruses are not alive, they can mutate every time they infect a host, so their genetic material changes quickly. Consider the H5N1 strain of influenza (flu) virus, which infects birds and other animals. H5N1 has a very high mortality rate in humans, but human-to-human transmission has been rare to date. The virus replicates in pigs without causing symptoms. Pigs transmit the virus to other pigs—and apparently to humans too. A phylogenetic analysis of H5N1 isolated from pigs showed that the virus "jumped" from birds to pigs at least three times since 2005, and that one of the isolates had acquired the potential to be transmitted among humans. Our increased understanding of the evolutionary history of this virus is helping us develop strategies to prevent it from spreading to humans again.

Figure 12.23 A blue wildebeest in Africa.
Conservation biologists discovered that a patchy availability of preferred food was hampering gene flow among wildebeest populations. The biologists recommended restoring grasses in some areas that had been cleared, to reestablish gene flow among isolated wildebeests.

Alan Lucas/Shutterstock.

Every living thing is related if you just look back far enough in time.

Take-Home Message 12.8

Why do we study evolutionary history?

• Evolutionary biologists study phylogeny in order to understand how all species are connected by shared ancestry.

• Among other applications, phylogeny research can help us to prioritize efforts to preserve endangered species, and to understand the spread of infectious diseases.

Summary

 Section 12.1 Populations tend to change along with the selection pressures that operate on them. Our overuse of antibiotics exerts directional selection favoring resistant bacterial populations, which are now common in the environment. We are running out of effective antibiotics to use as human drugs.

 Section 12.2 Alleles, which arise by mutation, are the basis of differences in the forms of traits shared by a species. All alleles of all genes in a population constitute a **gene pool**. **Microevolution**, which is change in **allele frequency** in a gene pool, occurs constantly in natural poplations because processes that drive it are always operating in nature.

 Section 12.3 Natural selection can occur in patterns. In **directional selection**, forms of a trait at one end of a range of variation are most adaptive. An intermediate form of a trait is most adaptive in **stabilizing selection**. In **disruptive selection,** extreme forms of a trait are adaptive and midrange forms are selected against.

 Section 12.4 **Sexual selection** is a mode of natural selection in which the adaptive traits are those that make their bearers better at securing mates. Any mode of natural selection can maintain multiple alleles at relatively high frequency in a population.

 Section 12.5 Allele frequency can change due to chance alone. This **genetic drift**, which is most pronounced in small populations, can lead to the loss of genetic diversity and cause alleles to become **fixed**. Genetic diversity may be reduced in populations that are **inbred**, and also in those that undergo an evolutionary **bottleneck** or have been founded by a small group of individuals (the **founder effect**). **Gene flow** can stabilize or change allele frequency.

 Section 12.6 The details of **speciation** differ every time it occurs, but **reproductive isolation**, the end of gene flow between populations, is always a part of the process. With **allopatric speciation**, a geographic barrier arises and interrupts gene flow between populations. After gene flow ends, genetic divergences that occur independently in the separated populations result in separate species. Speciation can also occur in the absence of a barrier to gene flow. **Sympatric speciation** occurs by genetic divergence within a population.

 Section 12.7 **Macroevolution** refers to large-scale patterns of evolution. A lineage may change very little over evolutionary time. In some cases, a body structure used for a particular purpose in a lineage served a different purpose when it first evolved in an ancestor. A **key innovation** can result in an **adaptive radiation**, or rapid diversification of a lineage into several new species. **Coevolution** occurs when two species act as agents of selection upon one another. A lineage with no more living members is **extinct**.

 Section 12.8 Evolutionary biologists reconstruct evolutionary history (**phylogeny**) by looking for derived traits. A **clade** consists of an ancestor in which a derived trait evolved, together with all of its descendants. Making hypotheses about the evolutionary history of a clade is called **cladistics**. These hypotheses are often represented as **cladograms**, which are diagrams of evolutionary connections among a group of clades. Each line in a cladogram represents a lineage, and a point where one lineage branches into two represents a shared ancestor.

Reconstructing phylogeny, which is based on the premise that all organisms are connected by shared ancestry, helps us preserve endangered species. It is also useful for studying the spread of viruses and other agents of infectious diseases.

Self-Quiz

Answers in Appendix I

1. _____ is the original source of new alleles.
   a. Mutation
   b. Natural selection
   c. Genetic drift
   d. Gene flow
   e. All are original sources of new alleles

2. Which is required for evolution to occur in a population?
   a. genetic diversity
   b. selection pressure
   c. gene flow
   d. none of the above

3. Match the modes of natural selection with their best descriptions.
   _____ stabilizing
   _____ disruptive
   a. eliminates extreme forms of a trait
   b. eliminates midrange forms of a trait

4. Sexual selection frequently influences aspects of body form and can lead to _____ .
   a. a sexual dimorphism
   b. male aggression
   c. exaggerated traits
   d. all of the above

5. The persistence of the sickle allele at high frequency in a population is a case of _____ .
   a. bottlenecking
   b. balanced polymorphism
   c. the founder effect
   d. inbreeding

6. _____ among populations can keep them similar to one another.
   a. Genetic drift
   b. Gene flow
   c. Mutation
   d. Natural selection

7. The theory of natural selection does not explain _____ .
   a. genetic drift
   b. the founder effect
   c. gene flow
   d. how mutations arise
   e. inheritance
   f. any of the above

8. Which of the following is *not* part of how we define a species?
   a. Its individuals appear different from other species.
   b. It is reproductively isolated from other species.
   c. Its populations can interbreed.
   d. Fertile offspring are produced.

9. Which of the following statements is correct?
   a. Genetic drift occurs only in small populations.
   b. Inbreeding increases genetic diversity.
   c. Gene flow can introduce new alleles into a population.

10. After fire devastates all of the trees in a wide swath of forest, populations of a species of tree-dwelling frog on either side of the burned area diverge to become separate species. This is an example of _____ .
    a. allopatric speciation
    b. adaptive radiation
    c. sympatric speciation
    d. an evolutionary bottleneck

11. Sex in many birds is typically preceded by an elaborate courtship dance. If a male's movements are unrecognized by the female, she will not mate with him. This is an example of _____ .
    a. reproductive isolation
    b. natural selection
    c. sexual selection
    d. all of the above

12. _____ is a way of reconstructing evolutionary history based on derived traits.
    a. Natural selection
    b. Phylogeny
    c. Gene flow
    d. Cladistics

13. The evolution of wings helped the insect clade to be very successful. In this example, wings are a(n) _____ .
    a. derived trait
    b. adaptive trait
    c. key innovation
    d. all of the above

14. In evolutionary trees, each line represents a(n) _____ .
    a. lineage
    b. extinction
    c. point of divergence
    d. adaptive radiation

15. Match the evolution concepts.
    _____ gene flow
    _____ sexual selection
    _____ derived trait
    _____ extinct
    _____ genetic drift
    _____ natural selection
    _____ cladogram
    _____ adaptive radiation
    _____ phylogeny
    _____ coevolution

    a. can lead to interdependent species
    b. changes in a population's allele frequencies due to chance alone
    c. alleles enter or leave a population
    d. evolutionary history
    e. adaptive traits make their bearers better at securing mates
    f. burst of divergences from one lineage into many
    g. no more living members
    h. diagram of sets within sets
    i. present in a group, but not in any of the group's ancestors
    j. operates on variations in shared traits

Critical Thinking

1. Species have traditionally been characterized as "primitive" and "advanced." For example, mosses were considered to be primitive, and flowering plants advanced; crocodiles were primitive and mammals were advanced. Why do most biologists of today think it is incorrect to refer to any modern species as primitive?

2. Rama the cama, a llama–camel hybrid, was born in 1997. The idea was to breed an animal that has the camel's strength and endurance, and the llama's gentle disposition. However, instead of being large, strong, and sweet, Rama is smaller than expected and has a camel's short temper. The breeders plan to mate him with Kamilah, a female cama. What potential problems with this mating should the breeders anticipate?

3. Two species of antelope, one from Africa, the other from Asia, are put into the same enclosure in a zoo. To the zookeeper's surprise, individuals of the different species begin to mate and produce healthy, hybrid baby antelopes. Explain why a biologist might not view these offspring as evidence that the two species of antelope are in fact one.

4. Some human traits may have arisen by sexual selection. Over thousands of years, women attracted to charming, witty men perhaps prompted the development of human intellect beyond what was necessary for mere survival. Men attracted to women with juvenile features may have shifted the species as a whole to be less hairy and softer featured than any of our simian relatives. Can you think of a way to test these hypotheses?

13

EARLY LIFE FORMS AND THE VIRUSES

Application ❯ 13.1 The Human Microbiome

REMEMBER: Biologists divide all life into three domains: Bacteria, Archaea, and Eukarya (Section 1.4).

Figure 13.1 Microbes that can live in the human gut.

(A) Dr. Fred Hossler/Visuals Unlimited, Inc.; (B) CDC/Dr. Stan Erlandsen.

A. Bacteria (*Bilophila wadsworthia*). This species is most abundant in people who eat a lot of animal fat, and it may promote irritable bowel disease.

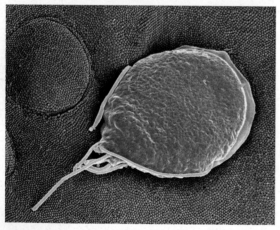

B. The flagellated protist *Giardia lamblia*. It helps fend off parasitic worms, but causes diarrhea and abdominal pain in some infected people.

microbiome Collection of microorganisms that inhabits a specific habitat, such as a human body.

pathogen Disease-causing agent.

The first forms of life lived as single cells and the overwhelming majority of modern organisms do the same. Consider that the single-celled organisms living in and on your body outnumber your cells by about ten to one. Even when healthy, a person is host to bacteria, archaea, protists, fungi, and viruses. The Human Microbiome Project, a collaborative endeavor that began in 2007, aims to identify the types of microorganisms we support—our **microbiome**—and how their presence affects our health and well-being. The project has already turned up some interesting findings.

Each person has a unique microbiome and its composition changes over time. The womb is sterile, so we acquire our first microorganism during birth. During normal childbirth, an infant becomes coated with bacteria from its mother's vagina. Some of these bacteria are swallowed and colonize the infant's digestive tract, where they help the infant digest milk sugars. Picking up maternal bacteria during birth may also kick-start the infant's immune system. Surgical delivery (delivery by a cesarean section or C-section) increases the risk of allergies and other immune disorders later in life. By one hypothesis, this heightened susceptibility to immune problems results from the failure of a surgical delivery to provide the infant with the types of bacteria that normally encourage immune development.

The current mix of species in your gut depends in part on your diet, and that species assortment may in turn affect your health. For example, some diets promote the growth of bacterial **pathogens**, or agents of disease. Consider that a diet rich in grain products selects for gut bacteria that can break down complex carbohydrates. An abundance of one such species, *Prevotella copri*, is associated with an increased risk for rheumatoid arthritis, a disorder in which the immune system attacks the joints. On the other hand, eating an excessive amount of animal fat selects for bacteria that can metabolize bile, an acidic substance essential to fat digestion. One bile-tolerant species, *Bilophila wadsworthia* (Figure 13.1A), increases the risk of irritable bowel disease (IBD). With IBD, chronic inflammation of the large intestine causes cramping, diarrhea, and weight loss. An estimated 14 million Americans have IBD and the prevalence of the disorder is on the rise worldwide. Increasing consumption of fat-rich, processed foods that encourage the growth of *B. wadsworthia* may be a factor in this increase.

The overwhelming majority of organisms that can live in the human gut are bacteria, but some protists also call the human gut home. The flagellated protozoan *Giardia lamblia* (Figure 13.1B) is one example. *G. lamblia* was first described in the late 1600s by Antoni van Leeuwenhoek, a pioneer in the use of microscopy. Van Leeuwenhoek noticed the random jerky movements of *G. lamblia* while examining a sample of his own feces. Today, *G. lamblia* infects an estimated 2 to 5 percent of adults in the United States. In nations where sanitation is poor, the adult infection rate can be as high as 30 percent. Unfiltered drinking water from streams or lakes and unpasteurized dairy products are common sources of infection.

G. lamblia infection sometimes cause IBD-like symptoms, but many infected people have no ill effects. Some even benefit from the protist's presence. Having *G. lamblia* in the gut reduces one's risk of infection by parasitic worms. Whether people infected by *G. lamblia* become ill depends both on their general health and the particular strain of *G. lamblia* that infects them. (In microbiology, the term "strain" refers to a genetically distinct subtype of a particular microorganism.)

13.2 On the Road to Life

REMEMBER: All organisms consist of complex organic compounds that they assembled from organic monomers (Section 2.6).

Conditions on the Early Earth Scientists estimate that Earth formed by about 4.6 billion years ago through the aggregation of dust and rock bits that were orbiting our sun. The composition of Earth's early atmosphere remains a matter of debate, but geologic evidence suggests our planet started out with little or no free oxygen (O_2). Had O_2 been present early on, we would see evidence of iron oxidation (rust formation) in Earth's most ancient rocks. However, these rocks show no sign of such oxidation. The apparent lack of O_2 interests scientists because it would have facilitated some proposed steps on the path to life. Had O_2 been present, oxidation reactions would have broken apart small organic compounds as quickly as they formed.

Liquid water is essential to life as we know it because molecules that carry out metabolic reactions have to be dissolved in water. At first, Earth's surface was molten rock, so all water was in the form of vapor. However, examination of crystals in ancient rocks indicates that by 4.3 billion years ago, Earth had cooled enough for water to pool on its surface.

Origin of the Building Blocks of Life Until the early 1800s, chemists thought that organic molecules possessed a special "vital force" and could only be made by living organisms. Then in 1825, a German chemist synthesized urea, a molecule abundant in urine. Later, another chemist made alanine, an amino acid. These synthetic reactions showed that nonliving mechanisms could yield organic molecules.

Today, there are three main hypotheses concerning the source of the organic building blocks for Earth's first life.

1. *Lightning fueled atmospheric reactions.* In 1950s, Stanley Miller and his colleagues tested the hypothesis that lightning-fueled atmospheric reactions could have produced simple organic compounds. They filled a reaction chamber with a mix of gases designed to simulate Earth's early atmosphere, then circulated the mixture while zapping it with sparks from electrodes (Figure 13.2). Within a week, this process produced simple organic compounds, including some amino acids present in living organisms.

2. *Delivery from space via meteorites.* The presence of amino acids, sugars, and nucleotide bases in meteorites that fell to Earth suggests an alternative origin for life's building blocks. Organic monomers that formed in interstellar clouds of ice, dust, and gases could have been delivered to Earth by meteorites. Keep in mind that during Earth's early years, meteorites fell to Earth thousands of times more frequently than they do today.

3. *Reactions at deep-sea hydrothermal vents.* Life's building blocks may also have formed in the sea, fueled by heat from hydrothermal vents. A **hydrothermal vent** is like an underwater geyser, a place where mineral-rich water heated by geothermal energy streams out through a rocky opening in the seafloor (Figure 13.3). Amino acids form spontaneously in a simulated vent environment.

Note that the three possible sources of organic monomers discussed above are not mutually exclusive. Most likely all three contributed to an accumulation of simple organic compounds in Earth's early seas.

Figure 13.2 Stanley Miller's experimental apparatus. It was used to test whether lightning-fueled reactions could have formed organic monomers in Earth's early atmosphere. Water vapor, hydrogen gas (H_2), methane (CH_4), and ammonia (NH_3) circulated in a glass chamber to simulate the atmosphere. Sparks provided by an electrode simulated lightning.

to vacuum pump — electrodes — spark discharge — CH_4 NH_3 H_2O H_2 gases — water out — condenser — water in — water droplets — water containing organic compounds — boiling water — liquid water in trap

Figure It Out: Why didn't Miller include oxygen (O_2) in the mix of gases?

Answer: Earth's early atmosphere lacked oxygen.

Figure 13.3 A hydrothermal vent on the seafloor. At such vents, mineral-rich water heated by geothermal energy streams out into cold ocean water.

Courtesy of the University of Washington.

hydrothermal vent Underwater opening from which mineral-rich water heated by geothermal energy streams out.

Simulations and experiments cannot prove how life or cells began, but they can show us what is plausible.

Origin of Metabolism Modern cells take up small organic molecules, concentrate them, and assemble them into larger organic polymers. Before there were cells, a nonbiological process that concentrated organic subunits would have increased the chance of polymer formation.

By one hypothesis, organic polymers began to form on clay-rich tidal flats. Clay particles have a slight negative charge, so positively charged molecules in seawater stick to them. At low tide, evaporation would have concentrated the subunits even more, and energy from sunlight might have induced the formation of polymers. Amino acids do form short chains under simulated tidal flat conditions.

The **iron–sulfur world hypothesis** proposes that early metabolic reactions took place in rocks around hydrothermal vents. Such rocks are porous, with many tiny chambers about the size of cells. Metabolism may have begun when iron sulfide in the rocks donated electrons to dissolved carbon monoxide (CO), setting in motion reactions that led to formation of larger organic compounds. In simulations of vent conditions, organic compounds such as pyruvate do form and accumulate. In addition, iron–sulfur clusters serve as cofactors in all modern organisms. The clusters function as electron donors in essential metabolic reactions. A universal requirement for iron–sulfur cofactors may be a legacy of life's rocky beginnings.

Origin of Genetic Material DNA is the genetic material in all modern cells. Cells pass copies of their DNA to descendant cells, which use instructions encoded in DNA to build proteins. Some of these proteins aid synthesis of new DNA, which is passed along to descendant cells, and so on. Protein synthesis depends on DNA, which is built by proteins. How did this cycle begin?

In the 1960s, Francis Crick and Leslie Orgel addressed this dilemma by proposing the **RNA world hypothesis**: Early on, RNA served a dual role, functioning both

Figure 13.4 Protocells.
Scientists test hypotheses about protocell formation by carrying out laboratory simulations and field experiments.

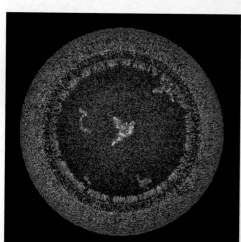

A. Illustration of a laboratory-produced protocell with a bilayer membrane of fatty acids and strands of RNA inside.

© Janet Iwasa.

B. Micrograph of a laboratory-formed protocell with RNA-coated clay (red) surrounded by fatty acids and alcohols.

From Hanczyc, Fujikawa, and Szostak, "Experimental Models of Primitive Cellular Compartments: Encapsulation, Growth, and Division"; www.sciencemag.org, *Science* 24 October 2003; 302;529, Fig. 2, p. 619. Reprinted with permission of the authors and AAAS.

C. Field-testing a hypothesis about protocell formation. David Deamer pours a mix of small organic molecules and phosphates into a pool heated by volcanic activity.

Photo by Tony Hoffman, courtesy of David Deamer.

as a genome and as a catalyst. Evidence that RNA can both store genetic information and function like an enzyme in protein synthesis supports this hypothesis. RNAs that function as enzymes (called ribozymes) are common in living cells. For example, some ribozymes cut noncoding bits (introns) out of newly formed RNAs (Section 7.3), and the rRNA in ribosomes speeds formation of peptide bonds during protein synthesis (Section 7.5).

If the earliest self-replicating genetic systems were RNA-based, why do all cells now have a genome of DNA? What was the selective advantage of DNA-based systems? The difference in stability of the two nucleic acids was probably a factor. Compared to a double-stranded DNA molecule, a single-stranded RNA breaks more easily and is more prone to replication errors. Thus, a switch from RNA to DNA would have made larger, more stable genomes possible.

Origin of Cell Membranes Self-replicating molecules and products of other early synthetic reactions would have floated away from one another unless something enclosed them. In modern cells, a plasma membrane serves this function. If the first reactions took place in tiny rock chambers, rock would have acted as a boundary. Over time, lipids produced by reactions inside such a chamber could have accumulated and lined the chamber wall, forming a protocell. A **protocell** is a membrane-enclosed collection of interacting molecules that can take up material and replicate. Scientists hypothesize that protocells were the ancestors of cellular life.

Researchers have combined organic molecules in the laboratory to yield synthetic protocells that have some lifelike properties. For example, some experiments have produced vesicle-like spheres in which a bilayer of fatty acids surrounds molecules of RNA (Figure 13.4A, B). These spheres "grow" by taking up and incorporating fatty acids and nucleotides from their surroundings. Mechanical force causes the spheres to divide into smaller spheres that have the same composition.

Biochemist David Deamer thinks that the conditions in hot, acidic pools near ancient volcanoes would have encouraged formation of protocells. To test this hypothesis, he carries out experiments both in the laboratory and in the field near currently active volcanoes (Figure 13.4C). His results show that conditions in these pools do favor formation of fatty acids that can self-assemble as vesicles.

Simulations and experiments cannot prove how life or cells began, but they can show us what is plausible. A variety of investigations by many researchers tell us this: Chemical and physical processes that operate today can produce simple organic compounds, concentrate them, and assemble them into protocells (Figure 13.5). Billions of years ago, the same processes may have led to the first life.

Figure 13.5 Proposed sequence for the evolution of cells. Scientists carry out experiments and simulations that test the feasibility of each step.

Take-Home Message 13.2

What do scientific studies reveal about the origin of life?

- Small organic subunits could have formed on the early Earth, or formed in space and fallen to Earth on meteorites.
- Complex organic molecules could have self-assembled from simpler ones.
- The first genetic material may have been RNA rather than DNA.
- Protocells—chemical-filled membranous sacs that grow and divide—may have been the ancestors of the first cells.

iron–sulfur world hypothesis Hypothesis that life began in rocks rich in iron sulfide near deep-sea hydrothermal vents.

protocell Membranous sac that contains interacting organic molecules; hypothesized to have formed prior to the earliest cells.

RNA world hypothesis Hypothesis that RNA served as the first material of inheritance.

Figure 13.6 Stromatolites.
The filament shown in the inset may be a chain of 3.5-billion-year-old fossil bacteria from an ancient stromatolite. The underlying photo shows modern stromatolites in Australia's Shark Bay. Each consists of living photosynthetic bacteria atop the remains of countless earlier generations of cells and the sediment that they trapped.

Background, Michael Aw/Lonely Planet Images/Getty Images; inset, Courtesy of John Fuerst, University of Queensland. Originally published in *Archives of Microbiology* vol. 175, p. 413–29, Lindsay MR, Webb RI, Strous M, Jetten MS, Butler MK, Forde RJ, Fuerst JA. Cell compartmentalisation in planctomycetes: Novel types of structural organization for the bacterial cell. *Arch. Microbiol.* 2001 Jun, 175(6):413–29.

13.3 Origin of the Three Domains

REMEMBER: Photosynthesis releases oxygen (Section 5.3), and aerobic respiration requires it (5.5). UV radiation can cause mutations (6.4).

Reign of the Prokaryotes The processes described in the previous section may have produced cellular life more than once. If so, all but one of those early lineages have become extinct. Studies of modern genomes tell us that all modern species descended from a common single-celled ancestor, a cell that lived perhaps as early as 4 billion years ago. Given what scientists know about relationships among modern species, most assume that this ancestor was prokaryotic, meaning it did not have a nucleus. Oxygen was scarce on the early Earth, so the ancestral cell must also have been anaerobic (capable of living without oxygen).

The two domains of prokaryotic cells, Bacteria and Archaea, diverged very early in the history of life. Shortly after this divergence, some bacteria began to capture and use light energy in photosynthetic pathways that did not produce oxygen. The photosynthetic bacteria grew in the sea as dense mats that trapped sediments. Over many years, cell growth and sediment deposition formed dome-shaped, layered structures called **stromatolites**, some of which were preserved as the earliest known fossils (Figure 13.6).

By 2.7 billion years ago, one lineage of bacteria began to carry out photosynthesis by the oxygen-releasing pathway. As a result of their activity, oxygen began to accumulate in the air and water. The rise in oxygen had two important consequences. First, oxygen created a new selective pressure, putting organisms that thrived in higher-oxygen conditions at an advantage. The pathway of aerobic respiration evolved and became widespread. This pathway requires oxygen, and it is far more efficient at releasing energy from organic molecules than other pathways. Second, ozone gas (O_3) formed and accumulated as the **ozone layer** in the upper atmosphere. The ozone layer prevents much of the sun's ultraviolet (UV) radiation from reaching Earth's surface. Such radiation can damage DNA and other biological molecules. Water screens out some UV radiation, but without the ozone layer to protect it, life could not have moved onto land.

Origin of Eukaryotes Eukaryotes first appear in the fossil record about 1.8 million years ago. All eukaryotes have a nucleus and an associated endomembrane system. These structures probably evolved from infoldings of the plasma membrane in a prokaryotic ancestor (Figure 13.7 **1**, **2**). No prokaryotes have a nuclear envelope, but some bacteria do have infoldings of their plasma membrane. Such folds provide multiple advantages. They increase the surface area that can hold membrane-associated enzymes and divide the cytoplasm into compartments inside which specific reactions can take place. If the membrane infoldings enclose a cell's DNA, they also protect the genetic material.

Mitochondria and chloroplasts resemble bacteria in their size and shape, and they replicate independently of the cell that holds them. Like bacteria, they have their own DNA in the form of a single circular chromosome. They also have at least two outer membranes, with the innermost membrane structurally similar to a bacterial plasma membrane. The **endosymbiont hypothesis** explains these similarities by proposing that mitochondria and chloroplasts are descended from bacteria that entered and lived inside a host cell. (*Endo–* means within; *symbiosis* means living together.) Endosymbionts that live inside a cell can be passed to the cell's descendants when the cell divides.

Evolution of mitochondria began when aerobic bacteria were taken up by or invaded an archaeal cell, then lived and replicated inside it **3**. When the host cell divided, it passed some endosymbionts along to its offspring. As the two species lived together over many generations, genes carried by both partners were free to mutate and to move between the host and its guests. Eventually, the host and endosymbionts could not live independently—the endosymbionts had become mitochondria **4**.

The endosymbiont hypothesis explains why gene sequence comparisons indicate that eukaryotes have both archaeal and bacterial ancestors. Eukaryotic genes that govern basic genetic processes (DNA replication, transcription, and translation) were passed down from an archaeal ancestor. By contrast, genes governing some metabolic processes came from the bacterial ancestors of mitochondria.

Although all eukaryotic lineages have mitochondria or organelles derived from them, only some have chloroplasts. Thus, biologists think that the two types of organelles were acquired independently. The first chloroplasts evolved from oxygen-producing bacteria that were engulfed by and lived in an early eukaryote **5**.

Figure 13.7 Evolution of eukaryotic organelles. The nuclear envelope and endomembrane components are derived from infoldings of the plasma membrane. Mitochondria and chloroplasts evolved from bacteria that lived inside a host cell.

Labels in figure:
ancestral archaeal cell — DNA
1 Portions of the plasma membrane fold inward.
3 Aerobic bacteria enter and live inside an archaeal cell.
2 Infoldings evolve into the nuclear envelope and the endomembrane system.
4 Over generations, the aerobic bacteria evolve into mitochondria.
5 Photosynthetic bacteria enter an early eukaryote and, over generations, evolve into chloroplasts.

Take-Home Message 13.3

How did the three domains arise?

- The first cells may have arisen as early as 4 billion years ago; they were anaerobic and prokaryotic.
- An early divergence separated ancestors of modern bacteria and archaea.
- Eukaryotes have a mixed ancestry. Their basic genetic apparatus is derived from archaea, but mitochondria and chloroplasts are descendants of bacteria.

endosymbiont hypothesis Hypothesis that mitochondria and chloroplasts evolved from free-living bacteria that entered and lived inside another cell.

ozone layer Atmospheric layer with a high concentration of ozone that prevents much UV radiation from reaching Earth's surface.

stromatolites Dome-shaped structures composed of layers of prokaryotic cells and sediments; form in shallow seas.

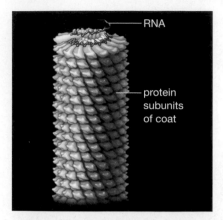

RNA

protein
subunits
of coat

A. Tobacco mosaic virus, a helical virus
that infects tobacco and related plants.

After Stephen L. Wolfe.

B. An adenovirus, a polyhedral virus
that infects animals. The 20-sided coat
encloses double-stranded DNA.

© Dr. Richard Feldmann/National Cancer Institute.

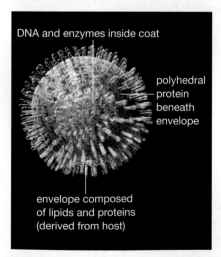

DNA and enzymes inside coat

polyhedral
protein
beneath
envelope

envelope composed
of lipids and proteins
(derived from host)

C. A herpesvirus, an enveloped virus that
infects animals. The envelope is derived
from a host cell.

© Russell Knightly/Science Source.

Figure 13.8 Examples of virus structure.

13.4 Viruses

REMEMBER: Viruses that infect bacteria played a role in the discovery that DNA is
the molecule of inheritance (Section 6.2).

In the late 1800s, biologists studying diseased tobacco plants discovered a previously unknown pathogen. It was smaller than even the smallest cells, and could not
be seen with a light microscope. The scientists called this unseen infectious agent a
virus, a term that means "poison" in Latin. Today, we define a **virus** as a noncellular
infectious particle that replicates only inside a living cell.

We do not know how viruses are related to cellular life. The fact that they can
only replicate inside cells suggests that they may have evolved from cells. Alternatively, viruses may be remnants of a time before cells.

Viral Structure and Replication A free viral particle (one that is not inside
a cell) always includes a viral genome enclosed within a protein coat. The viral
genome may be RNA or DNA, and it may be single-stranded or double-stranded.
The viral coat consists of many protein subunits that bond together in a repeating
pattern, producing a helical rod (Figure 13.8A) or many-sided (polyhedral) structure (Figure 13.8B). The coat protects the viral genetic material and plays a role in
infection. In all viruses, components of the viral coat bind to proteins at the surface
of a host cell. The coat may also enclose some viral enzymes that will act within the
host. In many animal-infecting viruses, the protein coat is enclosed within a **viral
envelope** (Figure 13.8C). The viral envelope is layer of cell membrane derived from
the host cell in which the viral particle formed.

Viral replication cycles vary in their details, but nearly all include the following steps. The virus first attaches to an appropriate host cell by binding to a specific
protein or proteins in the host's plasma membrane. Once a virus comes into contact
with and attaches to an the host cell, the viral genome, and in some cases other viral
components, enter into that cell.

A viral infection is like a cellular hijacking. Viral genes take over a host's cellular
machinery. They direct the cell to replicate viral DNA or RNA and to build viral
proteins. These viral components self-assemble to form new viral particles. The
particles may be released when the infected host cell bursts (lyses) or they may bud
from the host cell, taking some of its plasma membrane with them.

Bacteriophages **Bacteriophages**, sometimes called phages, are nonenveloped
viruses that infect bacteria. You learned earlier how Hershey and Chase used one
type of bacteriophage to identify DNA as the genetic material of all organisms (Section 6.2). This bacteriophage, called lambda, has a complex structure. A headlike
protein coat encloses the viral DNA. Other protein components allow the virus to
bind to a bacterium, pierce it, and inject viral DNA into it.

Bacteriophages replicate in bacteria by two pathways. Both pathways begin
when a bacteriophage attaches to a bacterial cell and injects its DNA (Figure 13.9).
In the lytic pathway, viral genes are expressed immediately ❶. The infected host
first produces viral components that self-assemble as virus particles. Then a viral-encoded enzyme breaks down the host's cell wall. Breakdown of the cell wall kills
the cell and releases viral particles into the environment.

In the lysogenic pathway, viral DNA becomes integrated into the host cell's
genome and viral genes are not immediately expressed, so the cell remains
healthy ❷. When the cell reproduces, viral DNA is copied and passed to the cell's

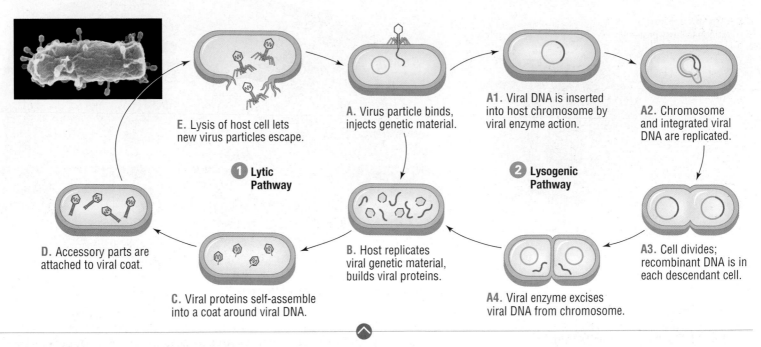

E. Lysis of host cell lets new virus particles escape.

A. Virus particle binds, injects genetic material.

A1. Viral DNA is inserted into host chromosome by viral enzyme action.

A2. Chromosome and integrated viral DNA are replicated.

1 Lytic Pathway

2 Lysogenic Pathway

D. Accessory parts are attached to viral coat.

B. Host replicates viral genetic material, builds viral proteins.

A3. Cell divides; recombinant DNA is in each descendant cell.

C. Viral proteins self-assemble into a coat around viral DNA.

A4. Viral enzyme excises viral DNA from chromosome.

Figure 13.9 The two bacteriophage replication pathways.

Left photo, Science Photo Library/Science Source.

descendants along with the host's genome. Like miniature time bombs, the viral DNA inside the new cells awaits a signal to enter the lytic pathway.

Some bacteriophages can only replicate by the lytic pathway. They always kill their host cell quickly and are not passed from one bacterial generation to the next. Others embark upon either the lytic or lysogenic pathway, depending on conditions in the host cell.

Plant Viruses Plant viruses are typically nonenveloped, with a helical structure and a genome of single-stranded RNA. The tobacco mosaic virus, illustrated in Figure 13.8A, is an example. Plant cells have a thick wall, so plants usually become infected only after insects, pruning, or some other mechanical injury creates a wound that allows the virus into a cell. Sucking insects such as aphids and whiteflies are the vector for many viral plant diseases. A **disease vector** is an organism that transmits a pathogen from one host to the next. Once a plant has become infected by a virus, little can be done to treat it. Thus, protecting crop plants from viruses depends mainly upon controlling insect vectors of viral diseases and breeding plants that are virus resistant.

Viruses and Human Health Some viruses have a beneficial effect on human health. For example, certain bacteriophages in the mucus that coats our airways and our gut help keep bacterial pathogens from infecting us.

Other viruses are themselves human pathogens. Most of these viruses produce mild symptoms and trouble us only briefly. For example, some rhinoviruses infect membranes of our upper respiratory system and cause common colds. Such a cold ends when the immune system eliminates all virus-infected cells. A minority of viral diseases are more persistent. Herpesviruses cause cold sores, genital herpes, mononucleosis, or chicken pox. Typically the initial infection causes symptoms for only a short time. However, the virus remains in the body in a latent state, and can reawaken later on. The herpes simplex virus 1 (HSV-1) can remain latent in nerve

A viral infection is like a cellular hijacking. Viral genes take over a host's cellular machinery.

bacteriophage Virus that infects bacteria.

disease vector Organism that carries a pathogen from one host to the next.

viral envelope A layer of cell membrane derived from the host cell in which an enveloped virus was produced.

virus A noncellular infectious particle with a protein coat and a genome of RNA or DNA; replicates only in living cells.

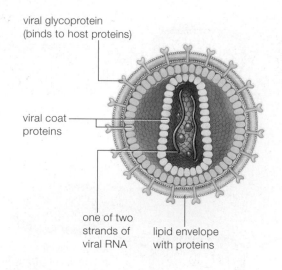

vial glycoprotein
(binds to host proteins)

viral coat
proteins

one of two
strands of
viral RNA

lipid envelope
with proteins

A. Structure of the virus.

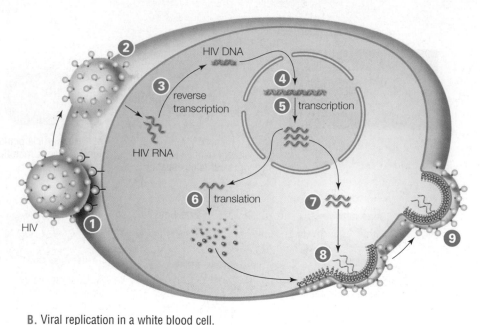

HIV DNA

②

③ reverse transcription

④

⑤ transcription

HIV RNA

⑥ translation

⑦

⑧

⑨

HIV

①

B. Viral replication in a white blood cell.

Figure 13.10 HIV, an enveloped RNA virus.

① Viral protein binds to proteins at the surface of a white blood cell.

② Viral RNA and enzymes enter the cell.

③ Viral reverse transcriptase uses viral RNA to make double-stranded viral DNA.

④ Viral DNA enters the nucleus and becomes integrated into the host genome.

⑤ Transcription produces viral RNA.

⑥ Some viral RNA is translated to produce viral proteins.

⑦ Other viral RNA forms the new viral genome.

⑧ Viral proteins and viral RNA self-assemble at the host plasma membrane.

⑨ New virus buds from the host cell, with an envelope of host plasma membrane.

Figure It Out: What is the product of reverse transcription of HIV RNA?
Answer: Double-stranded DNA

cells for years. When activated, the virus replicates and causes painful "cold sores" on the edge of the lips. Another type of herpesvirus causes genital herpes.

Some viruses can cause cancer. A few strains of human papillomavirus (HPV) can cause cancers of the cervix, penis, anus, or mouth. Infection by some hepatitis viruses increases the risk of liver cancer.

HIV—The AIDS Virus HIV (**human immunodeficiency virus**) is an enveloped RNA virus that replicates inside human white blood cells (Figure 13.10). It attaches to a cell via a glycoprotein that extends out beyond the viral envelope **①**. After attachment, the viral envelope fuses with the blood cell's plasma membrane, releasing viral enzymes and RNA into the cell **②**. A viral enzyme called reverse transcriptase uses viral RNA as a template to synthesize a double-stranded DNA **③**. This DNA enters the nucleus together with another viral enzyme that inserts the DNA into one of the host's chromosomes **④**. Once integrated, the viral DNA is replicated and transcribed along with the host genome **⑤**. Some of the resulting viral RNA is translated into viral proteins **⑥** and some becomes the genetic material of new HIV particles **⑦**. The particles self-assemble at the plasma membrane **⑧**. As the virus buds from the host cell, some of the host's plasma membrane becomes the viral envelope **⑨**. Each new virus can then infect another white blood cell. New HIV-infected cells are also produced when an infected cell replicates. The disease AIDS (acquired immune deficiency syndrome) arises as a result of HIV's detrimental effects on the immune system. We consider these effects in detail in Chapter 22.

The most common strain of HIV (HIV-1) evolved in west central Africa from a virus that infects nonhuman primates. In the mid-1960s, HIV-1 was introduced to Haiti, where it diversified and acquired distinctive mutations. By 1969, HIV-1 with Haiti-specific mutations reached the United States. It spread quietly until AIDS was identified as a threat in 1981. Today, more than 20 million people worldwide have died from AIDS. About 30 million are currently infected with HIV.

Drugs that fight HIV take aim at steps in viral replication. Some interfere with the way HIV binds to a host cell. Others impair reverse transcription or assembly of new virus particles. These antiviral drugs lower the number of HIV particles, so a person stays healthier. Lowering the concentration of HIV in body fluids also reduces the risk of passing the virus to others.

Ebola Like HIV, the Ebola virus is an enveloped RNA virus that emerged in Africa. It was identified in 1976. The virus infects fruit bats and nonhuman primates. New outbreaks in the human population arise when the virus gets into a person who has close contact with an infected animal, as by butchering it for food. The virus kills more than half of those of those it infects. Within three weeks of infection, a person develops flulike symptoms, followed by a rash, vomiting, diarrhea, and bleeding from the eyes, nose, mouth and other body openings. The virus is transmitted among people by direct contact with body fluids, so those who care for the sick must wear protective gear (Figure 13.11).

Until recently, all Ebola outbreaks had been confined to limited regions within Africa and had affected fewer than 500 people. However, an outbreak that began in Guinea in December of 2013 killed thousands and raised fears of a widespread epidemic. As of late 2014, there was no vaccine against Ebola and a limited supply of the experimental drugs that could be used to fight the disease.

New Flus Flus are caused by enveloped RNA viruses called influenza viruses. To keep up with ongoing mutations in influenza viruses, scientists create a new flu shot every year. The flu shot is a vaccine designed to protect against the influenza strains that scientists predict are most likely to pose a threat during the upcoming flu season. Unfortunately, determining which flu strains will be circulating in the future is not an exact science. Even after a flu shot, a person remains susceptible to a virus that differs from the strains targeted by the vaccine.

New influenza strains arise both through mutation and by **viral reassortment**, the swapping of genes between related viruses that infect a host at the same time (Figure 13.12). Consider what could happen if two influenza strains currently circulating underwent such a reassortment. The H5N1 strain is a bird flu that occasionally infects people who have direct contact with birds. When the virus does infect people, the death rate is high, about 60 percent. Fortunately, person-to-person transmission of the H5N1 virus is exceedingly rare. By contrast H1N1, commonly referred to as "swine flu," is easily passed between people, but only rarely deadly. The coexistence of these viruses raises the possibility of a potentially disastrous gene exchange. If H1N1 and H5N1 infected the same host simultaneously, the result could be a flu virus that is easily transmissible and deadly.

Figure 13.11 Fighting Ebola
Dr. Tom Frieden, head of U.S. Center for Disease Control, at an Ebola treatment center in Liberia during the 2014 outbreak. The micrograph on the right shows the virus, which has a threadlike structure.

Left, CDC/Sally Ezra; right, CDC/NIAID.

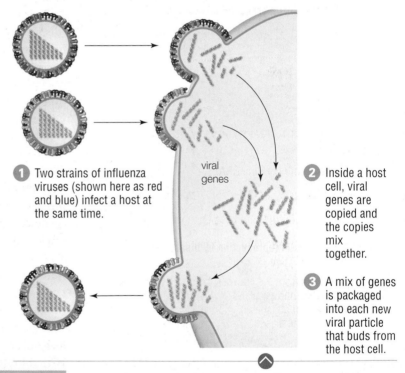

1 Two strains of influenza viruses (shown here as red and blue) infect a host at the same time.

viral genes

2 Inside a host cell, viral genes are copied and the copies mix together.

3 A mix of genes is packaged into each new viral particle that buds from the host cell.

Figure 13.12 Viral reassortment.
When a host cell is infected by two viruses of the same type, such as two influenza viruses, viral genes recombine to form viruses with new gene combinations.

Take-Home Message 13.4

What are viruses and how do they affect us?

- Viruses are noncellular particles that consist of genetic material wrapped in a protein coat. They replicate only inside living cells, and each type of virus infects and replicates inside a specific type of host.
- A virus harms and eventually kills a host cell. Viral genes direct the host cell's metabolic machinery to produce new viral particles.
- Viral genomes can be altered by mutation. Viruses with new combinations of genes also arise as a result of viral reassortment.

HIV (human immunodeficiency virus) Enveloped RNA virus that causes AIDS.

viral reassortment Two viruses of the same type infect an individual at the same time and swap genes.

13.5 Bacteria and Archaea

REMEMBER: All organisms are producers or consumers, and nutrients cycle between the two groups (Section 1.3).

Biologists have historically divided all life into two groups. Cells without a nucleus were prokaryotes and those with a nucleus were eukaryotes. More recently we learned that "prokaryotes" actually constitute two distinct lineages, now referred to as the domains Bacteria and Archaea. **Bacteria** are the more well-known and widespread group of cells that do not have a nucleus. **Archaea** are more closely related to eukaryotes than to bacteria, and many live in extreme habitats.

Structure and Function   Bacteria and archaea are small and, with rare exceptions, cannot be seen without a light microscope. Figure 13.13 shows a typical bacterial cell. It has no nucleus or membrane-enclosed organelles like those of eukaryotes. The prokaryotic chromosome (a ring of DNA) lies in the cytoplasm, as do the ribosomes. Nearly all bacteria and archaea have a porous cell wall around their plasma membrane. The wall gives the cell its shape, which may be spherical, spiral, or rod-shaped. A spherical cell is a coccus, a spiral-shaped one is a spirillum, and a rod-shaped one a bacillus. The cell depicted in Figure 13.13 is a bacillus. Like many bacteria, it has a capsule of secreted material around its cell wall.

Most bacteria can move from place to place. Some have one or more bacterial flagella that rotate like a propeller. Other bacteria glide along surfaces by using thin protein filaments called pili (singular, pilus) as grappling hooks. The pilus is extended out to a surface, sticks to it, then shortens, drawing the cell forward. Another type of pilus is used to draw cells together for gene transfers.

Reproduction and Gene Transfers   Bacteria and archaea have staggering reproductive potential. Division most commonly occurs by **binary fission**, a mechanism of asexual reproduction that yields two equal-sized, genetically identical descendant cells (Figure 13.14).   The process begins when the cell replicates its single chromosome, which is attached to the inside of the plasma membrane ❶. The DNA replica attaches to the plasma membrane adjacent to the parent molecule. Addition of new membrane and wall material elongates the cell and moves the two DNA molecules apart ❷. Then, membrane and cell wall material is deposited across the cell's midsection ❸, yielding two identical descendant cells ❹.

Bacteria and archaea do not reproduce sexually, but they can transfer genetic material among existing individuals. Three mechanisms permit such exchanges. With **transformation** (Figure 13.15A), a prokaryote takes up free DNA, such as that from a dead cell, from its environment. With **transduction** (Figure 13.15B), a virus picks up DNA from one host, then passes that DNA along to its next host. With **conjugation** (Figure 13.15C), one cell donates a small circle of DNA called a plasmid to another. A **plasmid** is a circle of double-stranded DNA with a few genes. Conjugation begins when a cell with a particular plasmid cell uses a special sex pilus to draw a cell without that plasmid close. The donor cell passes one strand of plasmid DNA to the recipient cell, then each cell makes the missing strand of DNA.

The ability of prokaryotic cells to acquire genetic information from other cells has important health implications. Suppose a gene for antibiotic resistance arises in one bacterial cell. Not only can this gene be passed on to that cell's descendants, but it can also be transferred to other existing cells. Such transfers speed the rate at which a gene spreads through a population.

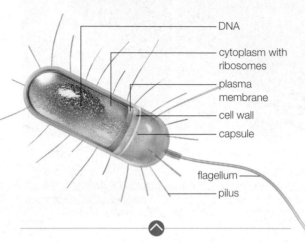

DNA
cytoplasm with ribosomes
plasma membrane
cell wall
capsule
flagellum
pilus

Figure 13.13 Generalized prokaryotic body plan.

Figure 13.14 Asexual reproduction by binary fission.

❶ A bacterium has one circular chromosome that attaches to the inside of the plasma membrane.

❷ The cell duplicates its chromosome, attaches the copy beside the original, and adds membrane and wall material between them.

❸ When the cell has just about doubled in size, a new membrane and wall are deposited across its midsection.

❹ Two genetically identical cells result.

Figure 13.15 Mechanisms of gene exchange between prokaryotic cells.

A. Transformation: taking up DNA from the environment.

B. Transduction: transfer of DNA by means of a virus.

C. Conjugation: direct transfer of a plasmid between cells.

Figure It Out: Which of these processes doubles the number of cells?

Answer: None of them. All are means of gene exchange, not reproduction.

archaea Lineage of prokaryotes most closely related to eukaryotes; many live in extreme environments.

autotroph Organism that uses carbon dioxide as its carbon source; obtains energy from light or breakdown of minerals.

bacteria Most diverse and well-known lineage of prokaryotes.

binary fission Method of asexual reproduction in which a prokaryote divides into two identical descendant cells.

conjugation Mechanism of gene transfer. One prokaryotic cell directly transfers a plasmid to another.

decomposer Organism that breaks down organic material into its inorganic subunits.

heterotroph Organism that obtains both carbon and energy by breaking down organic compounds.

plasmid Of many prokaryotes, a small ring of non-chromosomal DNA.

transduction Mechanism of gene transfer. A virus moves genes from one host cell to another.

transformation Mechanism of gene transfer. A prokaryotic cell takes up and uses DNA from its environment.

Metabolic Diversity A high degree of metabolic diversity contributes to the wide distribution of the prokaryotic lineages. Organisms obtain energy and nutrients from the environment in four different ways (Figure 13.16). All four nutritional modes occur among bacteria or archaea or in both.

Autotrophs are producers that build their own food using carbon dioxide (CO_2) as their carbon source. There are two subgroups: photoautotrophs and chemoautotrophs. Photoautotrophs are photosynthetic. They use the energy of light to assemble organic compounds from CO_2 and water. Many bacteria are photoautotrophs, as are plants and photosynthetic protists. Chemoautotrophs obtain energy by oxidizing (removing electrons from) inorganic molecules such as hydrogen sulfide or methane and use it to build organic compounds from CO_2. Chemoautotrophic bacteria and archaea are the main producers in dark environments such as the seafloor. So far, no eukaryotic chemoautotroph is known.

Heterotrophs cannot use inorganic sources of carbon. Instead, they obtain carbon by taking up organic molecules from their environment. As with autotrophs, there are two types. Photoheterotrophs harvest energy from light, and carbon from alcohols, fatty acids, or other small organic molecules. Heliobacteria that live in the soils of rice paddies are an example. Chemoheterotrophs obtain both energy and carbon by breaking down carbohydrates, lipids, and proteins. Most bacteria and some archaea are chemoheterotrophs, as are animals, fungi, and nonphotosynthetic protists. All pathogenic bacteria are chemoheterotrophs that extract the organic compounds they need to live from their host. Other prokaryotic chemoheterotrophs serve as **decomposers**, meaning they break down organic molecules into inorganic ones. By their actions, decomposers make nutrients that were tied up in wastes and remains accessible to producers.

Most eukaryotic organisms are aerobic, meaning they rely on aerobic respiration (Section 5.5) and thus need oxygen. By contrast, many bacteria and most archaea are anaerobes, which means they can tolerate an oxygen-free environment. Some are obligate anaerobes, meaning oxygen either slows their growth or kills them outright. Anaerobes are harmed by oxygen because oxidation reactions damage their biological molecules and, unlike aerobic cells, they do not have enzymes

Figure 13.16 Nutritional classification of organisms.

CARBON SOURCE	ENERGY SOURCE	
	Light	Chemicals
Inorganic source such as CO_2	**Photoautotrophs** bacteria, archaea, photosynthetic protists, plants	**Chemoautotrophs** bacteria, archaea
Organic source such as glucose	**Photoheterotrophs** bacteria, archaea	**Chemoheterotrophs** bacteria, archaea, fungi, animals, nonphotosynthetic protists

A. Thermally heated waters. Pigmented archaea color rocks in waters of this Nevada hot spring.

B. Highly salty waters. Pigmented extreme archaea color brine in this California lake.

C. The gut of many animals. Cows belch to expel methane produced by archaea in their digestive system.

Figure 13.17 Examples of archaeal habitats.

(A) © Savannah River Ecology Laboratory; (B) Courtesy of Benjamin Brunner; (C) Dr. John Brackenbury/Science Source.

that can repair that damage. We find obligate anaerobes in aquatic sediments and the animal gut. They can also infect deep wounds.

Many bacteria and some archaea can respond to adverse conditions by shutting down their metabolism and forming a dormant resting structure. Depending on the group and how the structure forms, it may be called a spore or a cyst. For example, some bacteria, including those that cause the diseases tetanus and anthrax, produce a resilient resting structure consisting of a stripped-down bacterial cell with a thick protective covering. This structure, called an endospore, can withstand heating, freezing, drying out, and exposure to ultraviolet radiation. Scientists have extracted bacterial endospores from the gut of a bee that had been encased in amber (fossilized tree sap) for at least 20 million years. When given nutrients and moisture, the endospores germinated and the cells within them became active once again.

Domain Archaea Archaea were discovered in the 1970s. Many thrive in seemingly hostile habitats. The **extreme thermophiles** live in very hot places. For example, archaea have been found in scalding hot water near deep-sea hydrothermal vents and in thermal springs (Figure 13.17A). The archaea that are **extreme halophiles** live in highly salty environments (Figure 13.17B).

Many archaea, including some extreme halophiles and thermophiles, are **methanogens**, chemoautotrophs that produce methane, an odorless flammable gas, as a by-product of their metabolic reactions. Methane-producing archaea abound in sewage, marsh sediments, and the animal gut (Figure 13.17C). About a third of the human population has significant numbers of methanogens in their intestine, so their flatulence (farts) contains methane.

As biologists continue to explore archaeal diversity, they are finding that these organisms are not restricted to extreme environments. They live alongside bacteria nearly everywhere. So far, scientists have not found any archaea that pose a major threat to human health. However, some that live in the mouth may encourage gum disease, and some that live in the gut may encourage weight gain.

Domain Bacteria Many bacteria play important roles in nutrient cycles. Photosynthesis evolved in many bacterial lineages, but only cyanobacteria (Figure 13.18A) use a pathway that produces oxygen as a by-product. Biologists infer that ancient cyanobacteria were the ancestors of modern chloroplasts. Thus, we have cyanobacteria and their chloroplast relatives to thank for nearly all the oxygen we breathe.

Some cyanobacteria also carry out **nitrogen fixation**, meaning they incorporate nitrogen from the air into ammonia (NH_3). Nitrogen fixation is an important ecological service provided only by bacteria. Photosynthetic eukaryotes need nitrogen, but they cannot use the gaseous form ($N\equiv N$) because they do not have an enzyme that can break the molecule's triple bond. They can, however, take up ammonia released by nitrogen-fixing bacteria.

Nitrogen-fixing bacteria of the genus *Rhizobium* live inside the roots of legumes, a group of plants that includes peas, alfalfa, and clover. The plants benefit from the presence of the bacteria, which provide them with ammonia. The bacteria benefit by living in the shelter of the roots and receiving sugar from the plant.

Bacteria also help cycle nutrients by acting as decomposers. Together with fungal decomposers, bacteria ensure that nutrients in wastes and remains of organisms return to the soil in a form that plants can use. Lactate-fermenting bacteria are among the decomposers. Sometimes these bacteria get into our food and spoil it, as when they cause milk to go sour. On the other hand, we use some lactate fermenters to make sauerkraut, pickles, cheese, and yogurt (Figure 13.18B). Lactate fermenters

nitrogen-fixing cell

Figure 13.18 Ecologically important bacteria.
A. Aquatic cyanobacteria. Cyanobacteria release oxygen as a by-product of photosynthesis. This species grows as long chains of cells connected by a secreted mucous sheath. Some specialized cells in the chain fix nitrogen.

B. Lactate-fermenting bacteria used to produce yogurt. Other lactate-fermenting bacteria live in the human gut or serve as decomposers in the soil.

(A) Michael Abbey/Visuals Unlimited, Inc.; (B) SciMAT/Science Source.

are also present in a healthy human gut and vagina. The acidity of the lactate these bacteria produce helps keep disease-causing organisms from taking hold. Other intestinal bacteria benefit us by producing essential vitamins or by breaking down materials we could not otherwise digest. For example, most of the vitamin K you need is produced by *Escherichia coli* bacteria in your large intestine.

Escherichia coli is the best-studied species of bacteria. Researchers often investigate genetic and metabolic processes in this species because it is easily grown in laboratories. *E. coli* is also used in industrial biotechnology. Recombinant *E. coli* now make hormones and other proteins for medical use.

When biotechnologists want to alter a plant's genome, they may turn to *Agrobacterium*. These soil bacteria have a plasmid that gives them the capacity to infect plants and cause a tumor. Scientists produce recombinant plants by inserting genes into the tumor-inducing plasmid, then infecting a plant with recombinant bacteria.

Bacteria cause many common diseases (Table 13.1). Some, such as whooping cough (pertussis) and tuberculosis, spread when a person with an active infection coughs or sneezes, distributing bacteria-laden droplets into the environment.

Impetigo, a skin disease, is caused by *Streptococcus* and *Staphylococcus* bacteria that infect outer skin layers. *Streptococcus* can also cause strep throat. Gonorrhea, syphilis, and chlamydia are bacterial diseases transmitted by sexual contact. Bacteria also enter our body in tainted food or water. Cholera, which kills about 100,000 people per year, spreads when bacteria-tainted feces contaminate drinking water. Lyme disease is a vector-borne bacterial disease. Ticks carry the bacteria that cause the disease between vertebrate hosts. Lyme disease may initially cause a bull's-eye-shaped rash at the site of the tick bite. Later, flulike symptoms occur.

Table 13.1 Examples of Bacterial Diseases

Disease	Description
Whooping cough	Childhood respiratory disease
Tuberculosis	Respiratory
Impetigo, boils	Blisters, sores on skin
Strep throat	Sore throat, can damage heart
Cholera	Diarrheal illness
Syphilis	Sexually transmitted disease
Gonorrhea	Sexually transmitted disease
Chlamydia	Sexually transmitted disease
Lyme disease	Rash, flulike symptoms, spread by ticks
Botulism, tetanus	Muscle paralysis by bacterial toxin

Take-Home Message 13.5

What are prokaryotes?

- Prokaryotes are cells that do not have a nucleus. They reproduce mainly by binary fission and they swap genes by conjugation and other processes.
- Archaea are the most recently discovered prokaryotic domain. Many archaea live in extremely hot or salty habitats.
- Bacteria benefit other organisms by releasing oxygen, fixing nitrogen, and serving as decomposers. We use them to produce foods and in biotechnology. Some bacteria cause human disease.

extreme halophile Organism that lives where the salt concentration is high.

extreme thermophile Organism that lives where the temperature is very high.

methanogen Organism that produces methane gas as a metabolic by-product.

nitrogen fixation Process of combining nitrogen gas with hydrogen to form ammonia.

Figure 13.19 Evolutionary tree for the eukaryotes.
Orange boxes indicate the protist lineages.

Figure 13.20 Euglena, a freshwater flagellated protozoan.
The species depicted here has chloroplasts that evolved from a green alga.

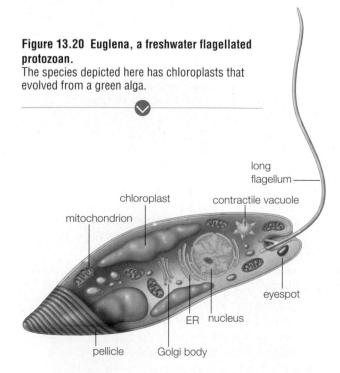

13.6 Protists

REMEMBER: Eukaryotic cells have a nucleus and other membrane-bound organelles and they use cilia and flagella to propel themselves (Section 3.5). Water moves across a plasma membrane by osmosis (4.5).

A diverse array of eukaryotic lineages are collectively referred to as **protists**. Protists were historically lumped together in a kingdom between prokaryotes and the "higher" forms of life (fungi, plants, and animals). Now that scientists have improved methods of comparing genomes, protists are being reclassified in ways that reflect their evolutionary relationships. Figure 13.19 shows where the protist lineages that we cover in this book fit in the eukaryote family tree. There are many additional protist lineages, but learning about a representative few will give you a good idea of the diversity of protist forms and of their importance in health and ecology. Notice that some of the protists are actually more closely related to plants, animals, or fungi than they are to other protists.

Being eukaryotes, all protist cells have a nucleus and a cytoskeleton with microtubules. Most also have mitochondria, endoplasmic reticulum, and Golgi bodies. All protists have multiple chromosomes, each consisting of DNA with proteins attached. Protists reproduce asexually by mitosis, sexually by meiosis, or both.

Most protist lineages include only single-celled species. However, colonial protists exist, and multicellularity evolved independently in several lineages. Cells of a **colonial organism** live together and behave in an integrated fashion, but still remain self-sufficient. Each retains the traits required to survive and reproduce on its own. By contrast, the cells of a **multicellular organism** have a division of labor and rely on one another for survival.

Flagellated Protozoans "Protozoans" is the general term for heterotrophic protists that live as single cells. **Flagellated protozoans** are single, unwalled cells that have one or more flagella. In protists, movement of the flagellum pulls a cell forward, rather than pushing the cell along as the flagellum of an animal sperm does. A pellicle, a layer of elastic proteins just beneath the plasma membrane, helps flagellated protozoans retain their shape.

Euglenoids have a single long flagella and most live in ponds and lakes (Figure 13.20). The interior of the cell has a higher solute concentration than the fresh water the cell lives in, so water tends to enter euglenoids by osmosis. Excess water collects in **contractile vacuoles**, organelles that can also contract and expel the water to the outside. Many euglenoids are heterotrophs, but some such as the one depicted in Figure 13.20 have chloroplasts that evolved from a green algae. Photosynthetic euglenoids are able to detect light using a an eyespot, an organelle near the base of their long flagellum.

Some flagellated protozoans live in the bodies of other organisms. *Giardia*, the intestinal parasite described in Section 13.1, is one example. *Trichomonas* (Figure 13.21A) is another. Its multiple flagella propel it through the human reproductive tract with jerky movements. *Trichomonas* causes trichomoniasis, a sexually transmitted disease commonly referred to as "trich." Trypanosomes are long, tapered cells with a single mitochondrion and a flagellum that is attached to the cell body by a membrane (Figure 13.21B). All trypanosomes are parasites of either plants or animals. Trypanosomes that cause human diseases such as sleeping sickness are transmitted by insect bites. Once inside a human host, trypanosomes live in the blood and other body fluids.

A. *Trichomonas*, a sexually transmitted pathogen.

flagellum attached to the cell body by a membrane

red blood cell

B. *Trypanosoma* in human blood.

Figure 13.21 Flagellated protozoans that parasitize humans.

(A) David M. Phillips/The Population Council/Science Source; (B) Oliver Meckes/Science Source.

ciliate Unwalled, single-celled protist with many cilia.

colonial organism Organism composed of many integrated cells, each capable of surviving and reproducing on its own.

contractile vacuole In freshwater protists, an organelle that collects and expels excess water.

flagellated protozoan Unwalled, single-celled protist that has one or more flagella.

foraminiferan Heterotrophic single-celled protist that secretes a calcium carbonate shell.

multicellular organism Organism composed of a variety of specialized cells, each unable to survive and reproduce on its own.

plankton Community of mostly microscopic drifting or swimming organisms.

protist General term for eukaryote that is not a fungus, plant, or animal.

Foraminifera **Foraminifera**, or forams, are single-celled predators that secrete a shell containing calcium carbonate. Including its shell, an individual foraminiferal cell can be as big as a grain of sand. Threadlike cytoplasmic extensions protrude through openings in the shell. Most forams live on the seafloor, where they probe the water and sediments for prey. Others are part of the marine **plankton**, a collection of tiny organisms that drift or swim in the open sea. Planktonic forams often have photosynthetic protists that live in their cytoplasm (Figure 13.22).

Foraminifera have lived and died in the oceans for more than 500 million years, so remains of countless cells have fallen to the seafloor. Over time, geologic processes transformed some accumulations of foraminiferal shells into chalk and limestone, two types of calcium carbonate–rich sedimentary rock. The giant blocks of limestone that were used to build the great pyramids of Egypt consist largely of the shells of ancient foraminifera.

Modern foraminifera play an important role in the global carbon cycle. By taking up carbon dioxide from seawater and incorporating it into their shells, foraminifera lower the ocean's carbon dioxide concentration and allow it to absorb more carbon dioxide from the air. Removing carbon dioxide from the air is important because the increasing concentration of atmospheric carbon dioxide is currently causing a global climate change.

200 µm

Figure 13.22 A planktonic foraminiferan.
The yellow dots are algae that live in its cytoplasm.

Courtesy of Allen W. H. Bé and David A. Caron.

Ciliates Ciliated protozoans, or **ciliates**, are unwalled cells with many cilia. Most ciliates are predators in seawater or fresh water. They feed on bacteria, algae, and one another. *Paramecium* is a freshwater ciliate commonly found in ponds (Figure 13.23). The cilia that cover its entire surface function in feeding and locomotion. They sweep water laden with bacteria, algae, and other food particles into an oral groove at the cell surface, and then to a gullet. Enzyme-filled vesicles digest food in the gullet.

Other ciliates live in the gut of mammalian grazers such as cattle and sheep. Like some bacteria, these ciliates help their host digest plant material. Only one species of ciliate (*Balantidium coli*) is a known human pathogen. It also infects pigs, and people become infected when pig feces containing a resting form of the ciliate taint drinking water. Infection causes nausea and diarrhea.

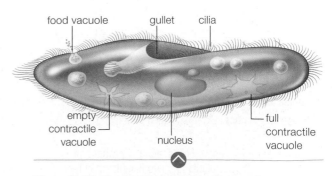

food vacuole   gullet   cilia

empty contractile vacuole   nucleus   full contractile vacuole

Figure 13.23 The freshwater ciliate *Paramecium*.

A. Photosynthetic dinoflagellate.

B. Bioluminescent dinoflagellates agitated by wave action give this shoreline an eerie glow.

Figure 13.24 Dinoflagellates.

(A) © Bob Andersen and D. J. Patterson; (B) Travelart/Alamy.

algal bloom Population explosion of single-celled aquatic organisms such as dinoflagellates.

apicomplexan Parasitic protist that enters and lives inside the cells of its host.

bioluminescence Light produced by a living organism.

dinoflagellates Single-celled, aquatic protist typically with cellulose plates and two flagella; may be heterotrophic or photosynthetic.

Dinoflagellates The name **dinoflagellate** means "whirling flagellate." These single-celled protists typically have two flagella, one at the cell's tip and the other running in a groove around the middle of the cell like a belt (Figure 13.24A). Combined action of the two flagella causes the cell to rotate as it moves forward. Most dinoflagellates deposit cellulose just beneath their plasma membrane, and these deposits form thick protective plates.

Some dinoflagellates live in fresh water and others in the oceans. Some prey on bacteria, others are parasites of animals, and still others have chloroplasts that evolved from red algae. A few species of photosynthetic dinoflagellates live inside the cells of reef-building corals. They supply their coral host, which is an invertebrate animal, with essential sugars. In exchange, the coral provides the dinoflagellates with nutrients, shelter, and the carbon dioxide necessary for photosynthesis. The coral cannot live without its protist helpers. If it loses them, it will starve.

In tropical seas, dinoflagellates are a common source of **bioluminescence**, which is light produced by a living organism (Figure 13.24B). Emitting light may protect a cell by startling a predator that was about to eat it. By another hypothesis, the flash of light acts like a car alarm. It attracts the attention of other organisms, including predators that pursue would-be eaters of dinoflagellate.

In nutrient-enriched water, free-living photosynthetic dinoflagellates or other aquatic protists sometimes undergo great increases in population size, a phenomenon known as an **algal bloom**. Algal blooms can harm other organisms. When the cells die, aerobic bacteria that feed on their remains use up all the oxygen in the water, so aquatic animals suffocate. In additions, some dinoflagellates produce toxins that can kill aquatic organisms directly and sicken people.

Apicomplexans **Apicomplexans** are parasitic protists that spend part of their life inside cells of their hosts. Their name refers to a complex of microtubules at their apical (top) end that allows them to enter a host cell. Apicomplexans infect a variety of animals, from worms and insects to humans. In most species, the life cycle is complicated, with multiple hosts and several forms. Consider *Plasmodium*, the apicomplexan that causes malaria (Figure 13.25). A female mosquito transmits the infectious form of *Plasmodium* (called a sporozoite) to a human when she bites ❶. The sporozoite travels through blood vessels to the liver, where it reproduces asexually ❷. Some of the resulting offspring, called merozoites, enter red blood cells, where they reproduce asexually to produce more merozoites ❸. Other merozoites enter into red blood cells and develop into immature gametes, or gametocytes ❹.

When a mosquito bites an infected person, it takes up gametocytes along with blood. The gametocytes mature in the mosquito's gut, then fuse to form zygotes ❺. Zygotes develop into new sporozoites that migrate to the insect's salivary glands, where they await transfer to a new vertebrate host ❻.

Malaria symptoms usually start a week or two after a mosquito bite, when infected liver cells rupture and release *Plasmodium* cells and cellular debris into the blood. Shaking, chills, a burning fever, and sweats result. After the first episode, symptoms may subside for weeks or even months. However, an ongoing infection damages the liver, spleen, kidneys, and brain. If untreated, malaria nearly always results in death. Malaria kills about half a million people each year, mainly in Africa.

The apicomplexan *Toxoplasma* also commonly infects humans. Most people never realize they are infected, but an infection can be deadly in someone with an impaired immune system, and an infection that begins during pregnancy can cause birth defects. Domestic cats that spend time ouside catching birds or rodents can serve as carriers of *Toxoplasma*; their feces may contain infectious cysts.

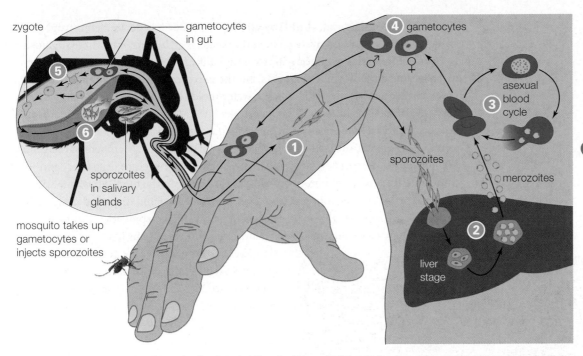

Figure 13.25 Life cycle of *Plasmodium*, the protist that causes malaria.

1 Infected mosquito bites a human. Sporozoites enter the blood, which carries them to the liver.

2 Sporozoites reproduce asexually in liver cells, then mature into merozoites. Merozoites leave the liver and enter the bloodstream, where they infect red blood cells.

3 Inside some red blood cells, merozoites reproduce asexually. These cells burst and release more merozoites into the bloodstream.

4 Inside other red blood cells, merozoites develop into male and female gametocytes.

5 A female mosquito bites and sucks blood from the infected person. Gametocytes in red blood cells enter her gut and mature into gametes, which fuse to form zygotes.

6 Zygotes develop into sporozoites that migrate to the mosquito's salivary glands.

Based on Fig. 1 from "Genetic linkage and association analyses for trait mapping in Plasmodium falciparum," by Xinzhuan Su, Karen Hayton & Thomas E. Wellems, *Nature Reviews Genetics 8*, 497–506 (July 2007).

Digging Into Data

How *Plasmodium* Summons Mosquitoes

Parasites sometimes alter their host's behavior in a way that increases their chances of transmission to another host. *Plasmodium* (the agent of malaria) would benefit by making its human host more attractive to hungry mosquitoes when immature gametes (gametocytes) are present in the host's blood. Such immature gametes are taken up by the mosquito along with blood, and they mature into gametes inside the mosquito's gut. Dr. Jacob Koella and his associates performed an experiment to see whether infection by *Plasmodium* makes a person more attractive to mosquitoes. The researchers recorded the response of mosquitoes to the odor of *Plasmodium*-infected children and uninfected children over the course of 12 trials on 12 separate days. They also recorded which stage of *Plasmodium* life cycle the infected children were carrying at the time. Figure 13.26 shows their results.

1. On average, which group of children was the most attractive to mosquitoes?
2. Did carrying noninfectious, asexual stage *Plasmodium* make children more attractive to mosquitoes than uninfected children?
3. Did the data support the hypothesis that the presence of infectious *Plasmodium* cells (gametocytes) makes an individual more attractive to mosquitoes?
4. Why would it be it beneficial for *Plasmodium* gametocytes to make a host attractive to mosquitoes?

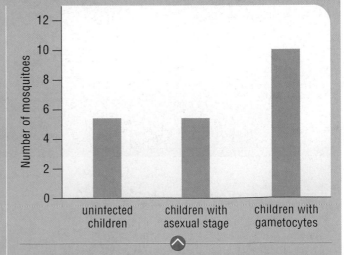

Figure 13.26 Number of mosquitoes (out of 100) attracted to uninfected children, children with asexual stages of *Plasmodium* (sporozoites, merozoites), and children with gametocytes. The bars show the average number of mosquitoes attracted to that category of child over the course of 12 separate trials.

Water Molds, Diatoms, and Brown Algae **Water molds** are heterotrophs known to scientists as oomyetes. The term means "egg fungus," and these organisms were once mistakenly grouped with the fungi. Like fungi, the water molds form a mesh of nutrient-absorbing filaments, but the two groups differ in many structural traits and are genetically distinct. Most water molds help decompose organic debris and dead organisms in aquatic habitats, but a few are parasites that have significant economic effects. Some grow as fuzzy white patches on fish in fish farms and aquariums. Others infect land plants, destroying crops and forests. Members of the genus *Phytophthora* are especially notorious. Their name means "plant destroyer" and they cause an estimated $5 billion in crop losses each year. In the mid-1800s, one species destroyed Irish potato crops, causing a famine that killed and displaced millions of people. Today, another *Phytophthora* species is causing an epidemic of sudden oak death in Oregon, Washington, and California. Millions of oaks have already died.

The closest relatives of water molds are two photosynthetic groups: diatoms and brown algae (Figure 13.27). Both have chloroplasts that include a brownish accessory pigment (fucoxanthin) that tints them olive green, golden, or dark brown.

Diatoms have a two-part silica shell, with upper and lower parts that fit together like a box with an overlapping lid. Some cells live individually, and others form chains ❶. Most diatoms float near the surface of seas or lakes, but some live in moist soil or in water droplets that cling to mosses in damp environments.

Diatom cells contain a large amount of oil. Oil is less dense than water, and its presence helps these photosynthetic cells stay afloat in sunlit waters. The oil also serves as a store of energy.

Like foraminifera, marine diatoms have lived and died in the oceans for many millions of years, and their remains form vast deposits on the seafloor. In some places, deposits of ancient diatom oil have been transformed into petroleum, which we extract to produce gasoline. In other places, diatom remains have been transformed into a silica-rich powder called diatomaceous earth. This material is quarried for use in filters, abrasive cleaners, and as an insecticide that is not harmful to vertebrates.

Brown algae are multicelled inhabitants of temperate or cool seas. In size, they range from microscopic filaments to giant kelps that stand 30 meters (100 feet) tall. Giant kelps form forestlike stands in coastal waters of the Pacific Northwest ❷. Like trees in a forest, kelps shelter a wide variety of other organisms. The Sargasso Sea in the North Atlantic Ocean is named for its abundance of *Sargassum*. This kelp forms vast, floating mats that can be up to 9 meters (30 feet) thick. The mats provide food and shelter to fish, sea turtles, and invertebrates.

Sargassum and other brown algae have commercial uses. Alginic acid from the cell walls of brown algae is used to produce algins, which serve as thickeners, emulsifiers, and suspension agents. Algins are used to manufacture ice cream, pudding, jelly beans, toothpaste, cosmetics, and other products.

Figure 13.27 Two related protist groups common in California's coastal waters.

❶ Microscopic diatoms have a silica shell.

❷ Kelp are large, multicelled brown alga.

brown alga Multicelled, photosynthetic protist with brown accessory pigments.

diatom Single-celled photosynthetic protist with brown accessory pigments and a two-part silica shell.

green alga Single-celled, colonial, or multicelled photosynthetic protist belonging to the group most closely related to land plants.

red alga Single-celled or multicelled photosynthetic protist with red accessory pigment.

water mold Heterotrophic protist that forms a mesh of nutrient-absorbing filaments.

Although some large brown algae have a plantlike form, this similarity is an example of morphological convergence, rather than evidence of shared ancestry. Brown algae evolved from a different single-celled ancestor than the lineage that includes red algae, green algae, and land plants.

Red Algae Some **red algae** are single cells, but most are multicelled forms that live in tropical seas. Most commonly they have a branching structure, but some form thin sheets (Figure 13.28). Coralline algae (red algae with cell walls hardened by calcium carbonate) are a component of tropical coral reefs. Red algae are tinted red to black by accessory pigments called phycobilins. These pigments absorb the blue-green light that penetrates deep into water. Phycobilins allow red algae to carry out photosynthesis at greater depths than other algae.

Red algae have many commercial uses. Nori, the sheets of seaweed used to wrap some sushi, is a red alga that is grown commercially. Agar and carrageenan are valuable products extracted from the cell walls of other red algae. Agar keeps baked goods and cosmetics moist, helps jellies set, and is used to make capsules that hold medicines. Carrageenan is added to soy milk, dairy foods, and the fluid that is sprayed on airplanes to prevent ice formation.

Green Algae **Green algae** include single-celled, colonial, and multicelled species (Figure 13.29). Most live in fresh water, but some are marine, and some grow on soil, trees, or other damp surfaces. A few single-celled species partner with a fungus to form a lichen.

The single-celled alga *Chlorella* is cultivated in ponds, dried, and sold in powdered or pill form as a nutritional supplement. *Chlorella* is also a promising candidate for biofuel production because it has a high oil content. Many ciliates, including some *Paramecium*, already get an energy boost from *Chlorella*. They feed on sugars produced by *Chlorella* cells that live inside them.

Figure 13.28 Branching and sheetlike red algae growing 75 meters (225 ft) beneath the sea surface in the Gulf of Mexico.

Image courtesy of FGB-NMS/UNCW-NURC.

Figure 13.29 Green algae.

(A) © Wim van Egmond/Visuals Unlimited; (B) Charles Kreb/Science Faction/SuperStock; (C) © Lawson Wood/Corbis.

A. Two desmids, a type of single-celled green alga that lives mainly in fresh water.

B. *Volvox*, a colonial, freshwater green alga. Each sphere is a colony made up of flagellated cells linked by thin cytoplasmic strands.

C. Sheets of sea lettuce (*Ulva*), a multicelled marine green alga. Although the sheets can be longer than your arm, they are thinner than a human hair.

Figure 13.30 *Amoeba proteus*, **a freshwater amoeba.**
Amoebas feed or shift position by extending lobes of
cytoplasm (pseudopods). This amoeba's food vacuoles
contain green algae engulfed by its pseudopods.

iStockphoto.com/micro_photo.

Red algae, green algae, and land plants share a variety of unique traits, including chloroplasts containing a particular type of chlorophyll and a cell wall of cellulose. These similarities are taken as evidence that these three groups share a common ancestor. Both red algae and green algae evolved from an ancestral protist that had chloroplasts descended from cyanobacteria. After the two algal lineages diverged, land plants evolved from one lineage of green algae.

Amoebas and Slime Molds The free-living amoebas and the slime molds are grouped together as amoebozoans. Members of this group do not have a cell wall, shell, or pellicle, so they continually change shape. A compact blob of a cell can extend lobes of cytoplasm called pseudopods (Section 3.5) to move about and to capture food.

Amoebas such as *Amoeba proteus* (Figure 13.30) always live and feed as solitary cells. Most amoebas are predators in freshwater habitats. Others live inside animals, and some cause human disease. Each year, about 50 million people suffer from amebic dysentery after drinking water contaminated by *Entamoeba histolytica* cysts. Inadequate sterilization of contact lenses or swimming with lenses can result in an eye infection by *Acanthamoeba*, an amoeba common in soil, standing water, and even tap water.

Slime molds are sometimes described as "social amoebas." There are two types, plasmodial slime molds and cellular slime molds.

Plasmodial slime molds spend most of their life cycle as a multinucleated mass called a plasmodium. The plasmodium forms when a diploid amoeba-like cell

Figure 13.31 Plasmodial slime mold on a log.
This multinucleated mass (the plasmodium) streams along
at a rate of about a millimeter an hour, engulfing any food it
encounters. As the plasmodium travels, it lays down a trail of
slime. If it later happens across its own trail, it will move off
in a different direction. In this way, the slime mold "remembers" where it has been and avoids revisiting areas where it
has already depleted its food supply.

Edward S. Ross.

**Figure 13.32 Life cycle
of the cellular slime
mold *Dictyostelium
discoideum*.**

Photo, Carolina Biological Supply
Company.

⑤ Spores give rise to amoeboid cells.

④ A fruiting body forms with resting spores atop a stalk.

Mature fruiting body

① Cells feed and multiply by mitosis.

② When food is scarce, cells aggregate.

③ The cells form a slug. It may start to develop as a fruiting body right away, or migrate about. In the slug, cells become prestalk (red) and prespore (tan) cells.

Migrating slug stage

divides its nucleus repeatedly by mitosis, but does not undergo cytoplasmic division. The resulting mass, which can be as big as a dinner plate, streams along the forest floor engulfing microbes and organic matter (Figure 13.31). When food supplies dwindle, a plasmodium develops into many spore-bearing fruiting bodies.

Cellular slime molds spend the bulk of their existence as individual haploid amoeboid (amoeba-like) cells. *Dictyostelium discoideum* is an example (Figure 13.32). Each cell eats bacteria and reproduces by mitosis ❶. When food runs out, thousands of cells aggregate to form a multicelled mass ❷. Environmental gradients in light and moisture induce the mass to crawl along as a cohesive unit often referred to as a "slug" ❸. When the slug reaches a suitable spot, its component cells differentiate to form a fruiting body. Some cells become a stalk, and others become spores atop it ❹. When a spore germinates, it releases a cell that starts the life cycle anew ❺.

Choanoflagellates   The **choanoflagellates** are the protist group with genes most similar to those of animals. Choanoflagellates also look a lot like the feeding cells of sponges, which are among the simplest animals. As a result, choanoflagellates are thought to be the closest living relatives of animals. Note that they are not considered ancestors of animals, but rather a group that shared a common single-celled ancestor with animals long ago.

Choanoflagellate means "collared flagellate." Each cell has a long flagellum surrounded by a ring (or collar) of tiny filaments reinforced with the protein actin (Figure 13.33A). Movement of the flagellum creates a current that draws water through the filaments. After tiny bits of food become entrapped, the cell extends pseudopods to capture them. The food is then digested within the cell body.

Most choanoflagellates live as single cells, but some form colonies (Figure 13.33B). The colonies arise when cells divide and the descendant cells stick together with the help of adhesion proteins. Choanoflagellate adhesion proteins are similar to those found in animals, and researchers have discovered that even solitary choanoflagellates have such proteins. By one hypothesis, the common ancestor of animals and choanoflagellates was a single-celled protist with adhesion proteins that helped it capture prey. Later, these proteins were put to use in a new context, helping cells stick together to form multicelled colonies. Later still, the proteins allowed animal cells to adhere to one another in multicelled bodies. This modification in the use of adhesion proteins is an example of how a trait that evolved to serve one function can later be modified and take on a different function.

A. Structure of a solitary choanoflagellate.

B. A colonial choanoflagellate.

Figure 13.33 Choanoflagellates, the modern protist group most closely related to animals.

(B) Courtesy of Damian Zanette.

Take-Home Message 13.6

What are protists?

- The protists are a diverse collection of eukaryotic lineages, some of which are only distantly related to one another.
- Most protists live as single cells, but there are colonial and multicelled species.
- Protists include photosynthesizers, predators, and decomposers in lakes, seas, and damp places on land. Protists also live inside other eukaryotes, including humans. Some of these protists are helpful, but others are parasites and pathogens.
- Green algae are the closest protist relatives of land plants, and choanoflagellates are the closest protist relatives of animals.

amoeba Solitary heterotrophic protist that feeds and moves by extending pseudopods.

cellular slime mold Heterotrophic protist that usually lives as a single-celled, amoeba-like predator. When conditions are unfavorable, cells aggregate into a cohesive group that can form a fruiting body.

choanoflagellates Heterotrophic protists with a collared flagellum; protist group most closely related to animals.

plasmodial slime mold Heterotrophic protist that moves and feeds as a multinucleated mass; forms a fruiting body when conditions are unfavorable.

Summary

Section 13.1 The human **microbiome**, the collection of cells that call our bodies home, includes a diverse variety of organisms. Diet and other factors affect the gut microbiome. Some organisms in the gut are helpful, but some can be **pathogens** (cause disease).

Section 13.2 Laboratory simulations provide indirect evidence that organic subunits can self-assemble under certain conditions. They also show how complex organic compounds and **protocells** may have formed on the early Earth. The **iron–sulfur world hypothesis** holds that these events took place near **hydrothermal vents**. The **RNA world hypothesis** holds that the first genome was RNA-based.

Section 13.3 Fossil **stromatolites** are evidence of early bacterial life. An early branching separated the bacteria from the archaea. Production of oxygen by some photosynthetic bacteria altered Earth's atmosphere and allowed formation of a protective **ozone layer**. Eukaryotes have a composite ancestry with both bacterial and archaeal genes. According to the **endosymbiont hypothesis**, mitochondria and chloroplasts evolved from bacteria.

Section 13.4 **Viruses** consist of RNA or DNA inside a protein coat. Some also have a **viral envelope**. Viruses replicate only in living cells, as when **bacteriophages** multiply in bacteria. Mutation and **viral reassortment** produce new types of viruses. Insects serve as **disease vectors** that spread some viral diseases. **HIV (human immunodeficiency virus)** is an enveloped RNA virus that infects human cells and causes AIDS.

Section 13.5 **Archaea** and **bacteria** are prokaryotic cells, meaning they lack a nucleus. They reproduce asexually by **binary fission**, and exchange genes through **transformation**, **transduction**, and **conjugation** (direct transfer of a **plasmid**).

Bacteria and archaea may be aerobic or anaerobic, and they show great nutritional diversity. **Autotrophs** use carbon dioxide as their carbon source. They include photoautotrophs such as cyanobacteria and chemoautotrophs such as the archaea that live near hydrothermal vents. By contrast, **heterotrophs** obtain carbon from organic compounds. Chemoheterotrophs that serve decomposers and pathogens obtain energy by breaking down organic compounds.

Archaea include heat-loving **extreme thermophiles** or salt-loving **extreme halophiles**. Others live in less extreme environments, such as the human gut. Some people have archaea that produce methane (**methanogens**) in their gut.

Bacteria play important ecological roles by serving as **decomposers**, releasing oxygen into the air, and carrying out **nitrogen fixation**. We use some bacterial species in biological research, biotechnology, and food production. Some bacteria benefit our health, as by living in our gut and producing vitamins. Others are pathogens.

Section 13.6 **Protists** are a diverse collection of lineages. Most lineages are single-celled, but some include **colonial organisms** or **multicellular organisms**. Nearly all protists live in water or in moist habitats, including host tissues.

Flagellated protozoans are single cells with a pellicle that helps hold their shape. Most are heterotrophs, but some euglenoids have chloroplasts. A **contractile vacuole** allows freshwater protozoans to expel excess water. Trypanosomes, *Giardia*, and *Trichomoniasis* are flagellated human pathogens.

Foraminifera are single-celled heterotrophs with calcium carbonate shells. They live on the seafloor or drift as marine **plankton**. Shells of foraminifera contribute to limestone and chalk.

Ciliates are single-celled heterotrophs that use cilia to move and feed. **Dinoflagellates** are single cells that move with a whirling motion and can be heterotrophs or photosynthetic. Some are **bioluminescent** and others cause **algal blooms**. **Apicomplexans**, such as the species that cause malaria, are parasites that spend part of their life in cells of their host.

Water molds are heterotrophs that grow as filaments. Some are pathogens of fish or plants. **Diatoms** are single-celled, silica-shelled, aquatic producers. **Brown algae** include the giant kelps, which are the largest protists.

Red algae can live at greater depths than other algae. Red algae share a common ancestor with **green algae**. Land plants evolved from a green alga.

Amoebas are shape-shifting cells that extend pseudopods to feed and move. They live in aquatic habitats and animal bodies. The related **cellular slime molds** spend part of their life as a single cell and part in a cohesive group that can migrate and differentiate to form a spore-bearing body. **Plasmodial slime molds** feed as a giant multinucleated mass, then form spores when food runs out. The **choanoflagellates** are the protists most closely related to animals.

Self-Quiz

Answers in Appendix I

1. A rise in atmospheric _____ allowed the formation of the ozone layer that screens out UV radiation from the sun.
   a. hydrogen   b. water   c. oxygen   d. ammonia

2. Stanley Miller's experiment demonstrated _____ .
   a. the great age of Earth
   b. that amino acids can assemble under some conditions
   c. that oxygen is necessary for life
   d. all of the above

3. The universal need for iron–sulfur cofactors is taken as evidence that metabolism may have begun _____ .
   a. on a meteorite
   b. on a mudflat
   c. on a rock near a hydrothermal vent

4. Mitochondria are most likely descendants of _____ .
   a. methanogenic archaea
   b. aerobic bacteria
   c. cyanobacteria
   d. green algae

5. The genetic material of HIV is _____ .
   a. protein   b. DNA   c. RNA   d. ATP

6. Viral transfer of genes between bacteria is called _____ .
   a. conjugation
   b. viral reassortment
   c. transduction
   d. transformation

7. All viruses have _____ .
   a. an envelope   b. ribosomes   c. DNA   d. a protein coat

8. Choanoflagellates are most closely related to _____ .
   a. bacteria   b. land plants   c. ciliates   d. animals

9. _____ take up carbon dioxide from seawater and use it to make a chalky shell.
   a. Ciliates   b. Diatoms   c. Foraminifera   d. Euglenoids

10. All _____ are parasitic eukaryotes that live in other cells.
    a. viruses
    b. apicomplexans
    c. euglenoids
    d. slime molds
    e. both a and b
    f. all are correct

11. Oil-rich remains of ancient _____ are the main source of the petroleum that we use to make gasoline.
    a. diatoms   b. ciliates   c. foraminifera   d. red algae

12. Some _____ live in corals and supply them with sugars.
    a. ciliates   b. viruses   c. kelps   d. dinoflagellates

13. Only certain _____ can carry out nitrogen fixation.
    a. green algae and land plants   c. bacteria
    b. diatoms   d. archaea

14. Genetic material of a _____ can be either DNA or RNA.
    a. bacteria   b. dinoflagellate   c. ciliate   d. virus

15. Match these terms with the appropriate definition.
    _____ green algae   a. protist population explosion
    _____ virus   b. social amoeba
    _____ bacteria   c. most diverse prokaryotes
    _____ brown algae   d. noncellular infectious agent
    _____ bioluminescence   e. include the largest protists
    _____ euglenoid   f. flagellate with chloroplasts
    _____ algal bloom   g. closest relative of plants
    _____ dinoflagellate   h. layered prokaryotes and sediment
    _____ slime mold   i. biologically produced light
    _____ stromatolite   j. whirling cell

Critical Thinking

1. Researchers looking for fossils of the earliest life forms face many hurdles. For example, few sedimentary rocks date back more than 3 billion years. Review what you learned about plate tectonics (Section 11.5). Explain why so few remaining samples of the earliest rocks remain.

2. The antibiotic penicillin acts by interfering with the production of new bacterial cell walls. Bacteria exposed to penicillin do not die immediately, but they cannot reproduce. Explain why.

3. Viruses that do not have a lipid envelope tend to remain infectious outside the body longer than enveloped viruses. "Naked" viruses are also less likely to be rendered harmless by soap and water. Can you explain why?

4. The apicomplexan that causes malaria had a photosynthetic ancestor and contains an organelle that evolved from its ancestral chloroplast. The organelle no longer functions in photosynthesis, but it does carry out some essential metabolic tasks. Why would targeting this organelle yield an antimalarial drug that would be likely to have minimal side effects?

Visual Question

1. The graphic below depicts the first steps in one of the two bacteriophage replication pathways. Which pathway is it and how can you tell?

14

PLANTS AND FUNGI

Application ❯ 14.1 Fungal Threats to Crops

REMEMBER: Plants are producers (Section 1.3) that make food by photosynthesis, whereas fungi are consumers that must obtain ready-made sugars.

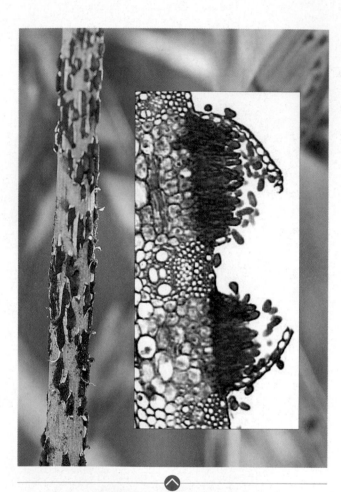

Figure 14.1 Wheat stem rust fungus, a current threat to world food supplies.
Wheat stem, with rust-colored fungal sporangia (spore-producing structures) on its surface. The inset micrograph shows spores (red) escaping from the sporangia.

Background, Photo by Yue Jin/USDA; inset, Courtesy of Charles Good, Ohio State University at Lima.

Plants are the main producers on land and thus serve as an important food source for both animals and fungi. As a result, we often find ourselves in competition with fungi for plant foods. For example, wheat is both an important agricultural crop and the host for wheat stem rust fungus (*Puccinia graminis*). Wheat stem rust fungus is an obligate plant parasite, meaning it can grow and reproduce only in a living plant.

Fungi disperse by releasing microscopic spores that travel on the wind. A wheat stem rust infection begins when a fungal spore lands on the leaf of a wheat plant. The spore germinates (becomes active), and a fungal filament grows into the plant. As fungal filaments extend through the plant's tissues, the filaments take up photosynthetic sugars that the plant would normally use to meet its own needs. As a result, an affected plant is stunted and produces little or no wheat. About a week after infection, tens of thousands of rust-colored spore sacs appear on the stem of the infected plant (Figure 14.1). Each spore can disperse and infect a new plant.

Outbreaks of wheat stem rust disease routinely destroyed wheat crops world-wide until the 1960s, when a plant breeding program headed by Norman Borlaug produced resistant strains of wheat. In 1970, Borlaug received the Nobel Peace Prize for his role in preventing food shortages that contribute to global instability. Planting rust-resistant wheats developed by Borlaug prevented outbreaks of wheat stem rust for decades. Then, in 1999, scientists discovered a new strain of wheat stem rust in Uganda. This strain, Ug99, has mutations that allow it to infect about 90 percent of the wheat varieties that were previously resistant to wheat stem rust.

Ug99 is now spreading. Fungal spores are microscopic, so they can lodge in crevices on dust particles. When winds lift these particles aloft, the spores go along for the ride. Dustborne fungal spores disperse long distances riding winds that swirl high above Earth's surface. As of 2014, windblown spores of Ug99 had reached Kenya, Ethiopia, and Sudan, crossed the Red Sea to Yemen, and from there crossed the Persian Gulf to Iran. Given the prevailing winds, India, the world's second largest wheat producer, is expected to be affected soon. Most likely, winds will eventually distribute Ug99 worldwide. Fungicides can minimize the damage, but are too expensive for farmers in developing nations.

The threat fungal diseases pose to crop plants is heightened by current agricultural practices. Farmers often plant a single variety of a crop in dense stands that cover an extensive area. The proximity of many genetically identical host plants allows a fungus to spread quickly through a field. In addition, farmers in different parts of the world often buy seeds from the same large companies and plant the same few varieties of a crop.

Protecting our food supply from the threat of plant diseases requires maintaining the genetic diversity of our crop plants and their wild relatives. Having many varieties of a crop increases the likelihood that at least one variety will be immune to any given disease. That variety can be planted or used to create new disease-resistant varieties, either through traditional plant breeding or by using genetic engineering to transfer disease-resistance genes. Maintaining the wild relatives of crop plants provides another potential source of disease-resistance genes. For example, researchers have discovered genes that confer Ug99 resistance in a wild grass related to wheat, and in a primitive wheat. Transferring these genes into widely used bread wheats could help ensure the safety of the world's wheat supply.

Figure 14.2 Traits of modern plant groups and relationships among them.

Photos, Courtesy of © Christine Evers.

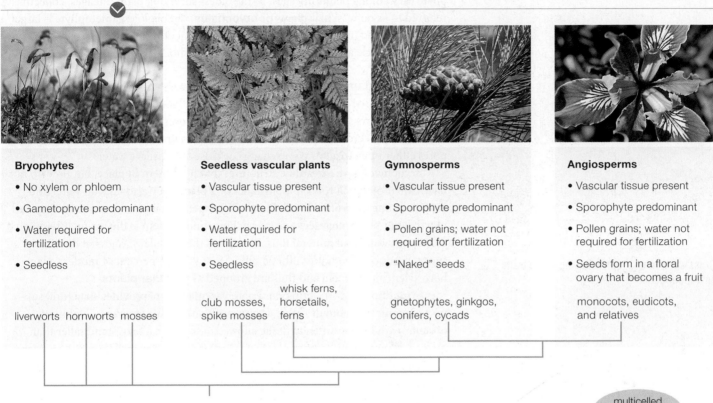

Bryophytes

- No xylem or phloem
- Gametophyte predominant
- Water required for fertilization
- Seedless

liverworts hornworts mosses

Seedless vascular plants

- Vascular tissue present
- Sporophyte predominant
- Water required for fertilization
- Seedless

club mosses,   whisk ferns,
spike mosses   horsetails,
               ferns

Gymnosperms

- Vascular tissue present
- Sporophyte predominant
- Pollen grains; water not required for fertilization
- "Naked" seeds

gnetophytes, ginkgos,
conifers, cycads

Angiosperms

- Vascular tissue present
- Sporophyte predominant
- Pollen grains; water not required for fertilization
- Seeds form in a floral ovary that becomes a fruit

monocots, eudicots,
and relatives

14.2 Plant Traits and Evolution

REMEMBER: Green algae are the closest relatives of land plants (Section 13.6).

Plants are a lineage of land-dwelling, multicelled, photosynthetic eukaryotes. They evolved from freshwater green algae (charophyte algae), and share many traits with this group. The defining trait that sets plants apart is a multicelled embryo that forms, develops within, and is nourished by the parent plant. For this reason, the clade of land plants is referred to as the embryophytes. Figure 14.2 summarizes the relationships among modern plants and their defining traits.

The story of plant evolution is one of adaptation to increasingly drier environments. An aquatic green alga lives surrounded by water, so it can absorb water and dissolved nutrients across its entire body surface. Water also buoys the alga's parts, thus helping it stand upright. Land plants, however, face the constant threat of drying out and must hold themselves upright. Changes in life cycle, structure, and the mechanisms of reproduction and dispersal adapted plants to life on land.

Life Cycle All plants have an **alternation of generations**, meaning their life cycle alternates between haploid and diploid multicelled bodies (Figure 14.3). The diploid generation, called the **sporophyte**, produces spores by meiosis ❶. A plant spore is a single diploid cell that undergoes mitosis and develops into the multicelled, haploid generation. The haploid **gametophyte** ❷ produces gametes by mitosis ❸. Gametes unite at fertilization to form a zygote ❹ that will develop into a diploid sporophyte ❺.

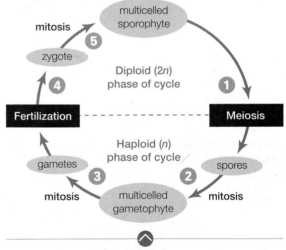

Figure 14.3 Plant life cycle (alternation of generations).

alternation of generations As in plants, a life cycle that alternates between a diploid spore-producing body and a haploid, gamete-producing one.

gametophyte Haploid gamete-forming body that forms in a plant life cycle.

plant Multicelled, typically photosynthetic organism; develops from an embryo that forms on the parent and is nourished by it.

sporophyte Diploid spore-forming body that forms in a plant life cycle.

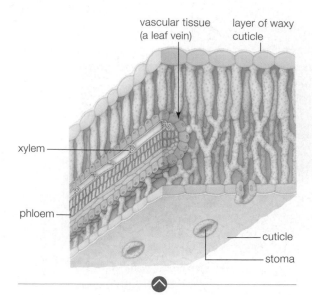

vascular tissue
(a leaf vein)

layer of waxy
cuticle

xylem

phloem

cuticle

stoma

Figure 14.4 Cross-section of a leaf, showing some traits that adapt plants to land.

bryophyte Member of a plant lineage that does not have vascular tissue; for example, a moss.

cuticle Secreted covering at a body surface. In plants it is waxy and helps conserve water.

lignin Compound that stiffens walls of some cells (including xylem) in vascular plants.

phloem Vascular tissue that distributes dissolved sugars.

pollen grain Immature male gametophyte of a seed plant.

seed Embryo sporophyte of a seed-bearing plant packaged with nutritive tissue inside a protective coat.

stomata Adjustable pores in a plant cuticle.

vascular plant A plant that has xylem and phloem.

xylem Vascular tissue that distributes water and dissolved mineral ions.

The relative size, complexity, and longevity of the sporophyte and gametophyte vary among plants lineages. In the oldest surviving plant lineages, collectively referred to as nonvascular plants or **bryophytes**, the haploid gametophyte is larger and longer-lived than the diploid sporophyte. In all other plants, sporophytes dominate the life cycle and gametophytes are tiny and short-lived.

Structural Adaptations to Life on Land The aboveground parts of most land plants are covered by a **cuticle** (Figure 14.4). This layer of waxy secretions slows evaporation across the body surface. Adjustable pores called **stomata** (singular, stoma) extend across the cuticle. Depending on environmental conditions, these pores either open to allow gas exchange or close to conserve water.

Bryophytes have threadlike structures that hold them in place, but they do not have true roots. Such roots not only anchor a plant, they also take up water and essential minerals from the soil. Plants with true roots have a vascular system with two types of tissues that serve as internal pipelines. **Xylem** is the vascular tissue that distributes water and mineral ions. **Phloem** is the vascular tissue that distributes sugars produced by photosynthetic cells. More than 90 percent of modern plants have xylem and phloem and thus are grouped as **vascular plants**.

In addition to distributing materials, vascular tissue provides structural support. Organic compounds called **lignins** stiffen the walls of xylem. Evolution of stems with lignin-stiffened tissue allowed vascular plants to stand taller than nonvascular ones and to branch. Most vascular plants also have leaves. Leaves are flattened, aboveground organs that increase the surface area available for intercepting sunlight and for gas exchange. They contain veins of vascular tissue.

Reproduction and Dispersal All bryophytes and some vascular plants produce sperm that have one or more flagella. To reach eggs, the sperm must swim through a film of water. As a result, these plants can reproduce only when their surroundings are damp. By contrast, seed-bearing vascular plants (seed plants) produce pollen grains. A **pollen grain** is a walled, immature male gametophyte. Winds or animals can carry pollen grains to a female gametophyte, thus fertilization can occur even in dry environments.

Bryophytes and seedless vascular plants disperse by releasing spores, but seed plants disperse by releasing seeds. A **seed** consists of an embryo sporophyte and food to support it, enclosed within a protective coat. There are two kinds of seed plants, gymnosperms and angiosperms. Angiosperms are the only plants that make flowers and disperse their seeds inside of fruits.

Take-Home Message 14.2

What adaptive traits allow plants to live on land and in dry places?

- Plants are multicelled, photosynthetic eukaryotes that protect and nourish their multicelled embryos on their body.
- Adaptations to life on land include a waxy cuticle with stomata, true roots, and vascular tissues that distribute materials and provide structural support.
- The plant life cycle alternates between two multicelled generations: a haploid gametophyte generation and a diploid sporophyte generation.
- Evolution of pollen grains and seeds gave seed plants the capacity to live in drier places than other plants.

14.3 Nonvascular Plants

REMEMBER: Gametes form by meiosis (Section 8.5).

Modern bryophytes (nonvascular plants) include 24,000 species belonging to three lineages: mosses, hornworts, and liverworts. Some mosses have tubes that conduct water and sugar, but none has the lignin-stiffened vascular pipelines that define vascular plants. As a result, nearly all nonvascular plants stand less than 20 centimeters (8 inches) high.

Mosses **Mosses** are the most diverse and familiar nonvascular plants. We will use the life cycle of one moss (*Polytrichum*) to illustrate a bryophyte life cycle (Figure 14.5). Like all bryophytes, this moss has a gametophyte-dominated life cycle. As in other mosses, the gametophyte has leaflike photosynthetic parts that grow from a central stalk ❶. Threadlike structures called **rhizoids** hold the gametophyte in place.

The moss sporophyte is not photosynthetic, so it must obtain nourishment from the gametophyte to which it is attached. The sporophyte consists of a stalk with a spore-producing organ (a sporangium) at its tip ❷. Meiosis of cells inside a spore chamber yields haploid spores ❸.

After dispersal by the wind, a spore germinates (becomes active) and grows into a gametophyte that has multicellular gamete-producing organs (gametangia. The moss we are using as our example has separate sexes, with each gametophyte producing either eggs or sperm ❹. Other bryophytes are bisexual. In either case, rain triggers the release of flagellated sperm that swim through a film of water to eggs ❺. Fertilization occurs inside the egg chamber and produces a zygote ❻ that develops into a new sporophyte ❼. Mosses also reproduce asexually by fragmentation when a bit of gametophyte breaks off and develops into a new plant.

There are about 14,000 named species of moss. Many of them can colonize rocky areas where the lack of soil prevents other plants from becoming established. Over time, decomposition of dead moss helps to create a layer of soil in which vascular plants can take root.

The 350 or so species of moisture-loving peat mosses (*Sphagnum*) are of great ecological and commercial importance. They are the main plants in peat bogs, a type of plant community that covers more than 350 million acres in high-latitude areas of Europe, Asia, and North America. Many peat bogs have existed for thousands of years, and layer upon layer of partially decayed plant remains have become compressed to form deposits of a carbon-rich material called peat. Blocks of peat can be cut, dried, and used as a clean-burning fuel. Freshly harvested peat moss is also an important commercial product. It is dried and added to planting mixes to help soil retain moisture.

Figure 14.5 Life cycle of the moss *Polytrichum*, shown at left.

❶ The leafy green part of a moss is the haploid gametophyte.

❷ The diploid sporophyte has a stalk and a capsule (sporangium). It is not photosynthetic.

❸ Haploid spores form by meiosis in the capsule, are released, and drift with the winds.

❹ Spores germinate and develop into male or female gametophytes with gametangia that produce eggs or sperm by mitosis.

❺ Sperm swim to eggs.

❻ Fertilization occurs in the egg chamber on the female gametophyte and produces a zygote.

❼ The zygote grows and develops into a sporophyte while remaining attached to and nourished by its female parent.

Bottom photo, Jane Burton/Bruce Coleman Ltd.

mosses Most diverse group of bryophytes (nonvascular plants). Low-growing plants that have flagellated sperm and disperse by producing spores.

rhizoid Threadlike structure that anchors some plants.

Figure 14.6 Liverwort.
Umbrella-like female gametangia of a liverwort (*Marchantia*). The attached sporophytes on their lower surface have spores (yellow) at their tips.

Dr. Annkatrin Rose, Appalachian State University.

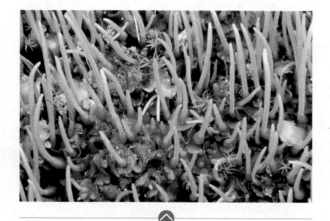

Figure 14.7 Hornworts.
Elongated photosynthetic sporophytes grow attached to the flat, ribbonlike gametophytes. Spores will form at the tip of these "horns."

age fotostock/SuperStock.

Figure It Out: Do spores of liverworts develop into sporophytes or gametophytes?

Answer: All plant spores develop into gametophytes.

Liverworts and Hornworts Liverworts and hornworts are less familiar nonvascular lineages. They often live alongside mosses in damp places.

Liverworts may be the most ancient of the surviving plant lineages. The oldest known fossils of land plants are spores that resemble those of modern liverworts. In addition, genetic comparisons put liverworts near the base of the plant family tree.

Marchantia is a widespread genus of liverworts that reproduces sexually by producing sperm and eggs in umbrella-like structures that grow from the surface of the flattened gametophyte (Figure 14.6). After fertilization, the sporophyte develops while still attached to the gametophyte. *Marchantia* species also reproduce asexually by producing small clumps of cells in cups on the gametophyte surface. The clumps are dispersed by rain and develop into new plants.

Hornworts have flattened, ribbonlike gametophytes. Fertilization of eggs that form on the gametophyte body produces a zygote that develops into a tall, horn-shaped sporophyte (Figure 14.7). Hornwort sporophytes, unlike those of mosses and liverworts, contain chloroplasts and grow continually. They can be several centimeters tall. These sporophyte traits, together with evidence from gene comparisons, suggest that hornworts are probably the nonvascular plants that are most closely related to vascular plants.

Take-Home Message 14.3

What are bryophytes?

- Bryophytes include three lineages of low-growing plants (mosses, hornworts, and liverworts). All have flagellated sperm and disperse by releasing spores.
- Bryophytes are the only modern plants in which the gametophyte dominates the life cycle and the sporophyte is dependent upon it.
- Liverworts are probably the oldest plant lineage. Hornworts are most likely the closest living relatives of vascular plants.

14.4 Seedless Vascular Plants

Seedless vascular plants include ferns, horsetails, and club mosses. Like bryophytes, these plants have flagellated sperm that require a film of water to swim to eggs. Also like bryophytes, they disperse by releasing spores.

Seedless vascular plants differ from bryophytes in other aspects of their life cycle and structure. The gametophyte is reduced in size and is relatively short-lived. Although the sporophyte develops on the gametophyte body, it survives on its own after the gametophyte dies. Lignin stiffens a sporophyte's body, and a system of vascular tissue distributes water, sugars, and minerals through it. These innovations in support and plumbing allow seedless vascular sporophytes to be large and structurally complex, with roots, stems, and leaves.

Ferns We begin our survey of the seedless vascular plants with the most diverse and familiar lineage, the **ferns**. Figure 14.8 shows the life cycle of a common North American fern. The leafy plant we envision when we think of a fern is a sporophyte ❶. In most ferns, roots and fronds (leaves) sprout from **rhizomes**, which are stems that grow along or just below the ground. Spores form by meiosis in capsules that cluster as **sori** (singular, sorus) on the underside of leaves ❷. When the spore

Figure 14.8 Life cycle of a common North American fern (*Woodwardia*).

1 The familiar leafy form is the diploid sporophyte.

2 Meiosis in cells on the underside of fronds produces haploid spores.

3 After their release, the spores germinate and grow into tiny gametophytes that produce eggs and sperm.

4 Sperm swim to eggs and fertilize them, forming a zygote.

5 The sporophyte begins its development attached to the gametophyte, but it continues to grow and live independently after the gametophyte dies.

Photo, A. & E. Bomford/Ardea, London.

Diploid (2*n*) phase

Haploid (*n*) phase

capsule pops open, wind disperses the spores. A spore develops into a gametophyte a few centimeters wide that forms eggs and sperm in chambers on its underside **3**. In the fern we are discussing, eggs and sperm form on the same gametophyte. In some other ferns, each gametophyte produces either sperm or eggs. In either case, sperm swim to eggs to fertilize them **4**. After fertilization, the resulting zygote develops into a new sporophyte, and its parental gametophyte dies **5**.

In many ferns, asexual reproduction occurs more frequently than sexual reproduction. New shoots and roots develop from a rhizome as it grows through the soil. Eventually the connection to the parent plant breaks, and that segment of rhizome becomes an independent plant.

Fern sporophytes vary enormously in their size and form. Some float on freshwater ponds and have fronds less than 1 centimeter wide. Many tropical ferns are **epiphytes**, plants that live attached to the trunk or branches of another plant but do not withdraw nutrients from it. The largest ferns are tree ferns that can be 25 meters (80 feet) tall.

Horsetails and Club Mosses Horsetails (*Equisetum*) are close relatives of ferns. They thrive along streams and roadsides, and in disturbed areas. A horsetail sporophyte has rhizomes and hollow stems with tiny nonphotosynthetic leaves at the joints (Figure 14.9A). Photosynthesis occurs in stems and leaflike branches. The stems contain silica, a gritty mineral that helps the plant fend off insects and snails. Before the invention of modern abrasive cleansers, silica-rich stems of some *Equisetum* species were used to scrub pots and polish metals. Depending on the species, spore-bearing structures form either at tips of photosynthetic stems or on specialized reproductive stems that do no have chlorophyll.

Club mosses are nonvascular plants common on the floor of temperate forests (Figure 14.9B). Their branching form makes them look like tiny pine trees. Roots

Figure 14.9 Seedless vascular plants.

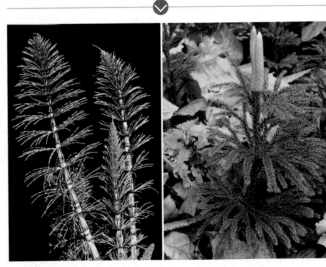

A. Horsetail   **B. Club moss**

(A) © William Ferguson; (B) © Martin LaBar, www.flickr.com/photos/martinlabar.

epiphyte Plant that grows on the trunk or branches of another plant but does not harm it.

ferns Most diverse lineage of seedless vascular plants.

rhizome Stem that grows horizontally along or just below the ground.

sorus Cluster of spore-forming chambers on a fern frond.

Figure 14.10 Coal forest.
An artist's depiction of a swamp forest during the Carboniferous period. An understory of ferns ❶ is shaded by tree-sized club mosses ❷ and horse-tails ❸.

Over millions of years, remains of such plants were transformed to coal (below).

Noraluca013/Shutterstock.com.

and upright stems with tiny leaves grow from a rhizome. Club moss spores have a waxy coating that makes them ignite easily. They were used to create flashes for early photography and are still used to produce special effects in theaters.

Ferns are currently the most diverse seedless vascular plants, but during the Carboniferous period (359–299 million years ago) relatives of modern club mosses and horsetails were the dominant plants in swamp forests (Figure 14.10). Some stood 40 meters (more than 130 feet) high. After forests of these plants formed, climates changed, and the sea level rose and fell many times. When the waters receded, the forests flourished. After the sea moved back in, submerged trees became buried in sediments that protected them from decomposition. As layers of sediments accumulated one on top of the other, their weight squeezed the water out of the saturated, undecayed remains, and the compaction generated heat. Over time, pressure and heat transformed the compacted organic remains into **coal**.

It took millions of years of photosynthesis, burial, and compaction to form coal. When you hear about annual production of coal or other fossil fuels, keep in mind that we do not really "produce" these materials, we only extract them. This is why fossil fuels are said to be nonrenewable sources of energy.

Take-Home Message 14.4

What are seedless vascular plants?

- Seedless vascular plants include ferns, club mosses, and horsetails.
- These plants disperse by releasing spores, and the life cycle is dominated by a sporophyte that has vascular tissue and lignin. The gametophyte is small and relatively short-lived.
- Like bryophytes, seedless vascular plants have flagellated sperm that must swim through a film of water to reach eggs.

coal Fossil fuel consisting primarily of the carbon-rich remains of seedless nonvascular plants.

14.5 Rise of the Seed Plants

Seed plants evolved from a lineage of seedless vascular plants about 400 million years ago. They survived alongside bryophytes and seedless nonvascular plants until the late Carboniferous, then rose to dominance as the climate became drier. Unique traits of seed plants give them a competitive advantage over seedless plants in places where water is scarce.

Gametophytes of seedless vascular plants are free-living, meaning they develop from spores that were released into the environment. By contrast, gametophytes of seed plants develop within the protection of spore-forming chambers (pollen sacs and ovules) on a sporophyte (Figure 14.11).

All nonvascular plants and most seedless vascular plants produce spores of just one type. By contrast, all seed plants produce two types of spores that differ in size. Meiosis of a cell inside a **pollen sac** produces four microspores **1**. Each **microspore** develops into a pollen grain, which is the sperm-producing gametophyte **2**. In **ovules**, meiosis and unequal cytoplasmic division produce three small cells that disintegrate and one large **megaspore** **3**. The megaspore develops into an egg-producing gametophyte **4**.

A seed plant releases its pollen grains, but holds onto its eggs. Wind or animals deliver pollen from one seed plant to the ovule of another, a process known as **pollination** **5**. Sperm of seed plants do not need to swim through a film of water to reach eggs, so these plants can reproduce in dry times.

After pollination, a pollen tube grows into the ovule and delivers a sperm, which in most seed plants is not flagellated, to the egg **6**. Fertilization produces a zygote within the ovule, which matures into a seed **7**. Releasing seeds puts seed plants at an advantage over plants that release spores. A seed contains a multicelled embryo sporophyte and stored food that the embryo can draw on during early development. By contrast, a plant spore is a single cell that does not have any food reserves. In addition, seeds often also have adaptations that aid in their long-distance dispersal. For example, the seed may have a winglike structure that helps it catch the wind.

Structural traits also gave seed plants an advantage. Some seed plants undergo secondary growth (growth in diameter) and produce wood. **Wood** is lignin-stiffened tissue that strengthens and protects older stems and roots. The giant nonvascular plants that lived in Carboniferous forests did not produce wood, and their trunks were supported by overlapping leaf bases. As a result, their trunks were softer and more flexible than those of woody seed plants, and they could not grow as tall as the tallest modern trees do.

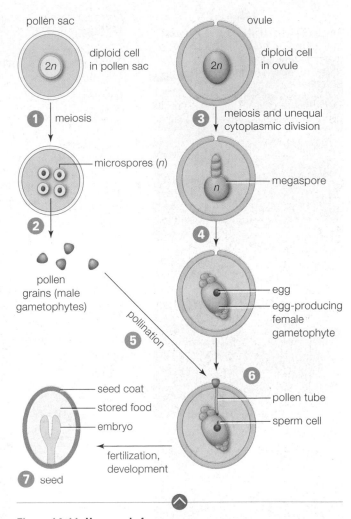

Figure 14.11 How seeds form.
This diagram shows the process in gymnosperms. The process is slightly different in angiosperms, as described in Section 14.7.

Figure It Out: Do pollen sacs form on sporophytes or on gametophytes?

Answer: Pollen sacs form on sporophytes. They produce spores that develop into male gametophytes.

megaspore In seed plants, a haploid cell that gives rise to a female gametophyte.

microspore In seed plants, a haploid cell that gives rise to a male gametophyte (pollen grain).

ovule Of seed plants, chamber inside which megaspores form and develop into female gametophytes; after fertilization this chamber becomes a seed.

pollen sac Of seed plants, chamber in which microspores form and develop into male gametophytes (pollen grains).

pollination Delivery of a pollen grain to the egg-bearing part of a seed plant.

wood Lignin-reinforced tissue produced by secondary growth of some seed plants.

Take-Home Message 14.5

What gave seed plants an adaptive advantage over spore-bearing plants?

- Seed plant spores develop into gametophytes while protected by the sporophyte body.
- Male gametophytes are pollen grains that can be carried to egg-bearing ovules even in dry environments.
- A seed contains a plant embryo and a food supply that it can draw upon during its early development.
- Some seed plants undergo secondary growth (they thicken) and become woody.

14.6 Gymnosperms

Gymnosperms are seed plants that produce seeds on the surface of ovules. Their seeds are said to be "naked," because unlike those of angiosperms, they are not inside a fruit. (*Gymnos* means naked and *sperma* is taken to mean seed.) However, many gymnosperms enclose their seeds in a fleshy or papery covering. There are four modern gymnosperm lineages.

Conifers **Conifers** (phylum Pinophyta) are the most diverse gymnosperms with more than 600 species. All are woody trees or shrubs with needlelike or scalelike leaves. Most conifers are evergreen, meaning they do not shed their leaves all at once. Evergreen conifers are the main plants in cool Northern Hemisphere forests. Their conical shape helps many conifers shed snow easily. Needlelike leaves with a heavy wax coating help them minimize water loss during a long winter when soil is frozen, or during a dry hot season. Conifers include the tallest trees in the Northern Hemisphere (Figure 14.12) and the longest-lived trees; some bristlecone pines are more than 4,000 years old.

A pine tree's life cycle is typical of conifers. The tree is a sporophyte and its cones are specialized spore-bearing structures. There are two types of cones: small, soft pollen cones and large, woody, ovulate cones (Figure 14.13). Both have scales (modified leaves) arranged around a central axis. Pollen made by a pollen cone is released and drifts on the wind. Ovulate cones produce a sticky substance that traps

ovule, which may become seed

Figure 14.13 Pine cones.
A. Pollen cone. **B.** Cross section of an ovulate cone.

(A) R. J. Erwin/Science Source; (B) © Stan Elems/ Visuals Unlimited.

Figure 14.12 Composite photo of one of the world's tallest trees, a coast redwood (*Sequoia sempervirens*) that stands 116 meters (379 feet) high in a Northern California forest.
Can you spot the three blue-shirted people climbing among its branches?

© James Balog/Aurora Photos.

pollen grains. After pollen is trapped, a tube grows into the ovulate cone scale and delivers a nonflagellated sperm cell to an egg in an ovule inside it. The pollen tube grows very slowly, so fertilization can occur as long as a year after pollination. After fertilization, the ovule develops into a seed.

Conifers are of great economic importance. For example, pines are the main source of lumber for building homes, and we use fir bark as mulch in our gardens. Some pines make a sticky resin that deters insects from boring into them. We use this resin to make turpentine, which is a paint thinner and solvent. We use oils from cedar in cleaning products, and we eat seeds, or "pine nuts," of pinyon pines.

Cycads and Ginkgos Cycads (phylum Cycadophyta) and ginkgos (phylum Ginkgophyta) are ancient lineages that were at their most diverse when dinosaurs walked the Earth. They are the only modern gymnosperms that have flagellated sperm. In both groups, a male plant produces pollen that is carried by the wind to a female plant. As in conifers, the ovule of the female plant secretes fluid that traps pollen grains. Sperm emerge from pollen grains, then swim through the secreted fluid to eggs in the plant's ovule.

About 130 species of cycads survive and they live mainly in tropical and subtropical regions. Cycads resemble palms or ferns but are not close relatives of either (Figure 14.14A). "Sago palms" commonly used in landscaping and as houseplants are actually cycads.

The only living ginkgo species is *Ginkgo biloba*, the maidenhair tree. It is native to China, but its pretty fan-shaped leaves (Figure 14.14B) and resistance to air pollution make it popular in many cities in the United States. Typically, male trees are planted because female trees make seeds with a fleshy covering that has a strong, unpleasant odor. Ginkgos are **deciduous plants**, meaning they shed all their leaves at the end of their growing season and spend the winter leafless and dormant.

Gnetophytes Gnetophytes (phylum Gnetophyta) include about 70 species. One of these, *Welwitschia mirabilis*, is an unusual-looking plant that lives in Africa's Namib desert (Figure 14.14C). It has a taproot, woody stem, and two long, straplike leaves that split lengthwise repeatedly, giving the plant a shaggy appearance. The leaves grow continually as a result of cell divisions at their base. *Welwitschia* is very long-lived, with some plants reaching an age of more than a thousand years. Members of the genus *Ephedra* are shrubs with jointed, photosynthetic branches and tiny, scalelike leaves (Figure 14.14D). Eurasian species of *Ephedra* contain large amounts of the amphetamine-like compound ephedrine and are used in traditional medicine. *Ephedra* species native to North America do not make this compound.

A. Cycad with palmlike leaves and fleshy seeds.

B. Ginkgo tree with fleshy seeds and fan-shaped seeds.

C. *Welwitschia* with seed cones and two long leaves.

D. *Ephedra* with photosynthetic stems and pollen cones.

Figure 14.14 Gymnosperm diversity.

(A, C) Fletcher and Baylis/Science Source; (B) left, KPG_Payless/Shutterstock.com; right, picturepartners/Shutterstock.com; (D) Christine Evers.

Take-Home Message 14.6

What are gymnosperms?

- Gymnosperms are one of the two lineages of seed-bearing vascular plants. They form seeds on the surface of cones or other spore-producing structures.
- Conifers, the most diverse gymnosperm group, are evergreen trees with needlelike leaves. They dominate cool, high-latitude forests.
- Cycads and ginkgos are two lineages with pollen grains that release flagellated sperm.
- Gnetophytes include diverse forms.

conifer Woody gymnosperm with needlelike leaves.

deciduous plant Plant that sheds all its leaves in preparation for a seasonal dormancy.

gymnosperm Seed plant that produces "naked" seeds (seeds that are not encased within a fruit).

Figure 14.15 Floral structure.

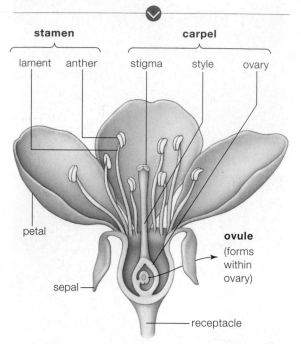

14.7 Angiosperms—Flowering Plants

REMEMBER: Coevolution refers to the joint evolution of two species as a result of their close ecological interaction (Section 12.7).

Floral Structure and Function **Angiosperms** are vascular seed plants, and the only plants that make flowers and fruits. A **flower** is a specialized reproductive shoot (Figure 14.15). Sepals, which usually have a green leaflike appearance, ring the base of a flower and enclose it until it opens. They surround a ring of petals, which are often brightly colored.

Stamens, a flower's pollen-producing parts, surround a **carpel** that captures pollen and produces eggs. Typically a stamen consists of a tall stalk, called the filament, topped by an **anther** that holds two pollen sacs.

A carpel has a sticky **stigma**, a region specialized for receiving pollen, at its tip. The stigma is located atop a stalk, called the **style**. At the base of the style is an **ovary**, a chamber containing one or more egg-producing ovules. After fertilization, an ovule matures into a seed and the ovary becomes the **fruit**. The name angiosperm refers to the fact that seeds form within the protection of the ovary. (*Angio*– means enclosed chamber, and *sperma*, seed.)

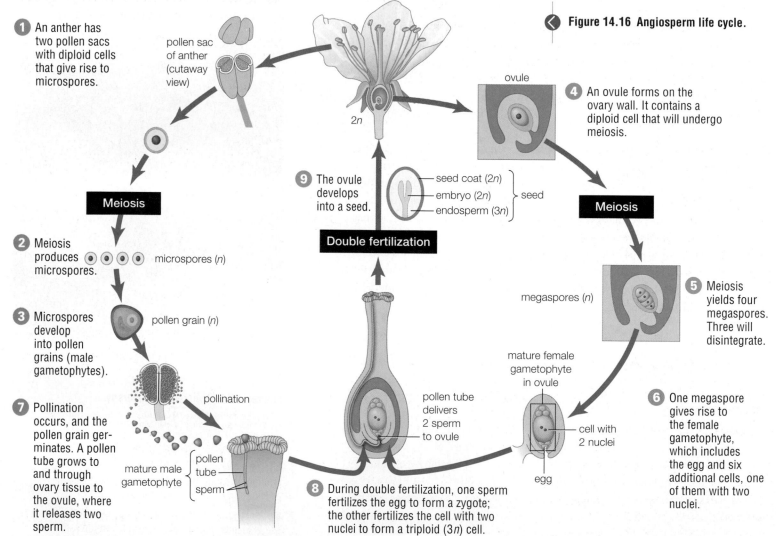

Figure 14.16 Angiosperm life cycle.

1 An anther has two pollen sacs with diploid cells that give rise to microspores.

pollen sac of anther (cutaway view)

Meiosis

2 Meiosis produces microspores.

microspores (*n*)

3 Microspores develop into pollen grains (male gametophytes).

pollen grain (*n*)

7 Pollination occurs, and the pollen grain germinates. A pollen tube grows to and through ovary tissue to the ovule, where it releases two sperm.

pollination

mature male gametophyte — pollen tube — sperm

8 During double fertilization, one sperm fertilizes the egg to form a zygote; the other fertilizes the cell with two nuclei to form a triploid (3*n*) cell.

Double fertilization

9 The ovule develops into a seed.

seed coat (2*n*)
embryo (2*n*) } seed
endosperm (3*n*)

2*n*

pollen tube delivers 2 sperm to ovule

4 An ovule forms on the ovary wall. It contains a diploid cell that will undergo meiosis.

ovule

Meiosis

5 Meiosis yields four megaspores. Three will disintegrate.

megaspores (*n*)

6 One megaspore gives rise to the female gametophyte, which includes the egg and six additional cells, one of them with two nuclei.

mature female gametophyte in ovule

cell with 2 nuclei

egg

A Flowering Plant Life Cycle Figure 14.16 shows a generalized life cycle for a flowering plant. A flower forms on the sporophyte body. Pollen sacs in the anthers hold diploid cells ❶ that produce microspores by meiosis ❷. The microspores develop into pollen grains (immature male gametophytes) ❸. Ovules form on the wall of an ovary at the base of a carpel ❹. Meiosis of cells in ovules yields haploid megaspores ❺. A megaspore develops into a female gametophyte consisting of a haploid egg, a cell with two nuclei, and a few other cells ❻.

Pollination occurs when a pollen grain arrives on a receptive stigma, the uppermost part of the carpel ❼. The pollen grain germinates, and a pollen tube grows through the style (the structure that elevates the stigma) to the ovary at the base of the carpel. Two nonflagellated sperm form inside the pollen tube as it grows.

Double fertilization occurs when a pollen tube delivers the two sperm into the ovule ❽. One sperm fertilizes the egg to create a zygote. The other sperm fuses with the cell that has two nuclei, forming a triploid (3n) cell. After double fertilization, the ovule matures into a seed ❾. The zygote develops into an embryo sporophyte and the triploid cell develops into **endosperm**, a nutritious tissue that will serve as a source of food for the developing embryo.

Keys to Angiosperm Diversity Flowering plants constitute 90 percent of all modern plant species. What accounts for angiosperm success? For one thing, they tend to grow faster than gymnosperms. Think of how a plant like a dandelion or a grass can grow from a seed and produce seeds of its own within a few months. In contrast, most gymnosperms take years to mature and produce seeds.

Evolution of flowers gave angiosperms an edge by facilitating animal-assisted pollination. After pollen-producing plants evolved, some insects began feeding on the plants' highly nutritious pollen. Plants gave up some pollen but gained a reproductive edge when insects unknowingly moved pollen, thus facilitating pollination. Many traits of flowering plants are adaptations that attract **pollinators**, animals that move pollen of one plant species onto female reproductive structures of the same species. Insects are the most common pollinators (Figure 14.17), but birds, bats, and other animals also serve in this role. Producing sugary nectar encourages more pollinator visits, which improves pollination rates and enhances seed production.

A variety of fruit structures help angiosperms disperse their seeds. Fruits ride the winds, stick to animal fur, or entice animals to eat them and release seeds in their feces. Gymnosperm seeds have less diverse dispersal mechanisms.

Major Groups The vast majority of flowering plants belong to one of two lineages. The 80,000 or so **monocots** include orchids, palms, lilies, and grasses, such as rye, wheat, corn, rice, sugarcane, and other important crop plants. **Eudicots** include most herbaceous (nonwoody) plants such as tomatoes, cabbages, roses, daisies, most flowering shrubs and trees, and cacti. Monocots and eudicots derive their group names from the number of seed leaves (cotyledons) in the embryo. Monocots have one seed leaf, and eudicots have two. The two lineages also differ in many structural details such as the arrangement of their vascular tissues and the number of flower petals. Although some eudicots undergo secondary growth (growth in width) and become woody, no monocots produce true wood. Section 27.2 describes the differences between eudicots and monocots in more detail.

Figure 14.17 A pollinator.
A bee unknowingly transfers pollen from flower to flower as it gathers pollen and sips nectar.

Courtesy of Christine Evers.

angiosperm Seed plant that produces flowers and fruits.

anther Part of the stamen that contains pollen sacs.

carpel Ovule-containing part of a flower.

double fertilization In flowering plants, one sperm fertilizes the egg, forming the zygote, and another fertilizes a diploid cell, forming what will become endosperm.

endosperm Nutritive tissue in an angiosperm seed.

eudicots Largest lineage of angiosperms; includes herbaceous plants, woody trees, and cacti.

flower Specialized reproductive shoot of a flowering plant.

fruit Mature ovary tissue that encloses a seed or seeds.

monocots Lineage of angiosperms that includes grasses, orchids, and palms.

ovary Of flowering plants, a floral chamber that holds one or more ovules.

pollinator Animal that moves pollen from one plant to another, thus facilitating pollination.

stamen Pollen-producing part of a flower. Consists of an anther that contains pollen sacs, atop a filament.

stigma Pollen-receiving part of a carpel.

style Elongated portion of a carpel that holds the stigma above the ovary.

A. Mechanized harvesting of wheat.

Photo USDA.

B. A field of cotton ready for harvest.

Photo by Scott Bauer, USDA/ARS.

Figure 14.18 Angiosperms as crops.

chytrid Fungus that produces flagellated spores.

club fungus Fungus that produces spores by meiosis in club-shaped cells.

fungus Spore-producing heterotroph that has cell walls of chitin and feeds by extracellular digestion and absorption.

hypha A single filament in a fungal mycelium.

mycelium Mass of threadlike filaments (hyphae) that compose the body of a multicelled fungus.

zygote fungus Fungus that usually grows as a mold; sexual reproduction yields a thick-walled zygospore.

Ecology and Human Uses of Angiosperms It would be nearly impossible to overestimate the importance of angiosperms. As the dominant plants in most land habitats, they provide food and shelter for land animals. They also supply many products that meet human needs (Figure 14.18).

Angiosperms provide nearly all of our food, directly or as feed for livestock. Cereal crops are the most widely planted. In the United States, more acreage is devoted to corn than to any other plant. Worldwide, rice feeds the greatest number of people. Wheat, barley, and sorghum are other widely grown grains. All are grasses. Legumes are the second most important source of human food. They can be paired with grains to provide all the amino acids the human body needs to build proteins. Soybeans, lentils, peas, and peanuts are examples of legumes.

Name a part of a plant, and humans eat it. In addition to the seeds of grains and legumes, we dine on leaves of lettuce and spinach, stems of asparagus, developing flowers of broccoli, modified roots of potatoes, carrots, and beets, and fruits of tomatoes, apples, and blueberries. Stamens of crocus flowers provide the spice saffron, and the bark of a tropical tree provides cinnamon.

Fibers used to make clothing come from two main sources, petroleum and plants. Plant fibers include cotton, flax, ramie, and hemp. We also use fibers from flowering plants to weave rugs, and in many places to thatch roofs. Oak and other hardwoods derived from angiosperms provide flooring and furniture.

We extract medicines and psychoactive drugs from plants. Aspirin is derived from a compound discovered in willows. Digitalis from foxglove strengthens a weak heartbeat. Coffee, tea, and tobacco are widely used plant-derived stimulants. Worldwide, cultivation of opium poppies (the source of heroin) and coca (the source of cocaine) have wide-reaching health, economic, and political effects.

Take-Home Message 14.7

What are angiosperms?

- Angiosperms are plants in which seeds develop inside ovaries that become fruits.
- They are the most diverse plant lineage. Adaptations that contributed to their success include a short life cycle, coevolution with insect pollinators, and a variety of fruit structures that aid in dispersal of seeds.
- There are two major angiosperm lineages: monocots and dicots.
- The overwhelming majority of plants grown as crops are angiosperms.

14.8 Fungal Traits and Diversity

REMEMBER: Glycogen is a complex carbohydrate (Section 2.7).

Yeasts, Molds, Mildews, and Mushrooms With this section, we begin our survey of another major lineage of eukaryotes, the fungi. Like plants, **fungi** have walled cells, spend their lives fixed in place, and produce haploid spores by meiosis. However, fungi are more closely related to animals than to plants. Like animals, fungi are heterotrophs and they store excess sugars as glycogen, rather than starch. Most are decomposers that feed on organic wastes and remains. A lesser number live on or in other living organisms. As a fungus grows in or over organic matter, it secretes digestive enzymes, then absorbs the resulting breakdown products.

A. Yeast (single-celled fungus).　　**B.** Mold growing on a grapefruit.　　**C.** Mildew growing on leaves.　　**D.** Mushrooms on a forest floor.

Fungi have a variety of growth habits. Yeasts live as single cells (Figure 14.19A). However, most fungi are multicelled. Molds, mildews, and mushrooms are familiar examples of multicelled fungi (Figure 14.19B–D). These fungi grow as a **mycelium** (plural, mycelia), a network of microscopic interwoven filaments. Each filament, or **hypha** (plural, hyphae), is a strand of walled cells arranged end to end (Figure 14.20). Fungal cell walls consist primarily of chitin, a complex carbohydrate also found in the body covering of crabs and insects.

Fungi do not have vascular tissue, but walls between cells in a hypha are porous, so materials flow among cells. As a result, nutrients or water taken up in one part of the mycelium can be shared with cells in other regions.

Some fungal bodies are enormous. The largest organism we know about is a soil fungus in Oregon. Its hyphae extend through the soil over an area of about 2,200 acres, and it is still growing. Given its growth rate and size, researchers estimate that this individual has been alive for 8,000 years.

Lineages and Life Cycles The oldest fungal fossils are about 500 million years old. They resemble **chytrids**, a group of modern fungi that are primarily aquatic and produce flagellated spores. The fossils and the existence of the aquatic, flagellated chytrids are taken as evidence that fungi, like plants and animals, evolved from an aquatic protist. By about 400,000 years ago, fungi had moved onto land.

Scientists estimate there are than a million species of fungus, although they have only named about 70,000. Most fungi are distributed among five major subgroups. In addition to the chytrids (phylum Chytridiomycota), there are zygote fungi (phylum Zygomycota), glomeromycete fungi (Glomeromycota), sac fungi (Ascomycota), and club fungi (Basidiomycota).

Most **zygote fungi** are molds that grow on or through organic matter as a mass of hyphae. As long as food is plentiful, the fungus reproduces asexually by making spores by mitosis at the tips of special hyphae. When food runs low, sexual reproduction occurs. As in all multicelled fungi, sexual reproduction begins with the fusion of haploid hyphae from two individuals of different mating strains. In zygote fungi, this fusion leads to formation of a thick-walled diploid spore called a zygospore (Figure 14.21). Meiosis of cells inside the zygospore yields haploid cells that develop into new mycelia.

Many sac fungi and club fungi produce spores on a multicelled fruiting body. A **club fungus** produces spores in club-shaped cells on a fruiting body such as a

Figure 14.19 Variety of fungal forms.

(A) Dr. John D. Cunningham/Visuals Unlimited, Inc.; (B) Photo by Scott Bauer/USDA; (C) Nigel Cattlin/Science Source; (D) Robert C. Simpson/Nature Stock.

Figure 14.20 Micrograph (top) and diagram (bottom) of club fungus hyphae.

one cell (part of one hypha of a club fungus mycelium)

Garry T. Cole, University of Texas, Austin/BPS.

pore in cross-wall

From Russell/Wolfe/Hertz/Starr, *Biology*, 3e, © Cengage Learning®.

Figure 14.21 Black bread mold, a zygote fungus.
The fungus usually grows as a mass of asexually reproducing hyphae. The inset micrograph shows a zygospore.

Background, © Micrograph J. D. Cunningham/Visuals Unlimited; inset, Ed Reschke.

Figure 14.22 Generalized life cycle for a club fungus.
Hyphae of different mating strains often grow through the same patch of soil.

1 Two haploid hyphal cells meet and their cytoplasm fuses, forming a dikaryotic (*n+n*) cell.

2 Mitotic cell divisions form a mycelium that produces a mushroom.

3 Spore-making cells form at the edges of the mushroom's gills.

4 Inside these dikaryotic cells, nuclei fuse, making the cells diploid (2*n*).

5 The diploid cells undergo meiosis, forming haploid (*n*) spores.

6 Spores are released and give rise to a new haploid mycelium.

After T. Rost, et al., *Botany*, Wiley, 1979.

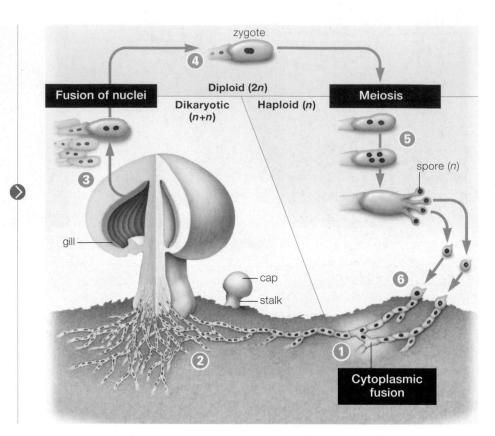

Figure It Out: Are cells that make up the stalk of a mushroom haploid, diploid, or dikaryotic?

Answer: Dikaryotic

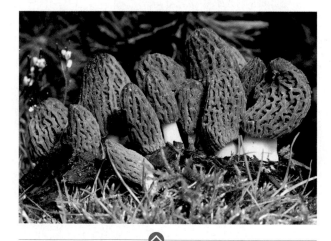

Figure 14.23 Morels, edible fruiting bodies of a sac fungus.

© unverdorben jr/Shutterstock.com.

common button mushroom (Figure 14.22). When haploid hyphae of two different club fungus individuals meet in the soil cytoplasmic fusion produces a dikaryotic cell **1**. Dikaryotic means "having two nuclei," being (*n+n*). Mitotic divisions produce a dikaryotic mycelium **2**. Multicelled club fungi spend most of their life cycle growing as a dikaryotic mycelium. When conditions favor reproduction, a sudden burst of hyphal growth produces a mushroom that emerges above the ground. A typical mushroom has a stalk and a cap, with thin sheets of tissue called gills on the cap's underside **3**. Fusion of haploid nuclei inside cells at the edges of the gills forms diploid zygotes **4**. Each zygote undergoes meiosis, producing haploid spores **5**. After a spore germinates, repeated mitotic divisions give rise to a new haploid mycelium **6**.

Some **sac fungi** form fruiting bodies that somewhat resemble those of club fungi (Figure 14.23). However, a sac fungus fruiting body does not have gills and sac fungus spores do not form in club-shaped cells. Instead, sac fungi form spores by meiosis inside a saclike cell. Hence the name "sac" fungus.

Take-Home Message 14.8

What are fungi?

- Fungi are heterotrophs that absorb nutrients from their environment. They live as single cells or as a multicelled mycelium and disperse by producing spores.
- Zygote fungi grow as molds that most often produce spores asexually.
- Mushrooms are spore-producing fruiting bodies of club fungi. Some sac fungi also produce large fruiting bodies that we eat.

sac fungus Fungus that produces spores by meiosis in saclike cells.

14.9 Ecological Roles of Fungi

REMEMBER: Fermentation is an anaerobic energy-releasing pathway (Section 5.6).

Decomposers Most fungi provide an important ecological service by breaking down complex compounds in organic wastes and remains. When a fungus secretes digestive enzymes onto these materials, some soluble nutrients escape into nearby soil or water. Plants and other producers can then take up these substances to meet their own nutrient needs. Bacteria also serve as decomposers, but they tend to grow mainly on surfaces. By contrast, fungal hyphae can extend deep into a dead log or another bulky food source and break it down from the inside (Figure 14.24).

Parasites Many sac fungi and club fungi are plant parasites. Powdery mildews (sac fungi) and rusts and smuts (club fungi) grow only in living plants. Wheat stem rust is an example. Hyphae of such fungi extend into cells of stems and leaves, where they suck up photosynthetically produced sugars. The resulting loss of nutrients stunts the plant, prevents it from producing seeds, and may eventually kill it. However, a plant usually does not die before the fungus has produced spores on the surface of its infected parts. Other pathogenic fungi produce toxins that kill plant tissues, then feed on the resulting remains. The club fungus *Armillaria* causes root rot by infecting trees and woody shrubs in forests worldwide. Once an infected tree dies, the fungus decomposes the stumps and logs left behind.

Animals are less vulnerable to fungal infections than plants and, among animals, those with a high body temperature are least susceptible. About 50,000 species of fungus can infect insects, whereas only a few hundred can infect mammals. Like

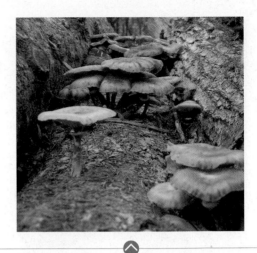

Figure 14.24 Fungi as decomposers.
Club fungus (*Armillaria*) fruiting bodies on pine logs. Fungal hyphae extend through the logs and digest them.

Chris Moody/Shutterstock.com.

Digging Into Data

Removing Fungus-Infected Stumps to Save Trees

The club fungus *Armillaria ostoyae* infects living trees and acts as a parasite, withdrawing nutrients from them. When the tree dies, the fungus continues to dine on its remains. Fungal hyphae grow out from the roots of infected trees and roots of dead stumps. If these hyphae contact roots of a healthy tree, they can invade and cause a new infection.

Canadian forest pathologists hypothesized that removing stumps after logging could help prevent tree deaths. To test this hypothesis, they carried out an experiment. In half of a forest, they removed stumps after logging. In a control area, they left stumps behind. For more than 20 years, they recorded tree deaths and whether *A. ostoyae* caused them. Figure 14.25 shows the results.

1. Which tree species was most often killed by *A. ostoyae* in control forests? Which was least affected by the fungus?
2. For the species most affected, what percentage of deaths did *A. ostoyae* cause in control and in experimental forests?
3. Do the overall data support the hypothesis that stump removal helps protect living trees from infection by *A. ostoyae*?

Figure 14.25 Results of a long-term study of how logging practices affect tree deaths caused by the fungus *A. ostoyae*.
In the experimental forest, whole trees—including stumps—were removed (brown bars). The control half of the forest was logged conventionally, with stumps left behind (blue bars).

After graph from www.pfc.forestry.ca.

Figure 14.26 White-nose syndrome.
Affected bats have white filaments of a parasitic sac fungus on their wings, ears, and muzzle.

USFWS/ScienceSource.

insects, amphibians do not maintain a high body temperature, and many species are currently threatened by a pathogenic chytrid inadvertently spread by humans.

In mammals, fungi seldom cause fatal infections. White-nose syndrome, a fungal disease currently decimating North American bats, is an exception (Figure 14.26). The first signs of this disease occurred in New York in 2006. By early 2014, white-nose syndrome had been discovered in 21 states and four Canadian provinces, and it had killed millions of North American bats. Scientists think the sac fungus that causes the disease was recently introduced to North America from Europe, where the bats have evolved resistance to it and are not sickened by it.

Human fungal infections most frequently involve body surfaces. Typically a fungus feeds on the outer layers of skin, secreting enzymes that dissolve keratin, the main skin protein. Infected areas become raised, red, and itchy. For example, several species of fungus infect skin between the toes and on the sole of the foot, causing "athlete's foot." Fungi also cause skin infections misleadingly known as "ringworm." No worm is involved. The ring-shaped lesion is caused by the growth of hyphae outward from the initial infection. Fungal vaginitis (a vaginal yeast infection) occurs when single-celled fungi that normally live in the vagina in low numbers undergo a population explosion. Symptoms of the infection include itching or burning sensations; a thick, odorless, whitish vaginal discharge; and pain during intercourse.

Fungi seldom cause systemwide disease in otherwise healthy people, but infections can be life-threatening in people whose immune system is impaired by AIDS, chemotherapy, or drugs that must be taken after an organ transplant.

Fungal Partnerships   Nearly all plants form mutually beneficial relationships, or **mutualisms**, with fungi. For example, many soil fungi take part in a **mycorrhiza** (plural, mycorrhizae), a mutually beneficial relationship with plant root cells (Figure 14.27). All **glomeromycete fungi** take part in a mycorrhiza in which hyphae penetrate the root cell wall and share space with the cell. An estimated 80 percent of vascular plants have a glomeromycete partner. Club fungi and sac fungi also take part in mycorrhizae. When they do, their hyphae grow into a root and between its cells. Hyphae of all mycorrhizal fungi functionally increase the absorptive surface area of their plant partner. Hyphae are thinner than even the smallest roots and can grow more easily between soil particles. The fungus shares water and nutrients taken up by its hyphae with root cells. In return, the plant supplies sugar to the fungus.

Fungal partners also enhance the nutrition of some animals. Chytrid fungi that live in the stomachs of grazing hoofed mammals such as cattle, deer, and moose

sporangium

plant root

hypha branching inside a root cell wall

10 μm

Figure 14.27 Mycorrhiza.
Specialized hyphae of glomeromycete fungus enter and branch inside the cell wall of a plant root cell.

© Dr. Mark Brundrett, The University of Western Australia.

Figure 14.28 Structure of a leaflike lichen.
Photo Gary Head.

photosynthetic cell

fungal hyphae

aid their hosts by breaking down otherwise indigestible cellulose. Similarly, fungal partners of leafcutter ants serve as an external digestive system. The ants gather bits of leaf to sustain the fungus that they cultivate in their nest. The ants cannot digest leaves, but they do eat the fungus.

Lichens are composite organisms consisting of a fungus and a single-celled photosynthetic species, either a green alga or a cyanobacterium. The fungus makes up most of the lichen's mass and shelters the photosynthetic species, which shares nutrients with the fungus. Lichens grow on many exposed surfaces (Figure 14.28). They are ecologically important as colonizers in places that are too hostile for other organisms. By releasing acids and retaining water that freezes and thaws, lichens help break down rocks and form soil.

Human Uses of Fungi Some fungi are important as human food crops. Mushroom farms produce many varieties of club fungi by inoculating sterilized compost with spores of the desired crop species. Mycorrhizal fungi, including chantrelles, morels, and truffles, cannot be cultivated, so they are gathered from the wild. Mushrooms taken from the wild should always be identified by an experienced mushroom forager before they are eaten because many edible species have poisonous look-alikes.

In addition to eating fungi, we make use of fungal fermentation reactions to produce food and drinks (Figure 14.29). A package of baker's yeast contains spores of a sac fungus (*Saccharomyces cerevisiae*). Set bread dough out to rise, and yeast cells carry out fermentation reactions that produce carbon dioxide, causing the dough to expand (rise). Another strain of *Saccharomyces* is used in the production of beer and wine.

Geneticists and biotechnologists also make use of the yeast *S. cerevisiae*. Like *E. coli* bacteria, *S. cerevisiae* grows readily in laboratories and it offers the added advantage of being eukaryotic like us. Checkpoint genes that regulate the eukaryotic cell cycle (Section 8.3) were first discovered in *S. cerevisiae*. This discovery was the first step toward our current understanding of how mutations of these genes cause human cancers. Genetically engineered *S. cerevisiae* and other yeasts are also now used to produce proteins that serve as vaccines or other medicines.

Some medicines and drugs are compounds first isolated from fungi. Most famously, the initial source of the antibiotic penicillin was the sac fungal mold *Penicillium*. Fungi have also yielded drugs used to lower blood pressure, reduce cholesterol levels, or to prevent rejection of transplanted organs. The hallucinogen LSD was first isolated from ergot, a club fungus that infects grains. So-called "magic mushrooms" contain compounds called psilocybins that induce a dreamlike state.

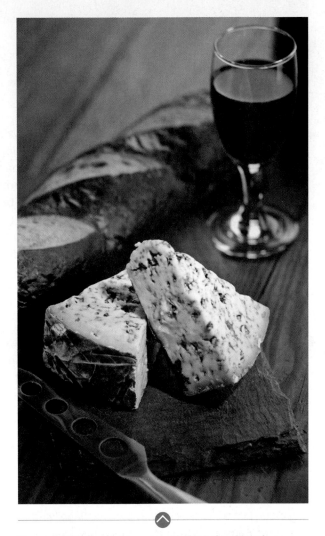

Figure 14.29 Products made with the help of fungi.
Sac fungal yeasts carry out fermentation necessary to make bread and wine. The blue material in "blue cheese" is hyphae of a sac fungal mold.

© CameronWhitman/iStockphoto.com.

Take-Home Message 14.9

How do fungi interact with other species?

- Mycorrhizal fungi live in or on plant roots in a mutually beneficial relationship. Fungi also live with single-celled photosynthetic cells as lichens.
- Fungi benefit other organisms when they feed on wastes and remains, releasing nutrients into the soil. They harm other organisms, including humans, by infecting and feeding on their tissues.
- We use fungi as food, as sources of drugs, and as a model for studying genetic processes.

glomeromycete fungus Soil fungus whose hyphae extend inside the cell wall of a plant root cell.

lichen Composite organism consisting of a fungus and a green alga or cyanobacterium.

mutualism Species interaction that benefits both species.

mycorrhiza Fungus–plant root partnership.

Summary

Section 14.1 Plants are food for both people and fungi. Winds spread spores of fungal pathogens, so outbreaks of fungal disease can often spread to crops over a wide area.

Section 14.2 **Plants** evolved from freshwater green algae. Their life cycle, an **alternation of generations**, includes two multicelled forms, a haploid **gametophyte** and a diploid **sporophyte**. The gametophyte dominates the life cycle of **bryophytes**, but the sporophyte dominates in **vascular plants**.

Key adaptations to dry habitats include a waterproof **cuticle** with **stomata**, and internal pipelines of **xylem** and **phloem**. Xylem reinforced by **lignin** helps vascular plants stand upright. **Seeds** and male gametophytes that can be dispersed without water (**pollen grains**) evolved in seed plants.

Section 14.3 Bryophytes include three lineages of low-growing plants: **mosses**, liverworts, and hornworts. The photosynthetic moss gametophyte is held in place by threadlike **rhizoids**. Flagellated sperm swim to eggs. The sporophyte remains attached to and often dependent upon the gametophyte even when mature.

sporophyte—
gametophyte—

Section 14.4 **Ferns** are seedless vascular plants. Sporophytes dominate their life cycle and produce spores in **sori**. Gametophytes produce flagellated sperm. Ferns grow from **rhizomes** (horizontal stems). Some live on trees as **epiphytes**. Other seedless vascular plants include club mosses and horsetails. **Coal** formed from the remains of ancient nonvascular seed plants.

Section 14.5 Seed-bearing vascular plants make two types of spores. **Microspores** give rise to pollen grains in a **pollen sac**. **Megaspores** form in **ovules**, and give rise to egg-producing female gametophytes. Even in the absence of water, winds or pollinators can move pollen, thus facilitating **pollination**. The seed is a mature ovule, with an embryo sporophyte and some nutritive tissue inside it. Some seed plants undergo secondary growth and produce **wood**.

Section 14.6 **Conifers**, cycads, ginkgos, and gnetophytes are among the **gymnosperms**. They are adapted to dry climates and bear seeds on exposed surfaces of spore-bearing structures. In conifers, these spore-bearing structures are distinctive cones. Conifers tend to be evergreen. Ginkgos are **deciduous plants**.

Section 14.7 **Angiosperms** are the dominant land plants. They alone have **flowers**. Pollen forms in **anthers**, the part of a **stamen** that holds pollen sacs. Many flowering plants coevolved with **pollinators** that deliver pollen to a receptive **stigma**. After pollination, a pollen tube grows through a **style** of the flower's **carpel** to the **ovary** at its base and **double fertilization** occurs. The ovary becomes a **fruit** containing one or more seeds. A flowering plant seed includes an embryo sporophyte and **endosperm**, a nutritious tissue. Most crops are angiosperms. There are two main lineages of flowering plants. **Monocots** include grasses and palm trees. **Eudicots** include most flowering trees and shrubs, as well as most herbaceous plants.

Section 14.8 **Fungi** are single-celled or multicelled heterotrophs. They secrete enzymes onto organic material and absorb the resulting breakdown products. Multicelled fungi grow as a **mycelium** composed of many filaments called **hyphae**. Fungi produce spores both sexually and asexually. A mushroom is a spore-producing body of a **club fungus**. Fungi of some **sac fungi** are also used as food. Bread mold is an example of a **zygote fungus**.

Section 14.9 Most fungi are decomposers. A **mycorrhiza** is a **mutualism**, as between a **glomeromycete fungus** and plant root cells. Fungi also partner with photosynthetic cells to form **lichens**. Some fungi infect plants or animals, causing disease. We eat fungi and use them to produce foods, drinks, and medicines.

Self-Quiz

Answers in Appendix I

1. Which of the following statements is not correct?
   a. Angiosperms produce pollen and seeds.
   b. Mosses are nonvascular plants.
   c. Ferns and angiosperms are vascular plants.
   d. Only gymnosperms produce fruits.

2. Which does *not* apply to all seed plants?
   a. vascular tissues   c. single spore type
   b. diploid dominance   d. all of the above

3. Bryophytes have independent _____ and dependent _____ .
   a. sporophytes; gametophytes   b. gametophytes; sporophytes

4. Ferns are classified as _____ plants.
   a. multicelled aquatic
   c. seedless vascular
   b. nonvascular seed
   d. seed-bearing vascular

5. The _____ produce flagellated sperm.
   a. ferns
   c. monocots
   b. conifers
   d. a and c

6. The _____ produced in the male cones of a conifer develop into pollen grains.
   a. ovules
   c. megaspores
   b. ovaries
   d. microspores

7. A seed is _____ .
   a. a female gametophyte
   c. a mature pollen tube
   b. a mature ovule
   d. an immature spore

8. Match the terms appropriately.
   _____ gymnosperm   a. gamete-producing body
   _____ sporophyte   b. help control water loss
   _____ horsetail   c. "naked" seeds
   _____ bryophyte   d. spore-producing body
   _____ gametophyte   e. nonvascular land plant
   _____ stomata   f. seedless vascular plant
   _____ angiosperm   g. flowering plant

9. All fungi _____ .
   a. are multicelled
   c. are heterotrophs
   b. form flagellated spores
   d. all of the above

10. Fungal decomposers derive nutrients from _____ .
    a. organic wastes and remains   c. living animals
    b. living plants   d. photosynthesis

11. A mushroom is _____ .
    a. the food-absorbing part of a fungus
    b. the only part of the fungal body not made of hyphae
    c. a reproductive structure that releases sexual spores
    d. the longest-lived part of the fungal life cycle

12. Human fungal infections most commonly involve _____ .
    a. the brain   c. the digestive system
    b. the heart   d. body surfaces

13. A _____ is a composite organism composed of a fungus and a single-celled photosynthetic species.
    a. mycorrhiza   c. decomposer
    b. lichen   d. ringworm

14. Cell walls of fungi are composed of _____ .
    a. cellulose   c. lignin
    b. keratin   d. chitin

15. Match the terms appropriately.
    _____ decomposer   a. filament made of walled cells
    _____ yeast   b. club fungus fruiting body
    _____ mushroom   c. fungus with flagellated spores
    _____ chytrid   d. mesh of fungal filaments
    _____ hypha   e. partners with a plant root
    _____ mycelium   f. single-celled fungus
    _____ glomeromycete fungus   g. breaks down organic matter

Critical Thinking

1. Early botanists admired ferns but found their life cycle perplexing. In the 1700s, they learned to propagate them by sowing what appeared to be tiny dustlike "seeds" from the undersides of fronds. Despite many attempts, the scientists could not find the pollen source, which they assumed must stimulate the "seeds" to develop. Imagine you could write to one of these botanists. Compose a note that would clear up their confusion.
2. Fungi play an important role in the breakdown of cellulose into its component simple sugars. What simple sugar is the final product of these reactions?
3. Fungal skin diseases are persistent, in part because fungi can penetrate deeper layers of skin than can ointments and creams. There are fewer antifungal drugs than antibacterial ones, and antifungals often have more severe side effects. Reflect on the evolutionary relationships among bacteria, fungi, and humans. Why it is harder to fight fungi than bacteria?

Visual Question

Match each term with a structure in the diagram below.
   _____ stigma
   _____ style
   _____ sepal
   _____ anther
   _____ filament
   _____ petal
   _____ ovule in ovary
   _____ ovary

Application ◆ 15.1 Medicines From the Sea

Animal life began in the sea, and Earth's oceans still hold the greatest store of animal diversity. In the sea, as on land, the majority of animals are **invertebrates**, meaning they do not have a backbone. Of all animals, only about 5 percent have a backbone and thus are described as **vertebrates**.

Many marine invertebrates produce compounds that are toxic to other organisms. Such toxins can protect an animal from predators, help it fend off pathogens, or assist in the capture of prey. Some invertebrate toxins also have effects in the human body, and thus can be useful as medicines. Consider the venom that some fish-eating cone snails use to subdue their prey (Figure 15.1). The snail's venom anesthetizes and paralyzes a fish, thus preventing the fish from struggling and possibly harming the snail as it captures and consumes it. Humans evolved from a fish ancestor and our nerves use the same chemical communication signals that fish nerves do. As a result, a toxin that acts on fish nerves also affects the function of the human nervous system. A person accidentally stung by a fish-eating cone snail may become numb at the site of venom injection, suffer from temporary paralysis, or even die.

A synthetic version of a peptide extracted from the venom of one cone snail is now used as a pain reliever. The drug, ziconotide (Prialt), is injected into the spinal cord to suppress pain that cannot be controlled by other means. Additional peptides isolated from cone snail venom are being tested as treatments for epilepsy, diabetes, and cancer.

Compounds derived from other marine invertebrates are also in use. AZT (azidothymidine), the first drug successfully used to treat AIDS, is a synthetic version of a molecule first discovered in a sponge. Other compounds made by sponges can be used to treat infections caused by herpesviruses. Sea whips, which are relatives of sea anemones, produce a variety of anti-inflammatory compounds, one of which is used in face creams.

Figure 15.1 Cone snail with its fish prey.
After administering a toxin that puts the fish into a stupor, the snail engulfs and devours it.
© K.S. Matz.

Finding a compound that might have medicinal value is only the first step in developing a new drug. For a compound to be used in clinical tests, researchers must obtain a sufficient amount of it. This can be difficult because many compounds of interest occur only at very low concentrations in animals. Consider the drug eribulin (Halaven), which is now used to treat some advanced-stage breast cancers. Sponges synthesize the compound on which eribulin is modeled, but only in tiny amounts. Obtaining enough of the sponge compound to test its efficiency as a cancer drug (300 milligrams) required processing more than one metric ton of sponge tissue.

Once the structure of a useful compound has been determined, chemists can usually manufacture that compound or one with similar structure and properties. Use of synthetically produced compounds prevents overharvesting of the species in which the compound was found. The breast cancer drug eribulin is a synthetic molecule with a slightly different structure than the molecule extracted from sponges.

15.2 Origins and Diversification

REMEMBER: Animals are eukaryotes with unwalled cells (Section 3.5). Choanoflagellates are protists closely related to animals (13.6). Homeotic genes regulate animal development (7.7). Tectonic plate movement causes supercontinents to form, rotate, and break up (11.5). Cutting off gene flow between populations can lead to speciation (12.6). With adaptive radiation, a lineage gives rise to many others (12.7).

Animals are multicelled consumers that take food into their body, where they digest it and absorb the released nutrients. An animal develops from an embryo (an early developmental stage) to an adult. Most animals reproduce sexually, some reproduce asexually, and some do both. Nearly all animals are motile (can move from place to place), during part or all of their life cycle.

Animal Origins The **colonial theory of animal origins** states that animals evolved from a heterotrophic protist that formed colonies. At first, all cells in the colony were identical. Each could survive and reproduce on its own. Later, mutations produced cells that specialized in some tasks and did not carry out others. Perhaps some cells captured food more efficiently, but did not make gametes, whereas other made gametes but did not catch food. The division of labor among interdependent cells made colonies more efficient, allowing them to obtain more food and produce more offspring. Over time, cells became increasingly interdependent and more specialized cell types evolved, producing the first animal.

Studies of modern protists support the colonial theory. Among choanoflagellates, the modern protists most closely related to animals, some species can live either as individual cells or as a colony. Colonies form when a cell undergoes mitosis and the resulting descendant cells stick together. Thus, cells of a choanoflagellate colony are, like the cells of an animal body, genetically identical.

Evidence of Early Animals We have no fossil evidence of the earliest animals. Most likely they were microscopic and, like the protist from which they evolved, had no hard parts that would fossilize easily. However, a collection of 570-million-year-old fossil organisms provides evidence of an early animal diversification. These fossils are collectively referred to as Ediacarans because they were found in Australia's Ediacara Hills. They include the remains of many small soft-bodied organisms that are widely thought to be marine invertebrates (Figure 15.2). Some of these fossil species may be early representatives of modern groups. Others belong to lineages that have no surviving members.

A great adaptive radiation of animals occurred during the Cambrian (541–485 million years ago). By the end of this period, all major animal lineages were present in the seas. Environmental factors encouraged diversification. During the Cambrian, global climate warmed and the amount of oxygen in the seas increased, making the environment more hospitable to animal life. (With rare exceptions, animals require oxygen.) Also during the Cambrian, the supercontinent Gondwana underwent a dramatic rotation. Movement of this landmass interrupted gene flow between populations, increasing the likelihood that speciation events would occur.

Biological factors also have encouraged animal diversification. After predatory animals arose, evolution of novel prey defenses, and predators able to overcome these, would have been favored. Duplications and divergence of homeotic genes probably facilitated such modifications. Changes in these genes can have dramatic effects on body plans.

Figure 15.2 Fossils of early animals.

(A) © DK Limited/Corbis; (B) Dr. Chip Clark.

A. Fossil of *Spriggina*, an Ediacaran that lived in the sea about 570 million years ago. It was a soft-bodied animal about 3 centimeters (1 inch) long. It is considered a possible ancestor of arthropods.

B. Fossil trilobite, an early arthropod. Trilobites arose during the Cambrian and thrived in the oceans for 270 million years before going extinct.

A great adaptive radiation of animals occurred during the Cambrian (542–488 million years ago).

animal A eukaryotic consumer that is made up of unwalled cells and develops through a series of stages. Most ingest food, reproduce sexually, and can move from place to place.

colonial theory of animal origins Well-accepted hypothesis that animals evolved from a colonial protist.

invertebrate Animal without a backbone.

vertebrate Animal with a backbone.

Sponges Cnidarians Flatworms Annelids Mollusks Roundworms Arthropods Echinoderms Chordates

5 Protostome Development

6 Deuterostome Development

3 Radial Symmetry

4 Bilateral Symmetry

2 Tissues

1 Multicellularity

Ancestral protist

Figure 15.3 Family tree for major animal phyla based on body form and genetic comparisons.
Vertebrates (animals with a backbone) are a subgroup of the chordates.

Figure It Out: What type of body symmetry do the arthropods have?

Answer: Bilateral symmetry

Major Groups and Evolutionary Trends Figure 15.3 shows relationships among the animal groups covered in this book. All animals are descended from the same multicelled ancestor **1**. The earliest animals were aggregations of cells, and sponges still show this level of organization. However, most animals have tissues **2**. A **tissue** consists of one or more types of cells that are organized in a specific pattern and that carry out a particular task. In the early animal lineages, embryos had two tissue layers: an outer ectoderm and an inner endoderm. In later lineages, cell movements produced a middle embryonic layer called mesoderm. Evolution of a three-layer embryo allowed an increase in structural complexity.

The structurally simplest animals such as sponges are asymmetrical, meaning you cannot divide their body into halves that are mirror images. Sea anemones and other cnidarians have **radial symmetry**: Body parts are repeated around a central axis, like the spokes of a wheel **3**. Radial animals have no front or back end. They attach to an underwater surface or drift along, so their food can arrive from any direction. Most animals have **bilateral symmetry**. They have a right and left half, with body parts repeated on either side of the body **4**. Bilateral animals have a distinctive "head end" that has a concentration of nerve cells.

There are two lineages of animals that have a three-layer embryo, and they differ somewhat in how they develop. In **protostomes**, the first opening that forms on an embryo becomes the mouth **5**. *Proto–* means first and *stoma* means opening. In **deuterostomes**, the second opening becomes the mouth **6**.

All animals take in and digest food, but the digestive process varies among groups. In sponges, digestion is intracellular. Cnidarians and flatworms digest food inside a saclike gut called a **gastrovascular cavity**. Food enters this cavity through the same opening that expels wastes, and the cavity also functions in gas exchange. Most bilateral animals have a tubular gut, or **complete digestive tract**, with an opening at either end. A tubular gut has advantages. Parts of the tube can be specialized for taking in food, digesting food, absorbing nutrients, or compacting waste. Unlike a saclike cavity, a tubular gut carries out all these tasks simultaneously.

A mass of tissues and organs surrounds a flatworm's gut (Figure 15.4A). However, most animals have a "tube within a tube" body plan, with a fluid-filled body cavity around the gut. Typically, this cavity is lined with tissue derived from mesoderm, and is called a **coelom**. Earthworms have this type of body plan (Figure 15.4B). Sheets of tissue (called mesentery) suspend the gut in the center of a coelom.

epidermis
(from ectoderm)

tissues, organs
(from mesoderm)

gut tissue
(from endoderm)

gut cavity

A. Flatworm, with no body cavity other than the gut.

coelom

epidermis
(from ectoderm)

mesentery

tissues, organs
(from mesoderm)

gut tissue
(from endoderm)

gut cavity

B. Annelid, with a mesoderm-lined, fluid-filled coelom. Sheets of mesentery hold organs in place.

pseudocoelom

epidermis
(from ectoderm)

tissues, organs
(from mesoderm)

gut tissue
(from endoderm)

gut cavity

C. Roundworm, with a pseudocoelom.

Figure 15.4 Variations in body plans among bilateral animals.
Diagrams show a cross-section through the body. Relative width of tissue layers is not to scale.

Figure It Out: Which of the animals shown above develops from an embryo with three tissue layers?

Answer: All three do.

A few invertebrates, such as roundworms, have a pseudocoelom, with only a partial mesodermal lining (Figure 15.4C).

Evolution of a fluid-filled body cavity provided a number of benefits. First, materials can diffuse through coelomic fluid to body cells. Second, muscles can redistribute the fluid to alter the shape of body parts in ways that allow movement. Finally, internal organs were no longer hemmed in by a mass of tissue so, over evolutionary time, they could become larger and move relative to one another.

Gases and nutrients diffuse quickly through the body of a small animal. However, diffusion alone cannot move substances through a large body fast enough to keep its cells alive. In most animals, a circulatory system speeds the distribution of substances through the body. In an **open circulatory system**, a heart pumps fluid out of vessels into internal spaces, from which it is taken up again by a heart. With a **closed circulatory system**, a heart or hearts propel blood through a continuous system of vessels. Materials carried by the blood diffuse out of vessels and into cells, and vice versa. A closed circulatory system allows for faster distribution of materials than an open one.

Segmentation is common in bilateral animals, meaning similar units are repeated along the length of the body. We clearly see body segments in annelids such as earthworms. Segmentation allows evolutionary innovations in body form. When many segments have organs that carry out the same function, some segments can become modified without endangering the animal's survival.

Take-Home Message 15.2

What are animals?

- Animals are multicelled heterotrophs that typically ingest food. Their cells are unwalled.
- Animals reproduce sexually and, in many cases, asexually. They go through a period of embryonic development, and most move about during at least part of the life cycle.
- Most animals have bilateral symmetry and are composed of tissues and organs.
- An animal's digestive cavity may be tubular or saclike, and the circulatory system (if present) may be open or closed.

bilateral symmetry Having right and left halves with similar parts, and a front and back that differ.

closed circulatory system System in which blood never leaves blood vessels and exchanges with cells take place across vessel walls.

coelom A body cavity completely lined by tissue derived from mesoderm.

complete digestive tract Tubular gut.

deuterostomes Animal lineage with a three-layer embryo in which the mouth is the second opening to form; includes echinoderms and chordates.

gastrovascular cavity Saclike gut.

open circulatory system System in which circulatory fluid leaves open-ended vessels and flows among tissues before returning to the heart.

protostomes Animal lineage with a three-layer embryo in which the first opening to form is the mouth; includes most bilateral invertebrates.

radial symmetry Having parts arranged around a central axis, like spokes around a wheel.

tissue One or more types of cells that are organized in a specific pattern and that carry out a specific task.

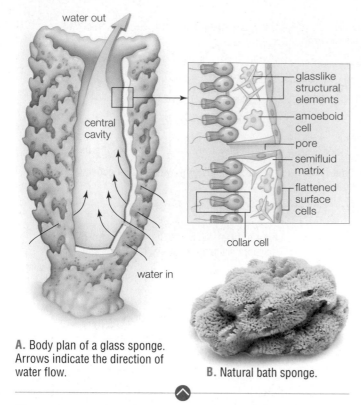

A. Body plan of a glass sponge. Arrows indicate the direction of water flow.

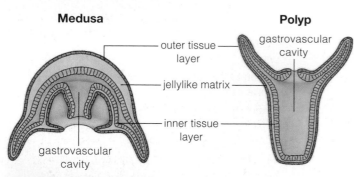

B. Natural bath sponge.

Figure 15.5 Sponges, animals without body symmetry or tissues.

(B) Image © ultimathule/Shutterstock.

Medusa   **Polyp**

outer tissue layer
gastrovascular cavity

jellylike matrix

inner tissue layer

gastrovascular cavity

A. Structure of the two body plans.

B. Jelly, a medusa.
© Boris Pamikov/Shutterstock.

C. Sea anemone, a polyp.
Ethan Daniels/Shutterstock.

Figure 15.6 Cnidarian body plans.

15.3 Invertebrate Diversity

Sponges Sponges (phylum Porifera) are aquatic animals with a hollow, asymmetrical body (Figure 15.5A). Adults are sessile, meaning they do not move about. Flattened cells cover a sponge's outer body surface, and flagellated collar cells line its internal cavities. A jellylike extracellular matrix lies between these cell layers. Movement of the collar cells' flagella causes water to flow in through many pores, then out through one or more larger openings.

In sponges, unlike most animals, digestion is intracellular. Collar cells filter bits of food from the water, engulf them by endocytosis, then break them down in vesicles. Amoeba-like cells in the matrix receive digested food from collar cells and distribute it to other cells. In many species, cells in the matrix also secrete fibrous proteins or glassy spikes (called spicules) that structurally support the body and discourage predators. Some protein-rich sponges are harvested from the sea, dried, cleaned, and bleached, then sold for bathing and cleaning (Figure 15.5B).

A typical sponge is a **hermaphrodite**: an individual that produces both eggs and sperm. Usually sperm are released into the water, and eggs are retained by the parent. After fertilization, the zygote develops into a ciliated larva. A **larva** (plural, larvae) is a free-living, sexually immature form in an animal life cycle. Sponge larvae exit the parental body, swim briefly, then settle and develop into adults.

Cnidarians Cnidarians (phylum Cnidaria) are aquatic, radially symmetrical animals with stinging tentacles. The two common body plans, medusa and polyp, both consist of two tissue layers with a secreted jellylike matrix between them (Figure 15.6A). A bell-shaped **medusa** (plural, medusae) swims or drifts about. Jellies (jellyfish) are medusae (Figure 15.6B). A **polyp**, such as a sea anemone, is tubular, and one end usually attaches to a surface (Figure 15.6C). In both polyps and medusae, tentacles surround the entrance to the gastrovascular cavity. This saclike area takes in and digests food, expels wastes, and functions in gas exchange.

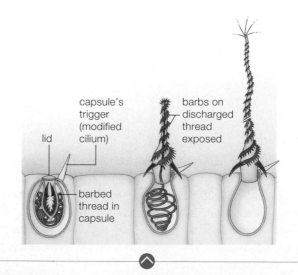

Figure 15.7 Action of a cnidarian's stinging cells (cnidocytes).

The phylum name Cnidaria is derived from *cnidos*, the Greek word for nettle, a kind of stinging plant. The name refers to unique stinging cells (cnidocytes) at the surface of the tentacles. The cells have a capsule-like organelle (a nematocyst) with coiled threadlike structure inside it (Figure 15.7). Touch causes the capsule to pop open, forcing the thread outward. The thread entangles prey or pierces it and delivers venom. Tentacles then push the prey through the mouth into the gastrovascular cavity, where gland cells secrete enzymes that digest it.

Reef-building corals are polyps that capture prey with their tentacles, but also rely on sugars made by photosynthetic protists (dinoflagellates) that live in their tissues. Each polyp secretes a hard, calcium carbonate–rich skeleton around its base. Over time, skeletal remains of many generations of polyps accumulate as a reef, with a layer of living polyps at its surface. Coral reefs are of great ecological importance because they provide food and shelter for many types of animals.

Flatworms **Flatworms** (phylum Platyhelminthes) are flat-bodied worms that do not have a body cavity. They are the simplest animals that develop from a three-layered embryo. Many flatworms live in the sea or in fresh water, but a few live in damp places on land. Still others live as parasites inside animals.

Planarians (Figure 15.8) are free-living flatworms common in ponds. Cilia on their surface allow them to glide along. A muscular tube (a pharynx) sucks food into a highly branched gastrovascular cavity. Nutrients and oxygen diffuse from the fine branches to body cells. Clusters of nerve cells in the head serve as a simple brain. The head also has chemical receptors and light-detecting eyespots.

Tapeworms are hermaphroditic flatworms that infect the vertebrate gut. The tapeworm has a segmented body, and it adds new segments (called proglottids) as it grows. Figure 15.9 shows the life cycle of the beef tapeworm, which can infect people who eat undercooked beef.

cnidarian Radially symmetrical invertebrate with two tissue layers; uses tentacles with stinging cells to capture food.

flatworm Bilaterally symmetrical invertebrate with organs but no body cavity; for example, a planarian or tapeworm.

hermaphrodite Animal that makes both eggs and sperm.

larva Preadult stage in some animal life cycles.

medusa Bell-shaped, free-swimming cnidarian body form.

polyp Tubular, typically sessile, cnidarian body form.

sponge Aquatic invertebrate that has no tissues or organs and filters food from the water.

Figure 15.8 Body plan of a planarian.

muscular tube that sucks up food and expels waste

branching gastro-vascular cavity

Figure 15.9 Life cycle of the beef tapeworm.

Right photo, Andrew Syred/Science Source.

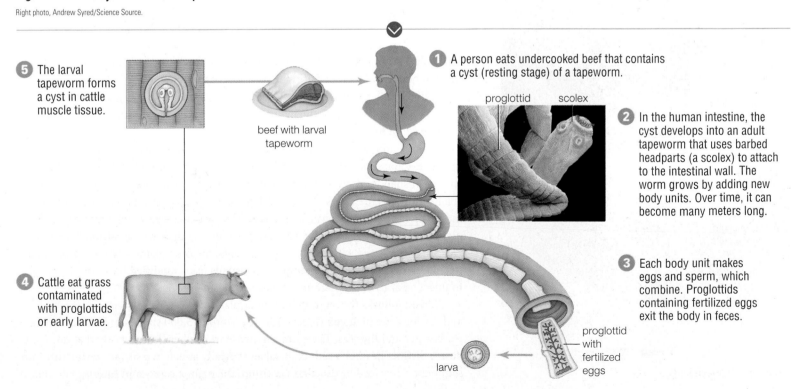

5 The larval tapeworm forms a cyst in cattle muscle tissue.

beef with larval tapeworm

1 A person eats undercooked beef that contains a cyst (resting stage) of a tapeworm.

proglottid   scolex

2 In the human intestine, the cyst develops into an adult tapeworm that uses barbed headparts (a scolex) to attach to the intestinal wall. The worm grows by adding new body units. Over time, it can become many meters long.

3 Each body unit makes eggs and sperm, which combine. Proglottids containing fertilized eggs exit the body in feces.

4 Cattle eat grass contaminated with proglottids or early larvae.

proglottid with fertilized eggs

larva

A. Body plan of an earthworm, an oligochaete.

anus
blood vessel
excretory organ
coelom
gut cavity
nerve cord
secretory region (clitellum)
one of five hearts
nerve cord
gut
brain
mouth

B. Sandworm (*Nereis*), a marine polychaete.

Darlyne A. Murawski, National Geographic Creative.

C. Blood-sucking leech swollen with blood.

J. A. L. Cooke/Oxford Scientific Films.

Figure 15.10 Annelids.

Flukes are parasites that have an unsegmented body with suckers. Some tropical freshwater snails carry blood flukes that infect human intestinal veins and cause schistosomiasis, a potentially fatal disease. People become infected when they wade in water containing larval flukes, which enter the body through a cut.

Annelids Annelids (phylum Annelida) are segmented worms with a coelom, a complete digestive system, and a closed circulatory system. The phylum name comes from the Greek word *annulus*, which means ringed. There are three major subgroups: oligochaetes, polychaetes, and leeches.

Earthworms are oligochaetes that live on land. Most have more than one hundred segments (Figure 15.10A). Each earthworm body segment has a fluid-filled coelomic chamber with paired excretory organs that remove waste from the fluid. An earthworm moves when muscles in the body wall change the shape of segments in a coordinated fashion. An earthworm eats its way through the soil, digesting the organic material that it takes in. The digestive tract extends through the coelom. A nerve cord runs the length of the body on the ventral (lower) side and connects to a simple brain. Five hearts pump blood through a closed system of blood vessels. Earthworms are hermaphrodites. Mucus produced by a secretory region called the clitellum glues worms together while they exchange sperm. Later, the clitellum secretes a silky case for the fertilized eggs that the worm deposits in the soil.

Most marine annelids are polychaetes, a lineage of annelids that typically have many bristles per segment. (*Poly–* means many and *chaete* means bristle.) Some polychaetes, including the sandworms often sold as bait, are active predators (Figure 15.10B). Others remain fixed in place and have feathery appendages specialized for filtering food from currents.

Leeches typically live in fresh water, although some are found in damp places on land. Most are scavengers and predators of invertebrates. An infamous few suck blood from vertebrates (Figure 15.10C). A protein in leech saliva keeps blood from clotting while the leech feeds. For this reason, doctors who reattach a severed finger or ear sometimes apply leeches to it. As the leeches feed, their saliva prevents unwanted clots from forming in blood vessels of the reattached body part.

Mollusks Mollusks (phylum Mollusca) have a small coelom and a soft, unsegmented body. *Mulluscus* is Latin for soft. The **mantle**, a skirtlike extension of the upper body wall, drapes over a visceral mass that contains the organs (Figure 15.11A). In most mollusks, the mantle secretes a hard, calcium-rich shell. A large, muscular foot functions in locomotion.

Among animals, mollusks are second only to arthropods in diversity. Most live in the seas, but some have adapted to life in fresh water or on land. With 60,000 species of snails and slugs, **gastropods** are the largest subgroup. Gastropod means "belly foot," and members of this group glide about on a broad muscular foot that makes up most of their lower body mass. The gastropod shell, when present, is one piece and usually coiled. Most gastropods scrape up algae with a **radula**, a tongue-like organ hardened with chitin, a tough polysaccharide. Some snails have adapted to a predatory lifestyle. For example, cone snails use a modified harpoonlike radula to inject a paralyzing venom into prey such as small fish.

Most mollusks that end up on dinner plates, including mussels, oysters, clams, and scallops, are **bivalves** (Figure 15.11B). About 15,000 species of bivalves live in fresh water and the seas. Their hinged, two-part shell encloses a body that does not have a distinct head or a radula. Bivalves typically attach to a surface or burrow into sediment. They feed by drawing water into the mantle cavity and filtering out bits of

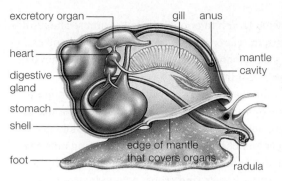

excretory organ

heart

digestive gland

stomach

shell

foot

gill   anus

mantle cavity

edge of mantle that covers organs

radula

A. Body plan of an aquatic snail (a gastropod).

B. Scallop, a bivalve with a two-part, hinged shell.

C. Octopus, a cephalopod.

Figure 15.11 Three types of mollusks.

(B) Frank Park/ANT Photo Library; (C) NURC/UNCW and NOAA/FGBNMS.

food. Being filter-feeders, bivalves occasionally take in toxins or pathogens that can sicken people who eat them. Risk of illness is greatest when bivalves are eaten raw or only partially cooked.

Cephalopods include cuttlefish, squids, nautiluses, and octopuses (Figure 15.11C). Cephalopod means "head-footed," and their foot has been modified into tentacles and/or arms that extend from the head. All cephalopods are predators and most have beaklike, biting mouthparts in addition to a radula. Cephalopods move by jet propulsion. They draw water into their mantle cavity, then force it out through a funnel-shaped siphon. Of all mollusks, only cephalopods have a closed circulatory system.

Cephalopods include the fastest (squids), largest (giant squid), and smartest (octopuses) invertebrates. Of all invertebrates, octopuses have the largest brain relative to body size, and show the most complex behavior.

Roundworms **Roundworms**, or nematodes (phylum Nematoda), are cylindrical, unsegmented worms with a pseudocoelom (Figure 15.12). They have a complete digestive system, excretory organs, and a nervous system, but no circulatory or respiratory organs. Like insects, roundworms secrete a protective cuticle, a body covering that is periodically molted (shed) and replaced as the worm grows.

Roundworms live in the seas, in fresh water, in damp soil, and inside other animals. Most of the nearly 20,000 species are free-living decomposers less than a millimeter long. Like the fruit fly, the free-living soil roundworm *Caenorhabditis elegans* is often used in scientific studies. It has the same tissue types as more complex organisms, but it is transparent and has fewer than 1,000 body cells. Such traits make it easy for scientists to monitor each cell's fate during development. Several scientists have made Nobel Prize–winning discoveries while studying *C. elegans*.

A few types of parasitic roundworms infect people. In the tropics, mosquitoes transmit parasitic roundworms that enter into human lymph vessels. The worms

Figure 15.12 Body plan of a free-living roundworm.

pharynx   intestine   eggs in uterus   gonad

pseudocoelom   muscular body wall   anus

annelid Segmented worm with a coelom, complete digestive system, and closed circulatory system.

bivalve Mollusk with a hinged two-part shell.

cephalopod Predatory mollusk with a closed circulatory system; moves by jet propulsion.

gastropod Mollusk that moves about on its enlarged foot.

mantle Skirtlike extension of tissue in mollusks; covers the mantle cavity and secretes the shell in species that have a shell.

mollusk Invertebrate with a reduced coelom and a mantle.

radula Tonguelike organ of many mollusks.

roundworm Unsegmented worm with a pseudocoelom and a cuticle that is molted as the animal grows.

A. A man with elephantiasis of his left leg. The swelling arises after roundworms damage lymph vessels.

B. Plant-infecting roundworm entering a root.

Figure 15.13 Parasitic roundworms.

(A) Courtesy of © Emily Howard Staub and The Carter Center; (B) William Wergin and Richard Sayre. Colorized by Stephen Ausmus.

Figure 15.14 Grasshopper body plan.
The body has three distinct regions. Compound eyes and a pair of antennae on the head provide sensory information.

From Russell/Wolfe/Hertz/Starr, *Biology*, 1e. © 2008 Cengage Learning®.

injure valves in these vessels, allowing lymph to pool in the lower limbs. The resulting condition is called elephantiasis, in reference to the fluid-filled, "elephant-like" legs (Figure 15.13A). The swelling is permanent because lymph vessels remain damaged even after the worms have been eliminated.

Pinworms (*Enterobius vermicularis*) are infectious roundworms common in the United States. Most infections occur in children. The worms, about the size of a staple, live in the rectum. At night, females crawl out and lay eggs on skin around the anus. Their movement produces an itching sensation, and scratching puts eggs onto fingertips and under fingernails. If swallowed, the eggs start a new infection.

Roundworms also affect our livestock, pets, and crops. A roundworm that infects pigs can also infect humans who eat undercooked pork, causing trichinosis. Dogs are susceptible to heartworms transmitted by mosquitoes. Cats usually become infected after eating an infected rodent that carries an infectious form of the worm. Roundworms can be important agricultural pests. Some roundworms suck nutrients from plant roots and others actually enter the plant (Figure 15.13B). Either way, a roundworm infection stunts the plant's growth and lowers crop yields.

Arthropods **Arthropods** (phylum Arthropoda) are invertebrates with jointed legs. They have a complete digestive system and an open circulatory system. Researchers have identified more than a million living species.

Arthropods secrete a cuticle hardened with chitin, the same material that hardens the mollusk radula. Arthropod cuticle is an external skeleton—an **exoskeleton**. It helps fend off predators and serves as a point of attachment for muscles. In land arthropods, the exoskeleton also helps conserve water and supports the animal's weight. A hard exoskeleton does not restrict growth, because, like roundworms, arthropods molt their cuticle after each growth spurt. A new cuticle forms under the old one, which is shed.

If an arthropod's cuticle were uniformly hard like a plaster cast, it would prevent movement. However, the cuticle thins at joints, where two hard body parts meet. "Arthropod" means jointed leg. Body parts move when the muscles that attach to the exoskeleton on either side of a joint contract.

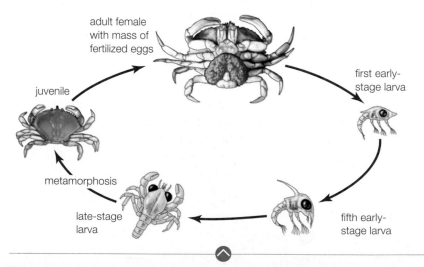

Figure 15.15 Development of a Dungeness crab (a crustacean arthropod).
Fertilized eggs develop into planktonic larvae that grow and molt. Metamorphosis (change in body form) occurs during the molt from a late-stage larva to a juvenile with the adult form. Additional growth and molting produces a sexually mature adult.

In early arthropods, body segments were distinct and all appendages were alike. In many groups, segments later fused into structural units such as a head, thorax, and abdomen (Figure 15.14). Specialized appendages such as wings developed on some segments.

Most arthopods have paired eyes. Insects and crustaceans have **compound eyes** that consist of many units, each with a lens. Such eyes excel at detecting movement. Many arthropods also have one or two pairs of **antennae**, sensory structures on the head that detect touch, odor, and vibrations.

The body plan of many arthropods changes during the life cycle. Individuals undergo **metamorphosis**: Tissues are remodeled as larvae develop into adults. For example, crab larvae swim near the ocean surface and filter food from the water, but adults are bottom-feeders (Figure 15.15). Each stage is specialized for a different lifestyle. Having such different body forms helps prevent adults and juveniles from competing with one another for resources.

Horseshoe crabs (Figure 15.16) are the oldest surviving arthropod lineage. They have lived in Earth's oceans for more than 400 million years. Their common name refers to the fused head and thorax segments that form a horseshoe-shaped region called the cephalothorax. A long spike (a telson) extends from the last abdominal segment and helps the animal right itself if it is turned over by the waves. Horseshoe crabs are bottom-feeders that live in shallow, nearshore waters. In the spring, they venture onto a sandy beach to mate and lay eggs in the damp sand. The billions of eggs they leave behind are an essential food source for migratory shorebirds.

antenna Of some arthropods, sensory structure on the head that detects touch and odors.

arthropod Invertebrate with jointed legs and a hardened exoskeleton that is periodically molted.

compound eye Eye that consists of many individual units, each with its own lens.

exoskeleton External skeleton.

metamorphosis Dramatic remodeling of body form during the transition from larva to adult.

Figure 15.16 Atlantic horseshoe crab (*Limulus*).

Ethan Daniels/Shutterstock.com.

Digging Into Data

Sustainable Use of Horseshoe Crabs

The Atlantic horseshoe crab, *Limulus polyphemus*, is an economically important species. The horseshoe crab's blue blood (right) is used to test for the presence of potentially deadly bacteria in injectable drugs and on medical implants. A liter of horseshoe crab blood sells for about $16,000. To keep horseshoe crab populations stable, blood is extracted from captured animals, which are then returned to the wild. Concerns about the survival of animals after bleeding led researchers to do an experiment. They compared survival of animals captured and maintained in a tank with that of animals captured, bled, and kept in a similar tank. Figure 15.17 shows the results.

1. In which trial did the most control crabs die? In which did the most bled crabs die?

2. Looking at the overall results, how did the mortality of the two groups differ?

3. Based on these results, would you conclude that bleeding harms horseshoe crabs more than capture alone does?

Mark Thiessen/National Geographic Creative

	Control Animals		Bled Animals	
Trial	Number of Crabs	Number That Died	Number of Crabs	Number That Died
1	10	0	10	0
2	10	0	10	3
3	30	0	30	0
4	30	0	30	0
5	30	1	30	6
6	30	0	30	0
7	30	0	30	2
8	30	0	30	5
Total	200	1	200	16

Figure 15.17 Mortality of young male horseshoe crabs kept in tanks for the two weeks after their capture.
Blood was drawn from half of the animals on the day of their capture. Control animals were handled, but not bled. The experiment was repeated eight times with eight different sets of horseshoe crabs.

Data: Walls, E., Berkson, J., *Fish. Bull.* 101:457–459 (2003).

abdomen cephalothorax

palps

A. Spider (tarantula). It kills prey with a venomous bite.

stinger

palps

B. Scorpion. It delivers venom through a stinger.

C. Tick. It sucks blood from vertebrates.

Figure 15.18 Arachnids. All have four pairs of legs.

(A) © Eric Isselée/Shutterstock.com; (B) © Frans Lemmens/The Image Bank/ GettyImages; (C) James Gathany/CDC.

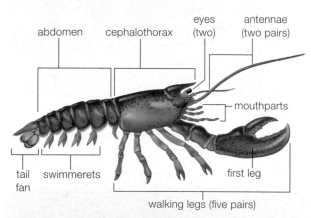

abdomen cephalothorax eyes (two) antennae (two pairs)

mouthparts

tail fan swimmerets first leg

walking legs (five pairs)

A. Body plan of an American lobster.

B. Krill swimming in Antarctic waters.

C. Barnacle filtering food from the water with its feathery legs.

Figure 15.19 Marine crustaceans.

(B) © David Tipling/Photographer's Choice/ Getty Images; (C) © Peter Parks/Image-quest Marine.

Despite their common name, horseshoe crabs are not true crabs, but rather close relatives of the arachnids. **Arachnids** have four pairs of walking legs, a pair of touch-sensitive appendages called palps, and no antennae. They include spiders, scorpions, ticks, and mites. Spiders and scorpions are venomous predators. Spiders deliver venom through fanglike mouthparts. They have a two-part body, with the cephalothorax (fused head and thorax) separated from the abdomen by a narrow "waist" (Figure 15.18A). The abdomen contains silk-making glands. Scorpions catch their prey with clawlike palps and subdue it with a venom-producing stinger on the final abdominal segment (Figure 15.18B). They eat insects, spiders, and even small lizards. Ticks suck blood from vertebrates (Figure 15.18C). Some serve as vectors for bacterial diseases, such as Lyme disease. Mites, the smallest arachnids, are usually less than a millimeter long. Many are scavengers, but some are parasites. Mites that burrow beneath the skin cause scabies in humans and mange in dogs. Some larval mites, commonly called chiggers, crawl into a hair follicle, secrete an enzyme that breaks down proteins, then suck up liquefied tissues.

Crustaceans are mostly marine arthropods that have two pairs of antennae. Decapod crustaceans are bottom-feeding scavengers with five pairs of walking legs. Many, such as shrimps, crabs, and lobsters (Figure 15.19A), are harvested as human food. Other animals also depend on crustaceans for food. Shrimplike crustaceans called krill abound in cool ocean waters (Figure 15.19B). Each is only a few centimeters long, but krill are so abundant and nutritious that a 100-ton blue whale can subsist almost entirely on the krill that it filters from seawater.

Barnacles are marine crustaceans that secrete a calcium-rich external shell (Figure 15.19C). Barnacle larvae swim, then settle and develop into adults that attach to a surface. Adults filter food from seawater with their feathery legs. Some barnacles are notable for the length of their penis, which can be eight times that of their body.

Isopods are a mostly marine group of crustaceans, but some live in damp land habitats. The species commonly known as pill bugs defend themselves from threats by rolling into a ball (right).

© Chris Howey/ Shutterstock.

Centipedes and millipedes constitute another arthropod lineage. These nocturnal ground dwellers have an elongated body with many similar segments (Figure 15.20). The head has paired antennae and two simple eyes. Centipedes are venomous predators with a flat, low-slung body and one pair of legs per segment. Most millipedes eat plant material. Their cylindrical body has two legs per segment and their cuticle is hardened with calcium carbonate.

Insects, which have three pairs of legs and one pair of antennae, are the most diverse arthropod group. Early insects were wingless, ground-dwelling scavengers that did not undergo metamorphosis. Modern bristletails and silverfish retain this type of body form and development. However, most modern insects have wings and undergo metamorphosis. With *incomplete* metamorphosis, an egg hatches into a nymph that differs somewhat in form from the adult. The nymph changes into the adult form over the course of several molts. Cockroaches, grasshoppers, and dragonflies develop in this way. With *complete* metamorphosis, a larva grows and molts without altering its form, then undergoes pupation. A pupa is a nonfeeding body in which larval tissues are remodeled into the adult form. Members of the most diverse insect orders such as flies, beetles, and the butterflies and moths all have two pairs of wings on the thorax and undergo complete metamorphosis (Figure 15.21).

Insects have important ecological roles. They serve as pollinators to many flowering plants (Figure 15.22A). They also serve as food for a variety of wildlife. Larval moths and butterflies (caterpillars) feed songbird nestlings. Aquatic larvae of dragonflies and mayflies serve as food for fish. Most amphibians and reptiles feed mainly on insects. In addition, insects dispose of wastes and remains. Flies and beetles are quick to find an animal corpse or a pile of feces (Figure 15.22B). They lay eggs in or on this organic material, and the larvae that hatch devour it. By their actions, these insects help distribute nutrients through the ecosystem. On the other hand, insects are our main competitors for plant foods. For example, Mediterranean fruit flies, or medflies, damage citrus crops (Figure 15.22C). Insects also transmit dangerous diseases. Mosquitoes spread malaria and carry parasitic roundworms between human hosts. Fleas can transmit plague and lice can transmit typhus. Bedbugs (Figure 15.22D) do not transmit disease, but their bites itch and infestations can have severe psychological and economic effects.

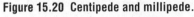

A. Centipede, a speedy predator.

B. Millipede, a scavenger of decaying plant material.

Figure 15.20 Centipede and millipede.

© Eric Isselée/Shutterstock.

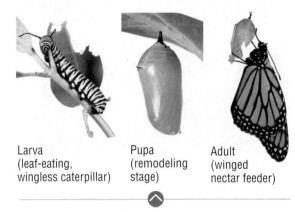

Larva
(leaf-eating,
wingless caterpillar)

Pupa
(remodeling
stage)

Adult
(winged
nectar feeder)

Figure 15.21 Complete metamorphosis in a butterfly.

Left and middle, © Jacob Hamblin/Shutterstock.com right, © Laurie Barr/Shutterstock.com.

A. Bee serving as a pollinator.

B. Dung beetle gathering feces.

C. Medfly, a threat to citrus.

D. Bedbug, a human parasite.

Figure 15.22 Ecological roles of insects.

(A) Photo by Jack Dykinga, USDA, ARS; (B) Gregory G. Dimijian, M.D./Science Source; (C) Photo by Scott Bauer/USDA; (D) CDC/Piotr Naskrecki.

arachnids Land-dwelling arthropods with four pairs of walking legs and no antennae; for example, a spider, scorpion, or tick.

crustaceans Mostly marine arthropods with two pairs of antennae; for example, a shrimp, crab, lobster, or barnacle.

insects Land-dwelling arthropods with a pair of antennae, three pairs of legs, and—in the most diverse groups—wings.

Figure 15.23 Echinoderms.

A. Sea star, and a close-up of its arm with little tube feet.

B. Sea urchin, which moves about on spines and a few tube feet. Sea urchins typically graze on algae.

C. Sea cucumber, with rows of tube feet along its elongated body. The hard parts have been reduced to microscopic plates embedded in a soft body.

(A) Herve Chaumeton/Agence Nature; (B) © Derek Holzapfel/photos.com; (C) Andrew David, NOAA/NMFS/SEFSC Panama City, Lance Horn, UNCW/NURC-Phantom II ROV operator.

chordates Animal phylum characterized by a notochord, dorsal nerve cord, pharyngeal gill slits, and a tail that extends beyond the anus. Includes invertebrate and vertebrate groups.

echinoderms Invertebrates with a water–vascular system and an endoskeleton made of hardened plates and spines.

endoskeleton Internal skeleton.

lancelets Invertebrate chordates that have a fishlike shape and retain their defining chordate traits into adulthood.

notochord Stiff rod of connective tissue that runs the length of the body in chordate larvae or embryos.

tunicates Invertebrate chordates that lose their defining chordate traits during the transition to adulthood.

water–vascular system Of echinoderms, a system of fluid-filled tubes and tube feet that function in locomotion.

Echinoderms Echinoderms (phylum Echinodermata) include about 6,000 marine invertebrates such as sea stars, sea urchins, and sea cucumbers (Figure 15.23). Their phylum name means spiny-skinned and refers to interlocking spines and plates of calcium carbonate embedded in the skin. The plates form an internal skeleton called an **endoskeleton**. Echinoderm adults are radially symmetrical.

Sea stars (sometimes called starfish) are the most familiar echinoderms. They do not have a brain, but they do have a decentralized nervous system. Eyespots at the tips of their arms detect light and movement. A typical sea star moves about on tiny, fluid-filled tube feet (Figure 15.23A). Tube feet are part of a **water–vascular system**, a system of fluid-filled tubes unique to echinoderms. The system includes a central ring and fluid-filled canals that extend into each arm. Side canals deliver coelomic fluid into muscular bulbs that function like the bulb on a medicine dropper. Contraction of a bulb forces fluid into the attached tube foot, extending the foot. A sea star glides along as coordinated contraction and relaxation of the bulbs redistribute fluid among hundreds of tube feet.

Most sea stars prey on bivalve mollusks. To feed, the sea star slides its stomach out through its mouth and into a bivalve's shell. The stomach secretes acid and enzymes that kill the mollusk and begin to digest it. Partially digested food enters the stomach and digestion is completed with the aid of digestive glands in the arms. A sea star's reproductive organs are also in its arms. Sexes are separate and eggs or sperm are released into the water. Fertilization produces an embryo that develops into a ciliated, bilaterally symmetrical larva. The larva swims about briefly, then develops into an adult. The bilateral symmetry of echinoderm larva, along with evidence from genetic studies, indicates that the ancestor of echinoderms was a bilateral animal. Developmental studies show that echinoderms have a deuterostome developmental pattern, and thus belong to the same lineage as chordates.

Sea stars and other echinoderms have a remarkable ability to regenerate lost body parts. If a sea star is cut into pieces, any portion with some of the central disk can regrow the missing body parts.

Take-Home Message 15.3

What are the traits of the main invertebrate groups?

- Sponges are filter-feeders that have specialized cells but no tissues. Cnidarians have two tissue layers and a radial body plan. Most other invertebrates have a bilateral body plan with three tissue layers that form organ systems.

- Three groups of "worms" differ in their traits and are not closely related. Flatworms do not have a coelom. Annelids have a coelom and a segmented body. Roundworms have a pseudocoelom and are the only worms that molt.

- Mollusks and arthropods are the two most diverse invertebrate phyla. Mollusks are soft-bodied, although many make a shell. Arthropods have a hardened exoskeleton and jointed legs. Insects, an arthropod subgroup, are the most diverse invertebrates and the only winged ones.

- Echinoderms are radial animals with bilateral larvae. They are on the same branch of the animal family tree as the chordates.

15.4 Introducing the Chordates

REMEMBER: Studies of similarities between genes of different organisms can be used to determine relationships among those organisms (Section 11.7).

Chordate Traits **Chordates** (phylum Chordata) are defined by four embryonic traits: (1) A **notochord**, a rod of stiff but flexible connective tissue, extends the length of the body and provides support. (2) A dorsal, hollow nerve cord parallels the notochord. (3) Gill slits open across the wall of the pharynx (throat region). (4) A muscular tail extends beyond the anus. Depending on the chordate group, some, all, or none of these traits persist in the adult.

Invertebrate Chordates There are two groups of invertebrate chordates, tunicates and lancelets. Both are marine and both filter food from currents of water that pass through gill slits in their pharynx.

Lancelets (subphylum Cephalochordata) have a fishlike shape and are 3 to 7 centimeters long (Figure 15.24). Lancelets retain all characteristic chordate traits as adults. The dorsal nerve cord extends into the head, where a single eyespot at its tip detects light. However, the head does not have a brain or any paired sensory organs similar to those of fishes.

Tunicates (subphylum Urochordata) are named for the secreted carbohydrate-rich covering or "tunic" that encloses the adult body (Figure 15.25A,B). Larval tunicates have all the typical chordate traits (Figure 15.25C). They swim about briefly, then undergo metamorphosis to the adult form. Of the four typical chordate traits, the adult retains only the pharynx with gill slits. Most adult tunicates attach to an undersea surface and filter food from the water. As water flows in an oral opening and past gill slits, bits of food stick to mucus on the gills and are then sent to a gut. Water leaves through another body opening.

Which invertebrate chordate is most closely related to vertebrates? An adult lancelet looks more like a fish than an adult tunicate does, but such superficial similarities are sometimes deceiving. Studies of developmental processes and gene sequences indicate that tunicates are the closest invertebrate relatives of vertebrates. Keep in mind that neither tunicates nor lancelets are ancestors of vertebrates. These groups share a recent common relative, but each has unique traits that put it on a separate branch of the animal family tree.

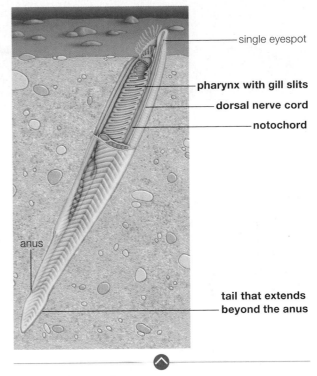

single eyespot

pharynx with gill slits

dorsal nerve cord

notochord

anus

tail that extends beyond the anus

Figure 15.24 Lancelet.
These marine invertebrates retain all chordate traits (labeled with bold text) into adulthood. Lancelets live in sandy sediments, and filter food from the water with their pharynx.

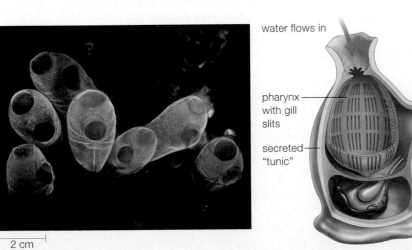

water flows in

water flows out

pharynx with gill slits

secreted "tunic"

2 cm

A. Adult tunicates.

B. Body plan of an adult tunicate.

Figure 15.25 Tunicates.
Larval tunicates are free-swimming and have all the characteristic chordate features. After metamorphosis, the adult retains only the pharynx with gill slits.

(A) Ethan Daniels/Shutterstock; (B,C) From Russell/Worlfe/Hertz/Starr, *Biology*, 1e © 2008 Cengage Learning®.

dorsal nerve cord   notochord   postanal tail

pharynx with gill slits

C. Body plan of a tunicate larva.

Figure 15.26 The chordate family tree.
The vast majority of the chordates are vertebrates.

Figure It Out: Do the cartilaginous fishes have a backbone?

Answer: Yes

bony fish Jawed fish with a skeleton composed mainly of bone.

cartilaginous fish Fish that has jaws, paired fins, and a skeleton made of cartilage; for example, a shark.

jawless fish Fish that has a skeleton of cartilage, but no jaws or paired fins; for example, a lamprey.

kidney Organ of the vertebrate urinary system that filters blood and adjusts its composition.

lung Saclike organ inside which blood exchanges gases with the air.

scales Hard, flattened elements that cover the skin of reptiles and some fishes.

tetrapod Vertebrate with four limbs.

vertebral column Backbone.

vertebrate Animal with a backbone; a fish, amphibian, reptile, bird, or mammal.

Vertebrate Traits and Trends

Vertebrates (subphylum Vertebrata) are the third major chordate subgroup. All vertebrates have a distinct head with a brain, and most have a pair of eyes. Their circulatory system is closed, with a single heart, and there is a complete digestive tract. Paired organs called **kidneys** filter blood, adjust its volume and solute composition, and eliminate wastes.

Figure 15.26 shows the chordate family tree and notes the innovations that define various lineages. Vertebrates ❶ are named for the **vertebral column** (backbone) that replaces the notochord as a vertebrate embryo develops. This flexible but sturdy structure encloses the spinal cord that develops from the nerve cord. The vertebral column is part of the vertebrate endoskeleton.

The next major innovation was jaws, which are hinged skeletal elements used in feeding ❷. Jaws evolved by modification of bony parts that supported the gill slits of early jawless fishes. Evolution of jaws opened up new feeding opportunities, allowing an adaptive radiation. The vast majority of modern fishes have jaws.

Some jawed fishes evolved internal air sacs ❸. In some species, such a sac functioned as a simple **lung**, a respiratory organ in which blood exchanges gases with the air. One lineage of fish with lungs also had bony paired fins ❹. Their descendants evolved bony paired limbs, becoming the first **tetrapods**, or four-legged walkers ❺. A special type of egg (the amniote egg) allowed one tetrapod group to lay eggs on land ❻. Members of that lineage, the amniotes (reptiles, birds, and mammals), are the most successful tetrapods on land.

Take-Home Message 15.4

What trends shaped chordate evolution?

- The earliest chordates were invertebrates, and two groups of invertebrate chordates (tunicates and lancelets) still exist. They are aquatic filter-feeders.
- Vertebrates evolved from an invertebrate chordate ancestor. All have a vertebral column, or backbone.
- Jaws, lungs, limbs, and waterproof eggs were key innovations that led to the adaptive radiation of vertebrates, first in the seas and then on the land.

15.5 Fishes and Amphibians

REMEMBER: Homologous structures are structures that are similar in different lineages because they evolved in a common ancestor. (Section 11.6).

We begin our survey of vertebrate diversity with fishes, the first vertebrate lineages to evolve and the most fully aquatic.

Jawless Fishes Modern **jawless fishes** have a smooth, elongated body without paired fins. Their skeletal elements are made of cartilage, the same type of connective tissue that supports your nose and ears. Figure 15.27 shows the distinctive mouth of one jawless fish, a lamprey. As adults, most lampreys feed on other fish. Lacking jaws, a lamprey cannot bite. Instead, it attaches to a fish using an oral disk ringed by horny teeth made of the protein keratin. Once attached, the lamprey secretes enzymes and scrapes up bits of its host's flesh with a tooth-covered tongue. The host fish often dies from blood loss or a resulting infection.

Jawed Fishes Jaws evolved from gill arches, which are skeletal elements that support a fish's gills (Figure 15. 28). Most modern jawed fishes have paired fins and **scales**: hard, flattened structures that grow from and often cover the skin. Scales and an internal skeleton make a fish denser than water and prone to sinking. Highly active swimmers have fins with a shape that helps lift them, something like the way that wings help lift up an airplane. Friction slows movement through water, so speedy swimmers typically have a streamlined body that reduces friction.

There are two groups of jawed fishes: cartilaginous fishes and bony fishes (Figure 15.29). As their name implies, **cartilaginous fishes** are jawed fishes with a skeleton made of cartilage. They include 850 species, with sharks being the best known. Some sharks are predators that swim in upper ocean waters. Others strain plankton from the water or suck up food from the seafloor. Human surfers and swimmers resemble typical prey of some predatory sharks, and rare attacks by a few species give the group as a whole an undeserved bad reputation. Worldwide, shark attacks kill about 25 people a year. For comparison, dogs kill about 30 people each year in the United States alone.

In **bony fishes**, an embryonic skeleton of cartilage becomes transformed to an adult skeleton consisting mainly of bone. Both cartilaginous fishes and bony fishes have paired fins, but only bony fishes can move those fins. Bony fishes also differ from other fishes in having their gill slits hidden beneath a gill cover. In jawless and cartilaginous fish, gill slits are visible at the body surface.

Figure 15.29 Jawed fishes.

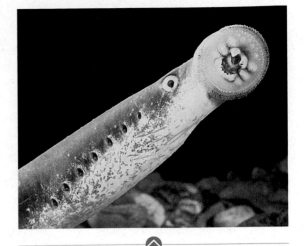

Figure 15.27 Lamprey, a jawless fish.
A lamprey has no paired fins and its gill slits are visible at the body surface. It attaches to other fishes with its oral disk and scrapes at their flesh.

Heather Angel/Natural Visions.

Jawless fish with skeletal elements supporting gills.

supporting structure for gill slits

gill slits

jaw, derived from support structure

Early jawed fish.

Figure 15.28 Evolution of jaws from gill supports.

A. Cartilaginous fish with jaws and paired fins. This species of shark is a swift predator.

© Luiz A. Rocha/Shutterstock.

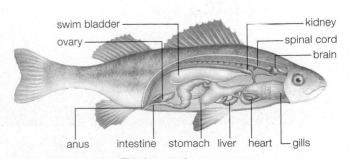

swim bladder — kidney
ovary — spinal cord
— brain

anus intestine stomach liver heart gills

B. Bony fish body plan. This is a perch.

After E. Solomon, L. Berg, and D.W. Martin, *Biology*, Seventh Edition, © Cengage Learning®.

There are two lineages of bony fishes. **Ray-finned fishes** have flexible fins supported by thin rays derived from skin. A gas-filled swim bladder helps them adjust their buoyancy. With about 30,000 species, ray-finned fishes constitute nearly half of the vertebrates. They include most familiar freshwater fish (Figure 15.30A) as well as marine species such as tuna, halibut, and cod.

Modern **lobe-finned fishes** include coelacanths (Section 12.7) and lungfishes (Figure 15.30B). These fish have thick, fleshy pelvic and pectoral fins with internal bony supports. As their name suggests, lungfishes have one or two lungs in addition to their gills. The fish inflates its lungs by gulping air, then air in the lungs exchanges gases with blood. Having lungs in addition to gills allows lungfishes to survive in low-oxygen waters.

Early Tetrapods   Genome comparisons indicate that, of the two modern lobe-finned groups, lungfishes are closest to tetrapods. Fossils reveal how the skeleton became modified as fishes adapted to swimming evolved into four-legged walkers (Figure 15.31). The bones inside a lobe-finned fish's pelvic and pectoral fins are homologous (Section 11.6) with amphibian limb bones. However, the transition to land was not only a matter of skeletal changes. Fishes have a two-chamber heart: one chamber receives blood, the other pumps it out. In amphibians, the heart became divided into three chambers: one to receive blood, one to pump blood to the lungs, and one to pump blood to the body. This change increased the rate of blood flow to the body and the efficiency of gas exchange. Changes in the inner ear improved the ability to detect airborne sounds, and eyelids prevented the eyes from drying out.

What was the advantage of living on land? An ability to survive out of water is useful in seasonally dry places. Individuals on land were also safe from aquatic predators and had access to a new food—insects—which had recently evolved.

A. Goldfish (carp), a ray-finned bony fish. Its fins consist of a web of skin supported by thin spines.

pelvic fin   pectoral fin

B. Lungfish, a lobe-finned bony fish. Its thick, fleshy fins have sturdy bony supports inside them.

Figure 15.30 Two types of fins in bony fishes.

(A) © iStockphoto.com/GlobalP; (B) © Wernher Krutein/photovault.com.

1 Fish (*Eusthenopteron*) with fins and no ribs.
© P. E. Ahlberg.

2 Fish (*Tiktaalik*) with ribs and modified fins.
Illústration by © Kalliopi Monoyios.

3 Early amphibian (*Icthyostega*) with ribs and limbs.
© P. E. Ahlberg.

Figure 15.31 Fossil species from the late Devonian show how vertebrates made the transition from water to land.

Figure 15.32 Amphibians. All have a scaleless body.

(A) Photo by James Bettaso, US Fish and Wildlife Service; (B) Stephen Dalton/Science Source; (C) © iStockphoto.com/Tommounsey.

A. Salamander, with equal-sized front and back limbs.

B. Frog, with long, muscular legs and short forelimbs.

C. Tadpole, a swimming frog larva with gills and a tail.

Modern Amphibians **Amphibians** are scaleless tetrapods that spend time on land, but require water to breed. Fertilization is typically external. Eggs and sperm are released into water through a **cloaca**, a body opening that also serves as the exit for urinary and digestive wastes. (Sharks, reptiles, birds, and egg-laying mammals also have a cloaca.) Amphibian larvae are aquatic and have gills. Most species lose their gills and develop lungs during the transition to adulthood. The adults are active predators with a three-chambered heart.

Of all amphibians, the 530 or so species of salamanders and newts most closely resemble early tetrapods in body form. Their forelimbs and back limbs are similarly sized, and they have a long tail (Figure 15.32A). Frogs and toads belong to the most diverse amphibian lineage, with more than 5,000 species. Long, muscular hindlimbs allow the tailless adults to swim, hop, and make spectacular leaps (Figure 15.32B). The much smaller forelimbs help absorb the impact of landings.

Larvae of salamanders look like small adults, except for the presence of gills. In contrast, frog and toad larvae differ markedly from adults. The larvae, commonly called tadpoles, have gills and a tail but no limbs (Figure 15.32C).

Many amphibian populations are in decline. Shrinking or deteriorating habitats are part of the problem. People commonly fill low-lying areas where pooling of seasonal rains could provide breeding grounds for amphibians. Climate change causes additional harm by fostering the spread of pathogens and parasites, and increasing the amount of UV radiation that reaches Earth's surface. Amphibians are also highly sensitive to water pollution. Their thin skin, unprotected by scales, allows waste carbon dioxide to diffuse out of their body. Unfortunately, it also allows chemical pollutants to enter.

Take-Home Message 15.5

What are the traits of fishes and amphibians?

- Lampreys are modern fishes that do not have jaws or paired fins.
- Cartilaginous fishes such as sharks have jaws and immobile paired fins. Like lampreys, they have a skeleton made of cartilage.
- Bony fishes are the most diverse fishes. They have paired movable fins.
- The lobe-finned fishes, a subgroup of the bony fishes, are the closest living relatives of the tetrapods.
- Amphibians are tetrapods with a three-chambered heart and a scaleless body. Fertilization typically takes place in water. Gilled larvae typically develop into adults that have lungs.

amphibian Tetrapod with a three-chambered heart and scaleless skin that typically develops in water, then lives on land as a carnivore with lungs.

cloaca Of some vertebrates, body opening that releases urinary and digestive waste, and also functions in reproduction.

lobe-finned fish Bony fish that has bony supports inside its fins.

ray-finned fish Bony fish with fins supported by thin rays derived from skin.

A. Hognose snakes hatching from leathery amniote eggs.

B. Komodo dragon, the largest lizard.

C. Turtle in a defensive posture.

D. Crocodile with its fish prey.

Figure 15.33 Diversity of nonbird reptiles.

(A) Z. Leszczynski/Animals Animals; (B) Soren Egeberg Photography/Shutterstock.com;
(C) Joel Sartore/National Geographic Creative; (D) Johan Swanepoel/Shutterstock.com.

15.6 Escape From Water—Amniotes

REMEMBER: A clade includes all descendants of the ancestor in which the unique trait that defines the clade arose (Section 12.8). Dinosaurs perished in a mass extinction about 66 million years ago (11.1).

Amniote Innovations **Amniotes**, a vertebrate group adapted to life on dry land, branched off from an amphibian ancestor during the Carboniferous. Amniotes have lungs throughout their life, and their skin is rich in keratin, a protein that makes it waterproof. A pair of well-developed kidneys help conserve water, and fertilization usually takes place inside the female's body. Sexes are typically separate and fixed for life. Amniotes produce eggs in which an embryo develops bathed in fluid, a trait that allows them to develop on dry land (Figure 15.33A).

An early branching of the amniote lineage separated ancestors of mammals from the common ancestor of all modern reptiles. You probably do not think of birds as reptiles, but from an evolutionary perspective, they are. To a biologist, the **reptiles** (clade Reptilia) include modern lizards, snakes, turtles, crocodilians, and birds, as well as the extinct dinosaurs.

The ability to regulate internal body temperature evolved in some early amniotes, including the ancestors of birds and of mammals. Amphibians, turtles, lizards, and snakes are **ectotherms**, which means "heated from outside." Ectotherms adjust their internal temperature by altering their behavior. They bask on a warm rock to heat up, or retreat into a burrow to cool off. In contrast, birds and mammals are **endotherms** that maintain their body temperature by varying their production of metabolic heat. Endotherms use energy staying warm, so they require more food than ectotherms. A bird or mammal requires far more calories than a lizard or snake of the same weight. However, because endotherms warm themselves, they can remain active at lower temperatures than ectotherms.

Nonbird Reptiles Lizards are the most diverse reptiles. Most are predators, although iguanas eat plants. The largest lizard, the Komodo dragon (Figure 15.33B), grows up to 3 meters (10 feet) long. It has a venomous bite, and trails its prey for hours or days until the poisoned animal collapses. Lizards typically lay eggs that develop outside the body, but some species give birth to live young. In live-bearers, eggs develop inside the mother, but they are not nourished by her tissues.

Snakes evolved from lizards during the Cretaceous and some snakes such as pythons retain bony remnants of ancestral hind limbs. All snakes are predators and all have teeth, but only some have fangs. Rattlesnakes and other fanged species bite and subdue prey with venom made in modified salivary (saliva-making) glands. Other snakes are constrictors that suffocate a prey animal by wrapping around it so tightly that it cannot expand its chest to inhale. Like lizards, most snakes lay eggs, but some hold eggs inside the body and give birth to live young.

Turtles have a bony, protective shell attached to their backbone (Figure 15.33C). Modern turtles do not have teeth. Instead, a thick layer of the protein keratin covers their jaws and forms a horny beak. Most turtles live either in the sea or in freshwater. Those that live entirely on land are commonly called tortoises.

Crocodilians live in or near water and include crocodiles, alligators, and caimans. They are predators with powerful jaws, a long snout, and sharp teeth (Figure 15.33D). Like birds, all crocodilians have a highly efficient four-chambered heart. They are the closest living relatives of birds.

The Jurassic and Cretaceous periods (201–66 million years ago) are sometimes referred to as the "Age of Reptiles." During this time the dinosaurs underwent a great adaptive radiation and became the dominant animals on land. They became extinct at the end of the Cretaceous, most likely as a result of an asteroid impact. Birds, which had earlier evolved from feathered dinosaurs by about 160 million years ago, survived this extinction event.

Birds **Birds** are the only modern animals with feathers, which are modified scales. The bird body is adapted to flight. Bird wings are homologous to our arms. Contraction of one set of muscles produces a powerful downstroke that lifts the bird (Figure 15.34). A less powerful set of muscles contracts to raise the wing. Birds have the most efficient respiratory system of any vertebrate. This system provides the oxygen necessary to produce a steady supply of ATP that can power flight. A four-chambered heart pumps oxygen-poor blood to the lungs and oxygen-rich blood to the rest of the body in separate circuits. Flying also requires good eyesight and a great deal of coordination. Compared to a lizard of a similar body mass, a bird has much larger eyes and a larger brain. Most birds are surprisingly lightweight. Air cavities inside a bird's bones keep its body weight low, as does the lack of a bladder (an organ that stores urinary waste in many other vertebrates). Rather than heavy, bony teeth, bird jaws are covered with a lightweight beak made of keratin. The structure of the beak adapts a bird to feed on a particular type of food.

In birds, as in other reptiles, fertilization is internal. Unlike most male reptiles, male birds do not typically have a penis. Thus, to inseminate a female, a male bird must press his cloaca against hers, in a maneuver poetically described as a cloacal kiss. A female bird lays an egg that has four membranes characteristic of **amniote eggs** (Figure 15.35A). Nutrients from the egg's yolk and water from the egg white (albumen) sustain the developing embryo. Like some turtles and all crocodilians, birds encase their eggs within a rigid shell hardened by calcium carbonate. Birds are endotherms, so their egg must be kept warm in order for the embryo to develop. In nearly all bird species, one or both parents incubate the eggs until they are ready to hatch. Unlike other reptiles, many birds hatch in a relatively undeveloped state and require extensive parental care before they can live on their own (Figure 15.35B).

Many birds make a seasonal migration. Migration typically involves spring travel to a breeding site where insects are abundant in the summer, then an autumn flight to a site where the bird spends the winter. Such flights can require remarkable endurance. One shorebird monitored by researchers flew from Alaska to New Zealand, a distance of 11,500 kilometers (7,145 miles), without stopping to feed or rest.

Mammals **Mammals** are the only amniotes in which females nourish their offspring with milk secreted from mammary glands. The group name is derived from the Latin *mamma*, meaning breast. Mammals are also the only animals that have hair or fur. Both are modifications of scales. Like birds, mammals are endotherms. A coat of fur or head of hair helps them maintain their core temperature. Mammals

Figure 15.34 Bird in flight.
Birds flap their wings to fly. The downstroke provides lift.
Eric Isselée/Shutterstock.com.

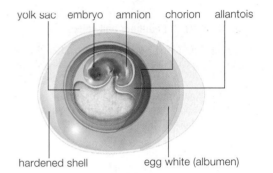

yolk sac   embryo   amnion   chorion   allantois

hardened shell   egg white (albumen)

A. Developing embryo inside an egg.

B. Newly hatched parrots.
Jane Burton/naturepl.com.

Figure 15.35 Bird development.

---

amniote Vertebrate that produces amniote eggs; a reptile, bird, or mammal.

amniote egg Egg with four membranes that allows an embryo to develop away from water.

bird Modern amniote with feathers.

ectotherm Animal that gains heat from the environment; commonly called "cold-blooded."

endotherm Animal that produces its own heat; commonly called "warm-blooded."

mammal Vertebrate that nourishes its young with milk from mammary glands.

reptile Amniote subgroup that includes lizards, snakes, turtles, crocodilians, and birds.

Figure 15.36 Mammal mothers and young.

A. Egg-laying mammal, the platypus. Her young hatched from eggs laid outside her body. They lick milk that oozes from her skin.

B. Marsupial, a kangaroo. Her young developed to an early stage in her body, then climbed into a pouch on her belly to nurse and complete development.

C. Placental mammal, a bear. Her young developed to a late stage inside her body. After birth, they suck milk from nipples on her chest and belly.

(A) Jean Phillipe Varin/Jacana/Science Source; (B) Craig Dingle/iStockphoto.com;
(C) Sergey Gorshkov/Nature Picture Library.

have a single lower jawbone and most have more than one type of tooth. By contrast, reptiles have a hinged two-part jaw, and all teeth are similarly shaped. Having a variety of different kinds of teeth allows mammals to eat more types of foods than most other vertebrates.

Mammals evolved early in the Jurassic, and early mouselike species coexisted with dinosaurs. By 130 million years ago, three lineages had evolved: **monotremes** (egg-laying mammals), **marsupials** (the pouched mammals), and **placental mammals** (mammals in which an organ called the placenta provides nutrients to developing offspring). Figure 15.36 shows examples of each group.

Only three species of monotremes survive, and most pouched mammals live in Australia or New Zealand. In contrast, placental mammals occur worldwide. What gives placental mammals their competitive edge? They have a higher metabolic rate, better body temperature control, and a more efficient way to nourish embryos. Compared to other mammals, placental mammals develop to a far more advanced stage inside their mother's body.

Rats and bats are the most diverse mammals. About half the 4,000 species of placental mammals are rodents and, of those, about half are rats. The next most diverse group is the bats, with about 375 species. Bats are the only flying mammals. Although some may look like flying mice, bats are more closely related to carnivores such as wolves and foxes than to rodents.

marsupial Mammal in which offspring complete development in a pouch on the mother's body.

monotreme Egg-laying mammal.

placental mammal Mammal in which developing offspring are nourished within the mother's body by way of a placenta.

primate Mammalian group with grasping hands; includes lemurs, tarsiers, monkeys, apes, and humans.

Take-Home Message 15.6

What are amniotes?

- Amniotes are vertebrates that are adapted to life on land by their unique eggs, waterproof skin, and highly efficient kidneys.
- One amniote lineage includes snakes, lizards, turtles, and crocodiles as well as the birds. Birds are the only amniotes with feathers. Like mammals, birds are endotherms; metabolic heat maintains their internal temperature.
- Mammals are amniotes that nourish their young with milk. Most have hair or fur. The three lineages are egg-laying monotremes, pouched marsupials, and placental mammals. Placental mammals are the most widespread and diverse.

15.7 Human Evolution

REMEMBER: Radiometric dating can reveal the age of fossils (Section 11.4). Mutation (12.2) and founder effects cause allele differences among populations (12.5).

Primate Traits **Primates** are an order of placental mammals that includes humans, apes, monkeys, and their close relatives. Primates first evolved in tropical forests, and many of the group's characteristic traits arose as adaptations to life among the branches. Primate shoulders have an extensive range of motion that facilitates climbing (Figure 15.37). Unlike most mammals, a primate can extend its arms out to its sides, reach above its head, and rotate its forearm at the elbow. Both hands and feet are capable of grasping. Many types of mammals have claws, whereas the tips of primate fingers and toes typically have touch-sensitive pads protected by flattened nails.

Compared to other mammals, primates have a large brain for their body size. The regions of the brain devoted to vision and to information processing are expanded, and the area devoted to smell is reduced. Most mammals have widely spaced eyes set toward the side of the skull, but primate eyes tend to be at the front of the head. As a result, both eyes view the same area, each from a slightly different vantage point. The brain integrates the differing signals it receives from the two eyes to produce a three-dimensional image. A primate's excellent depth perception adapts it to a life spent leaping or swinging from limb to limb.

Most primates spend their life in a social group that includes adults of both sexes. Females usually give birth to only one or two young at a time and provide care for an extended period after birth.

Primate Origins and Diversification Figure 15.38 shows the relationships among modern primate subgroups. Primates most likely arose before the demise of

Figure 15.37 Adapted to climbing.
A female orangutan (an Asian ape) demonstrates her wide range of shoulder motion and the grasping ability of her hands and feet. Eyes situated at the front of her face provide excellent depth perception.

© Thomas Marent/ ardea.com.

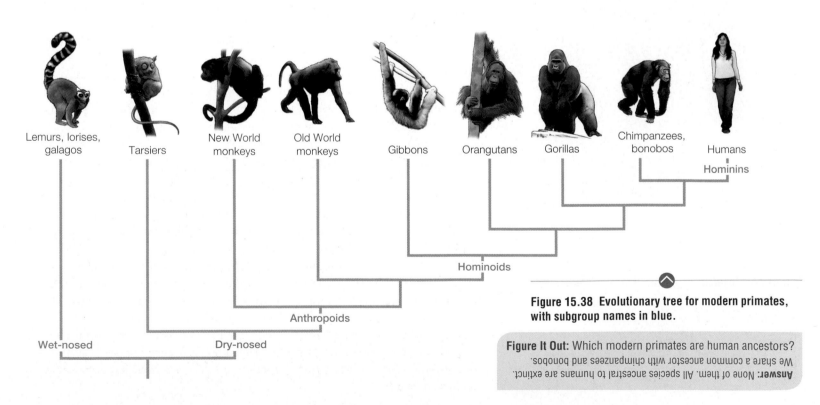

Lemurs, lorises, galagos | Tarsiers | New World monkeys | Old World monkeys | Gibbons | Orangutans | Gorillas | Chimpanzees, bonobos | Humans

Hominins

Hominoids

Anthropoids

Wet-nosed   Dry-nosed

Figure 15.38 Evolutionary tree for modern primates, with subgroup names in blue.

Figure It Out: Which modern primates are human ancestors?

Answer: None of them. All species ancestral to humans are extinct. We share a common ancestor with chimpanzees and bonobos.

A. Lemur, with a wet nose and cleft upper lip.

B. Tarsier, with a dry nose and uncleft upper lip.

C. Squirrel monkey (New World monkey), with a flat face and prehensile tail.

D. Baboon (Old World monkey), with a long nose and short tail.

E. Gorilla, the largest living great ape.

F. Chimpanzee, one of our two closest living relatives.

Figure 15.39 Diversity of modern, nonhuman primates.

(A) toos/iStockphoto.com; (B) roc8jas/iStockphoto.com; (C) primates.com; (D) © iStockphoto.com/JasonRWarren; (E) © Dallas Zoo, Robert Cabello; (F) Kenneth Garrett/National Geographic Image Collection.

the dinosaurs, sometime between 85 and 66 million years ago. Their closest living relatives are colugos, a group of nocturnal mammals that live in the forests of Southeast Asia and glide from tree to tree like flying squirrels.

Lemurs are members of the oldest existing primate lineage. Like dogs and most other mammals, lemurs are "wet-nosed"; they have a moist nose and an upper lip that attaches tightly to the underlying gum (Figure 15.39A). A dry nose and a movable, noncleft upper lip evolved in the common ancestor of tarsiers (Figure 15.39B) and all other modern primates. This innovation gave dry-nosed primates a wider range of facial expressions and allowed more complex vocalizations.

Anthropoids include monkeys, apes, and humans; "anthropoid" means humanlike. Nearly all anthropoids are diurnal (active during the day) and have good eyesight, including color vision.

New World monkeys (Figure 15.39C) climb through forests of Central and South America in search of fruits. They have a flat face and a nose with widely separated nostrils. A long tail helps them maintain balance. In many species, the tail is prehensile, meaning it grasps things.

Old World monkeys live in Africa, the Middle East, and Asia. They tend to be larger than New World monkeys, with a longer nose and closely set nostrils. Some are tree-climbing forest dwellers. Others, such as baboons (Figure 15.39D), spend most of their time on the ground in grasslands and deserts. Not all Old World monkeys have a tail, but in those that do the tail is typically short and never prehensile.

Tailless nonhuman primates are commonly called apes. About 15 species of small apes called gibbons inhabit Southeast Asian forests. Gibbons are sometimes referred to as "lesser apes," in comparison with the larger apes, or "great apes." The forest-dwelling orangutan of Sumatra and Borneo is the only surviving Asian great ape. All other great apes (gorillas, chimpanzees, and bonobos) are native to central Africa and spend most of their time on the ground. When walking, African apes lean forward and support their weight on their knuckles (Figure 15.39E). Gorillas, the largest living primates, live in forests and feed mainly on leaves. Chimpanzees (Figure 15.39F) and the bonobos are our closest living relatives. The chimpanzee/bonobo lineage and the lineage leading to humans diverged 6 to 13 million years ago.

Australopiths Humans and their closest extinct relatives are now grouped together as **hominins** (previously called hominids). The defining trait of hominins is **bipedalism**—habitual upright walking. Thus, researchers interested in human origins look for fossil evidence that a species walked upright. Some fossils of primates that may have been bipedal date back as long ago as 7 million years ago. However, the fossil record of these species is patchy and little is known about them.

The best-known early hominins are the **australopiths**, who lived in Africa from about 4 million to 1.2 million years ago. One australopith genus, *Australopithecus*, includes several species considered likely human ancestors. The fossil history of australopiths shows trends toward smaller teeth and improvements in the ability to walk upright, but little increase in brain size.

Fossil footprints in Tanzania document the passage of a bipedal species 3.6 million years ago (Figure 15.40A). The prints reveal that the walkers' feet had a pronounced arch and a big toe in line with the other toes—both adaptations to upright walking. The prints were probably made by *Australopithecus afarensis*, an australopith that lived in Tanzania and other parts of eastern Africa from 3.9 to 3 million years ago. An *A. afarensis* skeleton known as Lucy (Figure 15.40B) is the best-known representative of this species. *A. afarensis* is considered a possible human ancestor.

A. 3.6-million-year-old foot-prints of a bipedal species.

B. *A. afarensis* fossil (Lucy) from 3.5 million years ago.

A. Fossil of a male who lived in Kenya 1.5 million years ago.

B. Reconstruction based on a 700,000-year-old skull discovered in China.

Figure 15.40 Evidence of early hominins.

(A) Louise M. Robbins; (B) Dr. John D. Cunningham/Visuals Unlimited, Inc.

Figure 15.41 *Homo erectus*, **the first hominin species to leave Africa.**

(A) Science VU/NMK/Visuals Unlimited, Inc.; (B) Philippe Plailly & Atelier Daynes/Science Source.

Early Humans Humans are members of the genus *Homo*. Fossils of the oldest species in this genus, **Homo habilis**, date from 2.3 million years to 1.4 million years before the present. The species was named based on a fossil discovered in Kenya in 1964. The name *Homo habilis* means "handy man"; it refers to the stone tools found near the fossil. At the time of the fossil's discovery, the ability to make stone tools was considered a distinctly human trait. We now know that hominins used sharp stone edges to scrape meat from bones as early as 3.4 million years ago. Given the age of the bones, these early tool users were most likely australopiths. Classification of *H. habilis* continues to inspire debate. The species is australopith-like in its body proportions and brain size, so some scientists think it should be reclassified as a spe-cies of *Australopithecus*. Other scientists argue for the species' inclusion in the genus *Homo*, pointing out that the hands and arms are similar to those of modern humans.

Homo erectus, the first hominin with body proportions like our own, arose in Africa by about 2 million years ago. The species name means "upright man," and like us, *H. erectus* stood on legs that were longer than its arms. The most complete *H. erectus* fossil found thus far is a skeleton of a young male who lived about 1.5 million years ago in Kenya (Figure 15.41A). This individual, informally known as Turkana boy, was under age 14 when he died. However, he was already was 5 feet 2 inches (1.60 meters) tall and his brain was twice the size of a chimpanzee's. Foot-prints from the same region and time suggest that *H. erectus* had a gait like that of modern humans. As far as we know, *H. erectus* was the first hominin to venture out of Africa. By 1.75 million years ago, a population had become established in what is now the Eurasian country of Georgia. *H. erectus* went on to colonize Indonesia by 1.6 million years ago, and China by 1.15 million years ago (Figure 15.41B). At the same time, African populations continued to thrive.

anthropoids Monkeys, apes, and humans.

australopiths Informal name for a lineage of chimpanzee-sized hominins that lived in Africa between 4 million and 1.2 million years ago.

bipedalism Habitually walking upright.

hominin Humans and extinct humanlike species.

Homo erectus Human species that dispersed out of Africa.

Homo habilis Earliest named human species.

humans Members of the genus *Homo*.

Figure 15.42 Human dispersal routes.
Exposure of a temporary land bridge from Siberia to North America allowed entry into the Americas.

Photo, NASA; data, National Geographic.

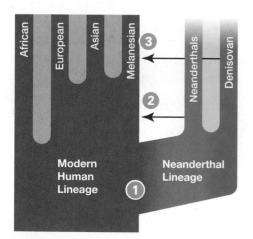

1. Perhaps as long as 700,000 years ago, a divergence separated the ancestors of modern humans from the lineage leading to Neanderthals and Denisovans.

2. About 60,000 years ago, Neanderthals living in the Middle East mated with modern humans who were venturing out of Africa.

3. About 40,000 years ago, ancestors of the modern humans who would later populate New Guinea and Australia mated with Denisovans.

Figure 15.43 One current model of human evolution.
The model is based on observed genetic similarities between some populations of modern humans and two fossil members of the genus *Homo*: Neanderthals and Denisovans.

Homo erectus is considered the most likely ancestor of our own species, and of Neanderthals, and some other more recently discovered hominins.

Homo Sapiens **Homo sapiens** is the species of anatomically modern humans. Compared to *H. erectus*, our species has a higher, rounder skull, a larger brain, and a flatter face with smaller jaw and teeth. We are also the only hominins with a chin, a protruding area of thickened bone in the middle of the lower jawbone.

To date, the oldest *H. sapiens* fossils discovered are two partial male skulls from Ethiopia, in East Africa. Known as Omo I and Omo II, the fossil individuals date to 195,000 years ago. Fossil remains of two adult males and a child who lived 160,000 years ago were also unearthed in this region. Other fossils show that *H. sapiens* reached South Africa by 115,000 years ago.

Genetic studies also support an African origin for our species. Modern Africans are more genetically diverse than people of any other region. This diversity indicates that African populations have existed for a very long time—long enough to accumulate a very large number of random mutations compared with other populations. Furthermore, most genetic variations seen in people native to regions outside of Africa are subsets of the variation found in Africa. This pattern is evidence that founder effects occurred after some people left their homeland to colonize the rest of the world. Over many generations, pioneers traveled along the coasts of Africa, then Eurasia and Australia. We know from fossilized feces deposited in an Oregon cave that they reached North America by about 14,000 years ago. Today, the distribution of specific mutations among different ethnic groups provides evidence of routes taken by the ancient travelers. By mapping the frequency of maternal and paternal genetic markers in modern peoples, geneticists have created a picture of how *Homo sapiens* dispersed across the world (Figure 15.42).

Neanderthals and Denisovans As early modern humans left Africa and expanded their range to Eurasia, they met and interbred with at least two other groups of hominins who had already become established there (Figure 15.43). These hominins had shared an ancestor with *H. sapiens* between 500,000 and 700,000 years ago 1. About 60,000 years ago, as some modern humans ventured into the Middle East, they met up with and interbred with Neanderthals (**Homo**

neanderthalensis) . As a result of this interbreeding, modern human populations native to regions outside of Africa have distinctive Neanderthal alleles that are rare among populations native to Africa. A second interbreeding event took place about 40,000 years ago somewhere in Asia. Here, humans interbred with Denisovans, a recently discovered, genetically distinct group of hominins ❸. As a result, we find unique Denisovan alleles in some modern Asians. Melanesians, such as people of Papua New Guinea, have an especially high percentage of Denisovan DNA.

Neanderthals have long been considered our closest extinct relatives. They first appeared about 230,000 years ago, and they left an extensive fossil record in the Middle East, Europe, and central Asia. The common conception of Neanderthals as brutish cavemen with poor posture arose from an early reconstruction based on a fossil of an individual deformed by arthritis. More recent reconstructions reveal Neanderthals were shorter than modern humans, but stood upright. They lived in regions where winters are cold, and a short, stocky body minimized the surface area available for heat loss. Modern Arctic peoples have a similar body shape. A recent reconstruction of a Neanderthal male, based on material from multiple fossils, stands about 164 centimeters (5′4″) tall (Figure 15.44). The Neanderthal braincase was longer and lower than that of modern humans, but their brain was as big as or bigger than ours. Their face had pronounced brow ridges, a large nose with widely spaced nostrils, and no chin.

Fossils of Neanderthal individuals who survived despite disabilities such as the loss of a limb testify to a compassionate social structure. Some simple burials suggest possible symbolic thought. Several lines of evidence suggest Neanderthals were able to speak. The most recent evidence of Neanderthals dates to about 40,000 years ago. Neanderthals may have been outcompeted by newly arrived *H. sapiens* or killed by diseases these migrants brought with them. Climate changes and volcanic eruptions may have contributed to their demise by altering the abundance of the animals they hunted. Meat made up the bulk of the Neanderthal diet.

Denisovans were first identified when researchers sequenced DNA extracted from a fossil pinky (small finger) bone discovered in Denisova Cave in Siberia. The researchers expected to find typical Neanderthal DNA. Instead, the finger's DNA contained unique sequences never seen before. As a result, the fossil was assigned to a group—the Denisovans—named for the site of its discovery. At this writing, Denisovans are not considered a distinct species, but rather a subgroup of Neanderthals.

Homo neanderthalensis   *Homo sapiens*

Figure 15.44 Skeletal comparison of Neanderthal and modern human males.
The Neanderthal skeleton is a reconstruction based on multiple fossils. Each color denotes a different fossil.

Courtesy of @ Blaine Maley, Washington University, St. Louis.

Take-Home Message 15.7

How did humans evolve?

- Humans, like other primates, have grasping hands and shoulders adapted to climbing. Unlike most primates, they walk upright.
- Australopiths are a group of upright-walking species that probably include human ancestors. They evolved in Africa by about 4 million years ago.
- The first humans, or members of the genus *Homo*, also arose in Africa. *Homo erectus* migrated out of Africa and into Europe and Asia.
- Modern humans (*Homo sapiens*) and the extinct Neanderthals are descendants of *Homo erectus*. By the currently favored model, modern humans evolved in Africa, then migrated worldwide.

Homo neanderthalensis Neanderthals. Closest extinct relatives of modern humans; had large brain, stocky body.

Homo sapiens Modern humans; evolved in Africa, then expanded their range worldwide.

Summary

Section 15.1 Invertebrates (animals without a backbone) are the most diverse animal group. They far outnumber **vertebrates**. Some compounds that invertebrates produce can be used as medicines.

Section 15.2 Animals are multicelled heterotrophs that ingest food and move about. The **colonial theory of animal origins** states that they evolved from a colonial protist. Animals with **tissues** have **radial symmetry** or **bilateral symmetry**. Most animals are either **deuterostomes** or **protostomes**, two groups that differ in their development. The animal gut is a saclike **gastrovascular cavity** or a tubelike **complete digestive tract**. Most bilateral animals have a **coelom** around the gut. Some have an **open circulatory system** or a more efficient **closed circulatory system**.

Section 15.3 Sponges are sessile filter-feeders with no body symmetry or tissues. Each individual is a **hermaphrodite**; it makes eggs and sperm. The sponge **larva** is free-swimming.

The radially symmetrical **cnidarians** have two body types: **medusa** (as in jellies) and **polyp** (as in sea anemones). Stinging cells help cnidarians capture prey.

Flatworms have simple organ systems, a gastrovascular cavity, and no coelom. Planarians are free-living flatworms, whereas the tapeworms and flukes are parasites.

Annelids are segmented worms that have a coelom, a complete digestive system, and a closed circulatory system. Earthworms are oligochaete annelids. The annelid lineage also includes leeches and polychaetes.

Mollusks have a **mantle**, a skirtlike extension of the body. In many groups, the mantle secretes a shell. **Gastropods** (such as snails) and **cephalopods** (such as octopuses) have a head with a **radula** used in feeding. **Bivalves** such as clams are filter-feeders.

Roundworms (nematodes) are unsegmented worms with a complete gut and an incompletely lined coelom. They may be free-living or parasitic.

Arthropods have a jointed **exoskeleton**. Horseshoe crabs are marine bottom-feeders. **Arachnids** include spiders, mites, ticks, and scorpions. Most **crustaceans**, such as lobsters, krill, and barnacles, are aquatic. Centipedes and millipedes have an elongated body and live on land. **Insects** have paired **antennae** and **compound eyes**, and most undergo **metamorphosis**. They are the most diverse animals and the only winged invertebrates.

Echinoderms such as sea stars have an **endoskeleton** of spines, spicules, or plates of calcium carbonate. A **water–vascular system** with tube feet allows adults to move about. Adult echinoderms are radial, but larva are bilateral, indicating the group has a bilateral ancestor.

Section 15.4 Four embryonic traits define the **chordates**: a **notochord**, a dorsal hollow nerve cord (which becomes a brain and spinal cord), a pharynx with gill slits, and a tail extending past the anus. Depending on the group, some or all of the features persist in adults.

Lancelets and **tunicates** are invertebrate chordates. Most chordates are vertebrates; they have a **vertebral column** of cartilage or bone. Jaws, **lungs**, limbs, and waterproof eggs are innovations that made the adaptive radiation of **vertebrates** possible. **Tetrapods** have four limbs. All vertebrates have a complete digestive system, closed circulatory system, and **kidneys**.

Section 15.5 The earliest fishes were **jawless fishes**. **Cartilaginous fishes** and **bony fishes** have jaws, **scales**, and paired fins. Jaws evolved from gill supports. The two bony fish lineages are the highly diverse **ray-finned fishes**, which includes most familiar fishes, and the **lobe-finned fishes**. **Amphibians** evolved from a lobe-finned fish and require water to reproduce. Existing amphibian groups include salamanders, frogs, and toads. Fertilization is external. Eggs and sperm exit the body through a **cloaca** that also expels digestive and urinary wastes.

Section 15.6 Amniotes were the first vertebrates that did not need external water for reproduction. Their skin and kidneys conserve water, and they produce **eggs** that have distinctive membranes. Mammals are one amniote lineage. **Reptiles**, including **birds**, are another. Most reptiles are **ectotherms**, but birds and mammals are **endotherms**; they maintain their temperature by metabolic production of heat). There are three lineages of **mammals**: egg-laying **monotremes**, pouched **marsupials**, and placental mammals. **Placental mammals** are the most diverse group.

Section 15.7 Primates are a lineage adapted to climbing and all have hands capable of grasping. Lemurs have a moist, doglike snout, but tarsiers and **anthropoids** (monkeys, apes, humans) have a dry nose and movable upper lip. Our closest living relatives are the chimpanzees and bonobos. **Bipedalism** defines the **hominins**, which include **humans** and their extinct close relatives. **Australopiths** were early hominins and some are considered likely human ancestors. The first named members of our genus, *Homo habilis*, resembled australopiths. *Homo erectus* had a larger brain and some populations became established outside of Africa. Modern humans (*Homo sapiens*) arose in Africa. As modern humans dispersed, some interbred with typical Neanderthals (*Homo neanderthalensis*) or with Denisovans. As a result, some human genomes contain alleles that can be traced to these groups.

Self-Quiz

Answers in Appendix I

1. True or false? Animal cells do not have walls.

2. The colonial theory of animal origins states that _____ .
   a. animals are more closely related to plants than to fungi
   b. animals evolved from a colonial protist
   c. most animals live in colonies
   d. all animals have a backbone

3. A body cavity that is fully lined with tissue derived from mesoderm is a _____ .
   a. pseudocoelom   c. coelom
   b. kidney   d. gastrovascular cavity

4. Most animal bodies have _____ symmetry.
   a. radial   b. bilateral   c. no

5. Earthworms are most closely related to _____ .
   a. insects   c. leeches
   b. tapeworms   d. roundworms

6. The _____ include the only winged invertebrates.
   a. flatworms   c. arthropods
   b. annelids   d. cnidarians

7. List the four distinguishing chordate traits. Which of these traits are retained by an adult tunicate?

8. Jaws evolved from _____ of jawless fishes.
   a. fins   b. gill supports   c. the backbone

9. All vertebrates are _____ but only some are _____ .
   a. tetrapods; mammals   c. amniotes; hominins
   b. chordates; amniotes   d. bipedal; australopiths

10. Amniote adaptations to land include _____ .
    a. waterproof skin   d. specialized eggs
    b. internal fertilization   e. a and c
    c. highly efficient kidneys f. all of the above

11. Birds and placental mammals _____ .
    a. are endotherms   c. have mammary glands
    b. lay eggs   d. have an open circulatory system

12. True or false? Bipedalism is the defining trait of primates.

13. *Homo erectus* is a likely ancestor of *H. sapiens* and _____ .
    a. *Homo habilis*   c. the great apes
    b. australopiths   d. Neanderthals

14. Match the organisms with the appropriate description.
    _____ sponges   a. most diverse vertebrates
    _____ cnidarians   b. no true tissues, no organs
    _____ flatworms   c. jointed exoskeleton
    _____ roundworms   d. mantle over body mass
    _____ annelids   e. segmented worms
    _____ arthropods   f. tube feet, spiny skin
    _____ mollusks   g. have specialized stinging cells
    _____ echinoderms   h. lay amniote eggs
    _____ amphibians   i. feed young secreted milk
    _____ fishes   j. unsegmented, molting worms
    _____ birds   k. first terrestrial tetrapods
    _____ mammals   l. tailless primates
    _____ apes   m. saclike gut, no coelom

15. Arrange the events in order, from most ancient to most recent.
    _____ 1   a. Cambrian explosion of diversity
    _____ 2   b. Origin of animals
    _____ 3   c. Tetrapods move onto the land
    _____ 4   d. Extinction of dinosaurs
    _____ 5   e. *Homo erectus* leaves Africa
    _____ 6   f. First jawed vertebrates evolve

Critical Thinking

1. In the summer of 2000, only 10 percent of the lobster population in Long Island Sound survived after a massive die-off. Many lobstermen in New York and Connecticut lost small businesses that their families had owned for generations. Some believe the die-off was the result of increased spraying of pesticides to control mosquitoes that carry West Nile virus. Fisherman saw no similar decrease in their catch. Explain why a chemical substance that targets mosquitoes might also harm lobsters but not fish.

2. Why is it more difficult to determine the sex of a newly hatched canary than a newborn puppy?

3. As human activities put more and more carbon dioxide (CO_2) into the air, the ocean takes up more CO_2 and becomes increasingly acidic. Increased ocean acidity makes it more difficult for marine animals to produce their calcium-rich hard parts. List three types of invertebrates that are likely to be adversely affected by an increase in ocean acidity.

4. One subgroup of ray-finned fishes, the teleosts, is exceptionally diverse. It includes about half of all vertebrate species. Early in their evolution, teleosts underwent a duplication of their entire genome. By one hypothesis, this event opened the way to their diversification. Explain why such a duplication might make the evolution of new traits more likely.

16 POPULATION ECOLOGY

Application ● 16.1 **A Honkin' Mess**

REMEMBER: Members of a population share a common gene pool (Section 12.2).

Visit a grassy park or a golf course and you may need to watch where you step. Wide expanses of grass with a nearby body of water attract large numbers of Canada geese, or *Branta canadensis* (Figure 16.1). These plant-eating birds produce slimy, green feces that can soil shoes, stain clothes, and discourage picnics. Goose feces also wash into ponds and waterways. The nutrients they add encourage bacteria and algae to grow, clouding the water and making swimming unappealing and possibly dangerous. Goose feces sometimes contain microbes that can sicken humans.

Geese can also pose a risk to air traffic. In January of 2009, both engines of a US Airways flight failed shortly after the plane took off from a New York City airport. Fortunately, the pilot was able to land the plane in the nearby Hudson River (Figure 16.1B), where boats unloaded all 155 people aboard. Afterwards, investigators from the Federal Aviation Agency found bits of feather, bone, and muscle in the plane's wing flaps and engines. Unique sequences in the DNA identified the tissue in both engines as Canada goose.

The number of Canada geese in the United States has increased dramatically. For example, Michigan had about 9,000 in 1970, and has 300,000 today. Controlling their number is challenging because several different Canada goose populations spend time in the United States. A **population** is a group of organisms of the same species who live in a specific location and breed with one another more often than they breed with members of other populations. In the past, nearly all Canada geese seen in the United States were migratory. They nested in northern Canada, flew to the United States to spend the winter, then returned to Canada. Most Canada geese still migrate, but some populations are permanent residents of the United States. The geese breed where they grew up, and the nonmigratory birds are generally descendants of geese deliberately introduced to a park or hunting preserve. During the winter, migratory birds often mingle with nonmigratory ones. For example, a bird that breeds in Canada and flies to Virginia for the winter finds itself alongside geese that have never left Virginia.

Life is more difficult for migratory geese than for nonmigratory ones. Flying to and from a northern breeding area takes lots of energy and is dangerous. A bird that does not migrate can devote more energy to producing young than a migratory one can. If the nonmigrant lives in a suburban or urban area, it also benefits from an unnatural abundance of food (grass) and an equally unnatural lack of predators. Not surprisingly, the greatest increases in Canada geese have been among nonmigratory birds living where humans are plentiful.

Migratory birds are protected under federal law and by international treaties. However, in 2006, increasing complaints about Canada geese led the U.S. Fish and Wildlife Service to exempt this species from some protections. The agency encouraged wildlife managers to look for ways to reduce nonmigratory Canada goose populations, without unduly harming migratory birds. To do so, these biologists need to know about the traits that characterize different goose populations, as well as how these populations interact with one another, and with other species.

These sorts of questions are the focus of the science of **ecology**, the study of interactions among organisms, and between organisms and their physical environment. Ecology is not the same as environmentalism, which is advocacy for protection of the environment. However, environmentalists often cite the results of ecological studies when drawing attention to environmental concerns.

A. Geese overrun a California park.

B. An airliner downed by a collision with geese.

Figure 16.1 Problems caused by Canada geese.

(A) Courtesy of Joel Peter; (B) AP Images/Steven Day.

A. Clumped distribution of hippopotamuses.

B. Near-uniform distribution of nesting seabirds.

C. Random distribution of dandelions.

Figure 16.2 Population distributions.

(A) Michael Poliza/National Geographic Creative; (B) © Eric and David Hosking/Corbis;
(C) Elizabeth A. Sellers/life.nbii.gov.

16.2 Characteristics of Populations

REMEMBER: Sampling error can lead to misleading results (Section 1.6).

When studying a population, ecologists collect information about its gene pool, reproductive traits, and behavior of its component individuals. They also look at **demographics**—vital statistics that describe the population.

Demographic Traits **Population size** refers to the number of individuals of a species in a population. **Population density** is the number of individuals in some specified area or volume of a habitat, such as the number of frogs per acre of rain forest or the number of amoebas per liter of pond water.

    Population distribution describes where individuals are relative to one another. Most populations have a clumped distribution, meaning members of the population are closer to one another than would be predicted by chance alone. Often, a patchy distribution of an essential resource draws individuals together, as when hippopotamuses gather in muddy river shallows (Figure 16.2A). Similarly, a cool, damp, north-facing slope may be covered with ferns, whereas an adjacent drier, south-facing slope has none. Limited dispersal ability increases the likelihood of a clumped distribution: The nut does not fall far from the tree. Asexual reproduction also results in clumping. It produces colonies of coral and vast stands of some trees. In addition, some animals are social; they benefit by living in a group.

    Competition for resources can produce a near-uniform distribution, with individuals more evenly spaced than would be expected by chance. Creosote bushes in deserts of the American Southwest grow in this pattern. Competition for water among the plants' root systems prevents them from growing near one another. Seabirds in breeding colonies often show a near-uniform distribution too. Each bird repels others that get within reach of its beak as it sits on its nest (Figure 16.2B).

    A random population distribution is rare in nature. Random distribution arises when resources are uniformly available, and proximity to others neither benefits nor harms individuals. For example, when wind-dispersed dandelion seeds land on the uniform environment of a suburban lawn, dandelions grow in a random pattern (Figure 16.2C).

    The scale of a study area and timing of the study can influence the observed pattern of distribution. For example, although seabirds may be spaced almost

demographics Statistics that describe a population.

ecology The study of interactions among organisms, and between organisms and their environment.

population A group of organisms of the same species who live in a specific location and breed with one another more often than they breed with members of other populations.

population density Number of members of a population in a given area.

population distribution The way in which members of a population are dispersed in their environment.

population size Number of individuals in a population.

uniformly at a nesting site, nesting sites are clustered along a shoreline. Also, these birds crowd together in the breeding season, but disperse at other times.

Age structure refers to the distribution of individuals among various age categories. A population's age structure affects its capacity for growth. A population with a large proportion of young individuals who have not begun to breed has a greater potential for growth than one composed mainly of older individuals.

Collecting Demographic Data Scientists often cannot directly count all the members of a population. Instead, they sample a population, then use data from that sample to estimate characteristics of the population as a whole. Plot sampling estimates the total number of individuals in an area on the basis of direct counts in part of the area. For example, to determine the number of daisies in a prairie or clams in a mudflat, ecologists begin by measuring the number of individuals in several 1-meter-by-1-meter-square plots. To calculate the total population size, researchers multiply the average number of individuals in the sample plots by the number of plots in the area the population inhabits. Size estimates from plot sampling are most accurate for organisms that do not move about and live in an area where conditions are uniform. Mark–recapture sampling is used to study mobile animals. Researchers capture animals, mark them, then release them. Sometime later, animals are captured again. The proportion of marked animals in the second sample is then taken to be representative of the proportion marked in the whole population:

$$\frac{\text{marked individuals in sampling at time 2}}{\text{total captured in sampling 2}} = \frac{\text{marked individuals in sampling at time 1}}{\text{total population size}}$$

Digging Into Data

Monitoring Iguana Populations

In 1989, Martin Wikelski started a long-term study of marine iguana populations in the Galápagos Islands. He marked the iguanas on two of the islands—Genovesa and Santa Fe—and collected data on how their body size, survival, and reproductive rates varied over time. The iguanas eat algae and have no predators, so deaths are usually the result of food shortages, disease, or old age. His studies showed that the iguana populations decline during El Niño events, when water surrounding the islands heats up.

In January 2001, an oil tanker ran aground and leaked a small amount of oil into the waters near Santa Fe. Figure 16.3 shows the number of marked iguanas that Wikelski and his team counted in their census of study populations just before the spill and about a year later.

1. Which island had more marked iguanas at the time of the first census?
2. How much did the population size on each island change between the first and second census?
3. Wikelski concluded that changes on Santa Fe were the result of the oil spill, rather than sea temperature or other climate factors common to both islands. How would the census numbers be different from those he observed if an adverse event had affected both islands?

Figure 16.3 Shifting numbers of marked marine iguanas on two Galápagos islands.
An oil spill occurred near Santa Fe just before the January 2001 census (blue bars). A second census was carried out in December 2001 (green bars).

Photo, Bruce Coleman/Photoshot.

Suppose, for example, that scientists capture, mark, and release 100 deer. Later, they return and once again capture 100 deer. Of these, 50 are marked. The proportion of marked deer in the second sample (50 percent) indicates that half of the deer in the population have been marked. Thus the 100 deer initially marked must have be members of a population of 200 deer.

Information about the traits of individuals in a sample plot or capture group can be used to infer properties of the population as a whole. For example, if half of the deer recaptured in a mark–recapture study are of reproductive age, half of the population is assumed to share this trait.

Any study that draws conclusions based on only a sample of a population is susceptible to sampling error (Section 1.6). The larger the sample, the more likely that conclusions drawn from that sample will be accurate.

Take-Home Message 16.2

How do scientists describe populations?

- Members of a population live in the same area and breed with one another.
- Populations can be described in terms of their size, density, and how their members are distributed through their environment.
- Population studies typically utilize sampling methods. Such methods run the risk of sampling error, which can be minimized by using a large sample size.

16.3 Population Growth

Exponential Growth A population grows when its birth rate exceeds its death rate. Ecologists measure births and deaths per individual, or per capita. For example, if the birth rate in a population of 2,000 mice is 1,000 young per month, then the birth rate is 1,000/2,000, or 0.5 per mouse per month. Subtract the per capita death rate from the per capita birth rate and you have the **per capita growth rate**. If the death rate in the population of 2,000 mice is 200 per month (0.1 per mouse per month), then the per capita growth rate is $0.5 - 0.1 = 0.4$ per mouse per month.

The **exponential model of population growth** describes how a population's size changes over time if its per capita growth rate is constant and its resources are unlimited. Under these theoretical conditions, the population growth in any interval (G) can be calculated as follows:

$$N \quad \times \quad r \quad = \quad G$$

| (number of individuals) | (per capita growth rate) | (population growth per unit time) |

Suppose we apply this to our population of 2,000 mice with their per capita growth rate of 0.4 per month. In the first month, the population grows by 2,000 mice × 0.4. This brings the size to 2,800. In the next month, 2,800 × 0.4, or 1,120 mice, are added, and so on (Figure 16.4A). At this growth rate, the number of mice would rise from 2,000 to more than 1 million in under two years! Plotting the size of this population against time produces a J-shaped, or "hockey stick," curve, characteristic of exponential growth (Figure 16.4B).

Exponential population growth is analogous to compounding of interest on a bank account that pays a fixed rate of return. Although the interest rate does not

Neil Burton/Shutterstock.com.

Starting Size of Population		Net Monthly Increase	New Size of Population
2,000	× r =	800	2,800
2,800	× r =	1,120	3,920
3,920	× r =	1,568	5,488
5,488	× r =	2,195	7,683
7,683	× r =	3,073	10,756
10,756	× r =	4,302	15,058
15,058	× r =	6,023	21,081
21,081	× r =	8,432	29,513
29,513	× r =	11,805	41,318
41,318	× r =	16,527	57,845
57,845	× r =	23,138	80,983
80,983	× r =	32,393	113,376
113,376	× r =	45,350	158,726
158,726	× r =	63,490	222,216
222,216	× r =	88,887	311,103
311,103	× r =	124,441	435,544
435,544	× r =	174,218	609,762
609,762	× r =	243,905	853,667
853,667	× r =	341,467	1,195,134

A. Increases in size over time. Note that the net increase becomes larger with each generation.

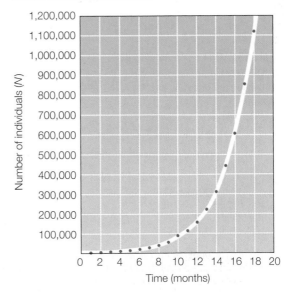

B. Graphing numbers over time yields a J-shaped curve.

Figure 16.4 Exponential growth in population of mice with a per capita growth rate (r) of 0.4 per mouse per month and an initial population size of 2,000.

No population can grow forever. Limiting factors put the brakes on population growth.

change, the amount of interest paid continually increases. Each year, the annual interest paid into the account adds to the size of the balance, and the following year's interest payment is based on that higher balance.

The exponential model of population growth assumes that resources are unlimited, so it cannot accurately predict the long-term growth of real populations. However, it does provide insight into the expected short-term growth of a population with plentiful resources. For example, when a few individuals of a species colonize a new habitat, the resulting population often grows exponentially for a period.

Carrying Capacity and Logistic Growth In the real world, the resources that organisms need to survive and reproduce are always limited. The **logistic model of population growth** addresses this limitation. With this model, the rate of population growth does not remain constant, but rather declines as population density increases. When there are few individuals relative to the amount of resources, the population grows exponentially (Figure 16.5 ❶).

As the number of individuals rises, **density-dependent limiting factors** such as competition begin to put the brakes on growth. As individuals become more and more crowded, they must compete with one another for food, hiding places, nesting sites, and other essential resources. Parasitism and disease also increase with population density, and both hinder population growth. As a result of density-dependent limiting factors, population growth begins to slow ❷.

Population growth continues until it eventually levels off at the environment's carrying capacity ❸. **Carrying capacity** is the maximum number of individuals of a species that a particular environment can sustain indefinitely. The carrying capacity for a species is not constant; rather, it depends on physical and biological factors that can change over time. For example, a prolonged drought can lower the carrying capacity for a plant species and for any animals that depend on that plant. The population size of one species can also affect the carrying capacity of another species with similar resource needs. For example, a grassland can support only so many grazers, so the presence of more than one grass-eating species lowers the carrying

Figure 16.5 One example of logistic growth: what happens when a few deer are introduced to a new habitat with finite resources.

❶ When the population is small, individuals have access to all the resources they require and the population grows exponentially.

❷ As population size grows, the growth rate begins to slow as density-dependent limiting factors begin to have an effect.

❸ Eventually, population size levels off. Population size plotted against time produces an S-shaped curve.

capacity for all of them. Human activities also affect carrying capacity. For example, human harvest of horseshoe crabs has decreased the carrying capacity for red knot sandpipers, a type of migratory bird. Horseshoe crab eggs are the sandpipers' main food during their long-distance migration.

Density-Independent Factors The logistic model of population growth describes what happens when density-dependent limiting factors influence population growth. However, other factors can affect population size, including harsh weather such as hurricanes, and natural disasters such as tsunamis or landslides. Such **density-independent limiting factors** increase the death rate in crowded and uncrowded populations alike; they do not arise as an effect of crowding.

In nature, density-dependent and density-independent factors often interact to determine the fate of a population. Consider what happened after the 1944 introduction of 29 reindeer to St. Matthew Island, an uninhabited island off the coast of Alaska. When biologist David Klein first visited the island in 1957, he found 1,350 well-fed reindeer munching on lichens (Figure 16.6). When Klein returned in 1963, he counted 6,000 reindeer. The population had soared far above the island's carrying capacity. Although a population can temporarily exceed the carrying capacity of its environment, the high density cannot be sustained. Klein observed that density-dependent negative effects were already apparent. For example, the average body size of the reindeer had decreased.

When Klein returned in 1966, only 42 reindeer survived. The single male had abnormal antlers, and was thus unlikely to breed. There were no fawns. Klein figured out that thousands of reindeer had starved to death during the winter of 1963–1964. That winter was unusually harsh, with low temperatures, high winds, and 140 inches of snow. Most reindeer, already in poor condition as a result of increased competition, starved when deep snow covered their food. A population decline had been expected—a population that exceeds its carrying capacity usually shrinks and falls below that capacity—but bad weather magnified the extent of the crash. By the 1980s, there were no reindeer on the island.

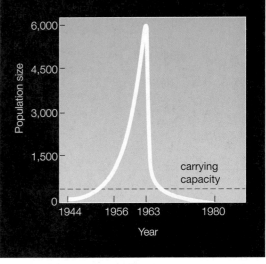

Figure 16.6 Changes in the size of a reindeer population. The reindeer were introduced to an island off the coast of Alaska in 1944.

Top, © Jacques Langevin/Corbis Sygma.

Take-Home Message 16.3

What factors affect population growth?

- The exponential model of population growth describes the growth of a population with unlimited resources. In this idealized circumstance, per capita growth rate remains constant, but the population grows faster and faster.
- The logistic model of population growth describes the growth of a population affected by density-dependent limiting factors. Such a population grows exponentially at first, then its growth rate declines as a result of competition for resources, infectious disease, and other negative effects of crowding.
- A population undergoing logistic growth levels off at the environment's carrying capacity for that species.
- Density-independent factors such as harsh weather are not addressed by models for population growth, but they too affect natural populations.

carrying capacity Maximum number of individuals of a species that a specific environment can sustain.

density-dependent limiting factor Factor whose negative effect on growth is felt most in dense populations; for example, infectious disease or competition for food.

density-independent limiting factor Factor that limits growth in populations regardless of their density; for example a natural disaster or harsh weather.

logistic model of population growth Model for growth of a population limited by density-dependent factors; numbers increase exponentially at first, then the growth rate slows and population size levels off at carrying capacity.

Table 16.1 Annual Plant* Life Table

Age Interval (days)	Survivorship (number surviving at start of interval)	Number Dying During Interval	Death Rate (number dying/number surviving)
0–63	996	328	0.329
63–124	668	373	0.558
124–184	295	105	0.356
184–215	190	14	0.074
215–264	176	4	0.023
264–278	172	5	0.029
278–292	167	8	0.048
292–306	159	5	0.031
306–320	154	7	0.045
320–334	147	42	0.286
334–348	105	83	0.790
348–362	22	22	1.000
362–	0	0	0
		996	

* *Phlox drummondii;* data from W. J. Leverich and D. A. Levin, 1979.

16.4 Life History Patterns

REMEMBER: Natural selection shapes the traits of a population (Section 12.3).

Biotic Potential The exponential growth rate for a population under ideal conditions is its **biotic potential**. This is the theoretical value that would hold if shelter, food, and other essential resources were unlimited and there were no predators or pathogens. Populations seldom reach their biotic potential because of limiting factors. Biotic potential is determined by **life history traits**, which are a set of heritable traits such as rate of development, age at first reproduction, number of breeding events, and life span. In this section, we look at how such life history traits vary and the evolutionary basis for this variation.

Describing Life Histories Each species has a characteristic life span, but only a few individuals survive to the maximum age possible. Death is more likely at some ages than others, as is reproduction. To gather information about age-specific risk of death, researchers focus on a **cohort**, a group of individuals that are all born at about the same time. Members of the cohort are tracked from birth until the last member of the cohort dies. Mortality data from a cohort study can be summarized in a life table. Table 16.1 provides an example of a life table.

A **survivorship curve** plots how many members of a cohort remain alive over time, providing information about age-specific death rates. Ecologists describe three generalized types of survivorship curves. A type I curve is convex (bulges outward), indicating survivorship is high until late in life (Figure 16.7A). This pattern is characteristic of humans and other large mammals that produce one or two young and care for them. A diagonal type II curve indicates that the death rate of the population does not vary much with age (Figure 16.7B). A type II curve is characteristic of lizards, small mammals, and large birds. In these groups, old individuals are as likely to die of disease or predation as young ones. A type III curve is concave (bulges inward), indicating that the death rate for a population is highest early in life (Figure 16.7C). Marine animals that release eggs into water have this type of curve, as do plants that release enormous numbers of tiny seeds.

Figure 16.7 Types of survivorship curves.

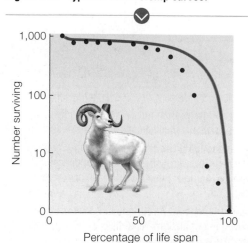

A. Type I curve. Mortality is highest very late in life. Data are for Dall sheep (*Ovis dalli*).

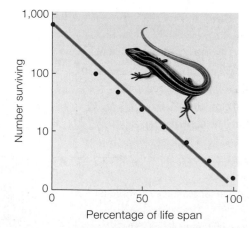

B. Type II curve. Mortality does not vary with age. Data are for a small lizard (*Eumeces fasciatus*).

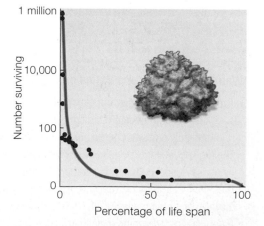

C. Type III curve. Mortality is highest early in life. Data are for a desert shrub (*Cleome droserifolia*).

Figure It Out: Which type of curve best fits the plant mortality data shown in Table 16.1?

Answer: Type III. Mortality is highest early in life.

Opportunistic life history	Equilibrial life history
shorter development	longer development
early reproduction	later reproduction
fewer breeding episodes, many young per episode	more breeding episodes, few young per episode
less parental investment per young	more parental investment per young
higher early mortality, shorter life span	low early mortality, longer life span

Figure 16.8 Opportunistic versus equilibrial life histories.
A, B. Dandelions and flies are opportunists.
C, D. Whales and coconut palms are equilibrial species.

(A) blueeyes/Shutterstock.com; (B) © Richard Baker; (C) © holbox/Shutterstock; (D) Florida Fish and Wildlife Conservation Commission/NOAA.

Evolution of Life Histories When producing offspring, an individual uses resources that it could otherwise use to grow and maintain itself. Species differ in the way in which they distribute their parental investment among offspring and over the course of a lifetime. Some invest little in each of many offspring, others invest a lot in only a few offspring. Some reproduce once, others many times. In studying this variation, ecologists have come to recognize two general life history strategies (Figure 16.8). Both strategies maximize the number of offspring that will be produced and survive, but each does so under different environmental conditions.

When a species lives where conditions vary in an unpredictable manner, its populations seldom reach their carrying capacity. As a result, there is usually little competition for resources among members of the same species. Such conditions favor an **opportunistic life history**, in which individuals produce as many offspring as possible, as quickly as possible. Opportunistic species (also called *r*-selected species) tend to have a short generation time and small body size. Because parental investment is spread across many offspring, each offspring receives a relatively small share. Opportunistic species usually have a type III survivorship curve, with mortality heaviest early in life. Annual plants such as dandelions (Figure 16.8A) have an opportunistic life history. They mature within weeks, produce many tiny seeds, then die. Flies are opportunistic animals. A female fly can lay hundreds of small eggs (Figure 16.8B) in a temporary food source such as a rotting tomato or a pile of feces.

When a species lives in a stable environment, its populations are often near their carrying capacity for that environment. Under these circumstances, competition for resources can be fierce and an **equilibrial life history**, in which parents produce a few, high-quality offspring, is adaptive. Equilibrial species (also called *K*-selected species) tend to have a large body and a long generation time. Consider a coconut palm, which grows for years before beginning to produce a few coconuts at a time (Figure 16.8C). Large mammals such as whales have this sort of life history too. They take years to reach adult body size and begin reproducing. When mature, a female whale produces only one large calf at a time (Figure 16.8C), and she continues to invest in the calf by nursing it after its birth. In both coconut palms and whales, a mature individual produces young for many years.

Some species such as century plants (a type of agave) and bamboo are large and long-lived, but reproduce only once. Atlantic eels and Pacific salmon are unusual among vertebrates in also having a one-shot reproductive strategy. Such a strategy can evolve when opportunities for reproduction are unlikely to be repeated. In the century plant and bamboo, climate conditions that favor reproduction occur only rarely. In the eels and salmon, physiological changes related to migration between fresh water and salt water make a repeat journey impossible.

biotic potential Maximum possible population growth under optimal conditions.

cohort Group of individuals born during the same interval.

equilibrial life history Life history favored in stable environments; individuals grow large, then invest a lot in each of a few offspring.

life history traits Set of traits related to growth, survival, and reproduction such as life span, age-specific mortality, age at first reproduction, and number of breeding events.

opportunistic life history Life history favored in unpredictable environments; individuals reproduce while young and invest little in each of many offspring.

survivorship curve Graph showing the decline in numbers of a cohort over time.

Guppy

A. Biologist David Reznick at his study site, a freshwater stream in Trinidad that is home to guppies and their predators.

B. Killifish. It preys on small guppies, thus selecting for individuals that grow quickly to large size before reproducing.

C. Pike cichlid. It preys on big guppies, thus selecting for individuals that reproduce early, while still young and small.

Figure 16.9 Effects of predation on guppy life history.

(A) Helen Rodd, inset David Reznick/University of California–Riverside, computer enhanced by Lisa Starr; (B,C) Hippocampus Bildarchiv.

Figure 16.10 A fisherman with a large Atlantic codfish. A human preference for big codfish selected for fish that matured while still young and small.

© Bruce Bornstein, www.captbluefin.com.

Predation and Life History Evolution Often, different populations within a species live in slightly different environments, and their life history traits reflect these differences. Consider one long-term study in which biologists David Reznick and John Endler documented the evolutionary effects of predation on the life history traits of guppies, a type of small freshwater fish. The study involved field-work in the mountains of Trinidad, an island in the southern Caribbean Sea. Here, guppies live in shallow freshwater streams (Figure 16.9A). Waterfalls in the streams function as barriers that keep guppies from moving from one part of the stream to another. As a result, guppy populations are genetically isolated. Waterfalls also restrict the movement of predatory fish, so different predators prey upon different guppy populations. Two kinds of guppy predators live in the streams. Killifish (Figure 16.9B) are relatively small and they prey on small, juvenile guppies but ignore full-grown adults. Pike cichlids (Figure 16.9C) are larger. They pursue large, mature guppies, while ignoring juveniles.

Reznick and Endler suspected that predation shapes guppy life history patterns through natural selection, and they devised an experiment to test this hypothesis. They found a pool that held guppies and adult guppy–eating pike cichlids, but no juvenile guppy–eating killifish. They left some of these guppies in place as a control group, and moved others to a pool that contained only killifish. They predicted that exposure to this previously unfamiliar predator would cause life history traits of the experimental population to evolve.

The results supported their hypothesis. Eleven years later, the guppy population at the experimental site had evolved. Compared to the control population, guppies in the experimental population grew faster, reproduced when larger, and produced bigger offspring. Selective pressure exerted by predation on the smallest fishes favored individuals that put their energy into growth rather than reproduction, until they reached the size at which they were too big to be eaten.

Evolution of life history traits in response to predation is not merely of theoretical interest. It has economic importance. Just as guppies evolved in response to predators, a population of Atlantic codfish (*Gadus morhua*) evolved in response to human fishing pressure. From the mid-1980s to early 1990s, the number of boats fishing for cod in the North Atlantic increased. As it did, the proportion of fishes that reproduced while young and small increased. Such individuals were at an advantage because both commercial fisherman and sports fishermen preferentially caught and kept larger fish (Figure 16.10). Fishing pressure continued to rise until 1992, when declining cod numbers caused the Canadian government to ban cod fishing in some areas. That ban, and later restrictions, came too late to stop the Atlantic cod population from crashing. In some areas, the population declined by 97 percent and has not recovered.

Take-Home Message 16.4

What factors shape life history patterns?

- Life history traits such as the age at which an organism first reproduces, the number of offspring produced at a time, and potential life span have a heritable basis.
- Rapid production of many offspring is adaptive when environmental factors keep population density low. Producing offspring that are fewer in number but better able to compete is adaptive if populations are often near carrying capacity.
- Predator preferences can influence the life history traits of their prey.

16.5 Human Populations

REMEMBER: Our species arose about 200,000 years ago in Africa (Section 15.7).

Population Size and Growth Rate For most of its history, the human population grew very slowly (Figure 16.11). The growth rate began to pick up about 10,000 years ago, and during the past two centuries, it soared. Three trends promoted the large increases. First, humans migrated into new habitats and expanded into new climate zones. Second, they developed technologies that increased the carrying capacity of existing habitats. Third, they sidestepped some limiting factors that typically restrain population growth.

Modern humans evolved in Africa approximately 200,000 years ago, and about 60,000 years ago they began to spread out across the globe. A large brain and the capacity to master a variety of skills gave humans an unmatched ability to live in a broad range of habitats. Humans learned how to start fires, build shelters, make clothing, manufacture tools, and cooperate in hunts. With the advent of language, knowledge of such skills did not die with the individual.

The invention of agriculture about 11,000 years ago provided a more dependable food supply than traditional hunting and gathering. A pivotal factor was the domestication of wild grasses, including species ancestral to wheat and rice. In the middle of the eighteenth century, people learned to harness energy in fossil fuels to operate machinery. This innovation opened the way to high-yielding mechanized agriculture and to improved food distribution systems. Food production was further enhanced in the early 1900s, when chemists discovered a way to convert gaseous nitrogen to ammonia. Previously, this process had been carried out primarily by nitrogen-fixing bacteria. Use of synthetic nitrogen fertilizers dramatically increased

Figure 16.11 Growth curve (red) for the world human population.
The gray box lists how long it took for our number to increase from 5 million to 7 billion.

Photo, NASA.

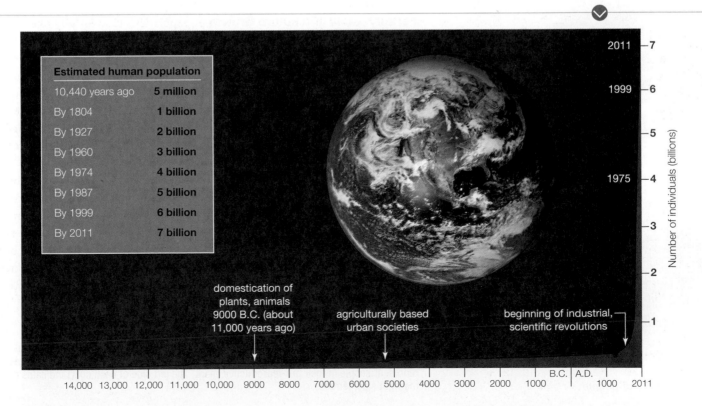

Estimated human population	
10,440 years ago	5 million
By 1804	1 billion
By 1927	2 billion
By 1960	3 billion
By 1974	4 billion
By 1987	5 billion
By 1999	6 billion
By 2011	7 billion

domestication of plants, animals 9000 B.C. (about 11,000 years ago)

agriculturally based urban societies

beginning of industrial, scientific revolutions

Number of individuals (billions)

14,000 13,000 12,000 11,000 10,000 9000 8000 7000 6000 5000 4000 3000 2000 1000 B.C. | A.D. 1000 2011

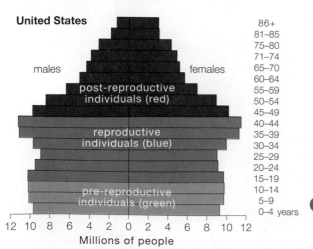

crop yields. Invention of synthetic pesticides in the mid-1900s also contributed to increased food production.

Disease has historically dampened human population growth. For example, during the mid-1300s, one-third of Europe's population was lost to a pandemic known as the Black Death. Beginning in the mid-1800s, an increased understanding of the link between microorganisms and illness led to improvements in food safety, sanitation, and medicine. People began to pasteurize foods and drinks, heating them to reduce the numbers of harmful bacteria. They also began to protect their drinking water. The first modern sewer system was constructed in London, England, in the late 1800s. By diverting wastewater downstream of the city's water source, the system lowered the incidence of waterborne diseases such as cholera and typhoid fever. Beginning in the early 1900s, new methods of sterilizing drinking water contributed to a further decline in these diseases in the most industrialized nations.

Advances in sanitation also lowered the death rate associated with medical treatment. In the mid-1800s, a German physician began urging doctors to wash their hands between patients. His advice was largely ignored until after his death, when Louis Pasteur popularized the idea that unseen organisms cause disease. Acceptance of this idea also revolutionized surgery, which had previously been carried out with little regard for cleanliness.

Vaccines and antibiotics also helped lower death rates. Vaccinations became widespread in developed countries during the 1800s. Antibiotics are a more recent development. Large-scale production of penicillin, the first antibiotic to be widely used, did not begin until the 1940s.

A worldwide decline in death rates without an equivalent drop in birth rates is responsible for the ongoing explosion in human population size. It took more than 100,000 years for the human population to reach 1 billion in number. Since then, the rate of increase has risen steadily. The population is now about 7 billion and it is expected to reach 9 billion by 2050.

Fertility Rates and Future Growth   A population's **total fertility rate** is the average number of children born to the females of a population during their reproductive years. In 1950, the worldwide total fertility rate averaged 6.5 children per woman. By 2011, it had declined to 2.5 but it remains above the replacement level, which is the number of children a woman must bear to ensure that two children grow to maturity and replace her and her partner. At present, this replacement level is 2.1 for developed countries and as high as 2.5 in some developing countries. (It is higher in developing countries because more female children die before reaching the age when they can reproduce.)

At present, China (with 1.3 billion people) and India (with 1.2 billion) hold 38 percent of the world population. The United States is the next largest, with 318 million. Compare the age structure, or distribution of individuals among age groups, for these three countries (Figure 16.12). Notice especially the size of the age groups that will be reproducing during the next fifteen years. The broader the base of an age structure diagram, the greater the proportion of young people, and the greater the expected growth. Government policies that favor couples who have only one child have narrowed China's pre-reproductive base.

Figure 16.12 Age structure diagrams for the world's three most populous countries. The width of each bar represents the number of individuals in a 5-year age group. Green bars represent people in their pre-reproductive years. The left side of each chart indicates males; the right side, females.

Even if every couple now alive decides to bear no more than two children, world population growth will not slow for many years, because more than one-third of the world population is now in the broad pre-reproductive base. About 1.9 billion people are about to enter the reproductive age bracket.

Effects of Industrial Development Industrial development affects the demographics of human populations. The most highly developed countries have the lowest fertility rate, the lowest infant mortality, and the highest life expectancy. The **demographic transition model** describes how birth and death rates change over the course of four stages of development. Living conditions are harshest during the preindustrial stage, before technological and medical advances become widespread. Birth and death rates are both high, so the growth rate is low. As industrialization begins and food production and health care improve, the death rate drops fast, but the birth rate declines more slowly. As a result, the population growth rate increases rapidly. India is in this stage. Once industrialization is in full swing, the birth rate moves closer to the death rate, and the population grows less rapidly. Mexico is currently in this stage. In the postindustrial stage, a population's growth rate becomes negative. The birth rate falls below the death rate, and population size slowly decreases. In some developed countries, the decreasing total fertility rate and increasing life expectancy have resulted in a high proportion of older adults.

Resource consumption rises with economic and industrial development. Ecological footprint analysis is one way to compare resource use. An **ecological footprint** is the amount of Earth's surface required to support a particular level of development and consumption in a sustainable fashion. Figure 16.13 shows per capita global footprint data for a few nations. Note that the average person in the United States has an ecological footprint nearly three times that of an average world citizen, and about nine times that of an average person living in India.

The average person in an industrialized nation uses far more nonrenewable resources than one in a less developed country. For example, the United States accounts for about 4.6 percent of the world's population, yet it uses about 25 percent of the world's minerals and energy supply. Billions of people living in India, China, and other less developed nations would like to own the same kinds of goods that people in developed countries enjoy. However, given current technology, Earth may not have the resources to make that possible. The World Resources Institute estimates that for everyone now alive to have an average American lifestyle would require the resources of four Earths. Finding ways to meet the wants and needs of expanding populations with limited resources will be a challenge.

> For everyone now alive to live like an average American would require the resources of four Earths.

Country	Hectares per Capita
United States	8.0
Canada	7.0
France	5.0
United Kingdom	4.9
Japan	4.7
Russian Federation	4.4
Mexico	3.0
Brazil	2.9
China	2.2
India	0.9
World Average	2.7

* Data from www.footprintnetwork.org

Figure 16.13 Ecological footprints.
A nation's ecological footprint estimates the amount of Earth's surface that a nation uses at its current level of development and consumption. It includes the area needed to secure food and manufacture products, as well as the natural area needed to take up the excess CO_2 produced by human activities.

Background photo, KonstantinChristian/Shutterstock.com.

Take-Home Message 16.5

What factors affect human population growth?

- Through expansion into new regions, invention of agriculture, and technological innovation, the human population has sidestepped environmental resistance to growth. Its size has been skyrocketing since the industrial revolution.
- Historically, death rates, and later birth rates, have fallen as nations have become more industrialized.
- Earth's resources are limited, so the current exponential growth of the human population is unsustainable.

demographic transition model Model describing the changes in human birth and death rates that occur as a region becomes industrialized.

ecological footprint Area of Earth's surface required to sustainably support a particular level of development and consumption.

total fertility rate Average number of children born to females of a population over the course of their lifetimes.

Summary

Section 16.1 A **population** is a group of individuals that live in a given area and tend to mate with one another. Human actions have influenced the growth of some Canada goose populations. The study of populations is one aspect of the field of biology known as **ecology**.

Section 16.2 Populations vary in their **demographics**, which are characteristics such as **population size**, **population density**, **population distribution**, and **age structure**. In most populations, individuals have a clumped distribution because of limited dispersal, a need for resources that are clumped, and/or the benefits of living in a social group.

Section 16.3 Birth and death rates determine how fast a population grows. The **exponential model of population growth** describes what happens with a constant, positive **per capita growth rate**. The population increases by a fixed percentage of the whole with each successive intervals, so a plot of numbers over time produces a J-shaped curve.

An essential resource that is in short supply is a **limiting factor** for growth. The **logistic model of population growth** model describes how population growth is affected by **density-dependent limiting factors**, such as disease or competition for resources. The population slowly increases in size, goes through a rapid growth phase, then stabilizes once carrying capacity is reached. **Carrying capacity** is the maximum number of individuals that can be sustained indefinitely by the resources available in the environment. Adverse weather and other **density-independent limiting factors** can affect any population regardless of its size.

Section 16.4 The maximum theoretical rate of population growth is the **biotic potential**. Limiting factors prevent populations from attaining this potential. Biotic potential is affected by aspects of an organism's **life history** such as age at maturity, number of reproductive events, offspring number per event, and life span. Life histories are often studied by following a **cohort**, a group of individuals that were born in the same time interval.

Three types of **survivorship curves** are common: a high death rate late in life, a constant rate at all ages, or a high rate early in life. Life histories have a genetic basis and are subject to natural selection. Depending on the environment, a population may be more successful if its individuals reproduce once, or many times. At low population density, an **opportunistic life history**, in which individuals make many offspring fast, is selectively advantageous. At a higher population density, an **equilibrial life**

history, in which individuals invest more in fewer, higher-quality offspring) is most favored. Predation can affect life history patterns because predators (including humans) act as selective agents that affect the traits of prey populations.

Section 16.5 The human population has now surpassed 7 billion. Expansion into new habitats and the invention of agriculture allowed early increases. Later, improved sanitation and technological innovations such as the invention of synthetic fertilizer raised the carrying capacity and minimized limiting factors that adversely affect other species.

A population's **total fertility rate** is the average number of children born to women during their reproductive years. The global total fertility rate is declining but it remains above the replacement rate. A population's age structure influences its growth. The pre-reproductive base of the human population is so large that the population is expected to grow for many years.

The **demographic transition model** describes how population growth rates have historically responded to industrialization. Developed nations have a lower growth rate, but larger **ecological footprint** than developing nations. Given current technology, Earth does not have enough resources to support the current population in the style of the most developed nations.

Self-Quiz

Answers in Appendix I

1. Most commonly, individuals of a population have a _____ distribution.
   a. clumped
   b. random
   c. nearly uniform
   d. none of the above

2. All members of a population _____ .
   a. are the same age
   b. reproduce
   c. belong to the same species
   d. all of the above

3. The exponential model of population growth assumes _____ .
   a. the death rate declines as population density increases
   b. per capita growth rate does not change
   c. industrialization causes a fall in birth rates
   d. resources are limited

4. Competition for resources and disease are _____ controls on population growth rates.
   a. density-independent
   b. density-dependent

5. For a given species, the maximum rate of population increase under ideal conditions is the _____ .
   a. biotic potential
   b. carrying capacity
   c. opportunistic strategy
   d. density control

6. An increase in the population of a prey species would most likely _____ the carrying capacity for that species' predators.
   a. increase
   b. decrease
   c. not affect
   d. stabilize

7. Members of a species with an _____ life history have many offspring and invest little in each one.
   a. equilibrial
   b. opportunistic

8. The logistic model of population growth takes into account _____ , but not _____ .
   a. density-dependent factors; density-independent factors
   b. density-independent factors; density-dependent factors

9. The human population is now about _____ .
   a. 7 billion   b. 7 million   c. 70 billion   d. 70 million

10. Compared to the less developed countries, the highly developed ones have a higher _____ .
    a. death rate
    b. birth rate
    c. total fertility rate
    d. ecological footprint

11. Carrying capacity _____ .
    a. is the same for all species that share a habitat
    b. is constant for a given species regardless of its habitat
    c. varies among both species and habitats
    d. is constant over time

12. The total fertility rate of a population is _____ .
    a. always higher than the replacement fertility rate
    b. the average number of offspring a female has in her lifetime
    c. the maximum number of children a woman could have if her resources were unlimited

13. If an exponentially growing population of 1,000 mice has a per capita growth rate (*r*) of 0.3 mice per month, how many mice will there be one month from now?
    a. 3,000
    b. 3,300
    c. 1,300
    d. 300

14. Human population growth was encouraged by _____ .
    a. invention of agriculture
    b. medical advances
    c. improved sanitation
    d. all of the above

15. Match each term with its most suitable description.
    ____ carrying capacity
    ____ logistic growth
    ____ exponential growth
    ____ demographic transition
    ____ limiting factor
    ____ cohort

    a. change in birth and death rates with industrialization
    b. group of individuals born during the same period of time
    c. population growth plots out as an S-shaped curve
    d. largest number of individuals sustainable by the resources in a given environment
    e. population growth plots out as a J-shaped curve
    f. essential resource that restricts population growth when scarce

Critical Thinking

1. Each summer, the giant saguaro cacti in deserts of the American Southwest produce tens of thousands of tiny black seeds apiece. Most die, but a few land in a sheltered spot and sprout the following spring. The saguaro is a CAM plant (Section 5.4) and it grows very slowly. After 15 years, a saguaro may be only knee high, and it will not flower until it is about age 30. The cactus may survive to age 200. Saguaros share their desert habitat with annuals such as poppies, which sprout just after the seasonal rains, produce seeds, and die in just a few weeks. How would you describe these two life history patterns? How could such different life histories both be adaptive in this environment?

2. A biologist catches 50 butterflies, marks them, then releases them. Later, the biologist returns and again catches 50 butterflies, 10 of them marked. What is the size of the population?

Visual Question

1. Consider the shape of the two age structure diagrams to the right. Which population will grow the fastest? Which includes the greater proportion of older people?

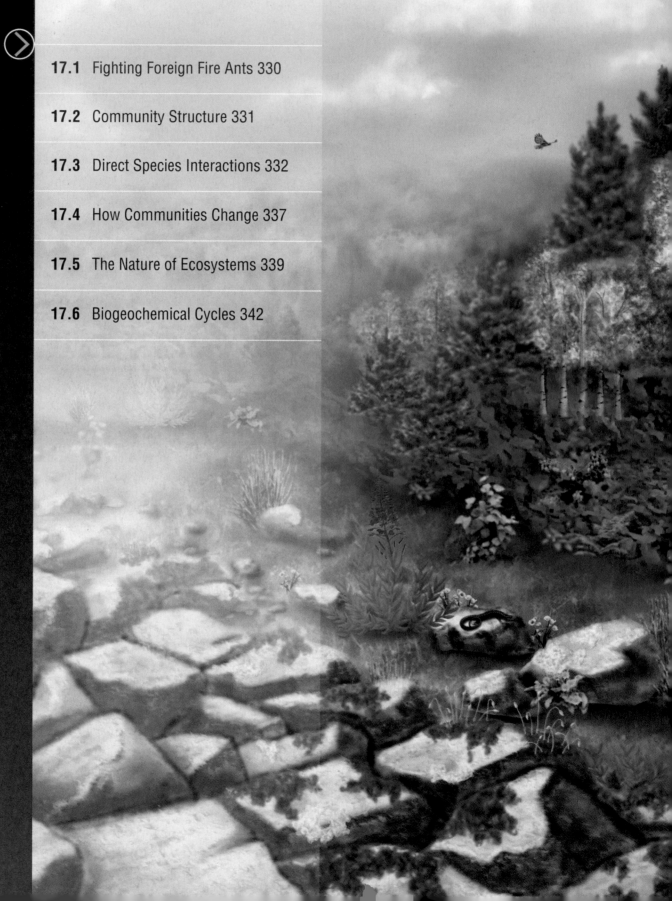

17

COMMUNITIES AND ECOSYSTEMS

Application ❯ 17.1 Fighting Foreign Fire Ants

A. One red imported fire ant worker. A colony can contain thousands of such workers, each with a stinger.

B. Red imported fire ants attack and kill eggs and hatchlings of ground-nesting birds such as quail.

C. A phorid fly attempting to lay its eggs in a red imported fire ant.

Figure 17.1 Red imported fire ant (*Solenopsis invicta*), an introduced threat to native species.

(A) Alex Wild/ Visuals Unlimited, Inc.; (B) © James Mueller; (C) Sanford Porter/USDA.

Red imported fire ants, or RIFAs (Figure 17.1A), are native to South America but, with a bit of human assistance, they have dramatically expanded their range. The ants first arrived in the southeastern United States in the 1930s, probably as stowaways on a cargo ship. They flourished in their new environment and eleven states now have well-established populations. In addition, the ants have used the United States as a springboard for dispersal to the Caribbean, Australia, New Zealand, and several Asian countries.

Red imported fire ants are considered pests because of their negative health and economic effects. The ants live in the ground and form moundlike nests. Accidentally step on one of these nests and large numbers of ants will rush out to bite and sting you. The venom injected by a RIFA's stinger causes an extremely painful burning sensation, followed by formation of a raised itchy bump at the site of the sting. As a result, a field or lawn colonized by red imported fire ants is inhospitable to humans, pets, and livestock. The ants' attraction to electricity also causes problems. For unknown reasons, large numbers of RIFAs sometimes congregate inside electrical motors, appliances, or switches, causing these devices to malfunction. Efforts to prevent the spread of red imported fire ants center on quarantines—prohibitions against moving soil that could contain the ants from affected areas to unaffected ones. Thus, arrival of RIFAs can spell disaster for commercial plant or sod growers.

Fire ants also have a negative impact on natural communities. A biological **community** includes all the species in a region. Competition with red imported fire ants typically causes a region's native ant populations to decline, and the resulting change in species composition can harm ant-eating animals. For example, the Texas horned lizard feeds mainly on native harvester ants, but it does not eat the red imported fire ants that have largely replaced its natural prey. The spread of RIFAs is also contributing to population declines in some birds, including bobwhite quails and vireos (a type of songbird). The ants decrease the abundance of insects that the birds would normally feed to their young. They also feed on birds' eggs and on nestlings. Ground-nesting birds are at special risk (Figure 17.1B). The presence of RIFAs can even affect native plants. The ants interfere with pollination by displacing or preying on native pollinators such as ground-nesting bees. The ants also interfere with the dispersal of native plants whose seeds are normally spread by native ant species that RIFAs have replaced.

Given all the difficulties RIFAs cause in the United States, you might wonder what happens in their native South America. The ants are not considered much of a problem there, in part because they are far less common. In South America, parasites and predators keep RIFA numbers far lower than those in affected parts of the United States. When some RIFAS left South America, they benefited by leaving their many natural enemies behind.

Biologists have now turned to some of those natural enemies to help them fight the RIFAs. Phorid flies are one such enemy. These flies lay their eggs in a RIFA's body (Figure 17.1C). After a fly egg hatches, it develops into a larva that grows inside the ant and feeds on the ant's soft tissues. When the fly larva is ready to undergo metamorphosis, it enters the ant's head and causes the head to fall off. The larva then develops into an adult fly inside remains of the head.

Phorid flies are not expected to kill off all *S. invicta* in affected areas. Rather, the hope is that the flies will reduce the density of the invader's colonies. When the flies are present, RIFAs try to avoid them and so spend less time foraging. The phorid flies did not evolve with our native ants, so they do not harm them.

17.2 Community Structure

Communities vary in size and often nest one inside another. For example, we find a community of microbial organisms inside the gut of a termite. That termite is part of a larger community of organisms that live on a fallen log. This community is in turn part of a still larger forest community.

Even communities that are similar in scale differ in their species diversity. There are two components to **species diversity**: The first, species richness, refers to the number of species that are present; the second is species evenness, or the relative abundance of each species. A pond that contains similar numbers of five species of fish has a higher species evenness, and thus a higher species diversity, than a pond with one abundant fish species and four rare ones.

Community structure is dynamic, which means that the array of species and their relative abundances change over time. Communities change over a long time span as they form and then age. Some change suddenly as a result of natural or human-induced disturbances.

Nonbiological Factors Factors related to geography and climate affect community structure. These factors include soil quality, sunlight intensity, rainfall, and temperature, which vary with latitude, elevation, and—for aquatic habitats—depth. Tropical regions receive the most sunlight energy and have the most even temperature. For most plants and animal groups, the number of species is greatest in the tropical regions near the equator, and declines as you move toward the poles. Tropical rain-forest communities are highly diverse (Figure 17.2), and forest communities in temperate regions are less so. Similarly, tropical reef communities are more diverse than comparable marine communities farther from the equator.

Biological Factors The evolutionary history and adaptations of the species in a community also influence community structure. Each species evolved in and is adapted to a specific **habitat**, the type of place where it typically occurs. All species of a community share the same habitat, but each has a unique ecological role that sets it apart. This role is the species' **niche**, which we describe in terms of the conditions, resources, and interactions necessary for survival and reproduction. Aspects of an animal's niche include temperatures it can tolerate, the kinds of foods it can eat, and the types of places where it can breed or hide. A description of a plant's niche would include details of its requirements for soil, water, light, and pollinators.

Species interactions also affect community structure. In some cases, the effect is indirect. For example, when songbirds eat caterpillars, the birds indirectly benefit the trees that the caterpillars feed on. Other interactions are direct; the actions of one species helps or harms another. Such direct interactions are the subject of the next section.

Figure 17.2 Tropical diversity.
Two of the 12 or so fruit-eating pigeon species that live in New Guinea's tropical rain forests (left). Top, a Victoria crowned pigeon, which is the size of a turkey. Below, the smaller superb crowned fruit pigeon.

Left, Donna Hutchins; right, top, © Martin Harvey, Gallo Images/Corbis; bottom, © Len Robinson, Frank Lane Picture Agency/Corbis.

Take-Home Message 17.2

What factors shape a community?

- Communities vary in their species diversity as a result of nonbiological factors such as differences in incoming sunlight, temperature, and soil quality.
- Biological factors such as species requirements for survival and reproduction, as well as interactions with other species, also influence community structure.

community All populations of all species in some area.

habitat The type of place in which a species lives.

niche The role of a species in its community.

species diversity The number of species and their relative abundance within a community.

Table 17.1 Interspecific Interactions

Type of Interaction	Direct Effect on Species 1	Direct Effect on Species 2
Commensalism	Benefits	None
Mutualism	Benefits	Benefits
Competition	Harmed	Harmed
Predation	Benefits	Harmed
Parasitism	Benefits	Harmed

Figure 17.3 Commensalism.
This tree provides orchids with an elevated perch from which they can capture sunlight. The presence of the orchids has no effect on the tree.

joloei/Shutterstock.

17.3 Direct Species Interactions

REMEMBER: Species interactions influence population size (Section 16.3) and can result in directional selection (12.3) and coevolution (12.7).

There are five types of direct interactions among species in a community: commensalism, mutualism, competition, predation, and parasitism (Table 17.1). Three of these—parasitism, commensalism, and mutualism—can be a **symbiosis**. Symbiosis means "living together." Symbiotic species, also known as symbionts, spend most or all of their life cycle in close association with each other. An endosymbiont is a species that lives inside its partner species.

Regardless of whether one species helps or hurts another, two species that interact closely may coevolve over generations. Recall that with coevolution, each species is a selective agent that shifts the range of variation in the other.

Commensalism and Mutualism **Commensalism** benefits one species and does not affect the other. For example, some orchids that live attached to a tree trunk or branch (Figure 17.3) benefit by having a perch in the sun, while the tree that provides this support is unaffected. As another example, commensal bacteria live in the gut of many animals. They benefit by having a warm, nutrient-rich place to live, and their presence neither helps nor harms their host.

Other gut bacteria assist their host by aiding in digestion or synthesizing vitamins. An interaction that benefits both species is a **mutualism**. Flowering plants and their pollinators are a familiar example. In some cases, coevolution of two species results in a mutual dependence. For example, there are several species of yucca plant, each pollinated by a single species of yucca moth, whose larvae develop on that plant species alone (Figure 17.4A). More often, mutualistic relationships are less exclusive. Most flowering plants have more than one pollinator, and most pollinators provide their service to more than one species of plant.

Photosynthetic organisms often supply sugars to their nonphotosynthetic partners, as when plants lure pollinators with nectar. In addition, many plants

 Figure 17.4 Mutualism.

(A) Bob and Miriam Francis/Tom Stack & Associates; (B) Sergey Uryadnikov/Shutterstock.com; (C) © Thomas W. Doeppner.

A. Yucca moth on a yucca flower. The yucca plant benefits by being pollinated. The moth benefits by laying eggs that develop within the yucca's fruit.

B. Lichen. It consists of a fungus and a green alga. The fungus supports and shelters the alga, which shares the sugar it makes with the fungus.

C. Anemonefish nestles among the tentacles of a sea anemone. In this mutually beneficial partnership, each species protects the other.

Figure 17.5 Active competition among scavengers.
After facing off over a carcass (left), an eagle attacked a fox with its talons (right). The fox then retreated, leaving the eagle to exploit the carcass.

© Pekka Komi.

make sugary fruits that attract seed-dispersing animals. Plants also provide sugars to mycorrhizal fungi and nitrogen-fixing bacteria. The plants' fungal or bacterial symbionts return the favor by supplying their host with other essential nutrients. Similarly, photosynthetic dinoflagellates provide sugars to corals, and photosynthetic bacteria or algae feed their fungal partner in a lichen (Figure 17.4B).

Mutualisms can also involve defense. A pink anemonefish will be eaten by a predator unless it has a sea anemone to hide in (Figure 17.4C). The anemone's tentacles are covered by stinging cells that do not affect the anemonefish, but help fend off predators that would eat the fish or its eggs. The anemone can survive on its own, but benefits by having a partner that chases away the few fish species that are able to feed on anemone tentacles.

From an evolutionary standpoint, mutualism is best considered as a case of reciprocal exploitation. Each participant increases its own fitness by extracting a resource, such as protection or food, from its partner. If taking part in the mutualism has a cost, selection will favor individuals who minimize that cost. Consider nectar production, which is energetically costly for a flower, but serves as a necessary payoff to pollinators. Natural selection will favor a flower that produces the minimum amount of nectar necessary to attract pollinators over one that expends additional energy to produce a more generous serving of nectar.

Interspecific Competition

Interspecific competition is competition between members of different species. Members of one species sometimes actively prevent members of another species from using a resource. For example, scavengers such as eagles and foxes sometimes fight over a carcass (Figure 17.5). Some plants also actively interfere with their competitors. For example, a sagebrush plant secretes chemicals into the soil, thus preventing potential competitors from growing nearby.

In other cases, competing species do not directly interfere with one another. Instead, all scramble for a share of a limited resource. For example, wolf spiders and carnivorous plants called sundews both capture and feed on insects (Figure 17.6). Although these organisms do not fight over food, they do compete. By catching and eating insects, each reduces the amount of food available to the other.

Figure 17.6 Scramble competition between members of different kingdoms.
Sundew plants (top) and wolf spiders (bottom) compete for food. Both feed on flies and other insects.

Top, scaners3d/Shutterstock.com; bottom, Cathy Keifer/Shutterstock.com.

commensalism Species interaction that benefits one species and has no effect on the other.

interspecific competition Two species compete for a limited resource and both are harmed by the interaction.

mutualism Species interaction that benefits both species.

symbiosis One species lives on or inside another in a commensal, mutualistic, or parasitic relationship.

Paramecium

A Time (days) **B** Time (days) **C** Time (days)

Figure 17.7 Competitive exclusion.
Two species of the ciliated protozoan *Paramecium*,
P. caudatum and *P. aurelia*, both feed on bacteria. When
grown in separate test tubes, each does well **A**, **B**.
When grown together **C**, one species drives the other to
extinction.

Left, Michael Abbey/Science Source.

Figure It Out: Which species of *Paramecium* was the
superior competitor?

Answer: *P. aurelia*

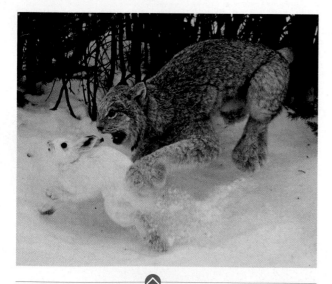

Figure 17.8 Predation.
Predators such as this lynx catch, kill, and devour their
prey, in this case a snowshoe hare.

Ed Cesar/Science Source.

Species compete most intensely when the supply of a shared resource is the
main limiting factor for both. In the early 1930s, G. F. Gause carried out a series of
experiments that led him to describe **competitive exclusion**: Whenever two species
require the same limited resource to survive or reproduce, the better competitor
will drive the less competitive species to extinction in that habitat. Gause studied
interactions between two species of ciliated protists (*Paramecium*) that eat the same
bacteria. He grew the species separately and together (Figure 17.7). In the mixed
cultures, population growth of one species always outpaced the other, which eventually died out.

Competitors whose resource needs are similar but not exactly the same can
coexist. However, competition among them suppresses their population growth. In
each species, competition with the other encourages directional selection. Those
individuals who differ most from competing species in terms of resource needs are
at a selective advantage. As a result of this selective pressure, the traits of competing species may evolve so there is less competition between them, an evolutionary
process referred to as **resource partitioning**.

Predator–Prey Interactions In **predation**, one free-living species captures, kills,
and eats another (Figure 17.8). Predators exerts selective pressure on prey, favoring
those with the best anti-predator defenses. Prey defenses in turn favor those predators best able to overcome them. As a result, predators and prey may engage in an
evolutionary arms race that continues over many generations.

You have already learned about some defensive adaptations. Many prey species
have hard or sharp parts that make them difficult to eat. Think of a snail's shell or
a porcupine's quills. Others have chemicals that taste bad or sicken predators. Most
defensive toxins in animals come from the plants they eat. For example, a monarch
butterfly caterpillar feeds on and takes up chemicals from the milkweed plant. If a
bird eats a monarch caterpillar or butterfly, plant-derived chemicals will sicken it.

Some well-defended prey have **warning coloration**, a conspicuous pattern or
color that predators learn to avoid. For example, stinging wasps and bees typically
have black and yellow stripes (Figure 17.9A). Their similar appearance is a type of
mimicry, an evolutionary pattern in which one species comes to resemble another.
In one type of mimicry, well-defended species benefit by looking alike. In another
type of mimicry, prey masquerade as a species that has a defense that they lack.
For example, some flies that cannot sting resemble stinging bees or wasps (Figure
17.9B). Such a fly benefits when predators avoid it after an encounter with the
better-defended look-alike species.

Some prey can startle an attacking predator. Section 1.5 described how eyespots
and a hissing sound protect some butterflies. A lizard's tail may detach from the
body and wiggle a bit as a distraction, while the lizard runs off. Skunks and some
beetles squirt foul-smelling, irritating repellents at potential predators.

A. The coloration of this yellow jacket wasp warns predators that it can sting.

B. This fly, which cannot sting, benefits by mimicking the color pattern of wasps.

Figure 17.9 Warning coloration and mimicry.

(A) © Kletr/Shutterstock.com; (B) Marco Uliana/Shutterstock.com.

Camouflage is a body shape, color pattern, or behavior that allows an individual to blend into its surroundings and avoid detection. Prey benefit when camouflage hides them from predators, and predators benefit when it hides them from prey (Figure 17.10).

In response to prey defenses, predators have evolved sharp teeth and claws that can pierce protective hard parts. Speedy prey select for faster predators. For example, the cheetah, the fastest land animal, can run 114 kilometers per hour (70 mph). Its preferred prey, Thomson's gazelles, run 80 kilometers per hour (50 mph).

Plants and Herbivores In **herbivory**, an animal feeds on a plant, which may or may not die as a result. Some plants can withstand loss of their parts and quickly grow replacements. For example, grasses are seldom killed by herbivores. They have a fast growth rate and store enough resources in their roots to replace the shoots lost to grazers. Other plants have traits that fend off herbivores. Physical deterrents include spines, thorns, and fibrous, difficult-to-chew leaves. Plants can also produce compounds that taste bad to herbivores or sicken them. Capsaicin, a compound that makes some peppers "hot," defends seeds against seed-eating mammals. Rodents find capsaicin-rich pepper fruits unpalatable, leaving them to be eaten by birds, which cannot taste capsaicin. A pepper benefits by deterring rodent seed eaters because rodents chew up and kill seeds, whereas birds excrete seeds alive and intact.

Parasites and Parasitoids **Parasites** withdraw nutrients from a living host. Some bacteria, protists, and fungi are parasites. Tapeworms and flukes are parasitic annelid worms. Some roundworms, insects, and crustaceans are parasites, as are all ticks (Figure 17.11A). Even a few plants are parasitic (Figure 17.11B).

Figure 17.11 Parasites.

(A) © Bill Hilton, Jr., Hilton Pond Center; (B) © The Samuel Roberts Noble Foundation, Inc.

A. Ticks sucking blood from a finch.

B. Dodder (*Cuscuta*). Roots that extend from the leafless golden stems withdraw water and nutrients from another plant.

Figure 17.10 Camouflage.
Frilly pink body parts of a flower mantis hide the mantis from its insect prey, which are attracted to the real flowers.

© Bob Jensen Photography.

camouflage Body shape, pattern, or behavior that helps a plant or animal blend into its surroundings.

competitive exclusion When two species compete for the same resource, the better competitor drives the weaker one to extinction in that habitat.

herbivory An animal feeds on a plant, which may or may not die as a result.

mimicry Two or more species come to resemble one another.

parasite A species that withdraws nutrients from another species (its host), usually without killing it.

predation One species (the predator) captures, kills, and feeds on another (its prey).

resource partitioning Evolutionary process whereby traits of competing species come to differ as a result of the selective pressure imposed by the competition.

warning coloration Distinctive color or pattern that makes a well-defended prey species easy to recognize.

A. Brood parasite. A cowbird chick with its foster parent. A female cowbird minimizes her cost of parental care by laying her eggs in the nests of other bird species.

B. Parasitoid. A commercially raised parasitoid wasp about to deposit a fertilized egg in an aphid. The wasp larva will devour the aphid from the inside.

Figure 17.12 Brood parasites and parasitoids.

(A) E. R. Degginger/Science Source; (B) © Peter J. Bryant/ Biological Photo Service.

Even when a parasite does not cause obvious symptoms, infection can weaken the host, making it more vulnerable to predation or less attractive to potential mates. Some parasitic infections cause sterility.

Adaptations to a parasitic lifestyle include traits that allow the parasite to locate hosts and to feed undetected. For example, ticks that feed on mammals or birds move toward a source of heat and carbon dioxide, which are likely signs of a potential host. A chemical in tick saliva acts as a local anesthetic, preventing the host from noticing the feeding tick. Parasites that live inside other organisms often have adaptations that help them evade a host's immune defenses.

Hosts' defenses against parasites include immune responses (a topic we consider in detail in Chapter 22) and behavioral responses. For example, many primates take turns removing ticks and other parasites from one another. Birds kill external parasites by preening their feathers, and their bill shape reflects this function, as well as its role in feeding. In pigeons, even a slight experimental modification of bill shape that has no effect on feeding can result in an increase in parasite numbers.

Brood parasites do not feed on their hosts, but rather steal parental care. A brood parasite tricks another animal into raising its young. For example, European cuckoos and North American cowbirds (Figure 17.12A) lay their eggs in the nests of other birds. When these eggs hatch, the foster parents care for the young as if they were their own. Freed from the constraints imposed by parental care, a female cowbird can lay as many as thirty eggs in a single season. Some other birds, fish, and insects are also brood parasites.

Parasitoids take the concept of forcing another to care for one's young to an even higher level. These insects lay their eggs inside other insects. The eggs hatch into larvae that devour the host from the inside, eventually killing it. The phorid flies that have been imported to attack red imported fire ants are parasitoids.

Parasites and parasitoids are often used in biological pest control. Such pest control provides an alternative to chemical pesticides, which typically kill or harm a wide variety of nontarget species. Species chosen for use as biological pest control agents target only a specific type of host. For example, the parasitoid wasp in Figure 17.12B is a biological control agent. It lays eggs only in aphids, which are sucking insects that damage many crop plants.

brood parasite An animal that tricks another species into raising its young, for example a cowbird.

parasitoid An insect that lays eggs in another insect, and whose young devour their host from the inside.

Take-Home Message 17.3

How do species interactions affect a community?

- In commensalism, one species benefits and the other is unaffected.
- In mutualism, two species exploit one another in a way that benefits both.
- Competition for resources has a negative effect on both competitors. If both depend on the same limited resource, the stronger competitor may drive the weaker one to local extinction, a process called competitive exclusion.
- Interspecific competition favors individuals of both species whose resource needs are most unlike those of the competing species. Over time, competition alters traits related to resource use and leads to resource partitioning.
- Predators benefit at the expense of their prey, and parasites benefit at the expense of their hosts. As a result, predators and parasites select for defensive traits in prey and hosts. These defenses in turn select for traits that help the predators and parasites overcome them.

17.4 How Communities Change

Ecological Succession Community structure—the kinds of species and their relative abundance in a habitat—changes constantly. In a process called **ecological succession**, the array of species gradually shifts over time as organisms alter their own habitat. One array of species is replaced by another, which is in turn replaced by another, and so on.

Primary succession occurs in habitats that lack soil and have few or no existing species. For example, a rocky area exposed by retreat of a glacier undergoes primary succession (Figure 17.13). At first, no multicellular organisms are present ❶. The community begins to change as pioneer species gain a foothold ❷. **Pioneer species** colonize new or vacated habitats. They often include lichens, mosses, and hardy annual plants with wind-dispersed seeds. As generations of pioneers live and die, they help build and improve the soil. In doing so, they set the stage for their own replacement. Seeds of shrubby species take root in the mats of pioneers ❸. Over time, organic wastes and remains build up and, by adding volume and nutrients to soil, this material allows tall trees to take hold ❹.

Secondary succession occurs after a natural or human disturbance removes the natural array of species, but not the soil. We observe this kind of succession after a fire destroys a forest or a plowed field is abandoned and wild species move in and take over.

The 1980 eruption of Mount Saint Helens, a volcano in Washington State, gave scientists an opportunity to observe succession in action (Figure 17.14). The eruption showered the area around the volcano with volcanic rock and ash, wiping out the existing plant life and covering the mature soil. Since then, plant life has colonized the area and succession is under way. Such field studies of succession have led to a change in the way the process is viewed. The concept of ecological succession was first developed in the late 1800s. At that time, it was viewed as a predictable and directional process that culminated in a "climax community," an array of species that persists over time and is reconstituted in the event of a disturbance. Physical factors such as climate, altitude, and soil type were thought to determine which species appeared in what order. Modern ecologists recognize that three types of factors affect succession: (1) physical factors such as climate, (2) chance events such as the order in which

ecological succession A gradual change in a community in which one array of species replaces another.

pioneer species Species that can colonize a new habitat.

primary succession Ecological succession occurs in an area where there was previously no soil.

secondary succession Ecological succession occurs in an area where a community previously existed and soil remains.

Figure 17.13 Artist's depiction of how primary succession in a previously glaciated area can lead to establishment of a forest community.

Figure 17.14 Ecological succession after a volcanic eruption.

(A) R. Barrick/ USGS; (B) USGS; (C) P. Frenzen, USDA Forest Service.

A. Mount Saint Helens erupted in 1980. Volcanic ash completely buried the community that had previously existed at the base of this volcano.

B. In less than a decade, numerous pioneer species had become established.

C. Twelve years after eruption, Douglas fir seedlings were taking hold in soils enriched by volcanic ash.

**Figure 17.15
A keystone species.**
The sea star *Pisaster*
lives along rocky
shores. Remove it and
the species diversity
of this community
declines.

lauraslens/Shutterstock.

**Figure 17.16 Exotic species that have become pests in
the United States.**
To learn about others, visit the National Invasive Species
Information Center online at www.invasivespeciesinfo.gov.

(A) Angelina Lax/Science Source; (B) © Greg Lasley Nature Photography, www.greglasley.net.

A. Kudzu native to Asia
is overgrowing trees
across the southeastern
United States.

B. Nutrias native to
South America are now
abundant in freshwater
marshes and riversides
of the Gulf States, the
Chesapeake Bay region,
and Oregon.

exotic species A species that has been introduced to
a new habitat and become established there.

keystone species A species that has a disproportionately large effect on community structure.

pioneer species arrive, and (3) the frequency and extent of disturbances. Because
the sequence of species arrivals and the frequency and extent of disturbances vary
in unpredictable ways, it is difficult to predict exactly how the composition of any
particular community will change in the future.

Adapted to Disturbance In communities that are repeatedly subjected to a
particular type of physical disturbance, individuals who withstand or benefit from
that disturbance have a selective advantage. For example, some plants in areas
subject to periodic fires produce seeds that germinate only after a fire has cleared
away potential competitors. Other plants have an ability to resprout quickly after
a fire. Because fire affects different species in different ways, the frequency of fire
influences competitive interactions. For example, when naturally occurring fires are
suppressed, plants adapted to periodic burning lose their competitive edge. The fire-
adapted plants can be overgrown by plants that devote all of their energy to growing
and reproducing, rather than investing in fire-related adaptations.

Species Losses or Additions A **keystone species** has a disproportionately
large effect on a community relative to its abundance. Robert Paine coined the term
to describe the results of his studies of a sea star (*Pisaster*) common along rocky
coastlines (Figure 17.15). When Paine experimentally removed the sea star from
some plots, species richness within those plots declined. The sea star feeds mainly
on mussels, and its presence prevents mussels from overgrowing other species.

Keystone species need not be predators. For example, the large rodents called
beavers can be a keystone species. A beaver cuts down trees by gnawing through
their trunk, then uses felled trees to build a dam. Construction of a beaver dam
creates a deep pool where a shallow stream would otherwise exist. By altering the
physical conditions in a section of the stream, the beaver affects the types of fish and
aquatic invertebrates that can live there.

Arrival of an **exotic species**—a species that was introduced to a new habitat
and became established there—can also alter community structure. In its new home,
an exotic species often has no coevolved parasites, pathogens, or predators to keep
it in check, so its numbers can soar. More than 4,500 exotic species have become
established in the United States. The red imported fire ants are one example. Kudzu
is another (Figure 17.16A). Native to Asia, this vine was introduced to the American
Southeast as a food for grazers and to control erosion, but it has become an invasive weed. Nutrias, large semiaquatic rodents from South America, are still another
(Figure 17.16B). Descendants of nutrias imported for their fur in the 1940s now live
wild in marshes where they threaten native plants and contribute to marsh erosion.

Take-Home Message 17.4

What causes changes in community structure?

- In ecological succession, one array of species changes the habitat in a way that allows
another array of species to take hold.
- Large and small disturbances shift community structure on an ongoing basis.
- A change in the presence or abundance of a keystone species has a great effect on
other species in a habitat.
- An exotic species that leaves behind the predators, parasites, and competitors it
evolved with can dramatically affect the structure of its adopted community.

17.5 The Nature of Ecosystems

REMEMBER: All life requires energy and nutrients (Section 1.3).

Overview of the Participants Organisms of a community interact with their environment as an **ecosystem**. In all ecosystems, there is a one-way flow of energy and a cycling of essential materials (Figure 17.17). An ecosystem's **producers** capture energy and use it to make their own food from nonbiological materials in the environment. Usually the producer's energy source is sunlight and the producers are plants and photosynthetic bacteria and protists. An ecosystem's **consumers** obtain energy and carbon by feeding on tissues, wastes, and remains of producers and one another. Herbivores, predators, and parasites are consumers that feed on living organisms. **Detritivores** such as crabs and earthworms eat tiny bits of organic matter, or detritus. Finally, wastes and remains of organisms are broken down into inorganic building blocks by bacterial, protist, and fungal **decomposers**.

Light energy captured by producers is converted to bond energy in organic molecules, which is then released by metabolic reactions that give off heat as a by-product. This is a one-way process because organisms cannot convert heat back into chemical bond energy.

Unlike energy, with its one-way flow, nutrients cycle within an ecosystem. The cycle begins when producers take up hydrogen, oxygen, and carbon from inorganic sources such as the air and water. They also take up dissolved nitrogen, phosphorus, and other necessary minerals. Nutrients move from producers into the consumers who eat them. Decomposition returns nutrients to the environment, from which producers take them up again.

Food Chains and Webs All organisms of an ecosystem take part in a hierarchy of feeding relationships referred to as **trophic levels** ("troph" means nourishment). When one organism eats another, energy (in the form of chemical bonds) and nutrients are transferred from the eaten to the eater. All organisms at the same trophic level are the same number of transfers away from the energy input into that system.

A **food chain** is one sequence of steps by which energy captured by primary producers moves to higher trophic levels. Consider one food chain in a tallgrass prairie (Figure 17.18). The main producers in this ecosystem—grasses and other plants—are at the first trophic level. Energy flows from the plants to grasshoppers, to sparrows, and finally to hawks. Grasshoppers are primary consumers and are at the second trophic level. Sparrows that eat grasshoppers are second-level consumers and at the third trophic level. Hawks that eat sparrows are third-level consumers and at the fourth trophic level.

consumer Organisms that obtains energy and nutrients from organisms or their remains.

decomposer Consumer that breaks organic remains into their inorganic building blocks.

detritivore Consumer that feeds on small bits of organic material (detritus).

ecosystem A community and its environment.

food chain Sequence of steps by which energy moves from one trophic level to the next.

producer Organism that captures light or chemical energy and makes its food from inorganic materials.

trophic level Position in a food chain.

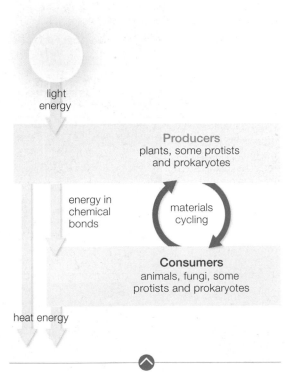

light energy

Producers
plants, some protists and prokaryotes

energy in chemical bonds

materials cycling

Consumers
animals, fungi, some protists and prokaryotes

heat energy

Figure 17.17 Generalized model for the one-way flow of energy (yellow arrows) and the cycling of materials (blue arrows) in an ecosystem.

Figure 17.18 One food chain in a tallgrass prairie.
Species at the first trophic level capture sunlight energy. Arrows represent the transfer of nutrients and energy from one trophic level to the next.

From left, © Van Vives; © D. A. Rintoul; © D. A. Rintoul; © Lloyd Spitalnik/ lloydspitalnikphotos.com.

First Trophic Level

Producer

Grass

Second Trophic Level

Primary Consumer

Grasshopper

Third Trophic Level

Second-Level Consumer

Sparrow

Fourth Trophic Level

Third-Level Consumer

Hawk

Higher Trophic Levels

A sampling of carnivores that feed on herbivores and one another

human (Inuk)

arctic wolf

arctic fox

gyrfalcon

snowy owl

ermine

Second Trophic Level

A sampling of primary consumers (herbivores) that eat plants

vole

arctic hare

lemming

mosquito   flea

Parasitic consumers feed at more than one trophic level.

First Trophic Level

Examples of primary producers (plants)

grasses, sedges

purple saxifrage

arctic willow

Detritivores and decomposers (nematodes, annelids, saprobic insects, protists, fungi, bacteria)

Figure 17.19 Arctic food web.
Arrows point from eaten to eater.

From left, top row, © Bryan & Cherry Alexander/Science Source; © Dave Mech; © Tom & Pat Leeson, Ardea London Ltd.; 2nd row, © Tom Wakefield/Bruce Coleman, Inc.; © Paul J. Fusco/Science Source; © E. R. Degginger/Science Source; 3rd row, © Hugo Wilcox/Minden Pictures; © Dave Mech; © Tom McHugh/Science Source; mosquito, Photo by James Gathany, Centers for Disease Control; flea, © Edward S. Ross; 4th row, © Jim Steinborn; © Jim Riley; © Matt Skalitzky; earthworm, © Peter Firus, flagstaffotos.com.au.

Food chains cross-connect with one another as a **food web**. Figure 17.19 shows some participants in an arctic food web. Nearly all food webs include two types of food chains. In grazing food chains, energy stored by producers flows next to herbivores, which tend to be relatively large animals. In a detrital food chain, energy in producers flows to detritivores, which tend to be smaller animals, and to decomposers. In most land ecosystems, detrital food chains predominate. For example, in an arctic ecosystem, grazers such as voles, lemmings, and hares eat some plant parts. However, far more plant matter becomes detritus that sustains soil-dwelling insects and decomposers such as bacteria and fungi. Detrital food chains and grazing food chains interconnect as the ecosystem's overall food web.

Figure 17.20 Satellite data showing net primary production for land and oceans.
Productivity is coded as red (highest) down through orange, yellow, green, blue, and purple (lowest).
NASA.

Primary Production and Inefficient Energy Transfers The flow of energy through an ecosystem begins with **primary production**: the capture and storage of energy by producers. Primary production, which is measured in terms of the amount of carbon taken up per unit area, varies seasonally and among habitats. On average, primary production is higher on land than it is in the oceans (Figure 17.20). However, because the oceans cover about 70 percent of Earth's surface, they contribute about half of Earth's total primary production.

An **energy pyramid** is a graphic representation of the proportion of the energy captured by producers that reaches higher trophic levels. Energy pyramids always have a large energy base, representing producers, and taper up. Figure 17.21 shows an energy pyramid for a freshwater ecosystem in Florida.

Only about 10 percent of the energy in tissues of organisms at one trophic level ends up in tissues of those at the next trophic level. Several factors limit the efficiency of transfers. All organisms lose energy as metabolic heat and this energy is not available to organisms at the next trophic level. Also, some energy gets stored in molecules that most consumers cannot break down. For example, most carnivores cannot access the energy tied up in bones, scales, hair, feathers, or fur. The inefficiency of energy transfers limits the possible length of food chains.

When people promote a vegetarian diet by touting the benefits of "eating lower on the food chain," they are referring to the inefficiency of energy transfers. A person eating a plant food involves only a single energy transfer. When a plant food is used to grow livestock, the animal uses some of the energy it obtains from plant food to sustain itself, loses some energy as heat, and invests some energy building inedible parts such as bones and hooves. Only a small percentage of the energy in the original plant food ends up as meat that a person can eat. Thus, feeding a population of meat-eaters requires far greater crop production than sustaining a population of vegetarians.

Oceans contribute about half of Earth's primary production.

Figure 17.21 Energy pyramid for a freshwater ecosystem.
Numbers are energy in kilocalories per square meter per year.

21	top carnivores
383	carnivores
3,369	herbivores
20,810	producers

Figure It Out: Approximately what percentage of the sunlight energy captured by producers was transferred to the herbivores that ate them?

Answer: 3,389/20,810 × 100 = 16 percent

Take-Home Message 17.5

How do organisms and their environment interact in ecosystems?

- Materials cycle between organisms and their environment, but energy flow is one-way because energy in chemical bonds is converted to heat.
- Energy and raw materials are taken up by producers, then flow to consumers.
- Food chains and food webs describe routes by which energy and nutrients move from one trophic level to another.
- Energy transfers between trophic levels are inefficient because energy is lost as metabolic heat, and some energy becomes tied up in materials that are not easily digested by consumers.

energy pyramid Diagram that illustrates the energy flow in an ecosystem.

food web System of cross-connecting food chains.

primary production The energy captured by an ecosystem's producers.

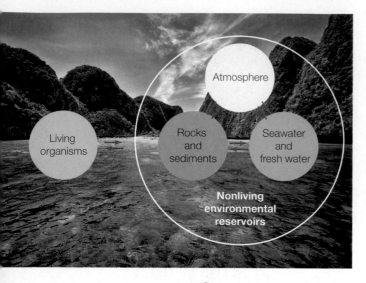

Figure 17.22 Generalized biogeochemical cycle.

Photo, Phaitoon Sutunyawatchai/Shutterstock.

Figure 17.23 The water cycle.
Water moves from the ocean into the atmosphere, onto land, and then back. The graphic below shows the volume of water in the various environmental reservoirs.

© Triff/Shutterstock.com.

Reservoir	Volume (10^3 km³)
Ocean	1,370,000
Polar ice, glaciers	29,000
Groundwater	4,000
Lakes, rivers	230
Atmosphere (water vapor)	14

17.6 Biogeochemical Cycles

REMEMBER: A triple bond holds the two nitrogen atoms in nitrogen gas together (Section 2.3). Movement of tectonic plates uplifts land (11.5). The ozone layer reduces the amount of UV radiation that reaches Earth's surface (13.3). A population explosion of single-cell producers in nutrient-enriched water is an algal bloom (13.6). Coal formed from ancient plants (14.4), and petroleum from diatoms (13.6).

In a **biogeochemical cycle**, an essential element moves from one or more environmental reservoirs, through the living components of an ecosystem, and then back to the reservoirs (Figure 17.22). Depending on the element, environmental reservoirs may include Earth's rocks and sediments, waters, and atmosphere.

Chemical and geologic processes move elements to, from, and among environmental reservoirs. Elements locked in rocks become part of the atmosphere as a result of volcanic activity. Movement of Earth's tectonic plates (Section 11.5) can uplift rocks, so an area that was once seafloor becomes part of a landmass. On land, erosion breaks down rocks, allowing the elements in them to enter rivers and flow to seas. Compared to the movement of elements within a community, movement of elements among nonbiological reservoirs is far slower. Processes such as erosion and uplifting operate over thousands or millions of years.

We focus on four biogeochemical cycles that move important elements: the water cycle, phosphorus cycle, nitrogen cycle, and carbon cycle.

The Water Cycle The **water cycle** moves water from oceans to the atmosphere, onto land and into freshwater ecosystems, and back to the oceans (Figure 17.23). Solar energy drives evaporation of water from the oceans and from freshwater reservoirs. Water that enters the lower atmosphere spends some time aloft as vapor, clouds, and ice crystals. By the process of precipitation, water falls from the atmosphere mainly as rain and snow. Oceans cover about 70 percent of Earth's surface, so most water evaporates from oceans and most precipitation falls on oceans. Most precipitation that falls on land runs into streams or seeps into the ground. Plant roots take up soil water, the water between soil particles. **Transpiration**, the evaporation of water from the aboveground parts of plants, returns most of this water to the atmosphere.

Our planet has a lot of water, but 97 percent of it is salt water. Of the 3 percent that is fresh water, most is frozen in glaciers. **Groundwater**, another freshwater reservoir, includes water in the soil, and water stored in porous rock layers called **aquifers**. About half of the population of the United States relies on aquifers for drinking water. Surface water (water in streams, rivers, lakes, and freshwater marshes) constitutes less than 1 percent of Earth's fresh water.

Movement of water results in the movement of soluble nutrients. Carbon, nitrogen, and phosphorus all have soluble forms that can be moved from place to place by flowing water. As water trickles through soil, it carries nutrient particles from topsoil into deeper soil layers. As a stream flows over limestone, water slowly dissolves the rock and carries carbonates back to the seas where the limestone formed. Flowing water can transport pollutants too; runoff from heavily fertilized lawns and agricultural fields carries dissolved phosphates and nitrates into streams and lakes.

The Phosphorus Cycle Atoms of phosphorus are highly reactive, so phosphorus does not occur naturally in its elemental form. Most of Earth's phosphorus is bonded to oxygen as phosphate (PO_4^{3-}), an ion that abounds in rocks and

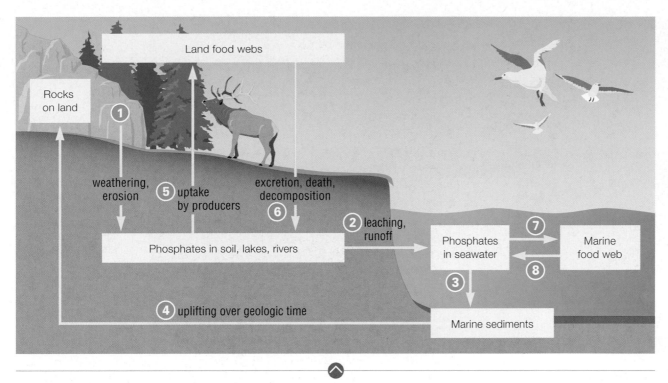

Figure 17.24 The phosphorus cycle.
In this cycle, most of the phosphorus moves in the form of phosphate ions. Earth's main phosphorus reservoir is rocks and sediments.

sediments. There is no commonly occurring gaseous form of phosphorus, so the atmosphere is not one of its reservoirs. In the **phosphorus cycle**, phosphate moves among Earth's rocks, soil, and water, and into and out of food webs (Figure 17.24). In the environmental portion of the cycle, weathering and erosion move phosphate ions from rocks into soil, lakes, and rivers ❶. Leaching and runoff deliver phosphate ions to the ocean ❷, where most of the phosphorus comes out of solution and settles as deposits along the edges of continents ❸. Over millions of years, movements of Earth's crust can uplift parts of the seafloor onto land ❹, where weather releases phosphates from the rocks. Dissolved phosphates enter streams and rivers, and the environmental portion of the phosphorus cycle starts over again.

All organisms require phosphorus to build ATP, nucleic acids, and phospholipids. The biological portion of the phosphorus cycle begins when producers take up phosphate. Roots of land plants take up dissolved phosphate from the soil water ❺. Land animals get phosphate by eating plants or one another. Phosphorus returns to the soil in wastes and remains ❻. In the seas, phosphorus enters food webs when producers take up dissolved phosphate from seawater ❼. As on land, wastes and remains replenish the supply ❽.

Lack of phosphates often limits plant growth, so most fertilizers include phosphate. Phosphorus-rich droppings from seabird or bat colonies can be harvested as a natural fertilizer, but most commercial fertilizer contains phosphorus derived from rock that has been mined and then chemically treated.

Access to phosphorus limits growth of algae and cyanobacteria too, so adding phosphorus to an aquatic habitat allows a population explosion of these organisms. The result is an algal bloom that clouds water and threatens other aquatic species. Humans encourage algal blooms by allowing phosphate-containing detergents, sewage, fertilizer runoff, and waste from livestock to pollute aquatic environments.

aquifer Porous rock layer that holds some groundwater.

biogeochemical cycle A nutrient moves among environmental reservoirs and into and out of food webs.

groundwater Water between soil particles and in aquifers.

phosphorus cycle Movement of phosphorus among rocks, water, soil, and living organisms.

transpiration Evaporation of water from a plant's aboveground parts.

water cycle Water moves from its main reservoir—the ocean—into the air, falls as rain and snow, and flows back to the ocean.

Figure 17.25 The nitrogen cycle. The main reservoir for nitrogen is the atmosphere. Activity of nitrogen-fixing bacteria converts gaseous nitrogen to forms that producers can use.

Figure It Out: What are the two forms of nitrogen that can be taken up by plants?
Answer: Ammonium and nitrate

carbon cycle Movement of carbon among rocks, water, the atmosphere, and living organisms.

nitrogen cycle Movement of nitrogen among the atmosphere, soil, and water, and into and out of food webs.

nitrogen fixation Conversion of nitrogen gas to ammonia.

The Nitrogen Cycle Earth's atmosphere, which is about 80 percent gaseous nitrogen (N_2), is the largest nitrogen reservoir. In the **nitrogen cycle**, nitrogen moves among the atmosphere, through reservoirs in soil and water, and into and out of food webs (Figure 17.25).

Plants cannot use gaseous nitrogen because they do not have an enzyme that can break the triple covalent bond between its two nitrogen atoms. Some bacteria do have such an enzyme, and these organisms carry out **nitrogen fixation**. They break the bonds in N_2, then use the nitrogen atoms to form ammonia, which dissolves and forms ammonium ions (NH_4^+) ❶. Plant roots take up ammonium from the soil ❷ and use it in metabolic reactions. Consumers get nitrogen by eating plants or one another.

Additional ammonium is added to the soil when bacterial and fungal decomposers break down the nitrogen-rich wastes and remains of organisms ❸. Other soil bacteria obtain energy by converting ammonium to nitrate (NO_3^-), a process called nitrification ❹. Like ammonium, nitrate can be taken up from the soil and used by producers ❺. Ecosystems lose nitrogen as some types of bacteria convert nitrate to gaseous forms that escape into the atmosphere ❻.

In the early 1900s, German scientists discovered a method of fixing atmospheric nitrogen and producing ammonium on an industrial scale. This process allowed the manufacture of synthetic nitrogen fertilizers that have boosted crop yields and help to feed the rapidly increasing human population. However, use of fertilizers, along with other human activities, has added large amounts of nitrogen-containing compounds to our water and air. Nitrates commonly pollute drinking water in agricultural areas, raising the risk of thyroid cancer. Nitrous oxide, a gas released by bacteria in overfertilized soil and by the burning of fossil fuel, is a greenhouse gas and contributes to destruction of the ozone layer (the layer that filters out dangerous ultraviolet radiation from the sun).

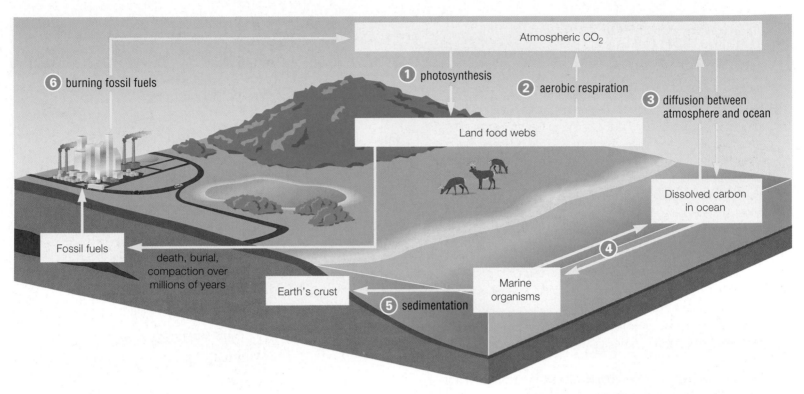

Figure 17.26 The carbon cycle.
Earth's crust is the largest carbon reservoir.

1 Carbon enters land food webs when plants take up carbon dioxide from the air and carry out photosynthesis.

2 Carbon returns to the atmosphere as carbon dioxide when plants and other land organisms carry out aerobic respiration.

3 Carbon diffuses between the atmosphere and the ocean. Carbon dioxide becomes bicarbonate when it dissolves in ocean water.

4 Marine producers take up bicarbonate for use in photosynthesis, and marine organisms release carbon dioxide produced by aerobic respiration.

5 Many marine organisms incorporate carbon into their shells. After they die, these shells become part of the sediments. Over time, these sediments become carbon-rich rocks such as limestone and chalk in Earth's crust.

6 Burning of fossil fuels derived from the ancient remains of plants adds additional carbon dioxide into the atmosphere.

The Carbon Cycle Carbon occurs abundantly in the atmosphere, combined with oxygen as carbon dioxide (CO_2). After water, carbon is the most abundant substance in living organisms. All molecules of life (carbohydrates, fats, lipids, and proteins) have a carbon backbone. In the **carbon cycle**, carbon moves among rocks, water, and the atmosphere, and into and out of food webs (Figure 17.26).

On land, plants take up and use carbon dioxide in photosynthesis **1**. Plants and most other land organisms release carbon dioxide back to the atmosphere when they carry out aerobic respiration **2**. Bicarbonate ions (HCO_3^-) form when carbon dioxide dissolves in water **3**. Aquatic producers take up bicarbonate and convert it to carbon dioxide for use in photosynthesis **4**. As on land, most aquatic organisms carry out aerobic respiration and release carbon dioxide.

Soil contains more than twice as much as carbon as the atmosphere. Soil carbon consists of organic wastes and remains along with living soil organisms. Over time, bacteria and fungi in the soil decompose organic material and release carbon dioxide into the atmosphere. The rate of decomposition and the carbon content of the soil varies with the regional climate. In the tropics, decomposition happens fast, so most carbon is stored in living plants, rather than in soil. By contrast, in temperate zone forests and grasslands, soil holds more carbon than the living plants. Soils hold the most carbon in the arctic, where low temperature hampers decomposition, and in peatbogs, where acidic, anaerobic conditions do the same.

Sedimentary rocks such as limestone constitute Earth's largest carbon reservoir. These rocks formed over millions of years by the compaction of the carbon-rich shells of marine organisms **5**. Plants do not take up dissolved carbon from the soil, so carbon in these rocks is not readily accessible to organisms in land ecosystems.

Deposits of fossil fuels formed over hundreds of millions of years from carbon-rich remains **6**. High pressure and temperature transformed the remains of ancient land plants to coal. A similar process transformed the remains of marine

Digging Into Data

Changes in Atmospheric Carbon Dioxide

To assess the impact of human activity on the carbon dioxide level in Earth's atmosphere, it helps to take a long view. One useful data set comes from deep core samples of Antarctic ice. The oldest ice core that has been fully analyzed dates back a bit more than 400,000 years. Air bubbles trapped in the ice provide information about the gas content in Earth's atmosphere at the time the ice formed. Combining ice core data with more recent direct measurements of atmospheric carbon dioxide—as in Figure 17.27—can help scientists put current changes in the atmospheric carbon dioxide into historical perspective.

1. What was the highest CO_2 level between 400,000 B.C. and 0 A.D.?
2. During this period, how many times did the CO_2 level exceed the CO_2 level observed in 1980?
3. The industrial revolution began around 1800. How did the CO_2 level change in the 800 years prior to this event? What about in the 175 years after it?
4. Which was greater, the rise in CO_2 level between 1800 and 1975 or the rise between 1980 and 2013?

Figure 17.27   Changes in atmospheric carbon dioxide (in parts per million). Data from 1980 on are direct measurements. Earlier data are based on ice cores.

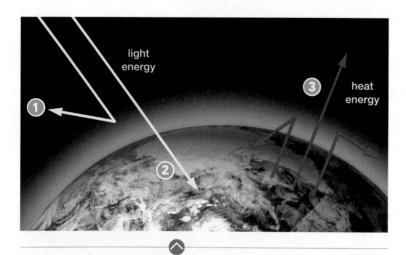

Figure 17.28   The greenhouse effect.

1 Light energy from the sun is reflected by Earth's atmosphere or surface.

2 More light energy reaches Earth's surface, and warms the surface.

3 Earth's warmed surface emits heat energy. Some of this energy escapes through the atmosphere into space. But some is absorbed and then emitted in all directions by greenhouse gases. The emitted heat warms Earth's surface and lower atmosphere.

diatoms to deposits of oil and natural gas. Until recently, the carbon in fossil fuels, like the carbon in rocks, had little impact on ecosystems. At the present time, our burning of this fuel adds billions of tons of CO_2 to the atmosphere every year.

The Greenhouse Effect and Global Climate Change   By burning fossil fuels and wood, humans have increased the amount of carbon dioxide and nitrogen oxides that enter the atmosphere. At the same time, we are cutting forests, thus decreasing the global uptake of carbon dioxide by plants. The result of these activities is a well-documented change in the composition of the atmosphere. For example, analysis of air samples taken at Mauna Loa Observatory in Hawaii show that between 1960 and the present the concentration of carbon dioxide rose from about 315 parts per million to nearly 400 parts per million. Analysis of air trapped in arctic ice and the composition of the shells of ancient marine organisms provide indirect evidence that the carbon dioxide content of Earth's atmosphere has not been this high for at least 15 million years.

These atmospheric changes can affect Earth's climate because carbon dioxide and nitrogen oxides are among the greenhouse gases. **Greenhouse gases** are atmospheric gases that slow the movement of heat from Earth to space (Figure 17.28). The process by which heat emitted by Earth's atmosphere warms its surface is called the **greenhouse effect**.

Figure 17.29 Global warming.
Variations in the annual average global temperature. The y-axis (vertical axis) is the deviation in degrees centigrade from the average temperature between 1901 and 2000. Blue bars indicate years that were cooler than this average, and red bars years that were warmer than this average.

Data from http://www.climate.gov/news-features/understanding-climate/climate-change-global-temperature.

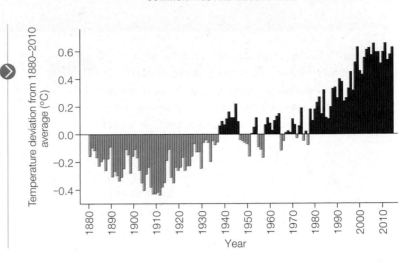

Given the greenhouse effect, we would predict that increases in atmospheric carbon dioxide and other greenhouse gases would raise the temperature of Earth's surface. Evidence from a variety of sources indicates that such global warming is under way (Figure 17.29). In the past 100 years, Earth's average temperature has risen by about 0.74°C (1.3°F), and the rate of warming is accelerating.

A rise of a degree or two in average temperature may not seem like a big deal, but it is enough to increase the rate of glacial melting, raise sea level, alter wind patterns, shift the distribution of rainfall and snowfall, and increase the frequency and severity of hurricanes. Scientists refer to the many climate-related effects of the rise in greenhouse gases as **global climate change**.

Earth's climate has varied greatly over its long history. During ice ages, much of the planet was covered by glaciers. Other periods were warmer than the present, and tropical plants and coral reefs thrived at what are now cool latitudes. Scientists can correlate some historical large-scale temperature changes with shifts in Earth's orbit, which varies in a regular fashion over 100,000 years, and Earth's tilt, which varies over 40,000 years. Changes in solar output and volcanic eruptions also influence Earth's temperature. However, most scientists see no evidence that any of these factors have a role in the current temperature rise.

In 2013, the Intergovernmental Panel on Climate Change reviewed the results of many scientific studies related to climate change. The panel included hundreds of scientists from all over the world. After reviewing all the data, the panel concluded that human activities, rather than any natural process, are the cause of the ongoing climate changes.

Human activities are the cause of ongoing climate change.

Take-Home Message 17.6

How do nutrients cycle between organisms and their environment?

- Water moves on a global scale from the ocean (its main reservoir), through the atmosphere, onto land, then back to the ocean.
- Phosphorus cycles between its main reservoir—rocks and sediments—and soils and water. Phosphorus enters food webs when producers take up dissolved phosphates. The atmosphere does not play a significant role in this cycle.
- Nitrogen moves from its main reservoir—the atmosphere—into soils and water, and into and out of food webs. Bacteria play a pivotal role in the nitrogen cycle by producing forms of nitrogen that producers can take up and use.
- Carbon's main reservoir is rocks, but most carbon enters food webs when producers take up dissolved carbon from water or carbon dioxide from the air.
- Human activities cause nutrient imbalances in land and aquatic habitats. Our activities also add excess greenhouse gases such as carbon dioxide and nitrogen oxides to the air. The rise in greenhouse gases is affecting global climate.

global climate change Wide-ranging changes in rainfall patterns, average temperature, and other climate factors that result from rising concentrations of greenhouse gases.

greenhouse effect Warming of Earth's lower atmosphere and surface as a result of heat trapped by greenhouse gases.

greenhouse gas Atmospheric gas that helps keep heat from escaping into space and thus warms the Earth.

Summary

Section 17.1 All species in a defined area are a **community**. Interactions among members of a community help keep populations in check. A species introduced to a new community without any natural enemies can increase in number and become a pest.

Section 17.2 Each species occupies a certain **habitat** and has a unique **niche**—the conditions and resources it requires, and the interactions it takes part in. The **species diversity** of a community is determined by nonbiological factors such as climate, as well as biological ones such as how species interact.

Section 17.3 Species interactions can result in coevolution. An association in which species live together is a **symbiosis**. In a **commensalism**, one species benefits and it neither helps nor harms the other. In a **mutualism**, two species exploit one another to their mutual benefit.

Interspecific competition harms both participants. **Competitive exclusion** occurs when species with identical resource needs share a habitat. **Resource partitioning** allows similar species to coexist.

Predation occurs when a free-living predator kills and eats its prey. In one type of **mimicry**, well-defended prey species have similar **warning coloration**. Less well-defended species also mimic well-defended ones. **Camouflage** can hide both predators and prey.

Herbivory may or may not kill a plant. **Parasites** withdraw nutrients from a host, usually without killing it. **Brood parasites** lay eggs in another's nest. **Parasitoids** are insects whose larvae develop inside and feed on a host, which they eventually kill.

Section 17.4 Ecological succession is the sequential replacement of arrays of species within a community. **Primary succession** occurs in habitats without soil. **Secondary succession** takes place in disturbed habitats. The first species in a community are **pioneer species**. Their presence may help other potential colonists. Community structure is not easily predicted. It is affected by physical factors, but also by random events and disturbances such as fires.

The presence of a **keystone species** has a large effect on community structure. Arrival of an **exotic species** can drastically alter a community.

Section 17.5 Producers in most **ecosystems** convert sunlight energy to chemical bond energy and take up nutrients that they and

the ecosystem's **consumers** require. A **food chain** is a path by which energy flows from one **trophic level** to another. Food chains intersect as **food webs**. In a typical land ecosystem, most energy in producers flows directly to **detritivores** and **decomposers**. Energy transfers are inefficient (energy is lost as heat and tied up in inedible parts), so most ecosystems support no more than a few trophic levels.

The rate of **primary production**—the capture and storage of energy by producers—varies with climate, season, and other factors. **Energy pyramids** show how available energy decreases as it is transferred from one trophic level to the next.

Section 17.6 In a **biogeochemical cycle**, water or a nutrient moves through the environment, then through organisms, then back to an environmental reservoir.

In the **water cycle**, water moves from the ocean into the atmosphere, falls on land, and flows back to the ocean, its main reservoir. **Transpiration** from plants releases water into the atmosphere. **Aquifers** and soil store **groundwater**, but most of Earth's fresh water is in the form of glacial ice.

In the **phosphorus cycle**, living things take up dissolved forms of phosphorus released by Earth's rocks and sediments. No gaseous form of phosphorus plays a role in this cycle.

The atmosphere is the main reservoir in the **nitrogen cycle**. Bacteria and the fertilizer industry carry out **nitrogen fixation** (convert atmospheric nitrogen to ammonium that plants can take up). Bacteria and fungi that act as decomposers also release ammonium. Nitrogen fertilizers commonly pollute water in agricultural regions.

The global **carbon cycle** moves carbon from its reservoirs in rocks and seawater, through its gaseous form (CO_2) in the atmosphere, and through living organisms.

Carbon dioxide and nitrogen oxides are **greenhouse gases**. Through the **greenhouse effect**, they trap heat in Earth's atmosphere and make life possible. An increase in greenhouse gases is causing **global climate change**.

Self-Quiz

Answers in Appendix I

1. A community consists of species _____ .
   a. that share a habitat
   b. that do not compete
   c. that share a gene pool
   d. with the same niche

2. Which of the following can be a symbiosis?
   a. interspecific competition
   b. predation
   c. commensalism
   d. all of the above

3. Match the species interaction with a suitable description.

_____ mutualism
_____ competition
_____ predation
_____ parasitism
_____ herbivory

a. A snake kills and eats a mouse.
b. A bee pollinates a flower while sipping floral nectar.
c. An owl and a wood duck both need a tree cavity to nest.
d. A mosquito sucks your blood.
e. A goat grazes on grass.

4. With interspecific competition, selection favors individuals of both species who are most _____ the competing species.
   a. similar to
   b. different from

5. Parasitoids are _____ that lay eggs in their hosts.
   a. birds   b. reptiles   c. insects   d. fish

6. The establishment of a biological community on a newly formed volcanic island is an example of _____ .
   a. primary succession
   b. secondary succession
   c. competitive exclusion
   d. resource partitioning

7. Match the terms with suitable descriptions.

_____ producer
_____ brood parasite
_____ decomposer
_____ detritivore
_____ exotic species

a. steals parental care
b. feeds on small bits of organic matter
c. degrades organic wastes and remains to inorganic forms
d. captures sunlight energy
e. new to a community

8. Match each substance with its largest environmental reservoir. One reservoir choice will be used more than once.

_____ carbon
_____ water
_____ phosphorus
_____ nitrogen

a. seawater
b. rocks and sediments
c. the atmosphere

9. Earth's largest reservoir of fresh water is _____ .
   a. lakes
   b. soil water
   c. glacial ice
   d. water in the bodies of living organisms

10. _____ convert nitrogen gas to a form producers can take up.
    a. Fungi
    b. Bacteria
    c. Mammals
    d. Mosses

11. Land plants take up _____ for photosynthesis from the air.
    a. carbon dioxide
    b. phosphate ions
    c. ammonium ions
    d. nitrogen gas

12. Addition of _____ to water encourages algal blooms.
    a. carbon dioxide
    b. phosphate ions
    c. salt
    d. bicarbonate ions

13. A biological control agent is _____ a pest species.
    a. the prey of
    b. a descendant of
    c. mutualistic with
    d. a natural enemy of

14. Greenhouse gases _____ .
    a. trap heat in the atmosphere
    b. are released by burning of fossil fuels
    c. may cause global climate change if they accumulate
    d. all of the above

15. A(n) _____ species is one that arrives early in succession.
    a. keystone   b. pioneer   c. commensal   d. exotic

Critical Thinking

1. With antibiotic resistance increasing (Section 12.1), researchers are looking for ways to reduce use of these drugs. Some cattle once fed antibiotic-laced food now receive food containing harmless bacteria that can live in the animal's gut. The idea is that if these bacteria are in place, then harmful bacteria with the same resource needs are less likely to thrive. Explain why this idea makes sense in terms of species interactions.

2. Figure 17.9 shows a stingless fly that mimics a stinging wasp. Researchers have found that in such mimicry systems, mimics benefit most when they are rare relative to the well-protected model. Can you explain why?

3. Ectotherms such as invertebrates, fish, and amphibians convert more of the energy in the food they eat into body tissues than do endotherms such as birds and mammals. What allows ectotherms to make more efficient use of food energy?

Visual Question

1. Consider this hypothetical energy pyramid for a forest. Which tier shows how much of the energy captured by the forest's trees ends up in herbivores?

18

THE BIOSPHERE AND HUMAN EFFECTS

Application ❯ 18.1 Going With the Flow

REMEMBER: Earthquakes result from the movement of tectonic plates (Section 11.5). Radioisotopes break down in a predictable manner (2.2).

A. March 16, 2011 distribution of atmospheric ^{137}Cs from the Fukushima atomic power plant. Red denotes the highest concentration of radiation. Release of radioactive material into the air began on March 12, 2011.

B. A biologist takes samples from kelp growing along the coast of California in July of 2014. Analysis of the samples showed that deep sea currents had not yet delivered ^{137}Cs from Fukushima to California's coastal waters.

Figure 18.1 Flow of radioactive pollutants from Japan.

(A) NOAA; (B) Photo by David Nelson.

Earth's air and seawater are in constant circulation and their movements facilitate the global distribution of energy and nutrients. Movements of air and seawater can also carry pollutants released in one region far from their source. Consider what happened after a powerful earthquake that occurred off the coast of Japan triggered a huge tsunami (tidal wave). Together, the earthquake and tsunami killed more than 20,000 people and decimated some of Japan's coastal cities. In the coastal town of Fukushima, damage to a nuclear power plant caused the release of radioactive material into the air and sea.

Prevailing winds carried radioisotopes accidentally released into the air at Fukushima eastward (Figure 18.1A). Rain deposited the vast majority of this radioactive material in the Pacific Ocean, but some remained aloft and continued farther east. Radioisotopes from Fukushima were first detected along the west coast of North America about 60 hours after their release, and in Europe about a week later. Within 18 days, some radioisotopes released at Fukushima had circled the globe.

Materials move more slowly in the ocean. Think about the millions of tons of debris that was dragged into the ocean by the tsunami. Most of this material sank, but some stayed afloat and was carried along by surface currents that flow east from Japan toward the Americas. Scientists have been monitoring the movement of the floating material and recording when objects that are clearly tsunami debris turn up along the west coast of North America. Several Japanese boats lost during the tsunami came ashore on the west coast of the United States in 2013. One arrived in Washington State carrying five live fishes and a variety of invertebrates native to the tropical Pacific.

Deep sea currents move material more slowly than winds or surface currents. Such currents are carrying radioisotope-contaminated water that was released into the sea at Fukushima eastward. The water contains cesium 137 (^{137}Cs), a radioisotope with a half-life of 30 years. Scientists expect deep marine currents to deliver water enriched with ^{137}Cs to the west coast of North America from about 2015 to perhaps as late as 2020. Although this water will certainly contain more ^{137}Cs than normal seawater, scientists think it is unlikely to pose a threat to human health.

Scientists are closely monitoring the movement of radioisotopes from Fukushima and their effects on natural communities. For example, one group of scientists is documenting the level of radiation in kelp forests along the western coasts of North America. They collect kelp samples several times a year (Figure 18.1B) and test the samples for radioisotopes released from Fukushima. Kelp was chosen because of its important role in the coastal ecosystem and because it tends to take up and concentrate radioactive material, making it easier to detect.

Bits of kelp tested shortly after the accident at Fukushima showed the presence of iodine 131, a radioisotope produced in nuclear reactors, but rare in nature. Researchers assume the ^{131}I traveled on the winds before falling into the sea and being taken up by the kelp. ^{131}I has a half-life of 8 days, so it is no longer a concern and there is no evidence that exposure to it did any long-term harm to the kelp.

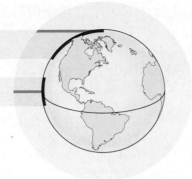

Figure 18.2 Variation in intensity of solar radiation with latitude.
For simplicity, we depict two equal parcels of incoming radiation on an equinox, a day when incoming rays are perpendicular to Earth's axis.

Rays that fall on high latitudes **A** pass through more atmosphere (blue) than those that fall near the equator **B**. Compare the length of the green lines. Atmosphere is not to scale.

Also, energy in the rays that fall at the high latitude is spread over a greater area than energy that falls on the equator. Compare the length of the red lines.

18.2 Factors That Affect Climate

REMEMBER: The biosphere includes all regions of Earth where life can exist (Section 1.2).

What lives where in the biosphere depends in large part on differences in regional climate. **Climate** refers to average weather conditions, such as temperature, humidity, wind speed, cloud cover, and rainfall, over a long interval. Air circulation patterns and ocean currents influence a region's climate, which in turn affects the types of species that can live there.

Air Circulation Patterns A region's latitude (how far north or south it is) determines how much light energy it receives. On any given day, equatorial regions receive more sunlight than higher latitudes for two reasons (Figure 18.2). First, sunlight traveling to high latitudes passes through more atmosphere to reach Earth's surface than sunlight traveling to the equator. Fine particles of dust, water vapor, and greenhouse gases absorb some solar radiation or reflect it back into space, so less light energy reaches the poles. Second, the energy in any incoming parcel of sunlight is spread out over a smaller surface area at the equator than at the higher latitudes. As a result of these factors, Earth's surface warms more at the equator than at the poles.

Latitudinal differences in surface warming, and the resulting effects on air and water, give rise to global patterns of air circulation and rainfall (Figure 18.3). At the equator, intense sunlight warms the air and causes evaporation of water from the ocean. As the air heats up, it expands and rises ❶. The same effect causes a hot-air balloon to inflate and rise when the air inside it is heated. As the equatorial air mass rises, it flows north and south and begins to cool. Cool air can hold less moisture than warm air, so moisture leaves the air as rain that supports tropical rain forests.

When the air reaches about 30° north and south latitude, it has cooled and dried out. Being cool, the air sinks downward ❷. When it descends, it draws moisture from the soil. As a result, deserts form at around 30° north and south latitude.

Air that continues flowing along Earth's surface toward the poles once again picks up heat and moisture. By a latitude of about 60° it has become warm and moist, and it rises, giving up moisture as rain ❸. In polar regions, cold air with little moisture descends ❹. Precipitation is sparse, and polar deserts form.

Landforms also affect rainfall. For example, moisture-laden air masses give up water as rain as they rise over coastal mountains. The dry region on the leeward side of such mountains is referred to as a **rain shadow**.

❹ At the poles, cold, dry air sinks and moves toward lower latitudes.

❸ Air rises again at 60° north and south, where air flowing poleward meets air coming from the poles.

❷ At around 30° north and south latitude, the air—now cooler and dry—sinks.

❶ Warm, moist air rises at the equator. As the air flows north and south, it cools and loses moisture as rain.

Figure 18.3 Air circulation patterns that result from latitudinal differences in the amount of solar radiation reaching Earth.

climate Average weather conditions in a region over a long time period.

rain shadow Dry region on the downwind side of a coastal mountain range.

Figure 18.4 Global ocean circulation.
Warm surface currents start moving from the equator toward the poles, but prevailing winds, Earth's rotation, gravity, the shape of ocean basins, and landforms influence the direction of flow. Water temperatures, which differ with latitude and depth, contribute to the regional differences in air temperature and rainfall.

NASA.

Ocean Circulation Latitudinal and seasonal variations in sunlight cause water to heat and cool. At the equator, where vast volumes of water warm and expand, the sea level is about 8 centimeters (3 inches) higher than at either pole. Existence of this slope causes sea surface water to move in response to gravity, from the equator toward the poles. As the water moves, it warms the air above it. At midlatitudes, oceans transfer 10 million billion calories of heat energy to the air every second!

Enormous volumes of water flow as ocean currents. The force of major winds, Earth's rotation, and topography determine the directional movement of these currents. Surface currents circulate clockwise in the Northern Hemisphere and counterclockwise in the Southern Hemisphere (Figure 18.4).

Swift, deep, and narrow currents of nutrient-poor water flow away from the equator along the east coast of continents. Along the east coast of North America, warm water flows north, as the Gulf Stream. Slower, shallower, broader currents of cold water parallel the west coast of continents and flow toward the equator.

biome Any of Earth's major land ecosystems, characterized by climate and main vegetation and found in several regions.

boreal forest Biome dominated by conifers that can withstand the cold winters at high northern latitudes.

temperate deciduous forest Biome dominated by broadleaf trees that grow in warm summers, then drop their leaves and go dormant during cold winters.

tropical rain forest Multilayered forest that occurs where warm temperatures and continual rains allow plant growth year-round. Most productive and species-rich biome.

Take-Home Message 18.2

What causes winds and ocean currents that affect climate?

- Longitudinal differences in the amount of solar radiation reaching Earth produce global air circulation patterns.
- Warmed water at the equator expands and flows "downhill" toward the poles. Wind, Earth's rotation, and topography affect the movement, establishing ocean currents that redistribute heat.
- The collective effects of air movements, ocean currents, and landforms determine regional temperature and moisture levels.

18.3 The Major Biomes

REMEMBER: Different mechanisms of fixing carbon adapt plants to different environmental conditions (Section 5.4). Deciduous plants lose their leaves seasonally (14.6).

Scientists classify the world's land ecosystems into a variety of **biomes**, each characterized by its climate and main type of vegetation. Figure 18.5 shows the distribution of the major biomes. A biome typically includes multiple nonadjacent regions.

Forest Biomes Evergreen broadleaf (angiosperm) trees dominate **tropical rain forests** of equatorial Asia, Africa, and South America (Figure 18.6A). Abundant rain, warm temperatures, and consistent day length allow plants to grow year-round. As a result of continual growth, rain forests take up more carbon dioxide and release more oxygen per unit area than any other biome. Thus, they are sometimes described as Earth's "lungs." Tropical rain forest is the most structurally complex biome, with many species of trees reaching 30 meters (100 feet) in height. Vines twine up the trees, and orchids and ferns grow on tree trunks and branches. Little light reaches the forest floor, so few plants grow there. Constant warmth encourages rapid decomposition and prevents leaf litter from accumulating. Thus, soil is relatively nutrient-poor. Per unit area, tropical rain forest holds more species of plants and animals than any other biome. It is the oldest biome, and its age is key to its diversity. Some modern rain forests have existed for more than 50 million years, which has provided ample time for many evolutionary branchings to occur.

Tropical regions where rains fall seasonally support tropical dry forests. The broadleaf trees that dominate such forests are smaller than rain forest trees and most lose their leaves and become dormant in the dry season.

Trees of **temperate deciduous forests** shed leaves and become dormant in preparation for winter, when conditions do not favor growth (Figure 18.6B). In spring, deciduous trees flower and put out new leaves. At the same time, leaves that were shed and accumulated during the prior autumn decay to form a rich soil. During the growing season, a somewhat open canopy allows sunlight to reach the ground, where shorter understory plants flourish.

Boreal forest, the conifer-dominated forest that sweeps across northern Asia, Europe, and North America (Figure 18.6C), is the most extensive biome. It is also referred to as taiga, which is Russian for "swamp forest," because rains that fall during the cool summers keep the soil soggy. Winters are dry and cold. The conifers are mainly spruce, fir, and pine. The conical shape of these trees helps them shed snow, and their needlelike leaves help minimize evaporative water loss during the winter when they cannot take up water from the frozen ground.

Equator

■ Tropical forest	☐ Savanna
■ Temperate deciduous forest	■ Chaparral
■ Boreal forest	▨ Desert
■ Temperate grassland	■ Tundra

Figure 18.5 Major land biomes.
Most biomes include areas on more than one continent.

From Russell/Wolfe/Hertz/Starr. *Biology*, 2e. © 2011 Brooks-Cole, a part of Cengage Learning®.

Figure 18.6 Examples of forest biomes.

(A) © Antonio Jorge Nunes/Shutterstock.com; (B) © James Randklev/Corbis; (C) © Serg Zastavkin, Used under license from Shutterstock.com.

A. Tropical rain forest.

B. Temperate deciduous forest in New England.

C. Boreal forest in Siberia during summer.

A. North American prairie, a temperate grassland, where bison are among the native grazers.

B. African savanna, a tropical grassland with scattered shrubs. It supports huge herds of grazing wildebeest.

C. Chaparral in California. Shrubby drought-resistant plants with small, leathery leaves predominate.

Figure 18.7 Biomes dominated by plants adapted to occasional fire.

(A) © Danny Barron; (B) Jonathan Scott/Planet Earth Pictures; (C) Jack Wilburn/Animals Animals.

Grasslands and Chaparral Grasslands are dominated by perennial grasses and other nonwoody plants that tolerate grazing, intervals of drought, and periodic fires—conditions that prevent trees and shrubs from taking over. Grasslands have highly fertile soil. They typically occur in the middle of continents and border shrublands or desert.

North America's temperate grasslands, called **prairies**, once covered much of the continent's interior, where summers are hot and winters are cold and snowy. The prairies supported herds of elk, pronghorn antelope, and bison (Figure 18.7A) that were prey to wolves. Today, these predators and prey are largely absent from most of their former range. Nearly all prairies have been plowed, and their rich soil now sustains production of wheat and other crops.

Savannas are tropical grasslands with a few scattered shrubs and trees. They lie between the tropical forests and hot deserts of Africa, India, and Australia. Temperatures are warm year-round, and there is a distinct rainy season. Africa's savannas are famous for their abundant wildlife. Herbivores include giraffes, zebras, elephants, a variety of antelopes, and immense herds of wildebeests (Figure 18.7b). Lions and hyenas eat the grazers.

Chaparral is a biome dominated by drought-resistant, fire-adapted shrubs that have small, leathery leaves. It occurs along the western coast of continents, between 30 and 40 degrees north or south latitude. Mild winters bring a moderate amount of rain, and the summer is hot and dry. Chaparral is California's most extensive ecosystem (Figure 18.7C). It also occurs in regions bordering the Mediterranean, as well as in Chile, Australia, and South Africa.

Deserts Low annual precipitation defines **desert**, a biome that covers about one-fifth of Earth's land surface. Many deserts are at about 30° north and south latitude, where global air circulation cause dry air to sink. Rain shadows also reduce rainfall. For example, the Himalayas prevent rain from falling in China's Gobi desert.

Lack of rainfall keeps the humidity in a desert low. With little water vapor to block rays, intense sunlight reaches and heats the ground. At night, the lack of insulating water vapor in the air allows the temperature to fall fast. As a result, deserts tend to have larger daily temperature shifts than other biomes.

Despite their harsh conditions, most deserts support some plant life (Figure 18.8A). Cacti are a family of plants native to Western Hemisphere deserts. All cacti are CAM plants (Section 5.4), which carry out gas exchange at night, when potential for evaporative water loss is lowest. When it does rain, cacti take up water and store it in their spongy tissues for use in drier times. Leaves modified as spines help cacti fend off animals that would like to tap this source of water.

Many deserts also contain woody shrubs such as mesquite and creosote. These plants have extensive root systems that take up the little water that is available. Mesquite roots can extend up to 60 meters (197 feet) beneath the soil surface. Desert shrubs often put out leaves only after a rain. Annual desert plants sprout and reproduce fast while soil remains moist after a rain.

Tundra **Tundra**, the biome with the coldest temperature, lies between the polar ice cap and the belts of boreal forests in the Northern Hemisphere. Most tundra is in northern Russia and Canada. Tundra is Earth's youngest biome, having appeared about 10,000 years ago when glaciers retreated at the end of the last ice age. Emergence of land from beneath these glaciers opened the way to a process of primary succession that has produced the currently existing communities. Snow blankets the arctic tundra for as many as nine months of the year. During the brief summer,

A. Mojave Desert after the winter rains. Annual plants sprout, flower, produce seeds, and die within weeks beneath slow-growing, drought-adapted, perennial cacti.

B. Arctic tundra during the summer, when shrubby plants grow under nearly continuous sunlight. Permafrost underlies the upper layer of defrosted soil.

Figure 18.8 Desert and tundra, biomes with extreme climates.

(A) © George H. Huey/Corbis; (B) © Darrell Gulin/Corbis.

plants grow fast under nearly continuous sunlight (Figure 18.8B). Lichens and shallow-rooted, low-growing plants are the producers for food webs that include voles, arctic hares, caribou, arctic foxes, wolves, and brown bears. Many migratory birds nest in the tundra during the summer, when the air is thick with insects.

Even in midsummer, only the surface layer of tundra soil thaws. Below that, a layer of permanently frozen soil called **permafrost** can be as thick as 500 meters (1,600 feet). Permafrost acts as a barrier that prevents drainage, so the soil above it remains perpetually waterlogged. Cool, anaerobic conditions slow decay, so organic remains accumulate, making the permafrost one of Earth's greatest stores of carbon. As global temperatures rise, the amount of frozen soil that melts each summer is increasing. With warmer temperatures, carbon stored in the soil is entering the atmosphere and encouraging further warming.

Tropical rain forest is Earth's oldest biome, and tundra is its youngest.

Take-Home Message 18.3

How do environmental factors affect life in the major biomes?

- Near the equator, year-round rains and warmth support highly productive, species-rich tropical rain forests. At higher latitudes, the trees of temperate deciduous forests drop their leaves and become dormant during winter. At still-higher latitudes, conifers dominate Northern Hemisphere boreal forests.
- Plants of African savannas and temperate grasslands are adapted to grazing and periodic fire. Shrubby chaparral plants are also adapted to fire.
- Deserts receive little rain, so perennial desert plants have traits that reduce their water loss or allow them to store water or tap into water deep in the soil. Annual desert plants complete their life cycle in a brief period after a rain.
- The most northerly and youngest biome is arctic tundra, where low plants are adapted to a short growing season and soil with a layer of permafrost.

chaparral Biome with cool, wet winters and hot dry summers; dominant plants are shrubs with small, leathery leaves.

desert Biome with little precipitation; its perennial plants are adapted to withstand drought.

permafrost Layer of permanently frozen soil in the Arctic.

prairie Temperate grassland biome of North America. Its grasses and other plants are adapted to recover after grazing and the occasional fire.

savanna Tropical biome dominated by grasses and other plants adapted to grazing, as well as a scattering of shrubs.

tundra Northernmost biome, dominated by low plants that grow over a layer of permafrost.

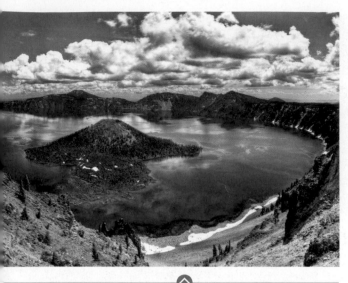

Figure 18.9 Low-nutrient lake.
Crater Lake in Oregon formed when a collapsed volcano began to fill with snowmelt about 7,700 years ago. From a geologic standpoint, it is a young lake, and its clear water is a sign of its low primary productivity.

© Lindsay Douglas/Shutterstock.

Figure 18.10 Examples of coastal ecosystems.

(A) © H. Mark Weldman Photography/Alamy; (B) © Douglas Peebles/Corbis.

A. South Carolina estuary dominated by cordgrass.

B. Florida wetland dominated by red mangroves.

18.4 Aquatic Ecosystems

REMEMBER: Photosynthetic protists are important producers in many aquatic ecosystems (Section 13.6). Succession changes community structure (17.4).

We can distinguish between types of aquatic ecosystems, just as we do biomes on land. Temperature, salinity, rate of water movement, and depth influence the composition of aquatic communities.

Freshwater Ecosystems A lake is a body of standing fresh water. All but the shallowest lakes have zones that differ in their physical characteristics and species composition. Near shore, where sunlight penetrates all the way to the lake bottom, rooted aquatic plants and algae that attach to the bottom are primary producers. A lake's open waters include an upper well-lit zone and, if a lake is deep or cloudy, a zone where light does not penetrate. Producers in the well-lit water include photosynthetic protists and bacteria. In the deeper dark zone, consumers feed on organic debris that drifts down from above.

A lake undergoes succession, meaning the community of lake organisms changes over time. A newly formed lake is deep, clear, and has few nutrients and a low primary productivity (Figure 18.9). As sediments accumulate, the lake becomes shallower. Nutrients accumulate and encourage growth of photosynthetic bacteria, diatoms, and other producers that cloud the water. These producers serve as food for tiny crustaceans, which are then eaten by fish.

Streams are flowing-water ecosystems. They typically originate from runoff or melting snow or ice. As they flow downslope, they grow and merge to form rivers. Properties of a stream or river vary along its length. The type of rocks a stream flows over can affect its solute concentration, as when limestone rocks dissolve and add calcium to the water. Shallow water that flows rapidly over rocks mixes with air and holds more oxygen than slower-moving, deeper water. Also, cold water holds more oxygen than warm water. As a result, different parts of a stream or river support species with different oxygen needs. For example, only cool, well-oxygenated water can support rainbow trout.

Marine Ecosystems An **estuary** is a mostly enclosed coastal region where seawater mixes with nutrient-rich fresh water from rivers and streams. Water inflow continually replenishes nutrients, so estuaries have a high primary productivity. Photosynthetic bacteria and protists that live on mudflats often account for a large portion of an estuary's primary production. Plants adapted to withstand changes in water level and salinity also serve as producers. For example, cordgrass (*Spartina*) is the dominant plant in the salt marshes of many estuaries along the Atlantic coast (Figure 18.10A). Cordgrass can withstand immersion during high tides, and specialized glands on its leaves excrete the salt that its roots take up with seawater.

Estuaries and tidal flats of tropical and subtropical latitudes often support nutrient-rich mangrove wetlands (Figure 18.10B). "Mangrove" is the common term for certain salt-tolerant woody plants. The plants have prop roots that extend out from their trunk and help the plant stay upright in the soft sediments. Specialized cells at the surface of some exposed roots allow gas exchange with air.

Along ocean shores, organisms are adapted to the mechanical force of the waves and to tidal changes. Many species are underwater during high tide, then exposed to the air when the tide is low. Along rocky shores, where waves prevent detritus from piling up, algae that cling to rocks are the producers in grazing food chains. In contrast, waves continually rearrange loose sediments along sandy shores, and make it difficult for algae to take hold. Here, detrital food chains start with organic debris from land or offshore.

Warm, shallow, well-lit tropical seas hold **coral reefs**, formations made primarily of calcium carbonate secreted by generations of corals, which are invertebrate animals (Section 15.3). Like tropical rain forests, tropical coral reefs are home to an extraordinary assortment of species (Figure 18.11A). The main producers in a coral reef community are photosynthetic dinoflagellates that live inside the reef-building corals. These protists provide their coral hosts with sugars. If an environmental change such as a shift in temperature stresses the coral, the coral expels its protist symbionts in a response called coral bleaching. If conditions improve fast, the symbiont population in the coral can recover. But when adverse conditions persist, the symbionts will not be restored, and the coral will die.

In the brightly lit waters of the open ocean, photosynthetic protists and bacteria are the primary producers, and grazing food chains predominate. Depending on the region, some light may penetrate as far as 1,000 meters (more than a half mile) beneath the sea surface. Below that, organisms live in darkness, and organic material that drifts down from above is the basis of detrital food chains.

On the seafloor, the greatest species richness occurs along the edges of continents. There are also some largely unexplored concentrations of biodiversity on seamounts and at hydrothermal vents. **Seamounts** are underwater mountains that stand 1,000 meters or more tall, but still lie below sea surface (Figure 18.11B). Seamounts attract large numbers of fishes and are home to many marine invertebrates. Like islands, they harbor many species that evolved there and live nowhere else. There are more than 30,000 seamounts, and scientists have just begun to document species that live on them.

Hot, mineral-rich water spews out from the ocean floor at **hydrothermal vents**. When mineral-rich hot water mixes with cold seawater, the minerals settle out as extensive deposits. Bacteria and archaea that can extract energy from these deposits serve as primary producers for food webs that include invertebrates such as tube worms and crabs (Figure 18.11C). As explained in Section 13.2, one hypothesis holds that life originated near hydrothermal vents.

A. Coral reef near Fiji.

B. Computer model of seamounts on the seafloor near the coast of Alaska. Patton Seamount, at the rear, stands 3.6 kilometers (about 2 miles) tall.

C. Hydrothermal vent community. Bacteria and archaea that extract energy from minerals are the producers here. Consumers include crabs and giant tube worms (close-up in inset photo) that grow meters long without ever eating. The worms rely on bacteria that live in their tissues to produce their food.

Figure 18.11 Regions of high biodiversity in the sea.

(A) © John Easley, www.johneasley.com. (B) NOAA; (C) NOAA/Photo courtesy of Cindy Van Dover, Duke University Marine Lab, inset © Peter Batson/imagequestmarine.com.

Take-Home Message 18.4

What factors shape aquatic ecosystems?

- Fast-flowing, cooler water holds more oxygen than warmer, slower-moving water, and the amount of available light decreases with the water's depth.
- In well-lit upper waters, producers such as aquatic plants, algae, and photosynthetic microbes are the base for grazing food chains. On coral reefs, the main producers are photosynthetic protists that live inside the coral's tissues.
- Detritus drifting down from above sustains most deep-water communities in lakes and oceans. However, hydrothermal vent communities on the ocean floor are sustained by energy that bacteria and archaea harvest from minerals.

coral reef In tropical sunlit seas, a formation composed of secretions of coral polyps that serves as home to many other species.

estuary A highly productive ecosystem where nutrient-rich water from a river mixes with seawater.

hydrothermal vent Place where hot, mineral-rich water streams out from an underwater opening in Earth's crust.

seamount An undersea mountain.

Figure 18.12 Driven to extinction by humans.
Artist's depiction of a dodo, a flightless bird, about a meter (3.3 feet) tall. It became extinct in the 1600s, after European sailors discovered its island home.

Figure 18.13 Species threatened by overharvesting.

(A) John Butler, NOAA; (B) © George Sanker/naturepl.com.

A. White abalone. There are now too few individuals left to reproduce in wild. The species was overharvested for use as human food.

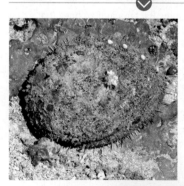

B. Black rhino. Since the 1960s, their number has been reduced from about 100,000 to about 5,000. They are killed for their horn, which is used in traditional Asian medicine.

18.5 Human Impact on the Biosphere

REMEMBER: Acidity is measured as pH (Section 2.5). An asteroid impact 66 million years ago caused extinction of the dinosaurs (11.1). The size of the human population has soared in the past century (16.5). Ultraviolet light is a mutagen (6.4). Increases in greenhouse gases are causing global climate change (17.6).

The rising size of the human population and its increasing industrialization have far-reaching effects on the biosphere. We begin by discussing how human activities affect individual species, then turn to their wider impacts.

Increased Species Extinctions Extinction, like speciation, is a natural process. Species arise and become extinct on an ongoing basis. The rate of extinction picks up dramatically during a **mass extinction**, an event in which many different kinds of organisms become extinct in a relatively short period. We are currently in the midst of such an event. Unlike previous extinction events, this one is not the inevitable result of a physical catastrophe such as an asteroid impact. Humans are the driving force behind the current rise in extinction rate.

There is indirect evidence that humans had a role in the prehistoric decline of many large animals, collectively referred to as megafauna. In North America, large herbivores such as camels, giant ground sloths, and mammoths and mastodons (relatives of elephants) became extinct after the arrival of humans about 15,000 years ago. Carnivores such as lions and saber-toothed cats also disappeared. By one hypothesis, humans directly caused declines in herbivorous species such as mammoths by hunting them. These declines then led to the extinction of their predators. Evidence that humans hunted some megafauna supports this hypothesis. For example, one 13,800-year-old mastodon rib bone found in Washington State had a spear point embedded in it.

More recent extinctions were certainly caused by humans. Consider what happened to the dodo (Figure 18.12) a large, flightless bird that lived on the island of Mauritius in the Indian Ocean. Dodos were plentiful in 1600, when Dutch sailors first arrived on the island, but 80 or so years later the birds were extinct. Sailors ate some, but destruction of nests and habitat by rats, cats, and pigs introduced by the sailors probably had a greater effect.

For the purposes of conservation, a species is considered extinct if repeated, extensive surveys of its known range repeatedly fail to turn up signs of any individuals. It is "extinct in the wild" if the only known members of the species are in captivity. Some species are difficult to find, so occasionally a population of a species previously thought to be extinct or extinct in the wild turns up. However, this is a rare occurrence.

An **endangered species** is one currently at a high risk of extinction in the wild. A **threatened species** is one that is likely to become endangered in the near future. Keep in mind that not all rare species are threatened or endangered. Some species have always been uncommon.

Overharvesting is one cause of current species declines. Consider the case of the white abalone, a gastropod mollusk native to kelp forests along the coast of California (Figure 18.13A). During the 1970s, harvest of this species for human food reduced the population to about 1 percent of its original size. In 2001, the abalone became the first invertebrate to be listed as endangered by the U.S. Fish and Wildlife Service. The white abalone's current population density is too low for effective reproduction in the wild. Abalones release their eggs and sperm into the water, a strategy that is effective only when many individuals live close to one another. In an attempt to save the species, white abalones are now being bred in captivity.

Species are overharvested not only as food, but also for the pet trade, and for use as ornaments or in traditional medicines. For example, black rhinos (Figure 18.13B) are threatened by poachers who kill them for their horn, which sells on the black market for nearly $30,000 per pound. The main market for rhino horn is Asia, where there is a mistaken belief that rhino horn has medicinal properties.

Each species requires a specific type of habitat, and any loss, degradation, or fragmentation of that habitat reduces population numbers. An **endemic species**, one that remains confined to the area in which it evolved, is more likely to go extinct than a species with a more widespread distribution. For example, giant pandas are endemic to China's bamboo forests and feed mainly on bamboo (Figure 18.14A). As these forests disappeared, so did pandas. Their population, which may once have been as high as 100,000 animals, is now reduced to about 1,600 animals in the wild.

Humans also degrade habitats in less direct ways. Consider what is happening to Edwards Aquifer in Texas. The aquifer consists of water-filled, underground limestone formations that supply drinking water to the fast-growing city of San Antonio. Excessive withdrawals of water from this aquifer, along with pollution of the water that recharges it, threaten species that live in the aquifer's depths. The Texas blind salamander (Figure 18.14B), which is endemic to this aquifer, is one such species.

Deliberate or accidental species introductions damage habitats too. Section 17.1 detailed the ways that red imported fire ants accidentally introduced to the United States from South America now threaten native species. Similarly, rats and domestic cats are decimating many of Hawaii's endemic bird species.

Decline or loss of one species sometimes endangers others. Consider buffalo clover, a plant that was abundant when many buffalo grazed in the American Midwest (Figure 18.14C). The clover thrived at the edges of woodlands, where buffalo enriched the soil with their droppings and helped to disperse the clover's seeds. Like most endangered species, buffalo clover faces a number of threats. In addition to the loss of buffalo, the clover is threatened by competition from introduced plants, attacks by introduced insects, and development of its habitat for housing.

The extent of the species declines is staggering. The International Union for Conservation of Nature (IUCN) reports that of the 48,677 named species they assessed, 36 percent were threatened or endangered. We do not know the level of threat for the vast majority of the approximately 1.8 million named species, or for the millions of species yet to be discovered.

endangered species A species that faces extinction in all or part of its range.

endemic species A species that evolved in one place and is found nowhere else.

mass extinction Event in which many species in many habitats become extinct in the same interval.

threatened species A species likely to become endangered in the near future.

Figure 18.14 Species currently threatened by habitat destruction and degradation.

(A) Hung Chung Chih/Shutterstock.com; (B) Joe Fries, U.S. Fish & Wildlife Service; (C) USDA Forest Service/Photo by Sarena Selbo.

A. Panda.

B. Texas blind salamander.

C. Buffalo clover.

Figure 18.15 Deforestation.
In Brazil, a tractor plows a field that once was forest. The soybeans grown in the field will be used to feed cattle and poultry.

Frontpage/Shutterstock.com.

Atlantic Ocean

Africa

Figure 18.16 Results of desertification.
Wind blows dust from the expanding Sahara Desert into the Atlantic Ocean.

© Geoeye Satellite Image.

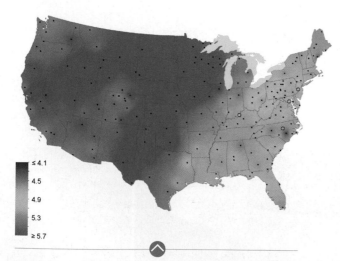

≤ 4.1
4.5
4.9
5.3
≥ 5.7

Figure 18.17 Acid rain.
Average precipitation acidities in the United States in 2012.

National Atmospheric Deposition Program/National Trends Network.

Deforestation and Desertification **Deforestation**, the removal of trees from an area, is an ongoing threat in many regions. Although the amount of forested land is currently stable or increasing in North America, Europe, and China, tropical forests continue to suffer heavy losses (Figure 18.15). Destruction of tropical rain forests is not merely a regional concern. As noted earlier, these forests have the highest primary production of any land biome, so their decline significantly affects carbon storage and oxygen production. In addition, these forests harbor the greatest number of species of any biome. Thus, destruction of an area of tropical rain forest endangers far more species than destruction of a comparable area of temperate zone or boreal forest.

Poor agricultural practices can encourage soil erosion and lead to a rapid shift from grassland or woodland to desert, a process called **desertification**. During the mid-1930s, large portions of prairie on North America's southern Great Plains were plowed, exposing the soil to the force of the region's constant winds. Coupled with a drought, the result was an economic and ecological disaster. Winds carried more than a billion tons of topsoil aloft, turning the region into what came to be known as the Dust Bowl. A similar situation now exists in Africa, where expansion of the Sahara desert causes clouds of dust to take flight over the Atlantic (Figure 18.16).

Drought encourages desertification, which results in more drought in a positive feedback cycle. Plants cannot thrive in a region where the topsoil has blown away. With less transpiration (evaporation from plant parts), less water enters the atmosphere, so local rainfall decreases.

The best way to prevent desertification is to avoid farming in areas subject to high winds and periodic drought. If these areas must be used, methods that do not repeatedly disturb the soil can minimize the risk of desertification.

Acid Rain A **pollutant** is a natural or man-made substance released into soil, air, or water in greater than natural amounts. It disrupts the physiological processes of organisms that evolved in its absence, or that are adapted to lower levels of it. **Acid rain** forms when air pollutants released by burning coal and other fossil fuels combine with water vapor and fall to Earth. The resulting rainfall can be ten times more acidic than normal (Figure 18.17). Acid rain that falls onto or drains into waterways, ponds, and lakes harms aquatic organisms. For example, heightened acidity impairs

Figure It Out: Is rain more acidic on the East Coast or the West Coast of the United States?

Answer: The East Coast

Digging Into Data

Accumulation and Transport of Radioisotopes by Tuna

Bluefin tuna are among the many fish species that were exposed to radioactive material released into the sea by the damaged Fukushima nuclear power plant. These tuna breed in the western Pacific, and some migrate to the Eastern Pacific when they are 1 to 2 years old. Daniel Madigan suspected that migrating bluefin tuna could transport radioactive material to the waters off the coast of California. To test this hypothesis, he and his collaborators analyzed the concentration of cesium radioisotopes (^{137}Cs and ^{134}Cs) in bluefin tuna and yellowfin tuna caught near San Diego in August 2011. Yellowfin tuna do not cross the Pacific; they spend their lives in California's coastal waters. The scientists compared these data with the radioisotope content of bluefin tuna that had been caught near San Diego in 2008. Figure 18.18 shows their results.

1. How did the radioisotope content of the bluefin tuna caught in August of 2011 compare to that of bluefins caught in 2008 (before the Fukushima accident)?
2. How did it compare to that of yellowfins caught in August of 2011?
3. Do these data support the hypothesis that bluefin tuna transported radioactive cesium from Fukushima in their body?
4. Bluefin tuna and yellowfin tuna are similarly sized top predators. Why was it important to compare species that were both the same size and at the same trophic level?

Tuna type, date sampled	Mean body mass (kg dry)	Mean age (years)	Mean radioisotope concentration (Bq per kilogram)	
			^{137}Cs	^{134}Cs
Bluefin, 2011 (n = 15)	1.5	1.5	6.3	4
Bluefin, 2008 (n = 5)	1.5	1.4	1.4	0
Yellowfin, 2011 (n = 5)	1.9	1.2	1.1	0

Figure 18.18 Radioactive cesium (Cs) concentration in two types of tuna caught along the coast of California.
^{137}Cs has a half-life of 30 years and some persists in the Pacific from weapons testing. ^{134}Cs has a half-life of 2 years and was undetectable in the Pacific prior to the accident at Fukushima. Becquerels (Bq) is a measure of radioactivity.

Adapted from the Proceedings of the National Academy of Sciences of the United States of America, Daniel J. Madigan, Zofia Baumann, and Nicholas S. Fisher, "Pacific bluefin tuna transport Fukushima-derived radionuclides from Japan to California."

---

the growth of diatoms, which are silica-shelled protists that serve as producers in many lakes. When acid rain falls on forests, it burns tree leaves and encourages loss of nutrient ions from the soil. As a result, trees become malnourished and more susceptible to disease.

In the United States, acid deposition peaked in the 1970s. Since then, federal regulations limiting sulfur dioxide emissions have helped reduce the acidity of precipitation. The world's greatest sulfur dioxide emitters are now China and India, where industrialization and coal use continue to increase.

Biological Accumulation and Magnification Two processes can cause the concentration of a chemical pollutant in an organism's body to rise far above its concentration in the environment. First, some chemical pollutants taken up by organisms accumulate in their tissues. As a result, older individuals contain more of the pollutant than younger ones. Second, by the process of **biological magnification**, a chemical pollutant passes through food chains, becoming increasingly concentrated in higher and higher trophic levels. For example, methylmercury released by coal-burning power plants is a toxic pollutant that can reach a very high concentration in large predatory fishes such as swordfish, bluefin tuna, and albacore (white) tuna. Methylmercury interferes with development of the nervous system, so pregnant women, nursing women, and children should avoid eating these fish.

The Trouble With Trash Historically, humans buried unwanted material in the ground or dumped it out at sea. Trash was out of sight, and also out of mind. We now understand that chemicals seeping from buried trash can contaminate

acid rain Rain containing sulfuric and/or nitric acid; forms when pollutants mix with water vapor in the atmosphere.

biological magnification A chemical pollutant becomes increasingly concentrated as it moves through a food chain.

deforestation Removal of all trees from a forested area.

desertification Conversion of grassland or woodlands to desertlike conditions.

pollutant A natural or man-made substance that is released into the environment in greater than natural amounts and that damages the health of organisms.

Figure 18.19 Death by plastic.
When scientists dissected this recently deceased Laysan albatross chick they found more than 300 pieces of plastic. The chick died after one piece punctured its gut wall.

Clair Fackler/ NOAA.

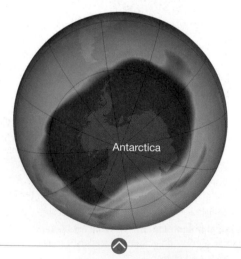

Antarctica

Figure 18.20 Ozone hole.
This graphic shows the September 2006 ozone hole, which was the largest ever recorded. Purple indicates the least ozone, with blue, green, and yellow indicating increasingly higher levels.

Check the current status of the ozone hole at NASA's website (http://ozonewatch.gsfc.nasa.gov/).

NASA Ozone Watch.

groundwater. The United States no longer dumps its trash at sea, but plastic and other garbage still enter our coastal waters. Foam cups and containers from fast-food outlets, foam peanuts, plastic shopping bags, plastic water bottles, and other material discarded as litter enter storm drains in coastal regions. From there it is carried to streams and rivers that can convey it to the sea. Waste plastic that enters the ocean poses a threat to marine life. For example, seabirds often eat floating bits of plastic or feed the plastic to chicks, with deadly results (Figure 18.19).

Ocean currents can carry bits of plastic for thousands of miles. These plastic bits can end up accumulating in some areas of the ocean. Consider the Great Pacific Garbage Patch, a region of the north central Pacific that the media often describes as an "island of trash." In fact, the plastic is not easily visible. Rather, the garbage patch is a region where a high concentration of confetti-like plastic particles swirl slowly around an area as large as the state of Texas.

Destruction of the Ozone Layer Between 17 and 27 kilometers (10.5 and 17 miles) above sea level, the concentration of ozone gas (O_3) is so great that scientists refer to this region as the **ozone layer**. The ozone layer benefits land organisms by absorbing most of the ultraviolet (UV) radiation in incoming sunlight. UV radiation, remember, can damage DNA and thus cause mutations.

In the mid-1970s, scientists noticed that the ozone layer was thinning. Its thickness had always varied a bit with the season, but now there was steady decline from year to year. By the mid-1980s, the spring ozone thinning over Antarctica was so pronounced that people were referring to the low-ozone region as an "ozone hole." Declining ozone quickly became an international concern. The ozone hole over the South Pole was a sign that the ozone layer was thinning.

Chlorofluorocarbons, or CFCs, were determined to be the main ozone destroyers. At the time, these odorless gases were widely used as propellants in aerosol cans, as coolants, and in solvents and plastic foam. In response to the potential threat posed by ozone thinning, countries worldwide agreed in 1987 to phase out the production of certain CFCs and other ozone-destroying chemicals. As a result of that agreement (the Montreal Protocol), the atmospheric concentrations of these pollutants is no longer rising dramatically.

Unfortunately, the ozone hole persisted (Figure 18.20). CFCs break down quite slowly, so scientists expect the existing pollutants to remain at a level that significantly impairs the ozone layer for many decades. In addition, our emission of other ozone-destroying chemicals continues to increase. The most notable of these chemicals is nitrous oxide (N_2O), which forms mainly as a result of the use of synthetic nitrogen fertilizers. Soil bacteria convert nitrogen from the fertilizer to nitrous oxide. Burning organic material, including fossil fuels, also contributes nitrous oxide to the atmosphere.

Global Climate Change The climate change resulting from our pouring greenhouse gases into the atmosphere has effects on ecosystems worldwide. For example, sea level is rising. An increase in average annual temperature elevates sea level by two mechanisms. First, water expands as it is heated. Second, heating melts glaciers and sea ice (Figure 18.21). Together, the processes have caused the sea level to rise about 20 centimeters (8 inches) in the past century. As a result, some coastal wetlands have disappeared underwater.

Until recently, ice sheets covered the Arctic Ocean year-round, making it difficult for ships to move to and from the Arctic landmass. Those ice sheets are now shrinking (Figure 18.21B). At the same time, ice on the Arctic landmass is melting.

Muir Glacier 1941

Muir Glacier 2004

A. Retreating glaciers in Alaska.

These changes will accelerate global warming because exposed soil absorbs more heat than ice, which is highly reflective. The changes will also make it easier for people to remove minerals and fossil fuels from the Arctic. With the world supply of fuel and minerals dwindling, pressure to exploit Arctic resources is rising. However, conservationists warn that extracting these resources will harm Arctic species, such as the polar bear, that are already threatened by other factors.

The temperature of the land and seas affects evaporation, winds, and currents. As a result, many weather patterns are expected to change as temperature rises. For example, warmer temperatures are correlated with extremes in rainfall patterns, with periods of drought interrupted by unusually heavy rains. Warmer seas are also expected to increase the intensity of hurricanes.

Climate change is already having widespread effects on biological systems. Arctic animals, such as polar bears and walruses, that would normally spend most of their time on ice are being forced onto land. In addition, temperature changes are cues for many temperate zone species, so warmer-than-normal springs cause deciduous trees to leaf out earlier, and spring-blooming flowers to blossom earlier. Animal migration times and breeding seasons are also shifting. Some species benefit from the warming, expanding their range to higher latitudes or elevations that were previously too cool to sustain them. Other species are harmed by the rising temperature. For example, warming of tropical waters stresses corals and has increased the frequency of coral bleaching.

1980

2012

B. Reduced permanent ice in the arctic. The bright white is the ice that persists all year. The larger light blue-gray area shows the maximum ice coverage in the winter. Graphics are based on data from satellites.

Figure 18.21 Evidence of a warming world.

(A left) National Snow and Ice Data center, W. O. Field; (A right) National Snow and Ice Data Center, B. F. Molnia; (B) NASA.

Take-Home Message 18.5

How do human activities affect the biosphere?

- We are increasing the rate of species extinctions by degrading, destroying, and fragmenting habitats, overharvesting species, and introducing exotic species.
- Human activities turn grasslands into deserts, strip woodlands of trees, and generate pollutants and trash that kill organisms and damage ecosystems.
- Some pollutants have global effects, as when CFCs cause thinning of the ozone layer, and increasing levels of greenhouse gases bring about climate change.

ozone layer Region of upper atmosphere with a high ozone concentration; acts as a sunscreen against UV radiation.

biodiversity Of a region, the genetic diversity within its species, variety of species, and variety of ecosystems.

conservation biology Field of applied biology that surveys biodiversity and seeks ways to maintain and use it.

ecological restoration Actively altering an area in an effort to restore or create a functional ecosystem.

indicator species A species that is particularly sensitive to environmental changes and can be monitored to assess whether an ecosystem is threatened.

The extinction of a species removes its unique assortment of traits from the world of life forever.

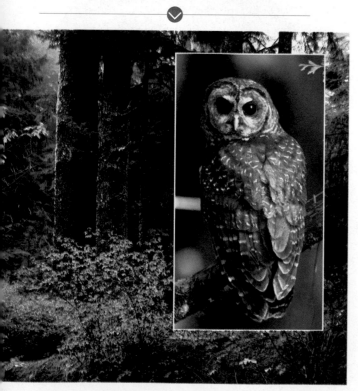

Figure 18.22 The Klamath–Siskiyou forest, one of North America's conservation hot spots. Endangered northern spotted owls (inset) nest in old-growth trees of this forest.

David Patte, USFWS; inset, USFWS.

18.6 Maintaining Biodiversity

REMEMBER: As a result of the lack of gene flow between species, each species has a unique assortment of traits (Section 12.5).

The Value of Biodiversity Every nation enjoys several forms of wealth: material wealth, cultural wealth, and biological wealth, or biodiversity. We measure a region's **biodiversity** at three levels: the genetic diversity within its species, its species diversity, and its ecosystem diversity.

Why should we protect biodiversity? From a purely selfish standpoint, doing so is an investment in our future. Healthy ecosystems are essential to the survival of our own species. Other organisms produce the oxygen we breathe and the food we eat. They also remove waste carbon dioxide from the air and decompose and detoxify other wastes. Plants take up rain and hold soil in place, preventing erosion and reducing the risk of flooding.

Compounds discovered in wild species often serve as medicines. Two widely used chemotherapy drugs, vincristine and vinblastine, were extracted from the rosy periwinkle, a low-growing plant native to Madagascar's rain forests. Cone snails that live in tropical seas are the source of a highly potent pain reliever.

Wild relatives of crop plants serve as reservoirs of genetic diversity that plant breeders can draw upon to protect and improve crops. Wild plants often have genes that make them more resistant to disease or adverse conditions than their domesticated relatives. Plant breeders can use traditional cross-breeding methods or biotechnology to introduce genes from wild species into domesticated ones, thus creating improved varieties.

A decline in some species can warn us that the natural support system we depend on is in trouble. **Indicator species** are particularly sensitive to environmental change, and can be monitored as indicators of environmental health. For example, a decline in lichens tells us that air quality is deteriorating. The loss of mayflies from a stream tells us that the quality of water in a stream is declining.

In addition, there are ethical reasons to preserve biodiversity. As we have emphasized many times, all living species are the result of an ongoing evolutionary process that stretches back billions of years. Each species has a unique combination of traits. The extinction of a species removes its unique collection of traits from the world of life forever.

Conservation Biology Biodiversity is currently in decline at all three levels and in regions worldwide. **Conservation biology** addresses these declines. The goals of this relatively new field of biology are to survey the range of biodiversity, and to find ways to maintain and use biodiversity in a manner that benefits human populations. The aim is to conserve as much biodiversity as possible by encouraging people to value it and use it in ways that do not destroy it.

With so many species and ecosystems at risk, conservation biologists must often make difficult choices about what areas to target for protection first. To prioritize their efforts, the biologists identify conservation hot spots, places that are richest in endemic species and under the greatest threat. The goal of prioritizing is to save representative examples of all of Earth's existing biomes. By focusing on hot spots, rather than individual species, scientists hope to maintain ecosystem processes that sustain biological diversity.

Conservation scientists of the World Wildlife Fund have defined 867 distinctive land ecoregions that they consider the top priority for conservation efforts. Each region has a large number of endemic species and is under threat. The Klamath–Siskiyou forest in southwestern Oregon and northwestern California is one hot spot (Figure 18.22). It is home to many rare conifers, and two endangered birds, the northern spotted owl and the marbled murrelet, nest in the forest's old growth. Endangered coho salmon breed in streams that run through it. Logging is the main threat to this region, but a newly introduced pathogen that infects conifers also poses a concern.

Protecting biological diversity can be a tricky proposition. Even in developed countries, people often oppose environmental protections because they fear such measures will have adverse economic consequences. However, taking care of the environment can make good economic sense. With a bit of planning, people can both preserve and profit from their biological wealth. For example, Costa Rica's Monteverde Cloud Forest Reserve protects more than 100 mammal species, 400 bird species, and 120 species of amphibians and reptiles. It is one of the few habitats left for the ocelot, puma, and jaguar (Figure 18.23). Each year, about 75,000 tourists visit the reserve, and the feeding, lodging, and guiding of these tourists provides much-needed employment to local people. These people also benefit from the ecological services that their forest continues to provide.

Ecological Restoration Sometimes, an ecosystem is so damaged, or there is so little of it left, that conservation alone is not enough to sustain biodiversity. **Ecological restoration** is work designed to bring about the renewal of a natural ecosystem that has been degraded or destroyed, fully or in part. Restoration work in Louisiana's coastal wetlands is an example. Louisiana contains more than 40 percent of the coastal wetlands in the United States. These marshes are an ecological and economic treasure, but they are in trouble. Dams and levees built upstream of the marshes hold back sediments that would normally replenish sediments lost to the sea. Channels cut through the marshes for oil exploration and production have encouraged erosion, and the rising sea level threatens to flood existing plants. Since the 1940s, Louisiana has lost an area of marshland the size of Rhode Island. Restoration efforts now under way aim to reverse some of those losses and protect remaining marsh species (Figure 18.24).

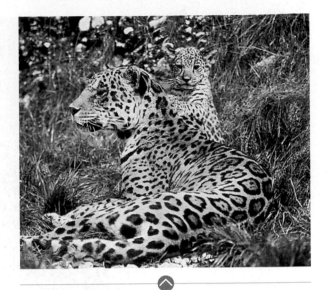

Figure 18.23 Jaguars, one of the many endangered species that live in Costa Rica's Monteverde Cloud Forest Reserve.

© Adolf Schmidecker/FPG/Getty Images.

Figure 18.24 Ecological restoration in Louisiana's Sabine National Wildlife Refuge.

(A) Diane Borden-Bilot, U.S. Fish and Wildlife Service; (B) U.S. Fish and Wildlife Service.

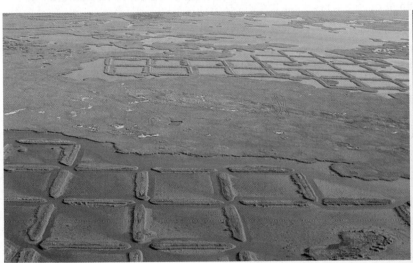

A. Where marsh has become open water, sediments are barged in and marsh grasses are planted on them.

B. Restoration protects native species such as the roseate spoonbills that nest in the marsh during the summer.

A. Right, Bingham copper mine near Salt Lake City, Utah. The mine is 4 kilometers (2.5 miles) wide and 1,200 meters (0.75 miles) deep.

B. Below, pelican covered with oil that accidentally leaked from a deepwater oil well in the Gulf of Mexico.

Figure 18.25 Environmental effects of resource extraction.

(A) © Lee Prince/Shutterstock.com; (B) Joel Sartore/National Geographic Creative.

Reducing Human Impacts Ultimately, the health of our planet depends on our ability to recognize that the principles of energy flow and of resource limitation that govern the survival of all systems of life do not change. We must take note of these principles and find a way to live within our limits. The goal is living sustainably, which means meeting the needs of the present generation without reducing the ability of future generations to meet their own needs.

Promoting sustainable living begins with acknowledging the environmental consequences of our own lifestyle. People in industrial nations use enormous quantities of resources, and the extraction and delivery of these resources have negative effects on biodiversity. In the United States, the size of the average family has declined since the 1950s, while the size of the average home has doubled. All the materials used to build and furnish those larger homes come from the environment. For example, an average new home contains about 500 pounds of copper in its wiring and plumbing. Where does copper come from? Like most other mineral elements used in manufacturing, it is generally mined from the ground (Figure 18.25A). Mining often generates pollution and produces ecological dead zones.

Nonrenewable mineral resources are also used in electronic devices such as phones, computers, televisions, and MP3 players. Constantly trading up to the newest device may be good for the ego and the economy, but it is bad for the environment. Reducing consumption by fixing existing products promotes sustainability, as does recycling. Recycling nonrenewable materials reduces the need for extraction of those resources from the environment, so it helps maintain healthy ecosystems.

Reducing energy use is another way to promote sustainability. Fossil fuels such as petroleum, natural gas, and coal supply most of the energy used by developed countries. You already know that use of these nonrenewable fuels contributes to global warming and acid rain. In addition, extraction and transportation of these fuels have negative impacts. Oil harms many species when it leaks from pipelines, ships, or wells (Figure 18.25B). Renewable energy sources do not produce

Figure 18.26 Volunteers restoring the Little Salmon River in Idaho so that salmon can migrate upstream to their breeding grounds.

Mountain Visions/NOAA.

greenhouse gases, but they have drawbacks too. For example, dams that generate renewable hydroelectric power may discourage endangered salmon from returning to streams above the dam to breed. Similarly, wind turbines can harm birds and bats. Panels used to collect solar energy are made using nonrenewable mineral resources, and manufacturing the panels generates pollutants.

In short, all commercially produced energy has some kind of negative environmental impact, so the best way to minimize that impact is to use less energy. Shop for energy-efficient appliances, avoid incandescent lightbulbs, and do not leave lights on in empty rooms. Walking, bicycling, and using public transportation are energy-efficient alternatives to driving. Shopping locally and purchasing locally produced goods also saves energy.

If you want to make a difference, learn about the threats to ecosystems in your own area. Are species threatened? If so, what are the threats? How can you help reduce those threats? Support efforts to preserve and restore local biodiversity. Many ecological restoration projects are supervised by trained biologists but carried out primarily through the efforts of volunteers (Figure 18.26).

Keep in mind that unthinking actions of billions of individuals are the greatest threat to biodiversity. Each of us may have little impact on our own, but our collective behavior, for good or for bad, will determine the future of the planet.

Take-Home Message 18.6

What is biodiversity and how can we sustain it?

- Biodiversity is the genetic diversity of individuals of a species, the variety of species, and the variety of ecosystems. Worldwide, biodiversity is declining at all of these levels.
- Conservation biologists are working to identify threatened regions with high biodiversity and prioritize which receive protection.
- Ecological restoration is the process of re-creating or renewing a diverse natural ecosystem that has been destroyed or degraded.
- Individuals can help maintain biodiversity by using resources in a sustainable fashion.

Summary

Section 18.1 Circulation of Earth's seas and atmosphere can distribute pollutants that were released in one part of the world across the globe.

 Section 18.2 Climate refers to average weather conditions over time. Variations in climate depend largely upon differences in the amount of solar radiation reaching different parts of Earth. The closer a region is to the equator, the more solar energy it receives. Warming of air and water at the equator sets in motion global patterns of air circulation and ocean currents. Circulating air and water distribute heat and moisture. Landforms also influence climate, as when coastal mountains cause a **rain shadow**.

 Section 18.3 Biomes are categories of major ecosystems on land. They are maintained largely by regional variations in climate and described mainly in terms of their dominant plant life.

Near the equator, high rainfall and mild temperatures support **tropical rain forests**, which are dominated by trees that remain leafy and active all year. This is Earth's most productive and oldest biome. In **temperate deciduous forests**, trees grow during warm summers, then lose their leaves and become dormant in winter. **Boreal forest**, dominated by conifers, is the most extensive biome. It is found only in the Northern Hemisphere where winters are cold and dry, and summers are cool and rainy.

Grasslands form at midlatitudes in the interior of continents and are dominated by plants adapted to grazing. **Prairie** is a North American grassland biome; **savanna** is an African one that also includes scattered shrubs. **Chaparral**, a biome dominated by shrubby plants with tough leaves, occurs in regions with cool, wet winters and hot, dry summers. Both grasses and chaparral plants are adapted to withstand periodic fires.

Deserts occur at latitudes about 30° north and south, where dry air descends and annual rainfall is sparse. Desert plants are adapted to withstand drought or complete their life cycle fast after a rain.

Tundra forms at high latitudes in the Northern Hemisphere. It is the youngest biome and has an underlying layer of **permafrost**.

Section 18.4 Aquatic ecosystems have gradients of sunlight penetration, water temperature, salinity, and dissolved gases. Lakes are standing bodies of water. Different communities of organisms live at different depths and distances from the shore.

Streams and rivers are flowing bodies of water. Physical characteristics of a stream or river that vary along its length influence the types of organisms that live in it. A semi-enclosed area where nutrient-rich water from a river mixes with seawater is an **estuary**, a highly productive habitat. Seashores may be rocky or sandy. Grazing food chains based on algae form on rocky shores. Detrital food chains dominate sandy shores.

Coral reefs are species-rich ecosystems found in well-lit tropical waters. Photosynthetic protists that live in the coral's tissues are the main producers in this ecosystem.

The open ocean's upper waters hold photosynthetic organisms that form the basis for grazing food chains. Deeper water communities usually subsist on material that drifts down from above. However, bacteria and archaea that can obtain energy from minerals serve as the producers at **hydrothermal vent** ecosystems. **Seamounts** are undersea mountains that have a high species richness.

 Section 18.5 Humans are causing a **mass extinction** by overharvesting, degrading and destroying habitats, and introducing exotic species. **Endangered species** are currently at risk of extinction; **threatened species** are likely to become endangered. **Endemic species** are more vulnerable to extinction than widely dispersed ones. Human activities can also threaten entire ecosystems, as when poor agricultural practices cause **desertification** or **deforestation** destroys a forest.

Pollutants such as those that cause **acid rain** threaten forest ecosystems. **Biological magnification** occurs when a pollutant is passed along a food chain. Trash that gets into fresh water can make its way into the oceans, where it degrades marine ecosystems. Ozone is a pollutant near the ground, but depletion of the **ozone layer** is a global threat caused by use of chemicals called CFCs.

Global warming caused by rising concentrations of greenhouse gases is melting polar ice and raising the sea level.

Section 18.6 Biodiversity includes the diversity of genes, species, and ecosystems. All three levels of biodiversity are declining. A decrease in biodiversity can harm humans. We rely on ecosystems to produce oxygen and decompose waste. We also benefit from many compounds produced by wild species, and by tapping their genetic diversity to enhance our crops. Loss of **indicator species** warns us that a habitat is being degraded.

Conservation biologists document the extent of biodiversity and look for ways to preserve it, while benefiting humans. They give priority to areas with high biodiversity that are most threatened. **Ecological restoration** is the work of actively renewing an ecosystem that has been damaged or destroyed. By living sustainably, we can maintain biodiversity and ensure that resources remain for future generations.

Self-Quiz

Answers in Appendix I

1. The most solar radiation reaches the ground at _____ .
   a. the equator
   b. the North Pole
   c. midlatitudes
   d. the South Pole

2. When air is heated, it _____ and can hold _____ water.
   a. sinks; less
   b. sinks; more
   c. rises; less
   d. rises; more

3. Most North American _____ has been converted to cropland.
   a. tundra
   b. prairie
   c. desert
   d. boreal forest

4. Plants in _____ are adapted to periodic fires.
   a. deserts
   b. boreal forests
   c. tropical rain forests
   d. chaparral

5. Permafrost underlies _____ .
   a. savanna   b. tundra   c. desert   d. prairie

6. The oldest and most productive biome is _____ .
   a. boreal forest
   b. tundra
   c. tropical rain forest
   d. desert

7. Bacteria and archaea that can obtain energy from minerals are the main producers at _____ .
   a. hydrothermal vents
   b. estuaries
   c. coral reefs
   d. seamounts

8. Which would hold more oxygen?
   a. a fast-moving, cool stream   b. a warm pond

9. Match the biome with the most suitable description.
   _____ tundra   a. fresh water and seawater mix
   _____ chaparral   b. low humidity, and little rainfall
   _____ desert   c. North American grassland
   _____ prairie   d. fire-adapted shrubs with tough leaves
   _____ estuary   e. low-growing plants over permafrost
   _____ boreal forest   f. most extensive biome
   _____ coral reef   g. main producers are protists
   _____ tropical rain   h. broadleaf trees are active all year
         forest

10. An _____ species has population levels so low it is at great risk of extinction in the near future.
    a. endemic
    b. endangered
    c. indicator
    d. exotic

11. An _____ species can be monitored to gauge the health of its environment.
    a. endemic
    b. endangered
    c. indicator
    d. exotic

12. The 1930s environmental disaster known as the Dust Bowl is an example of _____ .
    a. deforestation
    b. desertification
    c. ecological restoration
    d. species extinction

13. The ozone layer _____ .
    a. is getting thicker
    b. helps keep Earth warm
    c. is a layer in the deep ocean
    d. screens out UV radiation

14. Acid rain _____ .
    a. harms aquatic organisms
    b. kills trees
    c. is a type of pollution
    d. all of the above

15. As a result of biological magnification, _____ tend to have the highest concentration of pollutants in their bodies.
    a. plants
    b. algae
    c. predators
    d. herbivores

Critical Thinking

1. Like the salamander shown in Figure 18.14B, many animals that live in caves are blind but had sighted ancestors. By what process might a sighted animal species that colonized a cave lose its sight?

2. In one seaside community in New Jersey, the U.S. Fish and Wildlife Service suggested trapping and removing feral cats (domestic cats that live in the wild). The goal was to protect some endangered wild birds (plovers) that nested on the town's beaches. Many residents were angered by the proposal, arguing that the cats have as much right to be there as the birds. Do you agree? Why or why not?

3. Two arctic marine mammals that live in the same waters differ in the level of pollutants in their bodies. Bowhead whales have a lower pollutant load than ringed seals. What are some factors that might explain this difference?

4. The extent of damage done by acid rain depends to some extent on the type of rocks in the area where it falls. Acid rain that falls in a region with limestone or other calcium carbonate–rich rocks does less harm than acid rain that falls in a region without such rocks. How might the presence of limestone lessen the effects of an input of acid into an ecosystem? Hint: Review Section 2.5.

19 ANIMAL TISSUES AND ORGANS

Application ❯ 19.1 Growing Replacement Parts

REMEMBER: Differentiation is the process by which cells of a developing body turn on different genes (Section 7.7).

Figure 19.1 Regeneration in a sea star.
A sea star regrowing from a severed arm and a bit of central disk. Stem cells in the remaining bit of central disk are giving rise to replacement parts, including the new arms visible on the right.

Francois Michonneau/FLMNH.

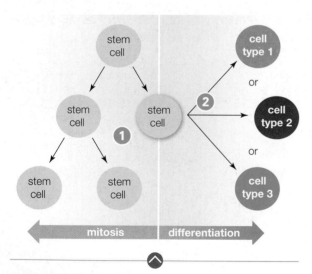

Figure 19.2 Characteristics of stem cells.
Stem cells can divide to form new stem cells or differentiate to form specialized cell types.

All animals can replace some tissues that are lost to injury, but invertebrates have the greatest capacity for regeneration. Slice a planarian (a type of freshwater worm) into 50 pieces and each piece can grow into a new worm. Similarly, some sea stars can regrow an entire body from a single arm and a bit of the central disk (Figure 19.1). No vertebrate can regenerate its body from a limb, but some salamanders can regrow a limb, and many salamanders and lizards can replace a lost tail. Mammals do not replace limbs or tails, although, like other animals, they routinely replace skin and blood cells.

In all animals, stem cells are the key to producing new or replacement tissues. **Stem cells** (Figure 19.2) are unspecialized cells that can either divide to produce more stem cells ❶ or differentiate into one of the specialized types of cells that characterize specific body parts ❷. All cells in an animal body "stem" from stem cells.

Human embryonic stem cells form soon after fertilization, when mitotic divisions produce a cluster of cells collectively smaller than the head of a pin. Each cell in this cluster is **pluripotent**, meaning it can develop into any of the cell types in a human body. Stem cells become increasingly specialized as development continues. For example, stem cells in your bone marrow can become blood cells, but not muscle cells or nerve cells.

Blood cells, skin cells, and other cell types that your body replaces on an ongoing basis arise continually from adult stem cells. This is why you can replace skin or blood lost to injury. However, human adults have few stem cells capable of replacing heart muscle or nervous tissue. Thus, these tissues are not replaced if they are damaged or lost. When the human spinal cord is cut, the result is a permanent paralysis. New nervous tissue does not form to replace the damaged part. Similarly, a heart attack weakens the heart by killing irreplaceable cardiac muscle cells.

Manipulating embryonic stem cells may provide a way to produce replacement muscle or nerve cells for people who need them. Clinical studies involving such replacement cells began recently. In 2011, researchers modified embryonic stem cells and used them to treat two patients blinded by eye diseases. A year later, the implanted cells had survived and produced moderate improvements in vision. Ongoing clinical trials will test this treatment in a larger group of patients and determine whether early stem cell treatment can prevent vision loss.

Embryonic stem cells must be derived from human embryos that have been donated or produced for this purpose. However, altering adult cells through biotechnology may allow us to grow replacement tissues using cells derived from adults. Adult cells that have been treated so they behave like embryonic stem cells are called induced pluripotent stem cells (iPSCs).

The first iPSCs were produced in 2006. If iPSCs are shown to be equivalent to embryonic cells, they could open the way to personalized tissue replacements. For example, a person who lost cardiac muscle to a heart attack could have his or her skin cells induced to become iPSCs. Those iPSCs could then be made to differentiate into cardiac muscle cells. Such cells would be less likely to be rejected than cells that were derived from another individual.

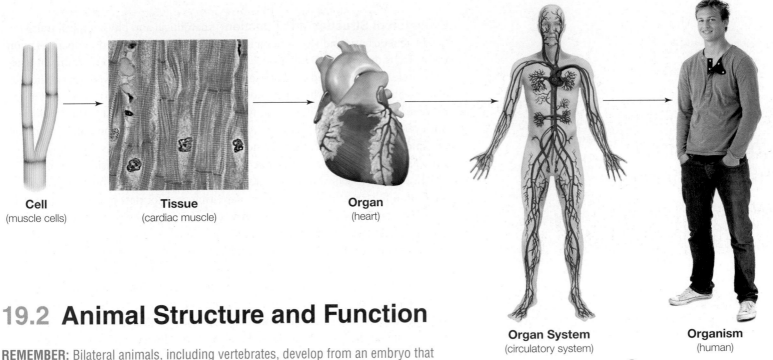

Cell
(muscle cells)

Tissue
(cardiac muscle)

Organ
(heart)

Organ System
(circulatory system)

Organism
(human)

Figure 19.3 Levels of organization in a typical animal body.

Tissue, Ed Reschke; Organism, Image © Yuri Arcurs/Shutterstock.com.

19.2 Animal Structure and Function

REMEMBER: Bilateral animals, including vertebrates, develop from an embryo that has three tissue layers (Section 15.2).

Organization and Integration All animals are multicelled, and nearly all have cells organized as tissues. A **tissue** consists of one or more cell types—and often an extracellular matrix—that collectively perform a specific task or tasks. Four types of tissue occur in all adult vertebrates:

1. Epithelial tissues cover body surfaces and line internal cavities.

2. Connective tissues hold body parts together and provide structural support.

3. Muscle tissues move the body and its parts.

4. Nervous tissues detect stimuli and relay information.

Each tissue is characterized by the types of cells it includes and their proportions. For example, nervous tissue includes neurons, a type of cell not found in muscle tissue or epithelial tissue.

Some animal tissues become organized into organs. An **organ** is a structural unit of two or more tissues organized in a specific way and capable of carrying out specific tasks. For example, a human heart is an organ (Figure 19.3) that includes all four tissue types. The heart has a wall made up mostly of cardiac muscle tissue, it is enclosed by a sheath of connective tissue, and its internal chambers are lined with epithelial tissue. Nervous tissue delivers signals to and from the heart.

In **organ systems**, two or more organs and other components interact physically, chemically, or both in a common task. For example, the force generated by a beating heart moves blood through a system of blood vessels that extends throughout the body.

As an animal goes about its activities, it gains and loses solutes and water, and is exposed to temperature variations. Each living cell engages in metabolic activities that keep it alive. At the same time, organs and organ systems interact to keep solute concentrations and temperature within ranges that cells can tolerate. Such interactions keep conditions in the internal environment within tolerable limits, a process we call **homeostasis**.

homeostasis Process of maintaining favorable conditions inside the body.

organ Structural unit that is composed of two or more tissues and adapted to carry out a particular task.

organ system Organs that interact closely in some task.

pluripotent Capable of developing into any type of cell in a multicelled body.

stem cell A cell that can divide and create more stem cells or differentiate to become a specialized cell type.

tissue A collection of one or more specific cell types that are organized in a way that suits them to a task.

Because evolution modifies existing structures, it does not always produce an optimal body plan.

Evolution of Structure and Function Anatomical and physiological traits have a genetic basis and vary among individuals, so natural selection operates upon them. Traits that help individuals survive and reproduce in their environment tend to be preferentially passed on. For example, vertebrates that evolved in water faced a new set of challenges when they moved onto land (Section 15.5). Gases can only move into and out of an animal's body by diffusing across a moist surface. That is not a problem for aquatic organisms, but on land, evaporation could dry out a moist external surface. The evolution of a new structural feature—lungs—allowed land animals to exchange gases with air across a moist surface deep inside their body.

Evolution typically does not produce entirely new tissues or organs. Instead, it modifies the structure and function of existing ones for new purposes, a process called exaptation. For example, lungs evolved from outpouchings of the gut, not from fish gills. As anyone who has remodeled a home knows, modifying an existing structure requires compromises unnecessary when you build something entirely new. Similarly, we see evidence of evolutionary compromises in many animal structural traits. For example, as a legacy of the lungs' ancient connection to the gut, the human throat opens to both the digestive tract and respiratory tract. As a result, food sometimes goes where air should and a person chokes. It would be safer if food and air entered the body through separate passages. However, because evolution modifies existing structures, it does not always produce an optimal body plan.

Take-Home Message 19.2

How does the structure of an animal body relate to its function?

• All animals are multicelled. Most have tissues, organs, and organ systems.
• Tissues, organs, and organ systems interact in homeostasis, the processes that keep conditions in the internal environment within the limits that cells tolerate.

19.3 Types of Animal Tissues

REMEMBER: Cell junctions hold cells together; cilia are movable structures that project from the surface of some eukaryotic cells (Section 3.5). DNA replication errors can lead to mutations that cause cancer (6.4).

Epithelial Tissues Look in a mirror and what you see is primarily **epithelial tissue**, or epithelium (plural, epithelia), a sheetlike tissue in which cells attach to one another by tight junctions. Your skin, hair, and nails consist of epithelial tissue or structures derived from it. Epithelium even covers the outer surface of your eyeballs. On the inside, epithelium lines your body's many tubes and cavities.

An epithelium secretes a layer of proteins called the basement membrane that glues it to underlying connective tissue (Figure 19.4). The opposite side of the epithelium faces a body cavity or the external environment. Blood vessels do not extend into epithelial tissue, so materials move to and from epithelial cells by diffusing through the underlying connective tissue and basement membrane.

We describe an epithelial tissue in terms of the shape of its cells and the number of cell layers. A simple epithelium is one cell thick, whereas a stratified epithelium has multiple cell layers. Cells in squamous epithelium are flattened or platelike. The thinnest type of epithelium, simple squamous epithelium, lines blood vessels and

Figure 19.4 General structure of an epithelium.

free surface of a simple epithelium

basement membrane (material secreted by epithelial cells)

underlying connective tissue

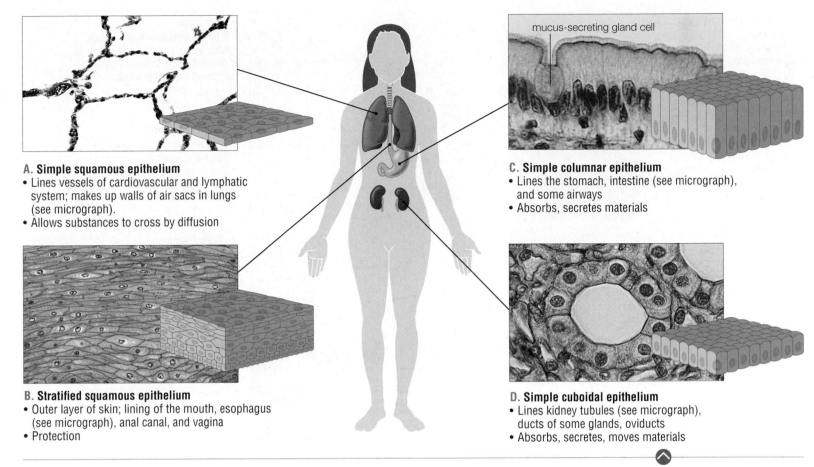

A. Simple squamous epithelium
- Lines vessels of cardiovascular and lymphatic system; makes up walls of air sacs in lungs (see micrograph).
- Allows substances to cross by diffusion

B. Stratified squamous epithelium
- Outer layer of skin; lining of the mouth, esophagus (see micrograph), anal canal, and vagina
- Protection

mucus-secreting gland cell

C. Simple columnar epithelium
- Lines the stomach, intestine (see micrograph), and some airways
- Absorbs, secretes materials

D. Simple cuboidal epithelium
- Lines kidney tubules (see micrograph), ducts of some glands, oviducts
- Absorbs, secretes, moves materials

Figure 19.5 Epithelial tissues.

(A, B) Dr. Gladden Willis/Visuals Unlimited, Inc.; (C) © Ed Reschke/Peter Arnold, Inc.; (D) © Don W. Fawcett.

the inner surface of the lungs (Figure 19.5A). Gases and nutrients diffuse across it easily, so it functions in exchange of materials. In contrast, stratified squamous epithelium (Figure 19.5B) has a protective function. It makes up the outer layer of your skin and the lining of your esophagus (the tube between your throat and stomach).

Cells in columnar epithelium are taller than they are wide (Figure 19.5C). Cells in cuboidal epithelium are short cylinders that look like cubes when viewed in cross-section (Figure 19.5D). Compared to squamous epithelial cells, columnar and cuboidal cells have a greater internal volume and can hold more metabolic machinery. These cells function in the absorption or secretion of substances. Their free surface often has cilia or microvilli. Microvilli are tiny immobile projections that increase a cell's surface area.

Some specialized epithelial cells secrete a substance that functions outside the cell. In most animals, these secretory cells cluster inside glands, which are organs that release substances onto the skin, or into a body cavity or a body fluid. **Exocrine glands** have ducts or tubes that deliver their secretions onto an internal or external surface. Exocrine secretions include mucus, saliva, tears, milk, digestive enzymes, and earwax. **Endocrine glands** have no ducts and they release signaling molecules called hormones into a body fluid. Most commonly, hormones enter the bloodstream, which then distributes them throughout the body.

Adults make few new muscle cells or nerve cells, but they continually replace epithelial cells. You shed and replaces about 0.7 kilogram (1.5 pounds) of skin each year. Similarly, the lining of your intestine is replaced every four to six days. All those cell divisions provide many opportunities for DNA replication errors that can lead to cancer. Thus, epithelium is the tissue most likely to become cancerous.

endocrine gland Ductless gland that secretes hormones into a body fluid.

epithelial tissue Sheetlike animal tissue that covers outer body surfaces and lines internal tubes and cavities.

exocrine gland Gland that secretes milk, sweat, saliva, or some other substance through a duct.

A. Loose connective tissue
- Underlies most epithelia
- Provides elastic support; stores fluid

B. Dense, irregular connective tissue
- Deep in skin and around intestine, in kidney capsule
- Binds parts together; provides support, protection

C. Dense, regular connective tissue
- Tendons and ligaments
- Stretchable attachment between parts

D. Cartilage
- Internal framework of airways, ears, nose; covers ends of bones
- Supports soft tissues; cushions, reduces friction at joints

E. Adipose tissue
- Beneath skin and around the heart and kidneys
- Stores energy-rich lipids; insulates; and cushions body parts

F. Bone tissue
- Bulk of most vertebrate skeletons
- Protects soft tissues; functions in movement; stores minerals; produces blood cells

G. Blood
- Connective tissue with fluid matrix (plasma) and cellular components
- Transports substances; functions in body defenses; helps maintain temperature

Figure 19.6 Connective tissues.

(A) © John Cunningham/ Visuals Unlimited; (B,C) Ed Reschke; (D,G) Science Source;
(E) © University of Cincinnati, Raymond Walters College, Biology; (F) Michael Abbey/
Science Source; center, Image © Yuri Arcurs, Used under license from Shutterstock.com.

Connective Tissues Connective tissues are the most abundant tissues in a vertebrate body. Their cells are separated from one another by a secreted extracellular matrix. **Loose connective tissue**, your most abundant tissue, holds organs in place and underlies epithelia. It consists of cells called **fibroblasts**, which are widely scattered in a matrix that they secreted (Figure 19.6A). Fibers of collagen, the body's most abundant protein, are a major component of this matrix. **Dense connective tissue** is stronger than loose connective tissue because its fibroblasts and collagen fibers are more tightly packed. In dense, irregular connective tissue, fibers are oriented every which way, as in Figure 19.6B. This tissue makes up deep skin layers, supports intestinal muscles, and forms capsules around kidneys. Dense, regular connective tissue has fibroblasts in rows between parallel, tightly packed bundles of fibers (Figure 19.6C). This organization helps keep the tissue from being torn when stretched. Tendons, which connect skeletal muscle to bone, and ligaments, which attach bones to one another, are dense, regular connective tissue.

All vertebrate skeletons include **cartilage**, a specialized connective tissue that has an extracellular matrix of collagen fibers and rubbery glycoproteins (Figure 19.6D). In human embryos, cartilage forms a model for the developing skeleton, then bone replaces most of it. Cartilage still supports your outer ears, nose, and trachea (windpipe). Pads of cartilage between bones serve as shock absorbers.

Many cells store some fat, but cells of **adipose tissue** bulge with so much fat that organelles, including the nucleus, get pushed to one side (Figure 19.6E). In addition to its energy-storage role, adipose tissue cushions body parts, and a layer of adipose tissue beneath the skin serves as insulation.

A. Skeletal muscle
- Interacts with bone to bring about movement, maintain posture
- Reflex activated, but also voluntarily controlled

B. Cardiac muscle
- Occurs only in the heart wall
- Contraction is not under voluntary control

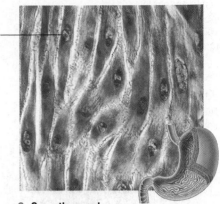

C. Smooth muscle
- Walls of digestive tract, arteries, reproductive tract, the bladder, other hollow organs
- Contraction is not under voluntary control

Figure 19.7 Muscle tissues.

(A,B) Ed Reschke; (C) Biophoto Associates/Science Source.

Bone tissue is a connective tissue in which a calcium-hardened extracellular matrix surrounds each cell (Figure 19.6F). This tissue is the main component of bones, the organs that interact with skeletal muscles to move a body. Bones also support and protect internal organs, and some produce blood cells.

Blood is considered a connective tissue because its cells and platelets descend from cells in bone (Figure 19.6G). Red blood cells transport oxygen. White blood cells defend the body against pathogens. Platelets are cell fragments that help blood clot. Plasma, the fluid portion of blood, consists mostly of water. It transports gases, proteins, nutrients, hormones, and other substances.

Muscle Tissues Cells of muscle tissues contract in response to signals from nervous tissue. **Skeletal muscle tissue** attaches to bones and moves the body or its parts. It consists of cylindrical cells called muscle fibers that have multiple nuclei. The fibers parallel one another and have a striated (striped) appearance under the microscope (Figure 19.7A). Each fiber contains stacks of many units of contraction, each with the same array of protein filaments that function in muscle contraction. Skeletal muscles contract reflexively, but we can also deliberately cause them to contract when we want to move a body part. That is why skeletal muscles are sometimes described as "voluntary" muscles.

Only the heart wall contains **cardiac muscle tissue** (Figure 19.7B). Its cells contract as a unit in response to signals that flow through gap junctions (Section 3.5) between their abutting plasma membranes. Each branching cardiac cell has a single nucleus. It has far more mitochondria than other types of muscle cells, because the heart's constant contractions require a steady supply of ATP. Cardiac muscle tissue also is striated, although less conspicuously so. Cardiac muscle and smooth muscle are both "involuntary" muscle; we cannot make them contract.

We find layers of **smooth muscle tissue** (Figure 19.7C) in the wall of many soft internal organs, including the stomach, bladder, and uterus. Among other tasks, smooth muscle contractions propel material through the gut, adjust the diameter of blood vessels, and constrict the pupil of the eye. Smooth muscle cells are unbranched with tapered ends and a single nucleus. Contractile units are not arranged in a repeating fashion, so smooth muscle is not striated. As in cardiac muscle, gap junctions relay signals between the cells.

adipose tissue Connective tissue with fat-storing cells.

blood Fluid connective tissue with cells that form inside bones.

bone tissue Connective tissue with cells surrounded by a mineral-hardened matrix of their own secretions.

cardiac muscle tissue Striated, involuntary muscle of the heart wall.

cartilage Connective tissue with cells surrounded by a rubbery matrix of their own secretions.

connective tissue Animal tissue with an extensive extracellular matrix; provides structural and functional support.

dense connective tissue Connective tissue with many fibroblasts and fibers in a random or a regular arrangement.

fibroblast Main cell type in soft connective tissue; secretes collagen and other components of extracellular matrix.

loose connective tissue Connective tissue with relatively few fibroblasts and fibers scattered in its matrix.

skeletal muscle tissue Striated, voluntary muscle that interacts with bone to move body parts.

smooth muscle tissue Involuntary muscle that lines blood vessels and hollow organs; not striated.

dendrites
(relay signals
to the cell body)

cell body (holds
nucleus, other
organelles)

axon (transmits electrical
signals along its length, and
releases chemical signals at
its endings)

neuroglial cell
wrapped around
the neuron's axon

Figure 19.8 Motor neuron and associated neuroglial cells.
The cell body and dendrites of a motor neuron are entirely in
the brain or spinal cord. The axon extends out to a skeletal
muscle. Neuroglial cells wrap around the axon of motor
neurons and act like insulation on a wire.

Nervous Tissue **Nervous tissue** allows a body to detect and respond to internal and external changes. It consists of neurons and neuroglia. **Neurons** are the signaling cells in nervous tissue. Neuron structure varies, but all neurons have a cell body that houses the nucleus and holds the bulk of the cytoplasm (Figure 19.8). Projecting out from the cell body are threadlike cytoplasmic extensions that function in receiving and sending signals. Dendrites are signal-receiving extensions; they convey electrical signals toward the cell body. An axon is a signal-sending extension; it conveys electrical signals away from the cell body and to its endings. When an electrical signal reaches an axon ending, the signal causes the ending to release a chemical signaling molecule. The released signaling molecules diffuse across a small gap to an adjacent neuron, muscle fiber, or gland cell, where they alter the behavior of the signal-receiving cell.

In addition to neurons, nervous tissue contains **neuroglia**, a diverse collection of cells that structurally and functionally support the neurons. Various types of neuroglia keep neurons positioned where they should be, provide them with nutrients, and insulate their signal-sending axon.

The bulk of your nervous tissue is in your brain and spinal cord, but bundles of some axons extend through your body as nerves. These axons are sometimes referred to as "nerve fibers."

Take-Home Message 19.3

What are the features of the four types of tissues?

- Sheetlike epithelial tissue covers the body surface and lines tubes and cavities. Tight junctions connect its cells, which have no matrix between them.
- Connective tissues consist of cells in a secreted extracellular matrix. Loose and dense connective tissues hold body parts in place. Adipose tissue stores fat. Bone and cartilage support body parts and function in movement. Blood is a transport medium with cellular components that were made inside bones.
- Muscle tissue consists of cells that contract. Skeletal muscle moves bones, and you can make it contract at will. Cardiac muscle (found in the wall of the heart) and smooth muscle (in walls of hollow organs) are involuntary, meaning you cannot control their contraction.
- Nervous tissue includes neurons. Neurons relay electrical signals along their plasma membrane and use chemicals to communicate with other cells.

19.4 Organs and Organ Systems

REMEMBER: UV light can damage DNA (Section 6.4). Amniotes have skin waterproofed by the protein keratin (15.6).

Skin is your largest organ and it has many functions. It contains sensory receptors that keep the brain informed of external conditions. It serves as a barrier to pathogens and it helps control internal temperature. In land vertebrates, skin also helps conserve water. In humans, reactions that produce vitamin D occur in the skin.

The outermost region of skin, the **epidermis**, is a stratified squamous epithelium (Figure 19.9) ①. Human epidermis consists mainly of keratinocytes, epithelial cells that synthesize the waterproofing protein keratin. Keratinocytes in deep

epidermis
stratified
squamous
epithelium

dermis
mainly dense
connective
tissue

hypodermis
mainly adipose
tissue and loose
connective tissue

hair

duct of sweat gland

blood vessel

pressure-sensitive
nerve ending

smooth muscle

sweat gland

hair follicle

oil gland

Figure 19.9 Structure of human skin, your largest organ.

Figure It Out: Is a sweat gland an endocrine gland or an exocrine gland?

Answer: An exocrine gland

epidermal layers continually divide by mitosis, so new cells displace the older ones toward the skin's surface. As cells move outward, they become flattened, lose their nucleus, and die. Dead keratinocytes accumulate at the skin surface, forming a tough layer that blocks the entry of many pathogens and helps conserve water.

Melanocytes, another type of epidermal cell, make pigments called melanins and donate them to keratinocytes. Variations in human skin color arise from differences in the distribution and activity of melanocytes, and in the type of melanin they produce. Melanin functions as a sunscreen, absorbing ultraviolet (UV) radiation that could damage DNA and other biological molecules. When skin is exposed to sunlight, melanocytes produce more of the brownish-black melanin, resulting in a protective "tan."

The **dermis** beneath the epidermis consists mostly of dense connective tissue ❷. Nervous tissue threads through the dermis, as do capillaries (small blood vessels). Sweat glands in the dermis are made up of epidermal cells that migrated into the dermis during development. Remember, glands are epithelial tissue.

Epithelial tissue embedded in the dermis also forms hair follicles. The base of a hair follicle holds living hair cells, the fastest-dividing cells in the human body. As these cells divide, they push cells above them up, lengthening the hair. The part of a hair that extends beyond the skin surface is the keratin-rich remains of dead cells. A smooth muscle attaches to each hair and can reflexively pull the hair upright in response to cold or fright. Secretions from oil glands next to each hair follicle keep hair smooth and shiny. Like a sweat gland, an oil gland is epidermal tissue embedded in the dermis.

Skin sits atop a layer of loose connective tissue and adipose tissue called the hypodermis ❸. This layer contains larger blood vessels that connect to the small ones running through the dermis. The depth of the hypodermis varies among body regions. The hypodermis beneath the skin of eyelids is thin, with few adipose cells. By contrast, the hypodermis of the buttocks is thickened by many adipose cells.

dermis Deep layer of skin; consists of connective tissue with nerves and blood vessels running through it.

epidermis Outermost, epithelial skin layer.

nervous tissue Animal tissue composed of neurons and supporting cells; detects stimuli and controls responses to them.

neuroglia In nervous tissue, cells that structurally and functionally support neurons.

neuron Main type of cell in nervous tissue; transmits electrical signals along its plasma membrane and communicates with other cells through chemical messages.

Digging Into Data

Growing Skin to Heal Wounds

Diabetes is a disorder in which the blood sugar level is not properly controlled. Among other effects, it reduces blood flow to the lower legs and feet. As a result, about 3 million diabetes patients have ulcers (open wounds that do not heal) on their feet. Each year, about 80,000 require amputations. Several companies provide cultured cell products designed to promote the healing of diabetic foot ulcers. Figure 19.10 shows the results of a clinical experiment that tested the effect of one such cultured skin product versus standard treatment for diabetic foot wounds. Patients were randomly assigned to either the experimental treatment group or the control group and their progress was monitored for 12 weeks.

1. What percentage of wounds had healed at 8 weeks when treated the standard way? When treated with cultured skin?
2. What percentage of wounds had healed at 12 weeks when treated the standard way? When treated with cultured skin?
3. How early was the healing difference between the control and treatment groups obvious?

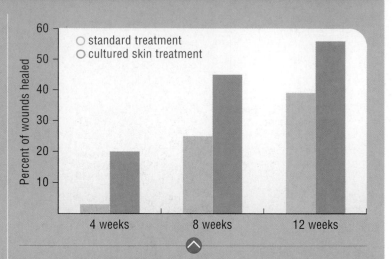

Figure 19.10 Use of cultured skin cells.
Results of a multicenter study of the effects of standard treatment versus use of a cultured cell product for diabetic foot ulcers. Bars show the percentage of foot ulcers that had completely healed.

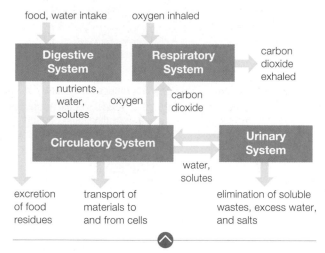

Figure 19.11 Flowchart of links between organ systems. This diagram shows some of the ways that organ systems interact to keep the body supplied with essential substances and eliminate unwanted wastes. Other organ systems that are not shown also take part in these tasks.

Organ Systems Organs interact as organ systems, and organ systems interact to keep an organism alive (Figure 19.11). All vertebrates have the same organ systems. Skin and structures derived from it such as hair, fur, hooves, claws, nails, and quills constitute a vertebrate's integumentary system (Figure 19.12 ❶). The nervous system ❷ is the body's main control center. The brain, spinal cord, and nerves are part of this system, as are sensory organs such as eyes. The endocrine system ❸ consists of hormone-secreting endocrine glands and cells. It works closely with the nervous system and, like that system, controls other organ systems. The muscular system ❹ consists of muscles that move the body and its parts. This system plays a role in regulating body temperature by generating heat. Bones are organs of the skeletal system ❺, which protects internal organs, serves as a point of attachment for skeletal muscles, stores minerals, and produces blood cells. The circulatory system ❻ consists of the heart and blood vessels. It cooperates with the respiratory, digestive, and urinary systems in delivering oxygen and nutrients to cells and clearing away wastes. It also helps regulate body temperature. The lymphatic system ❼ has vessels that move fluid (lymph) from tissues to blood. Organs involved in immunity are also part of this system. The respiratory system ❽ includes two lungs and the airways leading to them. It delivers oxygen from air to the blood, and expels carbon dioxide from blood into the air. The digestive system takes in food, breaks it down, delivers nutrients to the blood, and eliminates undigested wastes ❾. It includes organs of the gut (esophagus, stomach, intestine), as well as glandular organs such as the liver and pancreas, which supply substances that function in digestion. The urinary system consists of kidneys (organs that filter blood and create urine), the bladder, and ducts that deliver urine to the body surface for excretion ❿. It removes wastes from blood, and adjusts blood volume and solute composition.

Reproductive systems of both sexes include gamete-making organs (ovaries or testes) and ducts through which gametes travel ⓫.

1 **Integumentary System**

Protects body from injury, dehydration, pathogens; moderates temperature; excretes some wastes; detects external stimuli.

2 **Nervous System**

Detects external and internal stimuli; coordinates responses to stimuli; integrates body activities.

3 **Endocrine System**

Secretes hormones that control activity of other organ systems. (Male testes added.)

4 **Muscular System**

Moves the body and its parts; maintains posture; produces heat to maintain body temperature.

5 **Skeletal System**

Supports and protects body parts; site of muscle attachment; produces red blood cells; stores minerals.

6 **Circulatory System**

Distributes materials and heat through the body; helps maintain pH.

7 **Lymphatic System**

Collects and returns tissue fluid to the blood; defends the body against infection, cancers.

8 **Respiratory System**

Takes in oxygen necessary for aerobic respiration; expels carbon dioxide released by this pathway.

9 **Digestive System**

Takes in food and water; breaks food down and absorbs needed nutrients, then eliminates food residues.

10 **Urinary System**

Maintains volume and composition of blood; excretes excess fluid and wastes.

11 **Reproductive System**

Female: Produces eggs; nourishes and protects developing offspring. Male: Produces sperm and transfers them to a female.

Figure 19.12 Human organ systems.

Take-Home Message 19.4

What are the roles of organs and organ systems?

- The cells and tissues of an organ are organized to collectively carry out one or more essential tasks. For example, the skin—an organ with all four tissue types—protects the body, helps make vitamin D, detects events in the outside environment, and helps maintain body temperature.

- Organ systems are composed of interacting organs. All vertebrates have the same types of organ systems.

- Survival and reproduction depend on interactions among all organ systems.

Sensory Receptors
(monitor conditions)

Brain
(receives and integrates signal from sensory receptors, then signals muscles and glands)

Muscles and Glands
(act to alter conditions)

Figure 19.13 Components involved in homeostasis.

19.5 **Regulating Body Temperature**

REMEMBER: Homeostasis is one of the defining features of life (Section 1.3). Endotherms can adjust their production of metabolic heat to maintain a stable body temperature (15.6).

Homeostasis, again, is the process of keeping conditions in a body within the range that the body's cells can tolerate. In vertebrates, it involves interactions among sensory receptors, the brain, and muscles and glands (Figure 19.13). A sensory receptor is a cell or cell component that detects a specific stimulus. Sensory receptors involved in homeostasis function like internal watchmen that monitor the body for change. Information from sensory receptors throughout the body flows to the brain. The brain evaluates this information, then signals muscles and glands to take the actions necessary to adjust conditions in the internal environment.

Homeostasis often involves **negative feedback**, a mechanism in which a specific change stimulates a response that reverses that change (Figure 19.14).

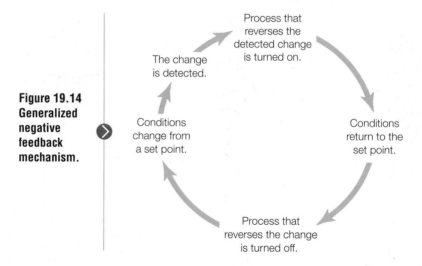

Figure 19.14 Generalized negative feedback mechanism.

The change is detected.

Process that reverses the detected change is turned on.

Conditions return to the set point.

Conditions change from a set point.

Process that reverses the change is turned off.

A heater with a thermostat provides a familiar nonbiological example of how negative feedback works. A person sets the thermostat to a desired temperature. When the temperature falls below this preset point, the heater turns on and emits heat. When the temperature rises to the desired level, the thermostat turns off the heater. A similar negative feedback mechanism maintains your core body temperature near 37°C (98.6°F). A brain region called the hypothalamus serves as your body's thermostat.

Consider what happens when you exercise on a hot day (Figure 19.15A). Muscle activity generates heat, so your body's internal temperature rises. Sensory receptors in your skin, spinal cord, and brain continually monitor temperature. When temperature rises, they send signals along nerves to the hypothalamus. In response, the hypothalamus sends out signals along nerves. As a result of these signals, muscle in the walls of blood vessels that lead to the skin relax. The diameter of the vessels increases, allowing more blood to flow from the body's hot interior to the skin, where it gives up heat to the environment. At the same time, sweat glands in the skin increase their output. Evaporation of sweat helps cool the body surface. As the rate of heat loss increases, body temperature returns to normal.

Figure 19.15 Negative feedback control of human body temperature.

1 Stimulus: Rise in body temperature

2 Sensory receptors alert the brain's temperature-regulating center

3 Response: Signals from brain cause sweating, and increased blood flow to the skin

4 Body temperature returns to normal

Normal body temperature

A. Response to an increase in body temperature.

1 Stimulus: Decline in body temperature

2 Sensory receptors alert the brain's temperature-regulating center

3 Response: Signals from brain cause decreased blood flow to the skin, shivering and non-shivering heat production

4 Body temperature returns to normal

B. Response to a decrease in body temperature.

Receptors in the skin also notify your brain when body temperature declines (Figure 19.15B). The brain then diverts blood flow away from the skin, lessening movement of heat to the body surface, where it would be lost to the surrounding air. With prolonged cold, the brain sends a signal to skeletal muscles, which respond by contracting ten to twenty times a second. This shivering response increases heat production by muscles. A specialized type of adipose tissue called brown fat can also contribute to increased heat production. Brown fat has an abundance of mitochondria, and in a process known as nonshivering heat production, these mitochondria reduce their output of ATP and instead release energy as heat.

Humans and other endotherms maintain tight control over their core body temperature. As a result, their body can be much hotter or colder than their surroundings (Figure 19.16). By contrast, ectotherms such as lizards cannot change their metabolic heat output to offset changes in the external temperature. Their core body temperature is generally about the same as that of their surroundings.

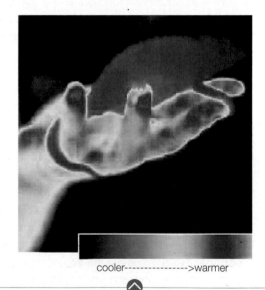

cooler---------------->warmer

Figure 19.16 Endotherm and ectotherm.
Photo of a chameleon (an ectothermic lizard) being held by a human (an endotherm). The photo was taken using heat-sensitive film. The lizard is about the same temperature as its surroundings, so it blends into the background. The human is warmer than its surroundings because he or she is producing a large amount of metabolic heat.

NASA/JPL-Caltech/L. Hermans.

Take-Home Message 19.5

How does your body maintain its temperature?

• Sensory receptors detect a change in body temperature and notify the brain. Signals from the brain then cause muscles and glands to react to this change.

• Temperature regulation involves negative feedback, in which a change triggers a response that reverses the change.

• Sweating and increasing the blood flow to the skin cool the body, whereas shivering and decreasing blood flow to the skin warm the body.

negative feedback A change causes a response that reverses the change.

Section 19.1 All animals have a capacity to replace some cells lost to injury. The replacement cells are derived from stem cells. **Stem**

cells can divide to form more stem cells or differentiate to become one or more specialized cell types. All body parts develop from embryonic stem cells, which are **pluripotent**. Replacement tissues derived from cultured embryonic stem cells or adult stem cells made to behave like embryonic ones (iPSCs) could one day help treat disorders resulting from the death of cells that are otherwise irreplaceable.

Section 19.2 Animal bodies are structurally and functionally organized on several levels. A **tissue** is a group of cells and intercellular substances that perform a common task. Tissues make up **organs**, which interact as **organ systems**. Activities at all levels contribute to **homeostasis**, the process of maintaining conditions inside the body within the range that body cells require.

Animal structure and function have a genetic basis and are shaped by natural selection. Evolution modifies existing structures, a mechanism that explains why some aspects of body plans can seem less than optimal.

Section 19.3 Most animals have four types of tissue. **Epithelial tissues** cover external body surfaces and line internal cavities and tubes. Tight junctions connect adjacent cells, and there is little matrix between cells. An epithelium has one free surface and one surface that is glued to an underlying tissue by its secretions. The free surface may have cilia or microvilli. **Endocrine glands** and **exocrine glands** are derived from epithelial tissue, as are hair and nails. Epithelial tissues are constantly replaced, and this constant turnover makes them especially vulnerable to cancer.

Connective tissues are the body's most abundant tissues. They consist of cells dispersed in a matrix of their own secretions. **Loose connective tissue**, which holds organs in place and underlies skin, and **dense connective tissue**, which makes up tendons and ligaments, have protein fibers, **fibroblasts**, and other cells in an extracellular matrix. Rubbery **cartilage**, fat-storing **adipose tissue**, mineralized **bone tissue**, and **blood** are specialized connective tissues.

Muscle tissues contract (shorten) when stimulated. **Skeletal muscle** moves body parts and is under voluntary control, whereas **smooth muscle** and **cardiac muscle** are not. Smooth muscle moves material through the digestive tract and adjusts the width of blood vessels. Cardiac muscle is the muscle of the heart.

Neurons serve as communication lines in **nervous tissue**. They relay electrical signals along their plasma membrane and send and receive chemical signals. Cells called **neuroglia** support the neurons.

Section 19.4 Skin is the body's largest organ. It includes a deep **dermis** layer consisting mainly of connective tissue. The outermost layer of skin, the **epidermis**, consists of epithelium. Keratinocytes in skin produce keratin, the main skin protein. Melanocytes produce melanin, the pigment that colors skin and protects against UV light. Skin functions in protection, temperature control, detection of environmental changes, vitamin D production, and defense.

An organ system consists of two or more organs interacting in tasks that keep individual cells as well as the whole body functioning. All vertebrates have the same types of organ systems.

Section 19.5 In vertebrates, homeostasis involves sensory receptors that detect changes, a brain that receives signals from receptors and coordinates responses, and muscles and glands that carry out responses. Homeostasis often involves **negative feedback** in which a change brings about a response that reverses the change. For example, in human temperature regulation, a rise in body temperature causes sweating and loss of heat from the skin, which cause body temperature to decline.

Self-Quiz

1. _____ tissues are sheetlike with one free surface.
   a. Epithelial   c. Nervous
   b. Connective   d. Muscle

2. _____ connect(s) epithelium to underlying connective tissue.
   a. Gap junctions   c. Keratinocytes
   b. Basement membrane   d. Plasma membrane

3. Most animals have glands derived from _____ tissue.
   a. epithelial   c. muscle
   b. connective   d. nervous

4. Only _____ cells have cilia or microvilli at their surface.
   a. epithelial   c. muscle
   b. connective   d. nervous

5. The most abundant protein in the human body is _____ , made by fibroblasts.
   a. collagen
   b. keratin
   c. melanin
   d. hemoglobin

6. _____ consists mainly of plasma.
   a. Loose connective tissue
   b. Blood
   c. Cartilage
   d. Bone

7. Your body converts excess carbohydrates and proteins to fats that accumulate in _____ .
   a. fibroblasts
   b. neurons
   c. adipose tissue cells
   d. melanocytes

8. _____ tissues are the body's most abundant and widely distributed tissue.
   a. Epithelial
   b. Connective
   c. Nervous
   d. Muscle

9. Cells of _____ can shorten (contract).
   a. epithelial tissue
   b. connective tissue
   c. muscle tissue
   d. nervous tissue

10. _____ muscle tissue has a striped appearance and is under voluntary control.
    a. Skeletal
    b. Smooth
    c. Cardiac
    d. a and c

11. _____ detects and integrates information about changes and controls responses to those changes.
    a. Epithelial tissue
    b. Connective tissue
    c. Muscle tissue
    d. Nervous tissue

12. Threadlike cytoplasmic extensions that extend out from the cell body of a _____ function in receiving and sending signals.
    a. neuron
    b. neuroglial cell
    c. fibroblast
    d. melanocyte

13. Functions of skin include _____ .
    a. defense against pathogens
    c. production of vitamin D
    b. helping to cool the body
    d. all of the above

14. With negative feedback, a change results in a response that _____ that change.
    a. reverses
    b. accelerates
    c. has no effect on

15. Match the terms with the most suitable description.
    _____ exocrine gland
    _____ endocrine gland
    _____ epidermis
    _____ dermis
    _____ smooth muscle
    _____ cardiac muscle
    _____ skeletal muscle
    _____ adipose tissue
    _____ blood
    _____ cartilage
    _____ simple squamous epithelium

    a. outermost skin layer
    b. secretes through duct
    c. in heart only
    d. support in ears and nose
    e. contracts, not striated
    f. deepest skin layer
    g. plasma, platelets, and cells
    h. ductless hormone secretor
    i. stores fat
    j. single layer of flat cells
    k. involved in voluntary movement

Critical Thinking

1. Leukemia is a cancer in which the body makes too many white blood cells. Affected people are sometimes treated with a bone marrow transplant. Radiation or chemotherapy drugs are used to kill the existing bone marrow, then a bone marrow transplant replaces it. Why would a bone marrow transplant affect the production of white blood cells?

2. Radiation and chemotherapy drugs preferentially kill cells that divide frequently, most notably cancer cells. These treatments also cause hair to fall out and disrupt gut function. Why?

3. Each level of biological organization has emergent properties that arise from the interaction of its component parts. For example, cells have a capacity for inheritance that the molecules making up the cell do not. Can you think of an emergent property of a tissue? Of an organ that contains that tissue?

Visual Question

1. The micrograph to the right shows cells from the lining of an airway that leads to the lungs. The gold cells are ciliated and the darker brown cells have microvilli and secrete mucus. Which of the four major types of tissue of tissue is this? How can you tell?

CNRI/Science Source.

20

HOW ANIMALS MOVE

Application ❯ 20.1 Bulking Up Muscles

REMEMBER: Mutations can alter protein function (Section 7.6). With incomplete dominance, the heterozygote has a phenotype intermediate between that of the two homozygotes (9.4).

A. Normal mouse (left) and a mouse in which the myostatin gene has been knocked out (inactivated).

B. Normal whippet (left) and a bully whippet (right). A bully whippet is homozygous for a mutation that prevents production of functional myostatin.

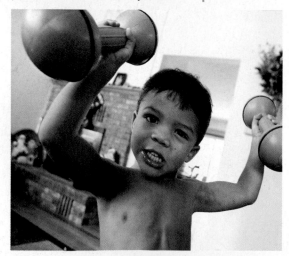

C. Liam Hoekstra, shown at age 3, lifting 5-pound (2.3-kilogram) dumbbells. At this time, he had about 40 percent more muscle than an average 3-year-old.

Figure 20.1 Effects of disrupted myostatin function. Myostatin inhibits muscle growth.

The more you use your muscles, the bulkier and more powerful they become. A mature skeletal muscle fiber cannot divide, but it can make more of the proteins involved in muscle contraction. Protein filaments inside a muscle fiber are continually built and broken down. Exercise tilts this process in favor of synthesis, so muscle cells get bigger and the muscle gets stronger. In addition, exercise encourages stem cells in skeletal muscle to divide and differentiate into muscle fibers.

Hormones affect muscle mass. Men tend to be more muscular than women because the male sex hormone testosterone encourages muscle cells to build more proteins. Human growth hormone has a similar effect. Taking synthetic versions of muscle-building (anabolic) hormones bulks up muscles, so some athletes use such hormones in the hope of boosting muscle size and improving their performance. However, most sport organizations consider the use of such drugs to be cheating and penalize athletes who use them. In addition, artificially elevating the level of anabolic hormones can have negative health effects.

Variations in some genes affect the ability to bulk up muscles. Consider the effects of myostatin, which is a regulatory protein that discourages muscle enlargement. When myostatin does not function normally, the result is myostatin-related muscle hypertrophy (excessive growth of muscle). Scientists can cause this condition in mice by knocking out the mouse myostatin gene (Figure 20.1A).

Myostatin-related muscle hypertrophy also occurs as a result of naturally occurring mutations. One such mutation occurs in some whippets, a type of dog bred for racing. "Bully whippets" are homozygous for a mutant myostatin allele that encodes a nonfunctional protein, so they are unusually muscular (Figure 20.1B). Because bully whippets do not conform to the appearance standard for their breed, they are not raced. However, whippets heterozygous for the mutant allele do race, and they tend to win races more often than normal whippets.

The only known case of complete myostatin deficiency in a human was reported in 2004. Physicians described a German boy who was born with highly muscular arms and thighs. By age 5, this boy could stand with his arms outstretched, holding a 3-kilogram (more than 5-pound) weight with each hand. Genetic testing revealed that the boy is homozygous for a mutation that causes premature termination of the myostatin RNA. As a result, he makes no myostatin.

Liam Hoekstra, an unusually strong American boy profiled in the 2009 documentary "The World's Strongest Toddler," also has myostatin-related muscle hypertrophy (Figure 20.1C). He makes myostatin, but his muscles do not respond to it.

Given what we know about myostatin's ability to suppress muscle growth, researchers are now seeking ways to inhibit myostatin production or interfere with myostatin's activity. Their goal is to come up with treatments for the muscle loss that occurs as a result of muscular dystrophy, cancer, and other disorders. Myostatin inhibitors could even help offset the progressive muscle loss that often accompanies aging. Will some athletes use such myostatin-inhibiting drugs to push their bodies beyond normal limits? No doubt they will. Nutritional supplements that purport to bind to myostatin and reduce its activity are already popular. In clinical tests, these supplements had no effect on strength, but hope for an extra edge keeps moving them off the shelf.

Figure 20.2 Hydrostatic skeleton of an earthworm.
Muscle contractions change the shape of an earthworm's body segments, just as squeezing a water-filled balloon alters its shape. In each body segment, muscles exert their force against the fluid trapped inside that segment. The fluid in all of the segments collectively constitutes a worm's hydrostatic skeleton.

Right, © Johnna Goodyear/Shutterstock.

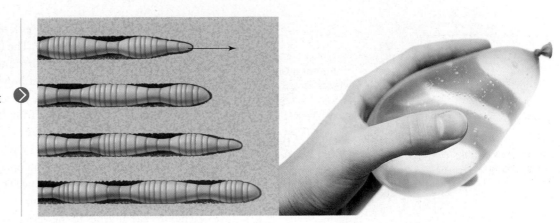

20.2 Skeletal Systems

REMEMBER: A coelom is a fluid-filled, lined body cavity (Section 15.2). The human skeleton is adapted to upright walking (15.7).

All animals move. During part or all of their life, they are capable of self-propelled movement from place to place. Most animals move about on a daily basis to find food and escape from predators.

Types of Skeletons Muscles bring about movement of a body or its parts by interacting with a skeleton. Soft-bodied invertebrates typically have a **hydrostatic skeleton**, an internal, fluid-filled chamber or chambers that muscles exert force against. For example, earthworms have a coelom divided into many fluid-filled chambers, one per segment. Muscles alter the shape of a body segment by squeezing its fluid-filled chamber (Figure 20.2). By analogy, think about how squeezing a water-filled balloon changes the balloon's shape. Coordinated changes in the shape of the earthworm's many body segments allow it to move through the soil.

In animals with an **exoskeleton**, a cuticle, shell, or other secreted hard external body part receives the force of a muscle contraction. For example, muscles attached to the cuticle of a fly's thorax change the shape of the thorax, thus causing the attached wings to flap up and down (Figure 20.3). The arthropod exoskeleton also protects the soft body tissues inside it. However, such a skeleton has some drawbacks. The exoskeleton consists of noncellular secreted material, so it cannot grow as the animal does. As an arthropod grows, it must periodically molt its old skeleton and replace it with a larger one. Repeatedly producing new exoskeleton uses resources that could otherwise be put to use in growth or reproduction. In addition, after an arthropod molts, it remains highly vulnerable to predators until its new exoskeleton hardens.

An **endoskeleton** is an internal framework of hardened elements to which muscles attach. Skeletal elements of an endoskeleton can grow with the animal. Echinoderms such as sea stars have an endoskeleton, as do all vertebrates. The term "vertebrate" refers to the **vertebral column**, or backbone, a skeletal feature common to all members of this group (Section 15.4). The vertebral column, along with the bones of the head and rib cage, constitute a vertebrate's axial skeleton. The appendicular skeleton consists of the pectoral (shoulder) girdle, the pelvic (hip) girdle, and limbs (or bony fins) attached to them (Figure 20.4).

thorax

Figure 20.3 Insect exoskeleton.
A fly's cuticle is an external skeleton. Muscles inside a fly's thorax pull on the cuticle of the thorax, thus changing its shape. The muscle-driven changes in thorax shape cause the fly's wings to pivot up and down.

Stephen Dalton/Science Source.

Figure 20.4 Features of the vertebrate endoskeleton.

▦ axial skeleton ▦ appendicular skeleton

rib cage vertebral column skull bones

pelvic girdle

pectoral girdle

endoskeleton Hard internal parts that muscles attach to and move.

exoskeleton Hard external parts that muscles attach to and move.

hydrostatic skeleton Fluid-filled chamber or chambers that muscles act on, redistributing the fluid.

vertebral column Backbone.

Skull
cranial bones

facial bones

Pectoral Girdle
clavicle (collarbone)

scapula
(shoulder blade)

Rib Cage
sternum
(breastbone)

ribs (12 pairs)

**Upper
Limb Bones**
humerus (upper
arm bone)

ulna (forearm
bone)

Vertebral Column
vertebrae

intervertebral
disk (cartilage)

radius (fore-
arm bone)

carpals
(wrist bones)

Pelvic Girdle
hip bones,
sacrum,
and coccyx

metacarpals
(palm bones)

phalanges
(finger bones)

**Lower Limb
Bones**

femur
(thighbone)

patella
(kneecap)

tibia (lower
leg bone)

fibula (lower
leg bone)

tarsals
(ankle bones)

metatarsals
(sole bones)

phalanges
(toe bones)

A. Bones of the human skeleton.

nutrient canal

location of
yellow marrow

compact
bone tissue

spongy
bone
tissue

B. Structure of a femur.

C. Spongy and compact bone tissue.

spongy
bone
tissue

compact
bone tissue

blood vessel

outer layer
of dense
connective tissue

Figure 20.5 Human skeletal anatomy.

The Human Skeleton For a closer look at vertebrate skeletal features, consider the human skeleton (Figure 20.5A). The flattened cranial bones of the skull fit together to form a braincase that surrounds and protects the brain. The skull's facial bones include cheekbones and other bones around the eyes, the bone that forms the bridge of the nose, and bones of the jaw.

The vertebral column extends from the base of the skull to the pelvic girdle, and consists of 23 bones called **vertebrae**. Thick pads of cartilage, called **intervertebral disks**, separate adjoining vertebrae. The spinal cord runs through the backbone and connects with the brain by way of an opening at the base of the skull. Viewed from the side, our backbone has an S shape. This curvature keeps our head and torso centered over our feet and is an adaptation to upright walking.

The rib cage attaches to the vertebral column. Both males and females have twelve pairs of ribs. Ribs and the breastbone, or sternum, form a protective cage around the heart and lungs.

The scapula (shoulder blade) and clavicle (collarbone) are bones of the human pectoral girdle. The upper arm has one bone, the humerus. The forearm has two bones, the radius and ulna. The wrist and palm each have multiple small bones, as does each finger.

The pelvic girdle protects internal organs and supports the weight of the upper body when you stand upright. It consists of two sets of fused bones, one set on each side of the body.

The largest bone in the body is the femur, or thighbone. It attaches to the lower leg bones, the tibia and fibula, at the knee. The knee is protected by the patella (kneecap). The smaller bone of the lower leg, the fibula, is not necessary for normal function. When a person has cancer of the jaw, reconstructive surgeons sometimes remove the cancerous jawbone and replace it with bone from the person's fibula. The ankle consists of multiple bones, as do the sole of the foot and each toe.

Bone Structure and Function A bone is an organ. An outer sheath of connective tissue covers the bone, and nerves and blood vessels extend through tiny passageways in the bone's interior. Bone tissue itself consists of bone cells in a secreted extracellular matrix of collagen hardened with calcium and phosphorus.

There are two types of bone. **Compact bone** makes up the outermost, weight-bearing part of a limb bone such as a femur (Figure 20.5B, C). Compact bone consists of many thin, concentric layers of extracellular matrix surrounding

Digging Into Data

Building Stronger Bones

Tiffany (left) was born with multiple fractures in her arms and legs. By age six, she had undergone surgery to correct more than 200 bone fractures. Her fragile, easily broken bones are symptoms of osteogenesis imperfecta (OI), a genetic disorder caused by a mutation in a gene for collagen. As bones develop, collagen forms a scaffold for deposition of mineralized bone tissue. This scaffold forms improperly in children with OI. Figure 20.6 shows the results of an experimental test of a new drug for OI. Bones of treated children, all less than two years old, were compared to bones of a control group of similarly affected, same-aged children who did not receive the drug.

1. An increase in vertebral area during the period of the study indicates bone growth. How many treated children showed such an increase?
2. How many of the untreated children showed an increase in vertebral area?
3. How did the rate of fractures in the two groups compare?
4. Do the results shown support the hypothesis that this drug can increase bone growth and reduce fractures in young children who have OI?

Treated child	Vertebral area in cm² (Initial)	(Final)	Fractures per year
1	14.7	16.7	1
2	15.5	16.9	1
3	6.7	16.5	6
4	7.3	11.8	0
5	13.6	14.6	6
6	9.3	15.6	1
7	15.3	15.9	0
8	9.9	13.0	4
9	10.5	13.4	4
Mean	11.4	14.9	2.6

Control child	Vertebral area in cm² (Initial)	(Final)	Fractures per year
1	18.2	13.7	4
2	16.5	12.9	7
3	16.4	11.3	8
4	13.5	7.7	5
5	16.2	16.1	8
6	18.9	17.0	6
Mean	16.6	13.1	6.3

Figure 20.6 Results of a clinical trial of a drug treatment for osteogenesis imperfecta (OI).
The drug being tested is a compound that reduces the rate of bone breakdown. Nine children with OI received the drug. Six others were untreated controls. Surface area of specific vertebrae was measured before and after treatment. Fractures occurring during the 12 months of the trial were also recorded.

Left, Courtesy of the family of Tiffany Manning.

---

canals through which nerves and blood vessels that service living bone cells run. The shaft and ends of a femur contain **spongy bone**, which has internal spaces. **Red marrow** fills spaces in spongy bone of limbs and some other bones, such as the sternum. Blood cells form mainly in red marrow. **Yellow marrow**, consisting mostly of adipose cells, fills the central cavity of adult limb bones.

Bones contain the body's main reserve of calcium and phosphorus. Ongoing mineral deposits to bone and removals from it help maintain the blood concentrations of calcium and phosphorus ions. Because calcium ions play a role in nerve cell function, muscle contraction, and other important processes, the calcium concentration in the blood is tightly controlled.

Some cells in bone continually secrete extracellular matrix and other bone cells break it down to release calcium. Until people reach their midtwenties, the rate of matrix deposition exceeds the rate of matrix breakdown. With increasing age, matrix production slows and bone density declines.

Osteoporosis is a disorder in which bone loss greatly outpaces bone formation, so bones become weaker and more likely to break. It most commonly occurs in postmenopausal women because they no longer produce sex hormones that encourage bone deposition. However, about 20 percent of osteoporosis cases occur in men. To reduce risk of osteoporosis, ensure that your diet provides plenty of calcium and vitamin D, which the body needs to absorb calcium from food.

Where Bones Meet—Skeletal Joints A **joint** is an area where bones come together. Connective tissue holds bones securely in place at **fibrous joints** such as those that hold cranial bones together. Pads or disks of cartilage connect bones at

compact bone Dense bone with concentric layers of matrix.

fibrous joint Joint where dense connective tissue holds bones firmly in place.

intervertebral disk Cartilage disk between two vertebrae.

joint Region where bones come together.

red marrow Bone marrow that makes blood cells.

spongy bone Lightweight bone with many internal spaces.

vertebrae Bones of the backbone (vertebral column).

yellow marrow Bone marrow that is mostly fat; fills cavity in most long bones.

femur

patella

cartilage

cruciate ligaments

menisci

tibia

fibula

Figure 20.7 Anatomy of the knee, a hinge-type synovial joint.
Bones are held in place by ligaments. The knee is also stabilized by wedges of cartilage called menisci. A torn ligament or damaged meniscus can impair knee function.

cartilaginous joint Joint where cartilage holds bones together and provides cushioning, as between vertebrae.

ligament Cord of dense connective tissue that holds bones in place at a synovial joint.

synovial joint Movable joint at which ligaments connect bones. For example, a shoulder, elbow, or knee.

tendon Strap of dense connective tissue that connects a skeletal muscle to bone.

cartilaginous joints. This flexible connection allows a bit of movement. Cartilaginous joints connect vertebrae to one another and connect some of the ribs to the sternum (breastbone). Movable joints such as those at the shoulder, elbow, wrist, hip, knee, and ankle are **synovial joints**. At these joints, the cartilage-covered ends of bones are separated by a small space. Cords of dense, regular connective tissue called **ligaments** hold bones of a synovial joint in place and form a capsule around the joint. The capsule's lining secretes a lubricating synovial fluid.

Synovial joints allow a variety of movements. Ball-and-socket joints at the shoulders and hips provide a wide range of rotational motion. At other synovial joints, including some in the wrists and ankles, bones glide past one another. Joints at the elbows and knees function like a hinged door; they allow the bones to move back and forth in one plane only.

Figure 20.7 shows the structure of the knee joint. In addition to ligaments, the knee is stabilized by wedges of cartilage called menisci (singular, meniscus). A bursa reduces friction between the patella (kneecap) and femur. A bursa is a fluid-filled sac that reduces friction between parts at a joint.

Joints are frequent sites of injury. A sprained ankle, the most common joint injury, occurs when one or more of the ligaments holding bones together at the ankle joint overstretches or tears. The sprain is usually treated immediately with rest, application of ice, compression with an elastic bandage, and elevation of the affected area. By contrast, a tear of the cruciate ligaments in the knee joint may require surgery. "Cruciate" means cross, and these short ligaments cross one another in the center of the knee joint. The cruciate ligaments stabilize the knee, and when they are torn completely, bones can shift so the knee gives out when a person tries to stand.

Arthritis is the chronic inflammation of a joint. The most common type of arthritis is osteoarthritis. It usually appears in older adults, whose cartilage has thinned at a frequently jarred joint or joints. Knees and hips are most often affected. The decrease in rubbery, protective cartilage sets the stage for damage to bones of the joint. Rheumatoid arthritis is an autoimmune disorder in which the immune system mistakenly attacks the fluid-secreting lining of synovial joints throughout the body. It can occur at any age.

Arthritis can be treated with drugs that relieve pain and minimize inflammation. Joints affected by osteoarthritis can also be replaced with artificial, or prosthetic, joints. Knee and hip replacements are now common and allow a person to resume normal activities.

Take-Home Message 20.2

What are the features of animal skeletons?

• Animal skeletons consist of hard structural elements or fluid-filled chambers that muscle contractions act upon.

• Humans have a bony endoskeleton with some features that facilitate upright walking.

• Bones are wrapped in connective tissue and have marrow in their interior. They are continually remodeled; minerals are removed and added as needed.

• Bones meet at joints. Bones remain in place at fibrous and cartilaginous joints, but can move relative to one another at synovial joints. Ligaments hold the bones of a synovial joint in place.

20.3 Functions of Skeletal Muscles

Skeletal muscles are sometimes referred to as voluntary muscles because they function in intentional movement. However, skeletal muscles also act to maintain posture and participate in reflexes. For example, a shivering reflex helps maintain core body temperature when external temperature plunges.

A sheath of dense, regular connective tissue encloses each muscle and extends beyond it as a straplike **tendon**. Most often, one or more tendons attach each end of the muscle to a bone. Figure 20.8 shows the two muscles of the upper arm: the biceps and the triceps. Biceps means "two-headed" in Latin and refers to the fact that the upper portion of the biceps attaches to the scapula (shoulder blade) by way of two tendons. At the opposite end of the muscle, a single tendon attaches the biceps to the radius in the forearm.

Keep in mind that muscles can pull on bones, but they cannot push them. To achieve the greatest range of motion, muscles work in opposition: Motion generated by contraction of one muscle is reversed by contraction of another. For example, the triceps in the upper arm opposes the biceps. When the biceps contracts, the triceps relaxes and the forearm is pulled toward the shoulder. Contraction of the triceps coupled with relaxation of the biceps reverses this movement, extending the arm.

You can feel the biceps contract if you extend one arm out, palm up, then place your other hand over the muscle and slowly bend your arm at the elbow. Although the biceps shortens only about a centimeter when it contracts, the forearm moves through a much greater distance. The elbow and many other joints function like a lever, a mechanism in which a rigid structure pivots about a fixed point (the bones are the rigid structures and the joints are the fixed points). Use of a lever allows a small force, such as that exerted by a contracting biceps, to overcome a larger one, such as the gravitational force acting on the forearm.

> Muscles can pull on bones, but they cannot push them.

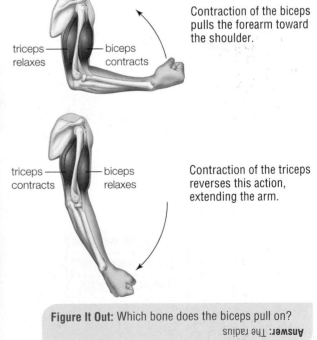

Contraction of the biceps pulls the forearm toward the shoulder.

Contraction of the triceps reverses this action, extending the arm.

Figure It Out: Which bone does the biceps pull on?

Answer: The radius

Figure 20.8 Opposing muscles of the upper arm.

actin Globular protein that makes up thin filaments in muscle fibers.

myofibrils Threadlike, cross-banded skeletal muscle components made up of sarcomeres.

myosin Motor protein that makes up thick filaments in muscle fibers.

sarcomere Contractile unit of skeletal and cardiac muscle.

skeletal muscle fiber Cylindrical, multinucleated cell that runs the length of a skeletal muscle.

sliding-filament model Explanation of how interactions among actin and myosin filaments shorten a sarcomere and bring about muscle contraction.

sphincter Ring of muscle that controls passage of material through a tubular organ or at a body opening.

Most skeletal muscles pull on bones, but some tug on other tissues. Some skeletal muscles move facial skin to bring about changes in expression. Others attach to and move the eyeball, or open and close eyelids. Some sphincters are also composed of skeletal muscle. A **sphincter** is a ring of muscle that controls passage of material through a tubular organ or at a body opening. Sphincters of skeletal muscle allow voluntary control of urination and defecation.

Some animals have boneless muscular organs capable of making complex movements. Your tongue is one example. Some tongue muscles have one end attached to the floor of the mouth and one free end. Other muscles are located entirely within the tongue and have no bony attachment at all. An elephant's trunk is another example of a boneless muscular organ.

Take-Home Message 20.3

How do skeletal muscles function?

- Most skeletal muscles interact with bones, to which they attach by means of a tendon.
- A skeletal muscle can pull on a bone, but cannot push against it.
- Skeletal muscles often work in opposition, with the action of one muscle opposing or reversing the action of another.

20.4 How Muscle Contracts

REMEMBER: Energy from ATP powers motor proteins (Section 3.5).

Muscle Components A skeletal muscle consists of connective tissue–wrapped bundles of muscle fibers (Figure 20.9). A **skeletal muscle fiber** is a multinucleated, cylindrical cell no wider than a hair. Each fiber runs the length of a muscle, so some fibers, such as those in your thigh, can be quite long. A skeletal muscle fiber has multiple nuclei because it formed by the fusion of many embryonic cells. The fiber also contains many mitochondria that supply the ATP for muscle contraction. The bulk of a muscle fiber is taken up by thousands of threadlike structures called **myofibrils**. A specialized type of endoplasmic reticulum called sarcoplasmic reticulum surrounds myofibrils. Between muscle contractions, calcium ions are actively transported into the sarcoplasmic reticulum and stored. When the muscle needs to contract, release of these stored ions will help bring about contraction.

A myofibril consists of many contractile units, called **sarcomeres**, attached end to end. When a myofibril is viewed under a microscope, dark-staining regions called Z lines define the ends of each sarcomere. The region between the Z lines contains thin and thick protein filaments arrayed parallel to one another and to the muscle's long axis. Two coiled strands of the globular protein **actin** are the main components of each thin filament. **Myosin**, a motor protein with a club-shaped head, makes up thick filaments. The myosin heads of the thick filaments are just a few nanometers away from the thin actin filaments, which have sites that can bind myosin.

Muscle bundles, muscle fibers, myofibrils, and thick and thin filaments of a sarcomere all have the same parallel orientation. As a result, they all pull in the same direction when a muscle contracts.

Figure 20.9 Structure of a vertebrate skeletal muscle.

Sliding Filaments The **sliding-filament model** explains how interactions between thick and thin filaments bring about muscle contraction. Neither actin nor myosin filaments change length, and the myosin filaments do not change position. Instead, myosin heads bind to actin filaments and slide them toward the center of a sarcomere. As actin filaments are pulled inward, the ends of the sarcomere are drawn closer, and the sarcomere shortens:

relaxed sarcomere   contracted sarcomere

Let's take a closer look at the molecular basis for sarcomere contraction (Figure 20.10). When a sarcomere is relaxed, myosin cannot bind to actin because other components of the thin filament block the myosin-binding sites. The myosin can, however, bind ATP ❶. A myosin molecule with bound ATP is in a low-energy state. Converting ATP to ADP and phosphate (P$_i$) energizes the myosin head in a manner analogous to stretching a spring ❷.

A muscle fiber contracts in response to a signal from the nervous system. More specifically, it responds to a signal from a motor neuron whose axon endings lie in close proximity to that fiber. In response to a signal from the motor neuron, calcium ions are released from the fiber's sarcoplasmic reticulum. The presence of calcium causes proteins of the thin filaments to shift, allowing myosin heads to bind to actin. Binding of myosin creates a cross-bridge (an attachment) between the thin and thick filaments and causes release of P$_i$ ❸. Release of P$_i$ from the myosin head triggers the power stroke ❹. Like a stretched spring returning to its original shape, a myosin head snaps back toward the sarcomere center. As the myosin head moves, it pulls the attached thin actin filament along and releases the bound ADP. As thin filaments are pulled along thick ones toward the sarcomere center, the sarcomere shortens.

A new molecule of ATP can now bind to the myosin head ❺. When it does, the cross-bridge between the thin and thick filaments breaks. The breaking of one cross-bridge does not allow a thin filament to slip backward because other cross-bridges hold the filament in place. During a contraction, each myosin head repeatedly binds, moves, and releases an adjacent thin filament.

Take-Home Message 20.4

How does skeletal muscle contract?

- A skeletal muscle shortens as a result of decreases in the length of its numerous component sarcomeres. Sarcomeres are the basic units of skeletal muscle contraction.
- The parallel orientation of a skeletal muscle's components directs the force of contraction toward a bone.
- Interactions between myosin and actin filaments in the many sarcomeres of a muscle cell collectively bring about muscle contraction.

❶ A relaxed sarcomere, in which myosin-binding sites on actin are blocked. Myosin heads have bound ATP and are in their low-energy state.

❷ The myosin heads hydrolyze ATP to ADP and P$_i$. The heads are now in a high-energy state, ready to interact with actin, but the myosin-binding sites on actin are blocked.

❸ When a nervous signal excites the muscle, myosin-binding sites open up. Myosin binds to actin, releasing bound phosphate and forming a cross-bridge between thin and thick filaments.

❹ Release of phosphate triggers a power stroke. Myosin heads contract and pull the bound thin filaments inward.

❺ When ATP binds to myosin, myosin releases its grip on actin and returns to its relaxed state, ready to act again.

Figure 20.10 Actin–myosin interaction during contraction. For simplicity we show only part of a sarcomere, with one thick filament and two thin ones.

Figure 20.11 Metabolic pathways by which muscle fibers produce the ATP to fuel their contraction.

Figure 20.12 Activity pattern and muscle fiber composition.

(A) Thomas Dressler/Gallo Images/Getty Images; (B) warmer/Shutterstock.com.

A. Cheetah, a fast-sprinting predator, whose leg muscles have a large proportion of white fibers.

B. Loris, a slow-moving primate, whose leg muscles have a large proportion of slow, red fibers. Creeping slowly along branches helps a loris escape attention of predators.

red muscle fiber Of skeletal muscle, a fiber that contains the oxygen-storing protein myoglobin and produces ATP primarily by aerobic respiration.

white muscle fiber Of skeletal muscle, a fiber that produces ATP primarily by lactate fermentation.

20.5 Fueling Muscle Contraction

REMEMBER: Aerobic respiration (Section 5.5) requires oxygen and is carried out in mitochondria. Lactate fermentation (5.6) does not require oxygen and takes place in the cytoplasm.

Three main pathways supply muscle with the ATP it needs to fuel contraction (Figure 20.11). At the onset of contraction, a muscle fiber taps into its store of creatine phosphate, a molecule that can transfer a phosphate to ADP to form ATP. Such transfers provide enough ATP to fuel muscle contraction for a few seconds, giving the muscle time to begin to increase its ATP output by other metabolic pathways.

We can divide muscle fibers into two categories based on the main pathway they use to make ATP. **Red muscle fibers** rely mainly on aerobic respiration, so they contain many mitochondria. The color of red fibers comes from myoglobin, a protein that, like hemoglobin, reversibly binds oxygen. During periods of rest, oxygen diffuses from the blood into red fibers, where it binds to myoglobin. During muscle activity, release of stored oxygen by myoglobin facilitates aerobic respiration. **White muscle fibers** make ATP mainly by lactate fermentation. They have relatively few mitochondria and do not contain myoglobin.

Muscle fibers can also be subdivided into fast fibers or slow fibers depending on how fast their myosin converts ATP to ADP. All white fibers are fast fibers. They contract rapidly but do not produce sustained contractions. The muscles that move your eye consist mainly of white fibers. Red fibers can be either fast or slow. Fast red fibers predominate in the human triceps muscle, which must often react quickly. Muscles involved in maintaining an upright posture, such as those in the back, consist mainly of slow red fibers.

The proportions of red fibers and white fibers in the limbs of different species reflect differences in how these animals use their limbs (Figure 20.12). Limb muscles of cheetahs, which are renowned for their sprinting ability, have mostly white fibers. By contrast, limb muscles of a loris, which is a stealthy, slow-moving primate, have mostly slow red fibers. Similarly, among human athletes, successful sprinters tend to have a higher-than-average percentage of fast, white fibers in their leg muscles whereas marathoners tend to have a high percentage of slow, red fibers.

Take-Home Message 20.5

What supplies the ATP for muscle contraction?

• Phosphate group transfer from creatine phosphate provides ATP quickly.

• With sustained muscle activity, ATP is produced by aerobic respiration (in red fibers) or lactate fermentation (in white fibers).

20.6 Exercise and Inactivity

When unrelenting stimulation keeps a skeletal muscle excited, muscle fatigue occurs, meaning the muscle's capacity to generate force declines despite ongoing stimulation. Engaging in aerobic exercise—exercise that is low intensity, but of long duration—can make skeletal muscles more resistant to fatigue. Aerobic exercise

Figure 20.13 Sitting, a health hazard.
Prolonged sitting causes metabolic changes in skeletal muscle that are associated with an increased risk of cardiovascular disease and diabetes.

© prodakszyn/Shutterstock.

increases a muscle's blood supply by boosting growth of new capillaries, increases the number of mitochondria and amount of myoglobin in existing red muscle fibers, and encourages conversion of white fibers to red ones. On the other hand, engaging in resistance exercise, such as weight lifting, encourages synthesis of additional actin and myosin filaments. The resulting increase in muscle mass allows for stronger contractions.

Prolonged sitting can be hazardous to your health (Figure 20.13). Sitting involves less muscle activity than standing or moving about, so sitters expend less energy. The more you sit, the fewer calories you burn and the more likely you are to become obese. Sitting has other negative metabolic effects too. When you sit, your leg muscles relax and they decrease their production of lipoprotein lipase (LPL). LPL is an enzyme that contributes to good health by facilitating uptake of lipids from the blood and production of HDL (the "good" cholesterol. When LPL declines, the risk of cardiovascular disease increases. In addition, relaxed muscles do not need to take up glucose to fuel ATP production, so prolonged sitting raises blood glucose level and increases the risk of diabetes.

There is increasing evidence that the health risks associated with prolonged muscle inactivity persist even if a person also gets regular exercise. In other words, exercising for an hour each morning, although it improves your health in some respects, does not cancel out the negative metabolic effects of sitting in place for several hours later in the day. The best way to prevent the health problems associated with inactivity is to avoid sitting for long intervals. When you must sit to carry out a task, get up every 20 minutes or so to move around.

Take-Home Message 20.6

What are the benefits of exercise?

- Exercise can increase the size of muscle fibers and reduce the tendency of muscles to become fatigued.
- Exercise increases the rate at which you burn calories, thus preventing obesity.
- Avoiding prolonged sitting prevents metabolic changes in muscle that can increase the risk of heart disease and diabetes.

Summary

Section 20.1 Exercise increases skeletal muscle mass, as do testosterone and growth hormone. The regulatory protein myostatin discourages muscle growth. Mutations that impair or eliminate myostatin function result in muscle hypertrophy.

Section 20.2 Muscles interact with skeletons. Earthworms and other soft-bodied invertebrates have a **hydrostatic skeleton** (fluid skeleton). Arthropods have an **exoskeleton** (external skeleton) and vertebrates have an **endoskeleton** (internal skeleton).

The human skeleton consists of a skull, a **vertebral column** (backbone), a rib cage, a pelvic girdle, a pectoral girdle, and paired limbs. The vertebral column consists of individual segments called **vertebrae**, with **intervertebral disks** between them. The spinal cord runs through the vertebral column and connects with the brain through a hole in the base of the skull. The shape of the human backbone is an evolutionary adaptation to upright walking.

Bones function in mineral storage, movement, and protection and support of soft organs. A long bone such as a femur (thighbone) has an outer layer of **compact bone** beneath a connective tissue sheath. Fatty **yellow marrow** fills most of the bone interior. **Spongy bone** with blood-producing **red marrow** fills the ends of the bone. Ongoing mineral deposits and removals help maintain blood levels of calcium and phosphorus, and also adjust bone strength. Bone density declines with age.

Bones meet at **joints**. **Fibrous joints** hold bones tightly in place and **cartilaginous joints** let them move a bit. **Synovial joints** allow the most motion. **Ligaments** connect bones at synovial joints.

Section 20.3 Most skeletal muscles move bones, to which they are connected by **tendons** of connective tissue. When skeletal muscles contract, tendons transmit force that makes the bones move. Muscles can only pull on bones, they cannot push them; thus many muscles work as opposing pairs. Skeletal muscles also allow you to change your facial expression and move your tongue. **Sphincters** of skeletal muscle provide voluntary control over urination and defecation.

Section 20.4 Skeletal muscles contract in response to signals from the nervous system. Internal organization of a skeletal muscle promotes directional contraction. A **skeletal muscle fiber** contains many long **myofibrils**. Each myofibril consists of **sarcomeres**, units of muscle contraction, lined up end to end along its length. Each sarcomere has parallel arrays of thin filaments composed of **actin** and thick filaments composed of

actin   myosin

myosin. The **sliding-filament model** describes how ATP-driven sliding of actin filaments along myosin filaments shortens the sarcomere. Muscle contraction occurs in response to a signal from a motor neuron. This signal triggers the release of calcium from a fiber's sarcoplasmic reticulum. The increase in calcium allows actin and myosin filaments to interact.

Section 20.5 Initial muscle contractions are fueled by dephosphorylation of creatine phosphate. For more prolonged activity, **red muscle fibers** (which contain myoglobin) rely on aerobic respiration, whereas **white muscle fibers** carry out lactate fermentation. The proportion of red and white fibers in limb muscles differs among species and reflects the manner in which an animal typically uses its limbs.

Section 20.6 Exercise increases the number of mitochondria and protein filaments in muscle fibers. Prolonged inactivity causes metabolic changes in muscle, and these changes contribute to an increased risk of cardiovascular disease and diabetes.

Self-Quiz

Answers in Appendix I

1. A hydrostatic skeleton consists of _____ .
   a. a fluid in an enclosed space
   b. hardened plates at the surface of a body
   c. internal hard parts
   d. none of the above

2. Bones are _____ .
   a. mineral reservoirs
   b. skeletal muscle's partners
   c. sites where blood cells form (some bones only)
   d. all of the above

3. The human backbone _____ .
   a. consists of vertebrae and intervertebral cartilage disks
   b. is S-shaped
   c. encloses the spinal cord
   d. all of the above

4. Bones move when _____ muscles contract.
   a. cardiac
   b. skeletal
   c. smooth
   d. all of the above

5. A ligament connects _____ .
   a. bones at a joint
   b. a muscle to a bone
   c. a muscle to a tendon
   d. a tendon to bone

6. Red marrow _____ .
   a. is the outermost layer of long bones
   b. fills the shaft of most long bones
   c. consists mainly of fat
   d. produces blood cells

7. Match each bone with its description.
   _____ femur   a. part of skull
   _____ radius   b. thighbone
   _____ vertebra   c. segment of backbone
   _____ sternum   d. breastbone
   _____ cranial bones e. forearm bone

8. The knee is a _____ .
   a. hinge-type synovial joint
   b. fibrous joint
   c. cartilaginous joint
   d. ball-and-socket type of synovial joint

9. A vertebrate skeletal muscle _____ .
   a. contracts in response to signals from a motor neuron
   b. pushes against and moves a bone
   c. is an involuntary muscle
   d. none of the above

10. A sarcomere shortens when _____ .
    a. thick filaments shorten
    b. thin filaments shorten
    c. both thick and thin filaments shorten
    d. none of the above

11. Release of calcium ions from a muscle fiber's _____ allows actin and myosin filaments to interact.
    a. multiple nuclei   c. mitochondria
    b. sarcoplasmic reticulum d. myofibrils

12. Binding of ATP to _____ activates it and prepares this protein to take part in muscle contraction.
    a. actin   c. collagen
    b. myosin   d. myostatin

13. A red muscle fiber _____ .
    a. contains an oxygen-storing protein
    b. produces ATP mainly by lactate fermentation
    c. has relatively few mitochondria
    d. is colored red by myosin

14. Prolonged muscle inactivity as in sitting causes a(n) _____ .
    a. increase in blood lipids
    b. increase in calories burned
    c. decrease in blood glucose
    d. all of the above

15. Match the words with their definitions.
    _____ myoglobin   a. long chain of sarcomeres
    _____ myostatin   b. makes ATP in white fibers
    _____ myosin   c. colors red fibers
    _____ myofibril   d. connects a muscle to a bone
    _____ collagen   e. ring of muscle
    _____ tendon   f. slows muscle protein production
    _____ sphincter   g. motor protein of sarcomere
    _____ osteoporosis   h. main protein in bone
    _____ arthritis   i. inflamed joint(s)
    _____ lactate fermentation j. decline in bone density

Critical Thinking

1. After death, a person no longer makes ATP, and calcium stored in the specialized endoplasmic reticulum of muscle fibers diffuses down its concentration gradient into the muscle cytoplasm. The result is rigor mortis—an unbreakable state of muscle contraction. Explain why the contraction occurs and why it is irreversible.

2. Athletes tend to have stronger bones than nonathletes, but different sports strengthen different bones. Volleyball, basketball, gymnastics, and soccer all thicken the femur, whereas swimming, skating, and bicycling have little effect on this bone. What does this tell you?

3. Bully whippets are homozygous for a deletion of two base pairs in the myostatin gene. The deletion changes an mRNA codon in the middle of the myostatin mRNA from UGU to UGA. Use your knowledge of the genetic code (Section 7.4) to determine the effect of this mutation on the structure of the resulting protein.

Visual Question

1. Is this sarcomere depicted below in a relaxed muscle or a muscle that is contracting? How can you tell?

21

CIRCULATION AND RESPIRATION

Application ❯ 21.1 **A Shocking Save**

The heart is the body's most durable muscle. It begins to beat during the first month of human development, and keeps going for a lifetime. A natural pacemaker in the heart wall generates electrical signals that stimulate contraction of heart muscle. In some people, this pacemaker malfunctions. Electrical signaling becomes disrupted, the heart stops beating, and blood flow halts. Such an event, called a sudden cardiac arrest, occurs in more than 400,000 people per year in the United States. In older people, sudden cardiac arrest usually occurs as a result of a heart attack; impaired blood flow to the heart causes the pacemaker to malfunction. In people under age 35, sudden cardiac arrest usually results from an inborn pacemaker defect.

About 10 seconds into a cardiac arrest, the brain becomes starved for oxygen and the person loses consciousness. The chance of surviving sudden cardiac arrest rises by 50 percent when the affected person receives cardiopulmonary resuscitation (CPR) within a few minutes of the event. With traditional CPR, a rescuer alternates between mouth-to-mouth respiration that forces air into an affected person's lungs and chest compressions that keep the victim's blood moving. Many people have an understandable reluctance to engage in mouth-to-mouth contact with a stranger. Fortunately, a new procedure called hands-only CPR or CCR (cardio-cerebral resuscitation) requires only chest compressions. As long as a person's airway is clear, chest compressions will move enough air into and out of a victim's lungs to oxygenate blood and will keep blood flowing to the person's heart and brain.

CPR cannot restart the heart. That task requires a defibrillator, a device that resets the natural pacemaker by delivering an electric shock to the chest. You have probably seen this procedure depicted in hospital dramas.

Matt Nader (Figure 21.1) learned about the importance of CPR and defibrillation when he experienced sudden cardiac arrest during a high school football game. He came off the field after a play, sat on the bench to talk to his coach, and felt a burning pain in his chest. His vision suddenly blurred, then he passed out. Nader's parents, who are physicians, were watching the game and rushed from their seats. They quickly determined that Matt did not have a pulse and, as they examined him, he stopped breathing.

Matt's parents began CPR on their son. At the same time, someone ran to get the school's automated external defibrillator (AED), a device about the size of a laptop computer. The AED provides simple voice commands that direct the user to attach electrodes to a person in distress. It then determines whether the person has a heartbeat and, if required, shocks the heart.

In Nader's case, the AED restarted his heart, quite possibly saving his life. Cardiologists determined that his sudden cardiac arrest had been caused by a genetic heart defect. To protect him, they implanted a small defibrillator inside his body. This device will provide a lifesaving shock if his heart stops again. After his recovery, Matt Nader appeared before the Texas State Legislature to testify about his experience and advocate for wider availability of AEDs in schools. Thanks in part to his efforts, Texas has passed a law requiring that all high schools have an AED available at athletic events and practices.

To find out where you can learn how to do CPR and use an AED, visit the website of the American Red Cross (www. redcross.org) or the American Heart Association (www.heart.org).

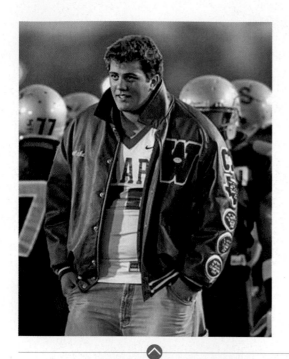

Figure 21.1 Surviving sudden cardiac arrest. Matt Nader (above) was a talented high school football player when he found out he had a heart defect. After his heart stopped during a game, CPR and quick defibrillation saved his life. Automated external defibrillators (AEDs) such as the one shown on the right can restart a heart and are designed to be simple enough to be used by a trained layperson.

(A) Courtesy of the family of Matt Nader; (B) Courtesy of ZOLL Medical Corporation.

21.2 How Substances Are Moved Through a Body

REMEMBER: Diffusion is the movement of a substance from an area where it is more concentrated to one where it is less concentrated (Section 4.5).

All animals must keep their cells supplied with oxygen and dispose of carbon dioxide. In cnidarians and flatworms, gas exchange with the environment occurs at the body surface and the lining of the gastrovascular cavity (Figure 21.2). In echinoderms and roundworms, gas exchange occurs at the body surface and at the surface of a coelom or pseudocoelom respectively. In all of the above groups, gases move through the body by diffusion through **interstitial fluid**, the fluid between cells.

Evolution of circulatory systems that facilitate distribution of gases and other materials through a larger, multilayered body opened the way to more complex animal body plans. A circulatory system is an organ system that speeds the distribution of materials within an animal body. It includes one or more **hearts** (muscular pumps) that propel fluid through vessels that extend through the body.

Open and Closed Circulatory Systems There are two types of circulatory systems. Arthropods and most mollusks have an **open circulatory system**, in which a heart or hearts pump fluid called **hemolymph** into vessels that open onto internal cavities (Figure 21.3A). Hemolymph leaves the vessels and enters these cavities, where it mixes with interstitial fluid and makes direct exchanges with cells. The hemolymph is then sucked back into the heart through pores in the heart wall.

By contrast, annelids such as earthworms, cephalopod mollusks such as octopuses, and all vertebrates have a **closed circulatory system**. In such a system a heart or hearts pump **blood** through a continuous series of vessels (Figure 21.3B). The system is "closed" because blood does not leave blood vessels to bathe tissues. Instead, transfers between blood and the cells of other tissues take place across the walls of the smallest-diameter blood vessels, which are called **capillaries**. A closed circulatory system distributes substances more quickly than an open one.

blood Fluid circulating in a closed circulatory system.

capillary Smallest-diameter blood vessel; site of exchanges of gases and other materials with the tissues.

closed circulatory system Circulatory system in which blood flows through a continuous network of vessels.

heart Muscular organ that pumps fluid through a body.

hemolymph Fluid circulating in an open circulatory system.

interstitial fluid Fluid between cells of a multicelled body.

open circulatory system Circulatory system in which the circulatory fluid (hemolymph) leaves vessels and flows among tissues before returning to the heart.

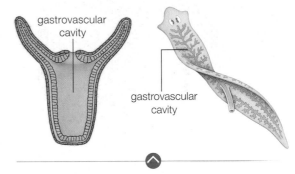

Figure 21.2 No circulatory system.
In sea anemones and flatworms, cells exchange gases with surrounding water and with water inside the gastrovascular cavity. There is no circulatory system to distribute materials through the body.

Figure 21.3 Open versus closed circulatory systems.

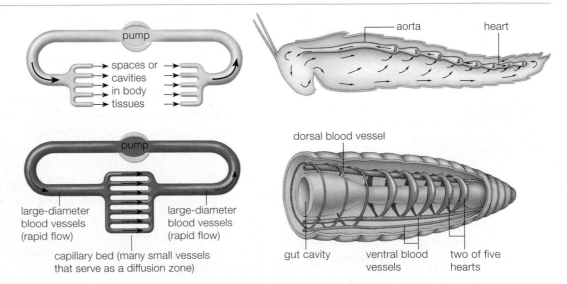

A. Open circulatory system.
A grasshopper's heart pumps hemolymph through a large vessel and out into tissue spaces. Hemolymph mingles with interstitial fluid, exchanges materials, and then reenters the heart through openings in its wall. Insect hemolymph is usually yellow to green in color.

B. Closed circulatory system.
An earthworm's hearts pump blood through vessels that extend through the body. Exchanges between blood and the tissues take place across the wall of the smallest vessels. Like vertebrate blood, earthworm blood contains hemoglobin and is red in color.

A. In fish, the heart has one atrium and one ventricle. Force of the ventricle's contraction propels blood through the single circuit.

B. In amphibians and most reptiles, the heart has three chambers: two atria and one ventricle. Blood flows in two partially separated circuits. Oxygenated and oxygen-poor blood mix a bit in the ventricle.

C. In crocodilians, birds, and mammals, the heart has four chambers: two atria and two ventricles. Oxygenated blood and oxygen-poor blood do not mix.

oxygen-poor blood   oxygen-rich blood

Figure 21.4 Variation among vertebrate circulatory systems.

Figure It Out: Which circuit delivers oxygenated blood to the muscle cells of a dog's legs?

Answer: The systemic circuit. It services all nonlung parts of the body.

Evolution of Vertebrate Cardiovascular Systems All vertebrates have a closed circulatory system, with a single heart and a network of blood vessels. These structures are collectively referred to as a cardiovascular system. The Greek word *kardia* means heart, and the Latin word *vasculum* means vessel.

The structure of the heart differs among vertebrate groups, as do the circuits through which blood flows. In most fishes, the heart has two main chambers, and blood flows in a single circuit (Figure 21.4A). One chamber of the heart, an **atrium** (plural, atria), receives blood. From there, the blood enters a **ventricle**, a chamber that pumps blood out of the heart. The pressure exerted by the contracting ventricle drives the blood through a series of vessels, into networks of capillaries inside each gill, through similar networks in body tissues and organs, and finally back to the heart. Traveling through capillaries reduces the pressure imparted to blood by the ventricle's contraction, so the blood is under less pressure when it leaves the gill capillaries, and even less as it travels back to the heart.

Adapting to life on land involved modifications of respiratory and circulatory systems. Amphibians and most reptiles have a three-chamber heart, with two atria emptying into one ventricle (Figure 21.4B). The heart of these animals propels blood through two circuits, increasing the speed of blood flow. The force of one contraction drives blood through the **pulmonary circuit**, to the lungs and then back to the heart. (The Latin word *pulmo* means lung.) A second contraction sends the newly oxygenated blood through the **systemic circuit**, which runs through capillaries in body tissues before returning to the heart. Oxygenated blood and oxygen-poor blood mix in the single ventricle.

Oxygen delivery improved with evolution of a four-chamber heart, which has two atria and two ventricles (Figure 21.4C). Crocodilians, birds, and mammals have a four-chamber heart. With two fully separated circuits, only oxygen-rich blood flows to tissues. Another advantage of a four-chamber heart is that blood pressure can be regulated independently in each circuit. One ventricle can contract forcefully to keep blood moving quickly through a long systemic circuit. At the same time, a less forceful contraction of the other ventricle sends blood to the lung where the higher blood pressure would damage delicate lung capillaries.

Crocodilians, birds, and mammals do not share an ancestor that had a heart with four chambers. Rather, such a heart evolved independently in each group. As a result, the structure of the heart differs somewhat among these groups.

Take-Home Message 21.2

What are features of animal circulatory systems?

• In an open circulatory system, fluid leaves vessels and seeps around tissues.

• Some invertebrates and all vertebrates have a closed circulatory system, in which blood flows through a continuous network of blood vessels. Blood flow through a closed circulatory system is faster than blood flow through an open circulatory system.

• In fishes, blood flows in one circuit. In other vertebrates, it flow in two circuits. The pulmonary circuit carries blood to and from lungs. The systemic circuit carries blood to and from other organs.

• Crocodilians, birds, and mammals have a four-chamber heart that prevents oxygen-rich blood from mixing with oxygen-poor blood.

21.3 Human Cardiovascular System

Like other mammals, humans have a four-chamber heart that pumps blood through two circuits (Figure 21.5). The short pulmonary circuit (gold boxes) carries blood to and from lungs. Blood enters this circuit when the heart's right ventricle pumps oxygen-poor blood into two pulmonary arteries ❶. An **artery** is a large-diameter blood vessel that carries swiftly moving blood away from the heart. Each pulmonary artery delivers blood to one lung. As blood flows through pulmonary capillaries (capillaries of the lung) ❷, it picks up oxygen and releases carbon dioxide. The newly oxygenated blood then returns to the heart through pulmonary veins ❸ that empty into the left atrium. A **vein** is a large vessel that carries blood back toward the heart.

Oxygenated blood next travels through the longer systemic circuit. The heart's left ventricle pumps blood into the body's largest artery, the aorta ❹. Some vessels that branch from the aorta deliver blood to the capillaries of the upper body ❺. Blood gives up oxygen and picks up carbon dioxide as it flows through these capillaries. It then returns to the heart's right atrium, through a large vein called the superior vena cava ❻. In this context, *superior* means uppermost. Other branches from the aorta deliver blood to capillaries in the abdomen and lower limbs ❼. The oxygen-poor blood that leaves these capillaries returns to the heart's right atrium via another large vein, called the inferior (meaning lower) vena cava ❽.

In a person at rest, a blood cell takes about a minute to make its way through both loops of the cardiovascular system. Over the course of its travels, the cell passes through the heart twice. Each time, contraction of a ventricle speeds it on its way.

Figure 21.5 The two circuits of the human cardiovascular system.
Pulmonary circulation is shown in the gold boxes and systemic circulation in the green boxes.

After Russell/Wolfe/Hertz/Starr. *Biology*, 2e. © 2011 Cengage Learning®.

Figure It Out: Which blood vessel returns oxygen-poor blood from capillaries in a person's legs to the heart?

Answer: Inferior vena cava

Systemic Circuit

capillaries of the head, neck, chest, and arms

Pulmonary Circuit

superior vena cava

aorta

pulmonary artery

pulmonary artery

pulmonary vein

pulmonary vein

Pulmonary Circuit

capillaries of right lung

inferior vena cava

capillaries of left lung

capillaries of the abdomen and lower limbs

Systemic Circuit

Take-Home Message 21.3

What are the components of a human cardiovascular system and how do they interact?

• The human cardiovascular system has a four-chamber heart that pumps blood through a network of blood vessels in two circuits.

• Oxygen-poor blood pumped out of the heart's right ventricle travels through the pulmonary circuit. Pulmonary arteries carry the blood to the lungs, then pulmonary veins return the now-oxygenated blood to the left atrium.

• Oxygen-rich blood pumped out of the left ventricle travels through the systemic circuit. The aorta gives off branches that supply capillary beds of the body. The superior vena cava and inferior vena cava return oxygen-poor blood to the right atrium.

artery Large-diameter blood vessel that carries blood away from the heart.

atrium Heart chamber that pumps blood into a ventricle.

pulmonary circuit Circuit through which blood flows from the heart to the lungs and back.

systemic circuit Circuit through which blood flows from the heart to the body tissues and back.

vein Large-diameter vessel that returns blood to the heart.

ventricle Heart chamber that pumps blood out of the heart and into an artery.

Contraction of the ventricles drives blood circulation. Atrial contraction only helps fill the ventricles.

21.4 The Human Heart

REMEMBER: Gap junctions are intercellular connections that allow ions to flow from one cell to another (Section 3.5).

The heart lies inside the thoracic cavity, between the lungs (Figure 21.6A). It is protected and anchored in place by pericardium, which is a double-layered sac of tough connective tissue. Fluid between the sac's two layers reduces the friction between them as the heart changes shape. Inside the sac, the heart wall consists mostly of cardiac muscle cells. Endothelium, a type of simple squamous epithelium, lines the heart's chambers, as well as all blood vessels.

A thick septum separates the heart's left and right sides (Figure 21.6B). Each side has two chambers: an atrium and a ventricle. The superior vena cava and inferior vena cava deliver oxygen-poor blood from the body to the right atrium. Pulmonary veins deliver oxygen-rich blood to the left atrium.

To flow from an atrium into a ventricle, or from a ventricle into an artery, blood must pass through a heart valve. These pressure-sensitive valves are like one-way doors that control the flow of blood through the heart. High fluid pressure forces a valve open, then the valve snaps shut when pressure declines. The "lup-dup" sound a beating heart makes arises from the closing of valves. Closing of atrioventricular (AV) valves between the atria and the ventricle produces the first heart sound. Simultaneous closing of valves at the entrances to the aorta and the pulmonary artery produces the second heart sound. With a heart murmur, heart valves do not close normally, so blood swirls through them, causing a swishing sound.

Figure 21.6 Location and structure of the human heart.
Arrows indicate the path of oxygenated blood (red) and oxygen-poor blood (blue).

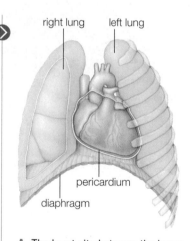

right lung   left lung

pericardium

diaphragm

A. The heart sits between the lungs.

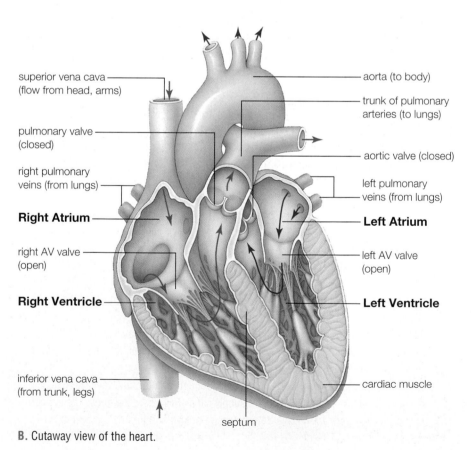

superior vena cava (flow from head, arms)

aorta (to body)

trunk of pulmonary arteries (to lungs)

pulmonary valve (closed)

aortic valve (closed)

right pulmonary veins (from lungs)

left pulmonary veins (from lungs)

Right Atrium

Left Atrium

right AV valve (open)

left AV valve (open)

Right Ventricle

Left Ventricle

inferior vena cava (from trunk, legs)

cardiac muscle

septum

B. Cutaway view of the heart.

The Cardiac Cycle With each heartbeat, the heart's chambers go through a sequence of relaxation and contraction known as the **cardiac cycle** (Figure 21.7). As we begin our observation of the cardiac cycle, both the atria and ventricles are relaxed. The pressure exerted by blood entering the atria has forced the AV valves open, and blood is filling the ventricles ❶. Contraction of the atria forces more blood into the ventricles ❷. Next, the ventricles begin contracting. The resulting rise in fluid pressure inside the ventricles causes the AV valves to shut. Pressure in the ventricles continues to rise until the aortic and pulmonary valves open, and blood rushes out of the ventricles and into the aorta and pulmonary arteries ❸. As fluid pressure in the ventricles declines, aortic and pulmonary valves close. At this point, the atria are starting to fill again ❹.

Contraction of ventricles is the driving force for blood circulation. Atrial contraction only helps fill ventricles. The structure of the cardiac chambers reflects this difference in function. Atria need only to generate enough force to squeeze blood into the ventricles, so they have relatively thin walls. Ventricle walls are much thicker. Contraction of muscle in the ventricle walls has to be strong enough to create a pressure wave that propels blood through an entire circuit. The left ventricle, which pumps blood throughout the body, has thicker walls than the right ventricle, which pumps blood only to the lungs and back.

Setting the Pace of Contractions The sinoatrial (SA) node, a clump of cells in the wall of the right atrium, is the **cardiac pacemaker** (Figure 21.8). In a healthy person at rest, the SA node sends out a muscle-exciting electrical signal about seventy times a minute. Gap junctions between adjacent muscle cells allow electrical signals generated by the SA node to spread across the atria, causing the atria to contract. At the same time, special noncontractile muscle fibers convey the excitatory signal to the atrioventricular (AV) node. This clump of cells is the only electric bridge to the ventricles, meaning the signal to contract can only reach the ventricles by way of the AV node. The time it takes for a signal to reach and cross this bridge allows blood arriving from the atria to fill the ventricles before the ventricles contract. From the AV node, the excitatory signal travels along conducting fibers in the septum between the heart's left and right halves. The fibers extend to the heart's lowest point and up the ventricle walls. As the excitatory signal spreads via gap junctions, ventricles contract from the bottom up, with a wringing motion.

As noted in Section 21.1, malfunctions of a heart's natural pacemaker can be dangerous or deadly. Implanting a battery-operated electrical pacemaker can regulate the heart rate of a person whose heart would otherwise beat too slow or too fast.

❶ Relaxed atria fill. Fluid pressure opens AV valves and blood flows into the relaxed ventricles.

❷ Atrial contraction squeezes more blood into the still-relaxed ventricles.

❹ As blood flows into the arteries, pressure in the ventricles declines and the aortic and pulmonary valves close.

❸ Ventricles start to contract and the rising pressure pushes the AV valves shut. A further rise in pressure causes the aortic and pulmonary valves to open.

Figure 21.7 The cardiac cycle.

Figure It Out: Which valves open in response to rising atrial pressure?

Answer: The AV valves

SA node (cardiac pacemaker)

AV node

fibers that relay signals

Figure 21.8 Components of the heart's signaling system. Signals from the SA node travel to the AV node and then on to the ventricles along junctional fibers.

Take-Home Message 21.4

How does the human heart function?

- The four-chamber heart is a muscular pump. Contraction of the thick-walled ventricles drives blood circulation. Atrial contraction fills the ventricles.
- The SA node is the cardiac pacemaker. Its spontaneous signals make cardiac muscle fibers of the heart wall contract in a coordinated fashion.

cardiac cycle Sequence of contraction and relaxation of heart chambers that occurs with each heartbeat.

cardiac pacemaker Group of heart cells (SA node) that emits rhythmic signals calling for muscle contraction.

plasma
(water,
plasma
proteins,
hormones,
nutrients,
dissolved
gases)

red   white
blood cell   blood cell   platelet

Figure 21.9 Components of blood.

Right, National Cancer Institute/Science Source.

**Figure 21.10
Blood clotting.**
Cells and platelets
become trapped
by threads of
fibrin derived from
a plasma protein
(fibrinogen).

Professor P. Motta/ Depart-
ment of Anatomy/ University
La Sapienca, Rome/Science
Source.

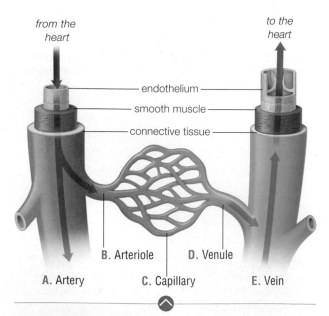

from the
heart

to the
heart

endothelium
smooth muscle
connective tissue

B. Arteriole   **D.** Venule

A. Artery   **C.** Capillary   **E.** Vein

Figure 21.11 Blood vessels.

From Russell/Wolfe/Hertz/Starr. *Biology*, 1e. © 2008, Cengage Learning®.

21.5 Blood and Blood Vessels

REMEMBER: Hemoglobin is an iron-containing protein that reversibly binds oxygen (Section 7.6). Stem cells give rise to specialized cells (19.1). Blood cells are made by stem cells in red bone marrow (20.2).

Components and Functions of Blood A human adult has about 5 liters of blood (a bit more than 5 quarts). Blood is—as the saying goes—thicker than water. It consists of plasma and cellular components (Figure 21.9). All cellular components descend from stem cells in red bone marrow.

Plasma, the fluid portion of blood, consists mostly of water with dissolved plasma proteins. More than half of these proteins are albumins, which are made by the liver. Albumins help transport steroid hormones, fat-soluble vitamins, and other lipids. The liver also makes plasma proteins that are essential to blood clotting. Another class of plasma proteins, the immunoglobulins, are made by white blood cells and function in immunity. Mineral ions, gases, sugars, amino acids, and water-soluble hormones and vitamins also travel in plasma.

Red blood cells (erythrocytes) are the most numerous blood cells. Stored hemoglobin gives them their red color. In mammals, red blood cells expel their nucleus and mitochondria before they enter the circulation. A circulating red blood cell is a flexible, hemoglobin-filled disk with a depression in the middle. The cell's flexibility allows it to slip easily through narrow blood vessels, and its thinness facilitates gas exchange. The mature cell will circulate for about four months.

White blood cells (leukocytes) function in housekeeping and defense. They engulf and digest cellular debris such as aged red blood cells and they detect and destroy pathogens. The next chapter describes the various types of white blood cells and their roles in immunity in more detail.

Platelets are bits of cytoplasm wrapped in plasma membrane. If an injury occurs, platelets release substances that initiate blood clotting. When small vessels are cut or torn, platelets clump together and temporarily fill the breach. They release substances that attract more platelets. During clot formation, enzymes convert fibrinogen, a soluble plasma protein, to insoluble threads of fibrin. Fibrin forms a mesh that traps cells and platelets (Figure 21.10). The entire mass is called a blood clot. Eventually, the clot retracts and forms a compact patch that seals the breach in the blood vessel.

High-Pressure Flow in Arteries Blood pumped out of ventricles enters arteries, which have a muscular wall reinforced with elastic tissue (Figure 21.11A). The elasticity of arteries helps keep blood moving, even between contractions. When a ventricle contracts, it forces blood into the arteries, causing their elastic walls to bulge. When the ventricle relaxes, artery walls rebound inward and squeeze the blood inside them a bit farther away from the heart. The bulging of an artery with each ventricular contraction is referred to as the pulse.

Blood pressure, the pressure exerted by blood against the walls of vessels that enclose it, is highest in arteries and declines as blood proceeds through a circuit. Blood pressure is usually measured in the brachial artery of the upper arm. Two pressures are recorded. **Systolic pressure**, the highest pressure of a cardiac cycle, occurs as the contracting ventricles force blood into the arteries. **Diastolic pressure**, the lowest blood pressure of a cardiac cycle, occurs when the ventricles are fully relaxed. Blood pressure is typically written as systolic value/diastolic value. It is measured in millimeters of mercury (mm Hg), a standard unit for measuring

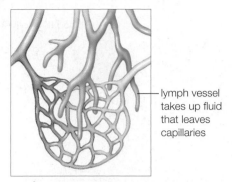

blood to venule

water

proteins remain in capillary

4 Near the venule, water enters blood by osmosis.

protein-free plasma

3 CO₂ diffuses and other wastes are transported into plasma.

2 O₂ diffuses, and nutrients such as glucose are transported out of blood and into interstitial fluid.

blood from arteriole

1 At the arterial end of a capillary, high blood pressure forces some protein-free plasma out between cells of the capillary wall.

cells surrounded by interstitial fluid

Figure 21.12 Capillary exchange.

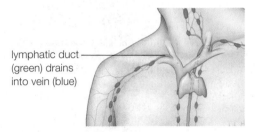

lymph vessel takes up fluid that leaves capillaries

A. Excess fluid that leaks out of capillaries enters adjacent lymph vessels.

lymphatic duct (green) drains into vein (blue)

B. Lymph vessels converge on lymph ducts that return lymph to large veins near the heart.

Figure 21.13 Interactions between the circulatory and lymphatic systems.

pressure. Normal blood pressure is about 120/80 mm Hg. In conversation, you would say this is a pressure of "120 over 80."

Adjusting Resistance at Arterioles Your body can alter the distribution of blood flow by adjusting the diameter of **arterioles**, vessels that receive blood from an artery and deliver it to capillaries (Figure 21.11B). Nervous or endocrine signals cause smooth muscle ringing an arteriole to constrict, so that the vessel narrows and less blood flows through it. Alternatively, signals can cause the smooth muscle to relax, so that the vessel dilates and blood flow increases. For example, after you eat, arterioles that supply blood to capillaries in your gut dilate, and arterioles supplying blood to capillaries that service skeletal muscles in your legs narrow. Conversely, when you run, arterioles in your legs dilate and those supplying the gut constrict.

Capillary Exchange and Function of the Lymph Vessels Flow rate slows dramatically when blood enters a network of capillaries, where blood exchanges fluids and solutes with cells (Figure 21.11C). Every cell is near at least one capillary. This proximity is essential because diffusion cannot move substances over long distances fast enough to keep cells alive.

Materials move between capillaries and body cells in several ways. A capillary wall is of a single layer of squamous epithelium. In most tissues, capillaries are leaky, with narrow spaces between the cells. At the arterial end of a capillary, pressure exerted by the beating heart forces plasma fluid rich in oxygen (O_2) and nutrients through these spaces into the surrounding interstitial fluid (Figure 21.12 **1**). Plasma proteins are too big to exit through the spaces between cells, so they remain in the vessel. Along the length of the capillary, oxygen diffuses from the blood into the interstitial fluid, while nutrients such as glucose are transported in the same direction **2**. Carbon dioxide (CO_2) diffuses into the capillary and other metabolic wastes are transported into it **3**. Near the venous end of the capillary bed, where blood pressure is lower, water moves by osmosis from the interstitial fluid into the protein-rich plasma **4**. With all the leaking and reentry of fluid, there is a small *net* outward flow from a capillary bed. The extra fluid enters vessels of the lymphatic system and becomes **lymph**. Lymph vessels converge and empty into lymphatic ducts that return fluid to veins near the heart (Figure 21.13).

arteriole Blood vessel that delivers blood from an artery to capillaries; site of adjustments to blood distribution.

blood pressure Pressure exerted by blood against the walls of blood vessels.

diastolic pressure Blood pressure when ventricles are relaxed.

lymph Fluid that has left capillaries and entered lymph vessels.

plasma Protein-rich fluid portion of blood.

platelet Cell fragment that functions in blood clotting.

red blood cell Hemoglobin-filled, oxygen-transporting cell.

systolic pressure Blood pressure when the ventricles are contracting.

white blood cell Blood cell that helps defend the body against pathogens and cleans up cellular debris.

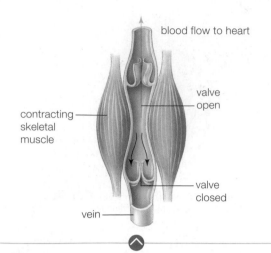

blood flow to heart

valve
open

contracting
skeletal
muscle

valve
closed

vein

Figure 21.14 Skeletal muscle's effect on a vein. When a skeletal muscle next to a vein contracts, the muscle's increased bulk pushes on the vein, forcing blood inside it through a one-way valve and in the direction of the heart.

Back to the Heart Blood from capillary beds enters **venules**, or "little veins," which merge into veins (Figure 21.11D, E). One-way, pressure-sensitive valves prevent blood in veins from flowing backward. The flexible vein wall can bulge greatly under pressure, so the veins can serve as a blood reservoir. When you stand, your veins can hold up to 70 percent of your total blood volume.

Like arteries and arterioles, veins have some smooth muscle in their wall. When blood needs to circulate faster, as during exercise, the muscle in vein walls contracts. The resulting increase in blood pressure drives more blood back to the heart. In addition, contractions of skeletal muscles help keep the blood moving. When these muscles contract, they bulge and press against nearby veins. The additional pressure on the blood in the veins helps move it toward the heart (Figure 21.14).

Take-Home Message 21.5

How do blood and blood vessels function?

- Blood consists mainly of plasma, a protein-rich fluid with solutes and dissolved gases. Red blood cells distribute oxygen. White blood cells defend the body. Platelets are cell fragments that help blood clot after an injury.
- Blood flows into arteries under high pressure. Elastic arterial walls help propel blood forward when ventricles are relaxed. Adjusting arteriole diameter alters how much blood flows to different body regions.
- Exchanges between the blood and cells occur at capillary beds. Venules and veins return blood to the heart.

21.6 Blood and Cardiovascular Disorders

REMEMBER: Diet affects the risk of cardiovascular disease (Section 2.1). Sickle-cell anemia is a genetic disorder caused by a mutation in the gene for hemoglobin (7.6).

Blood Disorders The most common blood disorder is anemia, in which red blood cells are in short supply or abnormal. As a result, oxygen delivery to cells falters, causing shortness of breath, fatigue, and chills. There are many causes for anemia. The most common is chronic blood loss, as from heavy menstrual periods or a bleeding ulcer, which can decrease the number of red blood cells. Sickle-cell anemia is a genetic disorder. A diet with too little iron can cause anemia too, because the lack of raw material slows production of iron-containing hemoglobin.

Problems can also arise when blood has an impaired or excessive tendency to clot. Hemophilia is a genetic disorder that impairs clotting (Section 9.7). It most commonly occurs because an enzyme that crosslinks strands of fibrin into a stable mesh is defective. Other disorders cause spontaneous clot formation in vessels. Such clots can stay in the vessel where they formed, or break loose and travel in the blood. In either case, the clot can clog a vessel and halt blood flow. A clot that blocks a blood vessel in the brain can cause a stroke, in which interrupted blood flow impairs brain cell function. Bursting of a blood vessel in the brain can also cause a stroke. Either way, a person who survives a stroke often has impairments caused by the death of brain cells.

Figure 21.15 Atherosclerosis.
Plaque accumulates in an artery making it more likely to rupture. If it does, a clot that forms at the rupture site further narrows the vessel.

Right, Biophoto Associates/Science Source.

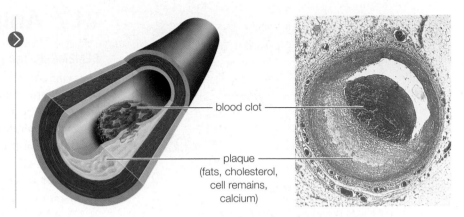

blood clot

plaque
(fats, cholesterol,
cell remains,
calcium)

Cardiovascular Disorders Disorders of the heart and blood vessels often involve **atherosclerosis**, commonly known as "hardening of the arteries." Affected vessels have plaques (lumps) on their inner wall (Figure 21.15). Plaque formation begins with accumulation of fat and cholesterol in a vessel wall. The plaque enlarges as the cells of the affected area becomes inflamed, fibrous scar tissue is laid down, and some cells become hardened by calcium. As a result of plaque formation, the artery becomes narrower and less elastic.

The cholesterol that ends up in plaques is delivered to artery walls by low-density lipoproteins (LDLs), so an excess of LDLs is associated with an increased risk of atherosclerosis. Conversely, high-density lipoproteins (HDLs), which carry cholesterol away from artery walls, contribute to artery health. Physicians monitor LDL and HDL levels to assess the risk of atherosclerosis.

Atherosclerosis raises the risk that a blood vessel in the brain, heart, or other organ will become clogged. Plaques make a vessel more likely to rupture, and the rupture may result in formation of a clot that cuts off blood flow completely. Interrupted blood flow to brain tissue causes a stroke, and interrupted blood flow to the heart causes a heart attack. In either cases, if the blockage is not removed fast, irreplaceable cells die. Drugs that dissolve clots restore blood flow and minimize cell death, but must be administered within an hour of the attack. Anyone suspected of having a stroke or heart attack should receive prompt medical attention.

Several methods are used to treat clogged coronary arteries. With coronary bypass surgery, a bit of blood vessel from elsewhere in the body is stitched to the aorta and to the coronary artery below a clogged region (Figure 21.16A). In laser angioplasty, laser beams vaporize plaques. In balloon angioplasty, doctors inflate a small balloon in a blocked artery to flatten the plaques. A wire mesh tube called a stent is then inserted to help keep the vessel open (Figure 21.16B).

Hypertension refers to chronically high blood pressure (above 140/90). It is sometimes described as a silent killer, because many affected people are unaware they have it. In some people, salt intake causes water retention that raises blood pressure, but the cause of hypertension is usually unknown. Regardless of its cause, hypertension makes the heart work harder than normal, causing it to become enlarged and less efficient. High blood pressure also damages blood vessels and encourages atherosclerosis.

vein from leg;
used to bypass
blockage

blocked
coronary artery

A. Coronary bypass surgery. Veins from another part of the body are used to divert blood past the blockages. This illustration shows a "double bypass," in which veins (green) from elsewhere in the body are placed to divert blood around two blocked coronary arteries.

plaque flattened by
balloon angioplasty

stent (metal mesh) placed
to keep artery open

B. Balloon angioplasty and the placement of a stent. After a balloonlike device is inflated in an artery to open it and flatten the plaque, a tube of metal (the stent) is inserted and left in place to keep the artery open.

Figure 21.16 Treating a blocked coronary (heart) artery. Such blockages are the main cause of heart attacks in older adults.

Take-Home Message 21.6

How do circulatory disorders affect health?

- Anemia causes fatigue and weakness by reducing oxygen delivery to tissues.
- A clot or an atherosclerotic plaque can clog a blood vessel and slow or halt blood flow, thus causing a stroke or heart attack.
- Hypertension puts a strain on the heart, raises the risk of atherosclerosis, and damages small blood vessels.

atherosclerosis Narrowing of an artery's interior due to lipid deposition and inflammation.

hypertension Chronically high blood pressure.

venule Blood vessel that connects a capillary to a vein.

Figure 21.17 Two sites of gas exchange.

① Cells of a respiratory surface exchange gases with the external environment and with fluid inside the body.

② Other body cells exchange gases with the internal environment (interstitial fluid).

In some invertebrates, gases simply diffuse between the two sites. In animals with a circulatory system, blood transports gases rapidly between the two exchange sites.

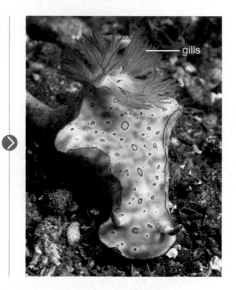

Figure 21.18 Sea slug with feathery external gills.

© Hal Beral VWPics/SuperStock.

21.7 Animal Respiration

REMEMBER: The process of aerobic respiration was described in Section 5.5.

Aerobic respiration is an energy-releasing pathway that requires oxygen (O_2) and produces carbon dioxide (CO_2). Here we focus on the physiological processes of **respiration**, which supply cells with oxygen from the environment, and deliver waste carbon dioxide to that environment.

Two Sites of Gas Exchange Gases enter and leave an animal's body across a thin, moist layer of epithelial cells called the **respiratory surface** (Figure 21.17). A typical respiratory surface is one or two cell layers thick. It must be thin because gases diffuse quickly only over very short distances. It is kept moist because gases cross a cell membrane quickly only if they first dissolve in fluid. Once dissolved, gases diffuse across the lipid bilayer of plasma membranes. The larger the area of a respiratory surface, the greater the speed of gas exchange.

A second exchange of gases takes place internally, at the plasma membrane of body cells. Here, oxygen diffuses from interstitial fluid into a cell, and carbon dioxide diffuses in the opposite direction. In most invertebrates and all vertebrates, a circulatory system enhances the movement of gases between these two sites of gas exchange.

Respiratory Systems Some invertebrates that live in aquatic or continually damp land environments have no special respiratory organs. They exchange gases across their entire body surface. In sea anemones and flatworms, gases diffuse from water into the body across the outer surface and the surface of their internal gastrovascular cavity. Similarly, an earthworm exchanges gases with the air all along the length of its body.

Other aquatic animals have **gills**, filamentous respiratory organs that increase the surface area available for gas exchange. As hemolymph or blood passes through the gills, it exchanges gases with water that surrounds them. Some sea slugs have external gills (Figure 21.18), as do all larval and some adult amphibians. Lobsters and clams have internal gills.

Figure 21.19 Fish respiration.

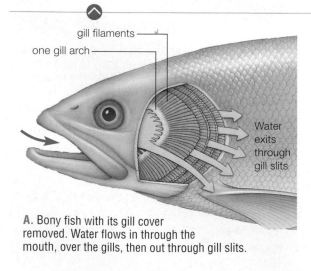

A. Bony fish with its gill cover removed. Water flows in through the mouth, over the gills, then out through gill slits.

B. Each gill has bony gill arches with many thin gill filaments attached.

C. Flow of blood and water in opposite directions maximizes diffusion of oxygen from the water into the blood.

Fishes have a pair of internal gills: rows of slits at the back of the mouth that extend to the body's surface (Figure 21.19A). Water flows into the mouth and pharynx, then over arrays of gill filaments (Figure 21.19B). Each filament has blood vessels that carry blood into and out of it (Figure 21.19C). Water flowing over gills and blood flowing through gill capillaries move in opposite directions. The result is a countercurrent exchange, a mechanism in which two fluids exchange substances while flowing in opposite directions. For the entire length of the capillary, the water next to the capillary holds more oxygen than the blood flowing inside it. As a result, oxygen continually diffuses from the water into the blood. The blood becomes increasingly oxygenated as it passes through the capillary.

The most diverse air-breathing land invertebrates are insects. They have a hard exoskeleton that helps conserve water but also prevents exchange of gases across the body surface. Insects have a **tracheal system** that consists of repeatedly branching, air-filled tubes that begin at spiracles, which are small openings across the body surface (Figure 21.20). Gas exchange occurs at the tips of the finest branches, where gases dissolve in interstitial fluid and diffuse through it to cells. The circulatory system does not play a role in most insect respiration.

Lungs are saclike internal respiratory organs. More than 450 million years ago, lungs evolved from outpouchings in the gut of some fishes. In oxygen-poor water, an ability to gulp air and exchange gases across a respiratory surface in lungs gave these fish an advantage. Later, an ability to breathe air allowed amphibians that descended from these fish to live on land.

Just about all land vertebrates have lungs. Frogs have small lungs that function in oxygen uptake, but most carbon dioxide diffuses outward across their skin. Toads, whose skin is usually dry, have larger lungs. In reptiles, birds, and mammals, the skin is waterproof, so the only respiratory surface is the lining of the lungs.

Birds have small, inelastic lungs that do not expand and contract when the bird breathes. Instead, large expandable air sacs attached to the lungs inflate and deflate (Figure 21.21). It takes two breaths to move air through this system. Oxygen-rich air flows through tiny tubes in the lung during both inhalations and exhalations. The lining of these tubes serves as the bird's respiratory surface and continual movement of air in one direction over this surface increases the efficiency of gas exchange.

By contrast, air flow in a mammalian lung is bidirectional. The route air travels to the respiratory surface is a dead end, so fresh air and oxygen-depleted air mix. As a result, mammalian lungs are less efficient at oxygen delivery than birds' lungs.

spiracle, entrance to the branching tracheal tubes

Figure 21.20 Tracheal tube respiratory system of an insect. Spiracles that open across the body surface are the entrance to chitin-reinforced tracheal tubes that carry air deep inside the body. These light-colored branching tubes are visible through the transparent cuticle of this butterfly larva.

© Andy Warren.

posterior air sacs lungs

anterior air sacs

Figure 21.21 Bird respiratory system with air sacs that keep air moving through lungs.

From Russell/Wolfe/Hertz/Starr. *Biology*, 1e. © 2008, Cengage Learning®.

Take-Home Message 21.7

What structural traits facilitate animal respiration?

- Respiration is a physiological process by which animals obtain oxygen and get rid of waste carbon dioxide by diffusion across a moist respiratory surface.
- In some invertebrates, the body surface serves as the respiratory surface.
- Gills increase the surface area for gas exchange in some invertebrates and vertebrates. In fish gills, opposing flow of water and blood increases the efficiency of gas exchange.
- A system of tracheal tubes delivers air deep inside an insect's body.
- Amphibians have small lungs but also exchange gases across their skin.
- Reptiles, birds, and mammals rely on lungs for respiration. Air sacs keep air flowing continually through a bird's lungs.

gills Folds or body extensions that increase the surface area for respiration.

lungs Internal saclike organs that serve as the respiratory surface in most land vertebrates and some fish.

respiration Physiological process by which gases enter and leave an animal body.

respiratory surface Moist surface across which gases are exchanged between animal cells and their environment.

tracheal system Tubes that deliver air from body surface to tissues of insects.

nasal cavity

pharynx (throat)

epiglottis

larynx (voice box)

trachea (windpipe)

left lung

bronchus

bronchiole

diaphragm

intercostal muscle

Figure 21.22 Human respiratory organs and associated structures. The diaphragm and intercostal muscles (muscles between the ribs) are muscles of respiration. The epiglottis is a flap of tissue that folds over the trachea and prevents food from entering it when you swallow.

Left, Martin Dohrn, Royal College of Surgeons/Science Source.

Resin cast of human lungs, with airways in white and blood vessels in red.

pulmonary capillaries associated with a cluster of alveoli

one alveolus (shown in cross section)

21.8 Human Respiratory Function

We turn now to the mammalian respiratory system, using the human respiratory system as our example.

From Airways to Alveoli   Take a deep breath. Now look at Figure 21.22 to get an idea of where the air traveled in your respiratory system. If you are healthy and sitting quietly, air probably entered through your nose, rather than your mouth. Air from the nostrils enters the nasal cavity, where it is warmed and moistened. The moist, warm air then flows into the **pharynx**, or throat.

Air continues to the **larynx**, a short airway commonly referred to as the voice box because it contains a pair of vocal cords (Figure 21.23). Each vocal cord consists of skeletal muscle covered by mucus-secreting epithelium. Contraction of the vocal cords changes the size of the **glottis**, the gap between them. When the glottis is wide open, air flows through it silently. When muscle contraction narrows the glottis, flow of air outward through the tighter gap makes vocal cords vibrate so they produce sounds. The tension on the cords and the position of the larynx determine the sound's pitch. To get a feel for how this works, place one finger on your "Adam's apple," the laryngeal cartilage that sticks out most at the front of your neck. Hum a low note, then a high one. You will feel the vibration of your vocal cords and how laryngeal muscles shift the position of your larynx.

Figure 21.23 The glottis and vocal cords from above.

Photos, Courtesy of Kay Elemetrics Corporation.

vocal cords

glottis (closed)

epiglottis

base of tongue

glottis closed

glottis open

Figure 21.24 How muscle actions alter the size of the thoracic cavity during one respiratory cycle.

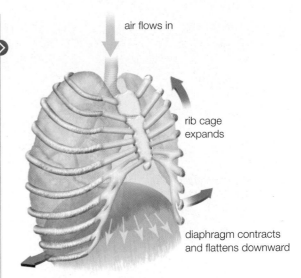

air flows in

rib cage
expands

diaphragm contracts
and flattens downward

A. Inhalation

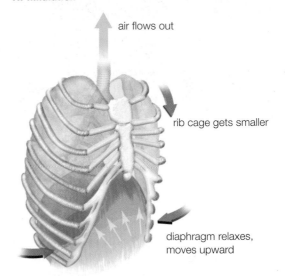

air flows out

rib cage gets smaller

diaphragm relaxes,
moves upward

B. Exhalation

When you breathe silently, the glottis is wide open and the **epiglottis**, a tissue flap at the entrance to the larynx, points upward. With the epiglottis in its upward position, air can enter the **trachea**, or windpipe. When you swallow, the epiglottis flops down and covers the larynx entrance, so food and fluids enter the esophagus. The esophagus connects the pharynx to the stomach.

The trachea branches into two airways, one to each lung. Each airway is a **bronchus** (plural, bronchi). Human lungs are cone-shaped organs in the thoracic cavity, one on each side of the heart. The rib cage encloses and protects the lungs. A two-layer-thick pleural membrane covers each lung's outer surface and lines the inner thoracic cavity wall.

Once inside a lung, air moves through finer and finer branchings of a "bronchial tree." The branches are called **bronchioles**. At the tips of the finest bronchioles are respiratory **alveoli** (singular, alveolus), little air sacs where gases are exchanged. Air in alveoli exchanges gases with blood flowing through pulmonary capillaries. Blood oxygenated in these capillaries returns to the heart, which then pumps it to the body's tissues.

How You Breathe The **diaphragm**, a broad sheet of smooth muscle beneath the lungs, partitions the human coelom into a thoracic cavity and an abdominal cavity. Of all smooth muscle, only the diaphragm can be controlled voluntarily. You can make it contract by deliberately inhaling. Movements of the diaphragm and the **intercostal muscles**, the skeletal muscles between the ribs, allow you to breathe.

A **respiratory cycle** includes one breath in (inhalation) and one breath out (exhalation). Inhalation is always active; muscle contractions drive it. Changes in the volume of the lungs and thoracic cavity cause the movement of air.

When you inhale, the diaphragm contracts and moves downward. At the same time, external intercostal muscles between the ribs contract, moving the rib cage upward and outward (Figure 21.24A). The thoracic cavity expands, and air is pulled into the lungs.

Exhalation is usually passive. When muscles that caused inhalation relax, the lungs passively recoil and lung volume decreases. This decrease in volume compresses alveolar sacs, and pushes air out of the lungs (Figure 21.24B).

Exhalation becomes active when you exercise vigorously or consciously attempt to expel more air. During active exhalation, internal intercostal muscles contract, pulling the thoracic wall inward and downward. At the same time, muscles of the abdominal wall contract. Abdominal pressure increases and pushes the diaphragm upward. As a result, the volume of the thoracic cavity decreases more than normal, and a bit more air is forced outward.

You do not have to think about breathing. Neurons in a part of your brain stem act as a pacemaker for respiration. When you rest, these neurons send out signals 10 to 14 times per minute. Nerves carry these signals to the diaphragm and intercostal muscles, causing contractions that result in inhalation. When you are more active, muscle cells increase their rate of aerobic respiration and produce more CO_2. This CO_2 enters blood, causing changes that are detected by receptors in arteries and the brain. In a homeostatic response to these changes, the brain alters the breathing pattern, so you breathe faster and more deeply.

alveoli Tiny, thin-walled air sacs that are the site of gas exchange in the lung.

bronchiole Small airway leading to alveoli.

bronchus Airway connecting the trachea to a lung.

diaphragm Dome-shaped muscle at the base of the thoracic cavity that alters the size of this cavity during breathing.

epiglottis Tissue flap that folds down to prevent food from entering the airways when you swallow.

glottis Opening formed when the vocal cords relax.

intercostal muscles Muscles between the ribs; help alter the size of the thoracic cavity during breathing.

larynx Short airway containing the vocal cords (voice box).

pharynx Throat; opens to airways and digestive tract.

respiratory cycle One inhalation and one exhalation.

trachea Major airway leading to the lungs; windpipe.

A

Figure 21.25 The site of gas exchange.
A. Cutaway view of a cluster of alveoli. Each alveolus is filled with air and surrounded by a network of pulmonary capillaries.

B. The respiratory surface consists of the wall of the alveolus, the wall of the capillary, and their fused basement membranes. Gases diffuse across this three-layered boundary.

Figure 21.26 Hemoglobin.
This protein consists of four globin molecules (blue and green). Each polypeptide associates with an iron-containing heme group (red) that can reversibly bind a molecule of O_2.

Exchanges at Alveoli Inhaled air exchanges gases with the blood at the alveoli. Alveolus means cavity in Latin, and each alveolus is a tiny sphere with a central air-filled cavity (Figure 21.25A). Alveolar epithelium contains squamous cells and secretory cells. The secretory cells release a substance that lubricates alveolar walls, preventing them from sticking together when the alveolus deflates. Pulmonary capillaries run across the outer surface of each alveolus. The basement membranes secreted by the alveolar and capillary cells fuse together as the respiratory membrane (Figure 21.25B). Gases diffuse quickly across this thin membrane between the air inside alveoli and the blood flowing through a capillary.

Transport of Gases Some oxygen can dissolve in plasma, but not enough to meet the body's needs. Nearly all the oxygen you inhale becomes bound to hemoglobin in your red blood cells. Hemoglobin is a protein that consists of four polypeptide chains (globin), each with an associated iron-containing cofactor called a heme (Figure 21.26). Oxygen binds reversibly to the iron in hemes. Hemoglobin tends to let go of oxygen in regions where the oxygen concentration is low, the temperature is warm, the pH is low, and the concentration of carbon dioxide is high. Such conditions occur in metabolically active tissues. After giving up oxygen to tissues, red blood cells return to the pulmonary capillaries to pick up more.

Carbon dioxide diffuses into the blood from body tissues, where its concentration exceeds that of the blood. On their return trip from the tissues to the lungs, red blood cells carry a small amount of carbon dioxide bound to hemoglobin. However, most carbon dioxide is transported to the lungs as bicarbonate ions (HCO_3^-). Enzymes inside red blood cells carry out a reaction that combines carbon dioxide with water to form these ions. Once formed, the bicarbonate diffuses out of the red blood cell and travels to the lungs in the plasma. In the pulmonary capillaries, bicarbonate is converted back to carbon dioxide. The carbon dioxide diffuses into the air in alveoli and is exhaled.

Respiratory Disorders Bronchitis is a common respiratory disorder in which the mucus-producing lining of the bronchi becomes inflamed. Inflamed epithelial cells secrete extra mucus that triggers coughing. Bacteria can colonize the mucus, leading to more inflammation, more mucus, and more coughing. Bronchitis often arises after an upper respiratory infection. Smoking irritates airways and can cause chronic bronchitis.

With asthma, an inhaled allergen or irritant triggers inflammation and constriction of airways, making breathing difficult. A tendency to have asthma is inherited, but avoiding potential irritants such as cigarette smoke and air pollutants can reduce the frequency of attacks. An acute asthma attack is treated with inhaled drugs that cause smooth muscle ringing the airways to relax.

With emphysema, tissue-destroying bacterial enzymes digest the thin, elastic alveolar wall. As these walls disappear, the area of the respiratory surface declines. Over time, the lungs become distended and inelastic, leaving the person constantly feeling short of breath. Some people inherit a genetic predisposition to emphysema. They do not have a functioning gene for an enzyme that inhibits bacterial attacks

Labels in Figure 21.25B:
- cells of alveolar wall
- cells of capillary wall
- fused basement membranes of both epithelial cell layers
- O_2
- CO_2
- B

Digging Into Data

Radon and Lung Cancer

Radon is a colorless, odorless gas emitted by many rocks and soils. It is formed by the radioactive decay of uranium and is itself radioactive. There is radon in the air almost everywhere, but routinely inhaling a lot of it raises the risk of lung cancer. Radon also seems to increase cancer risk far more in smokers than in nonsmokers.

Figure 21.27 is an estimate of how radon in homes affects risk of lung cancer mortality. Note that these data show only the death risk for radon-induced cancers. Smokers are also at risk from lung cancers that are caused by tobacco.

1. If 1,000 smokers were exposed to a radon level of 1.3 pCi/L over a lifetime (the average indoor radon level), how many would die of a radon-induced lung cancer?

2. How high would the radon level have to be to cause approximately the same number of cancers among 1,000 nonsmokers?

3. The risk of dying in a car crash is about 7 out of 1,000. Is a smoker in a home with an average radon level (1.3 pCi/L) more likely to die in a car crash or of radon-induced cancer?

Radon level (pCi/L)	Risk of cancer death from lifetime radon exposure	
	Never smoked	Current smokers
20	36 out of 1,000	260 out of 1,000
10	18 out of 1,000	150 out of 1,000
8	15 out of 1,000	120 out of 1,000
4	7 out of 1,000	62 out of 1,000
2	4 out of 1,000	32 out of 1,000
1.3	2 out of 1,000	20 out of 1,000
0.4	>1 out of 1,000	6 out of 1,000

Figure 21.27 Estimated risk of lung cancer death as a result of lifetime radon exposure.
Radon levels are measured in picocuries per liter (pCi/L). The U.S. Environmental Protection Agency considers a radon level above 4 pCi/L unsafe. To learn about testing for radon and what to do about high level, visit the EPA's radon information site at www.epa.gov/radon.

---

on alveoli. However, tobacco smoking is by far the main risk factor for emphysema (Figure 21.28).

Smoking tobacco or breathing secondhand tobacco smoke has a wide range of negative health effects. Tobacco smokers have more cardiovascular disease, with a risk of heart attack four to five times than of nonsmokers. Tobacco smoke also contains many carcinogens (cancer-causing chemicals). Lungs are most often affected; more than 80 percent of lung cancers occur in smokers. However, inhaling tobacco smoke also raises the risk of cancers of the mouth, larynx, esophagus, liver, pancreas, and colon. Smokers are more likely than nonsmokers to have diabetes or rheumatoid arthritis, and to go blind or become demented as they age.

A. Normal human lungs.

Take-Home Message 21.8

How does the human respiratory system function?

- Air enters through the nose or mouth, and flows through the pharynx, larynx (voice box), and trachea to the two bronchi that carry air into the lungs.
- Inhalation is always active. Contraction of the diaphragm and muscles of the rib cage increase the volume of the thoracic cavity and lungs, so air is sucked into the lungs. Exhalation is usually passive.
- Gas exchange takes place in alveoli. Oxygen diffuses from air inside alveoli into pulmonary capillaries. Carbon dioxide diffuses in the opposite direction.
- Most oxygen in blood is bound to hemoglobin in red blood cells. Most carbon dioxide is converted to bicarbonate, which travels in the plasma.

B. Lungs of a smoker with emphysema.

Figure 21.28 One effect of smoking on lungs.
© O. Aurbach/Visuals Unlimited.

Summary

 Section 21.1 When the heart stops pumping, blood flow halts and brain cells begin to die. CPR can keep some oxygenated blood moving to cells, but it cannot restart a heart. Reestablishing the normal rhythm requires a shock from a defibrillator.

Section 21.2 In some invertebrates, substances simply move to and from cells by diffusion through the **interstitial fluid**. However, most animals have a circulatory system, in which a **heart** pumps blood through blood vessels. In an **open circulatory system**, fluid called **hemolymph** leaves vessels and flows among tissues. In a **closed circulatory system**, fluid called **blood** stays inside vessels and exchanges take place at **capillaries**. A fish heart has one **atrium** and one **ventricle**, so blood flows in a single circuit. In other vertebrates, a **pulmonary circuit** services the lungs and a **systemic circuit** extends through the rest of the body.

Sections 21.3, 21.4 Each half of the human heart has two chambers: an upper atrium and a lower ventricle, with valves between them. During the **cardiac cycle**, electrical signals from the **cardiac pacemaker** (the SA node) trigger contraction of the atria, then the ventricles. The sounds we associate with a heartbeat are caused by closing of heart valves. Contraction of the ventricles drives blood flow away from the heart through **arteries**. Blood returns to the atria through **veins**.

 Sections 21.5, 21.6 Blood consists of plasma, blood cells, and platelets. **Plasma** is water with dissolved ions and molecules. **Red blood cells** are packed with hemoglobin and function mainly in the transport of oxygen. Many kinds of **white blood cells** function in housekeeping and defense. **Platelets** release substances that initiate blood clotting. All blood cells and platelets arise from stem cells in bone marrow.

Blood pressure, the pressure exerted by blood on the walls of blood vessels, is usually measured in arteries. It is recorded as **systolic pressure** over **diastolic pressure**. Chronic high blood pressure, or **hypertension**, threatens health, as does **atherosclerosis**, which narrows arteries. **Arterioles** adjust the distribution of blood through the body. Capillaries are diffusion zones for exchanges between blood and interstitial fluid. **Venules** carry blood to veins, which convey blood back to the heart and also function as a blood volume reservoir. Valves keep blood from moving backward in veins. **Lymph** is a mix of interstitial fluid and protein-free plasma that escaped from capillaries. Lymph vessels return lymph to veins.

 Section 21.7 **Respiration** is a physiological process by which gases enter and leave an animal body across a **respiratory surface**. In small, flattened animals, the body surface can serve as the respiratory surface. Sea slugs have external **gills** and fish have internal ones. Insects have a **tracheal system** that conveys air deep inside the body. Reptiles, birds, and mammals rely on their **lungs** for gas exchange. In birds, air sacs connected to lungs keep air flowing continually across the respiratory surface.

 Section 21.8 In humans, air flows through the nasal cavities, the **pharynx**, the **larynx**, **trachea**, **bronchi**, and **bronchioles**, which end at **alveoli** deep inside the lungs. When you swallow, the **epiglottis** covers the **glottis** so food does not enter the larynx.

A **respiratory cycle** is one inhalation and one exhalation. Inhalation requires energy. Contraction of the **diaphragm** and **intercostal muscles** between the ribs expand the chest cavity and pull air into the lungs. These events are reversed during exhalation, which is usually passive.

The walls of alveoli and of pulmonary capillaries combine as the respiratory membrane. Oxygen follows its concentration gradient from alveolar air spaces into the pulmonary capillaries, then into red blood cells, where it binds reversibly with hemoglobin. In capillary beds in the tissues, hemoglobin releases oxygen, which diffuses across interstitial fluid into cells. Carbon dioxide diffuses from cells into the interstitial fluid, then into the blood where it combines with water to form bicarbonate. In the lungs, bicarbonate is converted back to carbon dioxide and water. The carbon dioxide diffuses from the blood into air in alveoli, then is exhaled.

Bronchitis and emphysema are respiratory disorders that can be caused or worsened by smoking. Smoking also raises the risk of cancer, heart disease, and a variety of other disorders.

Self-Quiz

Answers in Appendix I

1. All vertebrates have _____ .
   a. an open circulatory system   c. a four-chamber heart
   b. a closed circulatory system   d. both b and c

2. The _____ serves as the cardiac pacemaker.
   a. AV node   c. AV valve
   b. SA node   d. brain

3. Blood pressure is highest in the _____ and lowest in the _____ .
   a. arteries; veins
   c. veins; arteries
   b. arterioles; venules
   d. capillaries; arterioles

4. Contraction of _____ is the main force driving the flow of blood away from the heart.
   a. the atria
   c. the ventricles
   b. arterioles
   d. skeletal muscle

5. At rest, the largest volume of blood is in the _____ .
   a. arteries
   c. veins
   b. capillaries
   d. arterioles

6. In the blood, most oxygen is transported _____ .
   a. in red blood cells
   c. bound to hemoglobin
   b. in white blood cells
   d. both a and c

7. The _____ circuit carries blood from the heart to the lungs, then back to the heart.
   a. pulmonary
   b. systemic

8. The heart chamber with the thickest wall pumps blood into the _____ .
   a. aorta
   c. pulmonary vein
   b. pulmonary artery
   d. superior vena cava

9. In human lungs, gas exchange occurs at the _____ .
   a. bronchi
   c. alveoli
   b. pericardium
   d. epiglottis

10. When you breathe quietly, inhalation is _____ and exhalation is _____ .
    a. passive; passive
    c. passive; active
    b. active; active
    d. active; passive

11. During inhalation, _____ .
    a. the thoracic cavity expands
    c. the diaphragm relaxes
    b. the glottis closes
    d. all of the above

12. The diaphragm is _____ muscle.
    a. smooth   b. skeletal   c. cardiac

13. Inhaled air flows from the larynx directly into _____ .
    a. the pharynx
    c. a bronchiole
    b. the trachea
    d. a bronchus

14. Match each disorder with its description.
    _____ anemia   a. heart stops beating
    _____ atherosclerosis   b. alveoli walls break down
    _____ hypertension   c. airways constrict
    _____ emphysema   d. impaired blood clotting
    _____ hemophilia   e. chronic high blood pressure
    _____ cardiac arrest   f. inflammation of bronchi
    _____ asthma   g. too few red blood cells
    _____ bronchitis   h. narrows arteries

15. Match the words with their descriptions.
    _____ plasma   a. receives blood from veins
    _____ alveolus   b. fluid component of blood
    _____ hemoglobin   c. site of gas exchange
    _____ veins   d. gap between vocal cords
    _____ trachea   e. drives blood flow from heart
    _____ glottis   f. windpipe
    _____ ventricle   g. blood volume reservoir
    _____ atrium   h. reversibly binds oxygen

Critical Thinking

1. Explain why a blood clot that forms in the leg, then breaks free as an embolus, is more likely to become stuck in a vessel in a lung than a vessel in the brain.

2. Carbon monoxide is a colorless, odorless gas produced by combustion, including burning of tobacco. If inhaled, it binds tightly to the oxygen-binding sites on hemoglobin, preventing oxygen from binding. Why does exposure to low levels of carbon monoxide produce symptoms like those of anemia?

Visual Question

1. Which of the graphics at the right depicts the point at which systolic blood pressure is measured, the point at which the blood pressure is highest?

Application ❯ 22.1 Frankie's Last Wish

REMEMBER: Viruses such as HPV cause a cell to make proteins that interfere with its own tumor suppressors; infection with HPV causes skin growths called warts, and some kinds are associated with neoplasms that form on the cervix; cancer occurs when the abnormally dividing cells of a malignant neoplasm physically and metabolically disrupt body tissues (Section 8.3). A virus is a noncellular infectious particle; its genes take over a host's cellular machinery, directing the cell to produce more viruses that are released when the infected host cell bursts (13.4).

Frankie McCullough had known for a few months that something was not quite right. She had not had an annual checkup in many years; after all, she was only 31 years old and had always been healthy. She had never doubted her own invincibility until the moment she saw the doctor's face change as he examined her cervix.

The cervix is the lowest part of the uterus, or womb. Cervical cells can become cancerous, but the process is usually slow. The cells pass through several precancerous stages that are detectable by routine Pap tests. Precancerous and even early-stage cancerous cells can be removed from the cervix before they spread to other parts of the body. However, plenty of women like Frankie do not take advantage of regular exams. They do not see a gynecologist unless they have pain or bleeding, which can be symptoms of advanced cervical cancer. Even with treatment, 85 percent of women who end up with this type of cancer will die within five years. Cervical cancer kills thousands of women each year in the United States, and many more in places where routine gynecological testing is not a common practice.

What causes cancer? At least in the case of cervical cancer, we know the answer to that question: Healthy cervical cells are transformed into cancerous ones by infection with human papillomavirus (HPV). HPV is a DNA virus that infects skin and mucous membranes. There are about 100 different strains of HPV; a few cause warts on the hands or feet, or in the mouth. About 30 others that infect the genital area sometimes cause genital warts, but usually there are no symptoms of infection. Genital HPV is spread very easily by sexual contact: A woman has a 50% chance of being infected with genital HPV within three years of becoming sexually active.

A genital HPV infection usually goes away on its own, but not always. A persistent infection with one of about 10 strains is the main risk factor for cervical cancer (Figure 22.1). Types 16 and 18 are particularly dangerous: At least one of these two HPV strains is found in most cervical cancers.

In 2006, the U.S. Food and Drug Administration (FDA) approved Gardasil, a vaccine against four types of genital HPV, including types 16 and 18. Cervarix, which was approved in 2009, targets types 16 and 18 only. Both vaccines consist of viral proteins that self-assemble into virus-like particles. These particles are not infectious (they contain no viral DNA), but the viral proteins they consist of elicit an immune response that can prevent HPV infection and the cervical cancer it causes. Immunization is most effective in girls who have not yet become sexually active, because they are least likely to have already become infected with HPV.

The HPV vaccine came too late for Frankie. Despite radiation treatments and chemotherapy, her cervical cancer spread quickly. She died in 2001, leaving a wish for other people: awareness. "If there is one thing I could tell a young woman to convince them to have a yearly exam, it would be not to assume that your youth will protect you. Cancer does not discriminate; it will attack at random, and early detection is the answer." Almost all women newly diagnosed with invasive cervical cancer have not had a Pap test in five years, and many have never had one.

Figure 22.1 HPV and cervical cancer.
Top, a Pap test reveals HPV-infected cervical cells among normal ones. Infected cells have enlarged, often multiple nuclei surrounded by a clear area. These changes sometimes lead to cervical cancer, which is treatable if detected early enough.

Bottom, Frankie McCullough (waving), who died of cervical cancer at age 32.

Top, Biomedical Imaging Unit, Southampton General Hospital/Science Source; bottom, In memory of Frankie McCullough.

Digging Into Data

Cervical Cancer Incidence in HPV-Positive Women

In 2003, Michelle Khan and her coworkers published results of their 10-year study correlating HPV status with cervical cancer incidence in women (Figure 22.2). All 20,514 participants were free of cervical cancer when the study began. The researchers tested the women's cervical cells for specific types of HPV at regular intervals.

1. At 110 months into the study, what percentage of women who were not infected with any type of cancer-causing HPV had cervical cancer? What percentage of women who were infected with HPV16 also had cancer?
2. In which group would women infected with both HPV16 and HPV18 fall?

Figure 22.2 Incidence of cervical cancer and HPV status. The data were grouped as follows: ● HPV16 positive; ○ HPV16 negative and HPV18 positive; ▼ All other cancer-causing HPV types combined; ▽ No cancer-causing HPV type detected.

22.2 Responding to Threats

REMEMBER: Receptor proteins trigger a cellular change in response to a stimulus (Section 3.3). Almost all prokaryotes have a cell wall, and many have pili and flagella (3.4). Some cells take in large particles by phagocytosis (4.6). Vertebrates are animals with a backbone (15.1). Interstitial fluid is the fluid between cells (21.2). White blood cells are a component of blood (21.5).

You continually cross paths with a tremendous array of viruses, bacteria, fungi, parasitic worms, and other agents of disease, but you need not lose sleep over this. Humans coevolved with these pathogens, so you have defenses that protect your body from them. The evolution of **immunity**, an organism's capacity to resist and combat infection, began before multicelled eukaryotes evolved from free-living cells. Mutations in the genes for membrane proteins introduced new molecular patterns that were unique in cells of a given type. As multicellularity evolved, so did mechanisms of identifying the patterns as self, or belonging to one's own body.

By about 1 billion years ago, nonself recognition had also evolved. Cells of all modern multicelled eukaryotes bear a set of receptors that collectively can recognize around 1,000 different nonself cues, which are called pathogen-associated molecular patterns (PAMPs). As their name suggests, PAMPs occur mainly on or in pathogens. They include proteins that make up bacterial flagella and pili, carbohydrates in bacterial cell walls, double-stranded RNA unique to some viruses, and so on. A PAMP is an example of **antigen**—any molecule or particle recognized by the body as nonself. When a cell's PAMP receptors bind to antigen, they trigger a set of immediate, general defense responses. PAMP receptors and the responses they trigger are part of **innate immunity**, a set of immediate, general defenses that help protect all multicelled organisms from infection. Vertebrates have an additional set of defenses carried out by interacting cells, tissues, and proteins. This **adaptive immunity** tailors immune defenses to a vast array of specific pathogens that an

> Humans coevolved with pathogens, so you have defenses that protect your body from them.

adaptive immunity In vertebrates, a set of immune defenses that can be tailored to specific pathogens as an organism encounters them during its lifetime.

antigen Molecule or particle that the immune system recognizes as nonself. Triggers an immune response.

immunity The body's ability to resist and fight infections.

innate immunity In all multicelled organisms, set of immediate, general defenses against infection.

Table 22.1 Features of Innate and Adaptive Immunity

	Innate Immunity	Adaptive Immunity
Response time	Immediate	About a week
How antigen is detected	Fixed set of receptors for pathogen-associated molecular patterns (PAMPs)	Antigen receptors produced by gene recombinations
Specificity	About 1,000 PAMPs	Billions of antigens
Persistence	None	Long-term

individual encounters during its lifetime. Innate and and adaptive immunity have different features (Table 22.1), but they function together.

The Defenders White blood cells participate in all immune responses. Many kinds circulate through the body in blood (Figure 22.3) and lymph; others populate the lymph nodes, spleen, and other tissues. All communicate with one another by secreting and responding to chemical signaling molecules. These molecules, which include proteins and polypeptides called cytokines, allow cells throughout the body to coordinate their activities during an immune response.

Different types of white blood cells are specialized for specific tasks. Those that are phagocytic (phagocytes) engulf and digest pathogens, dead cells, or other particles. Neutrophils, which circulate in blood, are the most abundant type. Phagocytic **macrophages** that migrate through tissues and interstitial fluid develop from monocytes that patrol the blood. **Dendritic cells**, which are also phagocytic, alert the adaptive immune system to the presence of antigen in solid tissue. Granules contain cytokines, local signaling molecules, destructive enzymes, and toxins such as hydrogen peroxide. A cell releases the contents of its granules (degranulates) in response to a trigger such as antigen binding. Neutrophils have granules, as do eosinophils that target multicelled parasites too big for phagocytosis. Basophils and mast cells degranulate in response to injury as well as antigen. Mast cells, unlike most other white blood cells, stay anchored in tissues. These cells also degranulate in response to signaling molecules secreted by cells of the endocrine and nervous systems. Lymphocytes are a special category of white blood cell with the collective capacity to target billions of specific antigens. **B cells** (B lymphocytes) make antibodies (more about these proteins in Section 22.4). **T cells** (T lymphocytes) play a central role in all adaptive immune responses. Cytotoxic T cells are specialized to kill infected or cancerous body cells. NK cells (natural killer cells) are lymphocytes that kill cancerous body cells undetectable by cytotoxic T cells.

neutrophil   monocyte   basophil   lymphocyte   eosinophil

Figure 22.3 Lineup of white blood cells.
These are a few types that circulate in blood. Staining reveals details such as lobed nuclei and cytoplasmic granules. Red blood cells (pink) are in the background.

© Antonio Zamora, www.scientificpsychic.com.

B cell B lymphocyte. Lymphocyte that can make antibodies.

dendritic cell Phagocytic white blood cell that alerts the immune system to the presence of antigen in solid tissues.

lysozyme Antibacterial enzyme in body secretions such as saliva and mucus.

macrophage Phagocytic white blood cell that patrols tissues and interstitial fluid.

normal flora Microorganisms that typically live on human surfaces, including the interior tubes and cavities of the digestive and respiratory tracts.

T cell T lymphocyte. Lymphocyte central to adaptive immunity; some kinds target infected or cancerous body cells.

Take-Home Message 22.2

How does immunity work?

- Innate immunity of multicelled organisms includes a system of immediate, general defenses that are triggered by a fixed number of antigens.
- Vertebrate adaptive immunity is a system of defenses that can target specific antigens.
- White blood cells are central to both systems; signaling molecules such as cytokines integrate their activities.

22.3 Innate Immunity Mechanisms

REMEMBER: All fluids outside of cells compose a body's internal environment (Section 1.3). Receptor proteins trigger a cellular change in response to a stimulus (3.3). Rows of tight junctions fasten the plasma membranes of adjacent cells and prevent body fluids from seeping between them (3.5). The rate of an enzymatic reaction typically increases with temperature until the enzyme denatures (4.4). Tonicity describes relative solute concentration of two fluids separated by a membrane (4.5). Phagocytosis is an endocytic pathway by which cells take in large particles (4.6). Lactate produced by fermentation reduces pH (5.6). A brain region called the hypothalamus serves as the body's thermostat (19.5). In atherosclerosis, fatty plaques form on the inner wall of blood vessels (21.6).

Normal Flora Your skin is in constant contact with the external environment, so it picks up many microorganisms. It normally teems with about 200 different kinds of yeast, protozoa, and bacteria—thousands to millions of microorganisms on every square inch of your external surfaces. Huge populations inhabit structures that open on the body's surface, including the eyes, nose, mouth, and anal and genital openings. Microorganisms that typically live on human surfaces, including the interior tubes and cavities of the digestive and respiratory tracts, are called **normal flora**. Body surfaces provide normal flora with a stable environment and nutrients. In return, their populations deter more dangerous species from colonizing (and potentially penetrating) body surfaces. Normal flora in the digestive tract help us digest food, and they also make essential nutrients such as vitamins K and B$_{12}$.

Normal flora are helpful only on body surfaces, however; many can cause or worsen disease when they invade tissues. Serious illnesses associated with normal flora include pneumonia; ulcers; colitis; whooping cough; meningitis; abscesses of the lung and brain; and cancers of the colon, stomach, and intestine. The bacterial agent of tetanus, *Clostridium tetani*, is considered a normal inhabitant of human intestines. The bacteria responsible for diphtheria, *Corynebacterium diphtheriae*, was normal skin flora before widespread use of the vaccine eradicated it. *Staphylococcus aureus*, a resident of human skin and linings of the mouth, nose, throat, and intestines, is also a leading cause of human bacterial disease.

Surface Barriers In contrast to body surfaces, the blood and interstitial fluid of healthy people are typically sterile (free of microorganisms). Physical, mechanical, and chemical barriers at body surfaces can prevent most microorganisms from entering the internal environment. The tough outer layer of vertebrate skin offers an example. Microorganisms flourish on skin's waterproof, oily surface, but they rarely penetrate its thick layer of dead cells (Figure 22.4A). The thinner epithelial tissues that line the body's interior tubes and cavities also have barrier defenses. Sticky mucus secreted by cells of these linings can trap microorganisms (Figure 22.4B). The mucus contains **lysozyme**, an enzyme that kills bacteria. In the sinuses and respiratory tract, the coordinated beating of cilia sweeps trapped microorganisms away before they have a chance to breach the delicate walls of these structures.

Your mouth is a particularly inviting habitat for microorganisms because it offers plenty of nutrients, warmth, moisture, and surfaces for colonization. Accordingly, it harbors huge populations of normal flora that can resist lysozyme in saliva. Microorganisms that get swallowed are typically killed in the stomach by gastric fluid, a potent brew of protein-digesting enzymes and acid. Any that survive passage to the small intestine are usually killed by salts secreted into intestinal fluid. Hardy

A. Skin's thick, waterproof layer of dead cells usually prevents normal skin flora from penetrating internal tissues.

B. *Staphylococcus aureus* bacteria (yellow) stuck in mucus secreted by goblet cells of nasal epithelia. Cilia on other cells sweep the bacteria-laden mucus toward the throat for disposal.

Figure 22.4 Examples of surface barriers.

(A) © Eye of Science/Science Source; (B) Juergen Berger/Science Source.

lipid
bilayer

pore

A. Complement activation. Top, activated complement proteins assemble into structures (purple) that insert themselves into a lipid bilayer. Bottom, the resulting pores cause the cell to burst.

B. Neutrophil nets. Two *Klebsiella* bacteria (purple) are ensnared in material that has been ejected by a neutrophil in lung tissue.

C. Phagocytosis. This macrophage is in the process of engulfing tuberculosis bacteria (red).

Figure 22.5 Some pathogen-busting mechanisms of innate immunity.

(A) Bottom, Robert R. Dourmashkin, courtesy of Clinical Research Centre, Harrow, England; (B) © 2010, Papayannopoulos et al. Originally published in *J. Cell Biol.* 191:677-691. doi: 10.1083/jcb.201006052 (Image by Volker Brinkman and Abdul Hakkim); (C) Science Source.

cells that reach the large intestine must compete with well-established populations of about 500 resident species specialized to live there. Any that displace normal flora are typically flushed out by diarrhea.

Urination's flushing action usually prevents pathogens from colonizing the urinary tract. Lactate produced by fermenting *Lactobacillus* bacteria helps keep the vaginal pH below the range of tolerance of other bacteria and most fungi.

Complement What happens if a microorganism slips by the body's surface barriers and enters the internal environment? Fast-acting, general response mechanisms can keep the invader from establishing a population in body tissues or fluids. These mechanisms include complement activation. **Complement** is a set of about 30 different kinds of proteins that circulate in inactive form throughout the body in blood and insterstitial fluid. Some complement proteins are activated when they bind directly to microorganisms. Others are activated by binding to antibodies clustered on the surface of a cell. Still others become activated when they encounter cytoplasmic or mitochondrial proteins leaking out of damaged body cells. In all cases, an activated complement protein activates other complement proteins, which activate other complement proteins, and so on. These cascading reactions quickly produce huge amounts of activated complement, which diffuses into surrounding tissues to form a gradient around an affected site.

Activated complement proteins form a coating that enhances the uptake of microorganisms, damaged cells, and debris by phagocytic white blood cells. The proteins can also assemble into structures that penetrate plasma membranes and form large channels through them (Figure 22.5A). Ions that flow through these channels disturb tonicity, causing the cells to burst. Normal body cells continuously produce proteins that inactivate complement, thus preventing a complement cascade from spreading too far into healthy tissue. Microorganisms do not make these inhibitory proteins, so they are singled out for destruction.

Phagocytosis Neutrophils, macrophages, and dendritic cells are mobile, and all can follow a chemical trail. Receptors for activated complement allow these phagocytes to follow a gradient of activated complement to its origin at an affected tissue, where they quickly engulf cells and particles coated with complement.

Neutrophils are among the first responders to injury or infection, collecting within minutes at a site of tissue damage. A neutrophil that engulfs a microorganism releases the contents of its granules both into the endocytic vesicle and to the exterior of the cell. Enzymes and toxins released into extracellular fluid destroy all cells in the vicinity—even healthy body cells. Neutrophils literally explode in response to a certain combination of signaling molecules and complement, in the process ejecting their nuclear DNA and associated proteins along with the contents of their granules. The mixture solidifies into a net, trapping pathogens near the released antimicrobial compounds (Figure 22.5B). Neutrophil nets are very effective at killing bacteria.

Macrophages in interstitial fluid engulf essentially everything except undamaged body cells (Figure 22.5C). However, these cells are more than just scavengers. Upon engulfing antigen, a macrophage secretes cytokines that alert other white blood cells to the presence of invading pathogens. Dendritic cells patrol tissues that contact the external environment, such as the lining of respiratory airways. Phagocytosis by dendritic cells plays a critical role in protecting the lungs from pathogens and harmful particles. However, the main function of dendritic cells is to present antigen to T cells (more on how this works in Section 22.5).

Inflammation and Fever Complement activation and cytokines released by phagocytic white blood cells typically trigger inflammation and fever. **Inflammation** is a fast, local response that simultaneously destroys affected tissues and jump-starts the healing process (Figure 22.6). Inflammation begins when basophils, mast cells, or neutrophils degranulate, releasing the contents of their granules into an affected tissue. Degranulation can occur in response to a number of stimuli, including PAMP receptor binding to antigen, for example on the surface of a bacterium ❶. A degranulating white blood cell releases cytokines. It also releases prostaglandins and histamines ❷, local signaling molecules that have two effects. First, they cause nearby arterioles to widen, so blood flow to the area increases. The increased flow speeds the arrival of more phagocytic white blood cells, which are attracted to the cytokines. Second, they cause spaces to open up between the cells making up the walls of nearby capillaries. Phagocytes arriving in the bloodstream can move quickly into tissues by squeezing through these spaces ❸. By the time this occurs, any invading cells have become coated with activated complement ❹, which makes them easy targets for the phagocytic cells ❺.

Symptoms of inflammation include redness and warmth that are outward indications of the area's increased blood flow. The increased permeability of capillary walls allows plasma proteins to escape into interstitial fluid. The fluid becomes hypertonic with respect to blood, and water follows by osmosis. The tissue swells with excess fluid, putting pressure on nerves and thus causing pain.

Inflammation continues as long as its triggers do. When these stimuli subside, for example after invading bacteria have been cleared from an infected tissue, macrophages produce compounds that suppress inflammation and promote tissue repair. If the stimulus persists, inflammation becomes chronic. Chronic inflammation is not a normal condition. It does not benefit the body; rather, it causes or contributes to several diseases, including asthma, Crohn's disease, rheumatoid arthritis, atherosclerosis, diabetes, and cancer.

Fever is a temporary rise in body temperature above the normal 37°C (98.6°F) that often occurs in response to infection or serious injury. Some cytokines stimulate brain cells to make and release prostaglandins, which act on the hypothalamus to raise the body's internal (core) temperature set point. As long as core temperature remains below the new set point, the hypothalamus sends out signals that cause blood vessels in the skin to constrict, which reduces heat loss from the skin. The signals also trigger an increase in the rate of heartbeat and respiration, as well as reflexive movements called shivering, or "chills," that increase the metabolic heat output of muscles. These responses all raise the body's internal temperature. If core temperature rises too much, sweating and flushing quickly lower it, thus maintaining the body's new temperature set point.

Fever enhances immune defenses by increasing the rate of enzyme activity, thus speeding up tissue repair and the formation and activity of phagocytic white blood cells. In addition, many pathogens multiply more slowly at the higher temperature, so white blood cells can get a head start in the proliferation race against them.

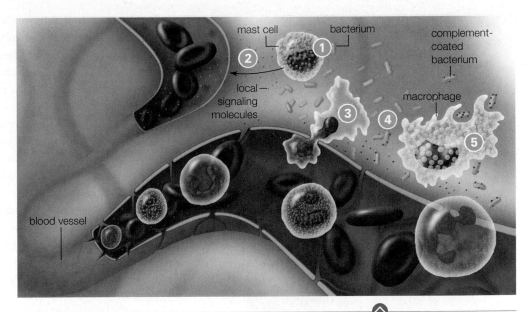

Figure 22.6 Example of inflammation as a response to bacterial infection.

❶ PAMP receptors on mast cells in the tissue recognize and bind to bacterial antigen.

❷ The mast cells release signaling molecules (blue dots) that cause arterioles to widen. The resulting increase in blood flow, which reddens and warms the tissue, hastens the arrival of phagocytes.

❸ The signaling molecules also increase capillary permeability, which allows phagocytes arriving in the bloodstream to move quickly through the vessel walls into the tissue. The tissue swells with fluid as plasma proteins seep out of the leaky capillaries.

❹ Meanwhile, bacterial antigens have triggered complement cascades, and invading bacteria have become coated with complement (purple dots).

❺ Phagocytes in the tissue recognize and engulf the complement-coated bacteria.

complement A set of proteins that circulate in inactive form in blood, and when activated play a role in immune responses.

fever A temporary, internally induced rise in core body temperature above the normal set point as a response to infection or tissue damage.

inflammation A local response to tissue damage or infection; characterized by redness, warmth, swelling, and pain.

A fever is a sign that the body is fighting something, so it should never be ignored. However, a fever of 40.6°C (105°F) or less does not necessarily require treatment in an otherwise healthy adult. Body temperature usually will not rise above that value, but if it does, immediate hospitalization is recommended. Brain damage or death can occur if the core temperature reaches 42°C (107.6°F).

Examples of Innate Responses   Acne is one effect of an innate response. This skin condition is caused in part by *Propionibacterium acnes*, a major bacterial constituent of normal flora. *P. acnes* feeds on sebum, a greasy mixture of lipids that lubricates hair and skin. Glands in the skin secrete sebum into hair follicles. During puberty, an increase in sex hormone production triggers an increase in sebum production, and the excess sebum combines with dead, shed skin cells to block the openings of hair follicles. *P. acnes* can survive on the surface of the skin, but far prefers anaerobic habitats such as the interior of blocked hair follicles. There, the cells multiply to tremendous numbers. Secretions of their flourishing populations leak into internal tissues of the follicles and initiate inflammation. The resulting pustules are called acne.

Another common innate response occurs in the mouth, which is normally inhabited by at least 700 species of microorganisms. These normal flora can accumulate in **dental plaque**, a thick biofilm of various bacteria and occasional archaea, their extracellular products, and saliva glycoproteins. Plaque sticks tenaciously to teeth (Figure 22.7). Some of the bacteria in plaque carry out lactate fermentation. The lactate they produce is acidic enough to dissolve minerals that make up the tooth, causing holes called cavities. In young, healthy people, tight junctions normally seal gum epithelium to teeth. The tight seal prevents oral microorganisms from entering gum tissue. As we age, the connective tissue beneath the epithelium thins, so the seal between gums and teeth weakens. Deep pockets form, and a nasty collection of anaerobic bacteria and archaea tends to accumulate in them. These microorganisms secrete destructive enzymes and acids that cause inflammation of the surrounding gum, a condition called periodontitis. Periodontal wounds are an open door to the circulatory system and its arteries, and all species of oral bacteria associated with periodontitis are also found in atherosclerotic plaque. Atherosclerosis is now known to be a disease of inflammation. What role oral microorganisms play in atherosclerosis is not yet clear, but one thing is certain: They contribute to the inflammation that fuels coronary artery disease.

A. Bacteria (*Streptococcus mutans*, gold, and *Porphyromonas gingivalis*, green), and yeast (red) in dental plaque.

B. Micrograph of toothbrush bristles scrubbing plaque on a tooth surface.

Figure 22.7 Dental plaque.

(A) © Dennis Kunkel Microscopy, Inc./Phototake; (B) www.zahnarztstuttgart.com.

Take-Home Message 22.3

What happens after antigen is detected inside the body?

- Surface barriers can usually prevent normal flora and other microorganisms from invading the internal environment.
- Antigen or tissue damage triggers complement activation. Activated complement recruits phagocytic white blood cells, coats cells and debris in an affected area, and triggers cell lysis.
- Phagocytic white blood cells release cytokines and antimicrobial molecules after engulfing antigen-bearing particles such as complement-coated microorganisms.
- Inflammation occurs when granular white blood cells release local signaling molecules that increase blood flow and attract phagocytes to a site of tissue damage or infection. Fever enhances immune defenses while slowing pathogen growth.

B cell

B cell receptor

A. An antibody molecule consists of four polypeptide chains joined in a Y-shaped configuration. The chains fold up so two two antigen-binding sites form at the tips of the "Y."

two binding sites for antigen

B. The antigen-binding sites of each antibody are unique. They bind only to antigen that has a complementary distribution of bumps, grooves, and charge.

C. Antibodies produced by a new B cell stay attached to its membrane as B cell receptors.

Figure 22.8 Antibody structure.

22.4 Antigen Receptors

REMEMBER: Plasma membranes incorporate receptor proteins, which trigger a change in the cell's activities in response to a stimulus (Section 3.3). Lysosomes are filled with powerful digestive enzymes (3.5). Cells use vesicles to take in or expel bulk materials (4.6). When a cell differentiates, it becomes specialized (6.1). Differentiation occurs as different cell lineages begin to express different subsets of their genes (7.7). The lymphatic system (19.4) includes organs involved in immunity as well as vessels that move fluid (lymph) from tissues to blood (21.5).

If innate immune mechanisms do not quickly rid the body of an invading pathogen, an infection may become established in internal tissues. By that time, long-lasting mechanisms of adaptive immunity have already begun to target the invaders specifically. These mechanisms are triggered by white blood cells that detect antigen via antigen receptors. Plasma membrane proteins that recognize PAMPs are one type of antigen receptor. Antibodies are another. **Antibodies** are Y-shaped antigen receptor proteins made by B cells. Each antibody molecule can bind to a particular antigen at the tips of its "Y" (Figure 22.8A). Only antigen with a complementary distribution of bumps, grooves, and charge can bind at these sites (Figure 22.8B).

Antibodies produced by a new B cell stay attached to the cell's plasma membrane as **B cell receptors**. The base of each receptor is embedded in the lipid bilayer of the cell's plasma membrane, and the two arms of the "Y" project into the extracellular environment (Figure 22.8C). Mature B cells secrete antibodies into the bloodstream. Many antibodies circulate in blood, and they can enter interstitial fluid during inflammation, but they do not kill pathogens directly. Instead, they activate complement and facilitate phagocytosis. Antibody binding also prevents some pathogens from attaching to body cells, and it neutralizes some toxic molecules.

Your T cells bear special antigen receptors called **T cell receptors**, or TCRs. Part of a T cell receptor recognizes antigen as nonself. Another part recognizes certain proteins in the plasma membrane of your body cells as self. These self-proteins are called **MHC markers** (left), after the major histocompatibility complex genes that encode them. MHC genes have thousands of alleles, so the cells of even closely related individuals rarely bear the same MHC markers.

an MHC marker

antibody Y-shaped antigen receptor protein made by B cells.
B cell receptor Antigen receptor on the surface of a B cell; an antibody that stays anchored in the B cell's plasma membrane.
dental plaque On teeth, a thick biofilm composed mainly of bacteria, their extracellular products, and saliva proteins.
MHC markers Self-proteins on the surface of vertebrate body cells.
T cell receptor (TCR) Antigen receptor on the surface of a T cell.

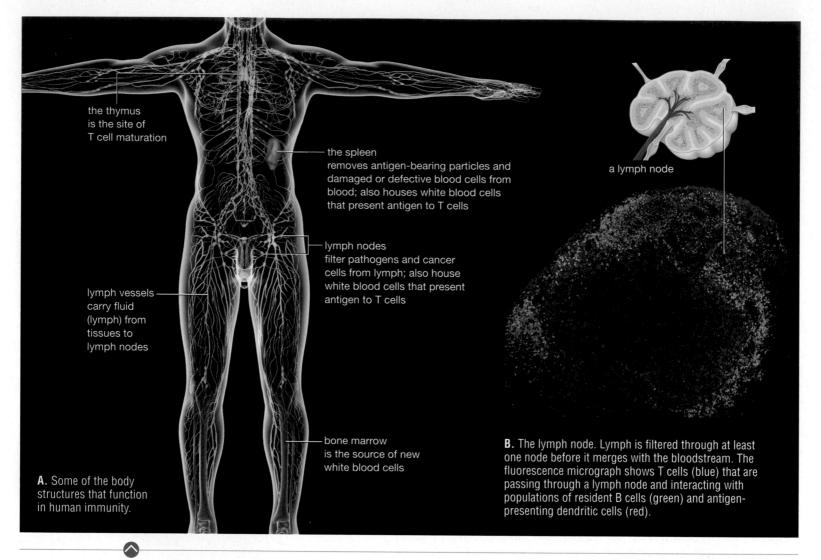

the thymus
is the site of
T cell maturation

the spleen
removes antigen-bearing particles and
damaged or defective blood cells from
blood; also houses white blood cells
that present antigen to T cells

lymph nodes
filter pathogens and cancer
cells from lymph; also house
white blood cells that present
antigen to T cells

lymph vessels
carry fluid
(lymph) from
tissues to
lymph nodes

a lymph node

bone marrow
is the source of new
white blood cells

A. Some of the body
structures that function
in human immunity.

B. The lymph node. Lymph is filtered through at least
one node before it merges with the bloodstream. The
fluorescence micrograph shows T cells (blue) that are
passing through a lymph node and interacting with
populations of resident B cells (green) and antigen-
presenting dendritic cells (red).

Figure 22.9 Battlegrounds of human immunity.

(A) layer 1 (lymphatic system), Somkiat/Colourbox; layer 2 (skeleton), Maya 2008/
Shutterstock.com; (B) bottom, Image courtesy of Dr. Fabien Garcon, The Babraham
Institute.

Humans can make billions of unique B and T cell receptors. This diversity
arises because the genes that encode these receptors occur in several segments, and
there are several different versions of each segment. The gene segments become
spliced together during B and T cell differentiation, but which version of each segment makes it into the antigen receptor gene of a particular cell is random. As a new
B cell or T cell differentiates, it ends up with one of about 2.5 billion potential combinations of gene segments. When the resulting gene is expressed, the cell produces
thousands of receptors, all of which can recognize the same, specific antigen.

Like all other blood cells, lymphocytes form in bone marrow (Figure 22.9A).
A new B cell is already making receptors before it even leaves the marrow. T cells
also form in bone marrow, but they mature in the thymus gland. There, they
encounter hormones that stimulate them to make receptors.

Antigen Processing A new lymphocyte is "naive," which means that no antigen
has bound to its receptors yet. B cell receptors can bind directly to antigen, but T cell
receptors cannot. T cell receptors recognize and bind only to antigen that has been
processed by an antigen-presenting cell. Macrophages, B cells, and dendritic cells do
the processing (Figure 22.10). First, one of these cells engulfs a bacterium or other

MHC marker

lysosome

particle that bears antigen 1. A vesicle that contains the antigen-bearing particle forms in the cell's cytoplasm and fuses with a lysosome 2. Lysosomal enzymes then proceed to break down the ingested particle into molecular bits. Lysosomes also contain MHC markers, and these bind to some of the antigen-bearing bits 3. The resulting antigen–MHC complexes become displayed at the cell's surface when the vesicles fuse with (and become part of) the plasma membrane 4. T cell receptors can recognize their specific antigen only when displayed together with MHC markers on the surface of an antigen-presenting cell.

Antigen-bearing particles in blood end up in the spleen; those in solid tissues or interstitial fluid end up in lymph nodes. Every day, billions of naive T cells filter through each lymph node and the spleen. As they do, they come into close contact with arrays of antigen-presenting cells that have taken up residence in these organs (Figure 22.9B). As you will see shortly, T cells with receptors that bind antigen–MHC complexes displayed by an antigen-presenting cell stimulate other white blood cells to divide and differentiate in an adaptive immune response.

During an infection, the lymph nodes swell because T cells accumulate inside them. When you are ill, you may notice your swollen lymph nodes as tender lumps under the jaw or elsewhere in your body.

T cell receptors only recognize antigen that has been processed by an antigen-presenting cell.

Take-Home Message 22.4

What are antigen receptors?

- Each B or T cell makes antigen receptors that can bind a specific antigen. Humans are capable of producing billions of unique antigen receptors.
- T cell receptors recognize self (MHC markers) and nonself (antigen).
- T cell receptors bind their specific antigen only in conjunction with MHC markers displayed by an antigen-presenting cell. The recognition process occurs in lymph nodes and the spleen.
- Antibodies are antigen receptors made only by B cells. Antibodies released into the circulatory system activate complement and facilitate phagocytosis.
- B cell receptors are antibodies that are not secreted; they remain attached to the B cell's plasma membrane.

Figure 22.11 Comparing the primary and secondary immune response.
A first exposure to an antigen triggers a primary immune response, in which effector cells form and fight the infection. Memory cells also form, but these are reserved for future battles.

If the same antigen enters the body at a later time, memory cells will carry out a secondary response that is faster and stronger than the primary response.

22.5 Adaptive Immune Responses

REMEMBER: Receptor proteins trigger a change in the cell's activities in response to a stimulus (Section 3.3). Most bacterial cell walls consist of peptides and polysaccharides (3.4).

Like a boxer's one-two punch, adaptive immunity has two separate arms: two types of responses that work together to eliminate diverse threats. Why two arms? Not all threats present themselves in the same way. For example, bacteria, fungi, or toxins can circulate in blood or interstitial fluid. These threats are intercepted quickly by phagocytes that initiate an **antibody-mediated immune response**. In this response, B cells produce antibodies specific to antigen detected in extracellular fluid. However, an antibody-mediated immune response is not the most effective way of countering other threats. Consider viruses, bacteria, fungi, and protists that reproduce inside body cells. These intracellular pathogens may be vulnerable to an antibody-mediated response only when they exit one cell to infect another. Intracellular pathogens are targeted by the **cell-mediated immune response**, in which cytotoxic T cells and NK cells detect and destroy infected or cancerous body cells.

Effector cells form during both antibody-mediated and cell-mediated immune responses. **Effector cells** are lymphocytes that act at once in a primary immune response. Some **memory cells** also form, and these long-lived lymphocytes are reserved for possible future encounters with the same threat. Memory cells can persist for decades after a primary response. If the same antigen is detected in the body later, memory cells carry out a faster, stronger secondary response (Figure 22.11).

Lymphocytes and phagocytic white blood cells interact to bring about the four defining characteristics of adaptive immunity:

Self/nonself recognition, based on the ability of T cell receptors to recognize self (in the form of MHC markers), and that of all antigen receptors to recognize nonself (in the form of antigen).

Specificity, which means that adaptive immune responses are tailored to combat specific antigens.

Diversity, which refers to the diversity of antigen receptors on a body's collection of lymphocytes. Having the ability to make billions of different antigen receptors, an individual has the potential to counter billions of different threats.

Memory, the capacity of the adaptive immune system to "remember" an antigen via memory cells. It takes about a week for B and T cells to respond in force the first time they encounter an antigen. If the same antigen shows up later, the response is faster and stronger.

Example of an Antibody-Mediated Response Suppose that you accidentally nick your finger. Being opportunists, some *Staphylococcus aureus* cells on your skin immediately enter the cut, invading your internal environment. Complement in interstitial fluid quickly attaches to carbohydrates in the bacterial cell walls, and complement activation cascades begin.

Within an hour, complement-coated bacteria tumbling along in lymph vessels reach a lymph node. There, they filter past millions of naive B cells. One of the naive B cells residing in that lymph node makes antigen receptors that recognize

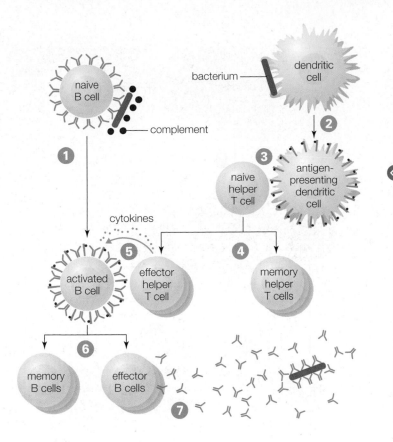

Figure 22.12 An example of an antibody-mediated immune response.

① The B cell receptors on a naive B cell bind to an antigen on the surface of a bacterium (red). The bacterium's complement coating (purple dots) triggers the B cell to engulf and digest it. Bacterial fragments bound to MHC markers become displayed at the surface of the B cell.

② A dendritic cell engulfs and digests the same kind of bacterium that the B cell encountered. Bacterial fragments bound to MHC markers become displayed at the surface of the dendritic cell.

③ The receptors of a naive helper T cell recognize bacterial antigen displayed by the dendritic cell. The two cells interact and disengage.

④ The T cell begins to divide. Its descendants differentiate into effector helper T cells and memory helper T cells.

⑤ The receptors of one of the effector helper T cells recognize and bind to antigen displayed by the B cell. Binding causes the T cell to secrete cytokines (blue dots).

⑥ The cytokines induce the B cell to undergo repeated mitotic divisions. Its many descendants differentiate into effector B cells and memory B cells.

⑦ The effector B cells begin making and secreting huge numbers of antibodies, all of which recognize the same antigen as the original B cell receptor. The new antibodies circulate throughout the body and bind to any remaining bacteria.

a polysaccharide in *S. aureus* cell walls (Figure 22.12). The B cell's receptors bind to one of the bacteria, and the complement coating stimulates the B cell to engulf it **①**. The B cell is now activated (it is no longer naive).

Meanwhile, more *S. aureus* cells have been secreting metabolic products into interstitial fluid around your cut. The secretions are attracting phagocytic white blood cells. A dendritic cell engulfs several bacteria, then migrates to the lymph node in your elbow. By the time it gets there, it has digested the bacteria and is displaying their fragments as antigens bound to MHC markers on its surface **②**. In the lymph node, one of your T cells recognizes and binds to the *S. aureus* antigen displayed by the dendritic cell **③**. This T cell is called a helper T cell because it helps other lymphocytes produce antibodies and kill pathogens.

The helper T cell and the dendritic cell interact in the lymph node for about 24 hours and then disengage. The helper T cell returns to the circulatory system and begins to divide, and a huge population of identical helper T cells forms. These clones mature as effector and memory cells **④**, each of which has receptors that recognize the same *S. aureus* antigen.

Let's now go back to that B cell in the lymph node. By now, it has digested the engulfed bacterium, and it is displaying bits of *S. aureus* together with MHC molecules on its plasma membrane. The new helper T cells recognize antigen displayed by the B cell. One of these helper T cells binds to the B cell. Like long-lost friends, the two cells stay together for a while and communicate **⑤**. One of the messages that is communicated consists of cytokines secreted by the helper T cell. The cytokines stimulate the B cell to undergo repeated mitotic divisions after the two cells disengage. A huge clonal population of descendant cells forms, and these B cells mature as effector and memory cells **⑥**. The effector B cells start releasing antibodies **⑦**. The antibodies are a secreted version of the original B cell's receptors, so they can recognize and bind to the same *S. aureus* cell wall polysaccharide.

antibody-mediated immune response Immune response in which antibodies targeting a specific antigen are produced.

cell-mediated immune response Immune response in which cytotoxic T cells and NK cells kill infected or cancerous body cells.

effector cell Antigen-sensitized lymphocyte that forms in an immune response and acts immediately.

memory cell Long-lived, antigen-sensitized lymphocyte that can act in a secondary immune response.

Figure 22.13 An example of a cell-mediated immune response.

1 A dendritic cell engulfs and digests a virus-infected cell. Bits of the virus bind to MHC markers, and the complexes become displayed at the dendritic cell's surface. The dendritic cell, now an antigen-presenting cell, migrates to a lymph node.

2 Receptors on a naive cytotoxic T cell bind to antigen displayed by the dendritic cell. The interaction activates the cytotoxic T cell.

3 Receptors on a naive helper T cell bind to antigen displayed by the dendritic cell. The interaction activates the helper T cell.

4 The activated helper T cell divides again and again. Its many descendants mature as effector and memory cells, each with T cell receptors that recognize the same antigen.

5 The effector helper T cells secrete cytokines.

6 The cytokines induce the activated cytotoxic cell to divide again and again. Its many descendants differentiate into effector and memory cells. Each cell bears T cell receptors that recognize the same antigen.

7 The new effector cytotoxic T cells circulate throughout the body. They kill any body cell that displays the viral antigen–MHC complexes on its surface.

Figure It Out: What do the red spots represent?

Answer: Viral antigen

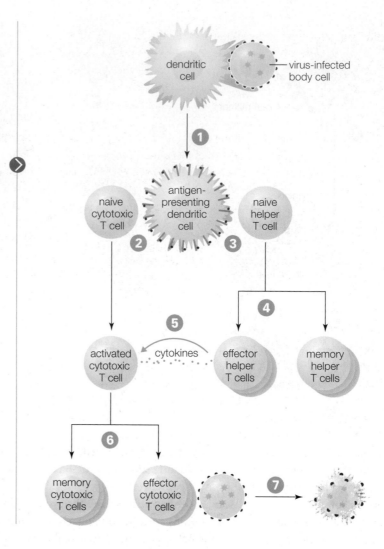

Tremendous numbers of antibodies now circulate throughout the body and bind to any *S. aureus* cells. An antibody coating prevents the bacteria from attaching to body cells while bringing them to the attention of phagocytes for quick disposal. Antibodies also glue the foreign cells together into clumps that are quickly removed from the circulatory system by the liver and spleen.

Example of a Cell-Mediated Response A cell-mediated immune response involves the production of cytotoxic T cells that recognize body cells bearing a specific antigen. This type of immune response does not involve antibodies. (Antibody-mediated immune responses target pathogens that circulate in blood and interstitial fluid, but they are not as effective against cancerous or infected body cells.) Most ailing body cells display antigens—molecules that are not found on healthy cells. For example, cancer cells display altered body proteins, and body cells infected with intracellular pathogens display polypeptides of the infecting agent. Both types of cell are killed by lymphocytes that act in cell-mediated responses.

A cell-mediated immune response often starts in interstitial fluid during inflammation, when a dendritic cell engulfs a sick body cell or the remains of one (Figure 22.13). The dendritic cell migrates to a lymph node or the spleen. As it does, molecules of the ingested cell become displayed together with MHC markers on its surface **1**. When the antigen-presenting dendritic cell reaches its destination, it comes into contact with millions of naive cytotoxic T cells ② and naive helper T cells ③. Some of the naive cells have T cell receptors that recognize molecules presented by the dendritic cell. These cells interact with the dendritic cell, then disengage. The T cells are now activated (they are no longer naive).

The activated helper T cells return to the circulatory system and begin to divide repeatedly. Their many descendants mature as effector and memory helper T cells ④. The new effector cells secrete cytokines ⑤. Any cytotoxic T cells that have been activated by interacting with an antigen-presenting cell (②) recognize these cytokines as a signal to divide repeatedly, and their many descendants mature as effector and memory cells ⑥.

All of the new effector cytotoxic T cells have receptors that recognize antigen displayed by that first ailing body cell. These cytotoxic T cells now circulate through-out blood and interstitial fluid, and bind to any other body cell displaying the same molecules together with MHC markers (Figure 22.14). Binding causes a cytotoxic T cell to release protein-digesting enzymes and small molecules called perforins. Perforins, like complement proteins, assemble into structures that penetrate a cell's plasma membrane and form large channels through it. The channels allow the enzymes to enter the body cell and cause it to burst or commit suicide ⑦.

As occurs in an antibody-mediated response, memory cells form in a primary cell-mediated response. If the same threat appears again at a later time, these memory cells will mount a faster, stronger secondary response.

Note that cytotoxic T cells must recognize MHC markers on the surface of an ailing cell in order to kill it. However, some infections and cancers alter body cells so much that part or all of their MHC markers are missing. NK cells are crucial for eliminating such cells from the body. Unlike cytotoxic T cells, NK cells can kill body cells that lack MHC markers. Cytokines secreted by effector helper T cells also stimulate NK cell division, and the resulting populations of NK cells recognize and attack body cells that have antibodies bound to them. They also recognize certain proteins displayed by body cells that are under stress. Stressed body cells with nor-mal MHC markers are ignored; only those with altered or missing MHC markers are killed. Because NK cells can operate early in immune defenses, they are often considered to be part of innate immunity. However, they also have features associ-ated with lymphocytes of adaptive immunity: activation by cytokines, for example, and memory.

Figure 22.14 Cytotoxic T cells killing a cancer cell.
The T cells are colored pink; the cancer cell, yellow.

Steve Gschmeissner/Science Source.

Take-Home Message 22.5

What happens during an adaptive immune response?

- Vertebrate adaptive immunity has four defining characteristics: self/nonself recogni-tion, specificity, diversity, and memory.
- Effector cells form during a primary adaptive response. Memory cells also form, but these are set aside for future encounters with the same threat. Memory cells can initi-ate a secondary response that is faster and stronger than a primary response.
- Antibody-mediated responses target antigen in blood or interstitial fluid. During an antibody-mediated response, B cells produce antibodies that bind to the antigenic particle and facilitate its removal from the body.
- Cell-mediated responses target infected or cancerous body cells. Cytotoxic T cells and NK cells that form during a cell-mediated immune response kill ailing body cells.

22.6 Immunity Gone Wrong

REMEMBER: A syndrome is a specific set of symptoms associated with a disorder (Section 9.6). Without treatment, people affected by SCID-X1 can survive only in germ-free isolation tents because they cannot fight infections (10.5). HIV is an enveloped RNA virus that replicates inside human white blood cells; the bird influenza virus H5N1 kills about 60% of the people it infects (13.4).

Despite built-in quality controls and redundancies in function, immunity does not always work as well as it should. Its complexity is part of the problem, because there are simply more opportunities for failure to occur in systems with many components. Even a small failure in immune function can have a major effect on health.

Overly Vigorous Responses An **allergen** is a normally harmless substance that stimulates an immune response in some people. Common allergens include drugs, foods, pollen, dust mite feces, fungal spores, and insect venom. Sensitivity to an allergen is an **allergy**. Some people are genetically predisposed to have allergies, but factors such as infections, emotional stress, exercise, and changes in air temperature can trigger or worsen them. Typically, a first exposure to an allergen stimulates an antibody-mediated response targeting the allergen. Some of the antibodies that are produced during this response become anchored to mast cells and basophils. Upon a later exposure, the anchored antibody binds to the allergen. Binding triggers the anchoring cell to degranulate, and prostaglandins and histamines it releases cause inflammation. If allergen is detected by mast cells in the lining of the respiratory tract, the resulting inflammation constricts the airways and causes mucus to be secreted; sneezing, stuffed-up sinuses, and a drippy nose result (Figure 22.15A). Antihistamines relieve these symptoms by dampening the effects of histamines. Other drugs can inhibit mast cell degranulation, thus preventing histamine release. Skin rashes and other contact allergies do not involve antibodies; they are caused by a cell-mediated response to an allergen (Figure 22.15B).

Immune defenses that eliminate a threat can also damage body tissues. Thus, mechanisms that limit these defenses are always in play. Consider that some complement proteins activate spontaneously, even in the absence of infection or tissue damage. Without inhibitory molecules that inactivate complement, complement cascades would occur constantly, with disastrous effects on body tissues.

Acute illnesses can arise when mechanisms that limit immune responses fail. Exposure to an allergen sometimes causes a rapid and severe allergic reaction called anaphylaxis. Huge amounts of inflammatory molecules, including histamines and prostaglandins, are released all at once in all parts of the body. Too much fluid leaks from blood into tissues, causing a sudden and dramatic drop in blood pressure (a reaction called shock). Rapidly swelling tissues constrict the airways and may block them. Anaphylaxis is rare but life-threatening and requires immediate treatment (Figure 22.15C). It may occur at any time upon exposure to even a tiny amount of allergen. Risks include any prior allergic reaction.

Severe episodes of asthma or septic shock occur when too many neutrophils degranulate at once. A "cytokine storm" occurs when too many white blood cells release cytokines at the same time. The cytokine overdose activates more white blood cells, which release more cytokines, and so on—a positive feedback loop that results in an exaggerated immune response. Massive inflammation and organ failure can occur, with potentially fatal results. Viruses such as H5N1 and Ebola have an unusually high mortality rate because they trigger cytokine storm.

A. Hay fever is caused by allergy to grass pollen. Symptoms such as sneezing and a runny nose are the outcome of inflammation of the mucous membranes in respiratory airways.

B. Direct contact with an allergen can cause a rash—an itchy, irritated patch of skin. In this case, the offending allergen was nickel (a metal) in a ring.

C. Anaphylaxis is a potentially lethal systemic response to an allergen.

Figure 22.15 Allergies.

(A) © Colin Hawkins/Cultura/Getty Images; (B) Biophoto Associates/ Science Source; (C) Hayley Witherell.

Table 22.2 Global HIV and AIDS Cases

Region	AIDS Cases	% of Adults Infected
Sub-Saharan Africa	22,100,000	4.7
Caribbean Islands	230,000	1.0
Central Asia/East Europe	1,300,000	0.7
North America	1,300,000	0.5
Latin America	1,400,000	0.4
South/Southeast Asia	3,700,000	0.3
Australia/New Zealand	48,000	0.2
Western/Central Europe	860,000	0.2
Middle East/North Africa	250,000	0.1
East Asia	880,000	0.1
Approx. worldwide total	32,100,000	0.8

Source: Joint United Nations Programme HIV/AIDS, 2012 report

Figure 22.16 In Cambodia, a mother lies dying of AIDS in front of her children.
This photo was taken in 2002. Today, the country has the highest incidence of AIDS in Southeast Asia.
© Masaru Goto/The World Bank.

People usually do not make antibodies to molecules that occur on their own healthy body cells, in part because the thymus has a built-in quality control mechanism that weeds out T cells with receptors that recognize one's own body proteins. If this mechanism fails, lymphocytes that do not discriminate between self and nonself may be produced. Such lymphocytes can mount an **autoimmune response**, which is an immune response that targets one's own tissues. Autoimmunity is beneficial when a cell-mediated response targets cancer cells, but in most other cases it is not. For example, a neurological disorder called multiple sclerosis occurs when self-reactive T cells attack nerves. Symptoms range from weakness and loss of balance to paralysis and blindness. Specific alleles for MHC markers increase susceptibility, but a bacterial or viral infection may trigger the disorder.

Immune Deficiency and AIDS Insufficient immune function—immune deficiency—renders an individual vulnerable to infections by opportunistic agents that are typically harmless to those in good health. Primary immune deficiencies, which are present at birth, are the outcome of mutations. Severe combined immunodeficiency (SCID) is an example. Secondary immune deficiency is the loss of immune function after exposure to a virus or other outside agent. **AIDS** (acquired immune deficiency syndrome), the most common secondary immune deficiency, occurs as a result of infection with HIV. Worldwide, more than 32 million individuals are infected with this virus (Figure 22.16 and Table 22.2).

A person newly infected with HIV appears to be in good health, perhaps fighting a cold or "the flu." But symptoms eventually emerge that foreshadow AIDS: fever, many enlarged lymph nodes, chronic fatigue and weight loss, and drenching night sweats. Then, infections caused by normally harmless microorganisms strike.

Even a small failure in immune function can have a major effect on health.

AIDS Acquired immune deficiency syndrome. A secondary immune deficiency that develops as the result of infection by HIV.

allergen A normally harmless substance that provokes an immune response in some people.

allergy Sensitivity to an allergen.

autoimmune response Immune response that targets one's own tissues.

When enough individuals refuse vaccination, outbreaks of preventable and sometimes fatal diseases occur.

Yeast infections of the mouth, esophagus, and vagina often occur, as well as a form of pneumonia caused by a fungus. Gastrointestinal inflammation due to infection by a yeast or virus causes diarrhea. Colored lesions that erupt are evidence of Kaposi's sarcoma, a type of cancer that is common among AIDS patients but rare among the general population. Other cancers are also common, as are infections by cancer-causing viruses such as Epstein–Barr virus. These medical problems are relatively uncommon in people with healthy immune systems.

HIV mainly infects macrophages, dendritic cells, and helper T cells. When virus particles enter the body, dendritic cells engulf them. The dendritic cells then migrate to lymph nodes, where they present processed HIV antigen to naive T cells. Armies of HIV-neutralizing antibodies and HIV-specific cytotoxic T cells form. This is a typical adaptive immune response, and it rids the body of most—but not all—of the virus. HIV persists in a few helper T cells in a few lymph nodes. For years or even decades, antibodies keep the level of HIV in the blood low, and cytotoxic T cells kill most of the HIV-infected cells. During this stage, infected people often have no symptoms of AIDS, but they can pass the virus to other people. Eventually, the level of virus-neutralizing antibodies plummets, and the production of T cells slows. Why this occurs is still a major topic of research, but its effect is certain: The immune system becomes progressively less effective at fighting the virus. The number of virus particles rises, and more and more helper T cells become infected. Lymph nodes begin to swell with infected T cells. Eventually, secondary infections and tumors kill the patient.

There is no way to rid the body of HIV, no cure for those already infected. Drugs that slow the disease's progress target processes unique to viral replication. A three-drug "cocktail" is currently the most successful AIDS therapy, and has changed the typical course of the disease from a short-term death sentence to a long-term, often manageable illness. Preventive use of a drug that consists of two reverse transcriptase inhibitors has recently been shown to greatly reduce the HIV infection rate in high-risk populations.

Most HIV infections are the result of having unprotected sex with an infected partner. Infected mothers can transmit HIV to a child; HIV also travels in tiny amounts of infected blood in the syringes shared by intravenous drug abusers, or by hospital patients in less developed countries. Many people have become infected via blood transfusions, but this transmission route is becoming rarer because blood is now tested prior to use for transfusions.

Most AIDS tests check blood, saliva, or urine for antibodies that bind to HIV antigens. These antibodies are detectable in 99 percent of infected people within three months of exposure to the virus.

Take-Home Message 22.6

What happens when immunity does not function as well as it should?

- An allergy is sensitivity to an allergen, a normally harmless substance that induces an immune response in some people. Severe allergic reactions and other overly vigorous immune responses can be life-threatening.
- Autoimmune diseases are caused by immune responses that inappropriately target normal body tissues.
- Immune deficiency, which can be inherited or triggered by environmental factors, causes an individual to be especially vulnerable to infections.

22.7 Vaccines

Immunization refers to procedures designed to induce immunity. In active immunization, a preparation that contains antigen—a **vaccine**—is injected or given orally. This immunization elicits a primary immune response, just as an infection would. A second immunization, or booster, elicits a secondary immune response for enhanced protection. In passive immunization, one individual receives antibodies that have been purified from the blood of another. Passive immunization offers immediate benefit for someone who has been exposed to a lethal agent such as tetanus, rabies, Ebola virus, or a venom or toxin. Because the antibodies were not made by the recipient's lymphocytes, effector and memory cells do not form, so benefits last only as long as the injected antibodies do.

The first vaccine was developed as a result of desperate attempts to survive epidemics of smallpox, a severe disease that kills up to one-third of the people it infects (Figure 22.17). Before 1880, no one knew what caused infectious diseases or how to protect anyone from getting them, but there were clues. In the case of smallpox, survivors seldom contracted the disease a second time. They were said to be immune—protected from infection. At the time, the idea of acquiring immunity to smallpox was extremely appealing. People had been risking their lives on it by poking into their skin bits of smallpox scabs or threads soaked in pus from smallpox sores. Some survived the crude practices, but many others did not.

By 1774, it was known that dairymaids usually did not get smallpox after they had contracted cowpox, a mild disease that affects humans as well as cattle. An English farmer collected pus from a cowpox sore on a cow's udder, and poked it into the arm of his pregnant wife and two small children. All survived the smallpox epidemic, but were from that time on subject to rock peltings by neighbors convinced they would turn into cows. Twenty years later, the English physician Edward Jenner injected liquid from a cowpox sore into the arm of a healthy boy. Six weeks later, Jenner injected the boy with liquid from a smallpox sore. Luckily, the boy did not get smallpox. Jenner's experiment showed directly that the agent of cowpox elicits immunity to smallpox. Jenner named his procedure "vaccination," after the Latin word for cowpox (*vaccinia*). Though it was still controversial, Jenner's vaccine spread quickly through Europe, then to the rest of the world. The last known case of naturally occurring smallpox was in 1977. Use of the vaccine eradicated the disease.

Today, we know that immunization with cowpox is effective against smallpox because the two diseases are caused by closely related viruses; antibodies produced during an infection with one of them recognize antigens of both. Vaccines for many other infectious diseases are the outcome of our increased understanding of immunity (Table 22.3). These vaccines have overwhelmingly reduced suffering and deaths, but public confidence is a necessary part of their success. When enough individuals refuse vaccination, outbreaks of preventable and sometimes fatal diseases occur.

Take-Home Message 22.7

How do vaccines work?

- Administering a vaccine elicits an immune response in the recipient that protects against future encounters with a disease-causing agent.
- Worldwide vaccination programs have greatly reduced suffering and deaths from many diseases. An effective vaccination program requires widespread participation.

Figure 22.17 Smallpox: viral cause and effect.
Worldwide use of the smallpox vaccine eradicated naturally occurring cases of smallpox; vaccinations for it ended in 1972.

Left, Eye of Science/Science Source; right, James Hicks, Centers for Disease Control and Prevention.

Table 22.3 Recommended Immunizations for Children

Vaccine	Age of Vaccination
HepB (hepatitis B)	Birth
HepB booster	1–2 and 6–18 months
Rotavirus (RV)	2, 4, and 6 months
DTaP (diphtheria, tetanus, pertussis)	2, 4, and 6 months
DTaP booster	15–18 months, 4–6 years, 11–12 years
HiB (*Haemophilus influenzae*)	2, 4, and 6 months
HiB booster	12–15 months
PCV13 (*Pneumococcus*)	2, 4, and 6 months
PCV13 booster	12–15 months
IPV (poliovirus)	2 and 4 months
IPV booster	6–18 months, 4–6 years
Influenza	Yearly after 6 months
MMR (measles, mumps, rubella)	12–15 months
MMR booster	4–6 years
Varicella (chicken pox)	12–15 months
Varicella booster	4–6 years
HepA (hepatitis A, 2 doses)	12–23 months
HPV (human papillomavirus, 3 doses)	11–12 years
Meningococcal	11–12 years
Meningococcal booster	16 years

Source: Centers for Disease Control and Prevention (CDC), 2014.

Summary

Section 22.1 Screenings, treatments, and vaccines for diseases such as cervical cancer are a direct outcome of our increasing understanding of the way the human body interacts with pathogens.

Section 22.2 The body's ability to resist and fight infections is called **immunity**. A microorganism or other antigen-bearing agent that breaches surface barriers triggers **innate immunity**, a set of general defenses that can prevent pathogens from becoming established in the body. **Adaptive immunity**, which can specifically target billions of different **antigens**, follows. Signaling molecules such as cytokines help coordinate the activities of white blood cells, including phagocytic **dendritic cells** and **macrophages**. Lymphocytes (**B cells**, **T cells**, and NK cells) are white blood cells with special roles in immune responses.

Section 22.3 Most **normal flora** that colonize body surfaces (including the linings of tubes and cavities) do not cause disease unless they penetrate inner tissues. Physical, mechanical, and chemical barriers (including **lysozyme**) fend off pathogens at body surfaces.

Microorganisms that breach surface barriers to enter the internal environment trigger general defense mechanisms of innate immunity. **Complement** can be activated by the presence of antigen or by tissue damage. White blood cells follow gradients of activated complement back to an affected tissue. Activated complement proteins also kill cells by puncturing their plasma membrane, and coat cells and particles to enhance their uptake by macrophages and other phagocytes (phagocytic white blood cells).

Phagocytes that engulf a microorganism or encounter damaged tissue release cytokines and local signaling molecules that initiate **inflammation**. Increased blood flow and capillary permeability speed delivery of more white blood cells to the affected area. Cytokines trigger **fever** that increases the metabolic rate and slows pathogen replication.

Normal flora that cause **dental plaque** can end up in atherosclerotic plaque, where they fuel inflammation associated with cardiovascular disease.

Section 22.4 **T cell receptors** are the basis of self/nonself discrimination; they recognize antigen displayed with **MHC markers** by an antigen-presenting cell. **B cell receptors** are **antibodies** that have not been released from a B cell. These antigen receptors collectively have the ability to recognize billions of specific antigens, a diversity that arises from random splicing of antigen receptor genes. Phagocytic white blood cells that engulf, process, and present antigen to T cells are central to all adaptive immune responses.

Section 22.5 B cells and T cells carry out adaptive immune responses. Four main characteristics of these responses are self/nonself recognition, sensitivity to specific threats, potential to intercept a tremendous number of diverse pathogens, and memory.

Antibodies that recognize a specific antigen are secreted by B cells during an **antibody-mediated immune response**. Antibody binding tags antigen-bearing particles for phagocytosis. Cytotoxic T cells that form in a **cell-mediated immune response** kill body cells that have been altered by infection or cancer. NK cells can kill ailing body cells that are missed by cytotoxic T cells. In both types of responses, **effector cells** form and target the antigen-bearing particles in a primary response. **Memory cells** that also form are reserved for a later encounter with the same antigen, in which case they trigger a faster, stronger secondary response.

Section 22.6 **Allergens** are normally harmless substances that induce immune responses in some people; sensitivity to an allergen is called **allergy**. A malfunction in the immune system or in its checks and balances can cause dangerous acute illnesses, or chronic and sometimes deadly disorders. In an **autoimmune response**, normal body cells are inappropriately recognized as foreign and attacked. Immune deficiency is a reduced capacity to mount an immune response. **AIDS** is caused by the human immunodeficiency virus (HIV). The virus infects white blood cells, so it eventually cripples the adaptive immune system.

Section 22.7 **Immunization** with **vaccines** designed to elicit immunity to specific diseases saves millions of lives each year as part of worldwide health initiatives. Vaccination programs require widespread participation in order to be effective.

Self-Quiz

Answers in Appendix I

1. _____ trigger immune responses.
   - a. Cytokines
   - b. Lysozymes
   - c. Antibodies
   - d. Antigens
   - e. Histamines
   - f. all of the above

2. Which of the following is *not* a surface barrier to infection?
   - a. resident bacterial populations
   - b. acidic gastric fluid
   - c. lysozyme in saliva
   - d. skin
   - e. complement activation
   - f. flushing action of diarrhea

3. Activated complement proteins can _____ .
   a. poke holes in cells
   b. promote inflammation
   c. attract macrophages
   d. all of the above

4. Which of the following is *not* part of innate immunity?
   a. phagocytic cells
   b. fever
   c. histamines
   d. cytokines
   e. inflammation
   f. complement activation
   g. presenting antigen
   h. all take part

5. Which of the following is *not* part of adaptive immunity?
   a. phagocytic cells
   b. antigen-presenting cells
   c. MHC markers
   d. cytokines
   e. antigen receptors
   f. complement activation
   g. antibodies
   h. all take part

6. _____ is/are characteristic of innate immunity, but not of adaptive immunity.
   a. Immediate responses
   b. The action of phagocytic cells
   c. Cytokine release
   d. Memory

7. Antibodies are _____ .
   a. antigen receptors
   b. made only by B cells
   c. proteins
   d. all of the above

8. _____ is/are characteristic of adaptive immunity, but not of innate immunity.
   a. Immediate responses
   b. The action of phagocytic cells
   c. Cytokine release
   d. Memory

9. A dendritic cell engulfs a bacterium, then presents bacterial bits on its surface along with a(n) _____ .
   a. MHC marker
   b. antibody
   c. T cell receptor
   d. antigen

10. Antibody-mediated responses are most effective against _____ .
    a. intracellular pathogens
    b. extracellular pathogens
    c. cancerous cells
    e. both a and c

11. Cell-mediated responses are most effective against _____ .
    a. intracellular pathogens
    b. extracellular pathogens
    c. cancerous cells
    d. both a and c

12. _____ are targets of cytotoxic T cells.
    a. Extracellular virus particles in blood
    b. Virus-infected body cells or tumor cells
    c. Bacterial cells in interstitial fluid
    d. Pollen grains in nasal mucus

13. Allergies occur when the body responds to _____ .
    a. pathogens
    b. toxins
    c. normally harmless substances
    d. all of the above

14. Match the immune cell with its main function.
    _____ dendritic cell
    _____ B cell
    _____ helper T cell
    _____ NK cell
    _____ cytotoxic T cell

    a. kills virus-infected body cells
    b. antigen-presenter
    c. activates other lymphocytes
    d. makes antibodies
    e. kills ailing body cells that lack MHC markers

15. Match the immunity concepts.
    _____ anaphylactic shock
    _____ immune memory
    _____ autoimmunity
    _____ inflammation
    _____ immune deficiency
    _____ antigen receptor
    _____ antigen processing

    a. recognizes antigen
    b. inadequate immune response
    c. general defense mechanism
    d. immune response against one's own body
    e. secondary response
    f. acute allergic reaction
    g. presenting antigen together with MHC markers

Critical Thinking

1. A tissue or organ or transplanted from one human into another is immediately attacked by the recipient's immune system. Which type of lymphocyte is responsible for this response, which is called transplant rejection?

2. A flu shot consists of a vaccine against several strains of influenza virus. This year, you get the shot and also "the flu." What happened? (There are at least three explanations.)

3. Monoclonal antibodies are produced by immunizing a mouse with a particular antigen, then removing its spleen. Individual B cells producing mouse antibodies specific for the antigen are isolated from the spleen, then fused with cancer cells. Each of the resulting hybrid cells can be cloned, which means it can be cultured indefinitely in the laboratory. The resulting cell lines produce and secrete antibodies that recognize the antigen to which the mouse was immunized. These antibodies are called monoclonal antibodies.

   Some monoclonal antibodies are used for passive immunization. They are effective, but only in the immediate term. Antibodies that are produced by one's own immune system can last up to about six months in the bloodstream, but monoclonals delivered in passive immunization often last for less than a week. Why the difference?

23

DIGESTION AND EXCRETION

Application ❯ 23.1 Causes and Effects of Obesity

Food provides the raw materials and energy we need to stay alive. The amount of energy that a human body can obtain from a particular food is measured in Calories (with a capital C). To maintain your weight, you must balance your energy intake with your energy output.

When the food you eat provides more energy than you need at the time, your body stores the excess in the bonds of organic compounds—mainly in fat molecules in adipose tissue. For most of human history, an ability to store energy as fat was purely advantageous. Accumulating fat when food is abundant increases the likelihood of survival if food later becomes scarce. However, most people in the United States now have more than enough food all of the time. As a result, about two-thirds of adults are overweight or obese.

Obesity is a disorder in which excess body fat impairs health. In a person who is obese, an excessive amount of fat surrounds and presses against internal organs, impairing the organs' function (Figure 23.1). For example, excessive fat in the abdomen impairs the diaphragm's ability to descend downward during inhalation, so breathing becomes more difficult. Obesity also impairs function at the cellular level. Adipose cells of people who are at a healthy weight hold a moderate amount of triglycerides. In obese people, adipose cells are overstuffed. Like cells stressed in other ways, the overstuffed cells respond by sending out chemical signals that summon up an inflammatory response. The resulting chronic inflammation harms organs throughout the body and increases the risk of cancer and diabetes.

Genetics plays a role in obesity. In terms of weight class (thin, average, overweight, or obese), people raised by adoptive parents resemble their biological parents more than their adoptive ones. Geneticists have even pinpointed some genes that affect the likelihood of obesity. The *fto* gene is one example. About 16 percent of people of European ancestry are homozygous for an *fto* allele that predisposes them to obesity. Compared to people without the high-risk allele, they tend to eat more food before they feel full and they are almost twice as likely to be obese.

Genetics can explain why one person is more likely than another to become overweight, but it cannot explain a national trend toward weight gain. Since 1980, the proportion of obese adults in the United States has doubled. There is no reason to think that the frequency of alleles that predispose people to obesity increased during this time. Decreasing exercise and increasing food intake are more likely causes of the rise in obesity.

Figure 23.1 Obesity—an unhealthy amount of fat.
Magnetic resonance images of an obese woman (left) and a woman of normal body weight (right). With obesity, the abdomen fills with fat that hampers the function of internal organs.

Marty Chobot/National Geographic Creative.

23.2 Two Types of Digestive Systems

REMEMBER: Animals are consumers (Section 1.3). They take food into their body, where they break it down and absorb the released nutrients (15.2).

In sponges, individual cells engulf food particles, break them down, then expel wastes. However, most animals have a digestive system, an organ system that carries out the following tasks:

1. **Ingestion:** Taking food into the chamber where digestion occurs. This chamber is an extension of the external environment, to which it opens. In animals with a digestive system, digestion is extracellular. Food does not enter individual cells; only the nutrients released from food do.

A. Saclike gastrovascular cavity of a planarian. There is a single opening to the cavity.

B. Complete digestive tract of an earthworm (an annelid). There are two openings, a mouth and an anus.

C. Diagram of an earthworm in cross section showing the "tube-within-a-tube" body plan.

2. **Digestion**: Breaking food down into components that can be absorbed into the body. Mechanical digestion smashes food into smaller and smaller fragments. Chemical digestion breaks large molecules in food into smaller ones.

3. **Nutrient absorption**: Movement of nutrient molecules into cells that line the digestive chamber.

4. **Elimination**: Expelling leftover material that was not digested and absorbed.

Flatworms and cnidarians have the simplest type of digestive system. Their saclike **gastrovascular cavity**, which functions in digestion and respiration, has a single opening. In flatworms, this opening is at the tip of a muscular pharynx (Figure 23.2A). As a result of the two-way traffic in a gastrovascular cavity, each load of food must be broken down, its nutrients absorbed, and any wastes eliminated before a new load of food can enter.

Most invertebrate groups and all vertebrates have a **complete digestive tract**, a tube that takes in food at one end and eliminates digestive wastes at the other. Earthworms have a complete digestive tract with a mouth and an anus (Figure 23.2B). An **anus** is a body opening that serves solely as the exit for digestive waste. The body plan of an animal with a complete digestive tract is sometimes described as a "tube within a tube." As Figure 23.2C illustrates, the coelom separates the outer tube (the body wall) from the inner tube (the digestive tract).

In many vertebrates, including amphibians, reptiles, and birds, digestive wastes leave the body through a cloaca (Figure 23.2D). A **cloaca** is a body opening that serves as the exit for digestive and urinary waste and also functions in reproduction.

Material moves through a complete digestive tract in one direction only, so new food can be taken in while previously ingested food is still being processed. The one-way flow also allows for regional specializations along the length of the tube. For example, both earthworms and pigeons have an expandable crop for storing food and a gizzard that specializes in smashing food.

D. Complete digestive tract of an bird. There are two openings, a mouth and a cloaca.

Figure 23.2 Animal digestive systems.

(B) From Russell/Wolfe/Hertz/Starr. *Biology*, 2e. © 2011, Cengage Learning®.

Take-Home Message 23.2

How do animals obtain the nutrients they need?

• Some invertebrates have a saclike gastrovascular cavity in which food is broken down and from which nutrients are absorbed.

• Most invertebrates and all vertebrates have a complete digestive tract. Some regions of this tubular system break down food and others absorb nutrients.

anus Body opening through which digestive wastes alone leave the body.

cloaca Body opening that functions in reproduction and serves as the exit for digestive and urinary wastes.

complete digestive tract Tubular gut with two openings.

digestion Breakdown of food into smaller bits and then into its component molecules.

elimination Expulsion of unabsorbed material from the digestive tract.

gastrovascular cavity Saclike cavity with one opening; functions in digestion and respiration.

ingestion Taking food into the digestive system.

nutrient absorption Movement of nutrient molecules from a digestive cavity into the body proper.

Figure 23.3 The human digestive system.

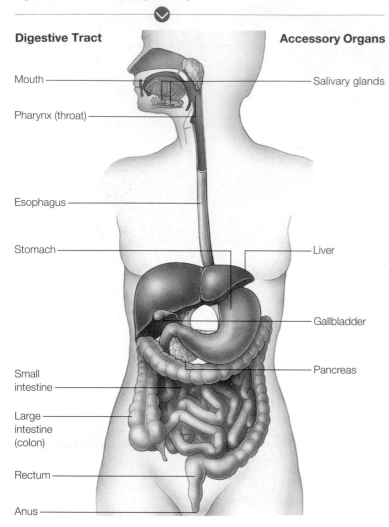

Digestive Tract

- Mouth
- Pharynx (throat)
- Esophagus
- Stomach
- Small intestine
- Large intestine (colon)
- Rectum
- Anus

Accessory Organs

- Salivary glands
- Liver
- Gallbladder
- Pancreas

23.3 Digestive Structure and Function

REMEMBER: Hydrolysis breaks large molecules into their smaller subunits (Section 2.6). Many bacteria live in the human gut (13.1). The pharynx is the entrance to the respiratory and digestive tracts (21.8).

The human digestive system includes a tubular digestive tract and the accessory organs that secrete substances into it (Figure 23.3).

In the Mouth Mechanical digestion begins in the mouth, where teeth tear food apart and smash it into smaller pieces. As a species, humans are omnivores, meaning our diet includes both plant and animal material, and the structure of our teeth adapts us to this varied diet. We have all four types of mammalian teeth, and all are equally well developed (Figure 23.4A). By comparison, carnivorous (meat-eating) mammals typically have enlarged canine teeth for piercing and tearing flesh (Figure 23.4B) and herbivores (plant eaters) tend to have reduced canine teeth (Figure 23.4C). Premolars of carnivores have a narrow, bladelike surface that helps them shear through meat. By contrast, premolars of herbivores and humans are broad and flat, the better for grinding plant material.

The tongue, a bundle of membrane-covered skeletal muscle, attaches to the floor of the mouth. Movements of the tongue help mix food with saliva from salivary glands. These exocrine glands open into the mouth beneath the tongue and on the inner surface of the cheeks beside the upper molars. Salivary amylase, an enzyme in saliva, begins the process of chemical digestion by breaking large polymers of starch into smaller disaccharides (two-sugar units). This reaction, like other digestive reactions, is an example of hydrolysis. Saliva also contains glycoproteins that combine with water to form mucus. The presence of mucus helps small food bits stick together in easy-to-swallow clumps.

Swallowing When food has been adequately chewed, the tongue pushes it to the back of the pharynx (throat), where its presence triggers a swallowing reflex. As

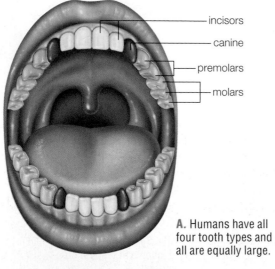

incisors
canine
premolars
molars

A. Humans have all four tooth types and all are equally large.

B. Carnivores have enlarged canine teeth and sharp premolars.

C. Herbivores have reduced canines, and large flat premolars and molars.

Figure 23.4 Variations in mammalian teeth. (B,C) From Russell/Wolfe/Hertz/Starr. *Biology,* 2e. © 2011, Cengage Learning®.

swallowing begins, the soft palate, a region of flexible tissue at the rear of the roof of the mouth, moves up to block the passage from the throat to the nasal cavity. Then the epiglottis (the flap of tissue at the entrance to the larynx) folds down and the vocal cords constrict. These actions block the route between the pharynx and larynx. With the passages to the nasal cavity and the larynx closed off, contraction of muscles in the pharynx wall can push food into the **esophagus**, the muscular tube that leads to the stomach.

If food or drink accidentally enters the larynx, a reflexive cough generally expels it. However, sometimes food gets through the larynx and obstructs the trachea. When this happens, a person chokes. A choking person cannot cough, speak, or breathe. To assist someone who is choking, the American Red Cross suggests first administering five sharp slaps to the victim's back between the shoulder blades, followed by five abdominal thrusts (Figure 23.5).

When swallowing proceeds normally and food enters the esophagus, waves of smooth muscle contraction, called peristalsis, move food to the stomach. The gastroesophageal sphincter connects the lower end of the esophagus with the stomach (Figure 23.6). (Gastro- or gastric refers to the stomach.) A **sphincter** is a ring of muscle that controls passage of material through a tube or a body opening. The gastroesophageal sphincter opens to allow food into the stomach, then closes to prevent stomach acid from splashing into the esophagus. In some people, this sphincter does not always shut properly, and acid reflux occurs; acidic fluids irritate the esophagus, causing the pain commonly called "heartburn." Occasional heartburn poses no threat to health and can be treated with over-the-counter antacids. Antacids act by making the fluid in the stomach less acidic. Chronic heartburn, which can raise the risk of esophageal cancer, should be investigated by a physician.

The Stomach The human **stomach** is a hollow, muscular organ that receives food from the esophagus, secretes substances that help break down food (especially proteins), and controls how quickly food passes into the next segment of the digestive tract (the small intestine). The stomach has a wall of smooth muscle layers and a lining of glandular epithelium. When the stomach is empty, this lining has

1. Slap the person between the shoulder blades 5 times.

2. Administer 5 abdominal thrusts. Stand behind the person and place one fist below the rib cage, just above the navel, with your thumb facing inward. Cover the fist with your other hand and thrust inward and upward with both fists.

3. Repeat alternating back blows and abdominal thrusts until the object blocking the airway is dislodged.

Figure 23.5 First aid for choking.

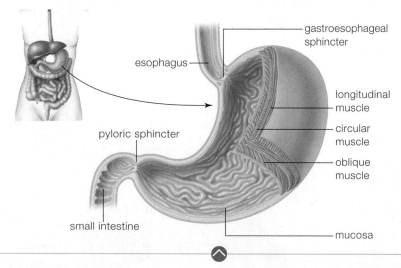

Figure 23.6 Human stomach.
The innermost glandular epithelium secretes gastric fluid. Three smooth muscle layers differ in their orientation, and their contractions cause a wringing action that mixes and moves food. Sphincters at either end of the stomach can close to keep acidic chyme inside the stomach.

esophagus Muscular tube that connects the pharynx (throat) to the stomach.

sphincter Ring of muscle that controls passage through a tubular organ or body opening.

stomach Muscular organ that receives food from the esophagus, mixes it with gastric fluid, and propels it into the small intestine.

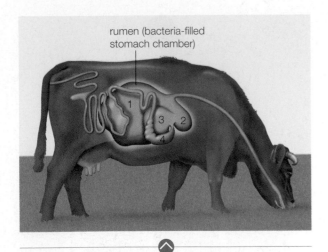

Figure 23.7 Digestive system of a cow, a ruminant.
The large stomach, highlighted in yellow, has four chambers. The largest chamber, the rumen, holds bacteria that can break down the cellulose in plant material.

rumen (bacteria-filled stomach chamber)

numerous folds. The folds disappear as the stomach fills with food. Arrival of food in the stomach also stimulates smooth muscle contractions and causes cells in the stomach lining to secrete **gastric fluid**. This fluid includes a strong acid (hydrochloric acid, or HCl), enzymes, and mucus. Stomach contractions mix food and gastric fluid together to form a semiliquid mass called **chyme**.

Protein digestion begins in the stomach. The acidity of gastric fluid denatures (unfolds) proteins in chyme. Then pepsin, an enzyme in gastric fluid, cuts the unfolded proteins into polypeptides—another hydrolysis reaction. Why doesn't this process also break down proteins in the stomach wall? Mucus secreted by the stomach lining protects the underlying tissue.

When something disrupts the stomach's protective mucus layer, gastric fluid and enzymes can erode the stomach lining, causing an ulcer. Continual use of aspirin, ibuprofen, and other nonsteroidal anti-inflammatory drugs increase the risk of an ulcer because one of their side effects is a decrease in the stomach's mucus production. Most ulcers occur after *Helicobacter pylori* bacteria make their way through gastric mucus and infect cells of the stomach lining. Antibiotics can halt the bacterial infection and allow the ulcer to heal.

Stomach-dwelling microbes play little role in human digestion, but they are essential to ruminants such as cattle, sheep, and goats. These hooved, plant-eating mammals have a large, multichambered stomach (Figure 23.7). The term "ruminant" refers to the largest chamber, the rumen, which is like a big fermenting vat. The rumen holds bacteria that, unlike animals, can break down the cellulose in plant walls. When a ruminant eats grass, the grass ferments for a while in the rumen before being moved back into the mouth as "cud" that is chewed and reswallowed. When cud is sufficiently broken down, it moves on through other stomach chambers and the remainder of the digestive tract. As a result of this process, ruminants can extract nutrients from plant material that most other animals cannot digest.

Digestion in the Small Intestine The **small intestine**, which receives food from the stomach, functions in digestion and is the main site of nutrient absorption. It is "small" only in terms of its diameter, which in humans is about 2.5 cm (1 inch). The small intestine is the longest portion of the human digestive tract.

The interior of the small intestine, like that of the stomach, is highly folded (Figure 23.8 ❶). Unlike the internal folds of the stomach, intestinal folds are

Figure 23.8 Structure of the small intestine.

(2–4) After Sherwood and others.

Figure It Out: Are microvilli multicelled or smaller than a cell?

Answer: Smaller than a cell

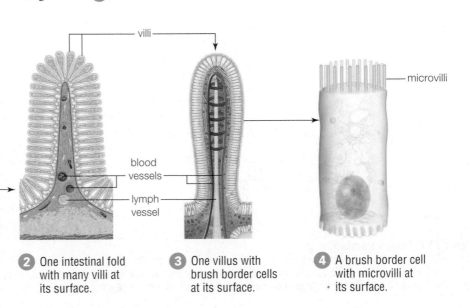

❶ Circular folds ring the interior of the small intestine. These folds are permanent; they remain even when the small intestine is filled with chyme.

❷ One intestinal fold with many villi at its surface.

❸ One villus with brush border cells at its surface.

❹ A brush border cell with microvilli at its surface.

villi

microvilli

blood vessels

lymph vessel

permanent. Each fold has many multicelled, millimeter-long projections called **villi** (singular, villus) ❷. The many villi that project from the intestinal lining give the lining a velvety appearance. Cells at the surface of each villus have even tinier projections called **microvilli** (singular, microvillus) ❸. The many microvilli at the surface of a cell make its outer edge look like a brush ❹, so these cells are sometimes called brush border cells. Collectively, the folds and projections of the small intestinal lining increase its surface area several hundredfold.

Chemical digestion of carbohydrates and proteins is completed in the small intestine, with help from the neighboring pancreas and liver (Figure 23.9). The **pancreas** is a large gland that secretes enzymes and a buffer into the small intestine. Pancreatic and small intestinal enzymes break carbohydrates down into monosaccharides (simple sugars) and polypeptides into amino acids.

Fat digestion occurs entirely in the small intestine. It requires enzymes and **bile**, a mix of salts, pigments, and cholesterol. Bile is made in the **liver**, a large organ that also stores glycogen and detoxifies substances. Between meals, the main bile duct from the liver to the small intestine is closed, so bile enters the **gallbladder**, which stores it. Eating a fatty meal stimulates the gallbladder to contract and squirt bile into the small intestine. Bile enhances fat digestion by helping to keep fat droplets from clumping together, a process called emulsification. Compared to bigger globules, tiny droplets present a greater surface area to fat-digesting pancreatic and intestinal enzymes. These enzymes break lipids into fatty acids and glycerol.

Sometimes components of bile accumulate as hard pellets called gallstones. Most gallstones do no harm, but some can block or become lodged in a duct. In such a case, the gallstones or the entire gallbladder can be removed surgically.

Absorption in the Small Intestine Figure 23.10 summarizes where nutrients are digested and the absorbable products of the digestive process. Absorption is the passage of solutes and water across the lining of the digestive tract and into the body proper. The process is facilitated by the large surface area of the small intestine, and by rings of smooth muscle in the intestinal wall. Contraction and relaxation of these muscles make chyme slosh back and forth, forcing it against the absorptive lining.

Products of carbohydrate digestion (monosaccharides) and protein digestion (amino acids) are actively transported into brush border cells. They diffuse across the cell, then are actively transported into the interstitial fluid inside the villus. From

> The small intestine is the longest portion of the human digestive tract.

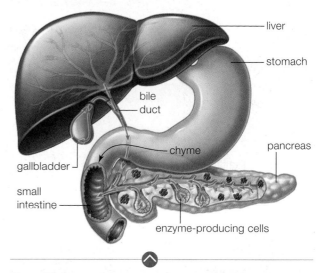

Figure 23.9 Organs that empty into the small intestine.
Chyme from the stomach joins bile and pancreatic enzymes, which enter through ducts.

From Russell/Wolfe/Hertz/Starr. *Biology*, 2e. © 2011, Cengage Learning®.

	Carbohydrates	Proteins	Lipids
Mouth	Salivary amylase begins digestion		
Stomach		Acid, pepsin begin digestion	
Small intestine	Pancreatic and intestinal enzymes complete digestion	Pancreatic and intestinal enzymes complete digestion	Pancreatic and intestinal enzymes carry out digestion with the assistance of bile
Absorbable products	Monosaccharides (simple sugars)	Amino acids	Fatty acids, glycerol

Figure 23.10 Nutrient breakdown locations and products.

bile Mix of salts, pigments, and cholesterol produced by the liver; aids in fat digestion.

chyme Mix of food and gastric fluid.

gallbladder Organ that receives bile from the liver and expels it into the small intestine.

gastric fluid Fluid secreted by the stomach lining; contains enzymes, acid, and mucus.

liver Organ that produces bile, stores glycogen, and detoxifies many substances.

microvilli Thin projections that increase the surface area of brush border cells.

pancreas Organ that secretes digestive enzymes into the small intestine and hormones into the blood.

small intestine Longest portion of the digestive tract, and the site of most digestion and absorption.

villi Multicelled projections from the lining of the small intestine.

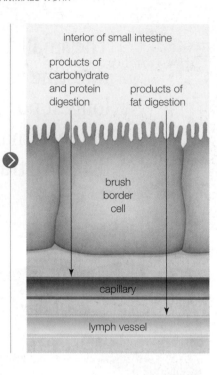

Figure 23.11 Absorption in the small intestine. Nutrients move across brush border cells into a villus, then enter the blood or lymph.

interior of small intestine

products of carbohydrate and protein digestion

products of fat digestion

brush border cell

capillary

lymph vessel

Figure 23.12 The large intestine.

(B) National Cancer Institute.

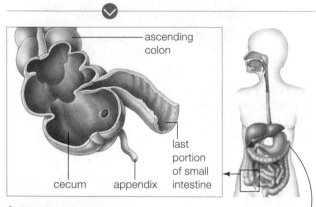

ascending colon

last portion of small intestine

cecum   appendix

A. Location of the appendix.

transverse colon

colon polyp

descending colon

B. Sketch and photo of polyps in the colon.

here, they enter a blood capillary (Figure 23.11). Products of fat digestion (fatty acids and glycerol) also cross brush border cells and enter the interstitial fluid. However, these molecules enter lymph vessels, which carry them to the blood.

All blood from capillaries in the small intestine flows through a vein (the hepatic portal vein) to the liver. Thus, blood with substances absorbed from the small intestine goes through the liver before returning to the heart and flowing to the rest of the body. The liver protects the body by detoxifying or removing from the blood dangerous substances that were ingested or formed by digestion.

Concentrating and Eliminating Wastes Contractions propel indigestible material, dead cells shed by the gut lining, and some water into the **large intestine**. As these materials travel through this organ, they become compacted as **feces**. The large intestine concentrates feces by actively pumping sodium ions across its wall, into the internal environment. Water follows by osmosis.

The initial portion of the large intestine is the cup-shaped cecum. A short, tubular **appendix** projects from the cecum (Figure 23.12A). Inflammation of the appendix, called appendicitis, requires prompt treatment. Removing an inflamed appendix prevents it from bursting and releasing bacteria into the abdominal cavity. Such ruptures can cause a life-threatening infection.

Beyond the cecum, the region of the large intestine known as the **colon** ascends the wall of the abdominal cavity, extends across that cavity, and descends. Contraction of smooth muscle in the colon wall mixes the colon's contents and propels them toward the rectum. The colon has a moderate pH and it moves materials along more slowly than other regions of the digestive tract. These factors encourage growth of bacteria such as *Escherichia coli*, which produces vitamin B_{12} that we absorb across our colon lining.

Some people are genetically predisposed to develop colon polyps, which are small growths on the colon wall (Figure 23.12B). Most polyps are benign, but some can become cancerous. If detected in time, colon cancer is highly curable. Blood in feces and dramatic changes in bowel habits may be symptoms of colon cancer and should be reported to a doctor. Also, anyone over the age of 50 should have a periodic colonoscopy, a procedure in which clinicians use a camera to examine the interior of the colon for polyps or cancer.

The colon connects to the **rectum**, the portion of the large intestine where feces are stored until they are expelled. When the rectum is full, stretching of its wall activates a defecation reflex that expels feces. We can override this reflex by contracting a sphincter made of skeletal muscle at the anus.

Take-Home Message 23.3

How is food processed by the mammalian digestive tract?

- Mechanical and chemical breakdown of food begins in the mouth. Saliva contains an enzyme that begins carbohydrate digestion.
- Acid and digestive enzymes begin the process of protein digestion in the stomach.
- In the small intestine, intestinal and pancreatic enzymes complete digestion of proteins and carbohydrates, and bile from the gallbladder assists in fat digestion.
- Most nutrients released by digestion are absorbed in the small intestine.
- The large intestine compacts undigested residues. The resulting feces are stored in the rectum before being expelled through the anus.

Figure 23.13 Fates of absorbed nutrients.

Photo, iStockphoto.com/fcafotodigital.

23.4 Human Nutrition

REMEMBER: Sections 2.7 to 2.9 described the structure of carbohydrates, fats, and proteins. Section 5.7 explained how your body releases energy stored in these organic molecules. Coenzymes (4.4) assist enzymes in these reactions.

Carbohydrates, fats, and proteins are called macronutrients because we require these substances in large amounts. Macronutrients function as sources of both energy and raw materials (Figure 23.13).

Carbohydrates Your body breaks down sugars and starches to release glucose, your cells' primary source of energy. When the amount of glucose absorbed exceeds the body's needs, the excess is stored. Blood that flows through capillaries in the small intestine carries glucose-rich blood to the liver, which stores glucose as glycogen. Liver and adipose cells use glucose to build fats.

Complex carbohydrates that you cannot digest are also essential. Such indigestible material is called dietary fiber. There are two types of fiber. Soluble fiber consists of polysaccharides that form a gel when mixed with water. A diet high in soluble fiber helps reduce the blood level of "bad cholesterol" (LDL). Insoluble fiber such as cellulose does not dissolve. It passes through the human digestive tract more or less intact. A diet high in insoluble fiber helps prevent constipation.

You may have noticed breads, pastas, and other carbohydrate-rich foods labeled as "gluten-free." Gluten is a protein found in wheat, barley, oats and some other grains. An estimated 1 percent of the population has celiac disease, a disorder in which gluten causes an autoimmune reaction that harms the small intestine's villi and interferes with nutrient absorption. Celiac disease is diagnosed by taking a biopsy of the intestine, and it is treated by eliminating gluten from the diet. Whether gluten also has negative effects in some people who do not have celiac disease remains unclear. Recent, well-controlled studies of people with self-diagnosed "gluten sensitivity," found no specific negative response to gluten. People who feel better when they eat a gluten-free diet may be benefiting by eliminating certain fermentable carbohydrates (collectively known as FOMAPs) also in grain-based products.

appendix Tubular projection from the first part of the large intestine.

colon Longest portion of the large intestine.

feces Unabsorbed food material and cellular waste that is produced by digestion.

large intestine Organ that concentrates and stores waste, then delivers it to the anus for excretion.

rectum Portion of the large intestine that stores feces until they are expelled.

A diet of plant foods can provide all essential amino acids.

Fats Fats serve as an energy source and as building blocks of cell membranes. They also help you take up fat-soluble vitamins. Your body can assemble most fats you need from carbohydrates. Fats that the body cannot make, such as linoleic acid, are **essential fatty acids**. Vegetable oils are a good source of these fatty acids.

By contrast, *trans* fatty acids, or "*trans* fats," which are used in some processed foods, should be avoided. A diet rich in these artificially modified fats increases the risk of heart disease. Most nutritionists also recommended moderating one's intake of foods rich in saturated fat in order to minimize the risk of cardiovascular disease. Such foods include eggs, red meat, and whole fat dairy products.

Proteins Your body uses amino acid components of dietary proteins for its own protein-building programs. Of the twenty common types of amino acids, eight are **essential amino acids** that you cannot synthesize and must obtain from food. Animal proteins (proteins in meats, eggs, and dairy products) contain all essential amino acids in the ratio that your body requires to build its own proteins. The ratio of amino acids is somewhat different in plant foods. Compared to meats, grains tend to have a lower percentage of the amino acid lysine, and legumes (peas and beans) tend to have a lower percentage of the sulfur-containing amino acids (methionine and cysteine). However, a vegan (plant only) diet that includes a variety of plant foods can provide adequate levels of all essential amino acids. A diet in which nonprocessed plant foods predominate has the added advantage of minimizing the intake of saturated fat and ensuring an adequate fiber intake.

Vitamins and Minerals **Vitamins** are organic substances that are required in small amounts for normal metabolism (Table 23.1). Vitamins A, E, D, and K are fat-soluble vitamins. Such vitamins are not affected by heat, so they are abundant in both cooked and fresh foods. Fat-soluble vitamins can be stored in the body's own

Table 23.1 Sources and Functions of Major Vitamins

Vitamin	Dietary Sources	Main Functions
Fat-soluble vitamins		
A	Yellow/orange fruits, green vegetables, milk, egg yolk, fish, liver	Used in synthesis of visual pigments, bone, teeth; maintains the skin
D	Fish liver oils, egg yolk, fortified milk; also synthesized in skin	Promotes bone health; enhances calcium absorption
E	Whole grains, dark green vegetables, vegetable oils	Counters effects of free radicals; helps maintain cell membranes; blocks breakdown of vitamins A, C in gut
K	Green leafy vegetables, cabbage	Acts in blood clotting; ATP formation via electron transfer chains
Water-soluble vitamins		
B_1 (thiamine)	Legumes, whole grains, green leafy vegetables, meats, eggs	Connective tissue formation; folate utilization; coenzyme action
B_2 (riboflavin)	Whole grains, poultry, fish, egg white, milk	Coenzyme in carbohydrate and amino acid metabolism
B_3 (niacin)	Green leafy vegetables, potatoes, peanuts, poultry, fish, meats	Coenzyme in carbohydrate metabolism
B_6	Spinach, tomatoes, potatoes, meats	Coenzyme in amino acid metabolism
Pantothenic acid	Meats, yeast, egg yolk	Coenzyme in glucose metabolism, fatty acid and steroid synthesis
Folate (folic acid)	Dark green vegetables, whole grains, yeast, lean meats; made by bacteria in large intestine	Coenzyme in nucleic acid and amino acid metabolism; required for normal nervous system development of embryo
B_{12}	Poultry, fish, red meat, dairy foods (not butter)	Coenzyme in nucleic acid metabolism
Biotin	Legumes, egg yolk	Coenzyme in fat, glycogen formation, and amino acid metabolism
C (ascorbic acid)	Fruits and vegetables, especially citrus, berries, cantaloupe, green pepper, cabbage, broccoli	Collagen synthesis; used in carbohydrate metabolism; structural role in bone, teeth, cartilage; may counter free radicals

fat, so they do not need to be replenished daily and deficiencies rarely occur. In fact, taking megadoses of fat-soluble vitamins can be harmful.

By contrast, water-soluble vitamins such as vitamin C and the B vitamins are not stored by the body. It is difficult to accumulate too much of these vitamins, because any excess is eliminated in the urine. Water-soluble vitamins must be replenished each day to prevent a deficiency. These vitamins are destroyed by heat, so uncooked foods are the best source.

Vitamin C is an antioxidant and it is needed to make collagen, the body's most abundant protein. A deficiency in this vitamin causes scurvy, a disorder in which skin and bones deteriorate and wounds are slow to heal. Many of the B vitamins are coenzymes. For example, the coenzyme NADH, which plays a role in aerobic respiration, is derived from vitamin B_3 (niacin).

Minerals are inorganic substances needed in small amounts for normal growth and metabolism. Calcium and phosphorus, the body's most abundant minerals, are components of teeth and bones. Sodium and potassium are important in nerve function, sulfur is a component of some proteins, chlorine is a component of the hydrochloric acid (HCl) in gastric fluid, and magnesium is a cofactor. Iodine, another essential mineral, is necessary to produce thyroid hormone, which has roles in development and metabolism.

Iron is an essential component of heme, the chemical group that binds to oxygen in hemoglobin. Heme is also a cofactor for many enzymes, including critical components of electron transfer chains. Worldwide, iron is the mineral most commonly deficient in the diet. Plant foods are low in iron, so people who avoid animal products are advised to take supplemental iron. Women are especially likely to be deficient in iron because they lose iron in blood when they menstruate.

USDA Dietary Recommendations Every five years, the United States government reviews nutrition research and issues updated dietary guidelines designed to promote health, prevent disease, and help people maintain a healthy weight. Figure 23.14 shows an example of their recommendations. Current guidelines recommend eating more vegetables and fruits, more whole grains, and more fat-free or low-fat milk products. People who are lactose intolerant or who wish to avoid animal products can substitute "milks" made from soybeans, rice, almonds, or other plants. The current guidelines also recommend avoiding prepared foods that are high in salt and added sugar.

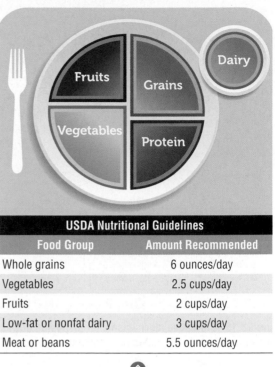

USDA Nutritional Guidelines	
Food Group	**Amount Recommended**
Whole grains	6 ounces/day
Vegetables	2.5 cups/day
Fruits	2 cups/day
Low-fat or nonfat dairy	3 cups/day
Meat or beans	5.5 ounces/day

Figure 23.14 Example of nutritional guidelines from the United States Department of Agriculture (USDA). Recommendations are for females between ages ten and thirty who get less than 30 minutes of vigorous exercise daily. Portions add up to a 2,000-kilocalorie daily intake.

You can generate your own healthy eating plan by visiting the USDA website: www.choosemyplate.gov.

USDA.

Take-Home Message 23.4

How do we use the nutrients in our food?

- Carbohydrates are broken down to glucose, the body's main energy source. Foods rich in complex carbohydrates also supply fiber and vitamins.
- Fats are burned for energy and used as building materials.
- Proteins are the source of amino acids that the body uses to build its own proteins.
- Vitamins are organic substances that are essential in small amounts for metabolism. Fat-soluble vitamins can be stored in the body, but water-soluble vitamins cannot.
- Minerals are inorganic substances required in small amounts for metabolic tasks.
- Added sugar and salt in foods and drinks do not meet any nutritional need and can in fact have negative health consequences.

essential amino acid Amino acid that the body cannot make and must obtain from food.

essential fatty acids Fatty acids that the body cannot make and must obtain from the diet.

mineral Inorganic substance that is required in small amounts for normal metabolism.

vitamin Organic substance required in small amounts for normal metabolism.

23.5 Fluid Regulation

REMEMBER: Osmosis is the movement of water across a membrane from a region of high solute concentration to a region of lower solute concentration (Section 4.5).

Fluid Homeostasis By weight, an animal consists mostly of water, with dissolved salts and other solutes. Extracellular fluid, the fluid outside cells, is the cells' environment. To keep the solute composition and volume of extracellular fluid within the ranges cells can tolerate, water and solute gains must equal water and solute losses. An animal loses water and solutes in excretions, exhalations, and secretions. It gains water by eating and drinking. In aquatic animals, water also moves into or out of the body by osmosis across the body surface.

Metabolic wastes enter and alter the composition of the extracellular fluid. Aerobic respiration produces waste carbon dioxide that diffuses out across the body surface or leaves with the help of respiratory organs. Protein breakdown produces ammonia (NH_3), a nitrogen-containing waste product that is toxic in large amounts. In most animals, special excretory organs rid the body of ammonia, other unwanted solutes, and any excess water.

Most aquatic invertebrates and many fish excrete ammonia directly but other animals convert it to another substance prior to excretion (Figure 23.15). Conversion of ammonia to uric acid is an energy-intensive process, but its saves a lot of water. Uric acid can be excreted as crystals with just a tiny bit of water, whereas ammonia must be dissolved in a large amount of water. Insects and birds excrete uric acid. Mammals excrete urea. Conversion of ammonia to urea requires a moderate energy investment and saves a moderate amount of water.

Fluid Regulation in Invertebrates Planarian flatworms, like other freshwater animals, have a higher solute concentration than the water around them, so water continually enters their body by osmosis. A system of branching tubes takes up body fluid at one end and expels excess water, ammonia, and other unwanted solutes at the other end (Figure 23.16A). The waste material exits through a pore in the body wall.

In insects, waste solutes are actively transported from hemolymph into a system of **Malpighian tubules**, which are excretory organs that extend from and empty into the gut (Figure 23.16B). Waste material delivered by Malpighian tubules joins with digestive wastes and both are expelled from the body through the anus.

Figure 23.15 Alternative mechanisms for disposing of waste amino groups formed by protein metabolism.

waste amino groups
(–NH₂)

↓

ammonia (NH₃)

Convert to uric acid	Excrete ammonia	Convert to urea
(highest energy cost, saves the most water)	(no added energy cost, highest water loss)	(moderate energy cost and water loss)

Figure 23.16 Examples of invertebrate excretory systems.

(B) left, © Stephen Dalton/Science Source; right, Susumu Nishinaga/Science Source.

A. Flatworm excretory system. Movement of cilia in porous cells at the tips of tubules draws interstitial fluid into tubules. Fluid exits tubules through a pore at the body surface.

ciliated, porous cell that draws fluid into tubule

pore through which fluid exits body

B. Insect Malpighian tubules (gold) are outpouchings of the gut (pink). Waste solutes are actively transported from hemolymph into a tubule. The tubules then deliver wastes to the gut for elimination through the anus.

Malpighian tubule

part of gut

Vertebrate Urinary System All vertebrates have a pair of kidneys. The **kidneys** filter water, excess ions, nitrogen-rich wastes, and other substances from blood. Then they adjust the composition of the filtrate and return nearly all water and nonwaste solutes to blood. The water and solutes that do not return to the blood are excreted from the body as **urine**.

The human **urinary system** consists of two kidneys, two ureters, a bladder, and a urethra (Figure 23.17). A human kidney is a bean-shaped organ about as large as an adult's fist, enclosed in a protective outer capsule of connective tissue. Urine flows from each kidney into a **ureter**, a tube that empties into a hollow, muscular **urinary bladder** where the urine is stored. When the bladder is full, a reflex action causes smooth muscle in the bladder wall to contract. At the same time, sphincters encircling the **urethra**, the tube that delivers urine to the body surface, relax. As a result of these two actions, urine flows out of the body. After age two or three, the brain can override this reflex and prevent urine from flowing through the urethra at inconvenient moments.

A human male's urethra runs the length of his penis, the organ that conveys both urine and sperm to the body surface. A sphincter cuts off urine flow during times of sexual excitement. In females, the urethra opens onto the body surface near the vagina. The female urethra is a shorter tube, so infectious organisms can more easily reach the urinary bladder. That is one reason women have bladder infections more often than men do.

Birds and other animals with a cloaca do not have a separate body opening that services the urinary tract. Thus, bird droppings contain a mix of feces and urinary waste. The chalky white component of bird droppings is uric acid crystals.

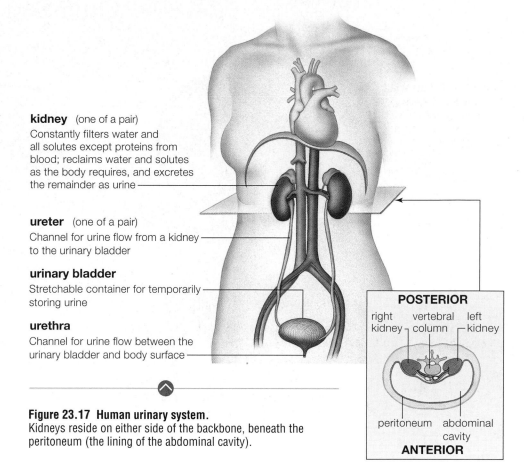

kidney (one of a pair)
Constantly filters water and all solutes except proteins from blood; reclaims water and solutes as the body requires, and excretes the remainder as urine

ureter (one of a pair)
Channel for urine flow from a kidney to the urinary bladder

urinary bladder
Stretchable container for temporarily storing urine

urethra
Channel for urine flow between the urinary bladder and body surface

POSTERIOR
right kidney   vertebral column   left kidney

peritoneum   abdominal cavity

ANTERIOR

Figure 23.17 Human urinary system.
Kidneys reside on either side of the backbone, beneath the peritoneum (the lining of the abdominal cavity).

Take-Home Message 23.5

How do animals regulate the composition of their body fluid?

• All animals must rid their body of metabolic wastes such as the ammonia formed by the breakdown of protein.

• Most animals have a urinary system that regulates the composition and volume of extracellular fluid.

• In vertebrates, kidneys filter the blood and form urine. In humans, urine from kidneys flows through ureters to a bladder, where it is stored until it flows out of the body through the urethra.

kidney Organ that filters blood and forms urine.

Malpighian tubules Of insects, organs that take up unwanted solutes from hemolymph and deliver them to the gut for excretion.

ureter Tube that carries urine from a kidney to the bladder.

urethra Tube through which urine from the bladder flows out of the body.

urinary bladder Hollow, muscular organ that stores urine.

urinary system Organ system that filters blood, and forms, stores, and expels urine.

urine Mix of water and soluble wastes formed and excreted by the urinary system.

Figure 23.18 How urine forms in the kidney. Nephron tubules interact with two sets of capillaries to form urine.

A. Cutaway view of a human kidney.

renal artery

renal vein

ureter

Filtration: Water and solutes filtered out of capillaries enter tubule

Reabsorption: Water and solutes leave tubules and return to the blood

Secretion: Unwanted solutes move from blood into filtrate

B. One nephron, the functional unit of a kidney.

urine

① Filtration (yellow arrow): At the start of a kidney tubule, blood pressure forces fluid out of a cluster of leaky capillaries and into the tubule.

② Reabsorption (green arrows): As filtrate continues through the tubule, water and solutes leave the tubule and return to the blood in adjacent capillaries.

③ Secretion (purple arrow): Some solutes are also transported in the opposite direction, from the blood into the filtrate inside the kidney tubule.

23.6 Kidney Function

REMEMBER: Blood pressure is the pressure imparted to the blood by contracting ventricles (Section 21.5).

How Urine Forms A kidney is a blood-cleansing organ. A renal artery carries blood to be cleansed to each kidney, and a renal vein transports cleansed blood away from it (Figure 23.18A). The Latin word *renal* means "relating to kidneys." Inside a kidney, blood flows into arterioles. The arterioles branch into capillaries that associate with kidney tubules as a **nephron**.

Nephrons cleanse the blood and form urine by three processes: filtration, reabsorption, and secretion (Figure 23.18B). Each nephron starts in the outer region of the kidney, where the wall of a kidney tubule cups around a cluster of capillaries. These capillaries are structurally specialized for filtering; they are much leakier than capillaries in most parts of the body. **Filtration** occurs when blood pressure forces water and solutes out through spaces between cells in the walls of the capillaries ①. Blood cells and proteins are too large to escape through these spaces, so they remain in the blood.

filtration In urine formation, blood pressure forces water and small solutes, but not blood cells or proteins, out across the walls of capillaries.

nephron Kidney tubule and associated capillaries; filters blood and forms urine.

reabsorption Water and solutes enter capillaries.

tubular secretion Substances are moved out of capillaries and into kidney tubules.

Digging Into Data

Pesticide Residues in Urine

Food that carries the USDA's organic label must be produced without pesticides such as malathion and chlorpyrifos, which are often used on conventionally grown crops. Chensheng Lu of Emory University used urine testing to find out how eating organic food affects the level of pesticide residues in a child's body (Figure 23.19). For fifteen days, the urine of twenty-three children (aged 3 to 11) was monitored for break-down products of pesticides. During the first five days, children ate their standard diet of conventionally grown foods. For the next five days, they ate organic versions of the same types of foods and drinks. For the final five days, the children returned to their standard diet.

1. During which phase of the experiment did the children's urine contain the lowest level of the malathion metabolite?
2. During which phase of the experiment was the maximum level of the chlorpyrifos metabolite detected?
3. Did switching to an organic diet lower the amount of pesticide residues excreted by the children?

4. Even during the conventional diet phases of this experiment, the pesticide metabolite levels remained far below those known to be harmful. Given these data, would you spend more to buy organically grown foods?

Study Phase	No. of Samples	Malathion Metabolite		Chlorpyrifos Metabolite	
		Mean (µg/liter)	Maximum (µg/liter)	Mean (µg/liter)	Maximum (µg/liter)
1. Conventional	87	2.9	96.5	7.2	31.1
2. Organic	116	0.3	7.4	1.7	17.1
3. Conventional	156	4.4	263.1	5.8	25.3

Figure 23.19 Effects of conventional and organic diet on metabolites (breakdown products) of malathion and chlorpyrifos in the urine of children. The difference in the average level of metabolites in the organic and inorganic phases of the study was statistically significant.

The solute-rich filtrate forced out of the capillaries enters the first part of the kidney tubule, where **reabsorption** begins. By this process, water and essential sol-utes that left the blood return to it as the filtrate flows through tubules . Essential solutes are actively transported out of the tubule and water follows by osmosis. About 99 percent of the water returns to the blood, along with all of the glucose and amino acids. Wastes such as urea and the breakdown product of hemoglobin remain in the filtrate. Degraded hemoglobin is what gives your urine its yellow color.

As the filtrate continues through the tubule, unwanted solutes from the blood are actively transported into it, a process called **tubular secretion** ❸. Secretion is essential to regulating the pH of the internal environment. Enzymes can function only within a limited pH range. An excess of hydrogen ions (H^+) can disrupt many essential metabolic processes. Kidneys help maintain the body's acid–base balance by regulating how much H^+ is secreted into the urine.

Metabolites (breakdown products) of foreign organic chemicals that have entered the bloodstream are also secreted into the urine. If you smoke cigarettes, use psychoactive drugs, take an antibiotic, or eat food that contains pesticides, your urine will contain breakdown products from these foreign substances. Urine-based drug tests are designed to detect such drug metabolites.

Feedback Control of Urine Formation The concentration of urine can vary. Sip soda all day and your urine will be dilute and light in color. Spend 8 hours or so without drinking, as when you sleep, and your urine becomes concentrated and darker in color. Your urine also becomes more concentrated when you lose a lot of water by sweating.

Stimulus

1 Water loss lowers the volume of blood and makes it more salty. The hypothalamus in the brain senses these changes and signals the adjacent pituitary gland to release ADH.

2 ADH travels through the blood to the kidney, where it affects kidney tubules. The ADH makes the tubules more permeable to water.

hypothalamus

Response

4 The hypothalamus senses the change in blood volume and concentration and stops calling for ADH secretion.

pituitary gland

3 More water is reabsorbed and less is lost in urine, so blood volume rises and blood becomes more dilute.

Figure 23.20 Negative feedback control of ADH secretion.
A negative feedback loop from kidneys to the brain helps adjust the volume of extracellular fluid. Nephrons in the kidneys reabsorb more water when we do not take in enough water or lose too much, as by profuse sweating.

Left, Evan Cerasoli.

Antidiuretic hormone (ADH) is a signaling molecule that increases water reabsorption, thus making the urine more concentrated (Figure 23.20). When you lose water, the resulting change in the volume and concentration of body fluid is detected by the hypothalamus, a brain region that is a control center for many homeostatic responses **1**. The hypothalamus stimulates the pituitary gland to secrete ADH. The ADH travels through the blood to cells of the kidney tubules, where it causes the tubules to become more permeable to water **2**. More water is reabsorbed, so less departs in the urine. Increased water reabsorption returns the body fluid's salt concentration and volume back to normal **3**. The hypothalamus senses the change and stops calling for ADH secretion **4**.

Alcohol inhibits ADH secretion. As a result, more fluid remains in the filtrate and the drinker needs to urinate more frequently. The excessive urination can lead to dehydration, which contributes to the symptoms of a hangover.

Impaired Kidney Function   Sometimes the substances dissolved in urine come out of solution and form hard kidney stones inside the kidney. About one million Americans receive treatment for kidney stones each year, with men four times as likely as women to be affected.

Most kidney stones do not cause symptoms, but some enter a ureter and cause extreme pain. When the ureter contracts around the stone, the ureter wall can be damaged, making the urine bloody. Usually the fluid pressure exerted on the stone by the urine pent up behind it forces the stone through the urinary tract and out of the body. If the kidney stone does not pass after a reasonable time, causes constant pain, or blocks flow of urine, it can be removed surgically or broken up by shock waves administered from outside the body. The best way to prevent kidney stones is to drink plenty of fluids.

Kidney function is measured as the rate of filtration. When the filtration rate falls by half, a person is in kidney failure. The vast majority of kidney failure occurs as a complication of diabetes mellitus or high blood pressure. These disorders

antidiuretic hormone Hormone produced by the pituitary gland; increases water reabsorption by the kidney.

kidney dialysis Procedure used to cleanse blood and restore proper solute concentrations in a person with impaired kidney function.

damage small blood vessels, including capillaries that filter blood into nephrons. Kidney failure can be fatal because wastes build up in the blood and interstitial fluid. The pH of body fluids rises, and changes in concentrations of other ions interfere with metabolism.

Kidney dialysis is used to restore proper solute balance in a person with kidney failure. "Dialysis" refers to the exchange of solutes between two solutions that are separated by a semipermeable membrane. With hemodialysis, a dialysis machine is connected to a patient's blood vessel (Figure 23.21A). The machine pumps the patient's blood through semipermeable tubes submerged in a warm solution of salts, glucose, and other substances. As the blood flows through the tubes, wastes dissolved in the blood diffuse out and solute concentrations return to normal levels. Cleansed, solute-balanced blood is returned to the patient's body. Typically, a person with kidney failure has hemodialysis three times a week at a dialysis center. Each treatment takes several hours. Peritoneal dialysis can be done at home. Each night, dialysis solution is pumped into a patient's abdominal cavity (Figure 23.21B). Wastes diffuse across the lining of this cavity and into the fluid, which is drained out the following morning.

When kidney failure is permanent, dialysis must be continued for the rest of a person's life, or until a kidney becomes available for transplant. The National Kidney Foundation estimates that every day, 17 people die of kidney failure while waiting for a transplant. Most kidneys used as transplants come from people who had arranged to be organ donors after their death. However, an increasing number of kidneys are removed from a living donor (Figure 23.22). A kidney transplant from a living donor has a better chance of success than one from a deceased person. One kidney is adequate to maintain good health, so the risks to a living donor are mainly related to the surgery—unless the donor's remaining kidney fails.

The benefits of organs from living donors, a lack of donated organs, and high dialysis costs have led some to suggest that people should be allowed to sell a kidney. Critics argue that it is unethical to tempt people to risk their health for money.

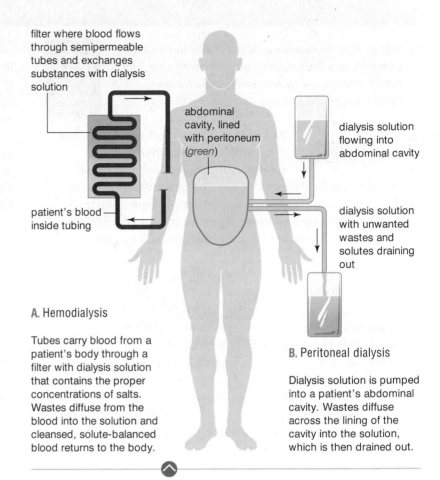

filter where blood flows through semipermeable tubes and exchanges substances with dialysis solution

abdominal cavity, lined with peritoneum (*green*)

patient's blood inside tubing

dialysis solution flowing into abdominal cavity

dialysis solution with unwanted wastes and solutes draining out

A. Hemodialysis

Tubes carry blood from a patient's body through a filter with dialysis solution that contains the proper concentrations of salts. Wastes diffuse from the blood into the solution and cleansed, solute-balanced blood returns to the body.

B. Peritoneal dialysis

Dialysis solution is pumped into a patient's abdominal cavity. Wastes diffuse across the lining of the cavity into the solution, which is then drained out.

Figure 23.21 Two types of kidney dialysis.

Take-Home Message 23.6

How does urine form?

- Filtration forces water and small solutes into the kidney tubules. The driving force for filtration is blood pressure.
- Most of the filtered water and solutes are reabsorbed into capillaries around the tubules.
- Wastes that are not absorbed end up in the urine, along with any excess solutes secreted into the filtrate.
- The hormone ADH is secreted by the brain and acts in the kidney to promote water reabsorption, thus concentrating the urine.

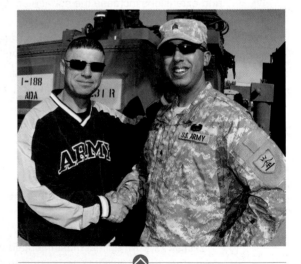

Figure 23.22 Living donor kidney transplantation. Spc. John Chase (left) with Sgt. Francisco Raatz (right), who donated a kidney to him. The two National Guardsmen had served together in Afghanistan before Spc. Chase went into kidney failure.

U.S. Army, Senior Master Sgt. David H. Lipp.

Summary

Section 23.1 An excessive caloric intake can lead to obesity, a condition in which accumulated fat threatens health. A tendency toward obesity can be inherited, but dietary habits and exercise also play a role.

 Section 23.2 Most animals have a digestive system that carries out **ingestion** and **digestion** of food, **nutrient absorption**, and **elimination** of digestive waste. The **gastrovascular cavity** of flatworms and cnidarians is saclike, with a single opening. Most animals have a **complete digestive tract**: a tube with two openings. Wastes leave the tract through an **anus** or a **cloaca**.

Section 23.3 Like other vertebrates, humans have a complete digestive tract. Chewing mixes food with saliva. An enzyme in saliva starts the process of starch digestion. The swallowing reflex moves food from the throat into the **esophagus**. Food continues through a **sphincter** into the **stomach**, a muscular sac with a lining that secretes **gastric fluid**. Protein digestion begins in the stomach.

The **small intestine** receives **chyme** (food mixed with gastric fluid) from the stomach, enzymes from the **pancreas**, and **bile** from the **gallbladder**. Bile, which assists in fat digestion, is made by the **liver**. Digestion of all nutrients is completed in the small intestine.

Most nutrients and fluid are absorbed into the internal environment across the wall of the small intestine. Intestinal folds are covered with multicelled projections (**villi**). Brush border cells at the villus surface have membrane extensions called **microvilli**. All of the folding and extensions increase the surface area for absorption.

The **large intestine** consists mainly of the **colon**, which absorbs minerals and water, and concentrates undigested residues as **feces**. The **rectum** stores feces until they exit through the anus. The **appendix** is a small projection near the start of the large intestine.

Section 23.4 The body burns carbohydrates as fuel and whole grains provide fiber. Fats are another source of energy. **Essential fatty acids** must be obtained from the diet, as must **essential amino acids**. **Vitamins** and **minerals** are necessary in small amounts for normal metabolism.

 Section 23.5 An animal must maintain the volume and composition of its fluid components within a narrow range. Most animals have excretory organs that rid the body of excess water and waste solutes. **Malpighian tubules** serve this function in insects. The human

urinary system interacts with other organ systems to balance the intake and output of solutes and water. The urinary system consists of a pair of blood-filtering **kidneys** that produce **urine**. The urine drains into a pair of **ureters**, is stored in a **urinary bladder**, and leaves the body through the **urethra**.

Section 23.6 Kidneys contain an enormous number of nephrons, or blood-filtering units. A **nephron** has a cup-shaped entrance around one set of blood capillaries and tubular regions that associate with another set of blood capillaries. Three processes contribute to urine formation by the nephrons:

Filtration. Blood pressure drives water and most solutes except proteins out of the capillaries at the start of a nephron. These enter the nephron's tubular parts.

Reabsorption. Water and solutes to be conserved move out of the tubular parts, then into the capillaries that thread around the nephron. A small volume of water and solutes remains inside the nephron as the forming urine. ADH (**antidiuretic hormone**) is a hormone that adjusts urine concentration by promoting water reabsorption. As a result, more water returns to the blood rather than leaving in the urine.

Tubular secretion. H^+ and some other solutes are transported from blood in capillaries into kidney tubules to be excreted.

Good kidney function is essential to life. **Kidney dialysis** or a transplant is an option for those with permanent kidney damage.

Self-Quiz

Answers in Appendix I

1. A digestive system functions in _____ .
   a. secreting enzymes
   b. absorbing nutrients
   c. eliminating wastes
   d. all of the above

2. Starch digestion begins in the _____ .
   a. mouth
   b. stomach
   c. small intestine
   d. colon

3. Digestion is completed and most nutrients are absorbed in the _____ .
   a. mouth
   b. stomach
   c. small intestine
   d. colon

4. Bile has roles in _____ digestion and absorption.
   a. carbohydrate
   b. fat
   c. protein
   d. amino acid

5. Bacteria that produce vitamin B$_{12}$ live in the _____ .
   a. stomach
   b. small intestine
   c. large intestine
   d. esophagus

6. Match each structure with its description.
   _____ gallbladder
   _____ large intestine
   _____ liver
   _____ small intestine
   _____ stomach
   _____ pancreas
   _____ gastrovascular cavity

   a. makes bile, stores glycogen
   b. compacts undigested residues
   c. secretes enzymes, buffer
   d. absorbs most nutrients
   e. secretes gastric fluid
   f. saclike gut
   g. stores, secretes bile

7. Essential fatty acids are _____ .
   a. *trans* fats
   b. saturated fats
   c. vitamins
   d. not made by the body

8. Microvilli _____ .
   a. increase the surface area of the stomach
   b. are the main type of cell in the lining of the small intestine
   c. project from the surface of cells in the small intestine
   d. produce bile

9. Iron is the _____ most often deficient in the diet.
   a. vitamin
   b. mineral
   c. essential fatty acid
   d. essential amino acid

10. Birds convert ammonia to _____ for excretion.
    a. proteins
    b. urea
    c. uric acid
    d. nucleic acid

11. Filtration moves _____ and small solutes into kidney tubules.
    a. water
    b. insoluble fiber
    c. large proteins
    d. red blood cells

12. Water loss triggers a(n) _____ in ADH secretion.
    a. increase
    b. decrease

13. Kidneys return water and small solutes to the blood by the process of _____ .
    a. filtration
    b. reabsorption
    c. tubular secretion
    d. both a and b

14. Kidneys adjust the blood acidity by increasing or decreasing the _____ of H$^+$.
    a. filtration
    b. reabsorption
    c. secretion
    d. both a and b

15. Match each structure with a function.
    _____ ureter
    _____ cloaca
    _____ urethra
    _____ nephron
    _____ Malpighian tubule

    a. exit for urine and feces
    b. delivers urine to body surface
    c. carries urine from kidney to bladder
    d. delivers solutes to insect gut
    e. many inside a kidney

Critical Thinking

1. Starch and sugar have the same number of calories per gram. However, not all vegetables are equally calorie dense. For example, a serving of boiled sweet potato provides about 1.2 calories per gram, while a serving of kale yields only 0.3 calories per gram. What could account for the difference in the calories your body obtains from these two foods?

2. Although rabbits cannot digest the cellulose in their all-plant diet, bacteria that live in their cecum can. However, to make full use of the nutrients released by bacterial action, a rabbit must produce and eat a special type of feces. Why must the rabbit reswallow this material to obtain the nutrients released by bacterial action?

3. Diabetes insipidus is a medical disorder in which a person produces an unusually large amount of highly dilute urine. Some cases are caused by a gene mutation, but in most people diabetes insipidus arises after a head injury. Explain how an injury to the head could affect kidney function.

4. Marine mammals and desert rodents both have highly efficient kidneys that produce only a tiny amount of very concentrated urine. What selective pressure shaped this trait in both animals?

Visual Question

1. Identify the organs illustrated in the diagram at the right.
2. Which of these organs are regions of the "tube-within-a-tube" digestive tract?

Application ◈ 24.1 Impacts of Concussions

REMEMBER: Neurons are the signaling cells of the nervous system (Section 19.3).

Brain tissue is surprisingly delicate. It is only about as firm as JELL-O or warm butter. The hard bones of the skull protect the brain, but there is a small fluid-filled gap between the brain and the skull. As a result, an impact to the head or a sudden stop can cause the brain to move within the skull. Such movement can tear, bruise, and stretch brain tissue. The mechanical stress also causes uncontrolled excitement of neurons in the brain and disrupts brain metabolism. The resulting mild traumatic brain injury is commonly called a concussion.

Symptoms of a concussion can include confusion, dizziness, blurred vision, increased sensitivity to light, headache, impaired short-term memory, difficulty concentrating, irritability, nausea, altered sleep patterns, and a temporary loss of consciousness. There is no treatment for concussion, other than physical and mental rest. In most cases, the brain heals itself within about 10 days. During recovery, it is especially important to avoid additional head trauma. Even a seemingly slight injury to the healing brain can result in increased swelling that can lead to permanent paralysis and, in some cases, death. The threat of this "second impact syndrome" makes it all the more important that an initial concussion be properly diagnosed. If a concussion is suspected, a person should halt physical activities and see a physician as soon as possible.

The Centers for Disease Control estimates that somewhere between 1.4 and 3.8 million concussions occur each year in the United States. Participating in contact sports raises the risk, but concussions also occur as a result of car accidents, workplace accidents, or simple falls. Military personnel deployed to combat zones often suffer concussions as a result of exposure to explosions. One recent study of U.S. soldiers who had served in Iraq found that about 15 percent had experienced symptoms of a concussion during their service.

Head injuries are an occupational hazard for professional football players (Figure 24.1). Physicians for the National Football League routinely diagnose between 100 and 150 concussions per season and many players sustain additional sub-concussion injuries. As a result, former football players have an elevated risk of degenerative brain disorders such as Parkinson's disease and Alzheimer's disease. Repeated head injuries can also result in chronic traumatic encephalopathy (CTE). With this incurable disorder, the brain shrinks as neurons die and a protein called tau accumulates. Symptoms can include memory loss, emotional problems, depression, suicidal impulses, and dementia. Consider the case of Dave Duerson, a former player for the Chicago Bears football team. In 2011, when Duerson was 50 years old, he killed himself with a shot to the chest. Before he ended his life, he sent text messages requesting that his brain be donated to a research center that studies sport-related brain injuries. A brain autopsy revealed that Duerson had, as he suspected, degenerative brain damage characteristic of CTE.

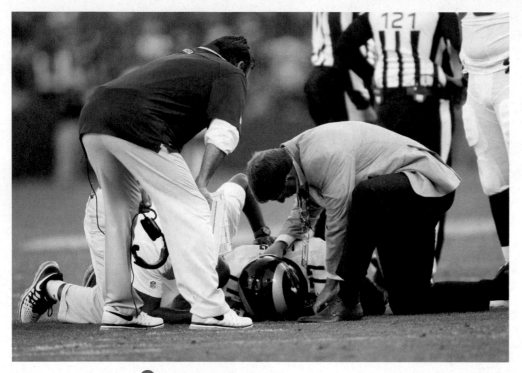

Figure 24.1 Football player with a concussion.
Jake Long of the St. Louis Rams football team is examined after a head-to-head collision with another player. He was diagnosed with a mild concussion.

Michael Zagaris/Getty Images Sport.

A. Nerve net (purple) of a sea anemone. There is no central organ that integrates signals.

B. Planarian nervous system. Two ganglia in the head serve as integrating centers. Nerve cords extend the length of the body along the ventral (lower) surface.

nerve cords with ganglia

C. Insect nervous system. A brain with hundreds of thousands of neurons integrates information. It connects to a ventral nerve cord that has a ganglion in each segment. The ganglia serve as local control centers.

Figure 24.2 Invertebrate nervous systems.

(C) From Russell/Wolfe/Hertz/Starr. *Biology*, 1e. © 2008 Cengage Learning®.

24.2 Animal Nervous Systems

REMEMBER: Cnidarians are aquatic, radial animals, but most animals have a bilateral body plan with a distinctive head end (Section 15.3). Humans are chordates (15.4).

For an animal body to function as an integrated whole, its cells must communicate with one another. Two organ systems—the nervous system and the endocrine system—facilitate long-distance communication among cells of an animal body. The endocrine system, which we discuss in the next chapter, coordinates activities by way of molecules called hormones that travel in the blood. Here, we focus on the nervous system. Neurons make up the communication lines of nervous systems. They transmit electrical signals along their plasma membrane and also send chemical messages to other cells. In most animals, nervous tissue also includes neuroglia, which are cells that structurally and functionally support neurons.

Invertebrate Nervous Systems In sea anemones and other cnidarians, neurons interconnect as a mesh called a **nerve net** (Figure 24.2A). Information flows in all directions among cells of the nerve net; there is no centralized, controlling organ that functions like a brain. The decentralized system allows these radially symmetrical, aquatic animals to respond to food and threats that arrive from all directions.

As animals with a bilateral body plan evolved, neurons that detect and process information about the external environment became concentrated at the body's anterior, or head, end. Consider the nervous system of a planarian flatworm (Figure 24.2B). A pair of ganglia in the head serve as integrating centers. Each **ganglion** (plural, ganglia) is a cluster of neuron cell bodies. A planarian's ganglia receive signals from eyespots and chemical-detecting cells on the head. They also connect to a pair of nerve cords that run along the animal's ventral (lower) surface. Nerves branch from the nerve cord and cross the body.

Insects and other arthropods also have sensory organs such as eyes and antennae concentrated at their head end. As in planarians, paired nerve cords run along the ventral surface (Figure 24.2C). In arthropods, however, the nerve cords connect to a simple brain. A **brain** is a central control organ of a nervous system. It receives and integrates sensory information, regulates internal processes, and sends out signals that bring about movement.

Vertebrate Nervous Systems A dorsal nerve cord (a nerve cord that runs along the back) is one of the defining features of chordate embryos. In vertebrates, the dorsal nerve cord evolved into a brain and spinal cord, which together constitute the animal's **central nervous system**. Nerves that extend from the central nervous

brain Central, integrating organ of a nervous system.
central nervous system Brain and spinal cord.
ganglion A cluster of neuron cell bodies.
nerve net Of some radial invertebrates, a mesh of neurons with no central organizing organ.

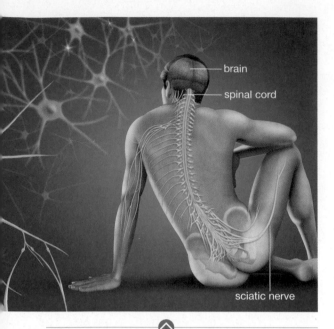

Figure 24.3 Human nervous system.
The brain and spinal cord are the central nervous system.

Nerves that emerge from the brain or spinal cord and extend to other body regions make up the peripheral nervous system. The sciatic nerve is an example.

system through the rest of the body constitute the **peripheral nervous system** (Figure 24.3).

The human peripheral nervous system consists of cranial nerves that connect to the brain and spinal nerves that connect to the spinal cord. Most cranial nerves, and all spinal nerves, carry signals both to and from the central nervous system. Consider the sciatic nerve, which runs from the spinal cord, through a buttock, and down the leg. If something touches your thigh, this nerve carries signals from receptors in your skin to the spinal cord. When you move your leg, the same nerve relays commands for movement from the spinal cord to muscles in the leg.

> ## Take-Home Message 24.2
>
> **How are animal nervous systems organized?**
>
> - The simplest nervous system is a nerve net, a mesh of interconnecting neurons with no central integrating region. Sea anemones have this type of system.
> - Most animals have a bilaterally symmetrical body with an integrating center at their head end. Nerves that run through the body carry signals to and from this center.
> - Vertebrates have a central nervous system (brain and spinal cord) and a peripheral nervous system (nerves that extend through the body and carry signals to and from the central nervous system).

24.3 Neuron Function

REMEMBER: Transport proteins allow ions to move across cell membranes and receptor proteins bind specific substances (Section 3.3).

Three Types of Neurons Neuron structure varies, but all neurons have a central cell body that contains the nucleus. Two types of cytoplasmic extensions project from the cell body. **Dendrites** are short, branched extensions that receive information. A neuron may have a few to many dendrites. An **axon** is a cytoplasmic extension that conveys an electrical signal along its length and releases chemical signals from its endings (the axon terminals). All neurons have only one axon.

In vertebrate nervous systems, information typically flows from sensory neurons, to interneurons, to motor neurons (Figure 24.4).

Figure 24.4 The three types of neurons.
Arrows indicate direction in which signals flow.

A. Sensory neurons become excited when receptor endings of their dendrites detect a specific stimulus, such as light or touch. Their axon terminals send chemical signals to interneurons or motor neurons.

B. Interneurons have dendrites that receive chemical signals from sensory neurons (as shown) or other interneurons. Their axon terminals send chemical signals to other interneurons or to motor neurons (as shown).

C. Motor neurons have dendrites that receive chemical signals from interneurons (as shown) or from sensory neurons. Their axon terminals send chemical signals to the muscle or gland that they control.

Sensory neurons are neurons that are excited by a specific environmental stimulus such as light or touch. Their dendrites have receptor endings that detect this stimulus. Their axon conveys information about the stimulus to an interneuron.

Interneurons are neurons that serve as relay stations. They both receive signals from and send signals to other neurons. Dendrites of interneurons are specialized for receiving chemical signals from other neurons.

Motor neurons control muscles and glands. Like interneurons, they have dendrites specialized for receiving signals from other neurons.

In vertebrates, most interneurons reside entirely within the central nervous system. By contrast, axons of sensory neurons and motor neurons extend through all regions of your body. A **nerve** is a bundle of such axons enclosed within a sheath of connective tissue (Figure 24.5A).

Neuroglia—Neuron Helpers In addition to neurons, nervous tissue includes a variety of cells collectively referred to as neuroglial cells or neuroglia. These cells do not directly take part in signaling, but they are integral to nervous system function. For example, neuroglia called Schwann cells wrap, one after another, around the axons of sensory and motor neurons. Collectively these neuroglia form a discontinuous covering called a **myelin sheath** around the axon (Figure 24.5B). The myelin sheath enhances nerve function by acting like the insulation on an electrical wire; its presence speeds the flow of electrical information along an axon. Another type of neuroglia performs the same task in the central nervous system.

Symptoms of multiple sclerosis illustrate the importance of the myelin sheath. With this disorder, the immune system mistakenly attacks the myelin-making cells of the central nervous system. As a result, transmission along the axons of interneurons slows, causing progressive weakness and fatigue, impaired balance, and vision problems. MS cannot be cured, although some treatments slow myelin loss.

Resting Potential The ability of neurons to communicate with one another and with other cells arises from properties of the neuron plasma membrane. A neuron's plasma membrane, like that of other cells, is composed mainly of a lipid bilayer that is impermeable to ions and large molecules. Transport proteins control movement of ions across the membrane. Also, like other cells, neurons have electrical and concentration gradients across their plasma membrane. Their cytoplasm has more negatively charged components than the interstitial fluid outside the cell.

Negatively charged proteins in the cytoplasm contribute to the gradient across the membrane. Being large and charged, these molecules cannot diffuse across the lipid bilayer of the cell membrane. The distributions of positively charged potassium ions (K^+) and positively charged sodium ions (Na^+) also play a role in the electrical gradient. There are more sodium ions in the fluid just outside a cell than there are inside it. The reverse is true for potassium ions. We can illustrate these ion concentration gradients as follows (the larger text represents higher concentrations; the green ball represents proteins):

A. Each nerve consists of many axons bundled together by connective tissue.

B. Each axon within the nerve has a discontinuous myelin sheath consisting of many Schwann cells that wrap around the axon.

Figure 24.5 Structure of a peripheral nerve.

axon Signal-sending cytoplasmic extension of a neuron.

dendrites Information-receiving cytoplasmic extension of a neuron.

interneuron Neuron that receives signals from, and sends signals to, other neurons.

motor neuron Neuron that controls a muscle or gland.

myelin sheath Of an axon, a discontinuous covering composed of multiple neuroglia that wrap around and insulate the axon.

nerve Many axons bundled together in connective tissue.

peripheral nervous system Nerves that extend through a vertebrate body and relay information to and from the central nervous system.

sensory neuron Neuron that is excited by a specific environmental stimulus.

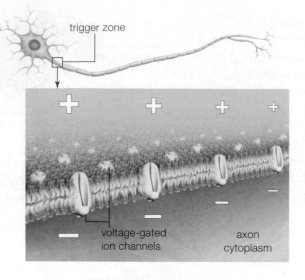

1 A neuron's trigger zone at resting potential. All voltage-gated ion channels (white) are closed.

voltage-gated ion channels

axon cytoplasm

2 At threshold potential, Na⁺ channels open and Na⁺ flows inward (blue arrows). The inside of the axon become increasingly positive.

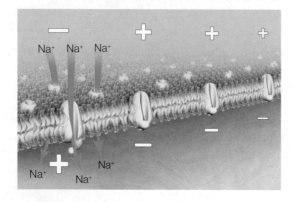

3 Na⁺ channels close and K⁺ channels open. K⁺ flows outward (red arrows).

Farther along the axon, diffusion of Na⁺ has driven the membrane to threshold and Na⁺ channels have begun to open.

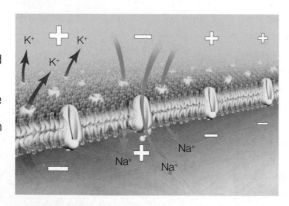

4 K⁺ gates at the original site close.

The action potential continues along as diffusion of Na⁺ triggers opening of gated Na⁺ channels farther and farther down the axon.

Above, a plot of a neuron's membrane potential over time. The spike is the action potential—a brief reversal in the polarity of the electric gradient across the membrane. Blue numbers correlate with numbered graphics (left), which show the events at one region of an axon membrane.

Figure 24.6 Action potential.

As in a battery, separation of charges constitutes potential energy. We describe potential energy as voltage, and the voltage across a cell's membrane is the membrane potential. The membrane potential of a neuron that is not being stimulated is its **resting potential**. It is usually about –70 millivolts (a millivolt [mV] is one-thousandth of a volt). The negative sign indicates the neuron's cytoplasm has more negative charges than the interstitial fluid around the neuron.

The Action Potential Neurons are "excitable" cells, meaning they undergo an action potential when properly stimulated. An **action potential** is a brief reversal of the electric gradient across the plasma membrane. If you plot the membrane potential at one point along an axon against time, the resulting graph will show a spike in voltage (Figure 24.6). The spike results from the ion flow through voltage-gated channel proteins. Such transport proteins have a gate that opens at a particular voltage, or membrane potential:

voltage-gated channel closed (no ions can pass through)

voltage-gated channel open (ions can pass through)

An action potential begins in a trigger zone, which is the region of the axon closest to the cell body. When a neuron is at rest, the trigger zone's gated channels for sodium and potassium are closed, and there are more negatively charged ions inside the axon than outside it **1**. A stimulus such as a signal from another neuron shifts membrane potential of the resting neuron. If the stimulus is large enough, the trigger zone reaches **threshold potential**, the potential at which gated sodium channels open.

Opening of gated sodium channels allows sodium ions to follow their concentration gradient and diffuse into the neuron ➋. Flow of these ions into the axon makes its cytoplasm more positive, causing more sodium channels to open. This accelerating inward flow of sodium ions is an example of positive feedback, in which a response amplifies the conditions that set it in motion.

The inward flow of sodium ions reverses the membrane potential. Within a millisecond, fluid inside the cell becomes more positively charged than the fluid outside. This brief reversal of potential across the membrane is the action potential.

Sodium continues to flow in until the heightened positive voltage triggers gated sodium channels to close and gated potassium channels to open ➌. As potassium ions flow out through the now-open channels, the axon's cytoplasm once again becomes more negatively charged than the interstitial fluid. This decline in membrane potential causes the gates on potassium channels to swing shut ➍.

An action potential is sometimes described as an all-or-nothing event because once the threshold potential is reached, an action potential always occurs and all action potentials are the same size. Peak voltage during an action potential is always the same—voltage always rises to the point where it triggers closing of gated sodium channels and opening of gated potassium channels. These events halt the voltage rise and bring the action potential in that region of the membrane to an end.

An action potential affects each region of the membrane only briefly as it moves along the axon from the trigger zone toward the axon terminals. After voltage-gated sodium channels in one region of an axon open, some of the sodium ions that rush inward diffuse into adjoining regions, where they lower the membrane potential to threshold level and set in motion a new action potential.

An action potential cannot move backward, because closing of the gate of a sodium channel inactivates that gate for a brief period. However, sodium gates in regions farther along the axon can and do swing open as these regions reach threshold. As gated sodium channels open in one region of the axon after another, the action potential moves steadily toward the axon terminals.

The Chemical Synapse Action potentials travel along an axon to its endings, but cannot jump from cell to cell. A **chemical synapse** is a communication point between a neuron and another cell (Figure 24.7 ➊). At the synapse, a narrow space called the synaptic cleft separates the axon terminal of a signal-sending neuron from an adjacent signal-receiving cell.

To signal another cell, a neuron releases a chemical signaling molecule called a **neurotransmitter**. Neurotransmitters convey information across the synaptic cleft. The axon endings of a neuron have neurotransmitter-filled vesicles in their cytoplasm ➋. Arrival of an action potential at an axon's endings causes exocytosis of these vesicles. The vesicles move to the plasma membrane and fuse with it. As fusion occurs, neurotransmitter is released into the synaptic cleft ➌.

The plasma membrane of a signal-receiving cell has receptor proteins that can bind a specific neurotransmitter ➍. Neurotransmitter molecules diffuse across the synaptic cleft and bind to these receptors. Some neurotransmitter receptors are also ion channels. In this case, binding of a neurotransmitter opens the channel through the receptors' interior. Depending on the type of receptor, ions diffuse into or out of the signal-receiving cell through the channel. Movement of these ions alters the membrane potential of the signal-receiving cell, pushing it either closer to its threshold potential or farther away from it.

Figure 24.7 Chemical synapse.

➊ An axon ending of one neuron send signals to another neuron at a chemical synapse.

➋ Neurotransmitter (green) is stored in vesicles inside an axon ending.

➌ Arrival of an action potential at the axon ending causes neurotransmitter release.

➍ Neurotransmitter binds to receptor in the membrane of the signal-receiving cell and opens a channel through the receptor. Ions flow through the channel and into the signal-receiving cell.

action potential Brief reversal of the charge difference across a neuron's plasma membrane.

chemical synapse Region where a neuron's axon terminal transmits signals to another cell.

neurotransmitter Chemical signal released by a neuron's axon terminals.

resting potential Membrane potential of a neuron at rest.

threshold potential Membrane potential at which voltage-gated sodium channels in a neuron axon open, causing an action potential.

A. Parkinson's disease affects about half a million people in the United States, including former heavyweight champion Muhammad Ali and actor Michael J. Fox.

Normal   With Alzheimer's

B. Brain PET scans. Red and yellow indicate regions with high metabolic activity in dopamine-secreting neurons.

Figure 24.8 Parkinson's disease.

(A) AP Images/Kenneth Lambert; (B) From Neuro Via Clinical Research Program, Minneapolis VA Medical Center.

endorphins Molecules that serves as the body's natural painkiller.

Different types of neurons make different neurotransmitters. Acetylcholine (ACh) is a neurotransmitter that acts on skeletal muscle, smooth muscle, the heart, many glands, and the brain. When motor neurons release acetylcholine into synapses with skeletal muscle cells, the muscle undergoes an action potential. The muscle then contracts. By contrast, release of ACH at a synapse with heart muscle inhibits muscle contraction. How can the same neurotransmitter have different effects? Skeletal muscle and heart muscle have different types of ACh receptors. Both bind ACh, but each allows a different ion to enter the muscle after ACh binds.

After neurotransmitter molecules have acted, they must be removed from synaptic clefts so that new signals can be sent. Some neurotransmitter molecules simply diffuse away. Membrane transport proteins pump others back into the neuron or into neighboring neuroglial cells. Enzymes secreted into the cleft break down neurotransmitter that remains in it. The importance of such enzymes is illustrated by the effects of nerve gases, which inactivate the enzyme that breaks down ACh. Exposure to a nerve gas causes ACh to accumulate in synaptic clefts, and the excess neurotransmitter results in confusion, headaches, and muscle paralysis. If nerve gas dosage is high enough, death follows.

Disrupted Synaptic Function Symptoms of some neurological disorders arise from an abnormal level of specific neurotransmitters. For example, Alzheimer's disease, the leading cause of dementia (loss of the ability to think), involves damage to brain neurons that release ACh. It begins with forgetfulness. As the disease progresses, a person becomes increasingly confused, cannot communicate, and eventually becomes incapable of living independently. Drugs that inhibit the enzymatic breakdown of ACh can slow the mental decline in some affected people.

The neurotransmitter dopamine affects motor control and reward-based learning. Damage to dopamine-secreting neurons in the area governing motor control results in Parkinson's disease (Figure 24.8). Tremors are an early symptom. Later, the sense of balance becomes impaired, and voluntary movement, including speech, becomes difficult. Because symptoms arise from a dopamine shortage, patients can be treated with a drug (levodopa) that the body converts to dopamine.

A lower than normal dopamine level in a different part of the brain can cause attention deficit hyperactivity disorder (ADHD). Affected people can have trouble concentrating and controlling impulses. Drugs used to treat ADHD increase dopamine availability in the brain. For example, Ritalin (methylphenidate) acts by preventing the reuptake of dopamine at a synapse.

Psychoactive Drugs Psychoactive drugs are chemicals that enter the brain and alter mood or perception by acting at synapses (Figure 24.9). Stimulants make users feel alert but also anxious, and they can interfere with fine motor control. Nicotine is a stimulant that, among other effects, binds to receptors for ACh in the brain and mimics the excitatory effects of ACh. By contrast, caffeine has a stimulating effect because it blocks receptors for a neurotransmitter that has an inhibitory effect in the brain. Cocaine and amphetamines produce their stimulating effects by blocking the reuptake of excitatory neurotransmitters from the synaptic cleft. Ecstasy is a type of amphetamine, as is methamphetamine.

Analgesics are psychoactive drugs that relieve pain. The opiate analgesics, such as morphine, codeine, heroin, and oxycodone, mimic **endorphins**, which are neurotransmitters that serve as the body's natural painkillers. Narcotic analgesics bind to endorphin receptors and elicit the same effects as endorphins. They dull pain and, when administered in larger doses, produce a rush of euphoria.

Depressants such as alcohol and barbiturates have a calming effect and they slow motor responses. Alcohol also stimulates the release of endorphins, so users typically experience a brief euphoria followed by depression.

Hallucinogens such as LSD (lysergic acid diethylamide) distort sensory perception and bring on a dreamlike state. Two related drugs, mescaline and psilocybin, have similar but weaker effects.

Marijuana, dried parts of the *Cannabis* plant, is classified as a hallucinogen because large doses can cause hallucinations. At more typical doses, users become relaxed, sleepy, uncoordinated, and inattentive. The main psychoactive ingredient in marijuana, THC (delta-9-tetrahydrocannabinol), mimics the function of the neurotransmitter anandamide. Another component of marijuana (cannabidiol) inhibits breakdown of anandamide. Anandamide increases appetite, decreases pain perception and anxiety, and has a role in deleting memories.

Use of many psychoactive drugs can lead to addiction. Addiction has many causes, but dopamine, a neurotransmitter that provides a feeling of pleasure, plays an important role in creating dependency. The brain releases a surge of dopamine in response to behaviors such as eating, which enhance survival, or engaging in sex, which enhances reproduction. Dopamine release helps individuals learn to repeat beneficial behaviors. Addictive drugs trigger dopamine release or prevent its reuptake, thus tapping into this ancient learning pathway. Drug users inadvertently teach themselves that the drug is essential to their well-being.

Continual use of a drug often results in tolerance, meaning the effectiveness of a given dose of the drug decreases over time. Although two beers make a nondrinker feel drunk, they have little effect on an alcoholic. Tolerance arises because

Figure 24.9 Three commonly used psychoactive drugs.
Nicotine in cigarettes is a stimulant. It binds to brain receptors for the neurotransmitter ACh and mimics ACh's excitatory effect in the brain.

Caffeine in coffee is also a stimulant. It binds to and prevents the function of brain receptors for adenosine, a neurotransmitter that slows activity of neurons in the brain.

Alcohol is a sedative. It binds to and enhances the function of receptors for GABA, a neurotransmitter that, like adenosine, has an inhibitory effect on brain activity.

Szasz-Fabian Jozsef/Shutterstock.com.

Digging Into Data

Prenatal Effects of Ecstasy

Animal studies are often used to assess effects of prenatal exposure to illicit drugs. For example, Jack Lipton used rats to study the behavioral effect of prenatal exposure to MDMA, the active ingredient in Ecstasy. He injected female rats with either MDMA or saline solution when they were 14 to 20 days pregnant. This is the period when their offspring's brains were forming. When those offspring were 21 days old, Lipton tested their response to a new environment. He placed each young rat in a new cage and used a photobeam system to record how much each rat moved around before settling down. Figure 24.10 shows his results.

1. Which rats moved around most (caused the most photobeam breaks) during the first 5 minutes in a new cage, those prenatally exposed to MDMA or the controls?

2. How many photobeam breaks did the MDMA-exposed rats make during their second 5 minutes in the new cage?

3. Which rats moved around the most during the last 5 minutes of the study?

4. Does this study support the hypothesis that MDMA affects a developing rat's brain?

Figure 24.10 Effect of prenatal exposure to MDMA on activity levels of 21-day-old rats placed in a new cage.
Movements were detected when the rat interrupted a photobeam. Rats were monitored at 5-minute intervals for a total of 20 minutes. Blue bars are results for rats whose mothers received saline; red bars are results for rats whose mothers received MDMA.

repeatedly taking a drug alters a user's body. Mechanisms of tolerance vary. Tolerance to a drug that causes release of a particular neurotransmitter, for example, may arise from decreased synthesis of the neurotransmitter or its receptors. Tolerance can also involve changes in how fast a neurotransmitter is cleared away.

> ## Take-Home Message 24.3
>
> **What are neurons and how do they transmit signals?**
>
> - Neurons are the signaling cells in animal nervous systems. They have dendrites that receive information and an axon that can signal other cells.
> - Sensory neurons are excited by environmental stimuli. Interneurons and motor neurons are excited by signals from other neurons.
> - When a neuron is sufficiently excited, action potentials move along its axon. Some neuroglia that wrap around the axon speed this process.
> - At a synapse, axon terminals of a signal-sending neuron release a neurotransmitter that binds to and affects the behavior of the signal-receiving cell.

24.4 The Central Nervous System

REMEMBER: An increase in brain size relative to body size is one of the trends that shaped human evolution (Section 15.7).

The brain and spinal cord are the organs of the central nervous system (CNS). Three membranes called the **meninges** enclose and protect these organs, which are bathed in clear **cerebrospinal fluid**. This fluid forms when water and small molecules are filtered out of the blood into brain cavities called ventricles.

A mechanism known as the **blood–brain barrier** controls the composition and concentration of cerebrospinal fluid. Cells that make up the walls of brain capillaries attach so tightly to one another that fluid cannot seep between them, as it does in most capillaries. As a result, molecules and ions in blood can move into cerebrospinal fluid only by traveling across cells of a capillary.

Two visibly different types of tissue occur in both the brain and spinal cord. **White matter** consists of bundles of myelin-sheathed axons. In the CNS, such bundles are called tracts, rather than nerves. The tracts carry information from one part of the central nervous system to another. **Gray matter** consists of cell bodies, dendrites, and supporting neuroglial cells. Thus, synapses of the central nervous system are located within the gray matter.

Regions of the Human Brain The average human brain weighs 1,240 grams, or 3 pounds. It contains about 120 billion interneurons, and neuroglia make up more than half of its volume. Figure 24.11 shows the structure of a human brain. Other vertebrate brains have the same functional regions, although the relative size of the regions and their arrangement differs.

The hindbrain sits atop the spinal cord. The medulla oblongata, the part of the hindbrain just above the cord, influences the strength of heartbeats and the rhythm of breathing. It also controls reflexes such as swallowing, coughing, vomiting, and sneezing. Just above the medulla oblongata is the pons, which also affects breathing. Pons means "bridge," and tracts extend through the pons to the midbrain.

A human brain contains about 120 billion interneurons.

corpus callosum	hypothalamus	thalamus	cerebrum
midbrain			
cerebellum			
pons			
medulla oblongata			

Forebrain	
Cerebrum	Localizes, processes sensory inputs; initiates, controls skeletal muscle activity; governs memory, emotions, abstract thought
Thalamus	Relays sensory signals to and from cerebral cortex; has a role in memory
Hypothalamus	With pituitary gland, functions in homeostatic control. Adjusts volume, composition, temperature of internal environment; governs behaviors that ensure homeostasis (e.g., thirst, hunger)
Midbrain	Relays sensory input to the forebrain
Hindbrain	
Pons	Bridges cerebrum and cerebellum, also connects spinal cord with forebrain. With the medulla oblongata, controls rate and depth of respiration
Cerebellum	Coordinates motor activity for moving limbs and maintaining posture, and for spatial orientation
Medulla oblongata	Relays signals between spinal cord and pons; functions in reflexes that affect heart rate, blood vessel diameter, and respiratory rate. Also involved in vomiting, coughing, other reflexive functions

Figure 24.11 Human brain.
Left, right half of a brain, with major structures labeled. Right, three main brain regions, major structures in each region, and their functions.

C. Yokochi and J. Rohen, *Photographic Anatomy of the Human Body*, 2nd Ed., Igaku-Shoin, Ltd., 1979.

The plum-sized **cerebellum** at the back of the hindbrain is densely packed with neurons, having more than all other brain regions combined. The cerebellum controls posture and coordinates voluntary movements. Excessive alcohol consumption disrupts coordination by impairing neurons in the cerebellum. That is why police often ask a person suspected of driving while drunk to walk a straight line.

In humans, the midbrain is the smallest of the three brain regions. It plays an important role in reward-based learning. The pons, medulla, and midbrain are collectively referred to as the brain stem.

The forebrain contains the **cerebrum**, the largest part of the human brain. The cerebrum receives sensory signals, integrates them, and initiates skeletal muscle movements. A fissure divides it into right and left hemispheres. The hemispheres are connected by the corpus callosum, a band of about 200 million axons that relay signals between them. Each hemisphere has a cerebral cortex, an outer layer of gray matter. The cerebral cortex is responsible for memory, emotions, language, and abstract thought.

Most sensory signals destined for the cerebrum pass through the adjacent thalamus, which sorts them and sends them to the proper region of the cerebral cortex. The **hypothalamus** ("under the thalamus") is the center for homeostatic control. It receives signals about the state of the body and regulates thirst, appetite, sex drive, and body temperature. It is also an endocrine gland that interacts with the adjacent pituitary gland.

blood–brain barrier Protective mechanism that prevents unwanted substances from entering the cerebrospinal fluid.

cerebellum Region of the hindbrain that coordinates voluntary movements.

cerebrospinal fluid Fluid that surrounds and fills cavities in the brain and spinal cord.

cerebrum Forebrain region that controls higher functions such as abstract thought and language.

gray matter Of the central nervous system, tissue that includes neuron cell bodies, dendrites, and axon terminals, as well as neuroglial cells.

hypothalamus Homeostatic control center in forebrain. Has both nervous and endocrine functions.

meninges Membranes that surround and protect the brain and spinal cord.

white matter Of the central nervous system, tissue that consists mainly of myelinated axons.

Figure 24.12 Lobes of the cerebrum.

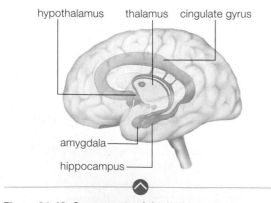

Figure 24.13 Components of the limbic system.

A Closer Look at the Cerebral Cortex The cerebral cortex is a 2-millimeter-thick, highly folded outer layer of gray matter. Prominent folds in the cortex are used as landmarks to define the cerebrum's frontal, parietal, temporal, and occipital lobes (Figure 24.12).

Much of each frontal lobe is devoted to areas that integrate information and bring about conscious actions. During the 1950s, more than 20,000 people had their frontal lobes deliberately damaged when they underwent frontal lobotomy. Doctors carried out this procedure to treat mental illness, personality disorders, and even severe headaches. Frontal lobotomies sometimes made patients calmer, but the procedure also permanently blunted emotions and impaired their ability to plan, to concentrate, and to behave appropriately in social situations.

The primary motor cortex near the rear of each frontal lobe controls skeletal muscles. Each hemisphere controls and receives signals from the opposite side of the body. For example, signals to move your right arm originate in the motor cortex of your left hemisphere. The primary somatosensory cortex of the parietal lobe receives sensory input from the skin and joints. When someone taps you on your left shoulder, signals that arrive in the primary somatosensory cortex of your right parietal lobe alert you to the tap. Another sensory area in the parietal lobe receives signals about taste. In the occipital lobe, the primary visual cortex integrates incoming signals from both eyes. The perception of sounds and odors arises in the primary sensory areas of the temporal lobe.

The two hemispheres differ somewhat in their function. About 90 percent of people are right-handed, and their left hemisphere is more active in controlling movement and in language. However, abilities of each hemisphere are flexible. When a stroke or injury damages one side of the brain, the other hemisphere often can take on new tasks. People can even function with a single hemisphere.

The Limbic System—Emotion and Memory The **limbic system** is a collection of structures that encircle the upper part of the brain stem (Figure 24.13). This system governs emotions, assists in memory, and correlates organ activities with self-gratifying behavior such as eating and sex. The limbic system is sometimes described as our emotional–visceral brain, to contrast it with the cerebral cortex.

Exactly how the structures of the limbic system give rise to different emotions is poorly understood, but we do know a bit about how its components function. For example, the hypothalamus summons up the physiological changes that accompany emotions. Signals from the hypothalamus make our heart pound and our palms sweat when we are fearful. The cingulate gyrus, an arch-shaped area above the corpus callosum, is activated when we experience physical or emotional pain, and it helps us learn to avoid such negative experiences. It also plays a role in forming emotional attachments and is active when a child experiences separation anxiety. The adjacent, almond-shaped amygdala becomes active when we are fearful, and it allows us to recognize emotions in the expressions of others. The amygdala is often overactive in people with panic disorders.

The hippocampus, a structure adjacent to the amygdala, plays an essential role in forming memory of facts and impressions. It helps you remember that a quarter is worth more than a dime, where your classes are held, how to find your way home, and what a lemon smells like. The essential role of the hippocampus in memory was first discovered in the 1950s after a man known as HM had both his hippocampi surgically removed to treat his seizures. The surgery alleviated HM's seizures, but also destroyed his ability to form new memories. Five minutes after meeting a person, HM was unable to remember that they had ever met.

The hippocampus is one of the first brain regions affected by Alzheimer's disease. Like HM, people in early-stage Alzheimer's usually have impaired short-term memory, but retain their memories of long-ago events. In many cases they also retain skill memories, such as the memory of how to play the piano or knit. Formation of such memories involves the cerebellum rather than the hippocampus.

The Spinal Cord The **spinal cord** runs through the vertebral column and connects the brain with peripheral nerves (Figure 24.14). Bones of the vertebral column enclose and protect the cord. Nerves that branch off from the spinal cord extend out through openings in the vertebrae. An opening at the base of the skull allows the cord to connect to the medulla oblongata.

Like the brain, the cord consists of gray matter and white matter. Gray matter fills an H-shaped region in the spinal cord's center. Surrounding this region is the cord's white matter.

An injury that disrupts the signal flow through the spinal cord can cause a permanent loss of sensation and paralysis. The effects are permanent because axons in the central nervous system, unlike those elsewhere in the body, are not repaired. Symptoms of a spinal cord injury depend on where the cord is damaged. Nerves carrying signals to and from the upper body originate higher in the cord than nerves that govern the lower body. An injury to the lower region of the cord often paralyzes the legs. An injury to the highest cord regions paralyzes all limbs, as well as muscles used in breathing. More than 240,000 Americans now live with a spinal cord injury. Researchers continue to investigate ways to restore function to such patients. One recent experiment involved transplanting neuroglial cells from a part of the brain involved in smell into spinal cord. The transplanted neuroglia encouraged nerve fibers to bridge the gap in the cord, restoring some function to the patient.

Doctors sometimes temporarily halt transmission of signals through the spinal cord to provide pain relief to a specific region of the body. Consider the most common way of lessening pain during childbirth—epidural anesthesia. During this procedure, a physician injects painkiller in the space between the meninges and spinal cord. The drug is administered in a region where it will partially numb the body from the waist down, without otherwise impairing perception or interfering with motor function.

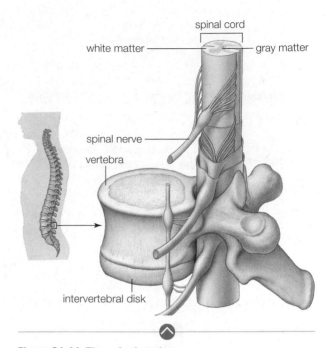

Figure 24.14 The spinal cord.
The cord runs through a cavity within the backbone. Damage to vertebrae or to the cartilage disks between them can put pressure on the spinal cord or the spinal nerves that extend out from the cord, causing pain and impairing function.

Take-Home Message 24.4

What are the functions of the central nervous system?

- The central nervous system consists of the brain and spinal cord.
- The brain stem regulates breathing, adjusts heart rate, and controls reflexes such as swallowing and coughing.
- The cerebellum coordinates voluntary movements.
- The cerebrum makes up the bulk of the forebrain. Its outer layer, the cerebral cortex, controls voluntary activity, sensory perception, abstract thought, and language and speech. The hypothalamus deep inside the forebrain is a center for homeostasis. It has both nervous and endocrine functions.
- Components of the limbic system, which resides deep inside the brain, play a role in emotion and in memory.
- The spinal cord carries information between peripheral nerves and the brain and also plays a role in some reflexes.

limbic system Collection of structures deep in the brain that have roles in emotion and memory.

spinal cord Portion of central nervous system that extends through the backbone and connects peripheral nerves with the brain.

24.5 **The Peripheral Nervous System**

The human peripheral nervous system includes 31 pairs of spinal nerves that connect to the spinal cord and 12 pairs of cranial nerves that connect to the brain. There are two functional divisions of this system (Figure 24.15). **Somatic nerves** carry signals from sensory neurons that monitor the external environment to the central nervous system, and carry commands for voluntary movements to skeletal muscles. The somatic nervous system allows you both to feel someone tickling your toes and to wiggle your toes in response.

Somatic nerves also play a role in some reflexes. A **reflex** is an involuntary response to a stimulus—a movement or other action that does not require thought. The stretch reflex, illustrated in Figure 24.16, is an example. This reflex causes a muscle to contract if some force stretches it. For example, stretching of the biceps excites sensory receptors called muscle spindles causing action potentials to travel along an axon to the spinal cord. Here they synapse with and excite motor neurons that control the biceps. In response, motor neurons send signals that cause the biceps to contract, steadying the arm. All reflexes involve the spinal cord or brain.

Autonomic nerves relay signals from the central nervous system to smooth muscle, cardiac muscle, and glands. They also relay signals about internal conditions from sensory receptors in internal organs to the central nervous system. Signals that travel along autonomic nerves both adjust your heart rate and tell your brain when your blood pressure drops too low.

There are two divisions of the autonomic nervous system: sympathetic and parasympathetic. **Parasympathetic neurons** are most active in times of relaxation. Signals from these neurons promote routine housekeeping tasks such as digestion

PERIPHERAL NERVOUS SYSTEM
Cranial and Spinal Nerves

Somatic Nerves	Autonomic Nerves	
	Nerves that carry signals to smooth muscle, cardiac muscle, and glands	
Nerves that carry signals to and from skeletal muscle, tendons, and the skin	**Sympathetic Division** Predominates in times of stress or danger; brings about fight–flight response	**Parasympathetic Division** Predominates under normal circumstances; facilitates resting and digesting

Figure 24.15 Components of the peripheral nervous system.

Figure 24.16 The stretch reflex, a spinal reflex.

Stimulus

1 Addition of fruit to a bowl stretches the biceps muscle of the arm holding the bowl.

2 Stretching out of the biceps stimulates muscle spindles, which are receptor endings of a sensory neuron. The stimulation triggers an action potential that travels along the sensory neuron's axon to the spinal cord.

white matter

gray matter

3 In the cord's gray matter, axon terminals of the sensory neuron synapse with a motor neuron, causing it to undergo an action potential.

4 The action potential travels along the axon of the motor neuron to the biceps muscle.

5 When the action potential arrives at the motor neuron's axon terminals, it triggers the neuron to release ACh. The ACh triggers the biceps muscle to contract.

axon terminals of motor neuron that synapse with a muscle fiber in biceps

muscle spindle (receptor endings of a sensory neuron)

Response

6 Stimulation of the biceps muscle causes it to contract, so the bowl is held steady.

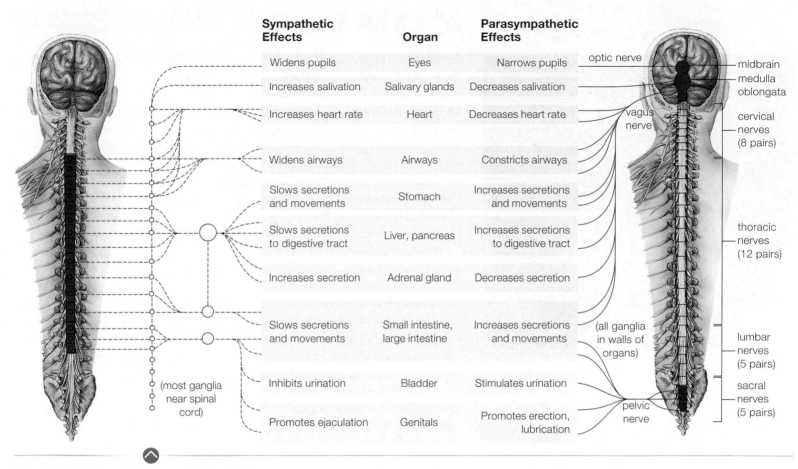

Sympathetic Effects	Organ	Parasympathetic Effects	
Widens pupils	Eyes	Narrows pupils	optic nerve
Increases salivation	Salivary glands	Decreases salivation	midbrain / medulla oblongata
Increases heart rate	Heart	Decreases heart rate	vagus nerve
Widens airways	Airways	Constricts airways	cervical nerves (8 pairs)
Slows secretions and movements	Stomach	Increases secretions and movements	
Slows secretions to digestive tract	Liver, pancreas	Increases secretions to digestive tract	thoracic nerves (12 pairs)
Increases secretion	Adrenal gland	Decreases secretion	
Slows secretions and movements	Small intestine, large intestine	Increases secretions and movements	(all ganglia in walls of organs)
Inhibits urination	Bladder	Stimulates urination	lumbar nerves (5 pairs)
Promotes ejaculation	Genitals	Promotes erection, lubrication	pelvic nerve / sacral nerves (5 pairs)

(most ganglia near spinal cord)

Figure 24.17 Autonomic nerves and their effects.
Autonomic signals travel to organs by a two-neuron path. The first neuron has its cell body in the brain or spinal region (indicated in red). This neuron synapses on a second neuron at a ganglion. Axons of these second neurons then synapse with the organ.

and urine formation. **Sympathetic neurons** are most active in times of excitement or danger. They promote what is called a "fight-or-flight" response. Increased sympathetic signals raise your heart rate and blood pressure, and make you sweat more and breathe faster. The signals put you in a state of intense arousal, so you are ready to fight or make a fast getaway.

Opposing sympathetic and parasympathetic signals govern most organs (Figure 24.17). Integration of the conflicting commands by cells in the signal-receiving organ determines the organ's response.

Take-Home Message 24.5

How does the peripheral nervous system function?

- Peripheral nerves connect the central nervous system with the rest of the body.
- The somatic part of this system controls skeletal muscles and conveys information about the external environment to the central nervous system.
- The autonomic part of this system carries information about internal conditions to the brain and sends signals to smooth muscle, cardiac muscle, and glands. Its parasympathetic division encourages housekeeping tasks, and its sympathetic division prepares a body for "fight or flight."

autonomic nerves Nerves that relay signals to and from internal organs and to glands.

parasympathetic neurons Of the autonomic nervous system, neurons that are most active at times of relaxation; govern housekeeping tasks.

reflex Automatic response to a stimulus.

somatic nerves Nerves that control skeletal muscle and relay signals from joints and skin.

sympathetic neurons Of the autonomic nervous system, neurons that are most active at times of excitement or danger.

A. Thermoreceptors in the skin of a vampire bat's nose help it locate veins filled with warm blood.

B. Pigeons are among the animals that can detect variations in Earth's magnetic field and use this information to navigate.

Figure 24.18 Animals with senses we lack.

(A) © Barry Mansell/naturepl.com; (B) @Andy Gehrig/Stockphoto.com.

---

pheromone Chemical signal used in communication between individuals of a species.

sensation Awareness of a stimulus.

sensory adaptation Diminishing response to a persistent stimulus.

sensory perception Meaning the brain assigns to a specific sensation.

sensory receptor Structure that is activated by a particular stimulus and whose activation results in action potentials in a sensory neuron.

24.6 The Senses

REMEMBER: The nervous signals discussed in this chapter involve receptor proteins (Section 3.3) and transport proteins (4.6).

Sensory Reception and Diversity The sensory portion of a vertebrate nervous system includes sensory neurons that become excited in the presence of specific stimuli, nerves that carry information about the stimulus to the brain, and brain regions that process this information.

Detection of sensory stimuli begins with the excitation of sensory receptors. A **sensory receptor** can be either the dendritic endings of a sensory neuron or a specialized epithelial cell that responds to a stimulus by exciting a sensory neuron. In either case, detection of a stimulus results in excitation of a sensory neuron.

We classify sensory receptors by the types of stimuli they respond to. Five types occur in most animals. Thermoreceptors respond to heat or to cold. Mechanoreceptors detect changes in pressure, position, or acceleration. Some detect shifts in a body's position or acceleration, others respond to touch or to stretching of a muscle, and still others respond to vibrations caused by pressure waves. Sound is a type of pressure wave, so auditory receptors are a type of mechanoreceptor. Pain receptors (nociceptors) detect injury. Chemoreceptors detect the presence of specific molecules in an environment. Photoreceptors respond to light energy.

Animals monitor the environment in different ways depending on the kinds and numbers of sensory receptors they have. Many animals have evolved sensory capacities we lack. For example, vampire bats use thermoreceptors on their nose to locate the blood vessels of their prey (Figure 24.18A); a vessel filled with blood is warmer than the skin that surrounds it. Humans detect only visible light, but insects and some other animals, including rodents, can also see ultraviolet light. Some aquatic animals have receptors that can detect electrical signals in water. Such receptors help some fish, amphibians, and even platypuses detect electrical signals produced by the nerves and hearts of their prey. Animals such as sea turtles, honeybees, and some birds have magnetoreceptors that aid their navigation by allowing them to detect variations in Earth's magnetic field (Figure 24.18B).

Sensation to Perception In animals that have a brain, processing of sensory signals gives rise to **sensation**: awareness of a stimulus. An animal's brain interprets action potentials on the basis of where they originate. This is why you may "see stars" if you press on your eyes in a dark room. The pressure causes action potentials to travel along optic nerves to the brain, which interprets all signals from this nerve as "light."

A strong stimulus causes a receptor to generate action potentials more often and longer than a weak signal does. The same receptors are stimulated by a whisper and a whoop. Your brain interprets the difference by variations in the frequency of the incoming signals. In addition, a strong stimulus recruits more sensory receptors, compared with a weak stimulus. A gentle tap on the arm activates fewer receptors than a slap.

Stimulus duration also affects how the stimulus is interpreted. In **sensory adaptation**, sensory neurons stop generating action potentials (or make fewer of them) despite continued stimulation. Walk into a house where an apple pie is in the oven and you will notice the sweet scent of baking apples immediately. Then, within a few minutes, the scent seems to lessen. The odor does not actually change in intensity, but chemoreceptors in your nose adapt to it.

Sensory perception arises when the brain assigns meaning to sensory signals. Consider what happens when you watch a person walking away from you. As the distance between the two of you increases, the image of the person on your eye becomes smaller and smaller. You perceive this change in sensation as evidence of increasing distance between you and the person, rather than a sign that the person is shrinking.

The Chemical Senses—Smell and Taste Smell (olfaction) and taste are chemical senses. Stimuli are detected by chemoreceptors that send signals when a specific chemical dissolved in the surrounding fluid binds to them. Binding of a chemical to an olfactory receptor causes action potentials to flow along axons to one of the brain's two olfactory bulbs. Here, axons synapse with interneurons that sort out components of a scent. From there, information flows along the olfactory tract to the cerebrum, where it is further processed (Figure 24.19).

Many animals use olfactory cues to find food and avoid predators. Many also communicate by use of pheromones. **Pheromones** are signaling molecules that are secreted by one individual and affect another member of the same species. For example, olfactory receptors on antennae of a male silk moth allow him to locate a pheromone-secreting female more than a kilometer upwind.

In reptiles and most mammals, a cluster of sensory cells forms a vomeronasal organ that detects pheromones. In primates, including humans, a reduced version of this organ is located in the nasal septum separating the two nostrils. Whether the human vomeronasal organ is functional, and what role, if any, it plays in human behavior, remains unclear.

Depending on the animal, chemoreceptors involved in taste can be on antennae, legs or tentacles, or inside the mouth. Humans have about 10,000 sensory organs called taste buds on the surface of the mouth, throat, and especially the upper part of the tongue (Figure 24.20). Each taste bud contains taste receptor cells and neurons. Receptor-covered microvilli of the taste receptor cells extend out through a pore, and thus come in contact with food molecules in saliva.

We perceive many different tastes, but all arise from a combination of five primary sensations: sweet (elicited by simple sugars), sour (acids), salty (NaCl or other salts), bitter (alkaloids), and umami (amino acids such as glutamate, which provides the savory taste typical of aged cheese and meat). Monosodium glutamate (MSG), a commonly used artificial flavor enhancer, stimulates the receptors responsible for the sensation of umami. All mammals generally have the same types of taste receptors, although cats, dolphins, and some other carnivores have lost the ability to detect sweetness.

Each taste receptor cell is most sensitive to one of the five primary tastes. Taste buds in all regions of the tongue include all five types of taste receptor cells. Thus, contrary to popular belief, the ability to detect a specific taste such as bitterness does not map to one specific region of the tongue.

Genetic differences in taste receptors can affect human taste. For example, the degree to which you find broccoli bitter depends largely on one gene that encodes a bitterness receptor. The dominant allele for this gene encodes a receptor that is activated by a compound in broccoli. To people with this allele, broccoli is somewhat to very bitter. A recessive allele encodes a receptor that does not respond to this compound. People homozygous for this allele do not find broccoli at all bitter.

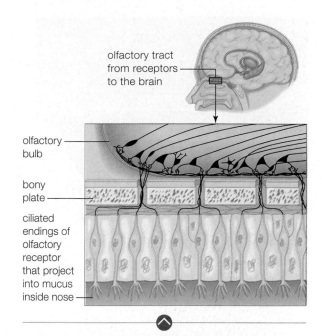

Figure 24.19 Pathway from sensory endings of olfactory receptors in the human nose to the cerebral cortex. Receptor axons pass through holes in a bony plate between the lining of the nasal cavities and the brain.

Figure 24.20 Taste receptors in the human tongue. The structures called circular papillae enclose epithelial tissue that contains taste buds. A human tongue has approximately 10,000 of these sensory organs, each of which has as many as 150 chemoreceptors.

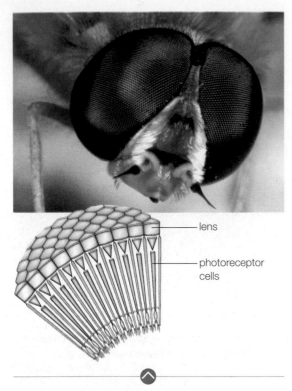

lens

photoreceptor
cells

Figure 24.21 Compound eyes of a fly.
Each eye contains multiple units, each with its own lens.

Top, Ablestock.com/photos.com.

Detecting Light Vision involves detection of light by photoreceptors, and processing of information from those receptors to form a mental image of objects in the environment. Some invertebrates such as earthworms have photoreceptors at their body surface. These animals use light as a cue to orient movements. However, they do not form a mental image of their surroundings as animals with eyes do.

Eyes are sensory organs with a dense array of photoreceptors. The most effective eyes have a **lens**, a transparent structure that bends light rays passing through it so the light falls on photoreceptors. Insects have compound eyes with many separate light-detecting units, each with its own lens (Figure 24.21). Compound eyes cannot detect visual details, but they are highly sensitive to movement. Cephalopod mollusks such as squids and octopuses have the most complex eyes of any invertebrate. Their camera eyes have an adjustable opening that allows light to enter a dark chamber. Each eye's single lens focuses incoming light onto a **retina**, a tissue densely packed with photoreceptors. The retina of a camera eye is analogous to the light-sensitive film in a film camera.

Vertebrates also have camera eyes, and because they are not closely related to cephalopod mollusks, camera eyes presumably evolved independently in the two lineages. This is an example of morphological convergence (Section 11.6).

The Human Eye A human eyeball sits inside a protective, cuplike, bony cavity called the orbit. Skeletal muscles that run from the rear of the eye to bones of the orbit move the eyeball.

Eyelids, eyelashes, and tears protect the delicate eye tissues. Periodic blinking is a reflex that spreads a film of tears over the eyeball's exposed surface. A protective mucous membrane, the conjunctiva, lines the inner surface of the eyelids and folds back to cover most of the eye's outer surface. Conjunctivitis, also called pinkeye, is an inflammation of this membrane caused by a viral or bacterial infection.

The eyeball is spherical and has a three-layered structure (Figure 24.22). The **cornea** that covers the front of each eye consists of a transparent protein called crystallin. A dense, white, fibrous sclera covers the rest of the eye's outer surface.

The eye's middle layer includes the choroid, iris, and ciliary body. The blood vessel–rich choroid is darkened by the brownish pigment melanin. This dark layer contributes to clear vision by reducing light reflection within the eyeball. Attached

Figure 24.22 Structure of the human eye.

sclera
choroid
retina
iris
lens
fovea
pupil
optic
disk
(blind
spot)
cornea
aqueous
humor
part of
optic
nerve
ciliary muscle
vitreous body

Figure 24.23 Pattern of retinal stimulation in the human eye.
Light rays that fall on the retina produce a pattern that is both upside down and inverted left to right.

to the choroid and suspended behind the cornea is a muscular, doughnut-shaped **iris**. It too contains melanin. Whether your eyes are blue, brown, or green depends on the amount of melanin in your iris.

Light enters the eye's interior through the **pupil**, an opening at the center of the iris. Muscles of the iris adjust pupil diameter in response to light conditions. In bright light the pupil contracts, so less light gets in. In low light the pupil widens, so more light enters the eye. Sympathetic stimulation also widens the pupil, presumably allowing a better look at the source of danger or excitement.

A ciliary body of muscle, fibers, and secretory cells attaches to the choroid and holds the lens in place behind the pupil. The stretchable, transparent lens is about 1 centimeter (1/2 inch) across and shaped like a bulging disk.

The eye has two internal chambers. The ciliary body produces the fluid that fills the anterior chamber. Called aqueous humor, this fluid bathes the iris and lens. A jellylike vitreous body fills the larger chamber behind the lens. The innermost layer of the eye, the retina, lines the back of this chamber. The retina contains the light-detecting photoreceptors.

The cornea and lens bend light rays coming from different points so they all converge at the back of the eye, on the retina. The image formed on the retina is an upside-down mirror image of the real world (Figure 24.23). However, the brain interprets this image so you perceive the world in its correct orientation.

When you see an object, you are perceiving light rays that are reflected from that object. The properties of light reflected from near and distant objects hit the eye at different angles, but adjustments of the lens ensure that all light rays become focused on the retina. Lens adjustments are carried out by a ciliary muscle that encircles the lens and attaches to it. When you focus on something close up, the ciliary muscle in each eye contracts, causing the lens to bulge outward. As a result, rays of light from the nearby object are bent so they are focused onto the retina (Figure 24.24A). When an object is farther away, light rays do not have to be bent as much to be focused on the retina. Ciliary muscles relax a bit, allowing the lens to flatten (Figure 24.24B).

The lens typically loses its flexibility as a person ages. That is why most people who are over age forty have impaired close vision and require reading glasses. Changes in the structure of proteins in the lens can result in a cataract, a clouding of the lens. Excessive exposure to ultraviolet radiation, smoking, use of steroids, and some diseases such as diabetes promote cataract formation. Typically, both eyes are affected. At first, a cataract scatters light and blurs vision. Eventually, the lens may become opaque, causing blindness. Cataract surgery restores normal vision by replacing a clouded lens.

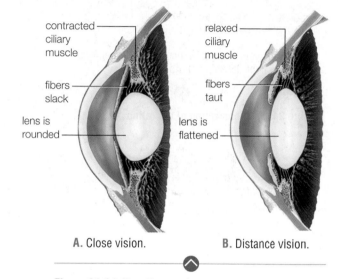

A. Close vision. **B.** Distance vision.

Figure 24.24 How the eye focuses.
The lens is encircled by ciliary muscle. Elastic fibers attach the muscle to the lens. The shape of the lens is adjusted by contracting or relaxing the ciliary muscle, increasing or decreasing the tension on the fibers, and changing the shape of the lens.

Bo Veisland/Science Source.

cornea Outermost layer at the front of the eye; bends light.

iris Ring of smooth muscle with pupil at its center; adjusts how much light enters the eye.

lens Structure that focuses light on an eye's photoreceptors.

pupil Opening through which light enters the eye.

retina Layer of eye that contains photoreceptors.

signal-processing cells

incoming rays of light

ganglion cells, whose axons constitute the optic nerve

A. Several layers of signal-processing interneurons overlie the light-detecting photoreceptors.

rod cell (detects dim light, changes in light intensity)

cone cell (provides color vision)

stacked, pigmented membrane

B. The two types of photoreceptors in the retina.

Figure 24.25 Organization of the retina.

(A, B) Based on www.occipita.cfa.cmu.edu.

Figure It Out: Which is closer to the eye's pupil, a rod cell or a ganglion cell?

Answer: A ganglion cell

At the Retina As previously noted, the cornea and lens bend light rays so they fall on the retina. The retina consists of several cell layers (Figure 24.25A). Rod cells and cone cells are the photoreceptors (Figure 24.25B). They lie beneath several layers of interneurons that process visual signals. **Rod cells** detect very dim light and respond to changes in light intensity that indicate motion. The sense of color and of acute daytime vision starts with light absorption by **cone cells**. There are three types, each having a different pigment. One cone cell pigment absorbs mainly red light, another absorbs mainly blue, and a third absorbs green. Normal human color vision requires all three kinds of cones. Individuals who lack one or more cone types are color blind (Section 9.7). They often have trouble distinguishing red from green in dim light, and some cannot do so even in bright light.

Signal integration and processing start in the retina. Signals flow from the rods and cones to adjacent cell layers that process the information and send signals to ganglion cells. Bundled axons of ganglion cells leave the retina as the beginning of the optic nerve. The part of the retina where the optic nerve exits lacks photoreceptors. It cannot respond to light and thus is called a "blind spot." We all have a blind spot in each eye, but usually do not notice it because information that is missed by one eye is provided by the other.

Hearing Hearing is the detection of sound, a form of mechanical energy. Sounds arise when a vibrating object causes pressure variations in air, water, or some other medium. When you clap your hands or shout, you create pressure waves that move

Figure 24.26 Anatomy of the human ear and how we hear.

(1) © Fabian/Corbis Sygma; (3) Medtronic Xomed; (4) top, Micrograph by Dr. Thomas R. Van DeWater, University of Miami Ear Institute.

inner ear
vestibular apparatus, cochlea

outer ear
pinna, auditory canal

middle ear
eardrum, ear bones

❶ The outer ear's flap and canal collect sound waves.

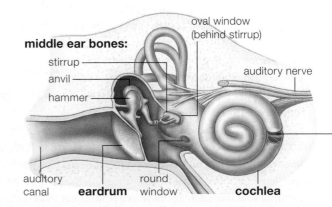

middle ear bones:
stirrup
anvil
hammer

oval window (behind stirrup)

auditory nerve

auditory canal

eardrum

round window

cochlea

❷ The eardrum and middle ear bones amplify sound.

through the air. The amplitude (height) of sound waves determines how loud a sound is. We measure loudness in decibels. A human ear can detect a 1-decibel difference between sounds. Normal conversation is about 60 decibels, a food blender operating at high speed produces a sound of about 90 decibels, and a chain saw makes a 100-decibel noise. A sound's frequency, the number of wave cycles per second, or hertz (Hz), determines its pitch. The more waves per second, the higher the pitch. Humans can hear sounds in the range of 20 to 20,000 Hz.

Water readily transfers vibrations to body tissues, so fishes do not need elaborate ears to detect sound waves. Air is less efficient at transferring sound waves to body tissues, so vertebrates that moved onto land faced a sensory challenge.

Certain features of mammalian ears are evolutionary adaptations that increase the efficiency of sound detection (Figure 24.26). For example, unlike amphibians and reptiles, most mammals have an outer ear that funnels sound inward **1**. The pinna, a skin-covered, sound-collecting, folded flap of cartilage, projects outward from the side of the head **1**. The outer ear also includes the auditory canal that carries sound to the middle ear.

The middle ear amplifies sound waves and transmits them to the inner ear **2**. Pressure waves funneled into the auditory canal cause the **eardrum**, a thin membrane, to vibrate. Behind the eardrum is an air-filled cavity with a set of small bones known as the hammer, anvil, and stirrup. These bones transmit the force of sound waves from the eardrum to the smaller surface of the oval window. This flexible membrane is the boundary between the middle ear and inner ear.

The inner ear contains the vestibular apparatus (discussed shortly), which functions in balance, and the cochlea. The pea-sized, fluid-filled **cochlea** resembles a coiled snail shell (the Greek word *koklias* means snail). Internal membranes divide the cochlea into three fluid-filled ducts **3**. Pressure from the stirrup on the oval window produces pressure waves in the fluid in these ducts. As these waves travel through the cochlear fluid, they cause the membranous walls of the ducts to vibrate.

The **organ of Corti**, the organ responsible for hearing, sits on a membrane (the basilar membrane) in one of the ducts **4**. It contains arrays of mechanoreceptors called hair cells. Specialized nonmotile cilia extend from the hair cells into an overlying membrane. When pressure waves move the membranes, the hair cells bend and undergo action potentials. These signals then travel along an auditory nerve into the brain.

cochlea Coiled structure in the inner ear that holds the sound-detecting organ of Corti.

cone cell Photoreceptor that provides sharp vision and allows detection of color.

eardrum Membrane of middle ear that vibrates in response to sound waves.

organ of Corti Sound-detecting organ in the cochlea.

rod cell Photoreceptor active in dim light; provides coarse perception of image and detects motion.

hair cells of organ of Corti

overlying membrane   basilar membrane

4 Pressure waves cause the basilar membrane beneath the organ of Corti to move upward. The movement pushes hair cells against an overlying membrane. The resulting action potentials travel along the auditory nerve to the brain.

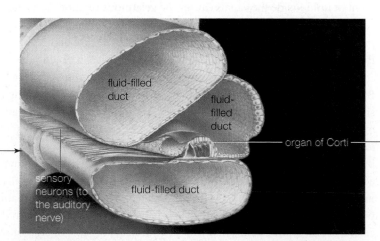

fluid-filled duct

fluid-filled duct

organ of Corti

sensory neurons (to the auditory nerve)

fluid-filled duct

3 One coil of the cochlea in cross-section. The organ of Corti detects pressure waves in fluid-filled ducts inside the cochlea.

Figure 24.27 Using echolocation to find prey.
A porpoise emits ultrasonic calls, then listens for the
echoes that bounce back from its fish prey.

The brain determines the volume and pitch of a sound
by assessing how many action potentials flow along the optic
nerve, and where in the cochlea those potentials come from. The
greater the volume of the sound, the more the hair cells bend.
The pitch of a sound determines where along the cochlea hair
cells bend the most. Higher-pitched sounds set up vibrations
near the base of the coil, whereas lower-pitched sounds cause
vibrations deeper inside the coil.

Many animals produce and respond to sounds at frequencies outside our range of hearing. Elephants, giraffes, and some
whales communicate through infrasound, which is pitched too
low for us to hear. Rodents and many insects communicate
using ultrasound, which is pitched above our range of hearing.
An ability to hear ultrasound also allows animals that prey on rodents to locate prey.
This is why dogs can hear the high pitch of a "dog whistle."

Porpoises and some bats can find their prey and navigate in the dark by using
ultrasound in echolocation. They emit ultrasound calls, then listen for the echoes
that bounce back from objects around them (Figure 24.27). The timing and other
properties of these echoes allow the animals to determine the position, size, and
speed of motion of these unseen objects.

Sense of Balance   Organs of equilibrium monitor a body's position relative to
gravity and its motion through space.

Many invertebrates have gravity-sensing organs called **statocysts**. A statocyst
is a hollow, fluid-filled sphere containing one or more dense particles that gravity
causes to sink to the lowest point in the sphere. Mechanoreceptors that line the walls
of the sphere become excited when the particle or particles press upon them.

The vertebrate organs of balance reside in the part of the inner ear called the
vestibular apparatus. The organs are located in the apparatus's three semicircular
canals, and in two sacs, called the saccule and utricle (Figure 24.28).

The semicircular canals function in the sense of rotational equilibrium. These
three canals are oriented at right angles to one another, so rotation of the head in
any combination of directions—front/back, up/down, or left/right—causes the fluid
inside them to move. At the base of each canal is a bulging area that contains a ridge
of tissue with an overlying gelatinous membrane. When you move your head, movement of fluid inside the canals causes the gelatinous membrane at the base of the
canal to bend. Cilia of hair cells (mechanoreceptors similar to those in the cochlea)
project into the gelatinous membrane. When the cilia are bent by the movement of
the membrane above them, the hair cells undergo action potentials.

The brain receives signals from semicircular canals on both sides of the head.
By comparing the number and frequency of action potentials coming from both
sides, the brain senses the angular movement and rotation of the head. Among
other things, your sense of rotational equilibrium allows you to keep your eyes
locked on an object even when you swivel your head or nod.

Organs in the saccule and utricle act in the sense of static or gravitational
equilibrium. These organs help the brain assess the head's position and how fast it is
moving in a straight line. They also act in keeping the head upright and maintaining
posture. Inside the saccule and utricle, a jellylike layer weighted with the mineral
calcite overlies hair cells. When you tilt your head, or start or stop moving, the
weighted mass shifts, bending hair cells and altering their rate of action potentials.
The brain also takes into account information from the eyes, and from receptors in

semicircular
canals

vestibular
nerve

saccule

utricle

gelatinous membrane
in a semicircular canal

hair cells with their cilia
embedded in membrane

sensory neurons

Figure 24.28 The vestibular apparatus.
Sensory organs within this fluid-filled structure function
in the sense of equilibrium.

the skin, muscles, and joints. Integration of this information provides awareness of the body's position and motion in space.

A stroke, an inner ear infection, or loose particles in the semicircular canals can cause vertigo, a sensation that the world is moving or spinning around. Vertigo also arises from conflicting sensory inputs, as when you stand at a height and look down. The vestibular apparatus reports that you are motionless, but your eyes report that your body is floating in space.

Mismatched signals also cause motion sickness. On a curvy road, passengers in a car experience changes in acceleration and direction that indicate "motion" to their vestibular apparatus. At the same time, signals from their eyes about objects inside the car tell their brain that the body is at rest. Driving can minimize motion sickness because the driver focuses on sights outside the car such as scenery rushing past, so the visual signals are consistent with vestibular signals.

The Somatosensory Cortex Thus far, the senses we have discussed involve receptors located in specific sensory organs. By contrast, the general senses involve receptors scattered throughout the body. Receptors in the skin give rise to sensations of touch, pain, warmth, and cold. Signals from the sensory neurons involved in these sensations travel along axons to the spinal cord, then along tracts in the spinal cord to the brain. The signals end up in the **somatosensory cortex**, a region of the cerebral cortex. This area contains a representation of the entire body, a kind of a map of its parts (Figure 24.29). However, the map is not to scale. Regions that are richly endowed with sensory receptors, such as the fingers and lips, are overrepresented in terms of area.

Signals from mechanoreceptors that detect stretching of tendons and muscles are also conveyed to the somatosensory cortex. As noted above, this information helps your brain keep track of where your body parts are located in space. The muscle spindle fibers that take part in the stretch reflex are one such type of receptor. The more a muscle stretches, the more frequently stretch receptors fire. Nearly all skeletal muscles and smooth muscles contain muscle spindles.

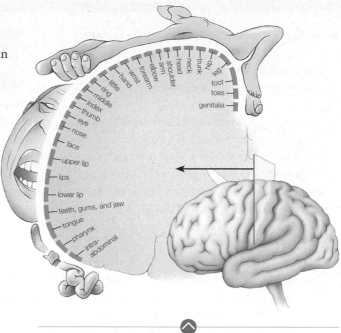

Figure 24.29 Representation of various body regions in the somatosensory cortex.
This brain region is a narrow strip of the cerebral cortex (shown here in yellow) that runs from the top of the head to just above each ear.

Left, After Penfield and Rasmussen, *The Cerebral Cortex of Man,* © 1950 Macmillan Library Reference. Renewed 1978 by Theodore Rasmussen.

Take-Home Message 24.6

How do sensory organs function?

- Different animals have different types of sensory receptors, and so have different sensory experiences.
- Receptors throughout the body give rise to somatic and visceral sensations.
- Smell and taste are chemical senses. They involve chemoreceptors that are activated by binding of a particular chemical.
- Human vision occurs when the various components of an eye focus light onto photoreceptors on the retina at the back of the eye.
- Hearing is the detection of pressure waves. Components of the ear collect, amplify, and sort out these waves, which are detected by mechanoreceptors. Mechanoreceptors in the inner ear also have a role in the sense of balance.
- Receptors in the skin send information about touch, temperature, and pain to the somatosensory cortex.

somatosensory cortex Region of the cerebral cortex that receives information about touch, temperature, and pain from receptors throughout the body.

statocyst Gravity-sensing organ of invertebrates.

vestibular apparatus Structure in the vertebrate inner ear that functions in the sense of balance.

Summary

Section 24.1 A concussion is a mild, traumatic brain injury. Most concussions impair brain function only briefly, but repeated head trauma can cause permanent damage to the brain.

Section 24.2 The simplest nervous system is a **nerve net**, a mesh of neurons with no central command. Most animals have a bilateral nervous system with a cluster of **ganglia** or a **brain** at the head end. The vertebrate nervous system is functionally divided into the **central nervous system** (brain and spinal cord) and the **peripheral nervous system** (nerves that connect the spinal cord and brain with the body).

Section 24.3 Neurons are electrically excitable cells that signal other cells by means of chemical messages. **Sensory neurons** detect stimuli. **Interneurons** relay signals between neurons. **Motor neurons** signal effectors (muscles and glands). Neuroglia support and assist neurons, as by insulating **axons** with a **myelin sheath**. A **nerve** is a bundle of many axons.

Signals reaching a neuron's **dendrites** can disturb the ion distribution across the plasma membrane. Stimulation may cause a cell that was at **resting potential** to reach **threshold potential**. The resulting **action potential** travels along the neuron's axon.

An action potential is an abrupt, brief reversal in the voltage across a neuron's plasma membrane. The reversal triggers an action potential at an adjacent membrane patch, then the next, and on to axon terminals. Action potentials are "all or nothing," meaning they occur only if a cell reaches threshold potential and they are always the same size.

Chemical synapses allow information to move between neurons, or between a motor neuron and a cell it controls. Arrival of an action potential triggers axon terminals of a neuron to release **neurotransmitters** that can bind to receptors on a signal-receiving cell. The response of that cell to any given signal depends in part on what other signals are arriving at the same time. Psychoactive drugs act by altering signaling at chemical synapses. Some mimic **endorphins**, which are natural painkillers.

Sections 24.4, 24.5 Organs of the central nervous system consist of **white matter** (myelin-sheathed axons) and **gray matter** (which includes cell bodies). These organs are enclosed in protective membranes (**meninges**) and surrounded by **cerebrospinal fluid**. The **blood–brain barrier** keeps unwanted substances from entering this fluid. The **spinal cord** relays signals from the body to the brain.

The hindbrain's medulla oblongata pons, along with the midbrain, constitute the brain stem. The brain stem regulates breathing and reflexes such as swallowing and coughing. The **cerebellum** in the hindbrain coordinates voluntary movements. The forebrain contains the **hypothalamus**, which regulates functions related to homeostasis. The **cerebrum** makes up the bulk of the forebrain. Its surface layer of gray matter, the cerebral cortex, governs language and abstract thought. The **limbic system** deep inside the forebrain functions in emotion and memory.

Nerves of the peripheral nervous system extend from cell bodies in the brain or spinal cord out through the body. They relay signals in both directions. The **somatic nerves** send signals to skeletal muscles and receive signals from receptors in joints and the skin. They plays a role in some **reflexes**, such as the stretch reflex. The **autonomic nerves** connect to internal organs. The two divisions of the autonomic system work in opposition. During a fight–flight situation, **sympathetic neurons** predominate. When you are in a relaxed state, **parasympathetic neurons** predominate.

Section 24.6 Sensory receptors respond to a specific stimulus. The brain evaluates information from sensory receptors based on which nerve delivers them, the rate of action potentials, and the number of axons firing in a given interval. Continued stimulation of a sensory receptor may lead to **sensory adaptation** (a diminished response). **Sensation** is detection of a stimulus, whereas **sensory perception** involves assigning meaning to a sensation.

Chemoreceptors involved in taste and smell detect specific chemicals. Human taste receptors are concentrated in taste buds on the tongue and walls of the mouth. Olfactory receptors line human nasal passages. Many animals also have chemoreceptors that detect social signaling molecules called **pheromones**.

In a human eye, light passes through the **cornea** at the eye's surface and enters its interior through the **pupil**, whose diameter is controlled by the **iris**. The **lens** focuses light onto photoreceptors (**rod cells** and **cone cells**) in the **retina**. Other cells of the retina integrate and process signals, which travel along optic nerves to the brain for final processing and interpretation.

The human outer ear collects sound waves. Sound waves vibrate the **eardrum**, and middle ear bones transmit these vibrations to the **cochlea** of the inner ear. The **organ of Corti** inside the cochlea contains mechanoreceptors that, when excited, send signals to the brain. The **vestibular apparatus** in the inner ear detects movement and changes in the body's position. In invertebrates, **statocysts** provide information about body position.

Sensory receptors in skin throughout the body relay information about touch, temperature, and pain to the **somatosensory cortex**, a region of the cerebral cortex that is like a map of the body.

Self-Quiz

1. The _____ of a neuron release(s) neurotransmitter.
   a. axon terminals
   b. cell body
   c. dendrites
   d. all of the above

2. _____ occur mainly in the brain and spinal cord.
   a. Sensory neurons
   b. Motor neurons
   c. Interneurons
   d. b and c

3. A(n) _____ has only a single axon.
   a. sensory neuron
   b. motor neuron
   c. interneuron
   d. all of the above

4. When you sit quietly, output from the _____ system prevails.
   a. sympathetic
   b. parasympathetic
   c. both
   d. neither

5. An action potential occurs when _____ .
   a. a neuron reaches threshold potential
   b. gated sodium channels close
   c. sodium–potassium pumps stop working
   d. gated potassium channels open

6. Skeletal muscles are controlled by the _____ system.
   a. sympathetic
   b. parasympathetic
   c. somatic
   d. both a and b

7. A _____ is an automatic response that does not require thought.

8. The two halves of the cerebrum _____ .
   a. have identical functions
   b. are part of the autonomic nervous system
   c. are connected by the corpus callosum
   d. consist mainly of motor neurons

9. The blood–brain barrier controls what enters _____ .
   a. blood
   b. cerebrospinal fluid
   c. peripheral nerves
   d. both a and b

10. Which of the following senses involve chemoreceptors?
    a. hearing
    b. smell
    c. touch
    d. taste
    e. both a and b
    f. both b and d

11. _____ is a reduced response to an ongoing stimulus.
    a. Propagation
    b. Perception
    c. Sensory adaptation
    d. Synaptic integration

12. Porpoises use _____ use to navigate and to locate prey.
    a. echolocation   b. statocysts   c. infrasound   d. thermoreceptors

13. In a vertebrate eye, photoreceptors are in the _____ .
    a. conjunctiva
    b. cornea
    c. lens
    d. retina

14. The compound eye of a fly has many _____ .
    a. pheromone receptors
    b. hair cells
    c. chemoreceptors
    d. lenses

15. Match each structure with its function.
    _____ rod cell
    _____ cochlea
    _____ cerebellum
    _____ brain stem
    _____ cerebral cortex
    _____ taste bud
    _____ myelin
    _____ neurotransmitter
    _____ blood–brain barrier
    _____ vomeronasal organ

    a. detects pheromones
    b. connects to spinal cord
    c. sorts out pressure waves
    d. protects brain and spinal cord from some toxins
    e. speeds signal transmission
    f. contains chemoreceptors
    g. secreted at synapse
    h. governs higher thought
    i. coordinates voluntary moves
    j. detects light

Critical Thinking

1. The blood–brain barrier in human newborns is not yet fully developed. Explain why this makes careful monitoring of diet and environmental chemical exposure particularly important.

2. After a leg injury, pain makes a person avoid putting too much weight on the affected leg. Shielding the injury gives it time to heal. An injured insect shows no such shielding response. Some have cited the lack of such a response as evidence that insects do not feel pain. But insects do produce substances similar to our natural painkillers. Is the presence of these compounds in insects sufficient evidence to conclude that they feel pain?

Visual Question

1. List some of the functions carried out by the region colored light pink in the graphic at the right.

489

25

ENDOCRINE CONTROL

Application ❯ 25.1 Endocrine Disrupters

We live in a world awash in synthetic chemicals. We drink from plastic bottles, wear synthetic fabrics, slather ourselves with synthetic skin products, and dose our food with synthetic pesticides. Enormous numbers of man-made compounds that are used to manufacture our computers and other electronic gadgets escape into the environment. What do we know about the safety of all these substances?

We have learned by sad experience that some synthetic chemicals are endocrine disrupters. Such substances interfere with the action of **animal hormones**, the signaling molecules secreted by endocrine cells. Consider the insecticide DDT, which has a structure similar to the female sex hormone estrogen. DDT was widely used until 1972, when it was banned in the United States because of its effects on nonpest species. In some birds, exposure to DDT caused eggshells to become so thin and brittle that the eggs broke before hatching. DDT was also a suspected carcinogen, and high levels of DDT in human milk was a matter of concern.

Many endocrine disrupters remain in use. For example, chemicals called phthalates (pronounced THAL-aytes) are commonly used to make plastics more flexible and to stabilize fragrances in scented products. Phthalates are not approved for use in human foods, but in 2011 a Taiwanese company illegally added them to a variety of foods and drinks. Examination of children who ingested the substances revealed changes in the thyroid gland. Results of this horrible unplanned experiment are consistent with epidemiological studies that implicate phthalates in disrupted thyroid function. Other studies have found a positive correlation between phthalate exposure and diabetes.

Phthalates also affect sex hormones. In adult men, a high concentration of phthalates is correlated with a low testosterone level and decreased sperm quality. Women who have a high phthalate level during pregnancy have an increased likelihood of giving birth to a son with an undescended testicle.

Phthalates were widely used in pacifiers and teething toys until 2007, when a U.S. law limiting the phthalate content of products intended for children under age twelve went into effect. However, opportunities for phthalate exposure remain. Vinyl, which releases phthalates into the air, is common in furniture, flooring, mattress covers, and shower curtains. Phthalates also remain in artificially scented goods ranging from air fresheners to personal care products.

Pediatrician Sheela Sathyanarayana found that the more scented powders, shampoos, and lotions a mother used on her infant, the higher the level of phthalates in the child's urine (Figure 25.1). The American Academy of Pediatrics recommends that parents choose unscented products for use on infants and young children. Similarly, many obstetricians recommend that pregnant women use personal care products that are labeled as organic or free of phthalates. Phthalates do not have to be listed as ingredients if they are a component of a fragrance.

Figure 25.1 A potential source of endocrine disruption. Phthalates, a class of chemicals common in artificially scented products such as bubble baths and shampoos, can alter the levels of some hormones.

Photo, © Jason L. Price/Shutterstock.com.

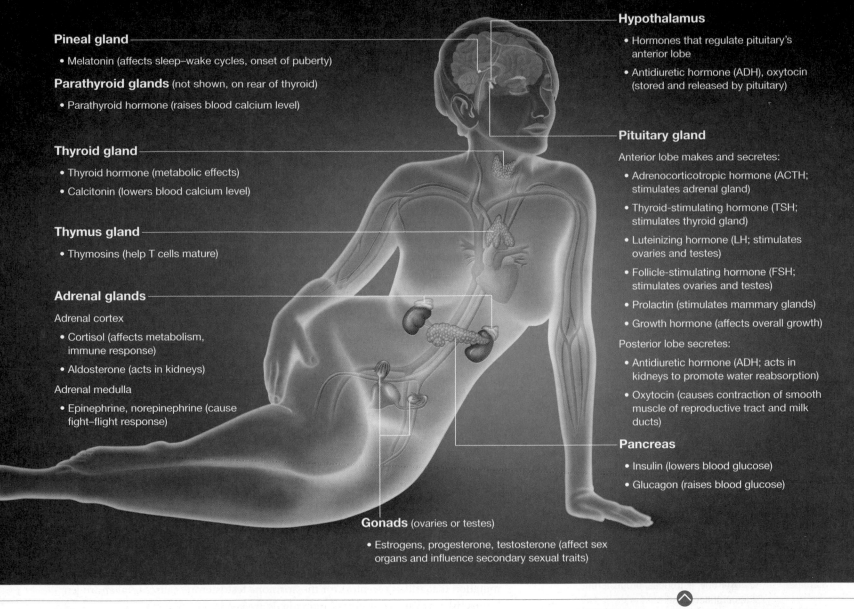

Pineal gland

• Melatonin (affects sleep–wake cycles, onset of puberty)

Parathyroid glands (not shown, on rear of thyroid)

• Parathyroid hormone (raises blood calcium level)

Thyroid gland

• Thyroid hormone (metabolic effects)
• Calcitonin (lowers blood calcium level)

Thymus gland

• Thymosins (help T cells mature)

Adrenal glands

Adrenal cortex

• Cortisol (affects metabolism, immune response)
• Aldosterone (acts in kidneys)

Adrenal medulla

• Epinephrine, norepinephrine (cause fight–flight response)

Hypothalamus

• Hormones that regulate pituitary's anterior lobe
• Antidiuretic hormone (ADH), oxytocin (stored and released by pituitary)

Pituitary gland

Anterior lobe makes and secretes:

• Adrenocorticotropic hormone (ACTH; stimulates adrenal gland)
• Thyroid-stimulating hormone (TSH; stimulates thyroid gland)
• Luteinizing hormone (LH; stimulates ovaries and testes)
• Follicle-stimulating hormone (FSH; stimulates ovaries and testes)
• Prolactin (stimulates mammary glands)
• Growth hormone (affects overall growth)

Posterior lobe secretes:

• Antidiuretic hormone (ADH; acts in kidneys to promote water reabsorption)
• Oxytocin (causes contraction of smooth muscle of reproductive tract and milk ducts)

Pancreas

• Insulin (lowers blood glucose)
• Glucagon (raises blood glucose)

Gonads (ovaries or testes)

• Estrogens, progesterone, testosterone (affect sex organs and influence secondary sexual traits)

Figure 25.2 Major glands of the human endocrine system and primary effects of the hormones they produce. The endocrine system also includes endocrine cells in many organs, such as the heart, kidneys, stomach, liver, small intestine, and skin.

25.2 Hormone Function

REMEMBER: Hydrophilic substances cannot diffuse across a lipid bilayer, but hydrophobic substances can (Sections 2.4, 3.3). Transcription is the first step in gene expression (7.2). Hormone-secreting cells are specialized epithelial cells (19.3).

An **endocrine gland** consists of many hormone-secreting epithelial cells embedded in connective tissue. Blood vessels in and near the gland take up the secreted hormone and distribute it through the body. Figure 25.2 provides an overview of the main glands of the human endocrine system. Other vertebrates have a similar system. Vertebrate endocrine and nervous systems are so closely linked that scientists sometimes refer to them collectively as the neuroendocrine system. Both endocrine cells and neurons receive signals from the hypothalamus in the forebrain (Section 24.4). Most organs respond to hormones as well as signals from the nervous system.

Hormones also play essential roles in the development and function of invertebrates. For example, an arthropod must molt its cuticle periodically as it grows, and

animal hormone Signaling molecule secreted by an endocrine gland or cell.

endocrine gland Gland that secretes hormones into the blood.

Figure 25.3 Hormone-induced molting in a crab.
This blue crab (right) has just shed its old exoskeleton (left) in response to a steroid hormone.

© Kevin Fleming/Corbis.

Table 25.1 Types and Examples of Hormones

Steroids	Testosterone, estrogens, progesterone, aldosterone, cortisol
Amines	Melatonin, epinephrine, thyroid hormone
Peptides	Glucagon, oxytocin, antidiuretic hormone, calcitonin, parathyroid hormone
Proteins	Growth hormone, insulin, prolactin, follicle-stimulating hormone, luteinizing hormone

amino acid–derived hormone An amine (modified amino acid), peptide, or protein that functions as a hormone.

second messenger Molecule that forms inside a cell when a hormone binds at the cell surface; sets in motion reactions that alter enzyme activity inside the cell.

steroid hormone Lipid-soluble hormone derived from cholesterol.

this process is under hormonal control (Figure 25.3). Chemicals that disrupt the action of hormones involved in molting are used as insecticides.

Types of Hormones Like neurotransmitters, hormones affect cells by binding to receptor proteins. A hormone's target is any cell with receptors for that hormone.

There are two categories of hormones, those derived from amino acids and those derived from cholesterol (Table 25.1). **Amino acid–derived hormones** include amine hormones (modified amino acids), peptide hormones (short chains of amino acids), and protein hormones (longer chains of amino acids). Peptide and protein hormones are polar, hydrophilic molecules, so they dissolve easily in blood, which is mostly water. Like other polar molecules, these hormones cannot diffuse across a lipid bilayer. They always bind to a receptor protein embedded in the plasma membrane of a target cell.

When a hormone binds to a receptor at the plasma membrane, a second messenger transmits the signal into the cell (Figure 25.4A). A **second messenger** is a molecule that forms inside a cell in response to an external signal. Formation of the second messenger sets in motion a chain of events that bring about the target cell's response to the signal. For example, binding of many protein hormones activates an enzyme that converts ATP to the second messenger cyclic AMP (cAMP). Formation of cAMP sets in motion a chain of enzyme activations and deactivations that affect the metabolism of the target cell.

Unlike protein and peptide hormones, **steroid hormones** can enter the nucleus and directly influence gene expression. These cholesterol-derived hormones are hydrophobic, so they can diffuse across the lipid bilayer of a cell membrane (Figure 25.4B). A steroid hormone diffuses into a target cell where it binds to receptors to form a hormone–receptor complex. The hormone–receptor complex functions in the target cell's nucleus. Here, it binds to the target cell's DNA and increases or decreases transcription of specific genes.

Hormone Receptors All hormone receptors are proteins, and mutations can alter their structure in a way that alters or prevents their function. For example, a mutation that affects receptors for the hormone testosterone causes total androgen insensitivity syndrome. Affected individuals are genetically male (XY genotype) and make testosterone, but without functional receptors for this hormone it is as if testosterone is not present. Such individuals form testes as embryos, but the testes do not descend into the scrotum, and genitals appear female. Total androgen insensitivity is often not diagnosed until affected individuals enter their teens. Being genetically male, they do not have female reproductive organs and do not menstruate.

Drugs that block a specific type of hormone receptors can be used to prevent unwanted hormonal effects. For example, some breast cancers have an increased number of estrogen receptors and these cancers grow in response to estrogen. The anticancer drug tamoxifen slows the growth of estrogen-sensitive breast cancers by binding to and blocking the cancer's estrogen receptors.

Variations in receptor structure affect how tissues respond to a hormone. The receptor for a hormone in one tissue may cause a different cellular response than a somewhat different receptor for the same hormone in another tissue. Most tissues have receptors for many hormones. The response called up by one hormone may oppose or reinforce that of another. For example, a skeletal muscle cell has receptors for glucagon, insulin, cortisol, epinephrine, estrogen, testosterone, growth hormone, somatostatin, and thyroid hormone, as well as others. The level of all of these hormones influences what happens in the cell.

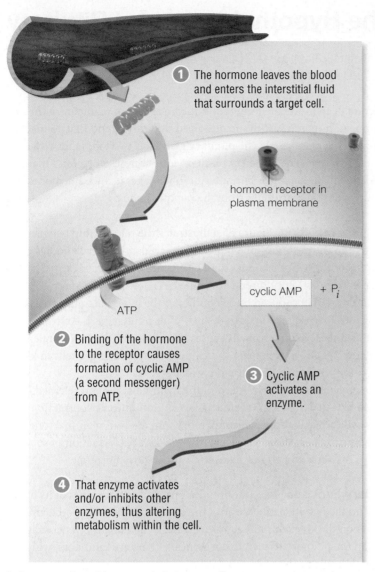

① The hormone leaves the blood and enters the interstitial fluid that surrounds a target cell.

hormone receptor in plasma membrane

cyclic AMP $+ P_i$

ATP

② Binding of the hormone to the receptor causes formation of cyclic AMP (a second messenger) from ATP.

③ Cyclic AMP activates an enzyme.

④ That enzyme activates and/or inhibits other enzymes, thus altering metabolism within the cell.

A. One type of peptide or protein hormone action.

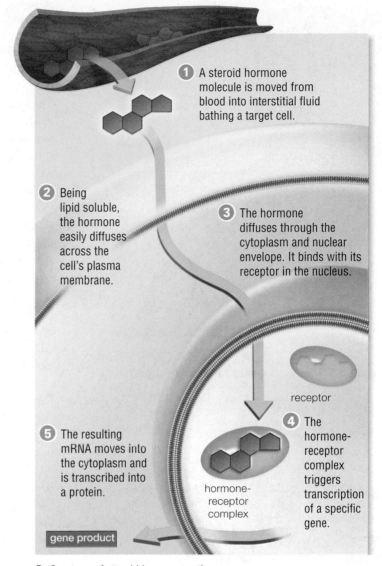

① A steroid hormone molecule is moved from blood into interstitial fluid bathing a target cell.

② Being lipid soluble, the hormone easily diffuses across the cell's plasma membrane.

③ The hormone diffuses through the cytoplasm and nuclear envelope. It binds with its receptor in the nucleus.

receptor

④ The hormone-receptor complex triggers transcription of a specific gene.

hormone-receptor complex

⑤ The resulting mRNA moves into the cytoplasm and is transcribed into a protein.

gene product

B. One type of steroid hormone action.

Figure 25.4 Two mechanisms of hormone action.

Figure It Out: Where in a cell are receptors for the peptide hormone insulin located?

Answer: In the cell's plasma membrane

Take-Home Message 25.2

How do hormones influence cells?

- Endocrine cells secrete hormones into the blood. Only cells with a receptor protein for that hormone are affected by it.
- Protein and peptide hormones bind to receptors in the plasma membrane. Binding results in formation of a second messenger that affects activity of the cell.
- Steroids diffuse across cell membranes and bind to intracellular receptors. Many receptor–hormone complexes bind to DNA and influence gene expression.
- Different types of cells have different receptors for the same hormone. The effect of a hormone varies with the properties of the receptor it binds to.
- Most cells have receptors for, and are influenced by, many different hormones.

Figure 25.5 The hypothalamus and pituitary gland.

Labels in figure: hypothalamus; anterior lobe of pituitary; posterior lobe of pituitary

25.3 The Hypothalamus and Pituitary

REMEMBER: Antidiuretic hormone (ADH) is one of the hormones that regulates the concentration of urine (Section 23.6).

The **hypothalamus** deep inside the forebrain is the body's main regulator of the internal environment. It interacts with the pea-sized **pituitary gland** that connects to it by a slender stalk (Figure 25.5). The pituitary gland has two lobes that differ somewhat in their function. The posterior pituitary lobe secretes hormones that were made inside the hypothalamus. The anterior lobe makes its own hormones but releases them in response to hormones from the hypothalamus.

Posterior Pituitary Function　Figure 25.6 illustrates the relationship between the hypothalamus and the posterior lobe of the pituitary gland. The cell bodies of specialized neurons in the hypothalamus produce hormones ❶. You learned earlier how one of these hormones—antidiuretic hormone (ADH)—affects kidneys and reduces urine output. ADH produced by cell bodies in the hypothalamus is transported through axons to axon terminals in the posterior pituitary ❷. Arrival of an action potential at these axon terminals causes the posterior pituitary to release ADH into the blood. When ADH reaches the kidney, it binds to target cells in kidney tubules and causes them to reabsorb more water ❸.

The hormone oxytocin is also produced in the hypothalamus and released by the posterior pituitary. In females, oxytocin triggers smooth muscle contractions that bring about childbirth. It also makes milk move into ducts of mammary glands when a female is nursing her young. Oxytocin also encourages formation of social bonds and for this reason is sometimes referred to as the "love hormone."

Anterior Pituitary Function　The anterior pituitary makes hormones of its own, but hormones from the hypothalamus control their secretion. Most hypothalamic hormones that act on the anterior pituitary are releasers, meaning they encourage their target cells in the pituitary to secrete hormones. Others are inhibitors, which reduce secretion of hormones by their target cells.

Most anterior pituitary hormones target other endocrine glands. Adrenocorticotropic hormone (ACTH) stimulates release of hormones by adrenal glands. Thyroid-stimulating hormone (TSH) stimulates secretion by the thyroid gland. Follicle-stimulating hormone (FSH) and luteinizing hormone (LH) encourage sex hormone secretion by gonads—a male's testes or a female's ovaries.

The anterior pituitary hormone prolactin targets mammary glands, the exocrine glands that secrete milk. Prolactin stimulates and sustains milk production after a woman has given birth.

Growth hormone (GH) is an anterior pituitary hormone with widespread effects throughout the body. It encourages the growth of bone and soft tissues during development, and it influences metabolism in adults.

Growth Disorders　Normally, a surge of GH production during teenage years results in a growth spurt, then GH production declines with age. Disorders that alter the amount of GH secreted or the timing of its secretion have dramatic effects on

Labels in figure: cell bodies of neurons in the hypothalamus; ❶ Cell bodies of neurons in the hypothalamus make ADH. axons; ❷ ADH is moved through axons into posterior pituitary. anterior pituitary; posterior pituitary; ADH; ❸ ADH released in the posterior pituitary enters the blood, and binds to target cells in kidney.

Figure 25.6 Production and secretion of ADH.
The hormone is produced in the hypothalamus, but stored in and released from the posterior pituitary gland.

B. Sultan Kosen, the tallest living man, with members of his medical team at the University of Virginia.

Kosen has both gigantism and acromegaly. He was diagnosed with a pituitary tumor at age 10. He continued to produce excess growth hormone until treatment halted his growth at 8 feet, 3 inches (2.5 meters).

A. A group of Ecuadorians with Laron syndrome, a heritable form of dwarfism caused by a mutated gene for the GH receptor. Jaime Guevara-Aguirre, at the rear in this photo, has studied the health of this group for more than 22 years.

body size and form. Pituitary dwarfism occurs when the body produces too little GH or receptors do not respond to it properly during childhood. Affected individuals are short but normally proportioned (Figure 25.7A). Pituitary dwarfism can be inherited, or arise as a result of a pituitary tumor or injury. Injections of recombinant human growth hormone (rhGH) made through genetic engineering (Section 10.4) increase the growth rate of children who have a naturally low GH level. However, such treatment is controversial. Many people object to the idea that short stature is a defect to be "cured."

Excessive GH secretion during childhood causes gigantism. Affected people have a normally proportioned body, but are unusually large (Figure 25.7B). When overproduction of GH begins or continues into adulthood, the result is acromegaly, a condition in which bones and cartilage thicken. Facial bones and bones of the hands and feet are the most obviously affected (Figure 25.7C). Both gigantism and acromegaly most often arise as the result of a benign pituitary tumor.

Injections of rhGH have been touted as a way to slow normal aging or boost athletic performance. However, such uses are not approved by regulatory agencies, have not been shown effective in clinical trials, and can have negative side effects, including increased risk of high blood pressure and diabetes.

age 16   age 52

C. A woman before and after she became affected by acromegaly. Excess growth hormone secretion thickens fingers, enlarges ears, lips, and nose, and makes the brow and chin protrude.

Figure 25.7 Growth hormone disorders.

(A) From *Sci Transl Med*, 16 February 2011: Vol 3, Issue 70, p70ra13. Reprinted with permission from AAAS; (B) © ZUMA Press Inc./Alamy; (C) Courtesy of Dr. William H. Daughaday, Washington University School of Medicine, from A. I. Mendelhoff and D. E. Smith, eds., *American Journal of Medicine*, 1956, 20:133.

Take-Home Message 25.3

How do the hypothalamus and pituitary function?

• The hypothalamus makes antidiuretic hormone (ADH) and oxytocin, which are secreted by the pituitary gland's posterior lobe. ADH promotes water reabsorption by kidneys. Oxytocin causes contraction of muscle during childbirth and nursing.

• The hypothalamus also makes releasers and inhibitors that control the pituitary's anterior lobe. Hormones secreted by the anterior pituitary govern secretions of the thyroid gland, adrenal glands, gonads, and mammary glands. The anterior pituitary also secretes growth hormone, which has effects on growth and development of tissues throughout the body.

growth hormone (GH) Anterior pituitary hormone that promotes growth and development throughout the body.

hypothalamus Forebrain region that controls processes related to homeostasis and has endocrine functions.

pituitary gland Pea-sized endocrine gland in the forebrain that interacts closely with the adjacent hypothalamus.

Thyroid hormone increases metabolic rate, so producing too little of it can cause weight gain.

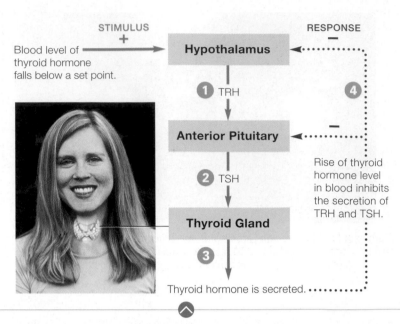

Figure 25.8 **The negative feedback loop that governs thyroid function.**

Left, Gary Head.

Figure 25.9 **A goiter (enlarged thyroid).**
In this case, it was caused by an iodine deficiency.

Scott Camazine/Science Source.

25.4 Thyroid and Parathyroid Glands

Thyroid Hormone The **thyroid gland** lies at the base of the neck, in front of the trachea. It secretes two molecules (triiodothyronine and thyroxine) that we will refer to collectively as thyroid hormone. Thyroid hormone acts throughout the body to increase resting metabolic rate, the rate at which the body uses energy to maintain itself. People who have a high metabolic rate release more energy as heat than those with a lower metabolic rate, and they can eat more food without gaining weight.

The anterior pituitary gland and hypothalamus regulate thyroid hormone secretion by way of a negative feedback loop (Figure 25.8). A low blood concentration of thyroid hormone causes the hypothalamus to secrete thyroid-releasing hormone (TRH) ❶. This releasing hormone causes the anterior pituitary to secrete thyroid-stimulating hormone (TSH) ❷. TSH in turn stimulates the secretion of thyroid hormone ❸. When the blood level of thyroid hormone rises, secretion of TRH and TSH declines ❹.

Thyroid hormone contains the mineral iodine, so a lack of iodine in the diet can cause thyroid hormone deficiency, or hypothyroidism. When iodine is in short supply, signals calling for thyroid hormone secretion cannot be turned off by the normal feedback mechanism. The ongoing stimulation of the thyroid causes it to enlarge. A visibly enlarged thyroid is called a goiter (Figure 25.9). A thyroid hormone deficiency also causes fatigue, confusion, increased sensitivity to cold temperature, and weight gain.

In the United States, where consumption of iodized salt has made dietary iodine deficiency rare, most thyroid disorders arise as a result of an autoimmune disorder. Depending on the type of cells the immune system attacks, the amount of thyroid hormone can increase or decrease. In either case, the thyroid may become enlarged. An excess of thyroid hormone causes nervousness and irritability, a chronic fever, and weight loss. Often, altered metabolism causes tissues behind the eyeball to swell, causing eyes to bulge in their sockets.

Thyroid hormone plays an important role in development of the human nervous system. If a mother lacks iodine during her pregnancy or has an immune disorder that lowers her production of thyroid hormone, development of her child's nervous system may be impaired, resulting in lowered intelligence. A low level of thyroid hormone during infancy or early childhood also stunts growth and impairs mental ability.

In amphibians, thyroid hormone regulates metamorphosis (Figure 25.10). Thus scientists can use tadpoles to test whether a chemical disrupts thyroid function. Commercially reared tadpoles are exposed to a chemical, then examined for developmental abnormalities as frogs. Such tests have shown that pollutants such as nitrates, which are common in drinking water, interfere with thyroid function.

Regulation of Calcium   **Parathyroid glands** are the main regulators of the blood's calcium level. There are four of these glands, each about the size of a grain of rice, on the thyroid's rear surface. When the calcium level in the blood declines, the glands release parathyroid hormone (PTH), which targets cells in bones and the kidneys. PTH increases the breakdown of bone, thus putting calcium ions into the bloodstream. In the kidneys, PTH increases calcium reabsorption, so less calcium is lost in the urine. It also increases production of an enzyme that activates vitamin D. This vitamin helps the intestine take up calcium from food.

Children who eat a diet deficient in vitamin D absorb too little calcium to build healthy new bone. Their low blood calcium also encourages secretion of PTH, which encourages the breakdown of existing bone. The resulting disorder is called rickets. Typical symptoms include bowed legs and pelvic deformities.

In adults, a benign parathyroid tumor sometimes causes excessive PTH secretion that leads to osteoporosis, a disorder in which bones weaken and become more likely to break. As with rickets, the high PTH level causes breakdown of existing bone and adds calcium to the blood. The elevated level of blood calcium raises the risk of kidney stones. The calcium comes out of solution in the kidney and forms deposits that can interfere with kidney and urinary function.

In many animals, a hormone called calcitonin plays an important role in calcium homeostasis. Calcitonin is produced by the thyroid and it opposes the effect of parathyroid hormone by encouraging bones to take up and incorporate calcium. In humans, calcitonin secretion occurs mainly during childhood, and adult blood calcium is regulated primarily by the parathyroids. Calcitonin supplements are sometimes used as a treatment for osteoporosis.

thyroid hormone required

Figure 25.10 Frog metamorphosis.
Thyroid hormone regulates the transition from a gilled tadpole with a tail to a tailless frog with lungs. Chemicals that block thyroid hormone action prevent this transition.

Eric Isselee/Shutterstock.com.

Take-Home Message 25.4

What are the functions of the thyroid and parathyroid glands?

- The thyroid gland regulates the metabolic rate of cells throughout the body. Normal thyroid function requires a diet with adequate iodine.
- Thyroid hormone also plays an essential role in human development and in amphibian metamorphosis.
- In human adults, the parathyroid glands are the main regulators of the level of calcium in the blood. Parathyroid hormone increases breakdown of bone and lessens calcium output in urine, thus increasing the blood calcium level. Calcitonin from the thyroid opposes the effects of parathyroid hormone in many animals.

parathyroid glands Four small endocrine glands in the neck whose hormone product increases the level of calcium in blood.

thyroid gland Endocrine gland at the base of the neck; produces thyroid hormone, which increases the metabolic rate and is essential for normal brain development, and calcitonin, which affects calcium level.

25.5 The Pancreas

REMEMBER: Animals store glucose as glycogen (Section 2.7). The pancreas is a multipurpose organ that produces both digestive enzymes and hormones (23.3).

The pancreas sits in the abdominal cavity, just behind the stomach (Figure 25.11). Most cells in this pancreas are devoted to production of digestive enzymes, but tiny clusters of hormone-producing cells are scattered throughout the organ.

Controlling Blood Glucose Food intake and intracellular demands for glucose vary, and both have the potential to shift the amount of glucose in the blood beyond its ideal range. However, opposing effects of two pancreatic hormones, glucagon and insulin, help smooth out fluctuations in blood glucose level. **Glucagon** is a hormone secreted by some cells of the pancreatic islets. It targets cells in the liver, where it causes the breakdown of glycogen into glucose. Thus, glucagon raises the level of glucose in blood. **Insulin** is a hormone secreted by other cells of the pancreatic islets. It causes its main targets—liver, fat, and skeletal muscle cells—to take up more glucose. Thus, insulin lowers the level of glucose in the blood.

When blood glucose level rises above a set point, the pancreas secretes less glucagon and more insulin (Figure 25.11 ❶–❺). As glucose is taken up and stored inside cells, blood glucose declines. In contrast, a decline in blood glucose below the set point increases glucagon secretion and slows insulin secretion ❻–❿. The resulting release of glucose from the liver causes the blood glucose level to rise.

Figure 25.11 Endocrine function of the pancreas.

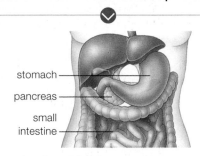

stomach
pancreas
small intestine

A. Above, location of the pancreas.

B. Right, how cells that secrete insulin and glucagon respond to a change in the level of glucose circulating in blood. These two hormones work antagonistically to maintain the glucose level within its normal range.

❶ After a meal, glucose enters blood faster than cells take it up, so its concentration in blood increases.

❷ In the pancreas, the increase in glucose causes a decrease in glucagon secretion.

❸ High glucose also stimulates insulin secretion.

❹ In response to insulin, adipose and muscle cells take up and store glucose; cells in the liver and muscle make more glycogen.

❺ As a result, insulin *lowers* the blood level of glucose.

❻ *Between* meals, the glucose level in blood declines.

❼ The decrease causes glucagon secretion.

❽ It also slows insulin secretion.

❾ In the liver, glucagon causes cells to break glycogen down into glucose, which enters the blood.

❿ As a result, glucagon raises the blood glucose level.

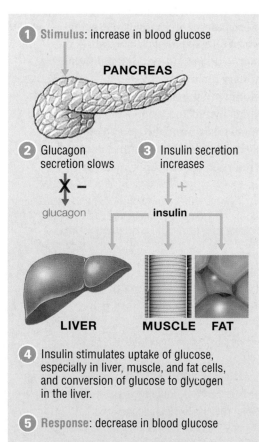

❶ **Stimulus:** increase in blood glucose

PANCREAS

❷ Glucagon secretion slows

❸ Insulin secretion increases

glucagon   **insulin**

LIVER   **MUSCLE**   **FAT**

❹ Insulin stimulates uptake of glucose, especially in liver, muscle, and fat cells, and conversion of glucose to glycogen in the liver.

❺ **Response:** decrease in blood glucose

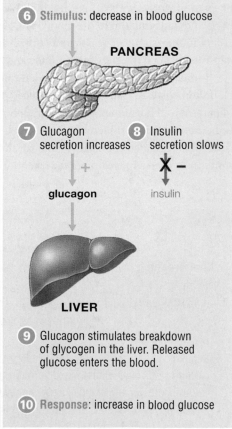

❻ **Stimulus:** decrease in blood glucose

PANCREAS

❼ Glucagon secretion increases

❽ Insulin secretion slows

glucagon   insulin

LIVER

❾ Glucagon stimulates breakdown of glycogen in the liver. Released glucose enters the blood.

❿ **Response:** increase in blood glucose

Table 25.2 Some Complications of Diabetes

Eyes	Changes in lens shape and vision; damage to blood vessels in retina; blindness
Skin	Increased susceptibility to bacterial and fungal infections; patches of discoloration; thickening of skin on the back of hands
Digestive system	Gum disease; delayed stomach emptying that causes heartburn, nausea, vomiting
Kidneys	Increased risk of kidney disease and kidney failure
Heart and blood vessels	Increased risk of heart attack, stroke, high blood pressure, and atherosclerosis
Hands and feet	Impaired sensations of pain; formation of calluses, foot ulcers; poor circulation in feet especially may lead to tissue death that can only be treated by amputation

Diabetes Mellitus Diabetes mellitus loosely translates as "passing honey-sweet water." People with this endocrine disorder produce sweet urine because their liver, fat, and muscle cells do not take up and store glucose as they should. The result is high blood sugar, or hyperglycemia. The disrupted metabolism affects cells throughout the body. Table 25.2 list the complications associated with diabetes.

There are two main types of diabetes. Type 1 diabetes accounts for 5 to 10 percent of diabetes cases. It occurs when an immune reaction, genetic defect, tumor, or toxin destroys insulin-producing cells of pancreatic islets so that little or no insulin is produced. Symptoms usually appear in childhood and adolescence. Affected people must monitor the amount of sugar in their blood and supply themselves with insulin via injections (Figure 25.12). In the more common type 2 diabetes, target cells stop responding to insulin. As a result, blood sugar levels remain high. It typically occurs in middle age, as insulin production declines. Genetics is a factor, but obesity increases the risk. Diet, exercise, and oral medication can control most cases of type 2 diabetes, but some affected people eventually need insulin injections.

The rate of type 2 diabetes is soaring worldwide. By one estimate, more than 150 million people are now affected. Western diets and sedentary lifestyles contribute to an increasing risk for the disease. Exposure to endocrine-disrupting chemicals may also play a role. Because of the many negative effects of diabetes, the increasing number of cases is currently a pressing concern of public health officials.

Figure 25.12 An insulin pump.
The device is programmed to inject insulin through a hollow tube that projects through the skin and into the body. The pump helps smooth out fluctuations in blood sugar, thus lowering risk of complications that arise from excessively low or high blood sugar.

© Elizabeth Musar; inset, © Manny Hernandez/Diabetes Hands Foundation, www.tudiabetes.com.

Take-Home Message 25.5

How do pancreatic hormones regulate blood glucose?

- The pancreas secretes two hormones that work in opposition to control the level of glucose in the blood.
- Insulin secreted in response to high blood glucose encourages target cells to take up and store glucose.
- Glucagon secreted in response to low blood glucose causes target cells to break down glycogen to glucose.

glucagon Pancreatic hormone that causes liver cells to break down glycogen and release glucose, thus raising the blood glucose level.

insulin Pancreatic hormone that causes cells to take up glucose from the blood, thus lowering the blood glucose level.

Digging Into Data

Effects of BPA on Insulin Function

Bisphenol A (BPA) is a chemical found in many plastic bottles and in the lining of food and beverage cans. It is also a suspected endocrine disrupter. A number of studies have found an association between a high level of BPA and an increased risk of type 2 diabetes. Angel Nadal suspected that BPA disrupts insulin metabolism by binding to and activating an estrogen receptor on pancreatic islet cells. Figure 25.13 shows the results of one of his experiments. Cultured cells from human pancreatic islets were exposed either to BPA or to DPN, a chemical that binds to and activates the estrogen receptor. The cells were then exposed to glucose and their insulin secretion was monitored.

1. Consider the two groups of cells that were exposed to glucose alone. How did the concentration of glucose affect their insulin secretion?
2. How did treating cells with DPN alter the response to glucose concentration?
3. How did treating cells with BPA alter the response to glucose concentration?
4. How were the effects of DPN and BPA similiar? How did they differ?
5. In another experiment, mouse islet cells lacking the estrogen receptor were exposed to BPA, and then to glucose. Insulin secretion of these cells did not differ from that of cells exposed to glucose alone. Explain how the combined data from these experiments support or refute the hypothesis that BPA alters human insulin secretion by binding to and activating the estrogen receptor.

Figure 25.13 Effects of BPA and DPN on glucose-stimulated insulin secretion by human pancreatic islet cells.
DPN is a chemical known to bind and activate estrogen receptors on pancreatic beta cells. A glucose concentration of 8 millimolar (mM) is equivalent to that of the blood after a meal.

After Soriano S, Alonso-Magdalena P, García-Arévalo M, Novials A, Muhammed SJ, et al. (2012) Rapid Insulinotropic Action of Low Doses of Bisphenol-A on Mouse and Human Islets of Langerhans: Role of Estrogen Receptor β. PLoS ONE 7(2): e31109. doi:10.1371/journal.pone.0031109.

25.6 The Adrenal Glands

REMEMBER: ADH is one of the hormones that regulates the concentration of urine (Section 23.6).

On top of each kidney is a grape-sized **adrenal gland**. (In Latin *ad–* means near, and *renal* refers to the kidney.) Each adrenal gland has two regions: an outer **adrenal cortex**, and an inner **adrenal medulla**. The two regions of the gland are controlled by different mechanisms, and they secrete different hormones.

The adrenal cortex secretes aldosterone and cortisol. Aldosterone promotes reabsorption of sodium and water in the kidney. Like ADH, it concentrates the urine. The hormone **cortisol** has wide-ranging effects on metabolism and immunity. Cortisol induces liver cells to break down their stored glycogen, and suppresses uptake of glucose by other cells. It prompts adipose cells to degrade fats, and skeletal muscles to degrade proteins. The breakdown products of fats and proteins serve as alternative energy sources (Section 5.7).

Under most circumstances, negative feedback loops with the anterior pituitary and hypothalamus maintain blood cortisol levels. Figure 25.14 shows what happens when the level of cortisol in blood decreases below its set point. This decline triggers secretion of corticotropin-releasing hormone (CRH) by the hypothalamus. CRH causes the anterior pituitary to secrete ACTH (adrenocorticotropin). ACTH in turn causes the adrenal cortex to secrete cortisol.

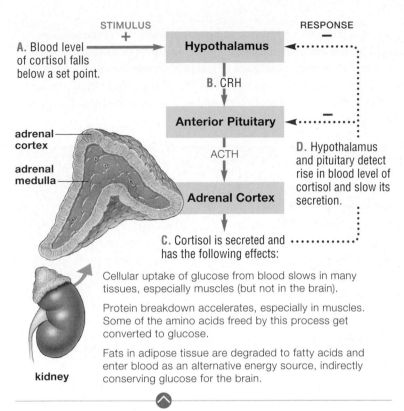

STIMULUS
+

A. Blood level of cortisol falls below a set point. ⟶ **Hypothalamus** ← RESPONSE −

B. CRH

Anterior Pituitary ← −

ACTH

D. Hypothalamus and pituitary detect rise in blood level of cortisol and slow its secretion.

Adrenal Cortex

C. Cortisol is secreted and has the following effects:

Cellular uptake of glucose from blood slows in many tissues, especially muscles (but not in the brain).

Protein breakdown accelerates, especially in muscles. Some of the amino acids freed by this process get converted to glucose.

Fats in adipose tissue are degraded to fatty acids and enter blood as an alternative energy source, indirectly conserving glucose for the brain.

adrenal cortex

adrenal medulla

kidney

Figure 25.14 Structure and function of the adrenal gland.
A negative feedback loop governs cortisol secretion. However, the nervous system can override the feedback controls during times of stress or danger.

With elevated cortisol caused by a pituitary tumor.

With normal cortisol after removal of the pituitary tumor.

Figure 25.15 Cushing's syndrome.
The woman shown developed Cushing's syndrome as a result of a pituitary tumor. Her cortisol level was restored to normal by treatment of this tumor.

Permission obtained from Blackwell Publishing © Holt RIG and Hanley NA (2006) *Essential Endocrinology & Diabetes*, edn 5.

Cortisol secretion rises dramatically with injury, illness, anxiety, or starvation. Under these circumstances, the nervous system overrides the feedback loop that stabilizes the cortisol level, and the cortisol level in the blood soars. At the same time, the adrenal medulla increases its output of norepinephrine and epinephrine. These hormones prepare the body for an emergency. They bring about dilation of the pupils, an increased respiratory rate, and a faster heart rate.

This stress response helps the body deal with an immediate threat by diverting resources from maintenance tasks to support a state of arousal. A stress response is adaptive for short periods of time, as when an animal is fleeing from a predator. However, long-term elevation of cortisol (as by chronic stress) is unhealthy because it interferes with immunity, memory, and sexual function. It also raises the risk of cardiovascular problems.

Cushing's syndrome is an endocrine disorder characterized by a chronically high level of cortisol. It results from an adrenal gland tumor, a pituitary tumor that causes excess secretion of ACTH, or ongoing use of synthetic steroids such as cortisone shots and prednisone. These drugs are given to relieve pain, inflammation, and other ailments. The body converts them to cortisol. People with Cushing's syndrome tend to put on fat around their torso and they have a puffy, rounded "moon face" (Figure 25.15). Blood pressure and blood glucose become unusually high. Impaired immunity makes infections common. Skin thins, bone density declines, and muscles shrink. Wounds may be slow to heal.

An abnormally low cortisol level, referred to as Addison's disease, is also unhealthy. It can be caused by some infections such as tuberculosis. However, in

adrenal cortex Outer portion of adrenal gland; secretes aldosterone and cortisol.

adrenal gland Endocrine gland located above kidney; secretes hormones with roles in urine formation and stress responses.

adrenal medulla Inner portion of the adrenal gland; secretes epinephrine and norepinephrine.

cortisol Adrenal cortex hormone that affects metabolism and immunity; secretions rise with stress.

developed countries, it usually results from autoimmune attacks on the adrenal glands. President John F. Kennedy had the autoimmune form of Addison's disease. Symptoms often include fatigue, weakness, depression, weight loss, and darkening of the skin. If cortisol level declines too much, blood sugar and blood pressure can fall to life-threatening levels. People with Addison's disease are treated with a synthetic form of cortisol.

Take-Home Message 25.6

What are the functions of the adrenal glands?

- The adrenal cortex secretes aldosterone and cortisol. Aldosterone makes the urine more concentrated. Cortisol affects metabolism and the stress response.
- The adrenal medulla releases epinephrine and norepinephrine, which prepare the body for excitement or danger.

25.7 Hormones and Reproductive Function

REMEMBER: Sexual reproduction involves formation of gametes (Section 8.5).

Gonads **Gonads** are an animal's primary reproductive organs—a female's ovaries or a male's testes. In addition to producing gametes (eggs or sperm), gonads secrete sex hormones, the steroid hormones essential for reproductive function. Sex hormones also affect secondary sexual traits, which are traits that differ between the sexes but do not play a direct role in reproduction.

Males and females produce the same sex hormones, but in very different proportions. Testes secrete mainly testosterone. **Testosterone** is a hormone that causes male genitals to form in embryos. During **puberty**, the period when sexual organs mature, testosterone triggers sperm formation and development of secondary sexual characteristics. In humans, these traits include facial hair and an enlarged larynx (voice box) that lowers the voice. Similarly, testosterone influences the development of a thick mane in male lions (Figure 25.16).

Ovaries produce mainly estrogens and progesterone. **Estrogens**, the primary female sex hormones, are responsible for maturation and maintenance of sex organs and for female sexual secondary traits. In humans, these traits include enlarged breasts and fat deposition at the hips. **Progesterone** prepares the body for pregnancy and maintains the uterus if a pregnancy does occur.

The hypothalamus and anterior pituitary control the secretion of sex hormones (Figure 25.17). In both males and females, the hypothalamus produces GnRH (gonadotropin-releasing hormone). This releaser causes the anterior pituitary to secrete follicle-stimulating hormone (FSH) and luteinizing hormone (LH). FSH and LH cause the gonads to secrete sex hormones. We discuss the reproductive role of sex hormones in more detail in Chapter 26.

The Pineal Gland The small, pinecone-shaped **pineal gland** lies deep inside the brain (Figure 25.18). It produces and secretes the hormone **melatonin**. However, these secretions halt when the retina is exposed to light, which causes signals to flow

Figure 25.16 A secondary sexual trait.
Testosterone secreted by testes stimulates the growth of a male lion's mane (left). A lioness (right) produces little testosterone, so she does not grow a mane.

Mattias Klum/National Geographic Creative.

Hypothalamus	The hypothalamus produces gonadotropin-releasing hormone that acts on the anterior pituitary.
GnRH	
Anterior Pituitary	The anterior pituitary makes follicle-stimulating hormone (FSH) and luteinizing hormone (LH), which both target cells in the testes or ovaries.
FSH, LH	
Gonads	The testes or ovaries produce sex hormones.
Sex hormones	

Figure 25.17 Control of sex hormone secretion.

along the optic nerve to the brain. The amount of light varies with time of day, and day length varies seasonally, so melatonin secretion rises and falls in daily and seasonal cycles.

In some mammals, mating behavior is seasonal and its timing is regulated by melatonin. For example, many temperate zone rodents breed only during summer when days are long, nights are short, and blood melatonin level is relatively low. During the winter, long nights allow the melatonin concentration to increase, and the high level of melatonin inhibits production of sex hormones.

In humans, melatonin secretion drops dramatically around the time of puberty. However, it remains unclear whether this decline is one of the causes of puberty or an effect of other hormonal changes.

In many animals, including humans, melatonin regulates sleep–wake cycles, in part by affecting body temperature. Just after sunrise, melatonin secretion decreases, body temperature rises, and we awaken. At night, melatonin secretion causes a decline in body temperature and we become sleepy. Because cycles of melatonin secretion that affect sleep–wake rhythms are set by exposure to light, travelers who fly across many time zones are advised to spend some time in the sun. The exposure to light can help them reset their internal clock and minimize the effects of jet lag.

Some people in latitudes where there are big seasonal shifts in day length have seasonal affective disorder (SAD), commonly called the "winter blues." During winter months, people with SAD tend to feel lethargic and depressed. Researchers have found that affected people secrete more melatonin in winter than in summer, whereas most people secrete the same amount year-round. Exposure to bright artificial light in the early morning is used to treat SAD.

Melatonin has a protective effect against some cancers. It directly inhibits division of cancer cells in animals. It also suppresses production of sex hormones, which encourage the growth of certain cancers. Some studies suggest that working night shifts or having poor sleep habits can disrupt melatonin secretion and increase the risk of cancer. For example, one study found female night-shift nurses have lower melatonin levels and a higher risk of breast cancer than their day-shift counterparts. Other studies suggest that a woman's risk of breast cancer decreases with the length of her average night's sleep. In men, night-shift work is associated with an increased risk of prostate cancer.

pineal gland

Figure 25.18 The human pineal gland.
Exposing the retina to light inhibits secretion of melatonin by this gland. Thus, melatonin secretion typically peaks at about 2 A.M., while we are asleep.

Right, diane39/iStockphoto.com.

Take-Home Message 25.7

What are the endocrine roles of the gonads and the pineal gland?

- A female's ovaries and a male's testes are gonads that make sex hormones as well as gametes. Males make mostly testosterone. Females make mostly estrogens and progesterone.

- In both sexes, sex hormone secretion is regulated by hormones secreted by the pituitary gland, in response to a releaser from the hypothalamus.

- The pineal gland deep inside the brain produces melatonin, a hormone that influences sleep–wake cycles and, in some animals, the timing of reproduction. Melatonin also has a protective effect against some cancers.

estrogens Sex hormones produced by a female's ovaries.

gonads Ovaries and testes; organs that produce gametes and secrete sex hormones.

melatonin Hormone produced by the pineal gland; affects onset of puberty, sleep–wake cycles.

pineal gland Endocrine gland in the forebrain that secretes melatonin; secretion declines when the eye is exposed to light.

progesterone Sex hormone produced by a female's ovaries.

puberty Period when reproductive organs mature and begin to function.

testosterone Sex hormone produced by a male's testes.

Section 25.1 Some synthetic chemicals such as phthalates and the insecticide DDT are endocrine disrupters, meaning they interfere with the function of **animal hormones**.

Section 25.2 All vertebrates have an **endocrine system** consisting of **endocrine glands** and cells that secrete hormones. Hormones enter the bloodstream and are carried to target cells elsewhere in the body. A cell is a target of a hormone only if it has receptors that can bind that hormone. There are two types of hormones: **amino-acid derived hormones** and steroid hormones. Peptide and protein hormones bind to membrane receptors at the cell surface. Binding commonly leads to formation of a **second messenger** molecule. The second messenger causes a series of enzyme activations in the cytoplasm. **Steroid hormones** are hydrophobic; they can enter a target cell and interact directly with DNA.

Section 25.3 The **hypothalamus**, a forebrain region, is structurally and functionally linked with the adjacent **pituitary gland**. The posterior pituitary releases antidiuretic hormone (ADH) and oxytocin made by cells in the hypothalamus. Oxytocin causes contraction of smooth muscle in milk ducts and the uterus. ADH acts on kidney tubules and promotes concentration of the urine.

The anterior pituitary makes and secretes its own hormones: Adrenocorticotropin tells the adrenal cortex to secrete cortisol; thyroid-stimulating hormone calls for thyroid hormone secretion; and follicle-stimulating hormone and luteinizing hormone stimulate hormone production by male and female gonads and have roles in gamete formation. Prolactin stimulates milk production by the mammary glands. **Growth hormone** has growth-promoting effects on cells throughout the body.

Section 25.4 A feedback loop to the anterior pituitary and hypothalamus governs the **thyroid gland** in the base of the neck. Iodine is required for synthesis of thyroid hormone, which increases the metabolic rate and plays an important role in development of the nervous system.

The four **parathyroid glands** located at the rear of the thyroid gland are the main regulators of calcium levels in the blood. They release parathyroid hormone in response to low calcium levels. Parathyroid hormone acts on cells in bone, the kidneys, and the intestine to raise calcium levels in the blood. The thyroid hormone calcitonin opposes the effect of parathyroid hormone.

Section 25.5 **Insulin** and **glucagon** are pancreatic hormones that regulate the level of glucose in the blood. Insulin stimulates glucose uptake by muscle and liver cells and thus lowers the blood glucose level. Glucagon stimulates the release of glucose, which increases blood levels. With the metabolic disorder diabetes mellitus, blood glucose remains chronically high as a result of either insufficient insulin or insulin receptors that do not respond to insulin.

Section 25.6 There is an **adrenal gland** atop each kidney. The **adrenal cortex** secretes aldosterone (which targets the kidney) and **cortisol** (the stress hormone). Cortisol secretion is governed by a negative feedback loop to the anterior pituitary gland and hypothalamus. In times of stress, the nervous system overrides feedback controls over cortisol secretion so cortisol level rises.

Norepinephrine and epinephrine released by neurons of the **adrenal medulla** influence organs during times of stress and excitement. They cause a fight–flight response.

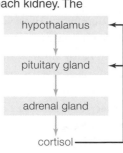

Section 25.7 **Gonads** (ovaries and testes) secrete **estrogens**, **progesterone**, and **testosterone**. These sex hormones are steroids that act in reproduction and in development of secondary sexual traits during **puberty**.

Melatonin secretion by the vertebrate **pineal gland** varies seasonally because exposure to light suppresses it. In humans, it affects the daily sleep–wake cycle. Disrupted melatonin secretion raises the risk of some cancers.

Self-Quiz

Answers in Appendix I

1. Peptide hormones are synthesized from _____ .
   a. amino acids   c. cholesterol
   b. proteins   d. nucleic acids

2. Antidiuretic hormone and oxytocin are hormones produced in the hypothalamus but released by the _____ .
   a. pineal gland   c. pancreas
   b. pituitary gland   d. thalamus

3. Protein hormones typically bind to receptors _____ .
   a. in the DNA   c. at the plasma membrane
   b. in mitochondria   d. both a and b

4. Match each pituitary hormone with its target.
   _____ antidiuretic hormone   a. gonads (ovaries, testes)
   _____ oxytocin   b. mammary glands, uterus
   _____ luteinizing hormone   c. kidneys
   _____ growth hormone   d. most body cells

5. Overproduction of _____ causes acromegaly and gigantism.
   a. melatonin   c. growth hormone
   b. insulin   d. cortisol

6. The _____ regulate(s) calcium levels in the blood.
   a. hypothalamus   c. pineal gland
   b. pancreas   d. parathyroid glands

7. _____ lowers blood sugar level; _____ raises it.
   a. Glucagon; insulin   c. Melatonin; insulin
   b. Insulin; glucagon   d. Cortisol; glucagon

8. A rise in the concentration of thyroid hormone in the blood slows the release of thyroid hormone. This is an example of _____ .
   a. positive feedback   b. negative feedback

9. The _____ produces digestive enzymes and hormones.
   a. pancreas   c. pineal gland
   b. hypothalamus   d. parathyroid gland

10. Thyroid hormone _____ .
    a. contains calcium that must be obtained in the diet
    b. is essential to amphibian metamorphosis
    c. is secreted in response to signals from the adrenal gland
    d. all of the above

11. A person with an overly active thyroid gland is more likely to be unusually _____ .
    a. heavy   c. cold
    b. anxious   d. both a and b

12. During stressful situations, the adrenal glands increase their output of _____ .
    a. cortisol   c. norepinephrine
    b. epinephrine   d. all of the above

13. The male sex hormone testosterone is secreted in response to secretion of hormones by the _____ .
    a. testes   c. pituitary gland
    b. ovaries   d. pancreas

14. Match each disorder with its symptom(s).
    _____ goiter   a. puffy face, slow to heal
    _____ pituitary dwarfism   b. very large hands, feet
    _____ Cushing's syndrome   c. enlarged thyroid
    _____ acromegaly   d. high blood sugar
    _____ diabetes   e. short but normally proportioned

15. Match the hormone source listed at left with the most suitable description at right.
    _____ adrenal cortex   a. makes gametes and hormones
    _____ thyroid gland   b. major control center
    _____ pineal gland   c. influences blood calcium level
    _____ parathyroid gland   d. stress increases secretions
    _____ pancreatic islet   e. light inhibits secretion
    _____ posterior pituitary   f. hormones require iodine
    _____ hypothalamus   g. regulates blood sugar level
    _____ testis   h. secretes hormones made by cell bodies in the hypothalamus

Critical Thinking

1. A woman affected by type 1 (insulin-dependent) diabetes miscalculates and injects herself with too much insulin. She soon begins to feel confused and sluggish. She calls for medical assistance and injects herself with the glucagon her doctor prescribed for such an emergency. When an ambulance arrives, she is given dextrose (a sugar) intravenously. How did the excess insulin cause her symptoms? How did the glucagon injection and the intravenous sugar help reverse them?

2. Women who are completely blind tend to undergo puberty at an earlier age than sighted women. They also are less likely to have breast cancer. By one hypothesis, blindness encourages early puberty and discourages breast cancer by its effect on melatonin secretion. Based on this information, would you expect the average melatonin level in blind women to be higher or lower than that of sighted women? Explain your answer.

3. Sex hormone secretion is governed by a negative feedback loop to the hypothalamus and pituitary, similar to that for thyroid hormone or cortisol. Because of this, a veterinarian can use a blood test to find out whether a female dog has been neutered. Dogs who still have their ovaries have a lower blood level of luteinizing hormone (LH) than dogs who have been neutered. Explain why removing a dog's ovaries would result in an elevated level of LH.

REPRODUCTION AND DEVELOPMENT

Application ⊕ 26.1 Assisted Reproduction

REMEMBER: During sexual reproduction, gametes (eggs and sperm) combine to produce a zygote (Section 8.5). Genetic screening methods detect alleles associated with inherited defects (9.9).

A couple who cannot conceive naturally or a woman who wishes to use sperm from a nonpartner donor may now choose to conceive through *in vitro* **fertilization** (**IVF**), an assisted reproductive technique that combines an egg and sperm outside the body. To obtain eggs for IVF, a woman is given hormones that encourage many eggs to mature at once. The mature eggs are removed from her ovaries via a hollow needle and are combined with sperm for fertilization (Figure 26.1A). After fertilization, each zygote undergoes mitotic divisions, forming a microscopic ball of cells that can be placed in a woman's womb to develop to term.

Louise Brown was the first child to be born after conception by IVF. Her birth in 1978 shocked scientists and the public alike. Many were appalled by the idea of what the media referred to as "test tube babies." Scientists expressed concern that this unnatural procedure would produce children with psychological and genetic defects. Ethicists and religious leaders warned about the dire societal implications of manipulating human embryos.

Despite these initial reservations, IVF has become widely accepted and practiced. Worldwide, the technique has resulted in the birth of more than 3 million children. The first test tube babies are now adults and have begun having children of their own. Louise Brown is among them. She now has two sons, both conceived naturally (Figure 26.1B).

Research into IVF opened the way to a variety of assisted reproductive technologies. If a man makes sperm, but does not ejaculate them or does not ejaculate enough to allow fertilization by normal means, intracytoplasmic sperm injection can put his sperm inside his partner's egg. A woman who does not make viable eggs but who wishes to carry a child can be implanted with an early embryo produced by IVF using an egg from an egg donor. A woman who produces viable eggs but cannot, or does not want to, carry them herself can have her egg fertilized by IVF. The resulting early embryo can then be implanted in a surrogate mother. Embryos can also be frozen for years before being used. With all IVF procedures, prospective parents can screen embryos for genetic defects before they are implanted.

The ability to "bank" human eggs—to freeze eggs and store them for later use—was a recent breakthrough. In the United States, egg banking first became an option in 2012. The capacity to store human eggs has dramatically cut the cost of obtaining donated eggs for use in IVF. It has also given women who face the loss of fertility as a result of a medical condition or aging a way to retain the option of reproducing later in life.

The United States Food and Drug Administration requires egg banks and sperm banks to screen donors for general health, test them for sexually transmitted diseases, take an extensive family history, and test semen for the presence of some pathogens. The FDA does not, however, mandate any type of genetic screening. To minimize genetic risks, egg banks and sperm banks now voluntarily screen for alleles most commonly associated with genetic disorders. However, even with extensive testing, they cannot eliminate the possibility of some less common harmful alleles slipping by. It is important to remember that unique allele combinations affecting health can arise whenever people mate, whether with a spouse or an anonymous donor. A shuffling of the genetic cards is integral to sexual reproduction.

A. Fertility specialist injecting a single sperm into an egg. The video screen shows a magnified view of the egg.

B. Louise Brown, the first child conceived by IVF, with her two sons, who were conceived naturally.

Figure 26.1 *In vitro* fertilization.

(A) © Heidi Specht, West Virginia University; (B) AP Images/Chris Radburn/PA Wire.

26.2 Modes of Reproduction

REMEMBER: Meiosis produces gametes that differ in their combination of alleles (Section 8.5). Amniotes are adapted to life on land by internal fertilization and by eggs in which membranes enclose the developing embryo. Placental mammals develop inside their mother's body, sustained by an organ called the placenta (15.6).

Asexual Reproduction   With **asexual reproduction**, a single individual produces offspring that are genetically identical to the parent and one another. As a result, the parent has all its genes represented in each offspring. Asexual reproduction can be advantageous in a stable environment where the allele combination that made a parent successful can be expected to do the same for its offspring. However, being locked into a particular allele combination can be disadvantageous in a changing environment.

Many invertebrates reproduce asexually. Some do this by fragmentation—a piece breaks off and grows into a new individual. In others, new individuals bud from existing ones (Figure 26.2A). In some insects, fish, amphibians, birds, and lizards, new individuals can develop from unfertilized eggs. Thus far, no mammal has been reported to reproduce asexually.

Sexual Reproduction   With **sexual reproduction**, meiosis and fertilization results in a unique combination of paternal and maternal alleles in each offspring. Genetic variety can be adaptive when offspring are likely to face different challenges than their parents. By reproducing sexually, a parent increases the likelihood that some of its offspring will inherit a combination of traits that allows them to succeed.

Some species switch between sexual and asexual reproduction. Plant-sucking insects called aphids provide one example. During the summer, a female aphid settles on a plant and gives birth to wingless daughters that develop inside her body from unfertilized eggs. An aphid's daughters live on the same plant as their mother, so they all experience more or less similar conditions. In autumn, males are produced and sexual reproduction yields genetically variable females. These females disperse and spend the winter in a resting state. In the spring, they seek out new plants on which to raise the new generation of genetically identical daughters.

Variations on Sexual Reproduction   Sexually reproducing animals that can produce both eggs and sperm are called **hermaphrodites**. Simultaneous hermaphrodites produce eggs and sperm at the same time. Tapeworms and some roundworms are simultaneous hermaphrodites that can fertilize themselves. Earthworms and slugs are simultaneous hermaphrodites too, but they require a partner (Figure 26.2B). During mating, each slug both donates sperm to and receives sperm from its partner. Some other mollusks and certain fishes are sequential hermaphrodites, meaning they switch from one sex to another during the course of a lifetime. More typically, vertebrates have separate sexes that remain fixed for life, with each individual making either eggs or sperm.

Fertilization may be external or internal. Fertilization is external in all amphibians and in most aquatic invertebrates and bony fishes. With external fertilization, large numbers of gametes are released into the water. Sperm swim to eggs, and development typically occurs in the environment. With internal fertilization, sperm are deposited directly into a female's reproductive tract (Figure 26.2C). Fertilization is internal in all cartilaginous fishes, some bony fishes, and most land animals, including insects, slugs, and the amniotes (reptiles, birds, and mammals).

A. Asexual reproduction in a hydra. A new individual (left) is budding from the body of its parent.

B. Sexual reproduction in banana slugs. Each slug both donates sperm to and receives sperm from its partner.

C. Sexual reproduction in elephants. The male is inserting his penis into the female. Eggs will be fertilized and the offspring will develop inside the mother's body, nourished by nutrients delivered by her bloodstream.

Figure 26.2 Animal reproduction.

(A) Biophoto Associates/Science Source; (B) Courtesy of Christine Evers; (C) © Gabriela Staebler Wildlife Photography.

asexual reproduction   Reproductive mode by which offspring arise from a single parent only.

hermaphrodite   Individual that can produce both eggs and sperm.

***in vitro* fertilization**   Procedure in which eggs are removed from a woman's ovary and fertilized outside her body.

sexual reproduction   Reproductive mode by which two parents produce offspring having a combination of their genes.

Figure 26.3 Live birth in a snake (adder).
Most snakes lay eggs, but adder eggs develop inside the mother's body. Developing young are nourished solely by egg yolk.

© Tony Phelps/naturepl.com.

blastula Hollow, fluid-filled ball of cells that forms early in animal development.

cleavage Mitotic division of cells in an early embryo.

ectoderm Outermost tissue layer of an animal embryo.

Egg Female gamete.

endoderm Innermost tissue layer of an animal embryo.

fertilization Egg and sperm combine and form a zygote.

gastrula Three-layered early embryo that forms when a blastula undergoes gastrulation.

gastrulation Animal developmental process by which cell movements produce a three-layered gastrula.

mesoderm Middle tissue layer of a three-layered animal embryo.

sperm Male gamete.

yolk Nutritious material in many animal eggs.

zygote Cell formed by fusion of gametes; first cell of a new individual.

After internal fertilization, a female either lays eggs or retains them inside her body while they develop. All birds and most insects lay eggs. Live birth occurs in some fishes, some lizards and snakes (Figure 26.3), and in all placental mammals.

A developing animal requires energy and nutrients. In most animals, yolk fills this need. **Yolk** is a thick fluid rich in proteins and lipids that is deposited in an egg as it forms. By contrast, placental mammals produce almost yolkless eggs. They nourish their embryos by means of a placenta, an organ that facilitates the exchange of substances between the maternal and embryonic bloodstreams.

Take-Home Message 26.2

How do animals reproduce?

• Some animals reproduce asexually, producing offspring identical to one another and to the parent. Most reproduce sexually and so produce genetically variable offspring.

• External fertilization is typical in aquatic animals, whereas in land-dwelling animals fertilization usually occurs inside a female's body.

• Fertilized eggs may be laid in the environment or develop inside the mother's body. In most animals, the egg's yolk provides the nutrients that sustain development.

26.3 Stages of Animal Development

REMEMBER: Differentiation is the process by which cells become specialized by turning on different genes (Section 6.1). Products of master genes affect expression of other genes (7.7). Most animals have an embryo with three tissue layers (15.2).

Sexual reproduction begins with formation of gametes. **Sperm** are male gametes and **eggs** are female gametes. **Fertilization** unites an egg and a sperm to produce a **zygote**, the first cell of the new individual. Developmental processes then transform a single-celled zygote into an adult animal. Figure 26.4 shows the developmental stages of one vertebrate, the leopard frog.

During **cleavage**, mitotic cell divisions increase the number of cells without increasing the zygote's original volume ❶. As a result, cells become more numerous but smaller. Cleavage also sets the stage for differentiation. Each cell produced by cleavage ends up with a somewhat different portion of the egg cytoplasm. The quantity and type of maternal mRNAs that a cell receives will determine which master genes it will turn on during early development.

Cleavage ends with the formation of a **blastula**, a hollow ball of cells with a fluid-filled center ❷. After a blastula forms, cell divisions slow and cells undergo a structural rearrangement called **gastrulation**. The resulting three-layered structure is the **gastrula** ❸. Its has three primary tissue layers: an outer layer of **ectoderm**, a middle tissue layer of **mesoderm**, and an inner tissue layer of **endoderm**.

After gastrulation, the embryo's tissues and organs begin to form ❹. Many organs incorporate tissues derived from more than one primary tissue layer. For example, the stomach's epithelial lining is derived from endoderm, and the smooth muscle that makes up the stomach wall develops from mesoderm.

In most animals, the individual that hatches from an egg or is born continues to grow and develop. In frogs, the egg hatches to release a larva (a tadpole). A larva is an immature animal that differs in form from the adult. A frog larva undergoes metamorphosis, a drastic remodeling of tissues, to take on its adult form ❺.

fluid-filled cavity

blastula

1 During cleavage, mitotic divisions divide a zygote's cytoplasm into multiple cells.

2 Cleavage produces a blastula.

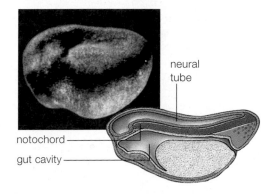

ectoderm

future gut cavity

ectoderm

mesoderm
endoderm

neural tube

notochord

gut cavity

3 During gastrulation, rearrangement of cells produces a gastrula that has three primary tissue layers: an outer ectoderm, a middle mesoderm, and an endoderm.

4 After gastrulation, organs begin to form.

Tadpole, a swimming larva with segmented muscles and a notochord extending into a tail.

Limbs grow and the tail is absorbed during metamorphosis to the adult form.

Sexually mature, four-legged adult leopard frog.

5 A frog egg hatches into a tadpole larva that develops into an adult.

Figure 26.4 Stages in the development of a vertebrate, the leopard frog.

(1–4) photos, Carolina Biological Supply Company; (5) left, center, © David M. Dennis/Tom Stack & Associates, Inc.; right, © John Shaw/Tom Stack & Associates.

Take-Home Message 26.3

How does development of a sexually reproducing animal proceed?

- Gametes unite at fertilization to form a zygote.
- Cleavage (a series of mitotic divisions) produces a blastula, a hollow ball of cells.
- Gastrulation, a series of cellular rearrangements, yields a three-layered gastrula. Cells of the gastrula go on to develop into specialized tissues and organs.
- In some animals, including frogs, a larva must undergo metamorphosis to take on the adult body form.

Figure 26.5 Components of the female reproductive system and their functions.

(A,C) From Russell/Wolfe/Hertz/Starr. *Biology*, 1e. © 2008 Brooks/Cole, a part of Cengage Learning, Inc.

A. Location of female reproductive organs, frontal view.

B. External sex organs, collectively referred to as the vulva.

26.4 Human Reproductive Function

REMEMBER: In animals, meiosis of germ cells produces gametes (Section 8.5). Hypothalamic and pituitary hormones control sex hormone production by gonads (25.7).

Human gametes form by meiosis of germ cells inside primary sex organs, or gonads. Sperm form inside male gonads, the **testes** (singular, testis). Eggs form inside female gonads, the **ovaries**. A male reproductive tract also has components that store sperm and deliver them into the female reproductive tract. A female's reproductive tract has components that receive sperm and sustain developing offspring.

Female Reproductive Anatomy A female's ovaries lie deep inside her pelvic cavity (Figure 26.5A, C). They are about the size and shape of almonds. In addition to producing eggs, ovaries secrete estrogens and progesterone, the main sex hormones in females. **Estrogens** trigger development of female secondary sexual characteristics and maintain the lining of the reproductive tract. **Progesterone** prepares the reproductive tract for pregnancy.

Adjacent to each ovary is an **oviduct**, a hollow tube that connects the ovary to the uterus. (Mammalian oviducts are sometimes referred to as Fallopian tubes.) An egg released by an ovary is drawn into an oviduct. Cilia in the oviduct lining then propel the egg along the length of the tube.

Both oviducts open into the **uterus**, a hollow, pear-shaped organ commonly called the womb. If fertilization occurs, an embryo forms and develops inside the uterus. A thick layer of smooth muscle makes up most of the uterine wall. The uterine lining (the endometrium) consists of glandular epithelium, connective tissues, and blood vessels. The lowest portion of the uterus narrows as the **cervix**, the region that connects to the vagina.

Ovary One of two female gonads. Makes eggs and secretes female sex hormones (estrogens and progesterone).

Oviduct One of a pair of ducts through which oocytes are propelled from an ovary to the uterus; usual site of fertilization.

Uterus Womb, chamber in which an embryo develops. Includes myometrium (smooth muscle layer) and endometrium (epithelial lining). Narrowed lower portion (the cervix) secretes mucus into the vagina.

Vagina Organ of sexual intercourse: birth canal.

Clitoris Highly sensitive erectile organ. Only the tip is externally visible; bulk of the organ extends internally on either side of the vagina.

Labium minus One of a pair of inner skin folds (the labia minora).

Labium majus One of a pair of fatty outer skin folds (the labia majora).

C. Side view, showing position of reproductive organs.

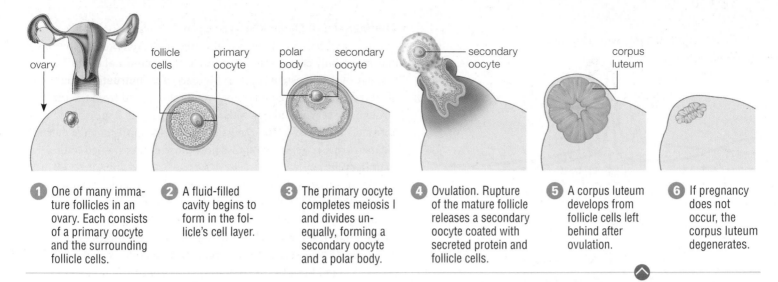

1. One of many immature follicles in an ovary. Each consists of a primary oocyte and the surrounding follicle cells.

2. A fluid-filled cavity begins to form in the follicle's cell layer.

3. The primary oocyte completes meiosis I and divides unequally, forming a secondary oocyte and a polar body.

4. Ovulation. Rupture of the mature follicle releases a secondary oocyte coated with secreted protein and follicle cells.

5. A corpus luteum develops from follicle cells left behind after ovulation.

6. If pregnancy does not occur, the corpus luteum degenerates.

Figure 26.6 The ovarian cycle.

The **vagina** is a muscular tube that extends from the cervix to the body's surface. It functions as the female organ of intercourse and also as the birth canal. Typically, when a girl is born, a membrane called the hymen partially covers the external entrance to her vagina. It is usually stretched open during the first episode of intercourse, if it has not previously been broken by other physical activity.

Genitals are the externally visible parts of the reproductive tract (Figure 26.5B). Female genitals include two pairs of liplike skin folds that enclose the openings of the vagina and urethra. Adipose tissue fills the labia majora, the thick outer folds. The thin inner folds are the labia minora. An erectile organ called the clitoris lies near the anterior junction of the labia minora. The clitoris and penis develop from the same embryonic tissue and both are highly sensitive to tactile stimulation.

Egg Production and Release A female's germ cells do not usually divide after she is born. At birth she has about 2 million primary oocytes, which are immature eggs that entered meiosis I but did not complete it. (An **oocyte** is an immature egg.) When a female reaches puberty, hormonal changes prompt her primary oocytes to begin to mature, one at a time, in an approximately 28-day ovarian cycle.

Figure 26.6 shows how an oocyte matures during the course of this cycle. A primary oocyte and the cells around it constitute an **ovarian follicle** 1. In the first part of the cycle, the oocyte enlarges and secretes a layer of proteins, and the cells around the oocyte divide repeatedly. As the follicle matures, a fluid-filled cavity opens around the oocyte 2. Often more than one follicle begins to develop, but typically only one becomes fully mature. In that follicle, the primary oocyte completes meiosis I and undergoes unequal cytoplasmic division 3. The result is a large secondary oocyte and a tiny **polar body**, which has no reproductive function and will later disintegrate. The secondary oocyte goes on to enter meiosis II, but does not complete it.

About two weeks after the onset of the cycle, **ovulation** occurs: The secondary oocyte and polar body are ejected into the adjacent oviduct 4. Once in the oviduct, an oocyte must meet up with sperm within 12 to 24 hours for fertilization to occur. The oocyte will not complete meiosis unless a sperm penetrates it.

Meanwhile, in the ovary, the cells of the ruptured follicle develop into a hormone-secreting structure called the **corpus luteum** 5. (The name means "yellow body" in Latin and refers to its yellowish color.) If pregnancy does not occur, the corpus luteum breaks down 6. Once it is gone, a new follicle can begin to mature.

cervix Narrowed region of the uterus that connects to vagina.

corpus luteum Hormone-secreting structure that forms from follicle cells after ovulation.

estrogen Sex hormone that causes development of female secondary sexual characteristics and maintains the reproductive tract.

oocyte Immature egg.

ovarian follicle Immature egg and surrounding cells.

ovary Female gonad; produces eggs.

oviduct Ciliated tube that connects an ovary to the uterus.

ovulation Release of a secondary oocyte by an ovary.

polar body Tiny cell that forms as a by-product of unequal cytoplasmic division during egg production.

progesterone Sex hormone that prepares the female reproductive tract for pregnancy.

testis Male gonad; produces sperm.

uterus Muscular chamber where offspring develop; womb.

vagina Female organ of intercourse and birth canal.

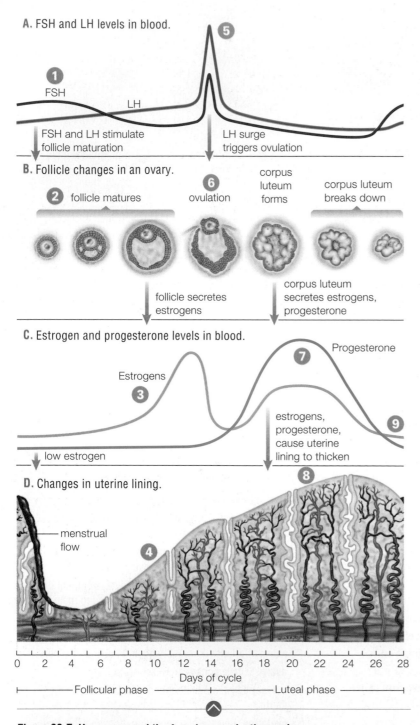

A. FSH and LH levels in blood.

⑤

①
FSH

LH

FSH and LH stimulate
follicle maturation

LH surge
triggers ovulation

B. Follicle changes in an ovary.

corpus
luteum
⑥   forms
② follicle matures   ovulation

corpus luteum
breaks down

follicle secretes
estrogens

corpus luteum
secretes estrogens,
progesterone

C. Estrogen and progesterone levels in blood.

Progesterone
⑦

Estrogens
③

estrogens,
progesterone,
cause uterine
lining to thicken   ⑨

low estrogen

D. Changes in uterine lining.

⑧

menstrual
flow
④

```
0   2   4   6   8   10  12  14  16  18  20  22  24  26  28
                   Days of cycle
├──────── Follicular phase ────────┼────── Luteal phase ──────┤
```

∧

Figure 26.7 Hormones and the female reproductive cycle.
Onset of menstrual flow is Day 1 of the approximately 28-day menstrual cycle.

A., B. Prompted by GnRH from the hypothalamus, the anterior pituitary secretes FSH and LH, which stimulate a follicle to grow and an oocyte to mature in an ovary. A midcycle surge of LH triggers ovulation and the formation of a corpus luteum. A decline in FSH after ovulation stops more follicles from maturing.

C., D. Early on, estrogens from a maturing follicle encourage repair and rebuilding of uterine lining. After ovulation, the corpus luteum secretes some estrogens and more progesterone to prime the uterus for pregnancy. If pregnancy occurs, the corpus luteum will persist, and will stimulate the maintenance of the uterine lining.

The Menstrual Cycle Cyclic events in the ovaries occur in synchrony with cyclic changes in the uterus. We refer to the approximately monthly changes in the uterus as the **menstrual cycle**. The first day of the menstrual cycle is the start of **menstruation**: the flow of bits of uterine lining and a small amount of blood from the uterus, through the cervix, and out of the vagina. Figure 26.7 shows how hormone levels and the thickness of the uterine lining change during the menstrual cycle. It also shows how those events are tied to the ovarian cycle.

At the onset of the menstrual cycle, gonadotropin-releasing hormone (GnRH) from the hypothalamus causes cells in the anterior pituitary to increase their secretion of two hormones ①. **Follicle-stimulating hormone** (**FSH**) stimulates an ovarian follicle to start maturing ②. **Luteinizing hormone** (**LH**) has a role in ovulation. (Both FSH and LH also function in sperm production in males.)

As a follicle matures, cells surrounding the oocyte secrete estrogens ③. The estrogens encourage the uterine lining to begin to thicken ④. A rising estrogen level also causes the pituitary to respond with an outpouring of LH ⑤. This LH surge encourages the primary oocyte to complete meiosis I and undergo cytoplasmic division. In addition, the LH surge causes the follicle to swell and burst ⑥. Thus, a midcycle surge of LH is the trigger for ovulation.

Immediately after ovulation, the estrogen level declines until the corpus luteum forms. The corpus luteum secretes some estrogens and a large amount of progesterone ⑦. Estrogens and progesterone stimulate the uterine lining to thicken and encourage blood vessels to grow through it ⑧. The uterus is now ready for a pregnancy.

If a pregnancy does not occur, the corpus luteum persists for about twelve days. During this period, the estrogen and progesterone it secretes prevents the hypothalamus from secreting FSH, so no other follicles start to mature.

When the corpus luteum does begin to break down, estrogen and progesterone levels fall ⑨. The hypothalamus senses this decline, and stimulates the pituitary to begin secreting FSH and LH once again. In the uterus, the decline in estrogens and progesterone causes the thickened lining to break down, and menstruation begins. Menstruation usually lasts for five to seven days.

A woman enters **menopause** when all the oocytes in her ovaries have either been released during menstrual cycles or have disintegrated as a result of normal aging. With no oocytes left to mature, production of estrogen and progesterone is dramatically diminished and menstrual cycles cease.

Vaginal bleeding between menstrual periods and bleeding after menopause require medical attention. Such bleeding can be a sign of uterine cancer, the most common cancer of the female reproductive tract. Ovarian cancer is less common, but causes more deaths. It does not alter menstrual cycles, so it often remains undetected until it has become well established or spread. There are no early symptoms of cervical cancer either, but this cancer can be detected by a Pap smear (a procedure in which cervical cells are sampled, stained, and examined under a microscope for changes indicating cancer).

Male Reproductive Anatomy Before a male is born, his testes descend into the scrotum, a pouch of skin and smooth muscle suspended below the bones of the pelvic girdle (Figure 26.8). A testis and the scrotum that enclose it are collectively called a testicle. When a man becomes cold or frightened, reflexive muscle contractions draw the testes closer to his body. When he feels warm, relaxation of smooth muscle in the scrotum allows his testes to hang lower, so the sperm-making cells do not overheat. These cells function best just below normal body temperature.

Immature sperm that form in the testes are moved into an **epididymis** (plural, epididymides), a coiled duct on top of the testis. As sperm pass through the epididymis, they mature and become motile. The last region of each epididymis stores mature sperm and is continuous with the first portion of a **vas deferens** (plural, vasa deferentia). In Latin, *vas* means vessel, and *deferens*, to carry away. A vas deferens is a duct that carries sperm away from an epididymis, and to a short ejaculatory duct. Ejaculatory ducts deliver sperm to the urethra, the duct that extends through a male's penis to an opening at the body surface.

The penis is the male organ of intercourse. It has a rounded head (the glans) at the end of a narrower shaft. Nerve endings in the glans make it highly sensitive to touch. Normally, when a man is not sexually excited, a retractable tube of skin called the foreskin covers the glans of his penis. In many cultures, males undergo circumcision, an elective surgical procedure that removes the foreskin.

Beneath the skin of the penis, connective tissue encloses three elongated cylinders of spongy tissue. When a male becomes sexually excited, signals from the nervous system cause blood to flow into the spongy tissue faster than it flows out. As fluid pressure rises, the normally limp penis becomes erect.

Sperm stored in the epididymides continue their journey toward the body surface only when a male reaches the peak of sexual excitement and ejaculates. During **ejaculation**, smooth muscle in the walls of the epididymides and vasa deferentia undergoes rhythmic contractions that propel sperm and accessory gland secretions out of the body as a thick, white fluid called semen.

ejaculation Smooth muscle contractions expel semen through the penis.

epididymis Duct that emerges from a testis and conveys sperm to a vas deferens.

follicle-stimulating hormone (FSH) Pituitary hormone that causes an ovarian follicle to mature in females and aids in sperm development in males.

luteinizing hormone (LH) Pituitary hormone that triggers ovulation in females and testosterone secretion in males.

menopause Of a human female, the end of fertility and of menstrual cycles.

menstrual cycle Approximately monthly cycle in which the uterine lining thickens in preparation for pregnancy, then is shed if pregnancy does not occur.

menstruation Flow of blood and bits of shed uterine lining out through the vagina.

vas deferens A long duct that conveys mature sperm from an epididymis toward the urethra.

Figure 26.8 Male reproductive anatomy.

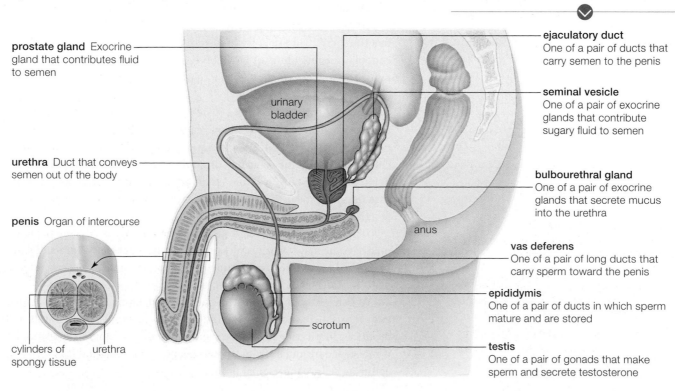

prostate gland Exocrine gland that contributes fluid to semen

urinary bladder

urethra Duct that conveys semen out of the body

penis Organ of intercourse

cylinders of spongy tissue

urethra

ejaculatory duct
One of a pair of ducts that carry semen to the penis

seminal vesicle
One of a pair of exocrine glands that contribute sugary fluid to semen

bulbourethral gland
One of a pair of exocrine glands that secrete mucus into the urethra

anus

vas deferens
One of a pair of long ducts that carry sperm toward the penis

epididymis
One of a pair of ducts in which sperm mature and are stored

scrotum

testis
One of a pair of gonads that make sperm and secrete testosterone

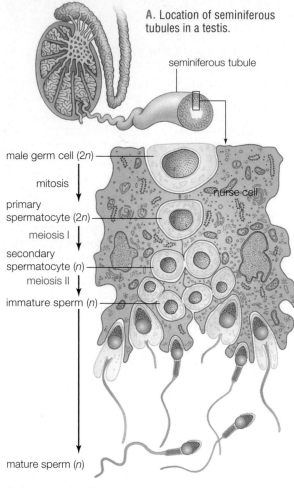

A. Location of seminiferous tubules in a testis.

seminiferous tubule

male germ cell (2n)

mitosis

nurse cell

primary
spermatocyte (2n)

meiosis I

secondary
spermatocyte (n)

meiosis II

immature sperm (n)

mature sperm (n)

B. Cross-section through one tubule.

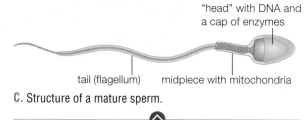

"head" with DNA and
a cap of enzymes

tail (flagellum)   midpiece with mitochondria

C. Structure of a mature sperm.

Figure 26.9 Sperm formation and structure.

Semen is a complex mixture of sperm, proteins, nutrients, ions, and signaling molecules. Sperm account for less than 5 percent of its volume. The other 95 percent is accessory gland secretions. **Seminal vesicles,** which are exocrine glands near the base of the bladder, secrete fructose-rich fluid into the vasa deferentia. Sperm use the fructose (a sugar) as their energy source. Another exocrine gland, the **prostate gland,** is the other major contributor to semen volume. It encircles the urethra and secretes fluid into it. Prostate gland secretions help raise the pH of the female reproductive tract, so it is more hospitable to sperm.

A young, healthy man has a prostate gland about the size of a walnut, but the gland often enlarges with age. Because the urethra runs through the prostate gland, prostate enlargement can narrow this duct and cause difficulty urinating. When this happens, medication, laser treatments, or surgery can alleviate symptoms. An enlarged prostate can also be a sign of cancer. Most prostate cancers grow relatively slowly and those in older men often do not require treatment. However, some prostate cancers grow quickly and can spread.

Testicular cancer is the most common cancer among men aged 15 to 34. To detect it, a man should examine his testes once a month, after a warm shower or bath, to look and feel for lumps, enlargement, or hardening.

How Sperm Form   Although smaller than a golf ball, each testis holds more than 100 meters of coiled **seminiferous tubules** (Figure 26.9A). Diploid male germ cells line the inner wall of each tubule (Figure 26.9B). These cells divide repeatedly by mitosis, and their offspring differentiate into primary spermatocytes. Primary spermatocytes undergo meiosis to form immature sperm. Large nurse cells inside seminiferous tubules support the developing sperm. Cells between seminiferous tubules secrete testosterone. Like egg production, sperm production is governed by the pituitary hormones LH and FSH. LH targets testosterone-secreting cells and encourages testosterone secretion. FSH targets nurse cells. In combination with testosterone, it causes nurse cells to produce chemicals essential to sperm development.

A mature sperm is a haploid cell with a "head" that is packed full of DNA and tipped by an enzyme-containing cap (Figure 26.9C). The other end has a flagellum that the sperm will use to swim toward an egg. The sperm does not have ribosomes, endoplasmic reticulum, or Golgi bodies. However, its midpiece contains many mitochondria that supply the ATP to power flagellum movement.

Sexual Intercourse   When a male is not sexually aroused, his penis remains soft and flexible because the arterioles that transport blood into cylinders of spongy tissue within the penis are somewhat constricted. With sexual excitement, these vessels widen and the veins that drain the penis narrow. As a result, inward flow of blood exceeds outward flow. As blood fills the interior of the penis, fluid pressure rises, and the organ enlarges and stiffens so that it can be inserted easily into a female's vagina.

The ability to achieve and sustain an erection peaks during the late teens. As a man ages, he may have episodes of erectile dysfunction. Viagra and similar drugs prescribed for erectile dysfunction cause blood vessels that carry blood into the penis to widen and deliver more blood.

When a woman becomes sexually excited, blood flow to the vaginal wall, labia, and clitoris increases. Glands in the cervix secrete mucus and glands on the labia produce a lubricating fluid. The vagina itself does not have any glandular tissue. It is moistened by mucus from the cervix and by plasma fluid that seeps out between epithelial cells of the vaginal lining.

During intercourse, body temperature, heart rate and breathing rate increase in both partners. The posterior pituitary increases its secretion of oxytocin. In the brain, the increased oxytocin inhibits signals from the amygdala, the part of the brain that controls fear and anxiety. In a reproductive tract, it encourages contraction of smooth muscles.

Continued mechanical stimulation of the penis or clitoris can lead to orgasm. During orgasm, endorphins flood the brain and evoke feelings of pleasure. At the same time, a surge of oxytocin causes strong, rhythmic contractions of smooth muscle in both the male and female reproductive tract. In males, orgasm is usually accompanied by ejaculation, an expulsion of semen through the penis.

A Sperm's Journey Ejaculation can put as many as 250 million sperm into the vagina. However, only a few hundred will survive the journey to the upper region of an oviduct, where fertilization typically occurs.

To travel from the vagina into the uterus, a sperm must swim through the cervical canal, a 3-centimeter-long tunnel through the center of the cervix. During most of a woman's reproductive cycle, a creamy acidic mucus fills the cervical canal and blocks the passage of sperm. As a woman's estrogen level increases in the days before ovulation, her cervical mucus becomes thinner and more alkaline, so sperm can swim through it. Even so, passing through the cervical canal remains difficult, and most sperm never reach the main chamber of the uterus.

Those sperm that do enter the uterus are propelled toward the oviducts by contractions of smooth muscle in the uterine wall. Sperm enter both oviducts, but because ovulation occurs in only one ovary at a time, about half of them will continue onward without any possibility of ever encountering an egg. An egg does produce chemicals that attract sperm, but this attraction only works at close range.

For a sperm, an oviduct is not simply a passageway to the egg. It is also the site of an important transformation. Here, a sperm acquires the capacity to fertilize an egg. The head of a freshly ejaculated sperm has a coating that interferes with the binding of the sperm to an egg. This coating is removed as the sperm passes through the initial portion of the oviduct.

> Ejaculation can put as many as 250 million sperm into the vagina.

Take-Home Message 26.4

What are the roles of human reproductive organs?

- Ovaries make eggs and secrete estrogens and progesterone. Oviducts connect the ovaries to the uterus, where offspring develop. The vagina is the organ of intercourse and also the birth canal.
- A woman is born with immature eggs. From puberty until menopause, these oocytes mature—one at a time—in an approximately monthly cycle. During each cycle, the uterine lining thickens in preparation for pregnancy. If pregnancy does not occur, the woman sheds the uterine lining (menstruates), and the cycle begins again.
- A male's testes produce sperm and secrete testosterone. Sperm and secretions from accessory glands form semen. Semen is propelled through a series of ducts and leaves the body through an opening in the penis.
- During intercourse the penis deposits sperm into the vagina. From here, sperm must pass through the cervix, continue through the uterus, and enter an oviduct. While in the oviduct, sperm become capable of fertilizing an egg.

prostate gland Exocrine gland that encircles a male's urethra; its secretions contribute to semen.

semen Sperm mixed with fluid secreted by exocrine glands.

seminal vesicles Exocrine glands that secrete fluid into vasa deferentia; main source of semen volume.

seminiferous tubule In a testis, a germ-cell containing tubule inside which sperm form.

A. Monitoring body temperature can help predict when a woman is about to ovulate.

B. Condom, a barrier to sperm. **C.** Birth control pills or hormone implants prevent ovulation.

Figure 26.10 Tools used to prevent to prevent pregnancy.

(A) Ever/iStockphoto.com; (B) Scott Camazine & Sue Trainor/Science Source; (C) SuperStock.

26.5 Reproductive Health

REMEMBER: HIV is a viral pathogen (Section 13.4) that impairs function of the immune system (22.6). HPV is a virus that causes cancer (22.1).

Contraception Methods used to prevent pregnancy are termed contraception. Table 26.1 lists common options and compares their effectiveness. The most effective option is abstinence—not having intercourse—which is 100 percent effective. Rhythm methods are forms of abstinence; a woman avoids sex during her fertile period. She calculates when she is ovulating by recording how long her menstrual cycles last, checking her temperature each morning, monitoring the thickness of her cervical mucus, or by checking some combination of these indicators (Figure 26.10A). However, cycles vary, so miscalculations are frequent and sperm deposited in the vagina up to three days before ovulation can survive to fertilize an egg.

Withdrawal, the removal of the penis from the vagina before ejaculation, can be ineffective because sperm can be in pre-ejaculation fluids. Similarly, rinsing out the vagina (douching) immediately after intercourse is unreliable because some sperm swim through the cervix within seconds of ejaculation.

Surgical methods are highly effective, but they are meant to make a person permanently sterile. Men may opt for a vasectomy. A doctor makes a small incision into the scrotum, then cuts and ties off each vas deferens. A tubal ligation blocks or cuts a woman's oviducts.

Physical and chemical barriers can prevent sperm from reaching an egg. Spermicidal foams and jellies poison sperm. They are not always reliable, but their use with a condom or diaphragm reduces the chance of pregnancy. A diaphragm is a flexible, dome-shaped device that covers the cervix. A cervical cap is a similar but smaller device. Condoms are thin sheaths worn over the penis or used to line the vagina during intercourse (Figure 26.10B).

An intrauterine device, or IUD, is inserted into the uterus by a physician. Some IUDs thicken cervical mucus so sperm cannot swim through it. Other IUDs shed copper, which keeps an early embryo from implanting in the uterus.

The birth control pill is the most common fertility control method in developed countries. "The Pill" is a mix of synthetic estrogens and progesterone-like hormones that prevents both maturation of oocytes and ovulation. When taken consistently, it is highly effective. A birth control patch is a small, flat adhesive patch applied to skin. It delivers the same mixture of hormones as an oral contraceptive, and it blocks ovulation the same way.

Hormone injections or implants prevent ovulation. Each injection acts for several months, whereas the implant lasts for three years. Both methods are effective, but may cause sporadic, heavy bleeding.

If a woman has unprotected sex or a condom breaks, she can turn to emergency contraception. Some so-called "morning-after pills" are available without a prescription to women over age 17. The most widely used emergency contraceptive, Plan B, delivers a large dose of synthetic progesterone that

Table 26.1 Mechanism and Effectiveness of Contraceptives

Method	Description	Pregnancy Rate*
Abstinence	No intercourse	0% per year
Rhythm method	Avoid intercourse in female's fertile period	25% per year
Withdrawal	End intercourse before male ejaculates	27% per year
Douche	Wash semen from vagina after intercourse	60% per year
Vasectomy	Cut or close off male's vasa deferentia	<1% per year
Tubal ligation	Cut or close off female's oviducts	<1% per year
Condom	Enclose penis, block sperm entry to vagina	15% per year
Diaphragm, cervical cap	Cover cervix, block sperm entry to uterus	16% per year
Spermicides	Kill sperm	29% per year
Intrauterine device	Prevent sperm entry to uterus or prevent implantation of embryo	<1% per year
Oral contraceptives	Prevent ovulation	<1% per year
Hormone patches, implants, or injections	Prevent ovulation	<1% per year
Emergency contraception pill	Prevent ovulation	15–25% per use**

* Percent of users who become pregnant with consistent, correct usage.
** Not meant for regular use. 15–25 percent of users become pregnant despite use of pill.

prevents ovulation and interferes with fertilization. It is most effective when taken immediately after intercourse, but has some effect up to three days later. Morning-after pills are not meant for regular use. They cause nausea, vomiting, abdominal pain, headache, and dizziness.

The drug mifepristone (RU-486) in combination with a prostaglandin is used to terminate an early unwanted pregnancy. The two chemicals interfere with pro-gesterone receptors in the uterus. Without progesterone's effects, the uterine lining cannot be maintained, and neither can pregnancy.

Infertility About 10 percent of couples in the United States are infertile, which means they do not have a successful pregnancy despite a year of trying. Infertility increases with age. In about half of cases, infertility arises in the female partner. The most common cause of female infertility is a hormonal disorder called polycystic ovarian syndrome. With this disorder, a follicle begins to mature, but ovulation does not occur; instead the follicle turns into a cyst (a fluid-filled sac).

Some sexually transmitted diseases can scar the oviducts and prevent sperm from reaching eggs. Endometriosis also interferes with oviduct function. With this disorder, endometrial tissue grows in regions of the pelvis outside the uterus. Like normal endometrium, the misplaced tissue grows and is shed in response to hormones. The result is pain and scarring that can distort oviducts. Endometriosis can also make intercourse painful, thus lessening its frequency and the associated opportunities for conception.

If fertilization does occur, scarring of oviducts can lead to an embryo implant-ing in the oviduct, causing a tubal pregnancy (Figure 26.11) that cannot develop to term and threatens the life of the mother. Benign tumors, endometriosis, and other uterine problems can also interfere with the ability of the embryo to implant.

1 cm

Figure 26.11 Tubal pregnancy.
Such pregnancies occur when an embryo implants in an oviduct, rather than the uterus. The embryo cannot develop to term and its presence can rupture the oviduct, causing bleeding, infection, and even maternal death.

Dr. E. Walker/Science Source.

Digging Into Data

Sperm Counts Down on the Farm

Contamination of water by agricultural chemicals has raised concerns that these chemicals may affect human health and reproduction. To determine the reproductive effects of chemical exposure, epidemiologist Shanna Swan and her colleagues studied sperm collected from men in four cities that differ in their proximity to agricultural fields (Figure 26.12). All of the men involved in the study were partners of women who had become pregnant and were visiting a prenatal clinic, so all were presumably fertile. Of the four cities, Columbia, Missouri, is located in the county with the most farmlands. New York City in New York is in an area with no agriculture.

1. In which cities did researchers record the highest and lowest sperm counts?
2. In which cities did samples show the highest and lowest sperm motility (ability to move)?
3. Aging, smoking, and sexually transmitted diseases adversely affect sperm. Do any of these variables explain the regional differences in sperm count?
4. Do these data support the hypothesis that living near farmlands can adversely affect male reproductive function?

	Location of clinic			
	Columbia, Missouri	Los Angeles, California	Minneapolis, Minnesota	New York, New York
Average age	30.7	29.8	32.2	36.1
Percent nonsmokers	79.5	70.5	85.8	81.6
Percent with history of STD	11.4	12.9	13.6	15.8
Sperm count (million/ml)	58.7	80.8	98.6	102.9
Percent motile sperm	48.2	54.5	52.1	56.4

Figure 26.12 Sperm study.
Data from a study of sperm collected from male partners of pregnant women who visited prenatal health clinics in one of four cities. STD stands for sexually transmitted disease.

A. Chlamydia-inflamed eyes of an infant who was infected by its mother during delivery.

B. Chancres caused by a syphilis infection that has spread throughout the body.

C. Genital warts (raised white areas) on the labia of a woman infected by HPV.

Figure 26.13 Consequences of STDs.
Use of a latex condom minimizes the risk of transmission of pathogens that can cause STDs.

(A) Western Ophthalmic Hospital/Science Source; (B) CNRI/Science Source; (C) CDC/Joe Millar.

Male infertility arises when a man makes abnormal sperm or too few sperm, or when sperm are not ejaculated properly. Disorders that lower testosterone can prevent sperm maturation. Some genetic disorders can interfere with sperm motility. In males, as in females, sexually transmitted disease can scar ducts of the reproductive system, causing blockages that keep sperm from leaving the body. Male infertility can also arise when the valve between the bladder and the urethra does not function properly, causing sperm to end up in the bladder rather than leaving the body through the urethra.

Sexually Transmitted Diseases Sexually transmitted diseases (STDs) are contagious diseases that spread through sexual contact. Women contract STDs more easily than men, and they have more complications. Women can also infect their offspring during childbirth, so sexually active women who intend to become pregnant should be tested for STDs. In both sexes, STDs can scar the reproductive tract and increase the risk of infertility.

To reduce the risk of STD transmission, physicians recommend use of a latex condom and a lubricant during sexual encounters. The condom serves as a barrier to the pathogen and the lubricant helps prevent small abrasions that would make it easier for the pathogen to enter the body.

Trichomoniasis, an STD commonly called "trich," is caused by the flagellated protozoan *Trichomonas vaginalis*. Infected women produce a yellowish discharge and have a sore, itchy vagina. Men are usually symptom-free but some studies suggest that untreated trichomoniasis is a risk factor for prostate cancer.

Chlamydia and gonorrhea are bacterial STDs that often occur together. Both can cause a discharge from the penis or vagina and pain during urination. However, these symptoms occur in only about half of infected women and even fewer infected men. Both diseases can also be passed from mother to child during childbirth, resulting in eye infections and other problems in the newborn (Figure 26.13A). An untreated gonorrhea infection can also harm joints and the heart.

The earliest symptom of syphilis, another bacterial STD, is flattened, painless ulcers called chancres (pronounced "shankers") at the site of initial infection. If the infection is untreated, bacteria spread and more widely distributed chancres appear (Figure 26.13B). A long-term infection can destroy organs throughout the body, causing joint pain, mental illness, blindness, and death.

Treatment with antibiotics can cure most bacterial STDS, but antibiotic resistance is on the rise. For example, a strain of gonorrhea recently discovered in Japan is resistant to all currently used antibiotics. Many bacteria strains that cause STDs resist commonly used drugs such as penicillin and tetracycline.

Some types of human papillomavirus (HPV) cause bumplike growths called warts on the hands or feet. Sexually transmitted strains of the virus can become established on the genitals (Figure 26.13C) or wherever else sexual behavior introduces them. In addition to warts, some sexually transmitted types of HPV cause cancer. HPV infection causes nearly all cervical cancers in women and most anal cancers in homosexual men. Oral sex can introduce HPV into the mouth and raise the risk of cancers of the mouth and throat. A recently approved vaccine can prevent HPV infection in both males and females, if given before viral exposure. No current treatment can cure an HPV infection, but the resulting warts can be removed by surgery, or by burning or freezing them.

The herpes simplex virus 2 causes genital herpes. Initial exposure to the virus commonly results in small sores at the site of infection. After these sores heal, the infection periodically becomes reactivated, causing tingling or itching, which may or may not be accompanied by visible sores. Antiviral drugs can promote healing of sores caused by an active infection and lessen the likelihood of viral reactivation, but they cannot eliminate the virus entirely.

Infection by HIV (human immunodeficiency virus) can cause AIDS (acquired immune deficiency syndrome), in which the immune system weakens and infectious organisms take hold. Section 13.4 described the HIV virus and Section 22.6 explained how its effects on the immune system lead to AIDS. With regard to sexual transmission, oral sex is least likely to pass on an infection. Unprotected anal sex is 5 times more dangerous than unprotected vaginal sex and 50 times more dangerous than oral sex. If you think you could have been exposed to HIV, get tested as soon as possible. Early treatment can prevent you from passing on the virus and developing AIDS.

Take-Home Message 26.5

What factors affect reproductive function?

- Contraceptives can prevent ovulation or fertilization.
- Female infertility arises from hormonal disorders, blocked reproductive ducts, or uterine problems. Infertile men may have a low sperm count, blocked reproductive ducts, or problems ejaculating.
- Sexually transmitted diseases scar the reproductive tract, increasing the risk of infertility. Some STDs can produce ill effects throughout the body.

26.6 Human Development

REMEMBER: Humans are chordates and their embryos have characteristic chordate traits (Section 15.4). They are also amniotes and their eggs have extraembryonic membranes typical of this group (15.6).

In humans, prenatal (pre-birth) development typically lasts for 38 weeks after fertilization. During that time, cell divisions transform a single-celled zygote into a newborn with about a trillion cells of many different types.

Fertilization Fertilization typically occurs in the upper region of the oviduct (Figure 26.14 **1**). A secondary oocyte released at ovulation still has some follicle cells around it and is encased within a layer of secreted proteins **2**. Sperm make their way between the follicle cells and bind to the protein layer. This binding triggers release of protein-digesting enzymes from the cap on the sperm's head. Although it only takes one sperm to fertilize an egg, it takes the presence of many to release enough enzyme to clear a way to the egg plasma membrane. This is why a man who has healthy sperm but a low sperm count can be functionally infertile.

Enzymes from many sperm clear a passage to the oocyte's plasma membrane. Receptors in this membrane bind a sperm's plasma membrane, and the two membranes fuse, allowing the sperm into the oocyte. Usually only one sperm enters because its entry triggers changes that prevent other sperm from binding.

Sperm entry prompts the secondary oocyte to complete meiosis II, forming a mature egg, along with a second polar body **3**. The egg nucleus and sperm nucleus combine to form the genetic material of the zygote **4**. Sperm mitochondria and

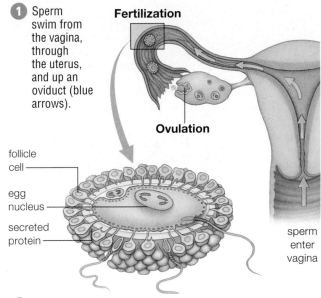

1 Sperm swim from the vagina, through the uterus, and up an oviduct (blue arrows).

follicle cell

egg nucleus

secreted protein

sperm enter vagina

2 Sperm surround a secondary oocyte and release enzymes that digest the protein layer that encloses it.

3 When a sperm enters the oocyte, the protein layer around the oocyte changes so other sperm cannot enter. Entry of a sperm stimulates completion of meiosis II in the oocyte's nucleus, producing a mature egg.

4 The sperm nucleus and egg nucleus combine to form the genetic material of the new individual.

nuclei fuse

Figure 26.14 Fertilization.

the flagellum enter the oocyte too, but they are typically broken down. Thus, mitochondria are inherited only from the mother.

Sometimes two eggs mature and are released at the same time, and each is fertilized by a different sperm. The result is fraternal twins. Such twins are no more similar than any other siblings.

From Cleavage to Implantation   Cleavage begins within a day of fertilization, as cilia propel the zygote through the oviduct toward the uterus (Figure 26.15). The zygote divides by mitosis to form two cells, which become four, which become eight, and so on ❶. Sometimes a cluster of four or eight cells splits in two and each cluster develops independently, resulting in identical twins: two individuals who have the same genotype. More typically, all of the cells adhere tightly to one another as the divisions continue.

By about 5 days after fertilization, the dividing cluster of cells has reached the uterus and formed a blastocyst, the mammalian version of the blastula ❷. The blastocyst consists of an outer layer of cells, a cavity filled with their fluid secretions, and an inner cell mass. Of 200 to 250 cells in the human blastocyst, only the thirty or so in the inner cell mass give rise to the embryo. The other cells give rise to the extraembryonic membranes.

Before a blastocyst can implant, it must escape from the protein coat that surrounds it. Once this coat has ruptured, the blastocyst attaches to the lining of the uterus and begins burrowing into it ❸.

As implantation continues, the inner cell mass develops into an embryonic disk with two flattened layers of cells ❹. At the same time, the extraembryonic membranes begin to form. A membrane called the **amnion** encloses a fluid-filled amniotic cavity between the embryonic disk and the blastocyst surface. Fluid in this cavity (amniotic fluid) acts as a buoyant cradle in which an embryo grows and moves freely, protected from temperature changes and mechanical impacts. As the amnion forms, other cells move around the inner wall of the blastocyst, forming the lining of a yolk sac. In humans, the yolk sac has no nutritional role.

As implantation continues, spaces in the uterine tissue around the blastocyst become filled with blood seeping in from ruptured maternal capillaries. An

Figure 26.15 Early embryonic development.
Graphics 3–5 depict a cross-section through the uterus.

❶ Days 1–4 Cleavage divides the zygote into many cells.

❷ Days 5–7 The blastocyst forms, expands, then sheds its protein coat.

❸ Days 8–9 Implantation begins. The blastocyst begins to burrow into the wall of the uterus.

❹ Days 10–11 The inner cell mass develops into an embryonic disk. Extraembryonic membranes form.

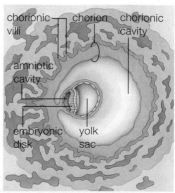

❺ Day 14 Chorionic villi (fingerlike extensions of the chorion) grow into blood-filled spaces in the lining of the uterus.

extraembryonic membrane called the **chorion** develops with many tiny fingerlike projections (chorionic villi) that extend into blood-filled maternal tissues ➎. The chorion will become part of the **placenta**, an organ that functions in exchanges of materials between a mother and her developing child. An outpouching from the yolk sac will become the fourth extraembryonic membrane, the allantois. In humans, the allantois forms blood vessels of the umbilical cord that connects the fetus to the placenta.

Implantation of a blastocyst prevents menstruation because the chorion secretes **human chorionic gonadotropin** (**HCG**). This hormone prevents degeneration of the corpus luteum. By the beginning of the third week, HCG can be detected in a mother's blood or urine. At-home pregnancy tests have a "dip stick" with a region that changes color when exposed to urine that contains HCG.

Embryonic and Fetal Development Gastrulation occurs on about day 15. Cells migrate inward along a depression that forms on the disk's surface ➏. Shortly thereafter, the embryonic disk develops two folds that will merge to form a neural tube, the precursor of the spinal cord and brain. Beneath the neural tube, mesoderm folds into another tube that develops into a notochord. The human notochord serves as the structural model for the developing backbone.

By the end of the third week, somites begin to appear on either side of the neural tube ➐. These paired segments of mesoderm will develop into the bones and skeletal muscles of the head and trunk.

The heart begins to beat a bit more than three weeks after fertilization. At first it is a linear tube of contractile cells that resembles a fish heart. The sound of its beating is still too faint to hear, even with a stethoscope. As development continues, this tubular heart bends back on itself, forming a heart with three chambers like that of an amphibian, and finally a heart with four chambers.

Bands of tissue called pharyngeal arches form at the onset of the fourth week ➑. These will later contribute to the pharynx, larynx, and the face, neck, mouth, and nose. In fishes, pharyngeal arches develop into gills, but a human embryo never has gills. It receives oxygen by way of the placenta.

At-home pregnancy tests detect the presence of HCG, a hormone produced early in pregnancy.

paired neural folds

future brain

pharyngeal arches

somites

➏ Day 16 Gastrulation has occurred and the neural tube is forming by the merging of two neural folds.

➐ Day 18–23 Bumps of mesoderm (somites) form. They will develop into muscle and bone of the head and trunk.

➑ Days 24–25 Pharyngeal arches appear. They will become structures in the head and neck.

amnion Extraembryonic membrane that encloses an amniote embryo and the amniotic fluid.

chorion Outermost extraembryonic membrane of amniotes; major component of the placenta in placental mammals.

human chorionic gonadotropin (**HCG**) Hormone produced by the chorion and later by the placenta; encourages maintenance of the uterus, thus sustaining a pregnancy.

placenta Of placental mammals, an organ composed of maternal and embryonic tissues that allows the exchange of substances between a developing individual and its mother.

Week 4

Weeks 5–6

yolk sac
connecting stalk
embryo

forebrain

future lens

pharyngeal
arches

developing heart

upper limb bud

somites

neural tube

lower limb
bud

tail

actual length

head growth exceeds
growth of other regions

retinal pigment

future external ear

upper limb differentiation
(hand plates develop, then
digital rays of future fingers;
wrist, elbow start forming)

umbilical cord formation
between weeks 4 and 8
(amnion expands, forms
tube that encloses the
connecting stalk and a
duct for blood vessels)

foot plate

actual length

Figure 26.16 Human embryonic and fetal development.

Top photos, © Lennart Nilsson/ Bonnierforlagen AB.

Figure It Out: What process eliminates the embryonic tail?

Answer: Apoptosis

When the fourth week ends, the embryo is 500 times its starting size, but it remains less than 1 centimeter long (Figure 26.16). A conspicuous tail accounts for about a sixth of the body's length. The pace of growth now slows as details of the embryo's organs begin to fill in. As development continues, limbs form. **Apoptosis**, a process by which cells self-destruct, sculpts individual digits from paddle-like extremities and shortens, then eliminates, the tail. Apoptosis is also responsible for the disappearance of a tadpole's tail during metamorphosis.

By the end of the eighth week, all organ systems have formed and we define the individual as a human **fetus**. A mother usually begins to feel reflexive fetal movements about five to six months after fertilization. Soft, fuzzy hair, the lanugo, covers

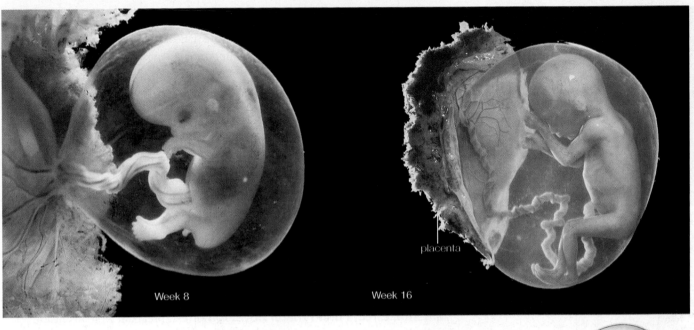

Week 8

Week 16

placenta

final week of embryonic period; embryo looks distinctly human compared to other vertebrate embryos

upper and lower limbs well formed; fingers and then toes have separated

primordial tissues of all internal, external structures now developed

tail has become stubby

actual length

Length: 16 centimeters
          (6.4 inches)
Weight: 200 grams
          (7 ounces)

WEEK 29
Length: 27.5 centimeters
          (11 inches)
Weight: 1,300 grams
          (46 ounces)

WEEK 38 (full term)
Length: 50 centimeters
          (20 inches)
Weight: 3,400 grams
          (7.5 pounds)

During fetal period, length measurement extends from crown to heel (for embryos, it is the longest measurable dimension, as from crown to rump).

the fetus. The fetal skin is wrinkled, reddish, and protected from abrasion by a thickened, cheeselike coating. Eyelids form and eyes open in the seventh month.

The optimal birthing time is about 38 weeks after fertilization. Technological advances have allowed us to keep premature infants alive at increasingly early stages. The survival of children born before 28 weeks (7 months) remains somewhat uncertain, mainly because the lungs have not fully developed. The prospects for survival increase after this. By 36 weeks, the survival rate is 95 percent. However, a fetus born between 36 and 38 weeks may still have some trouble breathing and maintaining a normal body temperature. Premature birth also increases the risk of a variety of birth defects.

apoptosis Self-destruction of a cell or cells.

fetus Developing human between 9 weeks after fertilization and birth.

A. Artist's depiction of the view inside the uterus, showing a fetus connected by an umbilical cord to the pancake-shaped placenta.

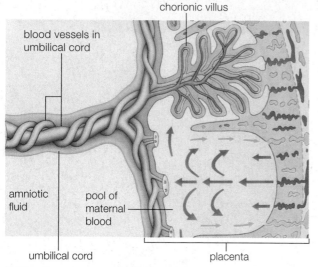

B. The placenta consists of maternal and fetal tissue. Fetal blood flowing in vessels of chorionic villi exchanges substances by diffusion with maternal blood around the villi. However, the bloodstreams do not mix.

Figure 26.17 Life-support system of a developing human.

Functions of the Placenta All exchanges of materials between an embryo and its mother take place by way of the placenta, a pancake-shaped, blood-engorged organ made of uterine lining and extraembryonic membranes (Figure 26.17).

The placenta begins forming early in pregnancy. By the third week, maternal blood has begun to pool in spaces in the lining of the uterus. Chorionic villi—the tiny fingerlike projections from the chorion—grow into the pools of maternal blood. Embryonic blood vessels extend through the umbilical cord to the placenta, and into the chorionic villi, where they are surrounded by the pooled maternal blood. The maternal and embryonic bloodstreams never mix. Instead, substances move between maternal and embryonic blood by diffusing across the walls of the embryonic vessels in the chorionic villi. Oxygen and nutrients diffuse from maternal blood into embryonic vessels in the villi. Wastes diffuse the other way, and the mother's body disposes of them.

The placenta also has a hormonal role. From the third month on, it produces large amounts of HCG, progesterone, and estrogens. These hormones encourage the ongoing maintenance of the uterine lining.

Maternal Effects on Prenatal Development A pregnant woman's health and behavior affect the development of the child she is carrying. For example, if a mother does not eat an adequate amount of iodine, her newborn may be affected by cretinism, a disorder that affects brain function and motor skills (Section 25.4). A maternal deficiency in folate (folic acid) puts her child at risk for neural tube defects. Such defects affect the brain, spine, or spinal cord.

Pathogens that cross the placenta can interfere with development. For example, infection by the rubella virus during the first eight weeks of pregnancy nearly always causes severe birth defects or miscarriage. A woman can avoid risks associated with rubella by getting vaccinated against the virus before pregnancy.

Alcohol crosses the placenta too. Alcohol use during pregnancy is the leading preventable cause of inborn mental impairment. Most doctors advise women who are pregnant or attempting to become pregnant to avoid alcohol.

Smoking cigarettes while pregnant interferes with delivery of oxygen across the placenta. In addition, levels of the addictive stimulant nicotine in amniotic fluid can be even higher than those in the mother's blood.

Environmental pollutants can also enter a woman's body and harm her child. Prenatal exposure to mercury impairs brain development, which is why pregnant women are advised to avoid eating large predatory fish such as albacore tuna. These fish often contain high levels of mercury.

Take-Home Message 26.6

How does a human develop?

- Cleavage produces a blastocyst that implants in the uterus. Projections from extra-embryonic membranes invade maternal tissue and form a placenta. Pathogens and toxins, as well as nutrients and gases, can cross the placenta.
- Gastrulation occurs two weeks after fertilization. A neural tube, notochord, and pharyngeal arches form later in the embryonic period.
- The fetal period begins at the end of the eighth week. Heartbeats are detected at about five months, and limb movements are felt at five to six months.

26.7 Birth and Milk Production

REMEMBER: Oxytocin and prolactin are pituitary hormones (Section 25.3).

Childbirth As a woman's pregnancy nears full term, her body prepares for childbirth. Her cervix, which has remained firm and has helped keep the fetus from slipping out of the uterus prematurely, becomes thinner, softer, and more flexible. This prepares it to stretch enough to allow the passage of the fetus.

The process of birth is known as labor, or parturition. Often a woman finds she is about to begin labor when "her water breaks." The amnion ruptures, and amniotic fluid drains out of the vagina. During labor, the cervix dilates and the fetus moves through it into the vagina, and then out of its mother's body (Figure 26.18).

Oxytocin stimulates smooth muscle contractions during labor. As the fetus nears full term, it usually becomes positioned so its head touches the cervix. Receptors in the cervix sense mechanical pressure and signal the hypothalamus, which calls for oxytocin secretion by the posterior pituitary. Binding of oxytocin to smooth muscle of the uterus increases the strength of contractions, which causes more mechanical pressure. This leads to more oxytocin secretion, and so on. The result is a positive feedback cycle—stretching of the cervix triggers more oxytocin secretion, which results in more stretching. This continues until the fetus has been expelled and there is no longer any mechanical pressure on the cervix.

Strong contractions help detach the placenta from the uterus and expel it as the afterbirth. Contractions also help to stop the bleeding where the placenta was attached to the wall of the uterus, by causing blood vessels at this ruptured attachment site to constrict. The umbilical cord is cut and tied off. A few days after shriveling up, the cord's stump has become the navel.

Nourishing the Newborn During pregnancy, prolactin secreted by the anterior pituitary triggers enlargement of mammary glands. After birth, an infant's sucking on a nipple triggers release of oxytocin. The oxytocin stimulates muscles around the milk glands to contract and force milk out through ducts. In addition to lactose and proteins, milk contains easily digested fats, vitamins and minerals, and enzymes that assist in digestion. Substances in milk encourage growth of beneficial bacteria and kill harmful ones. Milk also contains maternal antibodies that coat the lining of the newborn's throat and gut lining, lessening the risk of dangerous infections.

Women who are breast-feeding should keep in mind that toxins that enter their body may end up in their milk. Nicotine or alcohol in breast milk interferes with an infant's ability to sleep. Alcohol in milk may also interfere with a child's early motor development. Some viral pathogens, including HIV, are also transmitted in milk.

A. Fetus positioned for childbirth; its head is against the mother's cervix, which is dilating.

B. Muscle contractions stimulated by oxytocin force the fetus out through the vagina.

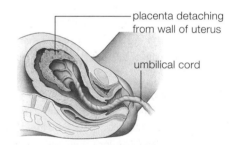

C. The placenta detaches from the wall of the uterus and is expelled.

Figure 26.18 Labor and childbirth.

Take-Home Message 26.7

What happens during birth and its aftermath?

- Hormonal changes prepare a woman's body to give birth and to nurse her newborn.
- During labor, oxytocin-stimulated contraction of uterine smooth muscle forces the fetus out of the body.
- Milk secreted by mammary glands provides nutrition and also has substances that protect the newborn from infection.

Summary

Section 26.1 *In vitro* **fertilization** has become a widely accepted solution for couples who have trouble conceiving naturally. Its success and acceptance opened the way for a variety of other assisted reproductive technologies.

Section 26.2 **Asexual reproduction** produces genetic copies of the parent. **Sexual reproduction** produces variable offspring. It is advantageous in places where conditions change from one generation to the next. Most animals reproduce sexually and have separate sexes, but some are **hermaphrodites**. Aquatic animals usually have external fertilization; they release gametes into the water. Most animals on land have internal fertilization, meaning gametes meet inside a female's body. In most animals, an egg's nutritious **yolk** supplies the energy and nutrients necessary for development.

Section 26.3 **Fertilization** unites an **egg** and a **sperm**, to produce a **zygote.** The zygote undergoes **cleavage**, which yields a **blastula**, a hollow, fluid-filled cluster of cells (right). During **gastrulation**, cells arrange themselves as a **gastrula** with three tissue layers: **ectoderm**, **mesoderm**, and **endoderm**. After gastrulation, the basic body plan is laid out and organ formation begins. In some animals, there is a larval stage, with a different body plan than the adult.

Section 26.4 **Ovaries**, a human female's gonads, produce eggs and sex hormones: **estrogens** and **progesterone**. A female is born with all the immature eggs (**oocytes**) she will ever have. After puberty, these eggs mature one at a time during an approximately monthly **menstrual cycle**. During this cycle, **follicle-stimulating hormone** (**FSH**) encourages maturation of an **ovarian follicle**. A primary oocyte completes meiosis I and undergoes unequal cytoplasmic division to form a large secondary oocyte and a tiny **polar body**. Near the midpoint of the menstrual cycle, a surge of **luteinizing hormone** (**LH**) triggers **ovulation**, which sends an oocyte down an **oviduct** toward the **uterus**. After ovulation, the remains of the ovarian follicle become a hormone-secreting **corpus luteum**. Hormones from the maturing follicle and the corpus luteum encourage thickening of the uterine lining. If pregnancy does not occur, the corpus luteum breaks down, and the uterine lining is shed during **menstruation**. Menstrual cycles cease permanently at **menopause**.

Testes hang beneath the pelvic girdle in a scrotum. Sperm form inside the **seminiferous tubules** of a testis, then enter an **epididymis**. Accessory glands such as the **seminal vesicles** and the **prostate gland**

produce other components of **semen**. These components are added as sperm pass through the **vasa deferentia**, which carry them to the urethra.

During intercourse, blood inflates and stiffens the penis. The male orgasm is usually accompanied by **ejaculation** of sperm into the **vagina**. Sperm must travel through the **cervix**, the entrance to the uterus, to reach the oviducts.

Section 26.5 Contraception refers to methods used to prevent pregnancy. Pregnancy can be avoided by abstaining from intercourse. The chance of pregnancy can be reduced by abstaining during a woman's fertile period, surgically severing reproductive ducts, placing a physical or chemical barrier at the entrance to the uterus, or administering synthetic female sex hormones to prevent ovulation.

Infertility can arise as a result of hormone disorders that affect sperm or egg production, blocked reproductive ducts, lack of sperm motility, or uterine problems that interfere with embryonic implantation or development.

Sexually transmitted diseases (STDs) are caused by protozoal, bacterial, and viral pathogens and are spread by unsafe sex. If untreated, STDs can cause sterility and some can be fatal.

Section 26.6 Fertilization usually occurs in an oviduct. Enzymes from many sperm digest the protein layer around the oocyte, allowing one sperm to bind to the oocyte plasma membrane. This sperm enters the oocyte and causes it to complete meiosis II. The sperm nucleus and egg nucleus unite. Cleavage produces a blastocyst that burrows into the uterus and begins producing a hormone (**human chorionic gonadotropin**) that prevents ovulation. The **amnion** encloses the developing embryo and the amniotic fluid. Another membrane, the **chorion,** combines with maternal tissue to form a **placenta** that passes oxygen and nutrients between mother and embryo.

Gastrulation occurs after the embryo has implanted. A neural tube forms (it will give rise to the brain and spinal cord), then somites appear. Pharyngeal arches develop, but the embryo never has functional gills.

By the end of the eighth week, apoptosis has eliminated the tail of the **fetus**, and all organ systems have formed.

Section 26.7 Hormones prepare a woman's body for pregnancy, labor, and nursing. During labor, oxytocin-induced uterine contractions expel the fetus and placenta. Milk secreted by mammary glands supplies nutrients and also contains antibodies that help a newborn resist infection. Prolactin stimulates milk production.

Self-Quiz

Answers in Appendix I

1. A(n) _____ produces both eggs and sperm.
   a. hermaphrodite   c. ovary
   b. asexual organism   d. testis

2. The end product of cleavage is a(n) _____ .
   a. blastula   b. gastrula   c. embryo   d. fetus

3. A(n) _____ undergoes gastrulation.
   a. gastrula   c. somite
   b. blastula   d. neural tube

4. Meiotic divisions of germ cells in the _____ give rise to sperm.
   a. seminiferous tubules   c. penis
   b. prostate gland   d. seminal vesicles

5. A male has an erection when _____ .
   a. muscles running the length of the penis contract
   b. germ cells release a surge of testosterone
   c. the posterior pituitary releases oxytocin
   d. spongy tissue inside the penis fills with blood

6. In humans, fertilization usually occurs in _____ .
   a. the uterus   b. the vagina   c. an oviduct   d. an ovary

7. During a menstrual cycle, a midcycle surge of _____ secreted by the pituitary triggers ovulation.
   a. estrogen   b. progesterone   c. LH   d. FSH

8. After ovulation, the corpus luteum secretes _____ .
   a. luteinizing hormone   c. progesterone
   b. follicle-stimulating hormone d. prolactin

9. A _____ implants in the lining of the human uterus.
   a. zygote   b. blastocyst   c. gastrula   d. fetus

10. Human milk contains _____ .
    a. antibodies   c. fats and proteins
    b. lactose   d. all of the above

11. During labor, secretion of oxytocin causes _____ .
    a. uterine contractions   c. ovulation
    b. rupture of the amnion   d. production of milk proteins

12. Birth control pills deliver synthetic _____ .
    a. estrogens and progesterone   c. follicle-stimulating hormone
    b. testosterone   d. luteinizing hormone

13. Number these events in human development correctly (1 to 6).
    _____ gastrulation
    _____ blastocyst forms
    _____ zygote forms
    _____ tail disappears
    _____ neural tube forms
    _____ implantation occurs

14. Which of the following STDs are caused by bacteria?
    a. chlamydia   d. trichomoniasis
    b. gonorrhea   e. a and b
    c. genital warts   f. all of the above

15. Match each human reproductive structure with its description.
    _____ testis   a. maternal and fetal tissues
    _____ vas deferens   b. adds fluid to sperm in semen
    _____ placenta   c. produces testosterone
    _____ vagina   d. produces estrogen and
    _____ ovary   progesterone
    _____ oviduct   e. conveys egg to uterus
    _____ prostate gland f. secretes milk
    _____ mammary gland   g. birth canal
                          h. conveys sperm toward urethra

Critical Thinking

1. Of all birds, the flightless kiwi has the highest proportion of yolk in its eggs and the longest incubation time. Explain why birds with the longest incubation times have the yolkiest eggs.

2. Fraternal twins are nonidentical siblings that form when two eggs are released and fertilized at the same time. Explain why an increased level of FSH raises the likelihood of fraternal twins.

3. Drinking alcohol decreases oxytocin production. What effect would you expect drinking to have on a woman's ability to nurse her infant?

Visual Question

1. The graphic below shows how the levels of two ovarian hormones change over the course of the female reproductive cycle. Which hormone does each color represent? (Day 1 is the start of the cycle, the onset of menstruation.)

Day 1   Day 28

27

PLANT FORM AND FUNCTION

Application ❯ 27.1 Leafy Cleanup Crews

REMEMBER: Roots of a vascular plant take up water and dissolved minerals from soil (Section 14.2).

From 1940 until the 1970s, the United States Army tested and disposed of weapons at J-Field, Aberdeen Proving Ground in Maryland. Obsolete chemical weapons and explosives were burned in open pits together with plastics, solvents, and other wastes. The soil became heavily contaminated. Toxic metals (such as lead, arsenic, and mercury) and organic compounds (including trichloroethylene, or TCE) ended up in groundwater seeping toward nearby marshes and the Chesapeake Bay. Concerned about the toxic seepage, the Army and EPA (Environmental Protection Agency) planted groves of fast-growing poplar trees at J-Field in 1996. Like other vascular plants, poplar trees have roots that take up water from the soil. Along with the water come nutrients and chemical contaminants, including TCE. Although TCE is toxic to animals, it does not harm plants. The poplars break down some of the toxin, but they release most of it into the atmosphere. Airborne TCE is the lesser of two evils: It breaks down much more quickly in air than it does in groundwater.

Using plants specifically to take up and concentrate or degrade environmental contaminants is called phytoremediation, and the practice is working splendidly at J-Field (Figure 27.1A). By 2001, the poplar trees had removed 60 pounds of TCE from the soil and groundwater. In addition, they had helped foster communities of microorganisms that were also breaking down toxins. Researchers estimate that the trees will have removed the majority of contaminants from J-Field by 2030.

Compared to other methods of cleaning up toxic waste sites, phytoremediation is usually less expensive—and it is more appealing to neighbors. Its success at J-Field has spurred similar cleanup strategies at other sites, including Ford Motor Company's Rouge Center in Dearborn, Michigan, where decades of steelmaking left soil contaminated with highly carcinogenic compounds. In 2000, the automaker funded development of a phytoremediation system that also boosted an initiative to restore the facility's native wildlife habitat (Figure 27.1B).

A. Poplar trees are removing toxic waste from groundwater at J-Field. Chemical weapons and explosives that were burned and dumped at the site (inset) heavily contaminated the soil.

B. This 10-acre green roof cleans contaminated storm water as it moves across Ford Motor Company's Rouge Center in Dearborn, Michigan.

Figure 27.1 Phytoremediation two ways.

(A) left, OPSEC Control Number #4 077-A-4; right, Billy Wrobel, 2004; (B) Reuters/Corbis.

Digging Into Data

TCE Uptake by Transgenic Plants

Researchers are designing transgenic plants with enhanced utility for phytoremediation. In 2007, Sharon Doty and her colleagues used *A. tumefaciens* (Section 10.4) to deliver a mammalian gene into poplar plants. The gene encodes cytochrome P450, an enzyme involved in the breakdown of many organic molecules, including TCE. Some of their results are shown in Figure 27.2.

1. On day 6, what was the difference between the TCE content of air around transgenic plants and that around vector control plants?

2. Assuming no other experiments were done, what two explanations are there for the results of this experiment? What other control might the researchers have used?

Figure 27.2 TCE uptake by transgenic poplar plants. Potted plants were kept in separate sealed containers with an initial 15,000 micrograms (µg) of TCE per cubic meter of air. Samples of the air in the containers were taken daily and measured for TCE content. Controls included a plant transgenic for a *Ti* plasmid with no inserted gene (vector control), and a bare-root transgenic plant (one that was not planted in soil).

27.2 Tissues in a Plant Body

REMEMBER: Cells of multicelled organisms are organized as tissues, organs, and organ systems (Section 1.2). Cellulose, a polysaccharide, is the main structural component of plants (2.7). Plant cells have walls; a cuticle of waxes and other hydrophobic compounds helps stems and leaves fend off insects and retain water (3.5). Most leaves and stems are studded with tiny, closable gaps called stomata (5.4). The evolution of lignin-stiffened tissue allowed vascular plants to branch and stand tall on land (14.2). Eudicots and monocots are the major angiosperm groups (14.7).

With more than 260,000 species (and counting), flowering plants dominate the plant kingdom. In this chapter, we focus mainly on the structure of eudicots and monocots. Like cells of most other multicelled organisms, those in plants are organized as tissues, organs, and organ systems. A vascular plant's body has two organ systems: shoots and roots (Figure 27.3). Stems, leaves, and reproductive organs such as flowers are shoots. Most shoots grow above the ground and most roots grow below it, but there are many exceptions to this general rule.

All plant parts are composed of ground, vascular, and dermal tissues. **Ground tissues** make up most of the soft internal parts of a plant, and include cells specialized for photosynthesis and for storage. **Vascular tissues**, which distribute water and nutrients to all parts of the plant body, have pipelines that thread through ground tissue. **Dermal tissues** cover and protect the plant's exposed surfaces.

Simple tissues have one cell type. There are three simple tissues: parenchyma, collenchyma, and sclerenchyma, each named after the cell type that makes it up (Table 27.1). Dermal and vascular tissues are complex tissues, which means they

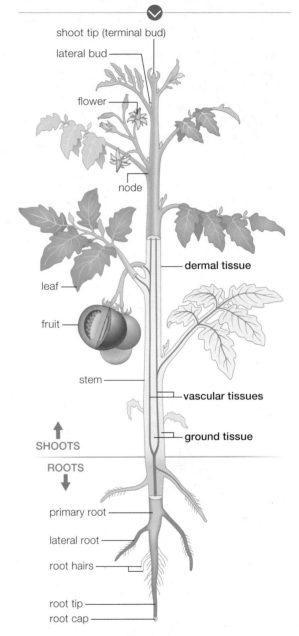

Figure 27.3 Body plan of a tomato plant.

shoot tip (terminal bud)
lateral bud
flower
node
dermal tissue
leaf
fruit
stem
vascular tissues
ground tissue
SHOOTS
ROOTS
primary root
lateral root
root hairs
root tip
root cap

Table 27.1 Simple and Complex Plant Tissues

Tissue Type	Main Components	Main Functions
Simple Tissues		
Parenchyma	Parenchyma cells	Photosynthesis, storage, secretion, tissue repair
Collenchyma	Collenchyma cells	Pliable structural support
Sclerenchyma	Sclerenchyma cells (fibers or sclereids)	Structural support
Complex Tissues		
Dermal		
Epidermis	Epidermal cells and their secretions	Secretion of cuticle; protection; control of gas exchange and water loss
Periderm	Cork cambium; cork cells; parenchyma	Forms protective cover on older stems and roots
Vascular		
Xylem	Tracheids, vessel elements; parenchyma cells; sclerenchyma cells	Water-conducting tubes; structural support
Phloem	Sieve elements, parenchyma cells; sclerenchyma cells	Sugar-conducting tubes and their supporting cells

dermal tissues Tissues that cover and protect the plant body.

ground tissues Tissues that make up most of the soft internal parts of the plant body.

vascular tissues Tissues that distribute water and nutrients through a vascular plant body.

collenchyma In plants, simple tissue composed of living cells with unevenly thickened walls; provides flexible support.

cotyledon Embryonic leaf of a flowering plant; also called a seed leaf.

epidermis Dermal tissue; outermost layer of a young plant.

mesophyll Photosynthetic parenchyma.

parenchyma In plants, simple tissue composed of living cells with functions that depend on location.

phloem Complex vascular tissue of plants; its living sieve elements compose sieve tubes that distribute sugars and other organic solutes.

sclerenchyma In plants, simple tissue composed of cells that die when mature; their tough walls structurally support plant parts. Includes fibers, sclereids.

xylem Complex vascular tissue of plants; its dead tracheids and vessel elements form tubes that distribute water and mineral ions.

consist of two or more cell types. Some plant cell types normally die at maturity. The rigid cell walls that remain after these cells die lend structural or functional support to the mature tissue. Figure 27.4 shows some locations of simple and complex tissues in a stem.

Parenchyma is a simple tissue that consists of parenchyma cells. The shape of these cells varies with their function, but all have a thin, flexible cell wall. Parenchyma has specialized roles that vary with its location in the plant. In stems and roots, for example, parenchyma cells store proteins, water, oils, and starch (the photo above left shows starch-packed organelles in parenchyma cells composing the ground tissue of a buttercup root). Chloroplast-containing parenchyma (below left) forms a special tissue called **mesophyll** that carries out photosynthesis. Other specialized functions of parenchyma include structural support and nectar secretion.

Collenchyma (left) is a simple tissue that supports rapidly growing plant parts such as young stems and leaf stalks. Collenchyma cells stay alive in mature tissue. A complex polysaccharide called pectin imparts flexibility to these cells' primary wall, which is thickened unevenly where three or more of the cells abut one another.

Variably shaped cells of **sclerenchyma** die when they are mature. Their thick cell walls, which contain a high proportion of durable compounds such as cellulose and lignin, lend sturdiness to plant parts and help them resist stretching and compression. Fibers are long, tapered sclerenchyma cells; they occur in bundles that support and protect vascular tissues in stems and leaves (the photo above left shows bundled fibers in vascular tissue of a sunflower stem). Fibers of some plants are used to make cloth, rope, paper, and other commercial products. Sclereids (below left) are sclerenchyma cells that strengthen hard seed coats, and they also make pear flesh gritty.

The first dermal tissue to form on a plant is **epidermis**, which in most species consists of a single layer of epidermal cells on the plant's outer surface. These cells secrete a waterproof, protective cuticle on their outward-facing cell walls. Hairs and other outgrowths of epidermal cells are also common (the photo above left shows some of these structures on the surface of a coleus leaf). Epidermis of leaves and young stems includes specialized cells such as those that form stomata. Plants close these tiny gaps to limit water loss from internal tissues. In older stems and roots, a complex dermal tissue called periderm replaces epidermis. The photo below left shows three types of cells making up periderm on a stem of elder.

Pipelines of complex vascular tissues distribute water and nutrients through all parts of the plant body. The two types of vascular tissue, xylem and phloem, are typically bundled together with sclerenchyma fibers in young plant parts. Both vascular tissues are composed of elongated conducting tubes. **Xylem**, the vascular tissue that conducts water and minerals dissolved in it, consists of two types of cells: vessel elements (Figure 27.5A) and tracheids (Figure 27.5B). Vessel elements and tracheids

parenchyma

collenchyma

sclerenchyma
(fibers)

xylem

phloem

epidermis

Figure 27.4 Locations of some tissues making up the stem of a eudicot (*Clematis*).

Dr. Keith Wheeler/Science Source.

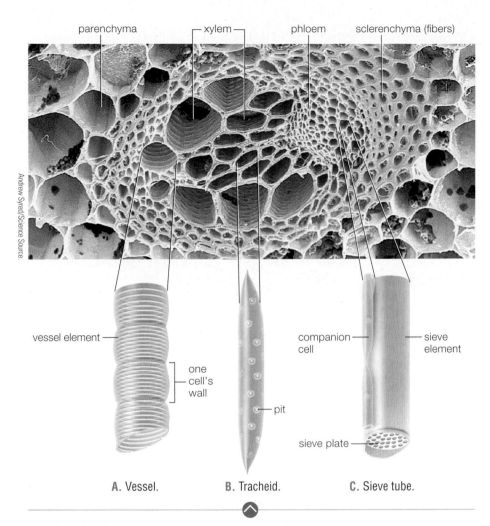

parenchyma xylem phloem sclerenchyma (fibers)

Andrew Syred/Science Source

vessel element

one cell's wall

pit

companion cell — sieve element

sieve plate

A. Vessel. **B.** Tracheid. **C.** Sieve tube.

Figure 27.5 Cells that make up conducting tubes of plant vascular tissues.

Forestry and Forest Products Research Institute, Japan.

are dead in mature tissue, but their stiff, waterproof walls remain to form interconnecting tubes that thread through the plant (left). The tubes conduct water, and their sturdy walls also lend structural support to plant parts. Pits in the walls allow water to move laterally between the tubes as well as vertically through them.

Phloem, the vascular tissue that conducts sugars and other organic solutes, consists of cells called sieve elements and their associated companion cells. Both types of cells are alive in mature tissue. Sieve elements connect end to end at perforated sieve plates (left), forming sieve tubes that conduct sugars to all parts of the plant (Figure 27.5C). Companion cells are a type of parenchyma. A companion cell provides each sieve element with metabolic support, and also transfers sugars into it.

Spike Walker/ Wellcome Images.

Eudicots and Monocots Monocots and eudicots have the same types of cells and tissues composing their parts, but the two lineages differ in many aspects of their organization and hence in the details of their structure (Figure 27.6). The names of the groups refer to one such difference, the number of seed leaves, or **cotyledons**, in their embryos. Monocot embryos have a single cotyledon; eudicot

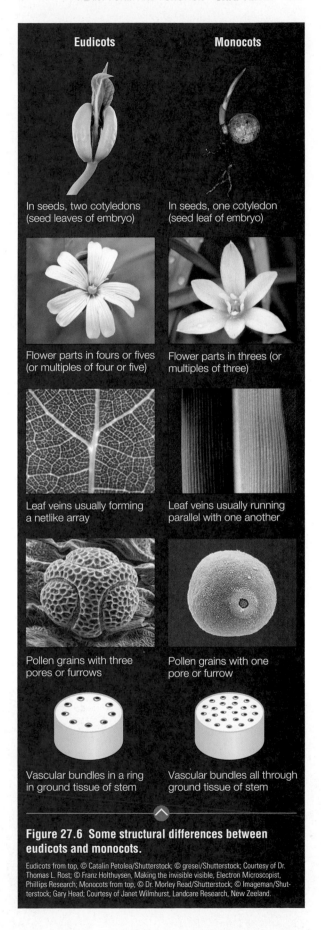

Eudicots | **Monocots**

In seeds, two cotyledons (seed leaves of embryo) | In seeds, one cotyledon (seed leaf of embryo)

Flower parts in fours or fives (or multiples of four or five) | Flower parts in threes (or multiples of three)

Leaf veins usually forming a netlike array | Leaf veins usually running parallel with one another

Pollen grains with three pores or furrows | Pollen grains with one pore or furrow

Vascular bundles in a ring in ground tissue of stem | Vascular bundles all through ground tissue of stem

Figure 27.6 Some structural differences between eudicots and monocots.

node A region of stem where new shoots and roots can form.

vascular bundle Multistranded bundle of xylem, phloem, and sclerenchyma fibers running through a stem or leaf.

embryos have two (Chapter 28 returns to development in plants). You will see in the next section how the tissue organization inside shoots and roots also differs among eudicots and monocots.

Take-Home Message 27.2

What is the basic structure of a flowering plant?

• Flowering plants typically have aboveground shoots and belowground roots. Both consist of ground, vascular, and dermal tissue systems.

• A plant's soft internal parts consist mainly of ground tissue. Vascular tissues that thread through ground tissue distribute water and nutrients. Dermal tissues cover and protect plant surfaces.

• Eudicots and monocots have the same types of tissues, but differ somewhat in their pattern of tissue organization.

27.3 Stems, Leaves, and Roots

REMEMBER: Leaves and stems are studded with stomata; when stomata close to conserve water on dry days, gas exchange comes to a halt (Section 5.4). Nitrogen-fixing *Rhizobium* bacteria live inside the roots of some plants (13.5). A mycorrhiza is a mutually beneficial relationship between a fungus and plant root cells (14.9).

Stems Stems form the basic structure for growth of a flowering plant, providing support and keeping leaves positioned for photosynthesis. They can grow above or below the soil, and in many species they are specialized for storage or asexual reproduction. Stems characteristically have **nodes**, which are regions that can give rise to new shoots or roots. Inside stems, xylem and phloem are organized in long, multistranded **vascular bundles**. The main function of vascular bundles is to conduct water, ions, and nutrients between different parts of the plant. Some components of the bundles—fibers and the lignin-reinforced walls of tracheids—also play an important role in supporting upright stems of land plants.

Vascular bundles extend through the ground tissue of all stems and leaves, but the arrangement of the bundles inside these plant parts differs between monocots and eudicots. The vascular bundles of monocot stems are typically distributed all throughout the ground tissue (Figure 27.7A). By contrast, the vascular bundles inside a typical eudicot stem are arranged in a characteristic ring (Figure 27.7B). The ring divides the stem's ground tissue into distinct regions of pith (inside the ring) and cortex (outside the ring).

Many plants have modified stem structures that function in storage and reproduction. Stolons are stems that branch from the main stem of the plant and grow horizontally on the ground or just under it. They are commonly called runners because in many plants they "run" along the surface of the soil. Stolons may look like roots, but they have nodes (roots do not have nodes). Roots and shoots that sprout from the nodes develop into new plants. The photo on the right shows new plants sprouting from nodes on strawberry plant stolons.

BJI/Blue Jean Images/Getty Images.

vilax/Shutterstock.com.

538

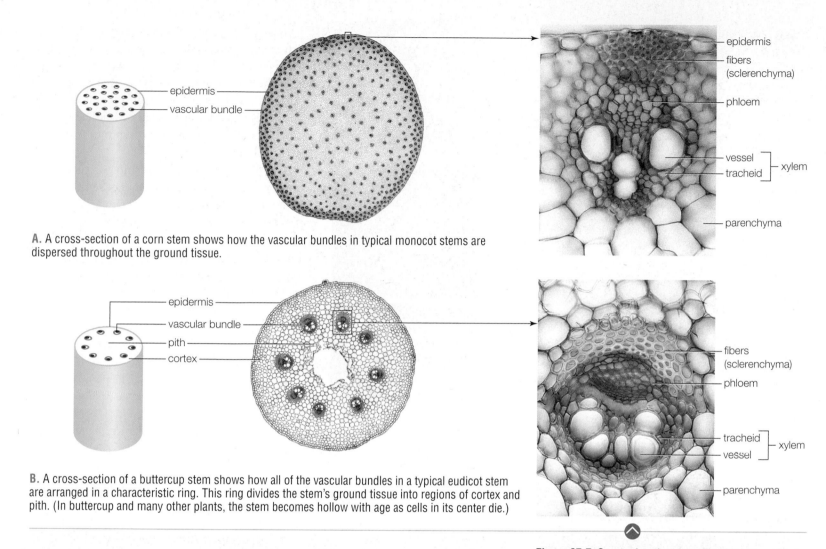

epidermis
vascular bundle

epidermis
fibers (sclerenchyma)
phloem
vessel
tracheid } xylem
parenchyma

A. A cross-section of a corn stem shows how the vascular bundles in typical monocot stems are dispersed throughout the ground tissue.

epidermis
vascular bundle
pith
cortex

fibers (sclerenchyma)
phloem
tracheid } xylem
vessel
parenchyma

B. A cross-section of a buttercup stem shows how all of the vascular bundles in a typical eudicot stem are arranged in a characteristic ring. This ring divides the stem's ground tissue into regions of cortex and pith. (In buttercup and many other plants, the stem becomes hollow with age as cells in its center die.)

Figure 27.7 Comparing the structure of a young shoot from A a monocot and B a eudicot.
Cells appear different colors in the sections because they have been stained with different dyes that bind to certain carbohydrates and lignin.

(A) middle, Dr. Keith Wheeler/Science Source; right, © Herve Conge/© ISM/Phototake; (B) middle and right, © ISM/Phototake.

A bulb is a short section of underground stem encased by overlapping layers of thickened, modified leaves called scales. The photo (left) shows clearly visible scales surrounding the stem at the center of an onion, which is the bulb of an *Allium cepa* plant. The scales develop from a basal plate at the base of the bulb, as do roots. Bulb scales contain starch and other substances that a plant holds in reserve during times when conditions in the environment are unfavorable for growth. When favorable conditions return, the plant then uses these stored substances to sustain rapid growth. The dry, paperlike outer scale of many bulbs serves as a protective covering.

A corm is a short, thickened underground stem. A corm is like a bulb in that it stores nutrients during times when conditions in the environment are unfavorable for growth, and roots grow from its basal plate. Unlike a bulb, however, a corm is solid rather than layered, and it has nodes. The photo on the left shows corms of crocus plants. New roots are developing from their basal plates, and new shoots are sprouting from their nodes.

leaf vein A vascular bundle in a leaf.

Ginger, irises, and some grasses have rhizomes, which are fleshy stems that typically grow under the soil and parallel to its surface. A rhizome is the main stem of the plant, and it also serves as the plant's primary storage tissue. Shoots that sprout from nodes grow aboveground for photosynthesis and flowering. The photo on the left shows shoots sprouting from rhizomes of turmeric plants.

Stem tubers are thick, fleshy storage structures that form on the stolons or rhizomes of some plant species. Most are underground and temporary. Like corms, stem tubers have nodes from which new shoots and roots can sprout. Unlike corms, they do not have a basal plate. The photo on the left shows how potatoes, which are stem tubers, grow on stolons. The "eyes" of a potato are nodes.

Many types of cacti and other succulents have cladodes, which are flattened, photosynthetic stems specialized to store water. New plants can form at nodes. The cladodes of some plants appear leaflike, but most are unmistakably fleshy. The photo (left) shows a spiky cladode of a prickly pear plant.

Leaves Leaves are the main organs of photosynthesis in most flowering plant species. They also function in gas exchange, and they are the major site of evaporative water loss. Typical leaves are thin, with a high surface-to-volume ratio. Leaves of most monocots are also long and narrow, with the base of the leaf wrapping around the stem and forming a sheath around it (Figure 27.8A). In most eudicots, a short

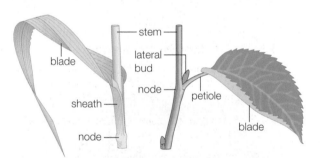

A. Typical leaves of monocots (left) and eudicots (right).

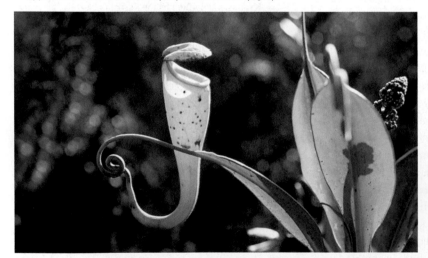

B. Example of specialized leaf form. Carnivorous plants of the genus *Nepenthes* grow in nitrogen-poor soil. They secrete acids and protein-digesting enzymes into fluid in a cup-shaped modified leaf. The enzymes release nitrogen from small animals (such as insects, frogs, and mice) that are attracted to odors from the fluid and then drown in it.

C. Example of leaf surface specialization: trichomes, which are outgrowths of epidermal cells. Glandular types such as the ones on this marijuana leaf secrete substances that deter plant-eating animals. Marijuana trichomes produce a chemical (tetrahydrocannabinol, or THC) that has a psychoactive effect in humans.

Figure 27.8 Leaf structure.

(B) Perennou Nuridsany/Science Source; (C) © Dan Legere Photography/www.danlegere.daportfolio.com.

1 epidermis

2 palisade mesophyll

3 spongy mesophyll

xylem

4 vascular tissue

phloem

1 epidermis

5 stomata

stalk called a petiole attaches the leaf to a stem. Eudicot leaves vary widely in their structure and some are highly specialized (Figure 27.8B).

Figure 27.9 illustrates the internal structure of a typical eudicot leaf. The bulk of the leaf consists of mesophyll—photosynthetic parenchyma—that lies between an upper and lower layer of epidermis ❶. Eudicot leaves are typically oriented perpendicular to the sun's rays, and have two layers of mesophyll. The uppermost layer is palisade mesophyll ❷. Cells in this layer have more chloroplasts than cells of the spongy mesophyll layer below. Cells of spongy mesophyll are more rounded than those of palisade mesophyll, and have larger air spaces between them ❸.

The vascular bundles of leaves are called **leaf veins**. Inside each vein, strands of xylem transport water and dissolved mineral ions to photosynthetic cells, while strands of phloem transport products of photosynthesis (sugars) away from them ❹. Layers of sclerenchyma around the vascular tissue stiffen a vein, and also provide support for the leaf's softer tissues. In most eudicot leaves, large veins branch into a network of minor veins.

The upper and lower surfaces of a leaf are sheets of epidermis one cell thick. Either or both surfaces may be smooth, sticky, or slimy, and have epidermal cell outgrowths such as hairs, scales, spikes, or hooks (Figure 27.8C). Epidermal cells secrete a translucent, waxy cuticle that slows water loss. A leaf's upper surface, which typically receives the most direct sunlight, may have a thicker cuticle than the lower surface, which tends to be shaded. The lower surface usually has more stomata ❺. Carbon dioxide needed for photosynthesis enters a leaf through open stomata, then diffuses through air spaces to mesophyll cells. Oxygen released by photosynthesis diffuses in the opposite direction.

Monocot and eudicot leaves have a somewhat different internal structure. Blades of grass and other monocot leaves that grow vertically can intercept light from all directions. Such leaves typically have a single layer of mesophyll. Unlike a eudicot's branching veins, monocot veins are similar in length and run parallel with the leaf's long axis (as shown in Figure 27.6).

Figure 27.9 Internal structure of a leaf.

The illustration above shows the organization of a typical eudicot leaf, with two layers of mesophyll. The micrograph below shows these layers in a barberry leaf.

❶ The leaf surface is epidermis with a secreted layer of cuticle.

❷ Palisade mesophyll has elongated cells.

❸ Spongy mesophyll has rounded cells.

❹ Inside leaf veins, vascular bundles of xylem (blue) and phloem (pink) transport materials to and from photosynthetic cells.

❺ Gas exchanges between air inside and outside of the leaf occur at stomata.

Last Refuge, Ltd./Phototake.

Figure It Out: What structure in this micrograph is the arrow pointing to?

Answer: A stoma

A. Fibrous root system.   B. Taproot system.

Figure 27.10 Types of root systems.
Typical monocots have a fibrous root system with many
adventitious roots; typical eudicots have a taproot system
with one main root.

(A) Don Nichols/Getty Images; (B) Caitlin Winner/Getty Images.

The roots of a
typical plant are at
least as extensive
as its shoots.

**Figure 27.11 Root hairs and a nodule on the young
root of an alfalfa plant.**
Both structures help the plant take up nutrients.

Ninjatacoshell.

Roots The roots of a typical plant are as extensive as its shoots, and often more
so. Roots branch from existing roots. They can also form on stems, leaves, or other
structures, in which case they are called adventitious roots.

When a seed germinates, the first structure to emerge from it is a root. In
typical monocots, this primary root is quickly replaced with a mat of adventitious
roots that arise from nodes on the developing stem. All of the roots in the resulting
fibrous root system are about equal in diameter—there is no dominant, central root
(Figure 27.10A). Roots in a fibrous system may or may not be branched. In typical eudicots, the primary root that emerges from a seed thickens and lengthens to
become a taproot. A taproot can give rise to other, lateral-branching roots, but even
so it remains the largest, dominant root. A eudicot taproot together with its smaller
lateral root branchings constitutes a taproot system (Figure 27.10B). In some eudicots, the taproot is eventually replaced by adventitious roots, so a taproot system
becomes a fibrous root system.

Roots anchor a plant in soil and often store organic molecules such as starch,
but their main function is to take up water—and the nutrients dissolved in it—from
soil. Sixteen elements are required for growth and maintenance of a plant body,
and only three of them (carbon, oxygen, and hydrogen) are abundantly available
in carbon dioxide and water molecules. The remaining elements occur in soil. Soil
consists mainly of mineral particles—sand, silt, and clay—that form by the weathering of rocks. Clay particles have a negative charge that attracts positively charged
mineral ions in soil water. Thus, clay-rich soil retains dissolved nutrients that might
otherwise trickle past roots too quickly to be absorbed. Larger sand and silt particles
intervene between tiny particles of clay. Soils with too much clay pack so tightly that
they exclude air—and the oxygen in it (cells in a plant's roots, like cells in its shoots,
require oxygen for aerobic respiration).

Soil also contains **humus**, which is decomposing organic material such as fallen
leaves and feces. Humus releases nutrients, and its negatively charged organic acids
can trap positively charged mineral ions in soil water. Humus also swells and shrinks
as it absorbs and releases water, and these changes in size open spaces for air to penetrate. Humus tends to accumulate in waterlogged soils because water excludes air,
and organic matter breaks down much more slowly in anaerobic conditions. Soils in
swamps, bogs, and other perpetually wet areas often contain more than 90 percent
humus. Very few types of plants can grow in these soils.

The dermal tissue on the surface of roots is a plant's absorptive interface with
soil. Epidermal cells on young roots often send out very thin extensions called **root
hairs** (Figure 27.11). These structures are tiny, but a huge number of them form.
Collectively, they greatly increase the root's surface area, thus maximizing its ability to absorb soil water. Most flowering plants also maximize their ability to take
up nutrients by taking part in mutualisms. In mycorrhizae, filaments of a fungus
(hyphae) form a velvety cloak around roots or penetrate their cells. The fungus takes
up some sugars and nitrogen-rich compounds from root cells. In return, the root
cells get some scarce minerals that the fungus is better able to absorb.

Legumes and some other plants can form mutualisms with nitrogen-fixing
Rhizobium bacteria. The roots of these plants release certain compounds into the
soil that are recognized by compatible nitrogen-fixing bacteria. The bacteria respond
by releasing signaling molecules that, in turn, trigger the roots to grow around and
encapsulate them inside swellings called **root nodules** (one is pictured in Figure
27.11). The association is beneficial to both parties. The plants require a lot of nitrogen, but cannot use nitrogen gas that is abundant in air. The bacteria in root nodules
convert this gas to ammonia, which is a form of nitrogen that the plant can use. In

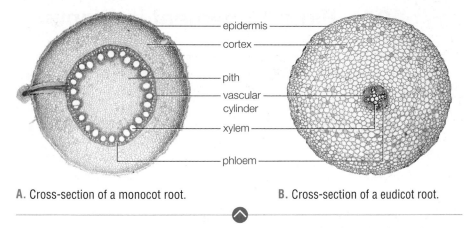

A. Cross-section of a monocot root.   B. Cross-section of a eudicot root.

Figure 27.12 Comparing the arrangement of tissues in monocot and eudicot roots.

(A) Michael Clayton/University of Wisconsin, Department of Botany; (B) © Dr. Brad Mogen/Visuals Unlimited Inc.

return for this valuable nutrient, the plant provides an oxygen-free environment for the anaerobic bacteria, and shares its photosynthetically produced sugars with them.

After soil water crosses a root's epidermis, it seeps into parenchyma that forms the root's region of cortex. The water moves from cell to cell of the cortex until it reaches the **vascular cylinder**, a column of vascular tissue running lengthwise through the center of the root. In a typical monocot root (Figure 27.12A), the vascular cylinder divides ground tissue into regions of pith (inside the ring) and cortex (outside the ring). By contrast, the vascular cylinder in a typical eudicot root contains very little pith or none at all. Most ground tissue in a eudicot root is cortex, and the vascular cylinder is mostly vascular tissue (Figure 27.12B).

In both monocots and eudicots, the outer boundary of the vascular cylinder is a sheet of endodermis, a single layer of cells that separates the root cortex from the vascular cylinder (Figure 27.13). Cells that make up endodermis secrete a waxy substance into their walls wherever they abut. This substance forms a waterproof band that prevents water from diffusing through endodermal cell walls into the vascular cylinder. Thus, water (and substances dissolved in it) can enter a vascular cylinder only by passing through the cytoplasm of an endodermal cell. The presence of endodermis allows a plant to regulate which solutes enter xylem for distribution to the rest of the plant body.

Figure 27.13 Root structure.

Figure It Out: Is this an illustration of a monocot root or a eudicot root?   **Answer:** Eudicot

Take-Home Message 27.3

How are plant tissues organized in shoots and roots?

- The arrangement of vascular bundles, which are multistranded bundles of vascular tissue, differs between eudicot and monocot stems. Many plants have modified stems that function in storage, reproduction, or both.
- Leaves are structurally adapted to intercept sunlight and exchange gases for photosynthesis. Components include veins, mesophyll, and cuticle-secreting epidermis.
- Root hairs and mutualisms with microorganisms maximize a root's ability to take up water and nutrients from soil.
- Water and substances dissolved in it must pass through endodermis of a root's vascular cylinder before entering xylem for distribution to the rest of the plant.

humus Decaying organic matter in soil.

root hairs Hairlike, absorptive extensions of a root epidermal cell; form on young roots.

root nodules On some plant roots, swellings that contain mutualistic nitrogen-fixing bacteria.

vascular cylinder Of a root, central column that contains vascular tissue and other supporting cells.

Figure 27.14 Cohesion–tension theory of water transport in vascular plants.

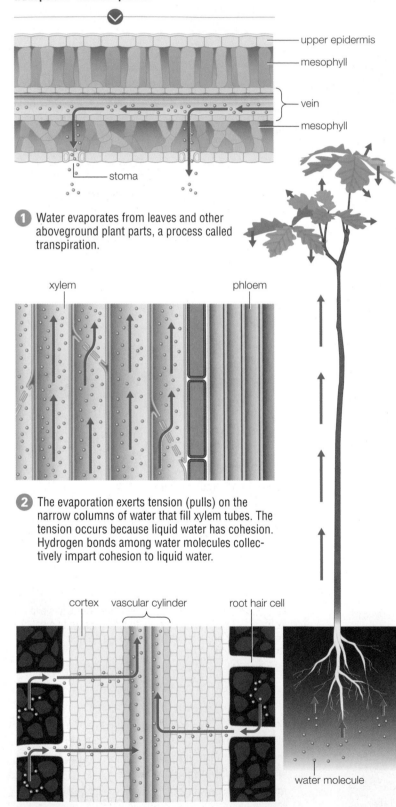

upper epidermis

mesophyll

vein

mesophyll

stoma

1 Water evaporates from leaves and other aboveground plant parts, a process called transpiration.

xylem   phloem

2 The evaporation exerts tension (pulls) on the narrow columns of water that fill xylem tubes. The tension occurs because liquid water has cohesion. Hydrogen bonds among water molecules collectively impart cohesion to liquid water.

cortex   vascular cylinder   root hair cell

water molecule

3 The tension inside xylem tubes extends from leaves to roots, where water molecules are being taken up from the soil.

27.4 Fluid Movement in Plants

REMEMBER: Water molecules resist separating from one another, and this resistance gives rise to cohesion; in evaporation, molecules escape the surface of a liquid and become vapor (Section 2.4). A waxy cuticle helps a plant retain water; plasmodesmata are open channels between adjacent plant cells (3.5). Tonicity describes relative solute concentration of two fluids separated by a membrane; osmosis is the movement of water across a membrane; turgor is pressure that a fluid exerts against a structure that contains it (4.5). Gases cannot diffuse across a plant cuticle, but oxygen produced by the light-dependent reactions of photosynthesis must escape the plant, and carbon dioxide needed for the Calvin–Benson cycle must enter it; thus, most leaves and stems are studded with stomata (5.4). Depending on environmental conditions, stomata open to allow gas exchange or close to conserve water (14.2).

Water Moves Through Xylem Water that enters a root travels to the rest of the plant inside tubes of xylem. These tubes consist of the stacked, interconnected walls of dead cells—vessel elements and tracheids (see Figure 27.5). How does water in xylem move all the way from roots to leaves? Tracheids and vessel elements that compose xylem tubes are dead, so these cells cannot be expending any energy to pump water upward against gravity. Rather, the upward movement of water in vascular plants occurs as a consequence of evaporation and cohesion. By the **cohesion–tension theory**, water in xylem is pulled upward by air's drying power, which creates a continuous negative pressure called tension. Figure 27.14 illustrates this mechanism, which begins with water evaporating from leaves and stems **1**. The evaporation of water from a plant's aboveground parts is called **transpiration**. Transpiration's effect on water inside a plant is a bit like what happens when you suck a drink through a straw. Transpiration exerts negative pressure (it pulls) on water. Because the water molecules are connected by hydrogen bonds, a pull on one tugs all of them. Thus, the negative pressure (tension) created by transpiration pulls on entire columns of water that fill xylem tubes **2**. The tension extends all the way from leaves that may be hundreds of feet in the air, down through stems, into young roots where water is being taken up from soil **3**. Water is pulled upward in continuous columns because xylem tubes are narrow, and water moving through a narrow conduit (such as a straw or a xylem tube) resists breaking into droplets. This phenomenon is partly an effect of water's cohesion, and partly because water molecules are attracted to hydrophilic materials (such as cellulose in the walls of xylem) making up the conduit.

A cuticle stops most water loss from aboveground parts, but only when stomata are closed. A pair of specialized epidermal cells borders each stoma (stoma is the singular form of stomata). When these two **guard cells** swell with water, they bend slightly so a gap (the stoma) forms between them (Figure 27.15). When the guard cells lose water, they collapse against one another, so the stoma between them closes. Stomata open or close based on an integration of cues that include humidity, light intensity, the level of carbon dioxide inside the leaf,

Figure 27.15 Stomata.
Whether a stoma is closed (left) or open (right) depends on how much water is plumping up the two guard cells bordering it.

Power and Syred/Science Source.

Figure 27.16 Some water-conserving structures in a pincushion leaf.
In this plant, leaf stomata are recessed beneath a ledge of cuticle. The ledge creates a small cup that traps moist air above the stoma.

Dr. Keith Wheeler/Science Source.

and hormonal signals from other parts of the plant. At sunrise, for example, the light causes guard cells to begin pumping potassium ions into their cytoplasm. Water follows the ions by osmosis, plumping the guard cells so the gap between them opens. Carbon dioxide from the air diffuses through the open stomata into the plant's tissues, and photosynthesis begins.

Open stomata allow water—a lot of it—to exit the plant. Even under conditions of high humidity, the interior of a leaf or stem contains more water than air. Thus, water vapor diffuses out of a stoma whenever it is open. Indentations, ledges, or other structures that reduce air movement next to a stoma decrease the rate of evaporation from it (Figure 27.16). Even so, around 95 percent of the water taken up from soil is lost by transpiration from open stomata.

Water is important for land plants, so why do they let almost all of it evaporate away? A cuticle is not only waterproof, it is also gasproof. When a plant's stomata are closed, it cannot exchange enough carbon dioxide and oxygen with the air to support critical metabolic processes. Opening and closing stomata allow a plant to balance its need for water with its need to exchange gases. A lot more water is lost through open stomata than gases are gained, however, and this is the reason for the fantastic amount of water loss by transpiration. During the day, each open stoma loses about 400 molecules of water for every molecule of carbon dioxide it takes in.

Sugars Flow Through Phloem Conducting tubes of phloem are called sieve tubes. Each sieve tube consists of living cells—a stack of sieve elements (see Figure 27.5). Unlike the cells that make up xylem tubes, sieve elements have no pits in their walls; fluid flows through their ends only. During differentiation of phloem, plasmodesmata connecting the sieve elements in a stack enlarge up to 100 times their original size, leaving very large pits in the cell walls. These pitted walls, which are called sieve plates after their appearance, separate stacked sieve elements in mature tissue (Figure 27.17). Also during differentiation, a sieve element loses most of its organelles, including the nucleus. How does it stay alive? Each sieve element has an associated companion cell that arises by division of the same parenchyma cell. The companion cell retains its nucleus and other components. Many plasmodesmata connect the cytoplasm of the two cells, so the companion cell can provide all metabolic functions necessary to sustain its paired sieve element.

A plant produces sugar molecules during photosynthesis. Some of these molecules are used by or stored in the cells that make them; the rest are conducted to other parts of the plant inside sieve tubes. The movement of sugars and other organic molecules through phloem is called **translocation**. Inside sieve tubes,

Figure 27.17 Sieve plates.
Left, sieve plates on the ends of two side-by-side sieve elements. Right, sieve plates (white arrows) in two columns of phloem. (Red stripes are rings of lignin in xylem vessel elements.)

Left, © J.C. Revy/ISM/Phototake; right, Spike Walker/Wellcome Images.

cohesion–tension theory Explanation of how transpiration creates tension that pulls a cohesive column of water upward through xylem.

guard cell One of a pair of cells that define a stoma.

translocation Movement of organic molecules through phloem.

transpiration Evaporation of water from above-ground plant parts.

Figure 27.18 Translocation of organic compounds in phloem from source to sink.
Water molecules are represented by blue balls; sugar molecules, by red balls. Each sieve tube is a stack of sieve elements that meet end to end at sieve plates. Sugars move into and out of a sieve element via its associated companion cell.

1 At a source region, sugars move into a companion cell, then into a sieve element.

2 The increase in solute concentration causes fluid in the sieve element to become hypertonic with respect to the surrounding cells. Water moves by osmosis from these cells into the sieve element, thus increasing pressure inside of it (turgor).

3 The pressure pushes the fluid through the sieve tube, toward a sink region where turgor is lower.

4 At a sink region, sugars move from sieve elements into sink cells. Water follows by osmosis.

fluid rich in sugars—mainly sucrose—flows from a **source** (a region of the plant where sugars are being produced or released from storage) to a **sink** (a region where they are being broken down or stored for later use). Photosynthetic tissues are typical source regions; tissues of developing shoots and fruits are typical sink regions.

Why do sugars flow from a source to a sink? A pressure gradient between the two regions drives the movement (Figure 27.18). Consider mesophyll cells in a leaf, which produce far more sugar than they use for their own metabolism. Photosynthesis in these cells sets up the leaf as a source region. Excess sugar molecules move from the cells into adjacent companion cells, then into associated sieve elements **1**. Sugar loading at a source region increases the solute concentration of a sieve tube's cytoplasm so that it becomes hypertonic with respect to that of the surrounding cells. Water follows the sugars by osmosis, moving into the sieve tube from the surrounding cells **2**. The rigid cell walls of mature sieve elements cannot expand very much, so the influx of water raises turgor in the tube. Pressure inside a sieve tube can be very high—up to five times higher than that of an automobile tire. The high fluid pressure pushes the sugar-rich cytoplasm from one sieve element to the next, toward a sink region where the turgor is lower **3**. Pressure inside a sieve tube decreases at a sink because sugars leave the tube in this region. Water follows, again by osmosis **4**, so turgor inside sieve elements decreases in these regions. This explanation of how turgor pushes sugar-rich fluid inside a sieve tube from source to sink is called the **pressure flow theory**.

Take-Home Message 27.4

How do fluids move through a plant body?

- A cuticle minimizes water loss from a plant's aboveground parts when stomata are closed. Open stomata allow plant tissues to exchange gases with air, but they also allow a lot of water to evaporate from the plant (transpiration).
- Xylem transports water. Transpiration pulls columns of water in xylem upward through a plant body, from roots to leaves.
- Phloem transports organic compounds. Sugars move into sieve tubes at sources, and exit at sinks. A pressure gradient pushes sugar-rich fluid inside a sieve tube from source to sink.

27.5 Plant Growth

REMEMBER: Stem cells are unspecialized cells that can either divide to produce more stem cells or differentiate into one of the specialized types of cells that characterize specific body parts (Section 19.1).

All plant tissues arise from the activity of **meristems**, which are regions of undifferentiated (stem) cells that divide continually during the growing season. Meristem cells are analogous to stem cells in animals: When they divide, some of their descendants remain undifferentiated, and others differentiate. In plants, the differentiating cells give rise to specialized tissues—vascular, ground, and epidermal tissues—as they enlarge and mature.

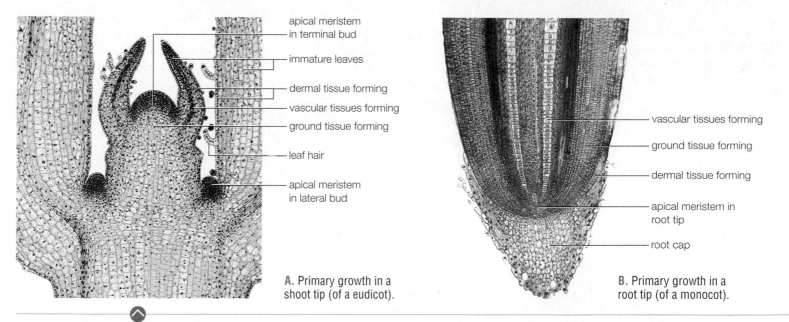

A. Primary growth in a shoot tip (of a eudicot).

B. Primary growth in a root tip (of a monocot).

Figure 27.19 Zones of primary growth in a shoot and a root.

(A) © Dale M. Benham, Ph.D., Nebraska Wesleyan University; (B) Biodisc/Visuals Unlimited.

A plant grows both by lengthening and by thickening. Roots and shoots lengthen during **primary growth**, and the plant produces soft parts such as leaves. Primary growth originates at **apical meristems** in the tips of shoots and roots. An apical meristem consists of a mass of rapidly dividing stem cells. Divisions push some of the cells toward the edge of the mass, and these displaced cells begin to differentiate. Lengthening occurs mainly because differentiating cells behind the mass of apical meristem elongate as they mature and take on their specialized function. The meristem remains at the tip of the lengthening structure, continually renewing itself and also producing cells that divide and differentiate behind it.

The tip of an actively lengthening shoot (Figure 27.19A) is called a terminal bud. As the shoot lengthens, masses of tissue near the sides of the apical meristem in the terminal bud bulge out and then develop into leaves. Other leaves form and mature in orderly tiers, one after the next, along the lengthening stem. As they do, lateral buds form in places where the leaves attach to the stem. Lateral buds also contain apical meristem. Hormonal signals can trigger the stem cells in a lateral bud to divide and give rise to a shoot—either a stem, a leaf, or a flower—that branches laterally from the main stem.

An actively lengthening root tip has a root cap, which is a dome-shaped mass of cells that protects the tip as the soft, young root grows through soil (Figure 27.19B). Cells of the root cap are shed continually as the root lengthens. Lateral roots also form, but not at nodes (roots have no nodes). Lateral roots originate with cells in the root's vascular cylinder. These cells divide in a direction perpendicular to the long axis of the root, then differentiate (a lateral root arising from the vascular cylinder is shown in Figure 27.12A).

Many plant species undergo **secondary growth**, during which their shoots and roots thicken and become woody. In eudicots and gymnosperms, the thickening originates at **lateral meristems**, which are cylindrical layers of meristem that run lengthwise through shoots and roots (Figure 27.20). Eudicots and gymosperms have two lateral meristems: vascular cambium and cork cambium. **Vascular cambium** produces secondary vascular tissue inside older stems and roots. When vascular

Figure 27.20 Lateral meristems.
In eudicots and gymnosperms, secondary growth (thickening) originates at two lateral meristems: vascular cambium and cork cambium.

apical meristem Meristem in the tip of a shoot or root; gives rise to primary growth (lengthening).

lateral meristem Vascular cambium or cork cambium; cylindrical sheet of meristem that runs lengthwise through shoots and roots; gives rise to secondary growth (thickening).

meristem Zone of undifferentiated plant cells; all plant growth arises from divisions of meristem cells.

pressure flow theory Explanation of how a difference in turgor between sieve elements in source and sink regions pushes sugar-rich fluid through a sieve tube.

primary growth Lengthening of young shoots and roots; originates at apical meristems.

secondary growth Thickening of older stems and roots; originates at lateral meristems.

sink Region of a plant where sugars are being used.

source Region of a plant where sugars are being produced or released from storage.

vascular cambium Lateral meristem that produces secondary xylem and phloem.

Figure 27.21
Secondary growth.
Divisions of cells on the inner surface of vascular cambium (dark orange) produce secondary xylem. As the inner core of xylem expands, it displaces the vascular cambium outward. Divisions of cells on the outer surface of the vascular cambium produce secondary phloem. Secondary growth (thickening) occurs simultaneously with primary growth (lengthening) at terminal and lateral buds.

© SeDmi/Shutterstock.com.

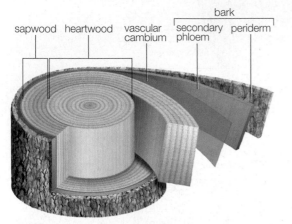

A. Structure and components of an older, woody eudicot stem. Wood is accumulated secondary xylem.

B. Sap oozes from xylem in a pine log. Still-functional secondary xylem is sapwood. The oldest xylem at the center of the structure (heartwood) no longer transports fluid because its tubes are plugged.

C. Tree rings are alternating bands of early and late wood. In temperate zone trees, one ring forms each year. Note the rings in **B**.

Figure 27.22 Wood.

(B) Brian Gordon Green/National Geographic Creative.

cambium cells divide, some of their descendants remain undifferentiated. The rest differentiate and give rise to secondary phloem on the outer surface of the vascular cambium, and to secondary xylem on the inner surface (Figure 27.21). Secondary xylem that has accumulated inside a cylinder of vascular cambium is called **wood**.

As wood accumulates, it displaces the vascular cambium toward the outer surface of the root or shoot. Displaced cells of the vascular cambium divide in a widening circle, so this meristem tissue keeps its cylindrical form. Another lateral meristem, cork cambium, eventually forms outside the cylinder of vascular cambium. **Cork cambium** produces **cork**, a tissue that consists of densely packed dead cells with thickened, waxy walls. Cork protects, insulates, and waterproofs the surface of a stem or root. It is part of periderm, the dermal tissue that replaces epidermis on the surfaces of older stems and roots. Periderm includes cork, cork cambium, and every other tissue between these two. **Bark** is the informal term for periderm and all other living and dead tissues outside the cylinder of vascular cambium (Figure 27.22A).

Over time, the secondary xylem at the center of an older woody stem or root can become plugged so that it no longer transports fluid. Sapwood, the region of still-functional secondary xylem, lies between heartwood and vascular cambium (Figure 27.22B). During spring in temperate zones, sugar-rich fluid (sap) travels through sapwood, from tree roots to buds. Sap collected from maple trees as it starts to flow in spring is boiled to make maple syrup.

In trees that are dormant in winter, vascular cambium gives rise to large-diameter, thin-walled xylem cells (early wood) during spring. Late wood, with small-diameter, thick-walled xylem cells, forms during summer. Thus, a cross-section reveals rings that consist of alternating bands of early and late wood. Each band is one growth ring, or "tree ring." In most temperate zone trees, one ring forms each year (Figure 27.22C). In tropical regions where weather does not vary seasonally, trees grow at the same rate all year and do not have rings.

Long-lived tree species such as redwoods and bristlecone pines add wood over centuries, one ring per year. Count an old tree's rings, and you have an idea

Digging Into Data

Tree Rings and Droughts

El Malpais National Monument, in west central New Mexico, has pockets of vegetation that have been surrounded by lava fields for about 3,000 years, so they have escaped wildfires, grazing animals, agricultural activity, and logging. Henri Grissino-Mayer generated a 2,129-year annual precipitation record using tree ring data from living and dead trees in this park (Figure 27.23).

1. Around 770 A.D., the Mayan civilization began to suffer a massive population loss, particularly in the southern lowlands of Mesoamerica. The El Malpais tree ring data show a drought during that time. Was it more or less severe than the Dust Bowl drought?
2. One of the worst population catastrophes ever recorded occurred in Mesoamerica between 1519 and 1600 A.D., when around 22 million people native to the region died. Which period between 137 B.C. and 1992 had the most severe drought? How long did that drought last?

Figure 27.23 Annual precipitation record for 2,129 years, inferred from compiled tree ring data in El Malpais National Monument, New Mexico. Data were averaged over 10-year intervals; graph correlates with other indicators of rainfall collected in all parts of North America. PDSI, Palmer Drought Severity Index: 0, normal rainfall; increasing numbers mean increasing excess of rainfall; decreasing numbers mean increasing severity of drought.

*A severe drought contributed to a series of catastrophic dust storms that turned the midwestern United States into a "dust bowl" between 1933 and 1939.

---

of its age. If you know the year in which the tree was cut, you can find out when a particular ring formed by counting the rings backward from the outer edge. Thickness and other features of the rings offer clues about the environmental conditions that prevailed during the years they formed. Consider how more fires tend to occur in warmer, drier conditions, so an increased frequency of fires can be evidence of a period of drought. Very old trees often bear the scars of many forest fires in their rings (Figure 27.24). In 2010, researchers used the rings and fire scars of ancient trees in California's Sequoia National Park to reconstruct a 3,000-year history of the region's climate. They discovered, for example, a dramatic increase in the number of forest fires between 800 A.D. and 1300 A.D. This 500-year period coincided with the Medieval Warm Period, an anomaly in climate that had previously been documented in some other parts of the world. Taken together, the evidence suggests this warming period was a global climate pattern that caused worldwide drought.

Take-Home Message 27.5

How does growth occur in plants?

- Primary growth (lengthening) originates at apical meristems in shoot and root tips.
- Secondary growth, which thickens older roots and shoots in many eudicots and gymnosperms, originates at lateral meristems (vascular cambium and cork cambium).
- Vascular cambium gives rise to secondary xylem and phloem. Cork cambium gives rise to periderm, a dermal tissue.
- The bark of woody plants consists of all living and dead tissues outside the ring of vascular cambium.

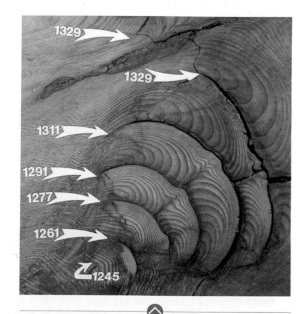

Figure 27.24 Fire scars. Scars that disrupt the normal, circular pattern of tree rings are evidence of past fires. These are fire scars in the wood of an ancient giant sequoia; numbers indicate the year a particular ring was laid down.

Photo by Tom Swetnam.

Figure It Out: How many forest fires affected this tree between the years 1245 and 1329? **Answer: Five**

Summary

Section 27.1 The ability of plants to take up substances from soil water is the basis for phytoremediation. This environmental cleanup method uses plants specifically to remove pollutants from a contaminated area.

Section 27.2 Most flowering plants have belowground roots and aboveground shoots (stems, leaves, and flowers). **Ground tissues** make up most of the soft internal tissues of a plant, and **dermal tissues** protect its surfaces. **Vascular tissues** conduct water and nutrients to all parts of the plant. Monocots and eudicots have the same tissues organized in different ways. For example, embryos of monocots have one **cotyledon**; those of eudicots have two.

Parenchyma, **collenchyma**, and **sclerenchyma** are simple tissues; each consists of only one type of cell. **Mesophyll** is photosynthetic parenchyma. Living cells in collenchyma have sturdy, flexible walls that support fast-growing plant parts. Cells in sclerenchyma die at maturity, but their lignin-reinforced walls remain and support the plant. Stomata open across **epidermis**, a dermal tissue that covers soft plant parts.

In vascular tissue, water and dissolved minerals flow through vessels of **xylem**, and sugars travel through vessels of **phloem**.

Section 27.3 **Vascular bundles** extending through stems conduct water and nutrients, and also help structurally support the plant. In most eudicot stems, vascular bundles form a ring that divides ground tissue into cortex and pith. In monocot stems, the vascular bundles are distributed throughout the ground tissue. Stems have **nodes**; roots do not. New shoots and roots can form at nodes. Some stems are specialized for storage, reproduction, or both.

Leaves, which are specialized for photosynthesis, contain photosynthetic mesophyll and vascular bundles (**leaf veins**) between upper and lower epidermis.

Roots are specialized for absorbing water from soil. Eudicots typically have a taproot system, and monocots have a fibrous root system. The availability of water and mineral ions in a particular soil depends on its proportions of sand, silt, and clay, and also on its **humus** content. Mycorrhizae as well as nitrogen-fixing bacteria in **root nodules** enhance nutrient uptake in many plants. **Root hairs** increase a root's surface area. Soil water absorbed by a root must cross endodermis to enter the **vascular cylinder** and be distributed in xylem to the rest of the plant.

Section 27.4 Water flows through xylem tubes from roots to shoot tips. These tubes consist of the walls of tracheids and vessel elements that formed in stacks and then died. The **cohesion–tension theory** explains how water moves upward through xylem:

Transpiration (the evaporation from aboveground plant parts, mainly at stomata) pulls water upward. This pull (tension) extends from leaves to roots because of water's cohesion inside the narrow tubes of xylem.

A cuticle helps a plant conserve water; stomata help it balance water conservation with gas exchange required for photosynthesis and aerobic respiration. A stoma closes when the pair of **guard cells** bordering it lose water and collapse against one another. It opens when the guard cells swell with water.

Sugars move through a plant by **translocation** in phloem's sieve tubes, which consist of stacked sieve elements separated by perforated sieve plates. By the **pressure flow theory**, the movement of sugar-rich fluid through a sieve tube is driven by a pressure gradient between **source** and **sink** regions.

Section 27.5 All plant tissues originate at **meristems**, which are regions of undifferentiated (stem) cells that divide rapidly during the growing season. **Primary growth** (lengthening) arises at **apical meristems** in the tips of young shoots and roots. **Secondary growth** (thickening) arises at **lateral meristems** (vascular cambium and cork cambium) in older stems and roots. **Vascular cambium** produces secondary xylem (**wood**) on its inner surface, and secondary phloem on its outer surface. **Cork cambium** gives rise to **cork**, which is part of periderm. **Bark** is an informal term for all tissue outside the ring of vascular cambium in a woody plant.

In many trees, one ring forms during each growing season. Tree rings hold information about environmental conditions that prevailed while the rings were forming.

Self-Quiz

Answers in Appendix I

1. Ground tissue consists mainly of _____ .
   a. pith
   b. the walls of dead cells
   c. parenchyma cells
   d. cork but not bark

2. Which of the following cell types remain alive in mature tissue? Choose all that apply.
   a. companion cells
   b. sieve elements
   c. tracheids
   d. vessel elements

3. Epidermis and periderm are _____ tissues.
   a. ground
   b. vascular
   c. dermal
   d. none of the above

4. An onion is a _____ (choose all that apply).
   a. root structure   c. bulb
   b. stem structure   d. corm

5. A vascular bundle in a leaf is called _____ .
   a. xylem   c. a vascular cylinder
   b. mesophyll   d. a vein

6. True or false? Lateral roots form at nodes on roots.

7. Typically, vascular tissue is organized as _____ in stems and as _____ in roots.
   a. multiple vascular bundles; one vascular cylinder
   b. one vascular bundle; multiple vascular cylinders
   c. one vascular cylinder; multiple vascular bundles
   d. multiple vascular cylinders; one vascular bundle

8. Roots and shoots lengthen through activity at _____ .
   a. apical meristems   c. vascular cambium
   b. lateral meristems   d. cork cambium

9. Plants obtain carbon from _____ .
   a. soil water   b. air   c. both a and b

10. Decomposing matter in soil is called _____ .
    a. clay   c. silt
    b. humus   d. sand

11. When guard cells swell, _____ .
    a. transpiration ceases   c. stomata open
    b. sugars enter phloem   d. root cells die

12. Water transport from roots to leaves occurs by _____ .
    a. a pressure gradient inside sieve tubes
    b. different solutes at source and sink regions
    c. the pumping force of xylem vessels
    d. transpiration, tension, and cohesion of water

13. Sugar transport from leaves to roots occurs by _____ .
    a. a pressure gradient inside sieve tubes
    b. different solutes at source and sink regions
    c. the pumping force of xylem vessels
    d. transpiration, tension, and cohesion of water

14. Tree rings occur _____ .
    a. when there are droughts during the time the rings form
    b. where environmental conditions influence xylem cell size
    c. if heartwood alternates with sapwood
    d. as epidermis replaces periderm

15. Match the plant parts with the best description.
    _____ mesophyll   a. separates cells in phloem tubes
    _____ cork   b. photosynthetic parenchyma
    _____ cotyledons   c. evaporation from stomata
    _____ parallel veins   d. characteristic of monocot leaves
    _____ vascular cylinder   e. only one in a monocot seed
    _____ wood   f. only in eudicots
    _____ transpiration   g. stem structure
    _____ corm   h. part of periderm
    _____ sieve plate   i. mass of secondary xylem
    _____ taproot   j. central column in roots

Critical Thinking

1. Oscar and Lucinda meet in a tropical rain forest and fall in love, and he carves their initials into the bark of a tree. They never do get together, though. Ten years later, still heartbroken, Oscar searches for the tree. Given what you know about primary and secondary growth, will he find the carved initials higher relative to ground level? If he goes berserk and chops down the tree, what kinds of growth rings will he see?

2. Manganese is a component of chlorophyll. How does a plant obtain this element? What effect would you expect a manganese deficiency to have on a plant?

3. Farmers sometimes plant clover or other legumes in fields during seasons when they are not growing crops. When a crop-planting season approaches, the "cover crop" is plowed under. Explain how this procedure enhances soil fertility.

4. Why do eudicot trees tend to be wider at the base than at the top?

5. Aboveground plant surfaces are typically covered with cuticle. Why do roots lack this protective coating?

Visual Question

1. Was the micrograph shown on the right taken from a section of stem or root? Monocot or eudicot?

© Mike Clayton/
University of Wisconsin
Department of Botany.

28

PLANT REPRODUCTION AND DEVELOPMENT

Application ❯ 28.1 **Plight of the Honeybee**

REMEMBER: Species with close ecological interactions may coevolve (Section 12.7). Angiosperms are vascular seed plants and the only plants that make flowers and fruits; many of their traits are adaptations that attract coevolved pollinators (14.7).

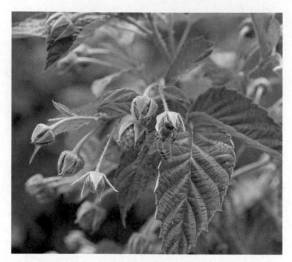

A. Honeybees are efficient pollinators of a variety of flowers, including those of berry plants.

B. Raspberry flowers can pollinate themselves, but the fruit that forms from a self-fertilized flower is of lower quality than that of a cross-pollinated flower. The two raspberries on the left formed from self-pollinated flowers. The one on the right formed from an insect-pollinated flower.

Figure 28.1 An example of the importance of insect pollinators.

(A) © Alan McConnaughey, www.flickr.com/photos/engrpiman; (B) Courtesy of James H. Cane, USDA-ARS Bee Biology and Systematics Lab, Utah State University, Logan, UT.

In the fall of 2006, commercial beekeepers worldwide began to notice something was amiss in their honeybee hives. All of the adult worker bees were missing from the hives, but there were no bee corpses to be found. The bees had suddenly (and quite unusually) abandoned their queen and many living larvae. The few young workers that remained were reluctant to feed on abundant pollen and honey stored in their own hive, or in other abandoned hives. A third of the colonies did not survive the following winter. By spring, the phenomenon had a name: colony collapse disorder (CCD). Since then, honeybee populations have continued to decline. In 2011, thirty percent of bee colonies in the United States and twenty percent of those in Europe collapsed.

A single honeybee visits hundreds, sometimes thousands, of flowers a day to find nectar and pollen. The insect uses "scent memory" to remember the location of these rewards. Then it finds its way back to the hive, navigating distances up to 8 kilometers (5 miles) to communicate the location of the flowers to other bees. Long-term exposure to low levels of insecticides might impair a bee's ability to carry out this mission, and many researchers suspect that it is an important factor in colony collapse disorder.

Synthetic insecticides called neonicotinoids are widely sprayed on crops, and they are a common component in seed treatments intended to protect the future plant from damage by crop pests. The chemicals are systemic, which means they are taken up by all tissues of a treated plant, including the nectar and pollen that honeybees collect from flowers as they forage. Neonicotinoids mimic the neurotoxic (nerve-killing) effect of nicotine, a natural insecticide made by plants such as tobacco. They are highly toxic to honeybees; even in doses that are too tiny to detect in the insects themselves, neonicotinoids increase honeybees' vulnerability to fungal infection. Dewdrops that form on plants grown from neonicotinoid-treated seeds contain enough of the insecticide to kill them. In sublethal doses, neonicotinoids impair honeybee learning and memory, so these insecticides are also likely to affect the bees' ability to navigate and communicate with hivemates. Another insecticide, coumaphos, makes the bees even more vulnerable to the effects of neonicotinoids. Coumaphos is commonly used to treat hives infested with parasitic varroa mites. Perhaps not coincidentally, these mites are often found in beehives affected by CCD.

Nearly all of our food crops are the fruits of angiosperms. Reproduction in many angiosperm species depends on honeybees and other **pollinators** that carry pollen from one plant to another (Figure 28.1A). Fruits do not form from unpollinated flowers, and in some species they only form when flowers receive pollen from another plant. Even species with flowers that can self-pollinate tend to make bigger fruits and more of them when they are cross-pollinated (Figure 28.1B). Honeybees are highly efficient pollinators, and they are the only ones that tolerate being trucked across the country in man-made hives, to wherever crops require pollination. The potential loss of their portable pollination service threatens the existence of nearly a third of our food supply, which is a reason CCD is currently in the spotlight. However, populations of bumblebees and other insect pollinators are also dwindling. Habitat loss may be the main factor, but diseases and insecticides that harm honeybees also harm other invertebrate pollinators.

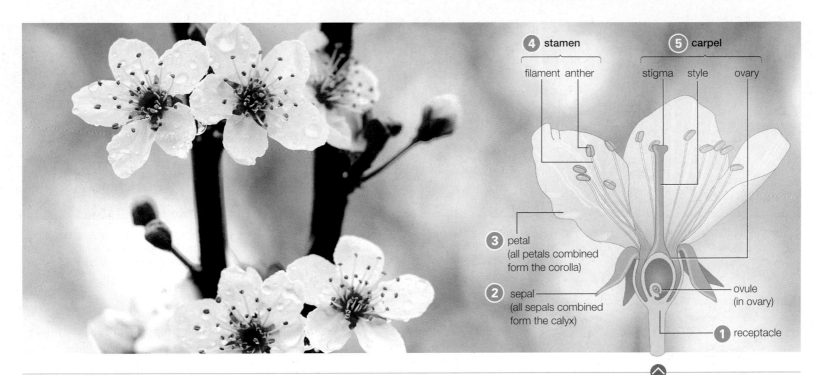

4 stamen

filament anther

5 carpel

stigma style ovary

3 petal
(all petals combined
form the corolla)

2 sepal
(all sepals combined
form the calyx)

ovule
(in ovary)

1 receptacle

Figure 28.2 Anatomy of a typical flower.
The structures that produce male gametophytes are
stamens, which consist of pollen-bearing anthers on
slender filaments. The structure that produces female
gametophytes is the carpel, which consists of ovary,
stigma, and style.

Left, © Martin Ruegner/Photographer's Choice RF/Getty Images.

28.2 Sexual Reproduction

REMEMBER: The independent evolution of similar body parts in different lineages
is morphological convergence (Section 11.6). In plants, the diploid sporophyte
produces spores by meiosis; the haploid gametophyte produces gametes by mitosis
(14.2). Monocots and eudicots differ in the details of their structure (27.2).

Flowers are the specialized reproductive structures of angiosperms (Figure 28.2).
The form and composition of flowers vary dramatically among species, but all
develop at the tips of special reproductive shoots. A flower's parts are modified
leaves that develop from a thickened region of stem called a receptacle **1**. A typical
flower has four spirals or rings (whorls) of modified leaves. The outermost whorl is
the calyx, a ring of leaflike sepals **2**. Sepals of most species are photosynthetic and
inconspicuous; they serve to protect the flower's more delicate parts. The whorl just
inside the calyx is the corolla (from the Latin *corona*, or crown), a ring of petals **3**.
Petals are usually the largest and most brightly colored parts of a flower.

Inside the corolla is a whorl of **stamens**, the reproductive organs that produce
the plant's male gametophytes. In most flowers, a stamen consists of a thin filament
with an **anther** at the tip **4**. A typical anther contains four pouches called pollen
sacs. **Pollen grains**, which are immature male gametophytes, form in pollen sacs.

The flower's innermost whorl consists of modified leaves folded and fused
into one or more carpels. **Carpels** are the reproductive organs that produce female
gametophytes **5**. The flowers of some species have one carpel; those of other species have several. A carpel—or a compound structure that consists of multiple fused
carpels—is commonly called a pistil. The upper region of a carpel is a sticky or hairy
stigma that is specialized to receive pollen grains. Typically, the stigma sits on top
of a slender stalk called a style. The lower, swollen region of a carpel is the **ovary**,
which contains one or more ovules. An **ovule** is a structure that bulges from the
inside of the ovary wall; a female gametophyte forms inside it.

anther Part of the stamen that produces pollen grains.

carpel Floral reproductive organ that produces the
female gametophyte; consists of an ovary, stigma,
and often a style.

flower Specialized reproductive structure of a flowering plant.

ovary In flowering plants, the enlarged base of a
carpel, inside which one or more ovules form.

ovule Of a seed-bearing plant, a structure in which a
female gametophyte forms.

petal Unit of a flower's corolla; often showy and
conspicuous.

pollen grain Immature male gametophyte of a seed
plant.

pollinator An animal that moves pollen from one
plant to another.

stamen Floral reproductive organ that consists of an
anther and, in most species, a filament.

stigma Upper part of a carpel; adapted to receive
pollen.

A. The solitary flower of the lady's slipper orchid is irregular.

B. A hyacinth flower is an elongated inflorescence.

C. A daisy is a composite flower with many individual florets.

D. A eucalyptus flower has no petals, so it is incomplete.

E. Some begonias have imperfect flowers: Female blossoms (with no stamens, left) form on the same plants as male blossoms (with no carpels, right).

Figure 28.3 Examples of structural variation in flowers.

(A) © Salawin Chanthapan/Shutterstock; (B–D) © Brian Johnston; (E) © Lauren Doyle.

A. Dandelion.　　**B.** Evening primrose.

Figure 28.4 UV-reflecting patterns in flower petals.
The top row shows flowers as we see them. The bottom row shows the same flowers photographed with a special filter that allows us to see reflected UV light, here represented by the color red. The bull's-eye patterns guide bees, which can see UV light, directly to the flower's reproductive parts.

Bjorn Rorslett/Science Source.

Flowers vary widely in form. Some are solitary blossoms (Figure 28.3A); others occur in clusters called inflorescences (Figure 28.3B). A composite flower is an inflorescence of many flowers ("florets") grouped as a single head (Figure 28.3C). Cherry blossoms (shown in Figure 28.2) and other regular flowers are symmetric around their center axis: If the flower were cut like a pie, the pieces would be roughly identical. Irregular flowers are not radially symmetric (the flower shown in Figure 28.3A is an example of an irregular flower).

A cherry blossom has all four sets of modified leaves (sepals, petals, stamens, and carpels), so it is called a "complete" flower. An "incomplete" flower lacks one or more of these structures (Figure 28.3D). Cherry blossoms are also "perfect" flowers, which means they have both stamens and carpels. Perfect flowers can be cross-pollinated (fertilized by pollen from other plants), or they can self-pollinate. Self-pollination can be adaptive in situations where plants are widely spaced. However, offspring that arise from self-pollinated flowers are often less vigorous than those of cross-pollinated plants. Accordingly, floral adaptations of many plant species encourage or even require cross-pollination. For example, a perfect flower may release pollen only after its stigma is no longer receptive to being fertilized. "Imperfect" flowers, which lack stamens or carpels (Figure 28.3E), cannot self-fertilize. In some species, stamen-bearing and carpel-bearing flowers form on different plants, so fertilization can only occur by cross-pollination.

Pollination, the arrival of pollen on a receptive stigma, is essential to sexual reproduction in flowering plants. Pollen is transferred from anther to stigma by a **pollination vector**, which can be an animal pollinator or an environmental agent such as wind. The diversity of flower form in part reflects a dependence on pollination vectors. Fragrance and other flower traits not directly related to reproduction are evolutionary adaptations that attract coevolved animal pollinators. An animal attracted to a particular flower often picks up pollen on a visit, then inadvertently transfers it to another flower on a later visit. The more specific the attraction, the more efficient the transfer of pollen among plants of the same species. Thus, a flower's traits typically reflect the sensory abilities and preferences of its coevolved pollinator. In an example of morphological convergence, the flowers of angiosperm species that depend on the same pollinator (bats, for example, or birds) often share a certain set of traits. Consider that bees have a keen sense of smell, and they can see UV light; bee-pollinated flowers tend to be fragrant, with a bull's-eye pattern

Digging Into Data

Who's the Pollinator?

Massonia depressa is a low-growing plant native to South African deserts. The dull, ground-level flowers of this monocot have tiny petals, a yeasty aroma, and a jellylike nectar—traits that led researchers to suspect they are pollinated by desert rodents. To test this hypothesis, the researchers trapped rodents in areas where *M. depressa* grows and checked them for pollen (Figure 28.5A). They also checked whether fruits and seeds form in the absence of rodents (Figure 28.5B).

1. How many rodents were captured? Of these, how many showed some evidence of ingesting *M. depressa* pollen?

2. Would this evidence alone be sufficient to conclude that rodents are the main pollinators for this plant?

3. How did the average number of seeds produced by caged plants compare with that of control plants?

4. Do these data support the hypothesis that rodents are required for pollination of *M. depressa*? Why or why not?

Type of rodent	Number caught	# with pollen on snout	# with pollen in feces
Namaqua rock rat	4	3	2
Cape spiny mouse	3	2	2
Hairy-footed gerbil	4	2	4
Cape short-eared gerbil	1	0	1
African pygmy mouse	1	0	0

A. Evidence of visits to *M. depressa* by rodents. The photo shows a gerbil pushing its head through the stamens of an *M. depressa* flower to reach the nectar. Note the pollen on the gerbil's snout.

	Mammals allowed access to plants	Mammals excluded from plants
Percent of plants that set fruit	30.4	4.3
Average number of fruits per plant	1.39	0.47
Average number of seeds per plant	20.0	1.95

B. Fruit and seed production of *M. depressa* with and without visits by mammals. Mammals were excluded from plants by wire cages with openings large enough for insects to pass through. 23 plants were tested in each group.

Figure 28.5 **Testing pollination of *M. depressa* by rodents.**
© Steven D. Johnson.

---

of UV-reflecting pigments in their petals (Figure 28.4). Bee-pollinated flowers also attract and communicate with individual bees via electric fields. Flowers pollinated by birds, which rely more on their excellent vision than their relatively poor sense of smell, are often bright red and unscented. Flowers pollinated by night-flying bats, which have a good sense of smell but poor vision, tend to be large and light-colored, with a strong nighttime fragrance. Not all flowers smell sweet; odors like dung or rotting flesh beckon beetles and flies.

Rewards offered by a flower reinforce a pollinator's memory of a visit, and encourage the animal to seek out other flowers of the same species. Many flowers exude a sweet fluid called **nectar** that is prized by many pollinators. Nectar is the only food for most adult butterflies, and it is the food of choice for hummingbirds. Honeybees convert nectar to honey, which helps feed the bees and their larvae through the winter. Bees also collect protein-rich pollen for food. Many beetles feed primarily on pollen. Pollen is the only reward for a pollinator of roses, poppies, and other flowers that produce no nectar.

Specializations of some flowers prevent pollination by all but one type of pollinator. For example, nectar or pollen at the bottom of a long floral tube may be accessible only to an insect with a matching feeding device. In some plants, stamens adapted to brush against a pollinator's body or lob pollen onto it will function only when triggered by that pollinator. Such relationships are to both species' mutual advantage: A flower that captivates the attention of an animal has a pollinator that spends its time seeking (and pollinating) only those flowers. Pollinators also benefit when they receive an exclusive supply of the flower's reward.

nectar Sweet fluid exuded by some flowers that attracts animal pollinators.

pollination The arrival of pollen on a receptive stigma.

pollination vector Environmental agent that moves pollen grains from one plant to another.

dormancy Period of temporarily suspended metabolism.

double fertilization Mode of fertilization in flowering plants in which one sperm cell fuses with the egg, and a second sperm cell fuses with the endosperm mother cell.

endosperm Nutritive tissue in the seeds of flowering plants.

germinate To resume metabolic activity after dormancy.

Pollen is transferred from anther to stigma by an animal pollinator or an environmental agent such as wind.

A New Generation Begins The life cycle of flowering plants (Figure 28.6) is dominated by the sporophyte, a diploid body that grows by mitotic divisions of a fertilized egg. Spores that form by meiosis inside sporophyte flowers develop into haploid gametophytes, which produce gametes.

The production of female gametes begins when a mass of tissue—the ovule— starts growing on the inner wall of an ovary ❶. One cell in the mass undergoes meiosis and cytoplasmic division, forming four haploid megaspores ❷. Three megaspores typically disintegrate, and the remaining one undergoes three rounds of mitosis without cytoplasmic division. The cell that results from these divisions has eight haploid nuclei ❸. The cytoplasm of this cell divides unevenly to form a seven-celled female gametophyte ❹. One of the cells, the endosperm mother cell, has two nuclei ($n + n$). Another cell is the egg.

The production of male gametes begins as masses of diploid, spore-producing cells form by mitosis inside anthers. Walls typically develop around the masses, so four pollen sacs form ❺. Each cell in the pollen sacs undergoes meiosis and cytoplasmic division to form four haploid microspores ❻. Mitosis and differentiation of a microspore produce a pollen grain ❼. A pollen grain consists of two cells, one inside the cytoplasm of the other, enclosed by a durable coat that will protect the cells on their journey to meet an egg. After the pollen grains form, they enter a period of suspended metabolism—**dormancy**—before being released from the anther when the pollen sacs split open ❽.

A pollen grain that lands on a receptive stigma **germinates**, which means it resumes metabolic activity after dormancy. The outer cell develops into a tubular outgrowth called a pollen tube ❾. The inner cell undergoes mitosis to produce two sperm cells (the male gametes) within the pollen tube. A pollen tube together with its contents of male gametes is the mature male gametophyte.

The pollen tube elongates at its tip, carrying the two sperm cells through the tissues of the style and ovary to the ovule. A pollen tube that reaches and penetrates the ovule releases the two sperm cells ❿. Flowering plants undergo **double fertilization**, in which one of the sperm cells delivered by the pollen tube fuses with (fertilizes) the egg and forms a diploid zygote; the other sperm cell fuses with the endosperm mother cell, forming a triploid ($3n$) cell. This cell gives rise to triploid **endosperm**, a nutritious tissue that forms only in seeds of flowering plants.

Take-Home Message 28.2

How does sexual reproduction occur in flowering plants?

- Flowers are the reproductive structures of angiosperm sporophytes. The diversity in flower form reflects angiosperm dependence on various pollination vectors.
- The parts of flowers that produce male gametophytes are stamens, which typically consist of a filament with an anther at the tip. Pollen forms inside the anthers.
- The parts of flowers that produce female gametophytes are carpels, which typically consist of stigma, style, and ovary. A gametophyte forms in an ovule inside the ovary.
- A pollen grain that germinates on a stigma develops into a male gametophyte, which consists of a pollen tube containing two sperm cells. The pollen tube grows into the carpel, enters an ovule, and releases the sperm cells.
- Double fertilization occurs when one of the sperm cells delivered by the pollen tube fuses with the egg, and the other fuses with the endosperm mother cell.

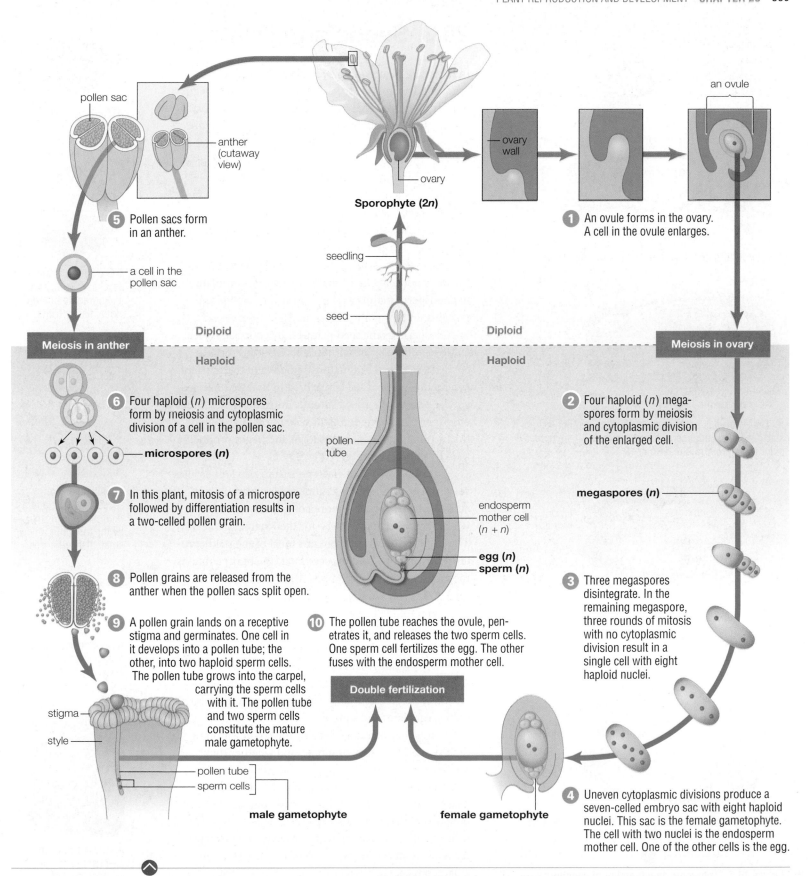

Figure 28.6 Life cycle of cherry, a eudicot.

Figure It Out: What structure gives rise to a pollen grain by mitosis?

Answer: A microspore

many ovules inside ovary wall

embryo

endosperm

A. After fertilization, an embryo forms inside each ovule.

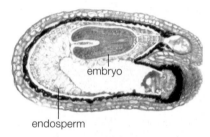

embryo

endosperm

B. The embryo is heart-shaped when its two cotyledons start forming. Endosperm tissue expands as the parent plant transfers nutrients into it.

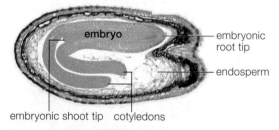

embryo

embryonic root tip

endosperm

embryonic shoot tip   cotyledons

C. In eudicots like *Capsella*, nutrients are transferred from endosperm into two cotyledons as the embryo matures. The developing embryo becomes shaped like a torpedo when the enlarging cotyledons bend.

embryo

seed coat

cotyledons

D. A tough seed coat forms around the mature embryo and its two enlarged cotyledons.

Figure 28.7 Embryonic development of shepherd's purse (*Capsella*), a eudicot.

(A,C,D) Michael Clayton, University of Wisconsin; (B) © David T. Webb.

28.3 **Seeds and Fruits**

In flowering plants, double fertilization produces a zygote and a triploid ($3n$) cell; both immediately begin mitotic divisions. The zygote develops into an embryo, and the triploid cell develops into endosperm (Figure 28.7). As the embryo approaches maturity, the outer cell layers of the ovule develop into a tough seed coat. The embryo sporophyte, its reserves of food, and the seed coat are now a mature ovule, a self-contained package called a **seed**. The seed may undergo dormancy until it receives signals that conditions in the environment are appropriate for germination.

As a seed is forming in an ovule, the parent plant transfers nutrients to it. These nutrients accumulate in endosperm mainly as starch with some lipids and proteins. In typical monocots, nutrient reserves stay in endosperm as the seed matures. By contrast, typical eudicot embryos transfer nutrients in endosperm to cotyledons as the seed matures, so most of the nutrients in a mature eudicot seed are stored in the two enlarged cotyledons. When a seed sprouts, nutrients stored in its endosperm or cotyledons will sustain rapid growth of the seedling until new leaves form and begin producing food by photosynthesis.

Nutrients stored in endosperm and cotyledons also sustain animals, including humans. For example, we cultivate many cereals—monocot grasses such as rice, wheat, rye, oats, and barley—for their nutritious seeds. The embryo (the germ) contains most of the seed's protein and vitamins, and the seed coat (the bran) contains most of the minerals and fiber. Milling removes the bran and germ, leaving only starch-packed endosperm.

Embryonic plants are nourished by nutrients in endosperm and cotyledons, but not by nutrients in fruits. A **fruit** is a seed-containing mature ovary, often with fleshy tissues that develop from the ovary wall. Apples, oranges, and grapes are familiar fruits, but so are many "vegetables" such as beans, peas, tomatoes, grains, eggplant, and squash.

Fruits may be categorized by the composition of their tissues, how they originate, and whether they are fleshy or dry (right). "True" fruits such as oranges consist only of the ovary wall and its contents (Figure 28.8A). "Accessory" fruits have tissues derived from other floral parts—petals, sepals, stamens, or receptacle—that expand along with the developing ovary. An apple is an accessory fruit; most of its flesh is an enlarged receptacle (Figure 28.8B). "Simple" fruits are derived from one ovary or a few fused ovaries. Cherries, pea pods, acorns, and shepherd's purse are examples.

A cherry is a true fruit, a simple fruit, and a fleshy fruit.

A blackberry is an accessory fruit, and an aggregate of individual fleshy fruits.

A pineapple is an accessory fruit, and a multiple fruit formed by the fusion of small, fleshy berries.

A pea pod is a true fruit, a simple fruit, and a dry fruit.

A strawberry is an accessory fruit, and an aggregate of individual dry fruits.

Figure 28.8 Parts of a fruit develop from parts of a flower.

Figure It Out: The flower that gave rise to the orange in **A** had how many carpels? Answer: 8

A. The tissues of an orange develop from the ovary wall.

B. The flesh of an apple is an enlarged receptacle.

By contrast, "aggregate" fruits such as blackberries are derived from separate ovaries of one flower that mature as a cluster. A "multiple" fruit such as a pineapple or a fig develops as a unit from several individually pollinated flowers.

Cherries, almonds, and olives are fleshy fruits, as are individual fruits of blackberries and other related species. Acorns, pea pods, and grains are dry fruits, as are the fruits of sunflowers, maples, and strawberries. Strawberries are not berries and their fruits are not fleshy. The red, juicy part of a strawberry is an enlarged receptacle with individual dry fruits on its surface. A "berry" is a fleshy fruit produced from one ovary. Grapes, tomatoes, citrus fruits, pumpkins, watermelons, and cucumbers are berries. Apples and pears are "pomes," fruits in which fleshy tissues derived from the receptacle enclose a core derived from the ovary wall.

The function of a fruit is to protect and disperse seeds. Dispersal increases reproductive success by minimizing competition for resources among parent and offspring. Just as flower structure is adapted to certain pollination vectors, so are fruits adapted to certain dispersal vectors: environmental agents of movement such as wind or water, or mobile organisms such as birds or insects. These adaptations are reflected in a fruit's form. For example, fruits dispersed by wind tend to be lightweight with breeze-catching specializations (Figure 28.9A). Fruits dispersed by water have water-repellent outer layers, and they float (Figure 28.9B). The fruits of many plants have specializations that facilitate dispersal by animals (Figure 28.9C). Some have hooks or spines that stick to the feathers, feet, fur, or clothing of more mobile species. Colorful, fleshy, or fragrant fruits attract birds and mammals that disperse seeds. The animal may eat the fruit and discard the seeds, or eat the seeds along with the fruit. Abrasion of the seed coat by teeth or by digestive enzymes in an animal's gut can help the seed germinate after it departs in feces.

A. Wind-dispersed fruits. Left, dry outgrowths of the ovary wall of a maple fruit form "wings" that catch the wind and spin the seeds away from the parent tree. Right, wind that lifts the hairy modified sepals of a dandelion fruit may carry the attached seed miles away from the parent plant.

B. Water-dispersed fruits. Left, fruits of sedges native to American marshlands have seeds encased in a bladderlike envelope that floats. Right, buoyant fruits of the coconut palm have tough, waterproof husks. They can float for thousands of miles in seawater.

C. Animal-dispersed fruits. Left, curved spines attach cocklebur fruits to the fur of animals (and clothing of humans) that brush past it. Right, the red, fleshy fruits of hawthorn plants are an important food source for cedar waxwings, which disperse the fruits' seeds in feces.

Figure 28.9 Some adaptations that aid fruit dispersal.

(A) left, R. Carr; right, © JupiterImages Corporation; (B) left, © T. M. Jones; right, Alex James Bramwell/Shutterstock; (C) left, © Robert H. Mohlenbrock © USDA-NRCS PLANTS Database; right, © Vishnevskiy Vasily/Shutterstock.

Take-Home Message 28.3

How do seeds and fruits develop?

- After double fertilization, the zygote develops into an embryo, and the endosperm becomes enriched with nutrients.
- A seed is a mature ovule that consists of an embryo sporophyte, its food reserves, and a seed coat.
- A fruit is a mature ovary, with or without accessory tissues that developed from other parts of the flower.
- A fruit protects and disperses seeds. Fruit specializations are adaptations to particular dispersal vectors.

fruit Mature ovary of a flowering plant, often with expanded accessory parts; encloses a seed or seeds.

seed Embryo sporophyte of a seed plant packaged with nutritive tissue inside a protective coat.

28.4 Early Development

REMEMBER: A form of a heritable trait that enhances an individual's fitness is called an adaptive trait, or evolutionary adaptation (Section 11.3). Monocots and eudicots differ in some details of their structure (27.2). Primary growth in plants (lengthening) originates at apical meristems in shoot and root tips (27.5).

An embryonic plant complete with shoot and root apical meristems forms as part of a seed. The embryonic shoot and root are separated by a tiny section of embryonic stem called hypocotyl (Figure 28.10).

As the seed matures, the embryo may enter a period of dormancy. A dormant embryo can rest in its protective seed coat for many years before it resumes metabolic activity and germinates. Seed dormancy is a climate-specific adaptation that allows germination to occur when conditions in the environment are most likely to support the growth of a seedling. For example, seeds of many annual plants native to cold winter regions are dispersed in autumn. If the seeds germinated immediately, the tender seedlings would not survive the upcoming winter. Instead, the seeds remain dormant until spring, when milder temperatures and longer day length favor the growth of seedlings. By contrast, the weather in regions near the equator does not vary by season. Seeds of most plants native to such regions do not enter dormancy; they germinate as soon as they mature.

Other than the presence of water, the triggers for germination differ by species. Some seed coats are so dense that they must be abraded or broken (by being chewed, for example) before germination can occur. Seeds of many cool-climate plants require exposure to freezing temperatures; those of some lettuce species, to bright light. In seeds of some species native to regions that have periodic wildfires, germination is inhibited by light and enhanced by smoke; in others, germination does not occur unless the seeds have been previously burned. Such requirements are evolutionary adaptations to life in a particular environment.

Germination typically begins with water seeping into a seed. Water causes the seed's internal tissues to swell, forcing the seed coat to rupture. Oxygen then diffuses into embryonic tissues. The water also provokes a series of events that result in enzymes breaking down stored starches into sugar subunits (we return to this topic in Section 28.6). Cells in the embryo's apical meristems use the sugars and the oxygen to run aerobic respiration. Energy released from the sugar breakdown fuels rapid divisions of the meristem cells, and the embryonic plant begins to grow. Germination ends when the embryonic root emerges from the seed coat.

After germination, the pattern of early growth varies. For example, in corn and other monocots, a rigid sheath called a coleoptile surrounds and protects the primary (first) shoot. In a typical monocot pattern of development (Figure 28.11), germination is followed by the emergence of the embryonic root and coleoptile from the seed coat **1**. The root grows down into the soil, becoming a primary root **2** as the coleoptile grows up **3**. When the coleoptile reaches the surface of the soil, it stops growing. The embryonic shoot develops into the primary (first) shoot as it emerges from the coleoptile **4**. The embryo's single cotyledon, which typically stays beneath the soil, functions mainly to transfer endosperm breakdown products to the seedling until leaves form on the shoot and begin photosynthesis **5**.

In eudicot seedlings, the primary shoot is not protected by a coleoptile. In a typical eudicot pattern of development (Figure 28.12), the embryonic root emerges from the seed **1**, followed by the hypocotyl. The hypocotyl bends into the shape of a hook as it lengthens **2**. The bent hypocotyl pulls the cotyledons upward through

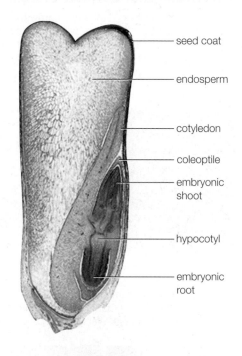

seed coat

endosperm

cotyledon

coleoptile

embryonic shoot

hypocotyl

embryonic root

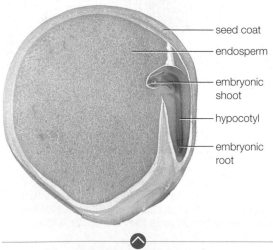

seed coat

endosperm

embryonic shoot

hypocotyl

embryonic root

Figure 28.10 Anatomy of a seed.
Top, corn, a monocot; bottom, pea, a eudicot.

As dormancy ends, cell divisions resume mainly at apical meristems of the embryonic shoot and root. These two structures are separated by hypocotyl, a section of embryonic stem.

In monocot grasses such as corn, the embryonic shoot is protected by a sheathlike coleoptile.

1 As a corn grain (seed) germinates, its embryonic root and coleoptile emerge from the seed coat.

2 The embryonic root develops into a primary root.

3 The coleoptile grows upward and opens a channel through the soil to the surface, where it stops growing.

4 The embryonic shoot develops into a primary shoot that emerges from the coleoptile.

5 The single cotyledon remains under the soil, transferring nutrients to the seedling until its new leaves can produce enough sugars by photosynthesis to sustain growth.

Figure 28.11 Early growth of corn, a typical monocot.

© Scott Sinklier/AgStock Images/Corbis.

1 A bean's seed coat splits and the embryonic root tip emerges.

2 The hypocotyl emerges from the seed and bends in the shape of a hook.

3 As the stem lengthens, it drags the two cotyledons upward toward the surface of the soil.

4 Exposure to light causes the stem to straighten.

5 Primary leaves emerge from between the cotyledons and begin photosynthesis.

6 Cotyledons typically undergo a period of photosynthesis before withering and falling off the stem.

Figure 28.12 Early growth of the common bean, a typical eudicot.

Nigel Cattlin/Visuals Unlimited.

the soil until it reaches the surface **3**. There, exposure to light causes the hypocotyl to straighten **4**. Primary leaves emerge from between the cotyledons and begin photosynthesis **5**. The cotyledons typically undergo a period of photosynthesis before shriveling up **6**. Eventually, the cotyledons fall off the lengthening stem, and the young plant's new leaves produce all of its food.

Take-Home Message 28.4

What happens during early plant development?

- After a seed germinates, nutrients stored in endosperm (of monocots) or cotyledons (of eudicots) support the seedling's growth until new leaves begin photosynthesis.
- A coleoptile protects the embryonic shoot of many monocot species.

A. Potatoes are tubers that grow on stolons. New roots and shoots sprout from their nodes, which we call "eyes." The new plants are clones, genetically identical with the parent.

B. Tiny new plants form at nodes on cladodes of many succulents. The new plants forming on this Mother of Thousands plant will break off the parent and fall to soil below, where they can take root.

C. In some plants, new individuals can form from mitotic divisions of cells in root tissue. The resulting clones are called root suckers. This photo shows suckers forming on an aspen root.

Figure 28.13 Examples of plant asexual reproduction.

(A) Image © Daniel Gale/Shutterstock.com; (B) Lisa Starr; (C) USDA/Photo by B. Campbell.

28.5 Asexual Reproduction

REMEMBER: Mitosis and cytoplasmic division are part of asexual reproduction (Section 8.2). A polyploid individual has three or more complete sets of chromosomes (9.8). New plant species can arise when the chromosome number multiplies (12.6). Stems have nodes, which can give rise to new shoots or roots; a corm is a short, thickened underground stem; cladodes are flattened, photosynthetic stems (27.3).

Most flowering plants can reproduce asexually, as when new roots and shoots grow from extensions or pieces of a parent plant. Each new plant is a clone, a genetic replica of its parent. For example, new roots and shoots can sprout from nodes on stems, and sometimes from roots (Figure 28.13). Entire forests of quaking aspen trees are actually stands of clones that grew from root suckers, which are shoots that sprout from shallow lateral roots. Suckers appear after aboveground parts of the aspens are damaged or removed. One stand of quaking aspens in Utah consists of about 47,000 shoots and stretches for 107 acres. Such clones are as close as any organism gets to being immortal. One of the oldest known plants is a clone: the one and only population of King's holly, which consists of several hundred stems along a river gully in Tasmania. Radiometric dating of the plant's fossilized leaf litter shows that the clone is at least 43,600 years old.

Agricultural Applications For thousands of years, we humans have been taking advantage of the natural capacity of many plants to reproduce asexually. For example, almost all houseplants, woody ornamentals, and orchard trees are clones grown from stem fragments (cuttings) of a parent plant. Propagating some plants from cuttings may be as simple as jamming a broken stem into the soil. This method harnesses the root- and shoot-forming ability of nodes on a stem. Other plants must be grafted. Grafting means inducing a cutting to fuse with and become supported by another plant. Often, the stem of a desired plant is grafted onto the roots of another.

Propagating a plant from cuttings ensures that offspring will have the same desirable traits as the parent. Consider familiar orchard apples, which are descendants of a wild species native to central Asia. Domesticated apple varieties are grafted because species in this genus do not breed true for fruit traits we value—color, flavor, size, sweetness, and texture (Figure 28.14). Most apple trees grown from seed produce unpalatable fruit. In the early 1800s, John Chapman (known as Johnny Appleseed) grew millions of apple trees from seed in the midwestern United States. He sold the trees to homesteading settlers, who made hard cider from the otherwise inedible apples. About one of every hundred trees produced sweet, tasty fruit; its lucky owner would graft the tree and patent it. An estimated 16,000 varieties of edible apples were discovered during this era. Few of them remain; only 15 varieties now account for 90 percent of the apples sold in U.S. grocery stores. All are clones of Chapman's original trees, and all are still grafted.

Grafting is also used to increase the hardiness of a desirable plant. In 1862, the plant louse *Phylloxera* was accidentally introduced into France via imported American grapevines. European grapevines had little resistance to this tiny insect, which attacks and kills the root systems of the vines. By 1900, *Phylloxera* had destroyed two-thirds of the vineyards in Europe, thus devastating the wine-making industry for decades. Today, French vintners routinely graft their grapevines onto the roots of *Phylloxera*-resistant American vines.

With **tissue culture propagation**, individual cells (typically stem cells from meristem) are coaxed to divide and form embryos in a laboratory. This technique can yield millions of genetically identical offspring from a single parent plant. It is frequently used in research aimed at improving food crops, and also to propagate rare ornamental plants such as orchids.

In some flowering plant species, fruits form even in the absence of fertilization; fruits that develop from unfertilized flowers have underdeveloped seeds or none at all. Plants with mutations that cause ovules or embryos to abort during development bear seedless fruit; such plants are necessarily sterile and must be propagated by grafting. Seedless grapes and navel oranges are like this. All commercially produced bananas are seedless because the plants that bear them are triploid ($3n$). During meiosis, three chromosomes cannot be divided equally between the two spindle poles; few viable gametes form (and few seeds form too). Grafting bananas and other monocots is notoriously difficult, so seedless banana plants are propagated by tissue culture or from shoots that sprout from nodes on their corms (Section 27.3).

Polyploid plants are common in nature, but it may take a long time for one to arise spontaneously. Plant breeders often use a microtubule poison called colchicine to artificially increase the frequency of polyploidy. Tetraploid ($4n$) plants produced by colchicine treatment are then crossed with diploid ($2n$) individuals. The resulting offspring are triploid and sterile: They make seedless fruit on their own or after pollination (but not fertilization) by a diploid plant. Seedless watermelons are produced commercially this way.

Take-Home Message 28.5

How do plants reproduce asexually?

- Many plants propagate themselves asexually when new shoots grow from a parent plant or pieces of it. Offspring are clones.
- Humans propagate plants asexually for agricultural or research purposes, for example by grafting or tissue culture.

Figure 28.14 Wild apples.
Color, flavor, size, sweetness, and texture vary in the fruits of wild apple trees (21 are shown in this photo). Apple trees are grafted because they do not breed true for these valuable traits.

Photo by Peggy Greb, USDA, ARS.

28.6 Plant Hormones

REMEMBER: Starch consists of coiled chains of glucose monomers (Section 2.7). Receptor proteins trigger a change in cellular activity in response to a stimulus (3.3). Feedback inhibition is a regulatory mechanism in which a change that results from an activity decreases or stops the activity (4.4). Gases diffuse directly through a lipid bilayer (4.5). Chlorophyll *a* is the main photosynthetic pigment in plants (5.2). Transcription factors affect whether and how fast a gene is transcribed (7.7). Endocrine cells secrete animal hormones (25.1). Stomata close to prevent water loss; phloem conducts organic compounds (27.4). All plant tissues originate with meristems (27.5).

In typical animals, most development occurs before adulthood. By contrast, most development in typical plants occurs at maturity. Plants, unlike animals, cannot move about to avoid unfavorable conditions. Instead, each plant adapts to its environment. It does so by adjusting development in response to cues such as temperature, gravity, night length, the availability of water and nutrients, and even the presence of pathogens or herbivores.

Developmental flexibility in plants depends on extensive coordination among individual cells. Cells in different tissues and even in different parts of a plant

A plant adjusts its development in response to environmental cues.

tissue culture propagation Laboratory method in which individual plant cells (typically from meristem) are induced to form embryos.

apical dominance Effect in which a lengthening shoot tip inhibits the growth of lateral buds.

auxin Plant hormone that causes cell enlargement; also has a central role in growth and development by coordinating the effects of other hormones.

cytokinin Plant hormone that promotes cell division in shoot apical meristem and cell differentiation in root apical meristem. Often opposes auxin's effects.

plant hormone Extracellular signaling molecule of plants that exerts its effect at very low concentration.

Table 28.1 Major Plant Hormones

Hormone	Effect
Abscisic acid (ABA)	
	Closes stomata in times of stress
	Inhibits shoot growth
	Inhibits seed germination
	Involved in environmental stress responses
	Involved in chloroplast movement
Auxin	
	Coordinates effects of other plant hormones during development of shoots and roots
	Role in apical dominance
	Causes cell enlargement
	Role in abscission
	Mediates tropisms
	Stimulates division of meristem cells
Cytokinin	
	Stimulates differentiation of cells in root apical meristem
	Stimulates cell division in shoot apical meristem
	Inhibits formation of lateral roots
Ethylene	
	Involved in breaking dormancy
	Stimulates flower organ development
	Stimulates fruit ripening
	Stimulates abscission
	Involved in stress responses
Gibberellin	
	Stimulates cell division, elongation in stems
	Mobilizes food reserves in germinating seeds
	Stimulates flowering in some plants

coordinate their activities by communicating with one another. As an example, a leaf being chewed by a caterpillar can signal other parts of the plant to produce appropriate caterpillar-deterring chemicals. Much of the cell-to-cell communication within plants involves hormones. A **plant hormone** is an extracellular signaling molecule that exerts its effect at very low concentrations (Table 28.1).

A cell's response to a hormone depends on the receptors it has, and, in plants, this response varies with the concentration of the hormone. The response also depends on the cell's integration of hormonal signals: One plant hormone can enhance or oppose another's effects on the same cell. Many hormones inhibit their own expression, a mechanism of negative feedback. Positive feedback loops in which a hormone promotes its own transcription are part of daily cycles of activity (we return to these cycles in Section 28.7).

Hormones operate in both plants and animals, but the origin, structure, and effects of these molecules differ between the two groups. For example, animals have organs specialized for hormone secretion; plants do not. All cells in a plant have the ability to make and release hormones. Animal hormones are defined by their ability to elicit an effect in a distant tissue. Similarly, a plant hormone may be released in a region of the body that is very far from cells in a targeted tissue, such as when cells in an actively growing root release cytokinins that keep shoot tips actively growing too. However, some plant hormones can have very local effects, such as when ethylene released from a cell in a ripening fruit affects the releasing cell as well as its neighbors.

Some molecules that act as hormones in plants act as hormones in animals, and vice versa. For example, plants make phytoestrogens that are structurally similar to estrogens, which are female sex hormones in mammals. The two types of compounds have similar hormonal effects in mammals, but phytoestrogens have nothing to do with sexual reproduction in plants. Rather, they function as part of plant defense mechanisms against fungal attack. The difference does not necessarily stem from differences in the structure of the hormones, but rather in the way the receptors for these compounds work in cells that bear them.

Auxin **Auxin** is a plant hormone that was first discovered for its ability to promote growth (Figure 28.15). It was later discovered to have a critical role in all aspects of plant development, starting with the first division of the zygote. Auxin is involved in polarity and tissue patterning in the embryo, formation of plant parts (primary leaves, shoot tips, stems, and roots), differentiation of vascular tissues, formation of lateral roots (and adventitious roots in some species), and, as you will see in the next section, shaping the plant body in response to environmental stimuli.

Researchers are still working out the mechanisms of many of these effects, but they do understand one way in which auxin promotes growth directly. During primary growth, the hormone causes young plant cells to expand by increasing the activity of transport proteins that pump hydrogen ions from cytoplasm into the cell wall. The resulting increase in acidity softens the wall. Turgor, the pressure exerted by fluid inside the softened wall, enlarges the cell irreversibly.

Auxin is present in almost all plant tissues, but is unevenly distributed through them. The hormone is made mainly in shoot apical meristems and in young leaves, then transported elsewhere. Some of the auxin produced in shoot tips is loaded into phloem, travels to roots, and then is unloaded into root cells. A cellular mechanism that distributes auxin directionally through adjacent cells dominates its movement over shorter distances. This mechanism is unique among plant hormones, and it is important because it establishes localized auxin concentration gradients. These gradients have a central role in coordinating the actions of other hormones.

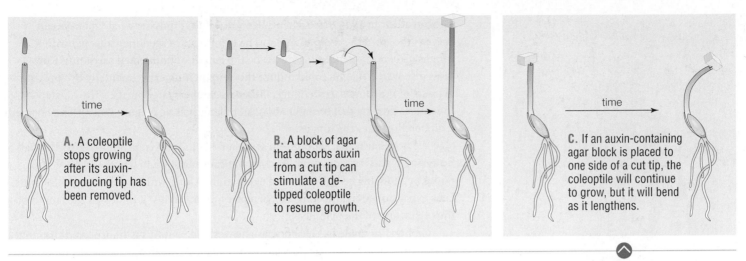

A. A coleoptile stops growing after its auxin-producing tip has been removed.

B. A block of agar that absorbs auxin from a cut tip can stimulate a de-tipped coleoptile to resume growth.

C. If an auxin-containing agar block is placed to one side of a cut tip, the coleoptile will continue to grow, but it will bend as it lengthens.

Figure 28.15 Experiments showing that a coleoptile lengthens in response to auxin produced in its tip.

Cytokinin The plant hormone **cytokinin** affects meristem cell division and differentiation, stimulates development of lateral buds, and inhibits development of lateral roots. Cytokinin and auxin influence one another's expression, a homeostatic mechanism that dynamically regulates their relative concentrations. They also influence the same cells. In shoot tips, for example, the two hormones act on shoot apical meristem cells to support cell division and to prevent differentiation. In most other contexts, cytokinin opposes auxin's effects. For example, in root tips, the two hormones work in opposition to maintain the balance of differentiating and undifferentiated cells in root apical meristem. In this context, auxin supports division of undifferentiated meristem cells, and cytokinin signals the cells to differentiate.

When a shoot is lengthening at the tip, its lateral buds are typically dormant, an effect called **apical dominance**. Auxin produced by apical meristem in the shoot's tip is transported down through cells of the stem, where it encourages the cells to lengthen. The presence of auxin in the stem also keeps the cytokinin level low. If a shoot's tip breaks off, its source of auxin disappears, so the cytokinin level rises in the stem (Figure 28.16A,B). The cytokinin moves into lateral buds and stimulates cell divisions of apical meristem inside them (Figure 28.16C). The newly active meristem cells produce auxin, which is then transported away from the bud tips and down the stem. The auxin flow causes the lateral buds to lengthen (Figure 28.16D).

Figure 28.16 Interaction of cytokinin and auxin in the release of apical dominance.

Above, Lisa Starr

A. Auxin is transported away from apical meristem and down through the stem. The presence of auxin in the stem keeps the level of cytokinin low.

B. Removing the tip ends auxin flow in the stem. As the auxin level declines, the cytokinin level rises.

C. The rise in the cytokinin level in the stem stimulates cell division in apical meristem of lateral buds. The cells begin to produce auxin.

D. Auxin is transported away from the lateral buds and down through the stem. This movement causes the lateral buds to lengthen.

Figure 28.17 Stem-lengthening effect of gibberellin.
The three tall cabbage plants to the right of the ladder were treated with gibberellin. The two short plants were not treated.

Sylvan H. Wittwer/ Visuals Unlimited.

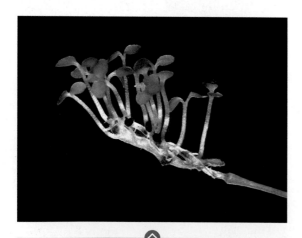

Figure 28.18 ABA prevents seed germination.
Part of the ABA signaling pathway in this thale cress plant has been knocked out. Its seeds germinated before they had a chance to disperse.

Taishi Umezawa, Kazuo Nakashima, Takuya Miyakawa, Takashi Kuromori, Masaru Tanokura, Kazuo Shinozaki, and Kazuko Yamaguchi-Shinozaki. Molecular Basis of the Core Regulatory Network in ABA Responses: Sensing, Signaling and Transport. Plant Cell Physiol (2010) 51(11): 1821–1839 first published online October 26, 2010. doi:10.1093/pcp/pcq156, by permission of Oxford University Press.

Gibberellin In 1926, researcher E. Kurosawa was studying what Japanese call *bakane*, the "foolish seedling" effect. The stems of rice seedlings infected with a fungus, *Gibberella fujikuroi*, grow twice the length of uninfected seedlings. Kurosawa discovered that he could induce this lengthening experimentally by applying extracts of the fungus to seedlings. Other researchers later purified the substance in the fungal extracts that brought about stem lengthening (Figure 28.17). They named it gibberellin, after the fungus.

We now know that **gibberellin** is a plant hormone that promotes growth in all flowering plants, among other functions. It causes a stem to lengthen between the nodes by inducing cell division and elongation. Along with auxin, gibberellin stimulates expansion along the long axis of a plant organ. Thus, it increases the length of the organ, and of the plant itself.

Gibberellin made by cells in young leaves and root tips is transported through phloem to the rest of the plant. Seeds also make this hormone. Seedless grapes tend to be smaller than seeded varieties because they have underdeveloped seeds that do not produce normal amounts of gibberellin. Farmers spray their seedless grape plants with synthetic gibberellin, which increases the size of the resulting fruit.

Gibberellin works by inhibiting inhibitors, thus removing the brakes on some cellular processes. For example, during germination of a barley seed, absorbed water causes cells of the embryo to release gibberellin. The hormone diffuses into a protein-rich layer of cells surrounding the endosperm. Gibberellin causes these cells to destroy a particular transcription factor—a molecule that suppresses transcription of the gene for amylase (an enzyme that breaks the bonds between glucose monomers in starch). The cells start producing amylase and releasing it into the endosperm's starchy interior, where the enzyme breaks down stored starch molecules. The embryo then begins to use the released glucose for aerobic respiration, which fuels rapid cell divisions of meristem cells in the embryonic shoot and root tips.

Abscisic Acid The plant hormone **abscisic acid** (ABA) was named because its discoverers thought it mainly mediated **abscission**, the process by which a plant sheds leaves or other parts. It was later discovered to have a much greater role in plant stress responses. Temperature extremes, lack of water, and other environmental stresses trigger an increase in ABA synthesis, release, and transport, so that the level of this hormone rises in the plant's tissues. In turn, the increase triggers expression of genes that help the plant survive the adverse conditions. For example, ABA is part of a response that causes a plant's stomata to close when water is scarce, thus preventing water loss by transpiration. It also has an important role in embryo maturation, seed and pollen germination, and fruit ripening; and like cytokinin it suppresses lateral root formation. ABA that accumulates in a seed as it forms, matures, and dries out prevents the seed from germinating too early. It exerts this effect mainly by inhibiting expression of genes involved in cell wall loosening and expansion—both critical processes for growth of an embryonic plant. ABA also inhibits expression of genes involved in gibberellin synthesis. Thus, a seed cannot germinate until its ABA level declines (Figure 28.18).

Ethylene The plant hormone **ethylene** is part of pathways that govern a wide range of metabolic and developmental processes in plants, including germination, growth, abscission, ripening, and stress responses. Ethylene is a gas that is soluble in water, and it freely crosses lipid bilayers. Two enzymes are involved in its synthesis, and each occurs in multiple versions encoded by slightly different genes. Expression of some of these genes is inhibited by ethylene (a negative feedback loop); expression petals drop   fruit forms   green fruit enlarges   fruit ripens   fruit is mature

Ethylene production

Days After Flower Opening

Figure 28.19 Ethylene production during strawberry formation and ripening.
Oscillations are normal daily cycles (see Section 28.7).

From left, (1) © Madlen/Shutterstock; (2) © Westend61/SuperStock; (3–4) © Anest/Shutterstock; (5–8) © Alena Brozova/Shutterstock; Graph data courtesy of Dr. Frans J.M. Harren and Dr. Simona M. Cristescu, Radboud University Nijmegen, The Netherlands.

of the others is enhanced by ethylene (a positive feedback loop). Negative feedback loops maintain a low level of ethylene that helps fine-tune ongoing metabolic and developmental processes such as growth and cell expansion. Positive feedback loops produce large amounts of ethylene required for intermittent processes such as germination, defense responses, abscission, and ripening (Figure 28.19). For example, when a fruit nears maturity, ABA triggers ethylene synthesis in a positive feedback loop. The resulting ethylene burst enhances transcription of genes whose products have several effects associated with ripening. The fruit changes color as (green) chlorophylls in its cells break down, and (red, orange, yellow, or purple) accessory pigments accumulate. Firm cell walls break down. Starch and acids are converted to sugars, and aromatic molecules are produced. The resulting color change, softening, and increased palatability attract animals that can disperse the fruit's seeds.

Because ethylene is a gas, it can diffuse from one fruit to initiate ripening in another. Humans use this property to artificially ripen several types of fruit. Hard, unripe fruit can be transported long distances with less damage than soft, ripe fruit. Upon arrival at its final destination, the fruit is exposed to synthetic ethylene, which jump-starts the positive feedback ripening loop.

Cells in different parts of a plant coordinate their activities by communicating with one another via hormones.

Take-Home Message 28.6

What coordinates growth and development in plants?

- Plant hormones are signaling molecules that coordinate activities among cells in different parts of the plant body. All cells in a plant can make them.
- Plant hormones are involved in all aspects of growth, development, and function. The different types have multiple functions and often affect the same cells.
- Auxin directly promotes cell enlargement during primary growth. It also coordinates the effects of other plant hormones involved in growth and development.
- Cytokinin and auxin act together and often antagonistically. The cytokinin–auxin balance controls cell division and differentiation in shoot and root apical meristem.
- Among other effects, gibberellin causes lengthening in plants by stimulating cell division and elongation along the long axis of stems and other organs.
- Abscisic acid is important in stress responses and stomata function.
- Ethylene produced in positive feedback loops is involved in intermittent processes such as fruit ripening.

abscisic acid Plant hormone involved in stress responses; inhibits germination.

abscission Process by which plant parts are shed.

ethylene Gaseous plant hormone involved in regulating growth and cell expansion. Participates in germination, abscission, ripening, and stress responses.

gibberellin Plant hormone that induces stem elongation and helps seeds break dormancy, among other effects.

28.7 Growth Responses

REMEMBER: A cytoskeleton of protein filaments is the basis of eukaryotic cell movement (Section 3.5). Pigments selectively absorb light of certain wavelengths (5.2). In the light-dependent reactions of photosynthesis, the flow of electrons through electron transfer chains drives ATP formation (5.3). In most eudicots, a short stalk called a petiole attaches the leaf to a stem (27.3).

Tropisms A plant responds to environmental stimuli by adjusting growth, thus optimizing its opportunities for photosynthesis, absorption of water and nutrients, reproduction, and other factors that affect its fitness. A directional adjustment of growth in response to environmental stimuli is called a **tropism**, and is typically mediated by hormones. Consider how, even if a seedling is turned upside down just after germination, its primary root and shoot will curve so the root grows down and the shoot grows up (Figure 28.20). This and any other growth response to gravity is a **gravitropism**.

A root or shoot "bends" because of differences in auxin concentration that occur after adjustments in the direction of local auxin transport. In shoots, auxin enhances cell elongation, so auxin that accumulates on one side of a shoot causes the shoot to bend away from that side. Auxin has the opposite effect in roots: Auxin that accumulates on one side of a growing root causes the root to bend toward that side. Gravity sensing in many organisms is based on organelles called statoliths. Plant statoliths are stuffed with dense grains of starch. They occur in root cap cells and also in specialized cells at the edge of vascular tissues in the stem. Starch grains are heavier than cytoplasm, so statoliths tend to sink to the lowest region of the cell, wherever that is (Figure 28.21). The shift causes the flow of auxin through cells in the root or shoot to be directed toward the down-facing side.

A **phototropism** is a growth response that orients parts of a plant in a direction influenced by light. For example, an elongating stem curves toward a light source, thus maximizing the amount of light intercepted by its photosynthetic cells. In shoots, phototropism occurs in response to blue light absorbed by nonphotosynthetic pigments called phototropins. In a shoot tip or coleoptile, light-energized

A. Regardless of how a corn seed is oriented in soil, the seedling's primary root always grows down, and its primary shoot always grows up.

B. These seedlings were rotated 90° counterclockwise after they germinated. The plants adjusted to the change by redistributing auxin, and the direction of growth shifted as a result.

C. Auxin transport was inhibited in these seedlings. They were also rotated 90° counterclockwise after germination, but the direction of growth did not shift.

Figure 28.20 Gravitropism.

(A) Michael Clayton, University of Wisconsin; (B, C) © Muday, GK and P. Haworth (1994) *Tomato root growth, gravitropism, and lateral development: Correlations with auxin transport.* Plant Physiology and Biochemistry 32, 193–203, with permission from Elsevier Science.

Figure 28.21 Gravity and statoliths.

Micrographs courtesy of Randy Moore from *How Roots Respond to Gravity*, M. L. Evans, R. Moore, and K. Hasenstein, Scientific American, December 1986.

— statoliths

A. This micrograph shows heavy, starch-packed statoliths settled on the bottom of gravity-sensing cells in a corn root cap.

Figure It Out: In which direction was this root rotated?

Answer: Counterclockwise

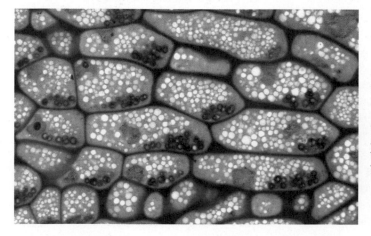

B. This micrograph was taken ten minutes after the root in **A** was rotated 90°. The statoliths are already settling to the new "bottom" of the cells.

A. Sunlight strikes only one side of a coleoptile. Auxin flow is directed toward the shaded side, so cells on that side lengthen more.

B. A shamrock plant adjusting its growth in response to a directional light source.

A. This is a section of a leaf of millet before exposure to an intense light source.

Figure 28.22 Phototropism.
Auxin-mediated differences in cell elongation between two sides of a coleoptile-protected primary shoot induce bending toward light. Red dots signify auxin molecules.

(B) © Cathlyn Melloan/Stone/Getty Images.

B. After 16 hours of exposure to an intense light source, chloroplasts have moved away from the external surfaces of the leaf, toward internal tissues that are more shielded from the light.

phototropins lift the brakes on local auxin transport. Thus, a stem exposed to a directional light source ends up with auxin flowing toward its shaded side, so cells on that side elongate more than cells on the illuminated side (Figure 28.22A). The difference causes the entire structure to bend toward the light as it lengthens (Figure 28.22B).

In another example of phototropism, chloroplasts are dragged from one position to another along actin filaments of the cytoskeleton depending on light intensity. In low light, chloroplasts are directed toward leaf surfaces and oriented perpendicular to the light source, thus maximizing their exposure to light for photosynthesis. In high-intensity light, chloroplasts move in the opposite direction—toward internal tissues of the leaf where they are more shielded (Figure 28.23), and oriented parallel to the light. Moving away from high intensity light minimizes thylakoid membrane damage, which can occur as a result of excess electrons accumulating in electron transfer chains of photosynthesis. Blue light drives the movement in both cases.

A plant's contact with an object may cause a change in the direction of its growth, a response called thigmotropism (Greek *thigma* means touch). We see thigmotropism when a vine's tendril coils around an object such as a wire or another

Figure 28.23 Movement of chloroplasts in response to intense light.
Mesophyll (**M**), bundle sheath cells (**B**), and vascular bundles (**V**).

Eri Maai, Shouu Shimada, Masahiro Yamada, Tatsuo Sugiyama, Hiroshi Miyake, Mitsutaka Taniguchi; *The avoidance and aggregative movements of mesophyll chloroplasts in C4 monocots in response to blue light and abscisic acid*; Journal of Experimental Botany, doi:10.1093/jxb/err008, by permission of Oxford University Press.

part of the plant (left). The mechanism that underlies thigmotropism is not well understood, but an immediate increase in cytoplasmic concentration of calcium that accompanies contact is likely to play a role. Plants (and animals) have plasma membrane transport proteins that flood cytoplasm with calcium ions upon mechanical disturbance of the membrane. Several gene products that can sense calcium ions are involved in the response to the flood of ions. These molecules trigger unequal growth rates of cells on opposite sides of the shoot tip, which causes the shoot to coil as it grows. A similar mechanism causes roots to grow away from contact, so they "feel" their way around rocks and other impassable objects in soil.

gravitropism In plants, directional growth response to gravity.

phototropism In plants, directional growth response to light.

tropism In plants, directional growth response to an environmental stimulus.

1 A.M.   6 A.M.   Noon   3 P.M.   10 P.M.   Midnight

Figure 28.24 A circadian rhythm: leaf movements by a young bean plant.
Physiologist Frank Salisbury kept this plant in darkness for twenty-four hours. Despite the lack of light cues, the leaves continued to unfold at sunrise (6 A.M.) and fold at sunset (6 P.M.).

Frank B. Salisbury.

Figure 28.25 Circadian cycles of gene expression for some genes involved in temperature stress responses.
Plants were kept in constant light; gray bars show night-time hours.

Molecules that help the plants tolerate cold nights and hot days are produced just before they are actually needed. The blue line shows the averaged expression of 46 genes involved in cold tolerance; the red line, averaged expression of 30 genes involved in heat tolerance.

Figure It Out: At what time of day is expression of cold tolerance genes lowest?   Answer: Early morning

Photoperiodic Responses   A **circadian rhythm** is a cycle of biological activity that repeats every twenty-four hours or so (circadian means "about a day"). For example, a bean plant holds its leaves horizontally during the day but folds them close to its stem at night. Bean plants exposed to constant light or constant darkness for a few days will continue to move their leaves in and out of the "sleep" position at the time of sunrise and sunset (Figure 28.24). Similar mechanisms cause flowers of some plants to open only at certain times of day. For example, flowers of plants pollinated mainly by night-flying bats open, secrete nectar, and release fragrance only at night. Periodically closing protects the delicate reproductive parts when the likelihood of pollination is typically lowest.

Circadian rhythms are driven by interconnected feedback loops involving transcription factors that regulate their own expression. These loops give rise to cycles in the levels of more than 30 percent of a plant cell's mRNAs (Figure 28.25). Cyclic shifts in gene expression underlie cyclic shifts in metabolism, for example between daytime starch-building reactions and nighttime starch-consuming reactions. Components of photosystems, ATP synthases, and other proteins used in photosynthesis are among the many gene products produced during the day and broken down at night. Similarly, many molecules used at night are broken down during the day.

At least six types of pigments provide light input into internal circadian "clocks," but the best known are phytochromes and cryptochromes. Phytochromes absorb red light; absorbing light with a wavelength of 660 nanometers (red) causes their structure to change from an inactive to an active form. Absorbing light with a wavelength of 730 nanometers (far-red, which predominates in shade) changes the molecule back to its inactive form. Cryptochromes absorb blue and UV light, thereby becoming activated and, interestingly, magnetic.

Except at the equator, the length of day varies with the season. Days are longer in summer than in winter, and the difference increases with latitude. These seasonal changes in light availability trigger seasonally appropriate responses in plants. **Photoperiodism** refers to an organism's response to changes in the length of day relative to night.

Flowering is a photoperiodic response in many species. Such species are termed long-day or short-day plants depending on the season in which they flower. These terms are somewhat misleading, as the main trigger for flowering is the length of night, not the length of day (Figure 28.26). Irises and other long-day plants flower only when the hours of darkness fall below a critical value, typically in summer. Chrysanthemums and other short-day plants flower only when the hours of darkness are greater than some critical value. Flowering is not photoperiodic in sunflowers, tomatoes, roses, and other day-neutral plants.

Some plants flower in spring only after exposure to a prolonged period of cold in the preceding winter, a response called **vernalization**. Researchers suspect that plant cells perceive temperature via their plasma membrane, because the lipid composition and calcium ion permeability of plasma membranes vary with temperature.

A. A flash of red light (660 nm) interrupting a long night causes plants to respond as if the night were short. Long-day plants flower; short-day plants do not.

B. A flash of far-red light (730 nm) cancels the effect of a red light flash. Short-day plants flower; long-day plants do not.

Figure 28.26 Experiments showing that long- or short-day plants flower in response to night length.
Each horizontal bar represents 24 hours. Blue indicates night length; yellow, day length.

Figure It Out: Which type of pigment detected the light flashes in these experiments?

Answer: A phytochrome

Regardless of the mechanism of detection, a "cold" signal influences gene expression in these plant species.

Yearly cycles of abscission and dormancy are photoperiodic responses too. Plants that drop their leaves before dormancy are typically native to regions too dry or too cold for optimal growth during part of the year. For example, deciduous trees of the northeastern United States lose their leaves in autumn, around September. The trees remain dormant during the months of harsh winter weather that would otherwise damage tender leaves and buds. Growth resumes in spring (April), when milder conditions return. On the opposite side of the world, many tree species native to tropical monsoon forests of south Asia lose their leaves during the summer dry season, which is between November and May. The region receives a lot of rain annually, but almost none of it falls during this period of the year. Dormancy offers the trees a way to survive the seasonal droughtlike conditions. New growth appears at the beginning of June, just in time to be supported by the ample water of monsoon rains.

Seasonal abscission of leaves (and fruits) is mediated by ethylene. Let's use a horse chestnut tree as an example. In this species, most root and shoot growth occurs between spring and early summer, from April to July. By midsummer, the tree is producing fruits and seeds. In August, the growing season is coming to a close, and nutrients are being routed to stems and roots for storage during the forthcoming period of dormancy. Ripe fruits and seeds (Figure 28.27A) release ethylene that diffuses into nearby twigs, petioles, and fruit stalks. The ethylene triggers cells in these zones to produce wall-digesting enzymes. The cells bulge and separate from one another as their walls soften. Tissue in the abscission zone weakens, and the structure above it drops. A scar often remains where the structure had been attached (Figure 28.27B).

A. Ripening chestnuts emit ethylene that stimulates abscission of the fruits and nearby leaves.

B. Horse chestnut trees take their name from abscission zones that leave horseshoe-shaped scars in this species.

Figure 28.27 Abscission as part of the normal life cycle of deciduous trees such as chestnuts.

(A) rhjelsand/iStockphoto.com; (B) © Adrian Chalkley.

Take-Home Message 28.7

How do environmental cues trigger growth responses in plants?

- Plants adjust the direction and rate of growth in response to environmental stimuli such as gravity, light, and contact. Hormones are typically part of these tropic responses.
- Plants respond to recurring cues from the environment with recurring cycles of activity. Pigments provide input into circadian cycles.
- The main environmental cue for flowering is the length of night relative to the length of day, which varies seasonally in most places. Low winter temperatures stimulate spring flowering in some plant species.
- The timing of seasonal abscission and dormancy is an evolutionary adaptation to recurring periods of environmental conditions that do not favor growth.

circadian rhythm A cycle of biological activity that repeats about every 24 hours.

photoperiodism Biological response to seasonal changes in the relative lengths of day and night.

vernalization Stimulation of flowering in spring by prolonged exposure to low temperature in winter.

Summary

Section 28.1 Colony collapse disorder (CCD) is killing honeybees. Declines in populations of bees and other **pollinators** negatively affect plant populations as well as other animal species that depend on the plants, including humans. Widely used neonicotinoid insecticides that are toxic to bees and increase their susceptibility to infection may contribute to CCD.

Section 28.2 **Flowers** are the reproductive structures of angiosperms. A whorl of sepals surrounds a whorl of **petals**, which in turn surround whorls of **stamens** and **carpels**. A carpel consists of a **stigma**, often a style, and an **ovary** inside which one or more **ovules** develop. A cell in an ovule gives rise to a female gametophyte, which contains an egg and an endosperm mother cell. A stamen consists of an **anther** on a thin filament. Cells in pollen sacs of anthers produce **pollen grains** that are released after a period of **dormancy**. Flower specializations (shape, pattern, color, and fragrance) are often adaptations to a particular **pollination vector**. Coevolved pollinators receive **nectar**, pollen, or another reward for visiting a flower. A pollen grain arrives on a receptive stigma at **pollination**. Upon **germination**, a pollen grain develops into a pollen tube that contains two sperm cells. The tube grows through the carpel and releases the sperm cells into the ovule. In **double fertilization**, one of the sperm cells fertilizes the egg, forming a zygote; the other fuses with the endosperm mother cell and gives rise to triploid **endosperm**.

Section 28.3 As a zygote develops into an embryo, endosperm collects nutrients from the parent plant, and the ovule's outer layers develop into a seed coat. A **seed** is a mature ovule: an embryo sporophyte and endosperm enclosed within a seed coat. Nutrients stored in endosperm or cotyledons make seeds a nutritious food source. As an embryo sporophyte develops, the ovary wall and sometimes other tissues mature into a **fruit**. Fruit specializations are adaptations to seed dispersal by specific vectors.

Section 28.4 Seeds often undergo a period of dormancy that does not end until species-specific environmental cues trigger germination. The embryonic root emerges from the seed coat upon germination; other patterns of early development vary.

Section 28.5 Many types of flowering plants can produce clonal offspring by reproducing asexually. Some are propagated commercially by grafting; the common laboratory technique of **tissue culture propagation** is used to clone some plants.

Section 28.6 Plants continue to develop through their lifetime. **Plant hormones** promote or arrest development in specific areas of a plant body by stimulating or inhibiting cell division, differentiation, or enlargement. Plant hormones work together or in opposition, and many have different effects in different regions of the plant body. **Auxin** promotes lengthening and also coordinates the effects of other hormones involved in growth. **Apical dominance** is a result of auxin transport through a growing shoot tip. **Cytokinin** acts together with auxin, often antagonistically, to balance growth with development in shoot and root tips. **Gibberellin** lengthens stems between nodes, and it is required for seed germination. **Abscisic acid** (ABA) plays a part in **abscission**, but has a greater role in seed germination and stress responses. Bursts of **ethylene** produced in positive feedback loops trigger special processes such as abscission and fruit ripening.

Section 28.7 In plants, a **tropism** is an adjustment in the direction of growth in response to environmental cues such as contact, gravity (**gravitropism**), or light (**phototropisms**). **Circadian rhythms** are driven by gene expression feedback loops that have input from nonphotosynthetic pigments. A **photoperiodism** is a response to change in the length of day relative to night. Seasonal cycles of abscission and dormancy are adaptations to seasonal changes in environmental conditions. Some plants require prolonged exposure to winter cold before they can flower (**vernalization**).

Self-Quiz

Answers in Appendix I

1. The arrival of pollen grains on a receptive stigma is called _____ .
   a. germination   c. pollination
   b. abscission   d. propagation

2. A pollinator may receive _____ when it visits a flower of a coevolved plant (choose all that apply).
   a. pollen   c. insecticides
   b. nectar   d. fruit

3. In flowers, the structure that produces male gametophytes is the _____ , and the one that produces female gametophytes is the _____ .
   a. pollen grain; flower   c. anther; stigma
   b. stamen; carpel   d. megaspore; microspore

4. The _____ of a flower contains one or more ovaries in which eggs develop, fertilization occurs, and seeds mature.
   a. pollen sac
   b. carpel
   c. receptacle
   d. sepal

5. Seeds are mature _____ ; fruits are mature _____ .
   a. ovaries; ovules
   b. ovules; stamens
   c. ovules; ovaries
   d. stamens; ovaries

6. Cotyledons develop as part of _____ .
   a. carpels
   b. accessory fruits
   c. embryo sporophytes
   d. flowers

7. In some species, exposure to _____ is a trigger for seed germination.
   a. light
   b. cold
   c. smoke
   d. all can be triggers

8. The three main parts of a mature eudicot seed are the _____ .
   a. pollen grain, egg, and seed coat
   b. embryo, endosperm, and seed coat
   c. megaspores, microspores, and ovule
   d. embryo, cotyledons, and seed coat

9. A new plant forms from a stem that broke off of the parent plant and fell to the ground. This is an example of _____ .
   a. tissue culture propagation
   b. exocytosis
   c. asexual reproduction
   d. nodal cloning

10. Banana plants produce seedless fruit because they are _____ .
    a. triploid
    b. monocots
    c. propagated by grafting
    d. treated with insecticides

11. Plant hormones _____ .
    a. often have multiple, overlapping effects
    b. are active in developing plant embryos
    c. are active in adult plants
    d. all of the above

12. Which of the following statements is false?
    a. Auxin and gibberellin promote stem elongation.
    b. Cytokinin promotes cell division in shoot tips.
    c. Abscisic acid promotes water loss and dormancy.
    d. Ethylene promotes fruit ripening and abscission.

13. In some plants, flowering is a _____ response.
    a. phototropic
    b. gravitropic
    c. photoperiodic
    d. thigmotropic

14. Match the response with its main trigger.
    _____ phototropism   a. contact with an object
    _____ gravitropism   b. blue light
    _____ thigmotropism   c. a long period of cold
    _____ photoperiodism   d. gravity
    _____ vernalization   e. night length

15. Match the observation with the hormone.
    _____ ethylene
    _____ cytokinin
    _____ auxin
    _____ gibberellin
    _____ abscisic acid

    a. Your cabbage plants bolt (they form elongated flowering stalks).
    b. The potted plant in your room is leaning toward the window.
    c. The last of your apples is getting really mushy.
    d. The seeds of your roommate's marijuana plant do not germinate no matter what he does to them.
    e. Lateral buds are sprouting.

Critical Thinking

1. The oat coleoptiles of the seedlings on the right have been modified: either cut or placed in a light-blocking tube. Which ones will still bend toward a directional light source?

2. Certain mutations in thale cress cause excess auxin production. Predict the impact of these mutations on the plants' phenotype.

3. Cattle in industrial dairy farms are typically given rBST, an animal hormone that increases a cow's milk production. There is concern that such hormones may have unforeseen effects on milk-drinking humans. Why is there much less concern about plant hormones in the plant foods we eat?

Visual Question

1. Is the seedling shown on the right a monocot or eudicot? How can you tell?

© Jubal Harshaw/ Shutterstock.

Answers to Self-Quizzes

Chapter 1

1.	b	1.2
2.	c	1.2
3.	d	1.3
4.	a	1.3
5.	c	1.3
6.	Animals	1.4
7.	c	1.3
8.	d	1.3
9.	a, d, e	1.2–1.4
10.	a, b	1.2, 1.4
11.	domains	1.4
12.	b	1.5
13.	b	1.5, 1.6
14.	c	1.2
	e	1.6
	b	1.4
	g	1.6
	d	1.5
	a	1.5
	f	1.3
15.	b	1.6

Chapter 2

1.	d	2.2
2.	Hydrogen	2.2
3.	a	2.3
4.	c	2.4
5.	c	2.4
6.	a	2.5
7.	e	2.7, 2.10
8.	c	2.8
9.	e	2.8
10.	b	2.9, 2.10
11.	d	2.9
12.	c	2.4
	b	2.2
	d	2.4
	a	2.2
	f	2.4
	e	2.2
	g	2.4
13.	a	2.10
14.	c	2.8
	e	2.7
	f	2.8
	d	2.10
	a	2.9
	b	2.10
15.	f	2.9
	a	2.8
	b	2.8
	c	2.10
	d	2.7
	e	2.10
	h	2.8
	g	2.9

Chapter 3

1.	c	3.2
2.	c	3.2, 3.4
3.	c	3.2
4.	c	3.2
5.	False	3.2
6.	b	3.3
7.	c	3.5
8.	c	3.3
9.	a	3.5
10.	a	3.5
11.	c, b, d, a	all 3.5
12.	d	3.5
13.	a	3.5
14.	b	3.5
15.	c	3.5
	f	3.5
	e	3.4, 3.5
	d	3.5
	a	3.5
	g	3.4, 3.5
	b	3.5

Chapter 4

1.	c	4.2
2.	b	4.2
3.	a	4.3
4.	c	4.3
5.	c	4.3, 4.4
6.	a	4.4
7.	d	4.4
8.	b	4.5, 4.6
9.	more, less	4.5
10.	c	4.5, 4.6
11.	a	4.5
12.	b	4.5
13.	b	4.6
14.	d	4.6
15.	c	4.3
	e	4.6
	f	4.2
	b	4.3
	a	4.4
	g	4.5
	h	4.6
	j	4.4
	i	4.4

Chapter 5

1.	b	5.1
2.	a	5.1, 5.2
3.	c	5.2
4.	b	5.3
5.	c	5.3
6.	c	5.3
7.	b	5.4
8.	a	5.4
9.	False	5.5
10.	c	5.5
11.	b	5.5
12.	d	5.6
13.	d	5.7
14.	a and d	5.3 and 5.5
15.	c	5.5
	a	5.5
	d	5.5
	f	5.2
	g	5.1
	e	5.4
	b	5.2, 5.3

Chapter 6

1.	d	6.1
2.	c	6.2
3.	c	6.2
4.	b	6.2
5.	b	6.2
6.	a	6.3
7.	b	6.3
8.	b	6.3
9.	a	6.4
10.	d	6.4
11.	c	6.4
12.	d	8.3, 8.4
13.	f	6.4
14.	d	6.4
15.	d	6.2
	c	6.1
	b	6.3
	a	6.4
	f	6.4
	e	6.2
	g	6.4

Chapter 7

1.	c	7.2
2.	c	7.2
3.	a	7.2
4.	b	7.2, 7.4
5.	a	7.3, 7.5
6.	a	7.4
7.	c	7.4
8.	d	7.5
9.	d	7.6
10.	b	7.7
11.	d	7.7
12.	d	7.7
13.	c	7.7
14.	True	7.7
15.	c	7.2, 7.3
	a	7.3
	d	7.5
	b	7.2, 7.5
16.	c	7.7
	g	7.6
	h	7.3
	e	7.4
	a	7.7
	f	7.7
	b	7.7
	d	7.5

Chapter 8

1.	d	8.2
2.	b	8.2
3.	a	8.2
4.	a	8.2
5.	c	8.2
6.	a	8.2
7.	b	8.4
8.	c	8.4
9.	b	8.5
10.	b	8.5
11.	d	8.5
12.	b	8.5
13.	b	8.5
14.	b, e, a, c, d	all 8.2
15.	c	8.2
	f	8.2
	a	8.3
	h	8.2
	g	8.5
	e	8.3
	b	8.5
	d	8.5

Chapter 9

1.	b	9.2
2.	a	9.2
3.	c	9.3
4.	c	9.3
5.	b	9.7
6.	False	9.4, 9.5
7.	c	9.4
8.	d	9.5
9.	b	9.6
10.	X from mom, Y from dad	9.7
11.	b	9.8
12.	True	9.8
13.	c	9.8, 9.9
14.	b	9.3
	d	9.3
	a	9.2
	c	9.2
15.	b	9.8
	a	9.6
	d	9.8
	e	9.2
	f	9.2
	c	9.7

Chapter 10

1.	c	10.2
2.	a	10.2
3.	b	10.2
4.	b	10.2
5.	e	10.2, 10.3
6.	b	10.3
7.	a	10.2
8.	b	10.3
9.	b	10.5
	d	10.2, 10.4
	e	10.5
	f	10.4
10.	e, a, d, b, c	10.2, 10.3
11.	c	10.4
12.	True	10.4
13.	b	10.5
14.	True	10.5
15.	c	10.3
	f	10.4
	d	10.5
	b	10.1
	a	10.4
	e	10.4

Chapter 11

1.	b	11.2
2.	c	11.2, 11.6
3.	d	11.3
4.	b	11.3
5.	b	11.4
6.	True	11.4
7.	c	11.4
8.	a	11.2, 11.4, 11.5
	c	11.5
	d	11.4, 11.5
	e	11.2, 11.4, 11.5
	f	11.1
9.	Gondwana	11.5
10.	c	11.1
11.	d	11.6
12.	d	11.6
13.	b	11.7
14.	j	11.3
	i	11.4
	e	11.3
	k	11.7
	f	11.4
	c	11.3
	b	11.3
	g	11.6
	h	11.6
	d	11.4
	a	11.7
15.	e	11.6

Chapter 12

1.	a	12.2
2.	d	12.5
3.	a, b	12.3
4.	d	12.4
5.	b	12.4
6.	b	12.5
7.	f	12.2, 12.3, 12.5
8.	a	12.6
9.	c	12.5, 12.6
10.	a	12.6
11.	d	12.4, 12.6
12.	d	12.6
13.	d	12.7, 12.8
14.	a	12.8
15.	c	12.5
	e	12.4
	i	12.8
	g	12.7
	b	12.5
	j	12.2, 12.3
	h	12.8
	f	12.7
	d	12.8
	a	12.7

Chapter 13

1.	c	13.3
2.	b	13.2
3.	c	13.2
4.	b	13.3
5.	c	13.4
6.	c	13.5
7.	d	13.4
8.	d	13.6
9.	c	13.6
10.	b	13.6
11.	a	13.6
12.	d	13.6
13.	c	13.5
14.	d	13.4
15.	g	13.6
	d	13.4
	c	13.5
	e	13.6
	i	13.6
	f	13.6
	a	13.6
	j	13.6
	b	13.6
	h	13.3

Chapter 14

1.	d	14.6, 14.7
2.	c	14.5
3.	b	14.3
4.	c	14.4
5.	a	14.4
6.	d	14.5, 14.6
7.	b	14.5–14.7
8.	c	14.6
	d	14.2
	f	14.4
	e	14.2
	a	14.2
	b	14.2
	g	14.7
9.	c	14.8
10.	a	14.9
11.	c	14.8
12.	d	14.9
13.	b	14.9
14.	d	14.8
15.	g	14.9
	f	14.8
	b	14.8
	c	14.8
	a	14.8
	d	14.8
	e	14.9

Chapter 15

1.	True	15.2
2.	b	15.2
3.	c	15.2
4.	b	15.2
5.	c	15.3
6.	c	15.3
7.	Notochord, nerve cord, pharynx with gill slits, tail that extends beyond anus. Pharynx with gill slits.	15.3
8.	b	15.5
9.	b	15.4, 15.6
10.	f	15.6
11.	a	15.6
12.	False	15.7
13.	d	15.7
14.	b	15.3
	g	15.3
	m	15.3
	j	15.3
	e	15.3
	c	15.3
	d	15.3
	f	15.3
	k	15.5
	a	15.5
	h	15.6
	i	15.6
	l	15.7
15.	1-b	15.2
	2-a	15.2
	3-f	15.4
	4-c	15.5
	5-d	15.6
	6-e	15.7

Chapter 16

1.	a	16.2
2.	c	16.1
3.	b	16.3
4.	b	16.3
5.	a	16.4
6.	a	16.3
7.	b	16.4
8.	a	16.3
9.	a	16.5
10.	d	16.5
11.	c	16.3
12.	b	16.5
13.	c	16.3
14.	d	16.5
15.	d	16.3
	c	16.3
	e	16.3
	a	16.5
	f	16.3
	b	16.4

Chapter 17

1.	a	17.1
2.	c	17.3
3.	b, c, a, d, e	17.3
4.	b	17.3
5.	c	17.3
6.	a	17.4
7.	d	17.5
	a	17.3
	c	17.5
	b	17.5
	e	17.4
8.	b, a, b, c	17.6
9.	c	17.6
10.	b	17.6
11.	a	17.6
12.	b	17.6
13.	d	17.3
14.	d	17.6
15.	b	17.4

Chapter 18

1.	a	18.2
2.	d	18.2
3.	b	18.3
4.	d	18.3
5.	b	18.3
6.	c	18.3
7.	a	18.4
8.	a	18.4
9.	e	18.3
	d	18.3
	b	18.3
	c	18.3
	a	18.4
	f	18.3
	g	18.4
	h	18.3
10.	b	18.5
11.	c	18.6
12.	b	18.5
13.	d	18.5
14.	d	18.5
15.	c	18.5

Chapter 19

1.	a	19.3
2.	b	19.3
3.	a	19.3
4.	a	19.3
5.	a	19.3
6.	b	19.3
7.	c	19.3
8.	b	19.3
9.	c	19.3
10.	a	19.3
11.	d	19.3
12.	a	19.3
13.	d	19.4
14.	a	19.5
15.	b, h, a, f, e, c, k, i, g, d, j	all 19.3

Chapter 20

1.	a	20.2
2.	d	20.2
3.	d	20.2
4.	b	20.4
5.	a	20.2
6.	a	20.2
7.	b, e, c, d, a	20.4
8.	d	20.4
9.	c	20.4
10.	b	20.4
11.	d	20.4
12.	a	20.4
13.	d	20.2
14.	b, e, c, d, a	all 20.2
15.	c	20.5
	f	20.1
	g	20.4
	a	20.4
	h	20.2
	d	20.3
	e	20.3
	j	20.2
	i	20.2
	b	20.5

Chapter 21

1.	b	21.2
2.	b	21.4
3.	a	21.3
4.	c	21.2
5.	c	21.5
6.	c	21.5
7.	a	21.2
8.	a	21.4
9.	c	21.8
10.	d	21.8
11.	a	21.8
12.	a	21.8
13.	b	21.8
14.	g	21.6
	h	21.6
	e	21.6
	b	21.8
	d	21.6
	a	21.1
	c	21.8
	f	21.8
15.	b	21.5
	c	21.8
	h	21.5, 21.6
	g	21.3
	f	21.8
	d	21.8
	e	21.2
	a	21.2

Chapter 22

1.	d	22.2
2.	e	22.3
3.	d	22.3
4.	g	22.3–22.5
5.	h	22.3–22.5
6.	a	22.2
7.	d	22.4
8.	d	22.2–22.5
9.	a	22.4
10.	b	22.5
11.	d	22.5
12.	b	22.5
13.	c	22.6
14.	b	22.5
	d	22.4, 22.5
	c	22.5
	e	22.5
	a	22.5
15.	f	22.6
	e	22.5
	d	22.6
	c	22.3
	b	22.6
	a	22.2, 22.4
	g	22.4

Chapter 23

1. d	23.2
2. a	23.3
3. c	23.3
4. b	23.3
5. c	23.3
6. g, b, a, d, e	23.3
c	23.2
f	23.1
7. d	23.4
8. c	23.3
9. b	23.4
10. c	23.5
11. a	23.6
12. a	23.6
13. b	23.6
14. c	23.6
15. c	23.5
a	23.2
b	23.5
e	23.6
d	23.5

Chapter 24

1. a	24.3
2. c	24.3
3. d	24.3
4. b	24.5
5. a	24.3
6. c	24.5
7. reflex	24.5
8. c	24.4
9. b	24.4
10. f	24.6
11. c	24.6
12. a	24.6
13. d	24.6
14. d	24.6
15. j	24.6
c	24.6
i	24.4
b	24.4
h	24.4
f	24.6
e	24.3
g	24.3
d	24.4
a	24.6

Chapter 25

1. a	25.2
2. b	25.3
3. c	25.2
4. c, b, a, d	25.3
5. c	25.3
6. d	25.4
7. b	25.4
8. b	25.4
9. a	25.5
10. b	25.4
11. b	25.4
12. d	25.6
13. c	25.7
14. c	25.4
e	25.3
a	25.6
b	25.3
d	25.5
15. d	25.6
f	25.4
e	25.7
c	25.4
g	25.5
h	25.3
b	25.3
a	25.7

Chapter 26

1. a	26.2
2. a	26.3
3. b	26.3
4. a	26.4
5. d	26.4
6. c	26.4
7. c	26.4
8. c	26.4
9. b	26.6
10. d	26.6
11. a	26.7
12. a	26.5
13. 4, 2, 1, 6, 5, 3	26.6
14. e	26.5
15. c	26.4
h	26.4
a	26.6
g	26.4
d	26.4
e	26.4
b	26.4
f	26.7

Chapter 27

1. c	27.2
2. a, b	27.2
3. c	27.2
4. b, c	27.3
5. d	27.3
6. False	27.3, 27.5
7. a	27.3
8. a	27.5
9. b	27.3, 27.4
10. b	27.3
11. c	27.4
12. d	27.4
13. a	27.4
14. b	27.5
15. b	27.2, 27.3
h	27.5
e	27.2
d	27.2, 27.3
j	27.3
i	27.5
c	27.4
g	27.3
a	27.2, 27.4
f	27.3

Chapter 28

1. c	28.2
2. a, b, c	28.1, 28.2
3. b	28.2
4. b	28.2, 28.3
5. c	28.3
6. c	28.3, 28.4
7. d	28.4
8. d	28.3, 28.4
9. c	28.5
10. a	28.5
11. d	28.6
12. c	28.6
13. c	28.7
14. b, d, a, e, c	all 28.7
15. c, e, b, a, d	all 28.6

Periodic Table of the Elements

The symbol for each element is an abbreviation of its name. Some symbols for elements are abbreviations for their Latin names. For instance, Pb (lead) is short for *plumbum*; the word "plumbing" is related—ancient Romans made their water pipes with lead.

Elements in each vertical column of the table behave in similar ways. For instance, all of the elements in the far right column of the table are inert gases; they do not interact with other atoms. In nature, such elements occur only as solitary atoms.

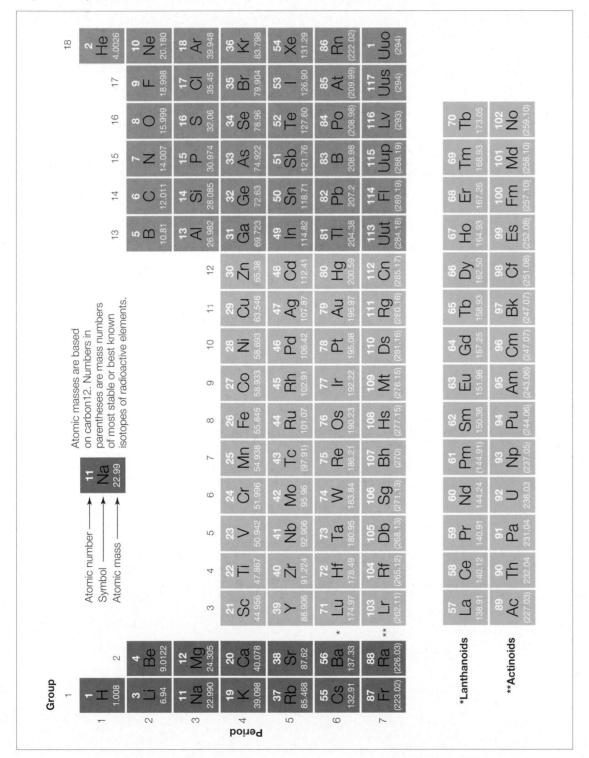

A Plain English Map of the Human Chromosomes

© 2002 Susan Offner/SK45176-02

Haploid set of human chromosomes. The banding patterns characteristic of each type of chromosome appear after staining with a reagent called Giemsa. The locations of some known genes are indicated. Also shown are locations that, when mutated, cause some of the genetic diseases discussed in the text.

Units of Measure

Length

1 kilometer (km) = 0.62 miles (mi)
1 meter (m) = 39.37 inches (in)
1 centimeter (cm) = 0.39 inches

To convert	multiply by	to obtain
inches	2.25	centimeters
feet	30.48	centimeters
centimeters	0.39	inches
millimeters	0.039	inches

Area

1 square kilometer = 0.386 square miles
1 square meter = 1.196 square yards
1 square centimeter = 0.155 square inches

Volume

1 cubic meter = 35.31 cubic feet
1 liter = 1.06 quarts
1 milliliter = 0.034 fluid ounces = 1/5 teaspoon

To convert	multiply by	to obtain
quarts	0.95	liters
fluid ounces	28.41	milliliters
liters	1.06	quarts
milliliters	0.03	fluid ounces

Weight

1 metric ton (mt) = 2,205 pounds (lb) = 1.1 tons (t)
1 kilogram (kg) = 2.205 pounds (lb)
1 gram (g) = 0.035 ounces (oz)

To convert	multiply by	to obtain
pounds	0.454	kilograms
pounds	454	grams
ounces	28.35	grams
kilograms	2.205	pounds
grams	0.035	ounces

Temperature

Celsius (°C) to Fahrenheit (°F): $°F = 1.8 \, (°C) + 32$

Fahrenheit (°F) to Celsius: $°C = \dfrac{(°F - 32)}{1.8}$

	°C	°F
Water boils	100	212
Human body temperature	37	98.6
Water freezes	0	32

APPENDIX IV

abscisic acid Plant hormone involved in stress responses; inhibits germination. **568**

abscission Process by which plant parts are shed. **568**

acid Substance that releases hydrogen ions in water. **32**

acid rain Rain containing sulfuric and/or nitric acid; forms when pollutants mix with water vapor in the atmosphere. **362**

actin Cytoskeletal protein that is the main component of thin filaments in muscle fibers. **396**

action potential Brief reversal of the charge difference across a cell's plasma membrane. **470**

activation energy Minimum amount of energy required to start a chemical reaction. **67**

active site Pocket in an enzyme where substrates bind and a chemical reaction occurs. **68**

active transport Energy-requiring mechanism in which a transport protein pumps a solute across a cell membrane against the solute's concentration gradient. **76**

adaptation *See* adaptive trait.

adaptive immunity In vertebrates, a set of immune defenses that can be tailored to specific pathogens as an organism encounters them during its lifetime. Characterized by self/nonself recognition, specificity, diversity, and memory. **425**

adaptive radiation Macroevolutionary pattern in which a burst of genetic divergences from a lineage gives rise to many new species. **227**

adaptive trait (adaptation) A form of a heritable trait that enhances an individual's fitness. **195**

adhesion protein Plasma membrane protein that helps cells stick together in animal tissues. Some types form adhering junctions and tight junctions. **51**

adipose tissue Connective tissue composed mainly of cells that specialized in the storage of fat. **378**

adrenal cortex Outer portion of adrenal gland; secretes the hormones aldosterone and cortisol. **502**

adrenal gland Endocrine gland located above kidney; secretes hormones with roles in urine formation and stress responses. **502**

adrenal medulla Inner portion of the adrenal gland; secretes epinephrine and norepinephrine. **502**

aerobic Involving or occurring in the presence of oxygen. **89**

aerobic respiration Oxygen-requiring metabolic pathway that breaks down sugars to produce ATP. Includes glycolysis, the Krebs cycle, and electron transfer phosphorylation. **89**

age structure Of a population, the distribution of individuals among various age groups. **316**

AIDS Acquired immune deficiency syndrome. A secondary immune deficiency that develops as the result of infection by HIV. **439**

alcoholic fermentation Anaerobic pathway that breaks down sugars and produces ATP, CO_2, and ethanol. **93**

algal bloom Population explosion of single-celled aquatic producers such as dinoflagellates. **252**

allele frequency Abundance of a particular allele among members of a population; expressed as a fraction of the total number of alleles. **214**

alleles Forms of a gene with slightly different DNA sequences; may encode slightly different versions of the gene's product. **140**

allergen A normally harmless substance that provokes an immune response in some people. **438**

allergy Sensitivity to an allergen. **438**

allopatric speciation Speciation pattern in which a physical barrier arises and ends gene flow between populations. **224**

alternation of generations As in plants, a life cycle that alternates between a diploid spore-producing body and a haploid, gamete-producing one. **263**

alveoli Tiny, thin-walled air sacs that are the site of gas exchange in the lung. **417**

amino acid Small organic compound that is a subunit of proteins. Consists of a carboxyl group, an amine group, and a characteristic side group (R), all typically bonded to the same carbon atom. **38**

amino acid–derived hormone An amine (modified amino acid), peptide, or protein that functions as a hormone. **494**

amnion Extraembryonic membrane that encloses an amniote embryo and the amniotic fluid. **524**

amniote Vertebrate that produces amniote eggs; a reptile, bird, or mammal. **302**

amniote egg Egg with four extraembryonic membranes; allows an embryo to develop away from water. **303**

amoeba Solitary heterotrophic protist that feeds and moves by extending pseudopods. **256**

amphibian Tetrapod with a three-chambered heart and scaleless skin. Larva are aquatic and have gills, but most are air-breathing as adults. **301**

anaerobic Occurring in the absence of oxygen. **89**

analogous structures Similar body structures that evolved separately in different lineages (by morphological convergence). **205**

anaphase Stage of mitosis during which sister chromatids separate and move toward opposite spindle poles. **134**

angiosperm Seed plant that produces flowers and fruits. **272**

animal A multicelled, eukaryotic consumer that is made up of unwalled cells and develops through a series of stages. Most ingest food, reproduce sexually, and can move from place to place. **8, 285**

animal hormone Signaling molecule secreted by an endocrine gland or cell. **492**

annelid Segmented worm with a coelom, complete digestive system, and closed circulatory system. For example, an earthworm. **290**

antenna Of some arthropods, sensory structure on the head that detects touch and odors. **293**

anther Of a flower, the part of the stamen that produces pollen grains in pollen sacs. **272, 555**

anthropoids Primate subgroup that includes monkeys, apes, and humans. **306**

antibody Y-shaped antigen receptor protein made by B cells. **431**

antibody-mediated immune response Immune response in which antibodies targeting a specific antigen are produced. Part of vertebrate adaptive immunity. **434**

anticodon In a tRNA, set of three nucleotides that base-pairs with an mRNA codon. **119**

antidiuretic hormone (ADH) Hormone that is released by the pituitary gland and acts in the kidney to increases water reabsorption. **460**

antigen Molecule or particle that the immune system recognizes as nonself. Its presence in the body triggers an immune response. **425**

anus Body opening through which digestive wastes alone leave the body. **447**

apical dominance Effect in which a lengthening shoot tip inhibits the growth of lateral buds. **567**

apical meristem Meristem in the tip of a shoot or root; gives rise to primary growth (lengthening) in a plant. **547**

apicomplexan Parasitic protist that enters and lives inside the cells of its host. For example, the parasite that causes malaria. **252**

apoptosis Self-destruction of a cell or cells. **526**

appendix Tubular projection from the initial portion of the large intestine. **452**

aquifer Porous rock layer that holds some groundwater. **342**

arachnids Land-dwelling arthropods with four pairs of walking legs and no antennae; for example, a spider, scorpion, or tick. **294**

archaea Group of single-celled organisms that lack a nucleus but are more closely related to eukaryotes than to bacteria. **8, 246**

arteriole Blood vessel that delivers blood from an artery to capillaries; site of adjustments to blood distribution. **411**

artery Large-diameter blood vessel that carries blood away from the heart. **407**

arthropod Invertebrate with jointed legs and a hardened exoskeleton that is periodically molted. **292**

asexual reproduction Reproductive mode of eukaryotes by which offspring arise from a single parent. **133, 511**

atherosclerosis Narrowing of an artery's interior due to lipid deposition and inflammation. **413**

atom Fundamental building block of all matter. Consists of varying numbers of protons, neutrons, and electrons. **5**

atomic number Number of protons in the atomic nucleus; determines the element. **25**

ATP Adenosine triphosphate. Nucleotide that consists of an adenine base, a ribose sugar, and three phosphate groups. Functions as a subunit of RNA and as a coenzyme in many reactions. Important energy carrier in cells. **41**

atrium Heart chamber that receives blood from veins and pumps blood into a ventricle. **406**

australopiths Informal name for chimpanzee-sized hominins that lived in Africa between 4 million and 1.2 million years ago. Some are considered likely ancestors of modern humans. **306**

autoimmune response Immune response that targets one's own tissues. **439**

autonomic nerves Nerves that relay signals to and from internal organs and to glands. **478**

autosome A chromosome that is the same in males and females. **107**

autotroph Organism that uses carbon dioxide as its carbon source, and obtains energy from light or breakdown of minerals. **247**

auxin Plant hormone that causes cell enlargement; also has a central role in growth and development by coordinating the effects of other hormones. **566**

axon Signal-sending cytoplasmic extension of a neuron. **468**

B cell B lymphocyte. Lymphocyte that can make antibodies during an antibody-mediated immune response. **426**

B cell receptor Antigen receptor on the surface of a B cell; an antibody that stays anchored in the B cell's plasma membrane. **431**

bacteria Singular, bacterium. The most diverse and well-known group of prokaryotes (organisms that lack a nucleus). **8, 246**

bacteriophage Virus that infects bacteria. **242**

bark In woody plants, informal term for all living and dead tissues outside the ring of vascular cambium. **548**

Barr body Condensed and inactivated X chromosome in a body cell of a female mammal (the other X chromosome is active). **125**

base Substance that accepts hydrogen ions in water. **32**

base-pair substitution Mutation in which a single base pair changes. **122**

bell curve Bell-shaped curve; typically results from graphing frequency versus distribution for a trait that varies continuously. **160**

bilateral symmetry Having right and left halves with similar parts, and a front and back that differ. **286**

bile A mixture of salts, pigments, and cholesterol that is produced by the liver and aids in fat digestion. **451**

binary fission Method of asexual reproduction in which a prokaryote divides into two identical descendant cells. **246**

biodiversity Scope of variation among living organisms. Of a region, the genetic diversity within its species, variety of species, and variety of ecosystems. **8, 366**

biofilm Community of microorganisms living within a shared mass of slime. **53**

biogeochemical cycle Cycle in which a nutrient moves among environmental reservoirs and into and out of food webs. **342**

biogeography Study of patterns in the geographic distribution of species and communities. **191**

biological magnification A chemical pollutant becomes increasingly concentrated as it moves through a food chain. **363**

biology The scientific study of life. **4**

bioluminescence Light produced by a living organism. **252**

biome Any of Earth's major land ecosystems, characterized by climate and main vegetation and found in several regions. **355**

biosphere All regions of Earth where organisms live. **5**

biotic potential Maximum possible population growth under optimal conditions. **320**

bipedalism Habitually walking upright. **306**

birds Common name for the lineage of feathered amniotes that descended from dinosaurs. **303**

bivalve Mollusk with a hinged two-part shell. **290**

blastula Hollow, fluid-filled ball of cells that forms early in animal development. **512**

blood Fluid connective tissue that circulates within a closed circulatory system. Consists of plasma and cellular components. **379, 405**

blood–brain barrier Protective mechanism that prevents unwanted substances from entering the cerebrospinal fluid. **474**

blood pressure Pressure exerted by blood against the walls of blood vessels. **410**

bone tissue Connective tissue with cells surrounded by a mineral-hardened matrix of their own secretions. **379**

bony fish Common name for a jawed fish with a skeleton composed mainly of bone. **299**

boreal forest At high northern latitudes, a biome dominated by conifers that withstand cold winters. **355**

bottleneck Reduction in population size so severe that it reduces genetic diversity. **220**

brain Central, integrating organ of a nervous system. **467**

bronchiole Small airway leading to alveoli. **417**

bronchus Airway connecting the trachea to a lung. **417**

brood parasite An animal that manipulates another species into raising its young, for example a cowbird. **336**

brown alga Multicelled, photosynthetic protist with brown accessory pigments; for example a kelp. **254**

bryophyte Member of a plant lineage that does not have vascular tissue; a moss, liverwort, or hornwort. **264**

buffer Set of chemicals that can keep the pH of a solution stable by alternately donating and accepting ions that contribute to pH. **32**

C3 plant Type of plant that uses only the Calvin–Benson cycle to fix carbon. **88**

C4 plant Type of plant that fixes carbon twice, in two cell types. **88**

Calvin–Benson cycle Cyclic carbon-fixing pathway that forms sugars from CO_2; light-independent reactions of photosynthesis. **87**

camouflage Evolved body shape, color pattern, or behavior that helps an organism blend into its surroundings. **335**

CAM plant Type of plant that fixes carbon twice, at different times of day. **88**

cancer Disease that occurs when a malignant neoplasm physically and functionally disrupts body tissues. **138**

capillary Smallest-diameter blood vessel; in a closed circulatory sytem, the site of exchanges of gases and other materials with tissues. **405**

carbohydrate Molecule that consists primarily of carbon, hydrogen, and oxygen atoms in a 1:2:1 ratio. Complex types (polysaccharides such as cellulose, starch, and glycogen) are polymers of monosaccharides. **34**

carbon cycle Movement of carbon among rocks, water, the atmosphere, and living organisms. **345**

carbon fixation Process by which carbon from an inorganic source such as carbon dioxide becomes incorporated (fixed) into an organic molecule. **87**

cardiac cycle Sequence of contraction and relaxation of heart chambers that occurs with each heartbeat. **409**

cardiac muscle tissue Striated, involuntary muscle that makes up the bulk of the heart wall. **379**

cardiac pacemaker Group of heart cells (SA node) that emits rhythmic electrical signals that stimulate muscle contraction. **409**

carpel Floral reproductive organ that produces the female gametophyte; consists of an ovary, stigma, and often a style. **272, 555**

carrying capacity Maximum number of individuals of a species that a specific environment can sustain. **318**

cartilage Connective tissue with cells surrounded by a rubbery matrix of their own secretions. **378**

cartilaginous fish Fish that has jaws, paired fins, and a skeleton made of cartilage; for example, a shark. **299**

cartilaginous joint Joint at which cartilage holds bones together and provides cushioning, as between vertebrae. **394**

cell Smallest unit of life; at minimum, consists of plasma membrane, cytoplasm, and DNA. **5**

cell cycle The collective series of intervals and events of a cell's life, from the time it forms until its cytoplasm divides. **133**

cell junction Structure that connects a cell to another cell or to extracellular matrix; e.g., tight junction, adhering junction, or gap junction (of animals); plasmodesmata (of plants). **58**

cell-mediated immune response Immune response in which cytotoxic T cells and NK cells kill infected or cancerous body cells. Part of vertebrate adaptive immunity. **434**

cell theory Theory that all organisms consist of one or more cells, which are the basic unit of life; all cells come from division of preexisting cells; and all cells pass DNA to offspring. **47**

cell wall Rigid but permeable layer of extracellular matrix that surrounds the plasma membrane of some cells. **52**

cellular slime mold Heterotrophic protist that usually lives as a single-celled, amoeba-like predator. With unfavorable conditions, cells aggregate into a cohesive group that can form a fruiting body. **257**

cellulose Tough, insoluble carbohydrate that is the major structural material in plants. **34**

central nervous system Brain and spinal cord. **467**

centromere Of a duplicated eukaryotic chromosome, constricted region where sister chromatids attach to each other. **106**

cephalopod Predatory mollusk with a closed circulatory system; moves by jet propulsion; for example, an octopus or squid. **291**

cerebellum Region of the hindbrain that coordinates voluntary movements. **475**

cerebrospinal fluid Fluid that surrounds and fills cavities in the brain and spinal cord. **474**

cerebrum Forebrain region that controls higher functions such as abstract thought and language. **475**

cervix Narrowed region of the uterus that connects to vagina. **514**

chaparral Biome with cool, wet winters and hot dry summers; dominant plants are shrubs with small, leathery leaves. **356**

charge Electrical property; opposite charges attract, and like charges repel. **25**

chemical bond An attractive force that arises between two atoms when their electrons interact; links atoms in molecules. *See* covalent bond, ionic bond. **28**

chemical synapse Region where a neuron's axon terminal transmits chemical signals to another cell. **471**

chlorophyll *a* Main photosynthetic pigment in plants. **84**

chloroplast Organelle of photosynthesis in the cells of plants and photosynthetic protists. Has two outer membranes enclosing semifluid stroma. Light-dependent reactions occur at its inner thylakoid membrane; light-independent reactions, in the stroma. **56**

choanoflagellates Heterotrophic protists with a collared flagellum; protist group most closely related to animals. **257**

chordates Animal phylum in which embryos have a notochord, dorsal nerve cord, pharyngeal gill slits, and a tail that extends beyond the anus. Includes invertebrate and vertebrate groups. **297**

chorion Outermost extraembryonic membrane of amniotes; major component of the placenta in placental mammals. **525**

chromosome Structure that consists of DNA together with associated proteins; carries part or all of a cell's genetic information. **106**

chromosome number The total number of chromosomes in a cell of a given species. **106**

chyme Mixture of food and gastric fluid. **450**

chytrid Fungus that produces flagellated spores. **275**

cilia Singular, cilium. Short, movable structures that project from the plasma membrane of some eukaryotic cells. **57**

ciliate Unwalled, single-celled protist with many cilia. **251**

circadian rhythm A cycle of biological activity that repeats about every 24 hours. **572**

clade A group whose members share one or more defining derived traits. **229**

cladistics Making hypotheses about evolutionary relationships among clades. **229**

cladogram Evolutionary tree diagram that summarizes hypothesized relationships among a group of clades. **229**

cleavage Mitotic division of cells in an early embryo. **512**

cleavage furrow In a dividing animal cell, the indentation where cytoplasmic division will occur. **136**

climate Average weather conditions in a region over a long time period. **353**

cloaca Of some vertebrates, a body opening that releases urinary and digestive waste, and also functions in reproduction. **301, 447**

clone Genetically identical copy of an organism. **100**

cloning vector A DNA molecule that can accept foreign DNA and be replicated inside a host cell. **176**

closed circulatory system System in which blood flows through a continuous network of vessels and exchanges with cells take place across vessel walls. **287, 405**

club fungus Fungus that produces spores by meiosis in club-shaped cells. **275**

cnidarian Radially symmetrical invertebrate with two tissue layers; uses tentacles with stinging cells to capture food. **288**

coal Fossil fuel consisting primarily of the carbon-rich remains of seedless nonvascular plants. **268**

cochlea Coiled structure in the inner ear that holds the sound-detecting organ of Corti. **485**

codominance Inheritance pattern in which the full and separate phenotypic effects of two alleles are apparent in heterozygous individuals. **155**

codon In an mRNA, a nucleotide base triplet that codes for an amino acid or stop signal during translation. **118**

coelom A body cavity completely lined by tissue derived from mesoderm. **286**

coenzyme An organic cofactor; e.g., NAD. **70**

coevolution The joint evolution of two closely interacting species; macroevolutionary pattern in which each species is a selective agent for traits of the other. **228**

cofactor A molecule or metal ion that associates with a protein and is necessary for its function. **70**

cohesion Property of a substance that arises from the tendency of its molecules to resist separating from one another. **31**

cohesion–tension theory Explanation of how transpiration creates a tension that pulls a cohesive

column of water upward through xylem, from roots to shoots. **544**

cohort Group of individuals born during the same interval. **320**

collenchyma In plants, simple tissue composed of living cells with unevenly thickened walls; provides flexible support. **536**

colon Longest portion of the large intestine. **452**

colonial organism Organism composed of many integrated cells, each capable of surviving and reproducing on its own. **250**

colonial theory of animal origins Well-accepted hypothesis that animals evolved from a colonial protist. **285**

commensalism Species interaction that benefits one species and has no effect on the other. **332**

community All populations of all species in a given area. **5, 330**

compact bone Dense bone with concentric layers of matrix. **392**

comparative morphology The scientific study of similarities and differences in body plans. **192**

competitive exclusion When two species compete for the same resource, the better competitor drives the weaker one to extinction in that habitat. **334**

complement A set of proteins that circulate in inactive form in blood; activated complement proteins attract phagocytic leukocytes, coat antigenic particles, and puncture lipid bilayers. **428**

complete digestive tract Tubular gut with two openings. **286, 447**

compound Molecule that has atoms of more than one element. **28**

compound eye Eye that consists of many individual units, each with its own lens. **293**

concentration Amount of solute per unit volume of solution. **30**

cone cell Photoreceptor that provides sharp vision and allows detection of color. **484**

conifer Woody gymnosperm with needlelike leaves. **270**

conjugation Mechanism of gene transfer in which one prokaryotic cell directly transfers a plasmid to another. **246**

connective tissue Animal tissue with an extensive extracellular matrix; provides structural and functional support. **378**

conservation biology Field of applied biology that surveys biodiversity and seeks ways to maintain and use it. **366**

consumer Organism that obtains energy and nutrients by feeding on the tissues, wastes, or remains of other organisms; a heterotroph. **6, 339**

continuous variation A range of small differences in a shared trait. **159**

contractile vacuole In freshwater protists, an organelle that collects and expels excess water. **250**

control group Group of individuals identical to an experimental group except for the independent variable under investigation. **13**

coral reef In tropical sunlit seas, a formation composed of the secretions of coral polyps; serves as home to many other species. **359**

cork Tissue that waterproofs, insulates, and protects the surfaces of woody stems and roots. **548**

cork cambium Lateral meristem that produces cork. **548**

cornea Outermost layer at the front of the eye; bends light. **482**

corpus luteum Hormone-secreting structure that forms from follicle cells after ovulation. **515**

cortisol Adrenal cortex hormone that affects metabolism and immunity; secretions rise with stress. **502**

cotyledon Embryonic leaf of a flowering plant; also called a seed leaf. Monocots have one; eudicots have two. **537**

covalent bond Type of chemical bond in which two atoms share a pair of electrons. **28**

critical thinking The act of evaluating information before accepting it. **12**

crossing over Process in which homologous chromosomes exchange corresponding segments during meiosis. **144**

crustaceans Lineage of mostly marine arthropods with two pairs of antennae; for example, a shrimp, crab, lobster, or barnacle. **294**

cuticle Secreted covering at a body surface. In plants it is waxy and helps conserve water. **58, 264**

cytokinin Plant hormone that promotes cell division in shoot apical meristem and cell differentiation in root apical meristem. Often opposes auxin's effects. **567**

cytoplasm Jellylike mixture of water and solutes enclosed by a cell's plasma membrane. **47**

cytoskeleton Network of protein filaments that support, organize, and move eukaryotic cells and their internal structures. *See* microtubule, microfilament, intermediate filament. **56**

data Results of an experiment or survey. **13**

deciduous plant Plant that sheds all its leaves in preparation for a seasonal dormancy. **271**

decomposer Organism that breaks down organic material into its inorganic subunits. **247, 339**

deforestation Removal of all trees from a forested area. **362**

deletion Mutation in which one or more nucleotides are lost from DNA. **123**

demographics Statistics that describe a population. **315**

demographic transition model Model describing how human birth and death rates change as a region becomes industrialized. **325**

denature To unravel the shape of a protein or other large biological molecule. **39**

dendrites Information-receiving cytoplasmic extension of a neuron. **468**

dendritic cell Phagocytic white blood cell that alerts the immune system to the presence of antigen in solid tissues. **426**

dense connective tissue Connective tissue with many fibroblasts and fibers in a random or a regular arrangement. **378**

density-dependent limiting factor Factor whose negative effect on growth is felt most in dense populations; for example, infectious disease or competition for food. **318**

density-independent limiting factor Factor that limits growth in populations regardless of their density; for example a natural disaster or harsh weather. **319**

dental plaque On teeth, a thick biofilm composed mainly of bacteria, their extracellular products, and saliva proteins. **430**

dermal tissues Tissues that cover and protect the plant body; e.g., epidermis, periderm. **535**

dermis Deep layer of skin; consists of connective tissue with nerves and blood vessels running through it. **381**

desert Biome with little precipitation; its perennial plants are adapted to withstand drought. **356**

desertification Conversion of grassland or woodlands to desertlike conditions. **362**

detritivore Consumer that feeds on small bits of organic material (detritus). **339**

deuterostomes Animal lineage with a three-layer embryo in which the mouth is the second opening to form; includes echinoderms and chordates. **286**

development Multistep process by which the first cell of a new multicelled organism gives rise to an adult. **6**

diaphragm Dome-shaped muscle at the base of the thoracic cavity that alters the size of this cavity during breathing. **417**

diastolic pressure Blood pressure when ventricles are relaxed. **410**

diatom Single-celled photosynthetic protist with brown accessory pigments and a two-part silica shell. **254**

differentiation Process by which cells become specialized during development; occurs as different cell lineages begin to use different subsets of their DNA. **101**

diffusion The spontaneous spreading of molecules or atoms. **73**

digestion Breakdown of food into smaller bits and then into its component molecules. **447**

dihybrid cross Cross between two individuals identically heterozygous for alleles of two genes; for example $AaBb \times AaBb$. **154**

dinoflagellates Single-celled, aquatic protist typically with cellulose plates and two flagella; may be heterotrophic or photosynthetic. **252**

diploid Having two of each type of chromosome characteristic of the species ($2n$). **107**

directional selection Mode of natural selection in which a phenotype at one end of a range of variation is favored. **215**

disease vector Organism that carries a pathogen from one host to the next. **243**

disruptive selection Mode of natural selection in which traits at the extremes of a range of variation are adaptive, and intermediate forms are not. **215**

DNA Deoxyribonucleic acid. Nucleic acid that carries hereditary information; consists of two chains of nucleotides twisted into a double helix. **6, 41**

DNA cloning Set of methods that uses living cells to mass-produce targeted DNA fragments. **176**

DNA fingerprinting *See* DNA profiling.

DNA library Collection of cells that host different fragments of foreign DNA, often representing an organism's entire genome. **176**

DNA polymerase DNA replication enzyme. Uses a DNA template to assemble a complementary strand of DNA. **108**

DNA profiling Identifying an individual by analyzing the unique parts of his or her DNA. **179**

DNA replication Process by which a cell duplicates its DNA before it divides. **108**

DNA sequence Order of nucleotides composing a strand of DNA. **106**

DNA sequencing *See* sequencing.

dominant Refers to an allele that masks the effect of a recessive allele paired with it on the homologous chromosome. **152**

dormancy Period of temporarily suspended metabolism. **558**

double fertilization In flowering plants, one sperm fertilizes the egg (to form the zygote), and a second sperm cell fuses with the endosperm mother cell (giving rise to endosperm). **273, 558**

eardrum Membrane at the entrance to the middle ear; vibrates in response to sound waves. **485**

echinoderms Invertebrates with a water–vascular system and an endoskeleton made of hardened plates and spines. **296**

ECM *See* extracellular matrix.

ecological footprint Area of Earth's surface required to sustainably support a particular level of development and consumption. **325**

ecological restoration Actively altering an area in an effort to restore or create a functional ecosystem. **367**

ecological succession A gradual change in a community in which one array of species replaces another. **337**

ecology The study of interactions among organisms, and between organisms and their environment. **314**

ecosystem A community interacting with its environment through a one-way flow of energy and cycling of materials. **5, 339**

ectoderm Outermost tissue layer of an animal embryo. **512**

ectotherm Animal that gains heat from the environment; commonly called "cold-blooded." **302**

effector cell Antigen-sensitized lymphocyte that forms in an immune response and acts immediately. **434**

egg Female gamete. **512**

ejaculation Smooth muscle contractions expel semen through the penis. **517**

electron Negatively charged subatomic particle. **25**

electron transfer chain Array of membrane-bound enzymes and other molecules that accept and give up electrons in sequence, thus releasing the energy of the electrons in small, usable steps. **72**

electron transfer phosphorylation Process in which electron flow through electron transfer chains sets up a hydrogen ion gradient that drives ATP formation. Occurs in photosynthesis and in aerobic respiration. **87**

electrophoresis Laboratory technique that separates DNA fragments by size. **178**

element A pure substance that consists only of atoms with the same number of protons. **25**

elimination Expulsion of unabsorbed material from the digestive tract. **447**

endangered species A species that faces extinction in all or part of its range. **360**

endemic species A species that evolved in one place and is found nowhere else. **361**

endocrine gland Ductless gland; secretes hormones that are distributed by the blood. **377, 493**

endocytosis Process by which a cell takes in a small amount of extracellular fluid (and its contents) by the ballooning inward of the plasma membrane. **76**

endoderm Innermost tissue layer of an animal embryo. **512**

endoplasmic reticulum (ER) Membrane-enclosed organelle that is a continuous system of sacs and tubes extending from the nuclear envelope. Smooth ER makes lipids and breaks down carbohydrates and fatty acids; ribosomes on the surface of rough ER make polypeptides that thread into its interior. **55**

endorphins Molecules that serves as a vertebrate body's natural painkillers. **472**

endoskeleton Internal skeleton; hard internal parts that muscles attach to and move. **296, 391**

endosperm Nutritive tissue in the seeds of angiosperms (flowering plants). **273, 558**

endosymbiont hypothesis Mitochondria and chloroplasts evolved from free-living bacteria that entered and lived inside another cell. **241**

endotherm Animal that produces its own heat; commonly called "warm-blooded." **302**

energy The capacity to do work. **65**

energy pyramid Diagram that illustrates the energy flow in an ecosystem. **341**

enzyme Organic molecule (protein or RNA) that speeds up a reaction without being changed by it. **34**

epidermis Outermost tissue layer of an animal or a young plant. **380, 536**

epididymis Duct that emerges from a testis and conveys sperm to a vas deferens. **517**

epigenetic Refers to heritable changes in gene expression that are not the result of changes in DNA sequence. **126**

epiglottis Tissue flap that folds down to prevent food from entering the airways during swallowing. **417**

epiphyte Plant that grows on the trunk or branches of another plant but does not harm it. **267**

epistasis Polygenic inheritance, in which a trait is influenced by multiple genes. **156**

epithelial tissue Sheetlike animal tissue that covers outer body surfaces and lines internal tubes and cavities. **376**

equilibrial life history Life history favored in stable environments; individuals grow large, then invest heavily in each of their few offspring. **321**

esophagus Muscular tube that connects the pharynx (throat) to the stomach. **449**

essential amino acid Amino acid that the body cannot make and must obtain from food. **454**

essential fatty acid Fatty acid that the body cannot make and must obtain from the diet. **454**

estrogen Sex hormone produced by a female's ovaries; causes development of female secondary sexual characteristics and maintains the reproductive tract. **504, 514**

estuary Highly productive, aquatic ecosystem where nutrient-rich water from a river mixes with seawater. **358**

ethylene Gaseous plant hormone involved in regulating growth and cell expansion. Participates in germination, abscission, ripening, and stress responses. **568**

eudicots Most diverse angiosperm lineage; characterized by two seed leaves; includes herbaceous plants, woody trees, and cacti. **273**

eugenics Idea of deliberately improving the genetic qualities of the human race. **185**

eukaryote Organism whose cells characteristically have a nucleus; a protist, fungus, plant, or animal. **8**

evaporation Transition of a liquid to a vapor. **31**

evolution Change in a line of descent. **193**

exocrine gland Gland that secretes milk, sweat, saliva, or some other substance through a duct. **377**

exocytosis Process by which a cell expels a vesicle's contents to extracellular fluid. **76**

exon Nucleotide sequence that remains in an RNA after post-transcriptional modification. **117**

exoskeleton External skeleton; hard external parts that muscles attach to and move. **292, 391**

exotic species A species that has been introduced to a new habitat and become established there. **338**

experiment A test designed to support or falsify a prediction. **13**

experimental group In an experiment, a group of individuals who have a certain characteristic or receive a certain treatment as compared with a control group. **13**

exponential model of population growth Model for population growth when resources are unlimited. The per capita growth rate remains constant as population size increases. **317**

extinct Refers to a species that no longer has any living members. **226**

extracellular matrix (ECM) Complex mixture of cell secretions, the composition and function of which vary by cell type. **58**

extreme halophile Organism that lives in a highly salty habitat. **248**

extreme thermophile Organism that lives in a high temperature habitat. **248**

facilitated diffusion Passive transport mechanism in which a solute follows its concentration gradient across a membrane by moving through a transport protein. **75**

fat Substance that consists mainly of triglycerides. Saturated types are composed mainly of triglycerides with three saturated fatty acid tails. **36**

fatty acid Organic compound with an acidic carboxyl group "head" and a long carbon chain "tail." *See* saturated fatty acid, unsaturated fatty acid. **36**

feces Mixture of unabsorbed food material and cellular waste that is excreted from the terminal end of the digestive tract. **452**

feedback inhibition Regulatory mechanism by which a change that results from some activity decreases or stops the activity. **72**

fermentation An anaerobic pathway that breaks down sugars to produce ATP. **92**

ferns Most diverse lineage of seedless vascular plants. **266**

fertilization Egg and sperm combine and form a zygote. **512**

fetus Developing human between 9 weeks after fertilization and birth. **526**

fever A temporary, internally induced rise in core body temperature above the normal set point as a response to infection or tissue damage. **429**

fibroblast Main cell type in soft connective tissue; secretes collagen and other components of extracellular matrix. **378**

fibrous joint Joint where dense connective tissue holds bones firmly in place. **393**

filtration In urine formation, blood pressure forces water and small solutes, but not blood cells or proteins, out across the walls of capillaries. **458**

first law of thermodynamics Energy cannot be created or destroyed. **65**

fitness Degree of adaptation to an environment, as measured by an individual's relative genetic contribution to future generations. **195**

fixed Refers to an allele for which all members of a population are homozygous. **220**

flagellated protozoan Unwalled, single-celled protist that has one or more flagella. **250**

flagellum Long, slender cellular structure used for movement. **53**

flatworm Soft-bodied, bilaterally symmetrical invertebrate with organs but no body cavity; for example, a planarian or tapeworm. **289**

flower Specialized reproductive structure of a flowering plant. **272, 555**

fluid mosaic Model of a cell membrane as a two-dimensional fluid of mixed composition. **50**

follicle-stimulating hormone (FSH) Pituitary hormone; causes ovarian follicle maturation in females and aids in sperm development in males. **516**

food chain Sequence of steps by which energy moves from one trophic level to the next. **339**

food web System of cross-connecting food chains. **340**

foraminiferan Heterotrophic single-celled protist that secretes a calcium carbonate shell. **251**

fossil Physical evidence of an organism that lived in the ancient past. **192**

founder effect After a small group of individuals found a new population, allele frequencies in the new population differ from those in the original population. **220**

free radical Atom with an unpaired electron; most are highly reactive and can damage biological molecules. **27**

fruit Mature ovary of a flowering plant, often with expanded accessory parts; encloses a seed or seeds. **272, 560**

fungus Single-celled or multicelled eukaryotic consumer that breaks down material outside its body, then absorbs nutrients released from the breakdown. **8, 274**

gallbladder Organ that receives bile from the liver, stores it, and expels it into the small intestine. **451**

gamete Mature, haploid reproductive cell; e.g., an egg or a sperm. **144**

gametophyte Haploid gamete-forming body that forms in a plant life cycle. **263**

ganglion A cluster of neuron cell bodies. **467**

gastric fluid Fluid secreted by the stomach lining; contains enzymes, acid, and mucus. **450**

gastropod Mollusk that moves about on an enlarged "foot" at its lower surface; for example a snail. **290**

gastrovascular cavity Saclike cavity with one opening; functions in digestion and respiration. **286, 447**

gastrula Three-layered early embryo that forms when a blastula undergoes gastrulation. **512**

gastrulation Animal developmental process by which cell movements produce a three-layered gastrula. **512**

gene A part of a chromosome that encodes an RNA or protein product in its DNA sequence. **115**

gene expression Process by which the information in a gene guides assembly of an RNA or protein product. Includes transcription and translation. **115**

gene flow The movement of alleles between populations. **221**

gene pool All the alleles of all the genes in a population; a pool of genetic resources. **214**

gene therapy Treating a genetic defect or disorder by transferring a normal or modified gene into the affected individual. **184**

genetic code Complete set of sixty-four mRNA codons. **118**

genetic drift Change in allele frequency due to chance alone. **220**

genetic engineering Laboratory process by which deliberate changes are introduced into an individual's genome. **181**

genetically modified organism (GMO) Organism whose genome has been modified by genetic engineering. **181**

genome An organism's complete set of genetic material. **176**

genomics The study of genomes. **179**

genotype The particular set of alleles that is carried in an individual's chromosomes. **152**

genus Plural, genera. A group of species that share a unique set of traits; first part of a species name. **10**

geologic time scale Chronology of Earth's history; correlates geologic and evolutionary events. **201**

germinate To resume metabolic activity after dormancy. **558**

gibberellin Plant hormone that induces stem elongation and helps seeds break dormancy, among other effects. **568**

gills Folds or body extensions that increase the surface area for respiration. **414**

gland *See* endocrine gland, exocrine gland.

global climate change Wide-ranging changes in rainfall patterns, average temperature, and other climate factors that result from rising concentrations of greenhouse gases. **347**

glomeromycete fungus Soil fungus whose hyphae extend inside the cell wall of a plant root cell. **278**

glottis Opening formed when the vocal cords relax. **416**

glucagon Pancreatic hormone that causes liver cells to break down glycogen and release glucose, thus raising the blood glucose level. **500**

glycolysis Set of reactions in which glucose is broken down to two pyruvate for a net yield of two ATP. Part of fermentation and aerobic respiration. **90**

GMO *See* genetically modified organism.

Golgi body Organelle that modifies polypeptides and lipids, then packages the finished products into vesicles. **55**

gonads Ovaries and testes; organs that produce gametes and secrete sex hormones. **504**

Gondwana Supercontinent that existed before Pangea, more than 500 million years ago. **201**

gravitropism In plants, directional growth response to gravity. **570**

gray matter Of the central nervous system, tissue that includes neuron cell bodies, dendrites, and axon terminals, as well as neuroglial cells. **474**

green algae Single-celled, colonial, or multicelled photosynthetic protists; algal lineage most closely related to land plants. **255**

greenhouse effect Warming of Earth's lower atmosphere and surface as a result of heat trapped by greenhouse gases. **346**

greenhouse gas Atmospheric gas that helps keep heat from escaping into space and thus warms the Earth. **346**

ground tissues Tissues that make up most of the soft internal parts of the plant body; all plant tissues other than vascular and dermal tissues. **535**

groundwater Water between soil particles and in aquifers. **342**

growth In multicelled species, an increase in the number, size, and volume of cells. **7**

growth hormone (GH) Anterior pituitary hormone that promotes growth and development throughout the body. **496**

guard cell In plants, one of a pair of specialized epidermal cells that define a stoma. **544**

gymnosperm Seed plant that produces "naked" seeds (seeds that are not encased within a fruit). **270**

habitat The type of place in which a species lives. **331**

half-life Characteristic time it takes for half of a quantity of a radioisotope to decay. **198**

haploid Having one of each type of chromosome characteristic of the species. **143**

heart Muscular organ that pumps fluid through a circulatory system. **405**

hemolymph Fluid circulating in an open circulatory system. **405**

herbivory An animal feeds on a plant, which may or may not die as a result. **335**

hermaphrodite Individual animal that makes both eggs and sperm. **288, 511**

heterotroph Organism that obtains both carbon and energy by breaking down organic compounds. **247**

heterozygous Having two different alleles of a gene. **152**

histone Type of protein that associates with eukaryotic DNA and structurally organizes chromosomes. **106**

HIV (human immunodeficiency virus) Enveloped RNA virus that causes AIDS. **244**

homeostasis Process in which an organism keeps its internal conditions within tolerable ranges by sensing and responding to change. **6, 375**

hominins Lineage of bipedal primates; includes humans and extinct humanlike species. **306**

Homo erectus Extinct human species with body proportions similar to modern humans; arose in and dispersed out of Africa. **307**

Homo habilis Earliest named human species; had australopith-like proportions and is known only from Africa. **307**

Homo neanderthalensis Neanderthals. Closest extinct relatives of modern humans; had large brain, stocky body. **308**

Homo sapiens Modern humans; evolved in Africa by about 200,000 years ago, then expanded their range worldwide. **308**

homologous chromosomes Chromosomes that have the same length, shape, and genes. In sexual reproducers, one member of a homologous pair is paternal and the other is maternal. **134**

homologous structures Body structures that are similar in different lineages because they evolved in a common ancestor. **204**

homozygous Having identical alleles of a gene. **152**

humans Members of the genus *Homo*. **307**

human chorionic gonadotropin (HCG) Hormone produced by the chorion and later by the placenta; encourages maintenance of the uterus, thus sustaining a pregnancy. **525**

humus Decaying organic matter in soil **542**

hydrogen bond Attraction between a covalently bonded hydrogen atom and another atom taking part in a separate covalent bond. **30**

hydrophilic Describes a substance that dissolves easily in water. **30**

hydrophobic Describes a substance that resists dissolving in water. **30**

hydrostatic skeleton Fluid-filled chamber or chambers that muscles act on to move body parts. **391**

hydrothermal vent Underwater opening from which mineral-rich water heated by geothermal energy streams out. **237, 359**

hypertension Chronically high blood pressure. **413**

hypertonic Describes a fluid that has a high overall solute concentration relative to another fluid. **73**

hypha A single filament in a fungal mycelium; consists of a chain of cells. **275**

hypothalamus Forebrain region that controls processes related to homeostasis; has both nervous and endocrine functions. **475, 496**

hypothesis Testable explanation of a natural phenomenon. **12**

hypotonic Describes a fluid that has a low solute concentration relative to another fluid. **73**

immunity The body's ability to resist and fight infections. **425**

immunization Any procedure designed to induce immunity to a specific disease; e.g., vaccination. **441**

inbreeding Mating among close relatives. **221**

incomplete dominance Inheritance pattern in which one allele is not fully dominant over another, so the heterozygous phenotype is an intermediate blend between the two homozygous phenotypes. **155**

indicator species A species that is particularly sensitive to environmental changes and can be monitored to assess the state of an ecosystem. **366**

inflammation A local response to tissue damage or infection; characterized by redness, warmth, swelling, and pain. **429**

ingestion Taking food into the digestive system. **446**

inheritance Transmission of DNA to offspring. **7**

innate immunity In all multicelled organisms, set of immediate, general defenses against infection. **425**

insects Land-dwelling arthropods with a pair of antennae, three pairs of legs, and—in the most diverse groups—wings. **295**

insertion Mutation in which one or more nucleotides become inserted into DNA. **123**

insulin Pancreatic hormone that causes cells to take up glucose from the blood, thus lowering the blood glucose level. **500**

intercostal muscles Muscles between the ribs; help alter the size of the thoracic cavity during breathing. **417**

intermediate filament Stable cytoskeletal element that structurally supports cell membranes and tissues; also forms external structures such as hair. **57**

interneuron Neuron that receives signals from, and sends signals to, other neurons. **469**

interphase In a eukaryotic cell cycle, the interval during which the cell grows, roughly doubles the number of its cytoplasmic components, and replicates its DNA in preparation for division. **133**

interspecific competition Two species compete for a resource in an interaction harmful to both. **333**

interstitial fluid Fluid between cells of a multicelled body. **405**

intervertebral disk Cartilage disk between two vertebrae. **392**

intron Nucleotide sequence that intervenes between exons and is removed during post-transcriptional modification. **117**

invertebrate Animal without a backbone. **284**

***in vitro* fertilization** Procedure in which eggs are removed from a woman's ovary and fertilized outside her body. **510**

ion Atom or molecule that carries a net charge. **27**

ionic bond Type of chemical bond in which a strong mutual attraction links ions of opposite charge. **28**

iris Ring of smooth muscle with pupil at its center; adjusts how much light enters the eye. **483**

iron–sulfur world hypothesis Hypothesis that life began in rocks rich in iron sulfide near deep-sea hydrothermal vents. **238**

isotonic Describes two fluids that have the same overall solute concentration. **73**

isotopes Forms of an element that differ in the number of neutrons their atoms carry. **25**

jawless fish Fish that has a skeleton of cartilage, but no jaws or paired fins; for example, a lamprey. **299**

joint Region where bones come together. **393**

karyotype Image of an individual's complement of chromosomes arranged by size, length, shape, and centromere location. **107**

key innovation An evolutionary adaptation that gives its bearer the opportunity to exploit a particular environment more efficiently or in a new way. **228**

keystone species A species that has a disproportionately large effect on community structure. **338**

kidney Organ of the vertebrate urinary system that filters blood, adjusts its composition, and forms urine. **298, 457**

kidney dialysis Procedure used to cleanse blood and restore proper solute concentrations in a person who has impaired kidney function. **461**

knockout An experiment in which a gene is deliberately inactivated in a living organism; also, an organism that has a knocked-out gene. **124**

Krebs cycle Cyclic pathway that, along with acetyl–CoA formation, breaks down pyruvate to carbon dioxide in aerobic respiration's second stage. **91**

lactate fermentation Anaerobic pathway that breaks down sugars and produces ATP and lactate. **93**

lancelets Invertebrate chordates that have a fishlike shape and retain their defining chordate traits into adulthood. **297**

large intestine Organ that concentrates and stores waste, then delivers it to the anus for excretion. **452**

larva Preadult stage in some animal life cycles. **288**

larynx Short airway containing the vocal cords (voice box). **416**

lateral meristem Cylindrical sheet of meristem that runs lengthwise through shoots and roots; gives rise to secondary growth (thickening) in a plant. *See* vascular cambium, cork cambium. **547**

law of nature Generalization that describes a consistent natural phenomenon that has an incomplete scientific explanation. **18**

leaf vein In vascular plants, a vascular bundle in a leaf. **541**

lens Structure that focuses light on an eye's photoreceptors. **482**

lichen Composite organism consisting of a fungus and a green alga or cyanobacterium. **279**

life history traits Characteristics related to growth, survival, and reproduction such as life span, age-specific mortality, age at first reproduction, and number of breeding events. **320**

ligament Cord of dense connective tissue that holds bones in place at a synovial joint. **394**

lignin Compound that stiffens walls of some cells (including xylem) in vascular plants. **264**

limbic system Collection of structures deep in the brain that have roles in emotion and memory. **476**

lineage Line of descent. **193**

lipid Fatty, oily, or waxy organic compound; e.g., a triglyceride, steroid, or wax. **36**

lipid bilayer Double layer of lipids arranged tail-to-tail; structural foundation of all cell membranes. **37**

liver Organ that produces bile, stores glycogen, and detoxifies many substances. **451**

lobe-finned fish Bony fish that has bony supports inside its fins. **300**

logistic model of population growth Model for growth of a population limited by density-dependent factors; numbers increase exponentially at first, then the growth rate slows and population size levels off at carrying capacity. **318**

loose connective tissue Connective tissue with relatively few fibroblasts and fibers scattered in its matrix. **378**

lungs Internal saclike organs inside which blood exchanges gases the the air; the respiratory organs in most land vertebrates and some fish. **298, 415**

luteinizing hormone (**LH**) Pituitary hormone that triggers ovulation in females and testosterone secretion in males. **516**

lymph Fluid that has left capillaries and entered lymph vessels. **411**

lysosome Enzyme-filled vesicle that breaks down cellular wastes and debris. **55**

lysozyme Antibacterial enzyme in body secretions such as saliva and mucus. **427**

macroevolution Evolutionary patterns and trends on a larger scale than microevolution. **226**

macrophage Phagocytic white blood cell that patrols tissues and interstitial fluid. **426**

Malpighian tubules Of insects, organs that take up unwanted solutes from hemolymph and deliver them to the gut for excretion. **456**

mammal Vertebrate that nourishes its young with milk from mammary glands. **303**

mantle Skirtlike extension of tissue in mollusks; covers the mantle cavity and secretes the shell if one is present. **290**

marsupial Mammal in which offspring complete development in a pouch on the mother's body. **304**

mass extinction Event in which many species in many habitats become extinct in the same interval. **360**

mass number Of an isotope, the total number of protons and neutrons in the atomic nucleus. **25**

master gene Gene encoding a product that affects the expression of many other genes. **124**

medusa Bell-shaped, free-swimming cnidarian body form. **288**

megaspore In seed plants, a haploid cell that gives rise to a female gametophyte. **269**

meiosis Nuclear division process that halves the chromosome number; the basis of sexual reproduction. **142**

melatonin Hormone produced by the pineal gland; affects onset of puberty, sleep–wake cycles. **504**

membrane potential *See* resting potential, action potential. **470**

memory cell Long-lived, antigen-sensitized lymphocyte that can act in a secondary immune response. **434**

meninges Membranes that surround and protect the brain and spinal cord. **474**

menopause Of a human female, the end of fertility and of menstrual cycles. **516**

menstrual cycle Approximately monthly cycle in which the uterine lining thickens in preparation for pregnancy, then is shed if pregnancy does not occur. **516**

menstruation Flow of blood and bits of shed uterine lining out through the vagina. **516**

meristem In a plant, zone of undifferentiated (stem) cells; all plant growth arises from divisions of these cells. *See* apical meristem, lateral meristem. **546**

mesoderm Middle tissue layer of a three-layered animal embryo. **512**

mesophyll Photosynthetic parenchyma. **536**

messenger RNA (**mRNA**) Type of RNA that carries a protein-building message. **115**

metabolic pathway Series of enzyme-mediated reactions by which cells build, remodel, or break down an organic molecule. **71**

metabolism All the enzyme-mediated chemical reactions by which cells acquire and use energy as they build and break down organic molecules. **34**

metamorphosis Dramatic remodeling of body form during the transition from larva to adult. **293**

metaphase Stage of mitosis at which all chromosomes are aligned in the middle of the cell. **134**

metastasis The process in which cells of a malignant neoplasm spread from one part of the body to another. **138**

methanogen Organism that produces methane gas as a metabolic by-product. **248**

MHC markers Self-proteins on the surface of vertebrate body cells; allow T cells to recognize a cell as belonging to the body. **431**

microbiome Collection of microorganisms that inhabits a specific habitat, such as a human body. **236**

microevolution Change in allele frequency. **214**

microfilament Cytoskeletal element composed of actin subunits. Reinforces cell membranes; functions in movement and muscle contraction. **57**

microspore In seed plants, a haploid cell that gives rise to a male gametophyte (pollen grain). **269**

microtubule Hollow cytoskeletal element composed of tubulin subunits. Involved in movement of a cell or its parts. **56**

microvilli Thin projections that increase the surface area of some cells such as those that line the small intestine. **451**

mimicry Evolutionary process whereby two or more species come to resemble one another. **334**

mineral Inorganic substance that is required in small amounts for normal metabolism. **455**

mitochondrion Double-membraned organelle that produces ATP by aerobic respiration in eukaryotes. **55**

mitosis Nuclear division mechanism that maintains the chromosome number. Basis of body growth and tissue repair in multicelled eukaryotes; also asexual reproduction in some eukaryotes. Has four stages: prophase, metaphase, anaphase, and telophase. **133**

model Analogous system used for testing a hypothesis. **12**

molecule Two or more atoms joined by chemical bonds. **5**

mollusk Invertebrate with a reduced coelom and a mantle. **290**

monocots Lineage of angiosperms that includes grasses, orchids, and palms. **273**

monohybrid cross Cross between two individuals identically heterozygous for alleles of one gene; for example *Aa* × *Aa*. **153**

monomer Molecule that is a subunit of polymers. **33**

monotreme Egg-laying mammal. **304**

morphological convergence Evolutionary pattern in which similar body parts evolve separately in different lineages. **205**

morphological divergence Evolutionary pattern in which a body part of an ancestor changes in its descendants. **204**

mosses Most diverse group of nonvascular plants; Low-growing plants that have flagellated sperm and disperse by producing spores. **265**

motor neuron Neuron that controls a muscle or gland. **469**

motor protein Type of energy-using protein that interacts with cytoskeletal elements to move the cell's parts or the whole cell. **57**

mRNA *See* messenger RNA.

multicellular organism Organism composed of a variety of specialized cells, each unable to survive and reproduce on its own. **250**

mutation Permanent change in the DNA sequence of a chromosome. *See* base-pair substitution, deletion, insertion. **109**

mutualism Species interaction that benefits both species. **278, 332**

mycelium Mass of threadlike filaments (hyphae) that compose the body of a multicelled fungus. **275**

mycorrhiza Fungus–plant root partnership. **278**

myelin sheath Of an axon, a discontinuous covering composed of multiple neuroglia that wrap around and insulate the axon. **469**

myofibrils Threadlike, cross-banded skeletal muscle components made up of sarcomeres. **396**

myosin A motor protein; makes up thick filaments in muscle fibers. **396**

natural selection Differential survival and reproduction of individuals of a population based on differences in shared, heritable traits. Driven by environmental pressures. **195**

nectar Sweet fluid exuded by some flowers that attracts animal pollinators. **557**

negative feedback A change causes a response that reverses the change. **384**

neoplasm An accumulation of abnormally dividing cells. **137**

nephron Kidney tubule and associated capillaries; filters blood and forms urine. **458**

nerve Many axons bundled together in connective tissue. **469**

nerve net Of some radial invertebrates, a mesh of neurons with no central organizing organ. **467**

nervous tissue Animal tissue composed of neurons and supporting cells; detects stimuli and controls responses to them. **380**

neuroglia In nervous tissue, cells that structurally and functionally support neurons. **380**

neuron Type of nervous tissue cell that transmits electrical signals along its plasma membrane and communicates with other cells through chemical messages. **380**

neurotransmitter Chemical signal released by a neuron's axon terminals. **471**

neutron Uncharged subatomic particle in the atomic nucleus. **25**

niche The role of a species in its community. **331**

nitrogen cycle Movement of nitrogen among the atmosphere, soil, and water, and into and out of food webs. **344**

nitrogen fixation Process of combining nitrogen gas with hydrogen to form ammonia. **248, 344**

node In plants, region of a stem where new shoots and roots can form. **538**

nondisjunction Failure of chromosomes to separate properly during nuclear division. **166**

normal flora Microorganisms that typically live on human surfaces, including the interior tubes and cavities of the digestive and respiratory tracts. **427**

notochord Stiff rod of connective tissue that runs the length of the body in chordate larvae or embryos. **297**

nuclear envelope A double membrane that constitutes the outer boundary of the nucleus. Nuclear pores in the membrane control the entry and exit of large molecules. **54**

nucleic acid Chain of nucleotides; DNA or RNA. **41**

nucleotide Small molecule with a deoxyribose or ribose sugar, a nitrogen-containing base, and one, two, or three phosphate groups; e.g., adenine, guanine, cytosine, thymine, uracil. Monomer of DNA or RNA; some have additional roles. **41**

nucleus Of an atom; core area occupied by protons and neutrons. **25** Of a eukaryotic cell; organelle with a double membrane that holds, protects, and controls access to the cell's DNA. **47**

nutrient Substance that an organism needs for growth and survival but cannot make for itself. **6**

nutrient absorption Movement of nutrients from a digestive cavity into the body proper. **447**

oncogene Gene that helps transform a normal cell into a tumor cell. **137**

oocyte Immature egg. **515**

open circulatory system Circulatory system in which the circulatory fluid (hemolymph) leaves open-ended vessels and flows among tissues before returning to the heart. **287, 405**

opportunistic life history Life history favored in unpredictable environments; individuals reproduce while young and invest a small amount in each of many offspring. **321**

organ Structural unit that is composed of two or more tissues and adapted to carry out a particular task. **375**

organ of Corti Sound-detecting organ in the cochlea. **485**

organ system Organs that interact closely in some task. **375**

organelle Structure that carries out a special metabolic function inside a cell. **47**

organic Describes a compound that consists mainly of carbon and hydrogen atoms. **33**

organism Individual that consists of one or more cells. **5**

osmosis Diffusion of water across a selectively permeable membrane; occurs when there is a difference in solute concentration between the fluids on either side of the membrane. **73**

ovarian follicle Immature egg and the cells surrounding it. **515**

ovary In flowering plants, the enlarged base of a carpel, inside which one or more ovules form. In animals, the egg-producing female gonad. **272, 514, 555**

oviduct Ciliated tube that connects an ovary to the uterus. **514**

ovulation Release of a secondary oocyte by an ovary. **515**

ovule Of seed plants, chamber inside which megaspores form and develop into female gametophytes; after fertilization, becomes a seed. **269, 555**

ozone layer Atmospheric layer with a high concentration of ozone that prevents much UV radiation from reaching Earth's surface. **240, 364**

pancreas Organ that secretes digestive enzymes into the small intestine and hormones (insulin and glucagon) into the blood. **451**

Pangea Supercontinent that began to form about 300 million years ago and broke up 100 million years later. **200**

parasite A species that withdraws nutrients from another species (its host), usually without killing it. **335**

parasitoid An insect that lays eggs in another insect, and whose young devour their host from the inside. **336**

parasympathetic neurons Of the autonomic nervous system, neurons that are most active at times of relaxation; govern housekeeping tasks. **478**

parathyroid glands Four small endocrine glands in the neck whose hormone product increases the level of calcium in blood. **499**

parenchyma In plants, simple tissue composed of living cells that have different functions depending on location; main component of ground tissue. **536**

passive transport Membrane-crossing mechanism that requires no energy input. **75**

pathogen Disease-causing agent. **236**

PCR Polymerase chain reaction. Laboratory method that mass-produces copies of a specific section of DNA. **177**

pedigree Chart of family connections that shows the appearance of a trait through generations. **160**

peptide bond A bond between the amine group of one amino acid and the carboxyl group of another; joins amino acids in proteins. **38**

per capita growth rate The number of individuals added during some interval divided by the initial population size. **317**

peripheral nervous system Nerves that extend through a vertebrate body and relay information to and from the central nervous system. **468**

permafrost Layer of permanently frozen soil in the Arctic. **357**

peroxisome Enzyme-filled vesicle that breaks down amino acids, fatty acids, and toxic substances. **55**

petal Unit of a flower's corolla; often showy and conspicuous. **555**

pH Measure of the amount of hydrogen ions in a fluid. **32**

phagocytosis "Cell eating"; an endocytic pathway by which a cell engulfs large particles such as microbes or cellular debris. **77**

pharynx Throat; opens to airways and digestive tract. **416**

phenotype An individual's observable traits. **152**

pheromone Chemical signal used in communication between members of a species. **481**

phloem Vascular tissue of plants; distributes sugars through its sieve tubes. Consists of sieve-tube members and companion cells. **264, 537**

phospholipid Lipid with a phosphate group in its hydrophilic head, and two nonpolar fatty acid tails; main constituent of eukaryotic cell membranes. **36**

phosphorus cycle Movement of phosphorus among rocks, water, soil, and living organisms. **343**

phosphorylation A chemical reaction in which a phosphate group is transferred from one molecule to another. **71**

photoperiodism Biological response to seasonal changes in the relative lengths of day and night. **572**

photosynthesis Metabolic pathway by which most autotrophs use light energy to make sugars from carbon dioxide and water. **6, 82**

phototropism In plants, directional growth response to light. **570**

phylogeny Evolutionary history of a species or group of species. **229**

pigment An organic molecule that can absorb light of certain wavelengths. Reflected light imparts a characteristic color. E.g., chlorophyll. **83**

pilus A protein filament that projects from the surface of some prokaryotic cells. **53**

pineal gland Endocrine gland in the forebrain that secretes melatonin; secretion declines when the eye is exposed to light. **504**

pioneer species Species that can colonize a new habitat. **337**

pituitary gland Pea-sized endocrine gland in the forebrain that interacts closely with the adjacent hypothalamus. **496**

placenta Of placental mammals, an organ composed of maternal and embryonic tissues that allows the exchange of substances between a developing individual and its mother. **525**

placental mammal Mammal in which developing offspring are nourished within the mother's body by way of a placenta. **304**

plankton Community of mostly microscopic drifting or swimming organisms. **251**

plant Multicelled, typically photosynthetic organism; develops from an embryo that forms on the parent and is nourished by it. **8, 263**

plant hormone Extracellular signaling molecule of plants that exerts its effect at very low concentration. E.g., auxin, gibberellin. **566**

plasma Protein-rich fluid portion of blood. **410**

plasma membrane Membrane that encloses a cell and separates it from the external environment. **47**

plasmid Of many prokaryotes, a small ring of nonchromosomal DNA. **246**

plasmodial slime mold Heterotrophic protist that moves and feeds as a multinucleated mass; forms a fruiting body when conditions are unfavorable. **256**

plate tectonics theory Theory that Earth's outermost layer of rock is cracked into plates, the slow movement of which conveys continents to new locations over geologic time. **201**

platelet Cell fragment that functions in blood clotting. **410**

pleiotropy Inheritance pattern in which a single gene affects multiple traits. **156**

pluripotent Capable of developing into any type of cell in a multicelled body. **374**

polar body Tiny cell that forms as a by-product of unequal cytoplasmic division during egg production. **515**

polarity Separation of charge into positive and negative regions. **28**

pollen grain Immature male gametophyte of a seed plant. **264, 555**

pollen sac Of seed plants, chamber in which microspores form and develop into male gametophytes (pollen grains). **269**

pollination Delivery of a pollen grain to the egg-bearing part of a seed plant. **269, 556**

pollination vector Environmental agent that moves pollen grains from one plant to another. **556**

pollinator Animal that moves pollen from one plant to another, thus facilitating pollination. **273, 554**

pollutant A natural or man-made substance that is released into the environment in greater than natural amounts and that damages the health of organisms. **362**

polymer Molecule that consists of multiple monomers. **33**

polymerase chain reaction See PCR.

polyp In cnidarians, a tubular, typically sessile, body form; for example a sea anemone. **288**

polyploid Having three or more ocomplete sets of chromosomes. **165**

population A group of organisms of the same species who live in a specific location and breed with one another more often than they breed with members of other populations. **5, 314**

population density Number of members of a population in a given area. **315**

population distribution The way in which members of a population are dispersed in their environment. **315**

population size Total number of individuals in a population. **315**

prairie Temperate grassland biome of North America. Its grasses and other plants are adapted to recover after grazing and the occasional fire. **356**

predation One species (the predator) captures, kills, and feeds on another (its prey). **334**

prediction Statement, based on a hypothesis, about a condition that should exist if the hypothesis is correct. **12**

pressure flow theory Explanation of how a difference in turgor between sieve elements in source and sink regions pushes sugar-rich fluid through a sieve tube. **546**

primary growth Lengthening of young shoots and roots; originates at apical meristems. **547**

primary production The energy captured by an ecosystem's producers. **341**

primary succession Ecological succession occurs in an area where there was previously no soil. **337**

primate Mammalian group with grasping hands; includes lemurs, tarsiers, monkeys, apes, and humans. **305**

primer Short, single strand of DNA or RNA that base-pairs with a specific DNA sequence in DNA synthesis. **108**

prion Infectious protein. **40**

probability Out of all possible outcomes of an event, the chance that a particular outcome will occur. **17**

probe Fragment of DNA or RNA labeled with a tracer; can hybridize with a nucleotide sequence of interest. **176**

producer Organism that makes its own food using energy and nonbiological raw materials from the environment. **6, 339**

product A molecule that is produced by a reaction. **66**

progesterone Sex hormone produced by a female's ovaries; prepares the reproductive tract for pregnancy. **504, 514**

prokaryote Informal name for a single-celled organism without a nucleus; a bacterium or archaeon. **8**

promoter In DNA, a nucleotide sequence to which RNA polymerase binds; site where transcription begins. **116**

prophase Stage of mitosis during which chromosomes condense and become attached to a newly forming spindle. **134**

prostate gland Exocrine gland that encircles a male's urethra; its secretions contribute to semen. **518**

protein Polymer of amino acids; an organic molecule that consists of one or more polypeptides. **38**

protist General term for eukaryote that is not a fungus, plant, or animal. **8, 250**

protocell Membranous sac that contains interacting organic molecules; hypothesized to have formed prior to the earliest cells. **239**

proton Positively charged subatomic particle that occurs in the nucleus of all atoms. **25**

protostomes Animal lineage with a three-layer embryo in which the first opening to form is the mouth; includes most bilateral invertebrates. **286**

pseudopod A temporary protrusion that helps some eukaryotic cells move and engulf prey. **57**

puberty Period when reproductive organs mature and begin to function. **504**

pulmonary circuit Circuit through which blood flows from the heart to the lungs and back. **406**

Punnett square Diagram used to predict the genetic and phenotypic outcomes of a cross. **153**

pupil Opening through which light enters the eye. **483**

radial symmetry Having parts arranged around a central axis, like spokes around a wheel. **286**

radioactive decay Process by which atoms of a radioisotope emit energy and subatomic particles when their nucleus spontaneously breaks up. **25**

radioisotope Isotope with an unstable nucleus. **25**

radiometric dating Method of estimating the age of a rock or fossil by measuring the content and proportions of a radioisotope and its daughter elements. **199**

radula Tonguelike organ of many mollusks. **290**

rain shadow Dry region on the downwind side of a coastal mountain range. **353**

ray-finned fish Bony fish with fins supported by thin rays derived from skin. **300**

reabsorption Water and solutes leave a kidney tubule and enter adjacent capillaries. **459**

reactant A molecule that enters a reaction and is changed by participating in it. **66**

reaction Process of molecular change. **34**

receptor protein Membrane protein that triggers a change in cell activity in response to a stimulus such as binding a certain substance. **51**

recessive Refers to an allele with an effect that is masked by a dominant allele on the homologous chromosome. **152**

recombinant DNA A DNA molecule that contains genetic material from more than one organism. **176**

rectum Portion of the large intestine that stores feces until they are expelled. **452**

red alga Single-celled or multicelled photosynthetic protist with a red accessory pigment. **255**

red blood cell Hemoglobin-filled cell that transports oxygen in the blood. **410**

red marrow Bone marrow that makes blood cells. **393**

red muscle fiber Of skeletal muscle, a fiber that contains the oxygen-storing protein myoglobin and produces ATP primarily by aerobic respiration. **398**

reflex Automatic response to a stimulus. **478**

reproduction Process by which parents produce offspring. *See* sexual reproduction, asexual reproduction. **7**

reproductive cloning A laboratory procedure that produces genetically identical animals; e.g., somatic cell nuclear transfer (SCNT). **101**

reproductive isolation The end of gene flow between populations. **222**

reptile Amniote subgroup that includes lizards, snakes, turtles, crocodilians, and birds. **302**

resource partitioning Evolutionary process whereby traits of competing species come to differ as a result of the selective pressure imposed by the competition. **334**

respiration Physiological process by which gases enter and leave an animal body. **414**

respiratory cycle One inhalation and one exhalation. **417**

respiratory surface Moist surface across which gases are exchanged between animal cells and their environment. **414**

resting potential Membrane potential of a neuron at rest. **470**

restriction enzyme Type of enzyme that cuts DNA at a specific nucleotide sequence. **175**

retina Layer of eye that contains photoreceptors. **482**

rhizoid Threadlike structure that anchors some plants. **265**

rhizome Stem that grows horizontally along or just below the ground. **266**

ribosomal RNA (rRNA) RNA that becomes part of ribosomes. **119**

ribosome Organelle of protein synthesis. An intact ribosome has two subunits, each composed of rRNA and proteins. **52**

RNA Ribonucleic acid. Nucleic acid with roles in gene expression; consists of a single-stranded chain of nucleotides. *See* messenger RNA, transfer RNA, ribosomal RNA. **41**

RNA polymerase Enzyme that carries out transcription (RNA synthesis). **116**

RNA world hypothesis Hypothesis that RNA served as the first material of inheritance. **238**

rod cell Photoreceptor active in dim light; provides coarse perception of image and detects motion. **484**

root hairs Hairlike, absorptive extensions of a root epidermal cell; form on young roots **542**

root nodules On some plant roots, swellings that contain mutualistic nitrogen-fixing bacteria. **542**

roundworm Unsegmented worm with a pseudocoelom and a cuticle that is molted as the animal grows. **291**

rRNA *See* ribosomal RNA.

sac fungus Fungus that produces spores by meiosis in saclike cells. **276**

salt Ionic compound that releases ions other than H^+ and OH^- when it dissolves in water. **30**

sampling error Difference between results derived from testing an entire group of events or individuals, and results derived from testing a subset of the group. **17**

sarcomere Contractile unit of skeletal and cardiac muscle. **396**

saturated fatty acid Fatty acid with only single bonds linking the carbons in its tail. **36**

savanna Tropical biome dominated by grasses and other plants adapted to grazing, as well as a scattering of shrubs. **356**

scales Hard, flattened elements that cover the skin of reptiles and some fishes. **299**

science Systematic study of the observable world. **12**

scientific method Making, testing, and evaluating hypotheses about the natural world. **13**

scientific theory Hypothesis that has not been disproven after many years of rigorous testing. **18**

sclerenchyma In plants, simple tissue composed of cells that die when mature; their tough walls structurally support plant parts. E.g., fibers, sclereids. **536**

seamount An undersea mountain. **359**

second law of thermodynamics Energy tends to disperse spontaneously. **65**

second messenger Molecule that forms inside a cell when a hormone binds at the cell surface; sets in motion reactions that alter enzyme activity inside the cell. **494**

secondary growth Thickening of older stems and roots; originates at lateral meristems. **547**

secondary succession Ecological succession occurs in an area where a community previously existed and soil remains. **337**

seed Embryo sporophyte of a seed plant packaged with nutritive tissue inside a protective coat. **264, 560**

semen Sperm mixed with fluid secreted by exocrine glands. **518**

seminal vesicles Exocrine glands that secrete fluid into vasa deferentia; main source of semen volume. **518**

seminiferous tubule In a testis, a germ-cell containing tubule inside which sperm form. **518**

sensation Awareness of a stimulus. **480**

sensory adaptation Diminishing response to a persistent stimulus. **480**

sensory neuron Neuron that is excited by a specific environmental stimulus. **469**

sensory perception Meaning the brain assigns to a specific sensation. **481**

sensory receptor Structure that is activated by a particular stimulus and whose activation results in action potentials in a sensory neuron. **480**

sequencing Laboratory method of determining the order of nucleotides in DNA. **178**

sex chromosome Member of a pair of chromosomes that differs between males and females. **107**

sexual reproduction Reproductive mode by which offspring arise from two parents and inherit genes from both. **141, 511**

sexual selection Mode of natural selection in which some individuals outreproduce others of a population because they are better at securing mates. **218**

shell model Model of electron distribution in an atom. **27**

short tandem repeat In chromosomal DNA, a sequence of a few nucleotides repeated multiple times in a row. Used in DNA profiling. **159**

single-nucleotide polymorphism (SNP) A one-nucleotide DNA sequence variation carried by a measurable percentage of a population. **174**

sink Region of a plant where sugars are being used. **546**

sister chromatids The two DNA molecules of a duplicated eukaryotic chromosome, attached at the centromere. **106**

skeletal muscle fiber Cylindrical, multinucleated cell that runs the length of a skeletal muscle. **396**

skeletal muscle tissue Striated, voluntary muscle that interacts with bone to move body parts. **379**

sliding-filament model Explanation of how interactions among actin and myosin filaments shorten a sarcomere and bring about muscle contraction. **397**

small intestine Longest portion of the digestive tract, and the site of most digestion and absorption. **450**

smooth muscle tissue Involuntary muscle that lines blood vessels and hollow organs; not striated. **379**

SNP *See* single-nucleotide polymorphism.

solute A dissolved substance. **30**

solution Uniform mixture of solute completely dissolved in a solvent. **30**

solvent Liquid in which other substances dissolve. **30**

somatic cell nuclear transfer (SCNT) Reproductive cloning method in which the DNA of an adult donor's body cell is transferred into an unfertilized egg. **100**

somatic nerves Nerves that control skeletal muscle and relay signals from joints and skin. **478**

somatosensory cortex Region of the cerebral cortex that receives information about touch, temperature, and pain from receptors throughout the body. **487**

sorus Cluster of spore-forming chambers on a fern frond. **266**

source Region of a plant where sugars are being produced or released from storage. **546**

speciation Evolutionary process in which new species arise; e.g., allopatric speciation, sympatric speciation. **222**

species Unique type of organism designated by genus name and specific epithet. Of sexual reproducers, often defined as one or more groups of individuals that can potentially interbreed, produce fertile offspring, and do not interbreed with other groups. **8**

species diversity The number of species and their relative abundance within a community. **331**

sperm Male gamete. **512**

sphincter Ring of muscle that controls passage of material through a tubular organ or at a body opening. **396, 449**

spinal cord Portion of central nervous system that extends through the backbone and connects peripheral nerves with the brain. **477**

spindle Temporary structure that moves chromosomes during nuclear division; consists of microtubules. **134**

sponge Aquatic invertebrate that has no tissues or organs and filters food from the water. **288**

spongy bone Lightweight bone with many internal spaces. **393**

sporophyte Diploid spore-forming body that forms in a plant life cycle. **263**

stabilizing selection Mode of natural selection in which an intermediate form of a trait is adaptive, and extreme forms are not. **215**

stamen Floral reproductive organ that consists of a pollen-producing anther and, in most species, a filament. **272, 555**

statistically significant Refers to a result that is statistically unlikely to have occurred by chance alone. **17**

statocyst Gravity-sensing organ of invertebrates. **486**

stem cell A cell that can divide and create more stem cells or differentiate to become a specialized cell type. **374**

steroid A type of lipid with four carbon rings and no fatty acid tails. **37**

steroid hormone Lipid-soluble hormone derived from cholesterol. **494**

stigma Upper part of a carpel; adapted to receive pollen. **272, 555**

stomach Muscular organ that receives food from the esophagus, mixes it with gastric fluid, and propels it into the small intestine. **449**

stomata Singular, stoma. Closable gaps formed by pairs of guard cells on aboveground plant surfaces. When open, they allow the plant to exchange gases with air. When closed, they limit water loss. **88, 264**

stroma The cytoplasm-like fluid between the thylakoid membrane and the two outer membranes of a chloroplast. Site of light-independent reactions of photosynthesis. **84**

stromatolites Dome-shaped structures composed of layers of prokaryotic cells and sediments; form in shallow seas. **240**

style Elongated portion of a carpel that elevates the stigma above the ovary. **272**

substrate Of an enzyme, a reactant that is specifically acted upon by the enzyme. **68**

surface-to-volume ratio A relationship in which the volume of an object increases with the cube of the diameter, and the surface area increases with the square. Limits cell size. **47**

survivorship curve Graph showing the decline in numbers of a cohort over time. **320**

symbiosis One species lives on or inside another in a commensal, mutualistic, or parasitic relationship. **332**

sympathetic neurons Of the autonomic nervous system, neurons that are most active at times of excitement or danger. **479**

sympatric speciation Divergence within a population leads to speciation; occurs in the absence of a physical barrier to gene flow. **224**

synovial joint Movable joint at which ligaments connect bones. For example, a shoulder, elbow, or knee. **394**

systemic circuit Circuit through which blood flows from the heart to the body tissues and back. **406**

systolic pressure Blood pressure when the ventricles are contracting. **410**

T cell T lymphocyte. Lymphocyte central to adaptive immunity; some kinds target infected or cancerous body cells. E.g., helper T cell, cytotoxic T cell. **426**

T cell receptor (TCR) Antigen receptor on the surface of a T cell; recognizes antigen only in conjunction with MCH markers. **431**

taiga *See* boreal forest.

taxon Plural, taxa. A rank in the classification of life; consists of a group of organisms that share a unique set of traits. **10**

taxonomy Practice of naming and classifying species. **8**

telophase Stage of mitosis during which chromosomes arrive at opposite ends of the cell. Two new nuclei form as the chromosomes loosen. **134**

temperate deciduous forest Biome dominated by broadleaf trees that grow in warm summers, then drop their leaves and become dormant during cold winters. **355**

temperature Measure of molecular motion. **31**

tendon Strap of dense connective tissue that connects a skeletal muscle to bone. **395**

testis Male gonad; produces sperm. **514**

testosterone Sex hormone produced by a male's testes. **504**

tetrapod Vertebrate with four limbs. **298**

theory *See* scientific theory.

threatened species A species likely to become endangered in the near future. **360**

threshold potential Membrane potential at which voltage-gated sodium channels in a neuron axon open, resulting in an action potential. **471**

thylakoid membrane A chloroplast's highly folded inner membrane system; forms a continuous compartment. Site of light-dependent reactions of photosynthesis. **84**

thyroid gland Endocrine gland at the base of the neck; produces thyroid hormone and calcitonin. **498**

tissue A collection of one or more specific cell types that are organized in a way that adapts them to a task. **286, 375**

tissue culture propagation Laboratory method in which individual plant cells (typically from meristem) are induced to form embryos. **564**

total fertility rate Average number of children born to females of a population over the course of their lifetimes. **324**

tracer A substance that can be traced via its detectable component. **26**

trachea Major airway leading to the lungs; the windpipe. **417**

tracheal system Of insects, branching tubes that deliver air from body surface to internal tissues. **415**

transcription RNA synthesis; process by which enzymes assemble an RNA using a strand of DNA as a template. Part of gene expression. **115**

transcription factor Regulatory protein that influences transcription by binding directly to DNA. **124**

transduction Mechanism of gene transfer. A virus moves genes from one host cell to another. **246**

transfer RNA (tRNA) RNA that delivers amino acids to a ribosome during translation. **119**

transformation Mechanism of gene transfer. A prokaryotic cell takes up and uses DNA from its environment. **246**

transgenic Refers to a genetically modified organism that carries a gene from a different species. **181**

translation Protein synthesis; process by which a polypeptide chain is assembled from amino acids in the order specified by an mRNA. **115**

translocation Movement of organic molecules through phloem. **545**

transpiration Evaporation of water from a plant's aboveground parts. **342, 544**

transport protein Membrane protein that passively or actively helps a specific ion or molecule move across the membrane. **51**

triglyceride A molecule with three fatty acid tails that is entirely hydrophobic; main component of fats. **36**

tRNA *See* transfer RNA.

trophic level Position in a food chain. **339**

tropical rain forest Multilayered forest biome that occurs where warm temperatures and continual rains allow plant growth year-round. Most productive and species-rich biome. **355**

tropism In plants, directional growth response to an environmental stimulus. **570**

tubular secretion Movement of substances out of capillaries and into kidney tubules for excretion in the urine. **459**

tumor A neoplasm that forms a lump. **137**

tundra Northernmost biome, dominated by low plants that grow over a layer of permafrost. **356**

tunicates Invertebrate chordates that lose their defining chordate traits during the transition to adulthood; adults have a secreted "tunic." **297**

turgor Pressure that a fluid exerts against a cell wall, membrane, or other structure that contains it. **74**

unsaturated fatty acid Fatty acid with one or more carbon–carbon double bonds in its tail. **36**

ureter Tube that carries urine from a kidney to the bladder. **457**

urethra Tube through which urine from the bladder flows out of the body. **457**

urinary bladder Hollow, muscular organ that stores urine. **457**

urinary system Organ system that filters blood, and forms, stores, and expels urine. **457**

urine Mix of water and soluble wastes formed and excreted by the urinary system. **457**

uterus Muscular chamber where offspring develop; womb. **514**

vaccine A preparation introduced into the body in order to elicit immunity to a specific antigen. **441**

vagina Female organ of intercourse and birth canal. **515**

variable In an experiment, a characteristic or event that differs among individuals or over time. **13**

vascular bundle In vascular plants, cylindrical bundle of xylem, phloem, and sclerenchyma fibers that runs through a stem or leaf. **538**

vascular cambium Lateral meristem that produces secondary xylem and phloem. **547**

vascular cylinder Central column of vascular tissue in a root of a vascular plant. **543**

vascular plant A plant that has xylem and phloem. **264**

vascular tissues Tissues that distribute water and nutrients through a vascular plant body; e.g., xylem, phloem. **535**

vas deferens A long duct that conveys mature sperm from an epididymis toward the urethra. **517**

vein Large-diameter vessel that returns blood to the heart. **407**

ventricle Heart chamber that pumps blood out of the heart and into an artery. **406**

venule Blood vessel that connects a capillary to a vein. **412**

vernalization Stimulation of flowering in spring by prolonged exposure to low temperature in winter. **572**

vertebra One of the bones of the backbone (vertebral column). **392**

vertebral column Backbone. **298, 391**

vertebrate Animal with a backbone. **284, 298**

vesicle Small, membrane-enclosed organelle; different kinds store, transport, or break down their contents. **54**

vestibular apparatus Structure in the vertebrate inner ear that functions in the sense of balance. **486**

villi Multicelled projections from the lining of the small intestine. **451**

viral envelope A layer of cell membrane derived from the host cell in which an enveloped virus was produced. **242**

viral reassortment Two viruses of the same type infect an individual at the same time and swap genes. **245**

virus A noncellular infectious particle with a protein coat and a genome of RNA or DNA; replicates only in living cells. **242**

vitamin Organic substance required in small amounts for normal metabolism. **454**

warning coloration Distinctive color or pattern that makes a well-defended prey species easy to recognize. **334**

water cycle Water moves from the ocean into the air, falls as rain and snow, and flows back to the ocean. **342**

water mold Heterotrophic protist that forms a mesh of nutrient-absorbing filaments; some are important plant pathogens. **254**

water–vascular system Of echinoderms, a system of fluid-filled tubes and tube feet that function in locomotion. **296**

wavelength Distance between the crests of two successive waves. **83**

wax Water-repellent substance that is a complex, varying mixture of lipids. **37**

white blood cell Blood cell that helps defend the body against pathogens and cleans up cellular debris. **410**

white matter Of the central nervous system, tissue that consists mainly of myelinated axons. **474**

white muscle fiber Of skeletal muscle, a fiber that produces ATP primarily by lactate fermentation. **398**

wood Lignin-stiffened secondary xylem of some seed plants. **269, 548**

xylem Complex vascular tissue of plants; its dead tracheids and vessel elements form tubes that distribute water and mineral ions. **264, 536**

yellow marrow Bone marrow that is mostly fat; fills cavity in most long bones. **393**

yolk Nutritious material in many animal eggs. **512**

zygote Diploid cell that forms when two gametes fuse; the first cell of a new individual. **144, 512**

zygote fungus Fungus that usually grows as a mold; sexual reproduction yields a thick-walled zygospore. **275**